Spatial Point Patterns

Methodology and Applications with R

CHAPMAN & HALL/CRC
Interdisciplinary Statistics Series

Series editors: N. Keiding, B.J.T. Morgan, C.K. Wikle, P. van der Heijden

Published titles

AGE-PERIOD-COHORT ANALYSIS: NEW MODELS, METHODS, AND EMPIRICAL APPLICATIONS Y. Yang and K. C. Land

ANALYSIS OF CAPTURE-RECAPTURE DATA R. S. McCrea and B. J. T. Morgan

AN INVARIANT APPROACH TO STATISTICAL ANALYSIS OF SHAPES S. Lele and J. Richtsmeier

ASTROSTATISTICS G. Babu and E. Feigelson

BAYESIAN ANALYSIS FOR POPULATION ECOLOGY R. King, B. J. T. Morgan, O. Gimenez, and S. P. Brooks

BAYESIAN DISEASE MAPPING: HIERARCHICAL MODELING IN SPATIAL EPIDEMIOLOGY, SECOND EDITION A. B. Lawson

BIOEQUIVALENCE AND STATISTICS IN CLINICAL PHARMACOLOGY S. Patterson and B. Jones

CLINICAL TRIALS IN ONCOLOGY, THIRD EDITION S. Green, J. Benedetti, A. Smith, and J. Crowley

CLUSTER RANDOMISED TRIALS R.J. Hayes and L.H. Moulton

CORRESPONDENCE ANALYSIS IN PRACTICE, SECOND EDITION M. Greenacre

DESIGN AND ANALYSIS OF QUALITY OF LIFE STUDIES IN CLINICAL TRIALS, SECOND EDITION D.L. Fairclough

DYNAMICAL SEARCH L. Pronzato, H. Wynn, and A. Zhigljavsky

FLEXIBLE IMPUTATION OF MISSING DATA S. van Buuren

GENERALIZED LATENT VARIABLE MODELING: MULTILEVEL, LONGITUDINAL, AND STRUCTURAL EQUATION MODELS A. Skrondal and S. Rabe-Hesketh

GRAPHICAL ANALYSIS OF MULTI-RESPONSE DATA K. Basford and J. Tukey

INTRODUCTION TO COMPUTATIONAL BIOLOGY: MAPS, SEQUENCES, AND GENOMES M. Waterman

MARKOV CHAIN MONTE CARLO IN PRACTICE W. Gilks, S. Richardson, and D. Spiegelhalter

MEASUREMENT ERROR AND MISCLASSIFICATION IN STATISTICS AND EPIDEMIOLOGY: IMPACTS AND BAYESIAN ADJUSTMENTS P. Gustafson

MEASUREMENT ERROR: MODELS, METHODS, AND APPLICATIONS J. P. Buonaccorsi

MEASUREMENT ERROR: MODELS, METHODS, AND APPLICATIONS J. P. Buonaccorsi

Published titles

MENDELIAN RANDOMIZATION: METHODS FOR USING GENETIC VARIANTS IN CAUSAL ESTIMATION S. Burgess and S.G. Thompson

META-ANALYSIS OF BINARY DATA USING PROFILE LIKELIHOOD D. Böhning, R. Kuhnert, and S. Rattanasiri

POWER ANALYSIS OF TRIALS WITH MULTILEVEL DATA M. Moerbeek and S. Teerenstra

SPATIAL POINT PATTERNS: METHODOLOGY AND APPLICATIONS WITH R A. Baddeley, E Rubak, and R. Turner

STATISTICAL ANALYSIS OF GENE EXPRESSION MICROARRAY DATA T. Speed

STATISTICAL ANALYSIS OF QUESTIONNAIRES: A UNIFIED APPROACH BASED ON R AND STATA F. Bartolucci, S. Bacci, and M. Gnaldi

STATISTICAL AND COMPUTATIONAL PHARMACOGENOMICS R. Wu and M. Lin

STATISTICS IN MUSICOLOGY J. Beran

STATISTICS OF MEDICAL IMAGING T. Lei

STATISTICAL CONCEPTS AND APPLICATIONS IN CLINICAL MEDICINE J. Aitchison, J.W. Kay, and I.J. Lauder

STATISTICAL AND PROBABILISTIC METHODS IN ACTUARIAL SCIENCE P.J. Boland

STATISTICAL DETECTION AND SURVEILLANCE OF GEOGRAPHIC CLUSTERS P. Rogerson and I. Yamada

STATISTICS FOR ENVIRONMENTAL BIOLOGY AND TOXICOLOGY A. Bailer and W. Piegorsch

STATISTICS FOR FISSION TRACK ANALYSIS R.F. Galbraith

VISUALIZING DATA PATTERNS WITH MICROMAPS D.B. Carr and L.W. Pickle

Chapman & Hall/CRC
Interdisciplinary Statistics Series

Spatial Point Patterns
Methodology and Applications with R

Adrian Baddeley
Curtin University
Australia

Ege Rubak
Aalborg University
Denmark

Rolf Turner
University of Auckland
New Zealand

CRC Press
Taylor & Francis Group
Boca Raton London New York

CRC Press is an imprint of the
Taylor & Francis Group, an **informa** business

A CHAPMAN & HALL BOOK

CRC Press
Taylor & Francis Group
6000 Broken Sound Parkway NW, Suite 300
Boca Raton, FL 33487-2742

© 2016 by Taylor & Francis Group, LLC
CRC Press is an imprint of Taylor & Francis Group, an Informa business

No claim to original U.S. Government works

Printed on acid-free paper
Version Date: 20151015

International Standard Book Number-13: 978-1-4822-1020-0 (Hardback)

Visit the Taylor & Francis Web site at
http://www.taylorandfrancis.com

and the CRC Press Web site at
http://www.crcpress.com

To Brian Ripley

Contents

Preface

In this book we aim to help researchers from a wide range of scientific fields to analyse their spatial data, when the data take the form of a map of point locations. The book is based on a lifetime of methodological research and scientific collaboration, and 25 years of software development, aimed at making the techniques accessible to everyone.

The `spatstat` package, which features prominently in this book, is free, open-source software in the R language. It provides a wide range of capabilities for spatial point pattern data, from basic data handling to advanced analytic tools.

Users of `spatstat` from all corners of the world, and from a wide range of sciences, have shared their scientific problems with us, generously contributed their data, written new software, identified bugs, offered suggestions for improvement, and posed challenging problems. We hope to repay this generosity by writing a book from the user's perspective: focused on practical needs, and explicitly answering the most frequently asked questions.

Although the authors were mathematically trained, this is not a book for mathematicians. There are many excellent textbooks for that purpose [484, 275, 277]. This book is written for scientific researchers and applied statisticians. It originated as a set of notes for a workshop on `spatstat`, and retains something of that style. Its focus is on the *statistical principles* of analysing spatial data, the *practicalities* of spatial data analysis, and the *scientific interpretation* of the results. It has been lovingly beaten into shape by the workshop participants, by the many users of `spatstat` who correspond with us, and by our scientific collaborators. They have challenged us to answer the real scientific questions, and to give clear, succinct explanations for all the concepts in the book.

Reviewing the recent literature, we found that a surprising number of methodological errors have gained currency — some of them serious enough to invalidate the entire data analysis. This has made us more determined that the book should correct those errors clearly and convincingly. We hope the reader will forgive us the occasional polemic. Technical details are given for backup when necessary, but are relegated to *starred sections* at the end of each chapter, where they can easily be skipped.

This book is dedicated to Brian D. Ripley, Emeritus Professor of Applied Statistics at Oxford University. The topic of this book owes a great deal to Brian Ripley, a pioneer and substantial contributor to statistical methods for spatial point patterns, to Markov chain Monte Carlo methods, and to statistical computing especially in S-PLUS™ and R, as well as to other areas of statistical science. The impact and importance of Ripley's scientific contributions deserve much greater recognition.

Acknowledgements

This book includes results from unpublished work by Ute Hahn (in Sections 16.8.3–16.8.5), by AB and Richard Gill (in Chapter 8), by AB, Andrew Hardegen, Tom Lawrence, Robin Milne, Gopalan Nair and Suman Rakshit (in Chapter 10), by AB, RT, Leanne Bischof and Ida-Maria Sintorn (in Chapter 16), and by AB, Greg McSwiggan, Gopalan Nair and Suman Rakshit (in Chapter 17). We gratefully acknowledge these contributors and thank them for their generosity.

Our special thanks go to Andrew Bevan, Achmad Choiruddin, Jean-François Coeurjolly, Ute Hahn, Kassel Hingee, Mahdieh Khanmohammadi, Jesper Møller, Tuomas Rajala, Farzaneh Safavi-manesh, Allan Trench, Rasmus Waagepetersen, and an anonymous reviewer, who read the typescript closely and gave us detailed feedback.

The spatstat package incorporates substantial amounts of code generously contributed by Kasper Klitgaard Berthelsen, Ute Hahn, Abdollah Jalilian, Marie-Colette van Lieshout, Tuomas Rajala, Dominic Schuhmacher and Rasmus Waagepetersen.

While every effort has been made to check the correctness and reliability of the software covered in this book, the software is provided without any warranty whatsoever, under the GNU Public Licence GPL 2.

Over the years, our collaborators and friends Ang Qi Wei, Eddy Campbell, Ya-Mei Chang, Jean-François Coeurjolly, David Dereudre, Peter Diggle, Richard Gill, Pavel Grabarnik, Eun-Jung Holden, Peter Kovesi, Marie-Colette van Lieshout, Jesper Møller, Tony Pakes, Jakob Gulddahl Rasmussen, Dominic Schuhmacher, Rohan Shah, Yong Song, Aila Särkkä, Eva Bjørn Vedel Jensen, and Rasmus Waagepetersen have made valuable contributions to the material now presented here. There were many generous donors of data, including S. Banerjee, M. Berman, T. Berntsen, R. Condit, P. Diggle, M. Drinkwater, S. Eglen, O. Flores, E.D. Ford, J. Franklin, N. Funwi-Gabga, D.J. Gerrard, A. Green, T. Griffin, U. Hahn, A.H. Hickman, S.P. Hubbell, J. Huntington, V. Isham, A. Jammalamadaka, C. Knox-Robinson, K. Kosic, P. Kovesi, R.A. Lamb, G.P. Leser, J. Mateu, E. Miranda, M. Numata, Y. Ogata, A. Penttinen, N. Picard, D. Stoyan, M. Tanemura, S. Voss, R. Waagepetersen, K.P. Watkins, and H. Wendrock. Valuable feedback on the book draft was also given by Hanne Lone Andersen, Daniel Esser, Robin Milne, Gopalan Nair, Suman Rakshit, and Torben Tvedebrink.

This book would not have been possible without support from the Centre for Exploration Targeting at the University of Western Australia; the Department of Mathematics and Statistics, Curtin University; CSIRO Division of Mathematics, Informatics and Statistics; Department of Mathematical Sciences, Aalborg University; Department of Statistics, University of Auckland; the Centre for Stochastic Geometry and Advanced Bioimaging, Aarhus University; School of Mathematics & Statistics, University of Western Australia; Department of Mathematics & Statistics, the University of New Brunswick; the Australian Research Council; the Danish Council for Independent Research, Natural Sciences; and the Natural Sciences and Engineering Research Council of Canada.

We are grateful to all of the readers who so generously contributed their time and effort. All three authors agree that any errors remaining should not in any way be blamed on the readers referred to, but are exclusively the fault of the other two authors.

Most of all we thank our wives and families for their unwavering support, patience and encouragement.

Part I

BASICS

1

Introduction

1.1 Point patterns

1.1.1 Points

A 'spatial point pattern' is a dataset giving the observed spatial locations of things or events. Examples include the locations of trees in a forest, gold deposits mapped in a geological survey, stars in a star cluster, road accidents, earthquake epicentres, mobile phone calls, animal sightings, or cases of a rare disease. The spatial arrangement of points is the main focus of investigation.

Interest in methods for analysing such data is rapidly expanding across many fields of science, notably in ecology, epidemiology, geoscience, astronomy, econometrics, and crime research.

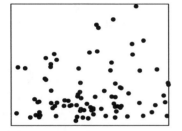 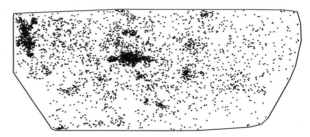

Figure 1.1. *Point pattern datasets with spatially varying density of points.* Left: *enterochromaffin-like cells in histological section of gastric mucosa (T. Bendtsen; see [484, pp. 2, 169]), interior of stomach towards top of picture.* Right: *sky positions of 4215 galaxies in the Shapley Supercluster (M. Drinkwater); survey region about 25 degrees across.*

One important task is to identify spatial trends in the density of points. The left panel of Figure 1.1 shows the locations of a particular type of cell seen in an optical microscope section of tissue from the stomach. The interior of the stomach is toward the top of the picture. There is an obvious gradient in the abundance of cells. The right panel of the Figure shows the sky positions of galaxies observed in an astronomical survey. There is a dramatic concentration of galaxies in two areas.

Statistical analysis of the spatial arrangement of points can reveal important features, such as a tendency for gold deposits to be found close to a major geological fault, or for cases of a disease to be more prevalent near a pollution source, or for bird nests to maintain a certain minimum separation from each other. Analysis of point pattern data has provided pivotal evidence for important research on everything from the transmission of cholera [621] to the behaviour of serial killers [381, 141] to the large-scale structure of the universe [445, 378].

Figure 1.2 shows the locations of larvae of the water strider *Limnoporus (Gerris) rufoscutellatus* (larval stage V) on the surface of a pond, observed in three different photographs. The research question is whether the larvae seem to be exhibiting territorial behaviour. If the larvae are territorial we would expect the spacing between points to be larger than if the points had been strewn completely at random. The human eye is not very good at judging this question: some kind of objective analysis is needed.

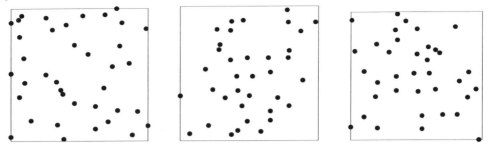

Figure 1.2. *Water strider larvae recorded from three photographs of the middle of a pond. Each frame is about 48 cm across. Data recorded by M. Nummelin, University of Helsinki, and kindly contributed by A. Penttinen.*

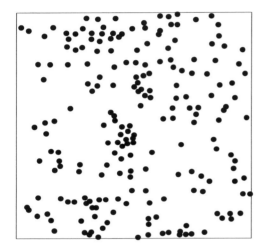

Figure 1.3. *Japanese black pine seedlings and saplings in a 10 metre square sampling region. Data recorded by M. Numata [503], and kindly supplied by Y. Ogata and M. Tanemura.*

Figure 1.3 depicts the locations of 204 Japanese black pine (*Pinus Thunbergii*) seedlings and saplings recorded in a 10×10 metre sampling region within a natural forest stand [503]. In mapping a snapshot of the forest, we hope to understand ecological processes, such as competition for resources (soil nutrients, light, water, growing space), and spatial variation in the landscape, such as variation in soil type or soil fertility.

A detailed analysis of these data [513, 516, 55] concluded that *both* these phenomena are present: there is spatial variation in the density of the forest, and also a tendency for trees to avoid growing close together, suggesting competition between neighbouring plants.

The spatial arrangement of points is often a surrogate for unobservable spatial variables (such as soil fertility or pollution exposure), or unrecorded historical events (such as territorial behaviour, forest succession, geological mineralisation history, or cosmological evolution). Conversely, spatial pattern affects other processes: the spatial pattern of individual organisms affects many aspects of an ecosystem.

There is no simple 'drag-and-drop' solution for statistical analysis of spatial point patterns, where we would simply instruct the computer to 'analyse' our data. It is a key principle of statistical methodology that the correct way to analyse data does not simply depend on the format of the data. It depends on how the data were obtained, and of course on the objectives of the analysis.

In particular we emphasise that 'point data' cannot necessarily be treated as a 'point pattern'. For example, measurements of soil acidity at a series of sampling locations in a field would not normally be treated as a point pattern: the sampling locations are artificial, and irrelevant to the study of soil acidity.

1.1.2   Points of several types

The points in a point pattern are often classified into different types. For example, trees may be classified into different species. Such a *multitype point pattern* introduces new scientific questions, and requires new kinds of statistical analysis, discussed in Chapter 14.

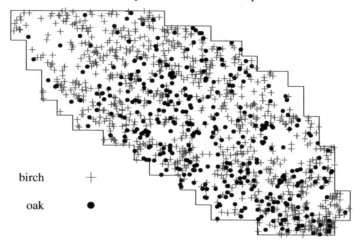

Figure 1.4. *Urkiola Woods data: locations of birch and oak trees in a secondary-growth forest in Urkiola Natural Park. Map about 220 metres across. Data collected by N.A. Laskurain, and kindly communicated by M. de la Cruz Rot.*

Figure 1.4 shows the locations of birch (*Betula celtiberica*) and oak (*Quercus robur*) trees in a secondary-growth forest in Urkiola Natural Park, Basque country, northern Spain. These are part of a more extensive dataset collected and analysed by Laskurain [404]. One important question is whether the two species have the same spatial distribution, or whether the relative proportions of the two species are spatially varying. In an extreme case a forest can be *segregated* into stands of different species, in the sense that there are regions where one species predominates over the other.

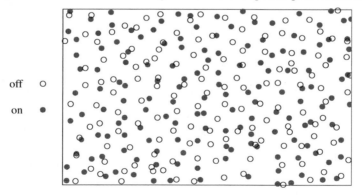

Figure 1.5. *Amacrine cells data: centres of 152 'on' cells and 142 'off' cells in a rectangular sampling frame about 1060 by 662 microns from the retina of a rabbit. Data from A. Hughes, and kindly supplied by P. Diggle.*

Figure 1.5 shows a point pattern of displaced amacrine cells in the retina of a rabbit, with cells categorised as either 'on' or 'off' according to their response to stimuli. Cells of the *same* type are regularly spaced apart: this is expected, because cells of the same type grow in the same layer or sheet of cells. A key research question is whether the two layers grew separately. This would be contradicted if cells of *opposite* type appeared to be correlated, that is, if the placement of the 'on' cells appeared to be affected by the location of the 'off' cells, or *vice versa* [223, 227].

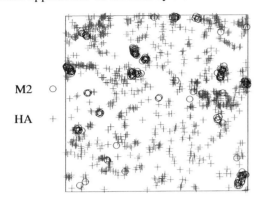

M2   ○

HA   +

Figure 1.6. *Locations of influenza virus proteins M2 and HA, mapped by immunogold labelling, on the surface of an infected cell. Field width is 3331 nanometres. Data kindly supplied by G.P. Leser and R.A. Lamb.*

Figure 1.6 shows the locations of influenza virus proteins on the surface of an infected cell [146]. The protein locations were mapped by immunogold labelling, that is, by growing antibodies to the specific proteins, attaching a gold particle to each antibody, and subsequently imaging the gold particles in electron microscopy. The research problem is to decide whether there is spatial association between the two proteins: this is important for the study of viral replication. While these data are superficially similar in structure to the amacrine cells data, the required analysis is completely different. Whereas the amacrine cells belong to two highly organised layers, it is appropriate to treat the individual influenza proteins (the individual circles and crosses in Figure 1.6) as individual, mobile, biochemical entities, each responding to its local environment.

Figure 1.7 shows the spatial locations of nests of two species of ants, *Messor wasmanni* and *Cataglyphis bicolor*, recorded by R.D. Harkness at a site in northern Greece, and described in [324]. The harvester ant *M. wasmanni* collects seeds for food and builds a nest composed mainly of seed husks. *C. bicolor* is a heat-tolerant desert foraging ant which eats dead insects and other arthropods. Interest focuses on whether there is evidence in the data for intra-species competition between *Messor* nests (i.e., competition for resources) and for preferential placement of *Cataglyphis* nests in the vicinity of *Messor* nests [359, 648], [594, sec. 5.3], [337, 293, 50], [52, sec. 11]. In this example, the role of the two species is not symmetric in the analysis, since one is potential food for the other.

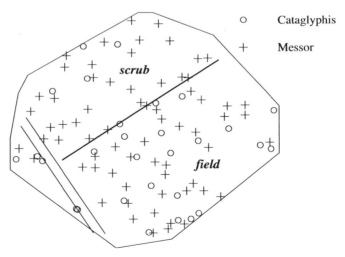

Figure 1.7. *Ants' nests of two species (*Messor wasmanni *and* Cataglyphis bicolor*) in a survey region about 425 feet (130 metres) across. Parallel lines show a walking track.*

1.1.3 Marked point patterns

The points in a point pattern may carry all kinds of attributes. A forest survey might record each tree's location, species, and diameter; a catalogue of stars may give their sky positions, masses, magnitudes, shapes, and colours; disease case locations may be linked to detailed clinical records. Auxiliary information attached to each point in the point pattern is called a **mark** and we speak of a **marked point pattern**.

The simplest example of a marked point pattern arises when the mark attached to each point is a single categorical value (such as a species label or disease status). Then we have a multitype point pattern as described above.

Figure 1.8 shows data from a forest survey in which the size of each tree was measured by its diameter at breast height (*dbh*). The locations and diameters of 584 Longleaf Pine (*Pinus palustris*) trees in a 200×200 metre region in southern Georgia (USA) were collected and analysed by Platt, Evans and Rathbun [551]. Each tree is represented in the Figure by a circle, centred at the tree location, with diameter proportional to *dbh*. Space-time models of forest succession have been used to account for the spatial pattern in this study [564]. In the upper right quarter of the survey there appears to be an area of smaller (and therefore probably younger) trees, suggesting that the forest may have been cleared in this area in previous decades.

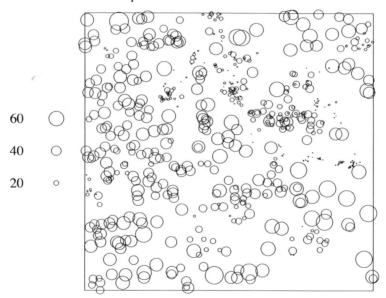

Figure 1.8. *Longleaf Pines data: tree locations and diameters in a 200 metre square survey region. Tree diameters are not to scale: legend shows diameters in centimetres.*

Figure 1.9 shows a longitudinal plane section through a filter made from bronze powder. The circles are the plane section profiles of bronze particles. The material was produced by sedimentation of bronze powder with varying grain diameter and subsequent sintering [91].

This example shows that physical objects, which are too large to be treated as ideal points at the scale of interest, can nevertheless be accommodated by recording the location of the centre of the object as a point, and treating the object's size and shape as attributes [519, chap. 9], [638].

In principle the mark attached to each point could be any type of information at all. It is often a multivariate observation: for example, in forest survey data, each tree could be marked by its species, its diameter, *and* a chemical analysis of its leaves. Each of these variables is called a *mark variable* and we speak of a *marked point pattern with multivariate marks* (Chapter 15).

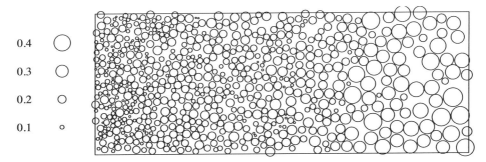

Figure 1.9. *Bronze particles in a filter, observed in longitudinal plane section,* 18×7 *mm. (R. Bernhardt, H. Wendrock; kindly contributed by U. Hahn.)*

1.1.4   Covariates

Our dataset may also include **covariates** — any data that we treat as explanatory, rather than as part of the 'response'.

One type of covariate is a *spatial function* $Z(u)$ defined at all spatial locations u. Figure 1.10 shows the locations of 3605 trees in a tropical rainforest, with a supplementary map of the terrain elevation (altitude) shown in the background. The covariate $Z(u)$ is the altitude at location u. Research questions for investigation include whether the forest density depends on the terrain, and whether, after accounting for the influence of terrain, there is evidence of spatial clustering of the trees. See Chapters 6, 9, and 12–14.

Analysis of forest survey data can involve several spatial covariates, giving information about terrain, hydrology (depth to water table, annual recharge flow), soil (soil type, acidity), understorey vegetation, and so on [684, 145].

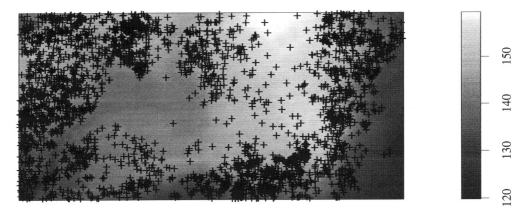

Figure 1.10. *Tropical rainforest data. Locations of* Beilschmiedia pendula *trees* (+) *and terrain elevation (greyscale) in a* 1000×500 *metre survey plot in Barro Colorado Island. Elevation is in metres above sea level. Part of a larger dataset containing the positions of hundreds of thousands of trees belonging to thousands of species [349, 167, 166].*

Another common type of covariate data is a *spatial pattern* such as another point pattern, a line segment pattern, or a collection of spatial features.   Figure 1.11 shows the result of an intensive geological survey of mineralisation in a 158×35 kilometre region in Queensland, Australia. It is a map of copper deposits (essentially pointlike at this scale) and geological lineaments (straight

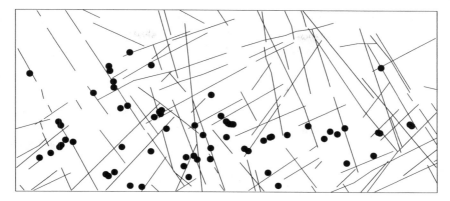

Figure 1.11. *Queensland copper data. Copper deposits (•) and geological lineaments (—) in a 158 × 35 km survey region. J. Huntington, CSIRO. Coordinates kindly provided by M. Berman and A. Green, CSIRO, Australia.*

lines). Lineaments are linear structures, mostly geological faults, which can easily be observed in aerial surveys. Copper deposits are hard to find, so the main question is whether the lineaments are 'predictive' for copper deposits, for example, whether copper is more likely to be found near lineaments. This kind of relationship is true for some minerals in some mineral provinces, typically because the mineralisation process involved liquids carrying the target mineral in solution which flowed from the deep earth to the surface through openings in the rock. Common practice is to define $Z(u)$ to be the distance from location u to the *nearest* lineament, and to use the function Z as the spatial covariate in statistical analysis [4, 5, 50, 55, 88, 89, 90, 139, 140, 269], [52, sec. 12].

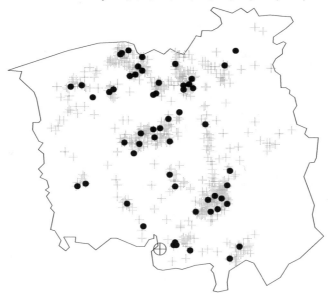

Figure 1.12. *Chorley-Ribble data. Spatial locations of cases of cancer of the larynx (•) and cancer of the lung (+) and a disused industrial incinerator (⊕). Survey area about 25 kilometres across.*

Figure 1.12 shows the Chorley-Ribble cancer data of Diggle [224] giving the residential locations of new cases of cancer of the larynx (58 cases) and cancer of the lung (978 cases) in the Chorley and South Ribble Health Authority of Lancashire, England, between 1974 and 1983. The location of a disused industrial incinerator is also given. The aim is to assess evidence for an increase in the incidence of laryngeal cancer close to the incinerator. The lung cancer cases serve as a surrogate for the spatially varying density of the susceptible population. Data analysis in [224, 236, 55, 331] concluded there is significant evidence of an incinerator effect.

1.1.5   Different spaces

In this book the points are usually locations in two-dimensional space, but they could also be locations in one dimension (such as road accidents recorded on a road network) or in three dimensions (such as cells observed in 3D microscopy) or in space-time (such as earthquake epicentre locations and times).

Figure 1.13. *Chicago crimes data. Crimes recorded over a two-week period in the vicinity of the University of Chicago. Survey area 1280 feet (390 metres) across. Extracted manually from a graphic published in the* Chicago Weekly News, *2002.*

Figure 1.13 shows the street address locations of crimes in the area of the University of Chicago over a two-week period, from a graphic displayed in the university newspaper, *Chicago Weekly News.* For such data it is clearly not appropriate to use statistical techniques designed for point patterns in two-dimensional space. The analysis needs to take into account the geometry of the street network; see Chapter 17.

1.1.6   Replicated patterns

The water striders data (Figure 1.2) were obtained from photographs taken at three different times and places. They can be regarded as independent repetitions or *replicates* of the same experiment. Replication plays an important role in classical statistics, because it enables us to separate different sources of variability in the experiment. In the case of the water striders, replication greatly strengthens the evidence for territorial behaviour [539].

Most of the current techniques for analysing spatial point patterns were developed, out of necessity, for dealing with a *single* observed point pattern in each study. When replicated point patterns are available, the power and scope of statistical techniques is vastly increased: see Chapter 16.

The point pattern of influenza virus proteins shown in Figure 1.6 is one of 41 point patterns obtained as the 'responses' in a designed experiment [146] involving two types of influenza virus, two choices of protein staining, and replicate observations.

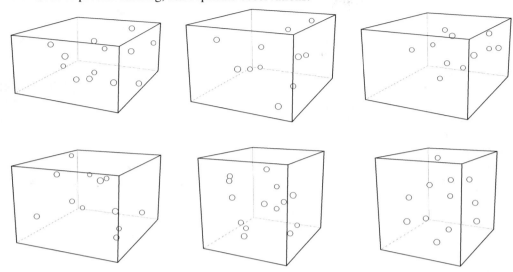

Figure 1.14. *Osteocyte lacunae data (subset of full dataset) from [66, 68]. Dimensions* $81 \times 100 \times d$ *μm (microns) where height d varies.*

Figure 1.14 shows replicated three-dimensional point patterns representing the positions of *osteocyte lacunae*, holes in bone which were occupied by osteocytes (bone-building cells) during life. Observations were made in several parts of four different skulls of Macaque monkeys using a confocal optical microscope [68].

1.2 Statistical methodology for point patterns

1.2.1 Summary statistics

A time-honoured approach to spatial point pattern data is to calculate a *summary statistic* that is intended to capture an important feature of the pattern.

For the water striders data, the feature of interest is the spacing between the water strider larvae. An appropriate summary statistic is the average distance from a larva to its *nearest neighbour*, the nearest other larva in the same pattern. This is appropriate because if the larvae are territorial we can expect them to try to increase the nearest-neighbour distance. The average nearest-neighbour distance is a numerical measure of the typical spacing between larvae.

For the three patterns in Figure 1.2 the average nearest-neighbour distances are 55, 49 and 54 *mm*, respectively. To interpret these values, we need a benchmark or reference value. Since our goal is to decide whether the water strider larvae are territorial or not, a suitable benchmark is the average nearest-neighbour distance that would be expected if the water striders are *not* territorial, that is, if the points were placed completely at random. This benchmark is slightly different for the three patterns in Figure 1.2, because they are slightly different in the number of points and the size of frame. A simple solution is to normalise the values, dividing the observed distance by the benchmark distance for each pattern. This ratio, called the *Clark-Evans index* [155], should be about 1 if the larvae are completely random, and greater than 1 if the larvae are territorial. The

Clark-Evans index[1] values for the water strider patterns are 1.304, 1.126 and 1.285, respectively, suggesting that the larvae are territorial. See Chapter 8 for further discussion.

A summary statistic can be useful if it is appropriate to the application, and is defined in a simple way, so that its values can easily be interpreted. However, by reducing a spatial point pattern to a single number, we discard a lot of information. This may weaken the evidence, to the point where it is impossible to exclude other explanations. For example, the Clark-Evans index is very sensitive to spatial inhomogeneity: index values greater than 1 can also be obtained if the points are scattered independently but unevenly over the study region. The analysis above does not necessarily support the conclusion that the water strider larvae are territorial, until we eliminate the possibility that the water striders have a preference for one side of the pond over another.

Summary *functions* are often used instead of numerical summaries. Figure 1.15 shows the estimated *pair correlation functions* for the three water strider patterns in Figure 1.2. For each value of distance r, the pair correlation $g(r)$ is the observed number of pairs of points in the pattern that are about r units apart, divided by the expected number that would be obtained if the points were completely random. Pairs of water striders separated by a distance of 3 centimetres (say) are much less common than would be expected if their spatial arrangement was completely random. See Chapter 7.

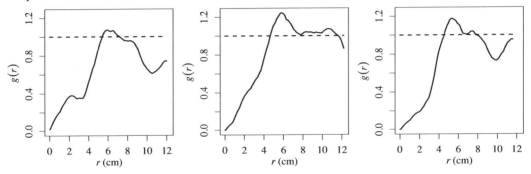

Figure 1.15. *Measuring correlation between points. Estimates of the pair correlation function (solid lines) for the three water strider patterns in Figure 1.2. Dashed horizontal line is the expected value if the patterns are completely random.*

To decide whether the deviations in Figure 1.15 are statistically significant, a standard technique is to generate synthetic point patterns which are completely random, compute the pair correlation function estimates for these synthetic patterns, and plot the envelopes of these functions (minimum and maximum values of pair correlation for each distance r). See Figure 1.16. This can be interpreted as a statistical test of significance, *with care* (see Chapter 10).

Often the main goal is to detect and quantify trends in the density of points. The enterochromaffin-like cells in gastric mucosa (left panel of Figure 1.1) clearly become less dense as we move towards the interior of the stomach (towards the top of the picture). Figure 1.17 shows two ways of quantifying this trend. In the left panel, the spatial region has been divided into equal squares, and the number of points falling in each square has been counted and plotted. The numbers trend downwards as we move upwards. The right panel shows an estimate of the spatially varying density of points using kernel smoothing.

1.2.2 Statistical modelling and inference

Summary statistics work best in simple situations. In more complex situations it becomes difficult to adjust the summary statistic to 'control' for the effects of other variables. For example,

---

[1] Using Donnelly's edge correction [244].

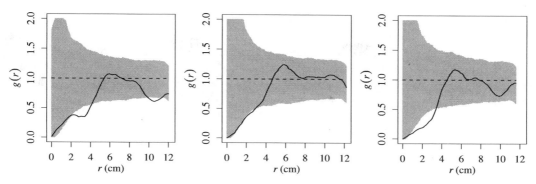

Figure 1.16. *Assessing statistical significance of correlation between points. Estimates of the pair correlation function (solid lines) for the three water strider patterns. Dashed horizontal line is the expected value if the patterns are completely random. Grey shading shows pointwise 5% significance bands.*

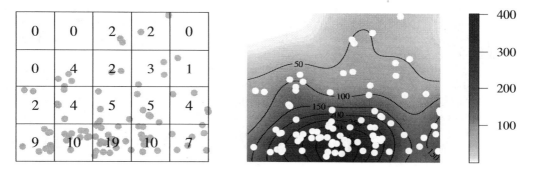

Figure 1.17. *Measuring spatial trend. Enterochromaffin-like cells in gastric mucosa (Figure 1.1) showing* (Left) *counts of points in each square of side length 0.2 units,* (Right) *kernel-smoothed density of points per unit area.*

the enterochromaffin-like cells (Figure 1.1) have spatially varying density. The Clark-Evans index cannot be used, unless we can think of a way of taking account of this spatial trend.

Analysing data using a summary statistic or summary function is ultimately unsatisfactory. If, for example, the conclusion from analysis of the water striders data is that there is insufficient evidence of territorial behaviour, then we have the lingering doubt that we may have discarded precious evidence by reducing the data to a single number. On the other hand if the conclusion is that there *is* evidence of territorial behaviour, then we have the lingering doubt that the summary statistic may have been 'fooled' by some other aspect of the data, such as the non-uniform density of points.

A more defensible approach to data analysis is to build a *statistical model*. This is a comprehensive description of the dataset, describing not only the averages, trends, and systematic relationships in the data, but also the variability of the data. It contains all the information necessary to *simulate* the data, i.e., to create computer-generated random outcomes of the model that should be similar to the observed data. For example, when we draw a straight line through a cloud of data points, a *regression model* tells us not only the position of the straight line, but also the scatter of the data points around this line. Given a regression model we can generate a new cloud of points scattered about the same line in the same way.

Statistical modelling is the best way to investigate relationships while taking account of other relationships, and while accounting for variability. By building a statistical model which includes all the variables that influence the data, we are able to *account for* (rather than 'adjust for') the effects of extraneous variables, and draw sound conclusions about the questions of interest.

A statistical model usually involves some *parameters* which control the strength of the relationships, the scale of variability, and so on. For example, a simple linear regression model says that the response variable y is related to the explanatory variable x by $y = \alpha + \beta x + e$ where β is the slope of the line, α is the intercept, and e is a random error with standard deviation σ. The numbers α, β, σ are the parameters of the regression model.

Fitting a model to data means selecting appropriate values of the model parameters so that the model is a good description of the data. For example, fitting a linear regression model means finding the 'line of best fit' (choosing the best values for the intercept α and slope β of the line) and also finding the 'standard deviation of best fit' (choosing the best value of σ to describe the scatter of data points around the line). The best-fit estimate of the model parameters usually turns out to be a sensible summary statistic in its own right.

Statistical modelling may seem like a very complex enterprise. Our correspondents often say *"I'm not interested in modelling my data; I only want to analyse it."* However, any kind of data analysis or data manipulation is *equivalent* to imposing assumptions. In taking the average of some numbers, we implicitly assume that the numbers all come from the same population. If we conclude that something is 'statistically significant', we have implicitly assumed a model, because the p-value is a probability according to a model.

The purpose of statistical modelling is to make these implicit assumptions explicit. By doing so, we are able to determine the best and most powerful way to analyse data, we can subject the assumptions to criticism, and we are more aware of the potential pitfalls of analysis. If we "only want to do data analysis" without statistical models, our results will be less informative and more vulnerable to critique.

Using a model is not a scientific weakness: it is a strength. In statistical usage, a model is always tentative; it is assumed for the sake of argument. In the famous words of George Box: "All models are wrong, but some are useful." We might even *want* a model to be wrong, that is, we might propose the model in order to refute it, by demonstrating that the data are not consistent with that model.

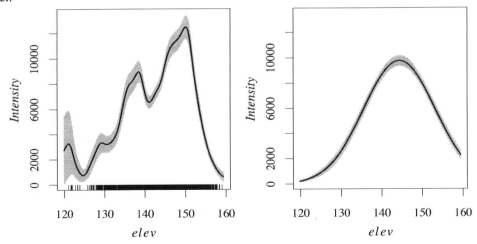

Figure 1.18. *Modelling dependence of spatial trend on a covariate.* Beilschmiedia *trees data (Figure 1.10). Estimated mean density of points assuming it is a function of terrain elevation.* Left: *nonparametric estimate assuming smooth function.* Right: *parametric estimate assuming log-quadratic function. Grey shading indicates pointwise 95% confidence intervals.*

It is often instructive to compare results obtained with different models. Figure 1.18 shows two different estimates of the density (in trees per square kilometre) of *Beilschmiedia* trees as a function of terrain elevation, based on the tropical rainforest data of Figure 1.10. The estimates are derived from two different statistical models, which assume that the *Beilschmiedia* tree locations are random, but may have a preference for higher or lower altitudes. The left panel is a nonparametric estimate, based on the assumption that this habitat preference is a smoothly varying function of terrain elevation. The right panel is a parametric estimate, based on a very specific model where the logarithm of forest density is a quadratic function of terrain elevation.

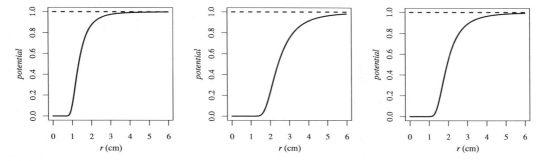

Figure 1.19. *Modelling dependence between points. Fitted 'soft core' interaction potentials for water striders.*

Figure 1.19 shows one of the results of fitting a *Gibbs model* to the water striders data. Such models were first developed in physics to explain the behaviour of gases. The points represent gas molecules; between each pair of molecules there is a force of repulsion, depending on the distance between them. In our analysis the parameters controlling the strength and scale of the repulsion force have been estimated from data. Each panel of Figure 1.19 shows the interaction probability factor $c(r) = e^{-U(r)}$ for a pair of points separated by a distance r, where $U(r)$ is the potential (total work required to push two points together from infinite distance to distance r). A value of $c(r) \approx 0$ means it is effectively forbidden for two points to be as close as r, while a value of $c(r) \approx 1$ indicates that points separated by the distance r are 'indifferent' to each other. These graphs suggest (qualitatively) that the water striders do exhibit territorial behaviour (and the model is quite suitable for this application). Figure 1.20 shows simulated realisations of the Gibbs model for the water striders data. Gibbs models are explained in Chapter 13.

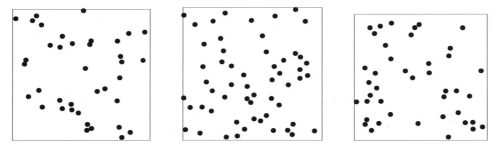

Figure 1.20. *Simulated realisations of the fitted soft core model for the water striders.*

1.2.3 Validation

The main concern with modelling is, of course, that the model could be wrong. Techniques for *validating* a statistical model make it possible to decide whether the fitted model is a good fit overall, to criticise each assumption of the model in turn, to understand the weaknesses of the analysis, and to detect anomalous data. Statistical modelling is a cyclic process in which tentative models are fitted to the data, subjected to criticism, and used to suggest improvements to the model.

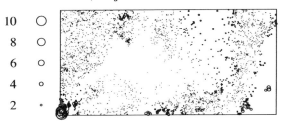 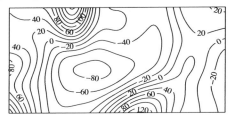

Figure 1.21. *Validation of trend model for* Beilschmiedia *trees.* Left: *influence of each point* (×1000). Right: *smoothed Pearson residual field.*

Figure 1.21 shows two tools for model validation, applied to the model in which *Beilschmiedia* density is a log-quadratic function of terrain elevation. The left panel shows the *influence* of each data point, a value measuring how much the fitted model would change if the point were deleted from the dataset. Circle diameter is proportional to the influence of the point. There is a cluster of relatively large circles at the bottom left of the plot, indicating that these data points have a disproportionate effect on the fitted model. Either these data points are anomalies, or the assumed model is not appropriate for the data.

The right panel of Figure 1.21 is a contour plot of the smoothed *residuals*. Analogous to the residuals from a regression model, the residuals from a point process model are the difference between observed and expected outcomes. The residuals are zero if the model fits perfectly. Substantial deviations from zero suggest that this model does not fit well, and indicate the locations where it does not fit.

1.3 About this book

Methods

This book describes modern statistical methodology and software for analysing spatial point patterns. It is aimed primarily at scientists, across a broad range of disciplines, who need to analyse their own point pattern data. It provides explanations and practical advice on data analysis for researchers who are results-oriented, together with guidance about the validity and applicability of different methods. Examples and code in the book, and online supplementary code, should make it easy for readers to begin analysing their own data.

The reader is assumed to have a general scientific training, including basic statistical methods. Using the book does not require advanced mathematical training, but mathematical formulae are supplied for completeness. The book is not biased toward any particular area of science, and aims to cover a wide range of techniques.

This book aims to explain the core principles of statistical methodology for spatial point patterns. 'Methodology' is more than just a collection of tools: it is a systematic, principled, coherent approach to analysing data. These principles guide the appropriate choice of tool for each situation, and guide the correct interpretation of the results of analysis. By explaining the core principles we also hope to make the methodology more comprehensible.

There has been a revolution in statistical methodology for spatial point patterns over the last decade. Classical tools popular since the 1980's, such as Ripley's K-function, have been extended and supplemented by new techniques. The classical *ad hoc* approach to analysing point patterns has been replaced by a systematic methodology which is much closer to mainstream statistical science [484, 277]. Using the new methodology, scientists can analyse their data in a more powerful and defensible way, while still relying on the same basic tools. This new methodology urgently needs to be communicated to the wider scientific community.

Rather than following the traditional structure of books on analysing spatial point patterns, this book is *structured around standard statistical concepts* as much as possible. This is both a peda-gogical decision (to make the book more easily comprehensible to people who are trained in basic statistical concepts) and a scientific goal (to bring spatial point process analysis closer to the statistical mainstream).

As statisticians, one of our greatest concerns is that statistical methods should be applied cor-rectly, and that the results should be interpreted correctly. This book draws attention to common errors and misunderstandings. Examples include confusion over the statistical significance of simu-lation envelopes, misunderstandings of the assumptions inherent in spatial logistic regression, inap-propriate use of techniques, and overemphasis on pseudoformal methods such as hypothesis testing. We hope the book will serve as an authoritative source of information about statistical methodology for spatial point patterns.

Software

All software featured in the book is free, open-source code written in the R language, and is eas-ily accessible online from the Comprehensive R Archive Network `cran.r-project.org` in the form of 'contributed packages'. The power of R makes it possible to write a new kind of statistics book: strongly governed by statistical principles, covering statistically advanced techniques, and yet focused on applications.

The main software package used in the book is `spatstat`, a contributed package for R written by the authors [51, 29]. Some other R packages are also mentioned. To the best of our knowledge, `spatstat` is one of the largest software packages available for spatial point pattern analysis, in any language. The result of 20 years' development by experts in spatial statistics, `spatstat` now has a wide range of capabilities (supported by more than 2000 functions). There is ample documentation in the help files installed with `spatstat`. This book complements the `spatstat` help files by focusing on the principles and applications of the techniques.

Most of this book is concerned with point patterns in two-dimensional (2D) space. The `spatstat` package was originally implemented for 2D point patterns. However, it is being extended progres-sively to 3D, space-time, and multidimensional space-time point patterns, and some material on these topics is included.

Contents of the book

Part I of the book gives an introduction to R software including `spatstat`, advice about collecting data, information about handling and manipulating data, and an accessible introduction to the basic concepts of point processes (Chapter 5).

Part II presents tools for Exploratory Data Analysis, including nonparametric estimation of in-tensity, correlation, and spacing. Intensity (Chapter 6) is the density of points per unit area, and can be used to study the dependence of the point pattern on an explanatory variable. Well-known mea-sures of correlation (Chapter 7) include Ripley's K-function, the pair correlation function, and the Fry plot; they can be used to detect clustering or regularity in a point pattern. Measures of spacing

(Chapter 8) are based mainly on the distance from a point to its nearest neighbour, or second-nearest neighbour, etc.

Part III discusses model-fitting and statistical inference for point patterns. The spatstat package has extensive facilities for fitting point process models to a point pattern dataset, for computing predictions from the model, and for validating the fitted model. The most fundamental models are Poisson point processes (Chapter 9). Fitting a Poisson model is equivalent to other popular techniques such as logistic regression and maximum entropy. Statistical hypothesis tests (Chapter 10) include Monte Carlo tests, envelope tests, goodness-of-fit tests, and analysis of deviance ('anova') for fitted models. Powerful diagnostics for validating a fitted model have recently been developed (Chapter 11). For point patterns which exhibit dependence between the points, Cox process and cluster process models (Chapter 12) and Gibbs point process models (Chapter 13) can be fitted. All the techniques described above can be extended to multitype point patterns (Chapter 14).

Part IV discusses point patterns with additional 'structure', such as complicated marks, space-time observations, three-dimensional and higher-dimensional spaces (Chapter 15), replicated observations (Chapter 16), and point patterns constrained to a network of lines (Chapter 17). Here the methodology and software is less developed and the coverage is briefer. Many interesting research challenges remain to be explored.

2

Software Essentials

This chapter introduces the R system (Section 2.1), additional packages for R (Section 2.2), and the `spatstat` package for analysing spatial point patterns (Section 2.3).

2.1 Introduction to R

The analyses of spatial point pattern data that we present in this book are all conducted in the R software system for statistical computing and graphics [353].

There are myriad reasons for choosing R, as an article in the *New York Times* explained [666]. It is free software, easy to install and to run on virtually all platforms. Basic commands are easy to learn. It is a powerful computing system, with all the advantages of deep-level computer science, but does not require the user to know a lot of computer science. The basic 'out-of-the-box' R system is already highly capable, and is easy to extend, partly thanks to its programming language, also called R. There are thousands of extension packages available, many of extremely high quality, produced by the very large developer/user community.

2.1.1 How to obtain and install R

R is free software with an open-source licence. It can be downloaded from `r-project.org`. For OS X® and Windows® you will most likely wish to download a 'binary' version of R, ready to install and use. The installation on your personal computer is easy and is accomplished by pointing at and clicking on the right icons. Many users of Linux operating systems will probably also wish to install a pre-compiled binary version of R. What needs to be done depends on the flavour of Linux: a web search for `install R for xxx` (where xxx represents your flavour of Linux) will set you on your way.

More experienced Linux users may want to download the 'source' which is bundled up in a file called `R-version.tar.gz` where *version* is the version number, something like `3.0.3`. Such source has to be compiled and installed. This is reasonably simple if you have the necessary compilers and attendant 'tools' on your computer.

The Documentation page on the R website lists many books, manuals, and online tutorials to help you learn to use R. We strongly recommend the manual *An Introduction to R*. Useful reference cards for R commands can be found on the webpage of Contributed Documentation.

R software is updated regularly, usually four times a year. The updates usually provide increased speed, new facilities, and bug fixes. It is advisable to 'stay current', updating the version of R when a new release appears. The `spatstat` package (on which much of this book is based) is also updated frequently, about once every two months, and again it pays to stay current.

For users who are new to R, we highly recommend installing the friendly user interface `RStudio`. It is free and open source software, and can be downloaded from `www.rstudio.com`.

2.1.2   **Running** R

Once R has been installed on a computer, the user can run it to start an *interactive session* in which the user will give commands and R will immediately execute them and report the results. How you start the session depends on the computer's operating system and on your preferences. For new users of R, the simplest advice is to double-click the R icon if it is available, and otherwise (for example on a Linux system) to type the single-letter command R into a system command line ('shell' or 'terminal') interface.

Basic installations of R on OS X and Windows have a simple point-and-click Graphical User Interface (GUI) which supports many tasks. Other GUIs are available as add-on packages, including JGR, RKWard, and Rcmdr ('R commander'). The contributed package rpanel enables users to build their own GUI control panels.

Many users prefer an 'integrated development environment' (IDE) which handles the editing, testing, and running of R code. The most popular of these are the highly recommended RStudio, and Tinn-R (for Windows only) and ESS (for the Emacs editor).

A sequence of R commands can be saved in a text file (created using a text editor or an IDE) and subsequently executed. In an interactive session, the user would type source("*filename*") to instruct R to execute the script.

R can also be run in a *non-interactive session*, when R is given a script of commands to execute, without interacting with the user. There are at least four ways of starting a non-interactive session, shown in Table 2.1. The manual **An Introduction to R** recommends the R CMD BATCH version.

```
R CMD BATCH filename
Rscript filename
R < filename
R -f filename
```

Table 2.1. *System commands (Mac™ or Windows command line, Linux shell commands) which will start a non-interactive session in* R, *where* filename *is the name of the text file containing* R *command input.*

An R session (interactive or non-interactive) starts in a particular folder ('directory') in the file system, and stores temporary data in that folder. At the end of a session, there is an option to save all the data and the complete history of the session in a *workspace* in the same folder. A new R session, if started in the same folder, 're-loads' the saved workspace and can thereby access the saved data and the session history, effectively taking up where the previous session left off. Use getwd to find out the current working directory, and setwd to change it.

In 'literate programming' [387] or 'dynamic report generation', the user writes a single document file containing a mixture of explanatory text and R commands. The file is processed automatically by executing the R commands, inserting the resulting output and graphics into the document file, and then typesetting the document. We strongly recommend this approach for research projects. The contributed R packages Sweave and knitr support dynamic report generation. This entire book was written as an Sweave document.

2.1.3   **How** R **commands are shown in this book**

In an R session, users can issue commands either by using a graphical point-and-click interface (if they have started their R session in such a way that such an interface is provided) or by typing commands line-by-line to a command interpreter. Typed commands are written in the R language [353]. Simple commands are very easy to learn, and users can gradually extend their skill to exploit progressively more of the powerful features of the language.

When it is waiting for a command, the R interpreter displays this prompt:

```
>
```

R can be used as a glorified calculator, and a typical series of R commands looks like this:

```
> 1+1
> 3 * (10 + 5)
> 2:5
> sqrt(2)
```

Note that you are not meant to type the > symbol; this is just the prompt for command input in R. To type the first command, just type 1+1.

In this book we will sometimes also print the response that R gives to a set of commands. In the example above, it would look like this:

```
> 1+1
[1] 2
> 3 * (10 + 5)
[1] 45
> 2:5
[1] 2 3 4 5
> sqrt(2)
[1] 1.414214
```

The annotation [1] indicates the first value in the output. This feature is useful when the output contains many values:

```
> (1:30)^2
 [1]   1   4   9  16  25  36  49  64  81 100 121 144 169 196 225 256 289 324
[19] 361 400 441 484 529 576 625 676 729 784 841 900
```

Data can be saved and recalled by 'assigning' the data to a name, using the assignment operator <-

```
> x <- 17
> x
[1] 17
```

In an interactive session, R will assume that your command is finished at the end of the line (i.e. when you press ENTER or RETURN) unless the command is obviously unfinished — for example if there is a parenthesis which has not yet been closed, or if the line ended with an arithmetic operator. In this case, R will print the *continuation character* '+' to indicate that it is expecting the previous command to be continued:

```
> folderol <- 1.2
> sin(folderol * folderol * folderol  * folderol  *
+     folderol  * folderol * folderol  * folderol * folderol)
[1] -0.9015787
```

For easier reading in the printed book, we have suppressed this behaviour by changing the continuation character to a blank space. This also makes it easier for users to copy-and-paste code directly from the e-book to the R command line.

Note that all R commands printed in this book that include the initial prompt '>' are capable of being executed in an R session. Commands shown without the initial prompt usually show the syntax or usage of a function but may not be immediately executable.

2.1.4   Getting help about R commands

For information about a specific function in R, the best resource is the installed help documentation. R has documentation for *every* function that is accessible to the user. The documentation specifies the arguments of the function, explains what they mean, and gives examples of their use. If you know the name of the function, then simply typing help(*functionname*) or ?*functionname* will display the help. For example, help(glm) or ?glm displays the help for the function glm, and help("[") shows the help for the subset operator.

If you are using the R command line interpreter, pressing the TAB key will show you the possible completions of a partially finished command. Typing gl and then TAB will give a list of all known functions and objects that begin with gl. Typing glm(with an open parenthesis, and then TAB, will list all the *arguments* of the function glm. Command-line completion is not available in some installations of R.

A web browser interface for the help documentation can be started by typing help.start(). This also gives access to online manuals, including **An Introduction to R**, and to a search engine for the help files (Search Engine and Keywords). The search engine can also be accessed directly by the R function help.search.

Other resources for searching include the RSiteSearch function in the base R system, and the function findFn from the contributed package sos (see Section 2.2.2 for details on how to install and use contributed packages).

Other sources of information include online manuals, books, tutorials, and the R-help mailing list. Suggestions for books may be found on the R website (look under Books). Under Manuals on the R web site and then contributed documentation you will find pointers to many freely available online books and tutorials. Under FAQs you can find answers to many questions about R.

It always helps to have a pretty good idea of just what you are looking for, and to have at least an inkling as to what the function you want might be called. If you are feeling lost, it is best to go to the manuals, books, and tutorials. If you go to **An Introduction to R** from the help.start() page, and use the web browser's Edit and then Find facility, and search for 'linear models' you will be taken immediately to a section called 'Linear models' giving ample information.

In order to make use of the R-help mailing list it is necessary to subscribe to this list before posting a question. See Mailing Lists on the R web site. *Please* read and follow the R-help posting guide, which explains what information you need to include when posting a question.

2.1.5   Vectors, lists, and indices

The power of R begins with the fact that any dataset can be stored and handled as a single entity, and referred to by a name which we choose, like x or MyTax2014.

If our data are just a sequence of numbers, they can be stored as a numeric *vector*.

```
> x <- c(3.2, 1.5, 3.6, 9.2, 2.4)
> x
[1] 3.2 1.5 3.6 9.2 2.4
> y <- 1:5
> y
[1] 1 2 3 4 5
```

Here c stands for 'concatenate' and simply collects all the numbers together into a vector. Many other ways to input data are explained in the standard introductions to R.

Standard 'calculator' commands can be applied to an entire vector of numbers at once (called a *vectorised* calculation):

```
> x + 5
```

```
[1]   8.2   6.5   8.6 14.2  7.4
> x/y
[1] 3.20 0.75 1.20 2.30 0.48
```

There are also vectors of character strings, vectors of logical values, and so on:

```
> month.name
 [1] "January"   "February"  "March"     "April"    "May"       "June"
 [7] "July"      "August"    "September" "October"  "November"  "December"
> x > 3
[1]  TRUE FALSE  TRUE  TRUE FALSE
```

To extract elements of a vector, use the subset operator [ ]. For example x[2] extracts the second element of x:

```
> x[2]
[1] 1.5
> month.name[8]
[1] "August"
```

If x is a vector, then x[s] extracts an element or subset of x. The subset index s can be:

- a positive integer: x[3] means the third element of x;

- a vector of positive integers indicating which elements to extract: x[c(2,4,6)] extracts the 2nd, 4th, and 6th elements of x;

- a vector of negative integers indicating which elements *not* to extract: x[-1] means all elements of x except the first one;

- a vector of logical values, of the same length as x, with each TRUE entry of s indicating that the corresponding entry of x should be extracted, and FALSE indicating that it should not be extracted. For example if x is numeric, then x[x > 3.1] extracts those elements of x which are greater than 3.1.

- a vector of logical values, of *shorter* length than x. The entries of s will be 'recycled' to obtain a logical vector of the same length as x, and the corresponding subset will be extracted. For example x[c(FALSE, TRUE)] extracts every second element of x.

- a vector of names: x[c("irving","melvin","clyde")] extracts those entries of x having the names "irving", "melvin" and "clyde", respectively, if there are any such entries, and yields a missing value NA otherwise.

It is helpful to remember that the R language does not recognise 'scalars'; everything is a vector. When we extract a single element from a vector x by typing x[3], the result is still a vector, which happens to have length 1.

In the examples above, the entries in the vector were *atomic* values — numbers, logical values, or character strings. A vector containing more complicated data types, or containing a mixture of different types of data, is called a *list*.

```
> b <- list(c("Tuesday", "Friday"), 3:7, c(FALSE,TRUE))
> b
```

```
[[1]]
[1] "Tuesday" "Friday"

[[2]]
[1] 3 4 5 6 7

[[3]]
[1] FALSE  TRUE
```

The key fact is that lists *are vectors*. Consequently all of the extraction procedures described above are applicable when x is a list. The subset operator [ ] applied to a list will always return a list (because a subset of a list is another list). If for example x is a list (having length at least 6) then x[c(2,4,6)] is a list of length 3, i.e. having 3 components. Notice in particular that if the length of the index vector s is 1, then what we get from x[s] is a *list* of length 1. For example, b[2] will return the list, of length 1, whose first (and only) entry is the vector 3:7.

If you want to get at the *object* constituting the second component of the list b then you need to use *double* brackets: b[[2]].

```
> b[2]
[[1]]
[1] 3 4 5 6 7
> b[[2]]
[1] 3 4 5 6 7
```

A list is like a carton of eggs: the difference between b[2] and b[[2]] is the difference between cutting down the egg carton so that it only holds one egg, and extracting the egg itself.

Like any vector, a list may have a sequence of *names* to distinguish its elements. Names may be attached when the list is created by giving the elements in name=value form:

```
> b <- list(Day="Tuesday", Number=3:7, Answers=c(FALSE,TRUE))
> names(b)
[1] "Day"     "Number"  "Answers"
> b
$Day
[1] "Tuesday"

$Number
[1] 3 4 5 6 7

$Answers
[1] FALSE  TRUE
```

Alternatively the names may be attached or changed afterward:

```
> names(b) <- c("Day", "Number", "Answers")
```

As for any vector, the subset operator can extract elements of a list by name:

```
> b[c("Day", "Answers")]
$Day
[1] "Tuesday"

$Answers
[1] FALSE  TRUE
```

```
> b[["Day"]]
[1] "Tuesday"
```

The operator $ can also be used to extract a *single* named entry, without needing the quotation marks:

```
> b$Day
[1] "Tuesday"
```

Note that the $ operator does 'partial argument matching', e.g., one could use b$D to extract the first component of b. (Using b[["D"]] would return NULL.) Partial argument matching saves keystrokes, but involves risks. If the list b also had a component named D then b$D would extract *that* component and *not* the component named Day. Unless you are very sure about the names of your list you would be well advised to type the complete name of the desired component.

2.1.6   Matrices and data frames

A matrix is a two-dimensional array of values indexed by row and column number: m[i,j] is the entry on row i and column j.

```
> m <- matrix(1:12, nrow=3, ncol=4)
> m
     [,1] [,2] [,3] [,4]
[1,]    1    4    7   10
[2,]    2    5    8   11
[3,]    3    6    9   12
> m[2,3]
[1] 8
```

A matrix contains atomic values (numbers, character strings, logical values) which are all of the same type. Notice that, when the vector $1:12$ was converted to the matrix m, the sequence of values was copied *column-by-column*. This is the ordering in which matrix data are stored internally in R. Whole rows or columns can easily be extracted:

```
> m[2,]
[1]  2  5  8 11
> m[,3]
[1] 7 8 9
```

Note the presence of the comma. If a single subset index is used, it will be interpreted using the column-by-column ordering:

```
> m[5]
[1] 5
```

To copy the values in row-by-row order, set byrow=TRUE:

```
> matrix(1:12, nrow=3, ncol=4, byrow=TRUE)
     [,1] [,2] [,3] [,4]
[1,]    1    2    3    4
[2,]    5    6    7    8
[3,]    9   10   11   12
```

A very important data type in R is a *data frame*. This is a two-dimensional array like a matrix, except that different columns may contain different types of data.

```
> d <- data.frame(i=7:10, month=month.name[7:10],
                  has.r=((1:4) > 2))
> d
   i      month has.r
1  7       July FALSE
2  8     August FALSE
3  9  September  TRUE
4 10    October  TRUE
```

We think of the different columns as containing different variables, while the rows correspond to different experimental subjects or experimental units on which these variables have been measured. The columns have *names*, which can be assigned when the data frame is created as above, and extracted or changed using `colnames`:

```
> colnames(d) <- c("month", "name", "has.r")
> d
  month      name has.r
1     7      July FALSE
2     8    August FALSE
3     9 September  TRUE
4    10   October  TRUE
```

A data frame can be indexed using two indices, in the same way as a matrix:

```
> d[2,3]
[1] FALSE
> d[,3]
[1] FALSE FALSE  TRUE  TRUE
```

However, because of the heterogeneous structure of the data, a data frame is actually stored as a `list`, in which the successive entries are successive columns of the data frame. We can also extract *columns* by using the subset operators for a list:

```
> d[[3]]
[1] FALSE FALSE  TRUE  TRUE
> d[c("month","name")]
  month      name
1     7      July
2     8    August
3     9 September
4    10   October
> d$month
[1]  7  8  9 10
```

A data frame can be viewed and edited in a spreadsheet-style interface using the R functions `View` and `edit`, for example `View(d)` or `dnew <- edit(d)`.

2.1.7 Objects, classes, and methods in R

The R system can accept vague instructions, like `plot(x)`, and take sensible action, appropriate to the type of data stored in `x`. But how is this possible? Essentially, datasets are tagged as belonging to different *classes* of data, and there are separate *methods* for plotting each class. We need to know a bit about how this works.

In computer science jargon, R is an 'object-oriented' language. All individual entities in R are *objects*. A vector, matrix, or data frame is an object. Datasets are stored as objects. A dataset with some kind of structure on it (e.g., a contingency table, a time series, a point pattern) is treated as a single object.

The famous `sunspots` dataset, installed with R, is a good example. It is a time series, giving the number of sunspots that were observed each month from January 1749 to December 1983. In addition to the counts themselves, the dataset `sunspots` includes information such as the start and finish times. This dataset is treated as a single entity — an object — which can be plotted, summarised, and assigned as the value of another variable.

```
> plot(sunspots)
> summary(sunspots)
> X <- sunspots
```

Each object in R is identified as belonging to a particular **class**. The `sunspots` dataset has class `"ts"` (time series), which can be discerned using the `class` function:

```
> class(sunspots)
[1] "ts"
```

The function `plot` is a so-called **generic** function, for which there may be different **methods** for different classes of data. The method for plotting an object of class `"ts"` is the function `plot.ts`.

When the user types the command `plot(sunspots)`, the R system recognises that the function `plot` is generic, inspects the object `sunspots`, identifies that it belongs to class `"ts"`, and looks for a function called `plot.ts`. This latter function is then called to do the actual work: the system executes `plot.ts(sunspots)`. In R jargon, the call to the generic `plot` has been 'dispatched' to the method `plot.ts`. The plot method for time series produces a display that is sensible for time series, with axes properly annotated.

> Tip: To find out how to customize the plot for an object of any class `"foo"`, consult `help(plot.foo)` rather than `help(plot)`.

What we have described here is the simplest arrangement of this kind, called 'S3 method dispatch'. There are more intricate arrangements including 'S4 method dispatch', which are also supported in R, but these are not used in `spatstat`.

To see a list of all methods available in R for a particular generic function such as `plot`:

```
> methods(plot)
```

To see a list of all methods that are available for the class `"ts"`:

```
> methods(class="ts")
```

Note that the output of `methods` depends on what packages are currently loaded.

2.1.8 The return value of a function

Every function in R returns a value. The return value may be NULL, or a single number, a list, or any kind of object.

If the function performs a complicated analysis of a dataset, the return value will typically be an object belonging to a special class. This is a convenient way to handle calculations that yield large or complicated output. The result of the calculations can then be stored as a single object, and printed or displayed or analysed using functions specially designed for that class, with minimal effort by the user.

When you type an R expression on the command line, the result of evaluating the expression is usually printed.

```
> 1+1
[1] 2
> sin(pi/3)
[1] 0.8660254
```

However, just to confuse matters, the result of a function may be tagged as *'invisible'* so that it is not printed.

```
> fun <- function(){invisible(42)}
> fun()
>
```

Invisibility is lost as soon as the value is manipulated in any way, for example by assigning it to an object and then printing the object:

```
> x <- fun()
> x
[1] 42
```

or simply by enclosing it in parentheses:

```
> (fun())
[1] 42
```

You can also use the name .Last.value to capture the return value of the last executed command.

```
> fun()
> .Last.value
[1] 42
> x <- .Last.value
> x
[1] 42
```

Graphics functions, such as methods for plot, typically return their results invisibly. Often the result is NULL and this is returned invisibly because it is more elegant. However, complicated graphics functions may return a result that specifies how the data were plotted: the positions of histogram bars, the numerical scale, and so on. This is returned invisibly to avoid deluging the user with unwanted information. When we want to annotate or modify a plot, the return value becomes useful.

Tip: To find out the format of the output returned by a particular function *fun*, type help(*fun*) and read the section headed 'Value'.

2.1.9 Factors

In statistics, a 'factor' is a categorical variable, that is, a variable whose values are discrete categories (e.g., 'red', 'green', 'blue', 'orange' or 'animal', 'vegetable', 'mineral'). The possible categories are called the *levels* of the factor.

In R factors are represented by objects of class `"factor"`. A factor dataset f contains values `f[1]`, `f[2]`, ..., `f[n]` which are treated as categorical values.

```
> col <- c("red", "green", "red", "blue",
            "blue", "green", "red")
> col
[1] "red"    "green" "red"    "blue"  "blue"  "green" "red"
> f <- factor(col)
> f
[1] red    green red    blue  blue  green red
Levels: blue green red
```

Factors are superficially similar to vectors of character strings, but their conceptual nature and practical use are very different. The R system treats factors and non-factors very differently.

A `factor` can be recognised by the way it is printed: there are no quotes around the factor values, and the printout ends with a list of all the possible levels. Alternatively one can use `is.factor`:

```
> is.factor(col)
[1] FALSE
> is.factor(f)
[1] TRUE
```

To display the possible levels of a factor f, type `levels(f)`. To control the order of the levels and the names of the levels use the arguments `levels` and `labels` in the call to `factor`:

```
> factor(col, levels=c("red", "green", "blue"), labels=c("r","g","b"))
[1] r g r b b g r
Levels: r g b
```

Other useful commands include `relevel` and `mergeLevels`.

A factor can represent a choice among different alternatives, or a division into groups, or a yes/no response. In a clinical trial of pain relief, the treatment administered to each volunteer could be either Aspirin, Paracetamol, or Placebo. This information could be recorded as a factor `treat` where `treat[i]` is the treatment administered to subject i. The factor `treat` would serve as an explanatory variable in the data analysis. In a study of social disadvantage, each postal district could be classified as Low, Medium, or High socioeconomic status. Factors commonly occur when a numeric variable is converted into a grouping variable by dividing the numerical range into a number of groups using the function `cut`:

```
> x <- c(1,7,4,10,2,14,15,4,9)
> cut(x, c(0,5,10,15))
[1] (0,5]   (5,10]  (0,5]   (5,10]  (0,5]   (10,15] (10,15] (0,5]   (5,10]
Levels: (0,5] (5,10] (10,15]
```

Posts to the `r-help` mailing list reveal there is a lot of confusion about factors in R, and some mistakes seem to recur. It is well worth the effort to gain a better understanding of factors, because they are so important and pervasive.

The key thing to understand is that a `factor` is represented in the software by *encoding* the categorical values as numbers from 1 to L, where L is the number of levels. The list of all possible levels is also stored, so that the original factor values can be reconstructed when presenting the data to the user. This trick makes computation faster, because it is easier to compare numbers than character strings, and may also save computer memory, because it takes less space to store numbers than long strings.

For example, the factor `f` described above is internally represented by the numerical codes,

```
> as.integer(f)
[1] 3 2 3 1 1 2 3
```

and the factor levels,

```
> levels(f)
[1] "blue"  "green" "red"
```

For example, the first entry in `as.integer(f)` is the number 3, meaning the third level of the factor, which is `"red"`, so the first entry in `f` is the categorical value `red`.

Usually we do not need to think about this internal mechanism. One source of confusion arises when the levels of the factor are actually numbers stored as character strings, such as `"1.02"` `"2.03"` `"3.04"`. If `x` is a factor with such levels, one might reasonably expect `as.numeric(x)` to be a numeric vector whose entries were the numbers resulting from converting the character strings to numeric values. Wrongo! In this setting `as.numeric(s)` returns the vector of integer indices used in the storage method described above. To get the vector of levels-converted-to-numbers that you really want, you can do `as.numeric(as.character(x))` or more efficiently `as.numeric(levels(x))[x]`.

```
> x <- factor(c("1.02","2.03","2.03",
                "1.02","3.04","1.02",
                "2.03","1.02","3.04","2.03"))
> x
 [1] 1.02 2.03 2.03 1.02 3.04 1.02 2.03 1.02 3.04 2.03
Levels: 1.02 2.03 3.04
> as.numeric(x)
 [1] 1 2 2 1 3 1 2 1 3 2
> as.numeric(as.character(x))
 [1] 1.02 2.03 2.03 1.02 3.04 1.02 2.03 1.02 3.04 2.03
> as.numeric(levels(x))[x]
 [1] 1.02 2.03 2.03 1.02 3.04 1.02 2.03 1.02 3.04 2.03
```

For more information about factors, start with the basic manual `An Introduction to R` (section 4) from the R web page `www.rproject.org`, under `manuals`. Other sources include [609, p. 20 ff.] and [588, sec. 4.8, p. 39 ff.].

2.1.10 Formulas

In statistics we often speak of a variable being 'explained by' another variable. A `formula` is a special idiom in the R language which is used to express this relationship. The formula `y ~ x` essentially means 'y depends on x', while `y ~ x + z` means 'y depends on x and on z'. A formula is recognizable by the presence of the tilde character '~'. The variable to the left of the tilde is explained by, predicted by, or dependent on the variables appearing to the right of the tilde.

Formulas have many potential uses; the most important ones are for *graphics* and for *statistical modelling.* In graphics, the formula y ~ x means that y should be plotted against x. In modelling, y ~ x means that y is the response variable and x is the predictor variable. The precise interpretation depends on the context.

A `formula` can be saved as an object in its own right:

```
> a <- (y ~ x + z)
> a
y ~ x + z
> class(a)
[1] "formula"
```

This object is just a symbolic expression. The variable names x, y, z appearing in the formula are just symbols, and do not have to correspond to any real data. Importantly, the operators +, -, *, /, ^, : are also just symbols, and do not have any particular meaning at this stage.

Typically the user will give a `formula` as one of the arguments to a function that performs graphical display or fits a statistical model:

```
plot(y ~ x)
lm(y ~ x)
```

The function will then interpret the formula in a way that is appropriate to the context. It is only at this stage that the variable names x and y will usually be expected to correspond to datasets in the R session, and the operators +, -, *, /, ^, : will be given a particular meaning (which is not the same as their usual meaning in arithmetic).

2.1.10.1 Formulas in graphics

A formula can be used to specify which variables should be plotted against each other in a graphic. Figure 2.1 illustrates how the formula y ~ x can be interpreted in different ways depending on the type of data in x. In this example we have used the standard datasets `stackloss` and `chickwts` supplied with the R system, which are automatically available in any R session. See `?stackloss` and `?chickwts` for information.

The left panel of Figure 2.1 was produced by the command

```
> plot(stack.loss ~ Water.Temp, data=stackloss)
```

in which the variables `stack.loss` and `Water.Temp` are both numeric vectors, which are columns of the data frame `stackloss`. The right panel was produced by the superficially similar command

```
> plot(weight ~ feed, data=chickwts)
```

in which `weight` and `feed` are columns of the data frame `chickwts`. The difference is that `feed` is a `factor` representing the type of food given to each chicken in the experiment. The sensible way to plot weight against feed type is using a boxplot for each group of observations.

Notice the argument `data` used in these examples. This tells `plot` that values for the variables appearing in the formula can be found in correspondingly-named columns of the data frame given by the `data` argument. This is a very handy convention which improves readability and supports well-organised code.

Apart from the generic function `plot` there are many other graphics functions which accept formula arguments and interpret them in some appropriate way. The help files for these functions explain how the formula will be interpreted in each case. For example,

```
> plot(stack.loss ~ Water.Temp + Air.Flow, data=stackloss)
```

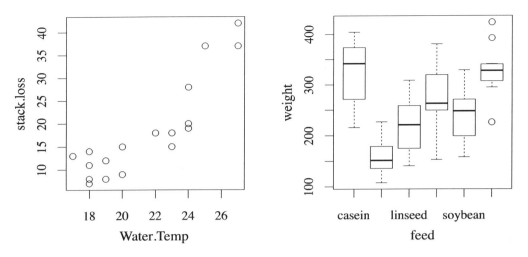

Figure 2.1. *Results for plotting y against x depend on the type of data in x.* Left: *x is a numeric vector.* Right: *x is a factor. In both cases, y is numeric.*

produces two[1] scatterplots (`stack.loss` against `Water.Temp` and `stack.loss` against `Air.Flow`) while

```
> library(scatterplot3d)
> with(stackloss,
      scatterplot3d(stack.loss ~ Water.Temp + Air.Flow))
```

produces a three-dimensional scatter plot of the three variables.

2.1.10.2   Formulas in statistical models

A much more important and powerful use of formulae in R is to specify a statistical model.

In the `stackloss` data, suppose we want to fit a simple linear regression of `stack.loss` on `Water.Temp`. This can be done simply by typing

```
> lm(stack.loss ~ Water.Temp, data=stackloss)
```

Here `lm` is the command to fit a Linear Model. The formula specifies the 'systematic' component of the model, which in this case is the linear relationship $y = \alpha + \beta x$, where y is stack loss, x is water temperature, and α, β are parameters (the intercept and slope of the regression line) which are to be estimated from the data. The output from this command is:

```
Call:
lm(formula = stack.loss ~ Water.Temp, data = stackloss)

Coefficients:
(Intercept)    Water.Temp
   -41.911         2.817
```

The fitted coefficients are displayed above: $\hat{\alpha} = -41.911$ and $\hat{\beta} = 2.817$.

Note that the coefficients α and β do not appear in the formula `stack.loss ~ Water.Temp`.

---

[1]Multiple plots are displayed fleetingly, one after the other, unless you type `par(ask=TRUE)` to pause them, or `par(mfrow=c(1,2))` to display them side by side.

Only the names of *variables* appear in the formula: response variables to the left of the tilde, and explanatory variables to the right.

The conciseness of model formulae makes them extremely powerful. Even very complicated models can be specified succinctly. Expressions for simple models are very easy to learn and more complex models can be worked up to with a modicum of effort.

The idea of specifying a model by a compact formula syntax is certainly not original or exclusive to R. The particular syntax used in R has its origins in a paper by Wilkinson and Rogers [702]. However, the language capabilities of R make it very easy to use formulae.

There is a common syntax for interpreting formulae which is used by *almost all*[2] of the important model-fitting functions in R. We will briefly explain some of the basics of this syntax.

In a model formula, the symbols +, -, *, /, ^ and : do not have their usual meaning as arithmetic operations. Instead they are operators which combine terms in a model.

The operator '+' is used to add terms to a model. To fit a multivariable linear regression of `stack.loss` on `Water.Temp` and `Air.Flow`, we simply type

```
> lm(stack.loss ~ Water.Temp + Air.Flow, data=stackloss)
```

The operator '-' is used to remove a term from a model. For example, a model is assumed by default to include a constant term (such as the intercept α in the linear regression described above). If an intercept is not wanted, it can be explicitly removed using -1. To fit a proportional regression ($y = \beta x$) instead of a linear regression ($y = \alpha + \beta x$) of `stack.loss` on `Water.Temp`, we type

```
> lm(stack.loss ~ Water.Temp - 1, data=stackloss)
```

One can also use +0 instead of -1 for the same purpose. The meaning of other model operators *, /, ^, and : is explained in Chapter 9.

The precise interpretation of a model formula also depends on the type of data involved. For example, with the `chickwts` dataset,

```
> lm(weight ~ feed - 1, data=chickwts)
Call:
lm(formula = weight ~ feed - 1, data = chickwts)

Coefficients:
    feedcasein  feedhorsebean     feedlinseed   feedmeatmeal     feedsoybean
         323.6          160.2           218.8          276.9           246.4
 feedsunflower
         328.9
```

The explanatory variable `feed` is a factor, so by fitting a linear model we are simply estimating the mean weight of chickens in each group. Further explanation is given in Chapter 9.

2.1.11 Graphics devices in R

R has quite a few different graphics devices, but here we only mention the most basic facts. The default graphics device used in an interactive session depends both on your operating system and whether or not you use RStudio. By default RStudio uses its own graphics device (`RStudioGD`) which makes it easy for the user to browse through previously generated plots and export them to PDF files, for example. However, for some interactive plotting tasks it may be preferable to use your system's default R graphics device. On Windows such a device is opened by the command `windows()`, on OS X the command `quartz()` is used, and on Linux the corresponding command is `x11()`.

---

[2]The `nls` function is one example where the syntax is slightly different.

2.2 Packages for R

2.2.1 Packages and libraries

A `package` is a collection of additional features that can be added to R. It is usually supplied as an archive file with extension `.zip` or `.tar.gz`, compacted from several files containing code in the R language and other languages, help files, datasets, vignettes (demonstration examples with explanation), and test examples.

A standard installation of R comes with several essential default packages, such as `stats` (which as we saw above contains the critically useful function `lm`) and `graphics`. These packages are 'loaded' (see below) automatically whenever R is started. In addition a number of recommended packages (e.g., `boot`, `MASS`, and `survival`) are installed automatically along with R but are *not* loaded until asked for.

Additional packages can easily be added to an existing installation of R as described below. A package only needs to be installed once (although if the version of R is upgraded, the packages may need to be re-installed). It is probably a good idea to do a complete re-install of your installed packages when you have upgraded R. The function `update.packages()` makes this easy.

When a package is installed, it is installed into a directory ('folder') which is usually referred to as a '`library`'. That is, a library is a collection of packages. To access the features of a particular package in an R session, the package must be *loaded* in the session, by typing `library(`*name*`)` where *name* is the name of the package, with or without quotes. The appropriate metaphor is that the `library` function 'checks the requested package out of the library'. Core packages such as `stats` are automatically 'checked out' or loaded at the start of an R session. However, other packages are not, because of the large overhead that would be involved.

2.2.2 Contributed packages

One of the great strengths of R is the large range of user-contributed packages which can be freely downloaded from the R archive site `cran.r-project.org` under 'Contributed Packages'. At the time of writing there are almost 6000 contributed packages, supporting everything from accelerometer data analysis to zooplankton image processing.

The technical quality of the contributed packages is very high. In particular, they have all passed a rigorous automatic checking process which verifies that the code works, the examples can be executed, the documentation is consistent with the code, and so on.

The scientific quality is also generally very high. The vast majority of contributed packages are written by specialists in the relevant field. They are all documented with help files, most of which are well written and which contain illuminating examples. The help files are often supported by detailed technical documentation, either installed in the package (usually in the form of a 'vignette') or published in the *Journal of Statistical Software* `jstatsoft.org`. For details on how to use vignettes, see Section 2.4.3. There is a large community of enthusiasic users who quickly detect problems and contribute new code.

Instructions on how to install a package are given at `cran.r-project.org`. Following is a brief explanation.

On a Windows or OS X system, start an R session. In the graphical interface menu bar, pull-down the appropriate menu item for installing packages, and select the package name from the list. If this menu item is not available (typically for internet security reasons), manually download packages from `cran.r-project.org` under `Contributed packages`: visit the page for each desired package, and download the Windows or OS X binaries. Then install these by starting an R session and using the appropriate menu item to install a local file.

On a Linux system, start R as superuser, and issue the command `install.packages("`*name*`")` where *"name"* is the name (in quotes) of the desired package. If this is the first time you have installed a package, you may may be asked to choose a directory to contain your library of R packages. If this is unsuccessful, manually download packages from `cran.r-project.org` under `Contributed packages`: visit the page for each desired package, download the Linux tar file *name*`.tar.gz`, and install it by the R command `install.packages("name.tar.gz")`.

A package may depend on other packages: the dependencies are stated on the package's download page. If you are installing a package manually, you will need to download all its dependencies manually, and install them before installing the desired package.

Example datasets are included in most contributed packages. The package documentation includes an entry for each installed dataset. In any package, a dataset named *foo* can be accessed by typing `data(foo)`. After this is done, the name *foo* will be recognised in R commands and the dataset will be retrieved.

```
> library(maptools)
> data(SplashDams)
> SplashDams[1:3,]
```

In some packages, including `spatstat`, installed datasets are *'lazy-loaded'*, which means that the `data(foo)` command is not necessary.

2.2.3 Contributed R packages for spatial point patterns

Amongst the thousands of contributed packages on CRAN, there are hundreds which analyse spatial data. A selective overview of these packages is provided in the Spatial Task View (follow the links to *Task Views — Spatial* on `cran.r-project.org`).

For spatial point pattern data, some useful contributed packages are listed in Table 2.2.

adehabitat	habitat selection analysis [138]
ads	spatial point pattern analysis [538]
aspace	'centrographic' analysis of point patterns
DCluster	detecting clusters in spatial count data [287]
ecespa	spatial point pattern analysis [206]
lgcp	log-Gaussian Cox processes [650, 651]
MASS	pioneering code for spatial statistics [670]
ppmlasso	modelling with Lasso penalty [568, 567]
SGCS	spatial graph techniques for detecting clusters
sp	classes and methods for spatial data [109, 111]
sparr	analysis of spatially varying relative risk [202]
spatgraphs	graphs constructed from spatial point patterns
spatial	the pioneering R package for spatial data [670]
spatialkernel	interpolation and segregation of point patterns
spatialsegregation	segregation of multitype point patterns
splancs	spatial and space-time point pattern analysis [590]

Table 2.2. *Contributed R packages with substantial functionality for spatial point patterns.*

There are also numerous small packages that provide a single tool for spatial point pattern analysis, such as a new statistical technique described in recent literature. The CRAN website search is useful for finding these packages using a keyword or an author's name.

The format in which point pattern data are represented will usually be different in different packages. The `sp` package offers a standard set of spatial data types in R, but many other packages (including `spatstat`) do not conform to these standards. Several packages conform to the `spatstat` formats instead. The very useful `maptools` package can convert between different data structures and file formats. Handling data files is discussed at greater length in Chapter 3.

2.3   Introduction to `spatstat`

2.3.1   The `spatstat` package

The analysis of spatial point patterns as discussed in this book is conducted predominantly using the `spatstat` package. This is a contributed R package for analysing spatial data, with a major focus on point patterns. Most of the available functionality is for *two-dimensional* spatial point patterns, but this is now being extended to three-dimensional space, space-time, and other domains such as networks.

Both this book and the `spatstat` package are oriented toward producing *complete statistical analyses* of spatial point pattern data. Such analyses include the creation, manipulation, and plotting of point patterns; exploratory data analysis (such as producing basic summaries of point patterns, kernel smoothing, the calculation and plotting of Ripley's *K*-function and other summary functions, and the calculation of quadrat counts and nearest neighbour distances); random generation and simulation of patterns; parametric model-fitting (Poisson models, logistic regression, Gibbs models, Cox models, cluster processes); formal statistical inference (confidence intervals, hypothesis tests, model selection); and informal diagnostics and model validation. In addition to providing the standard tools that would be expected in a package for spatial point pattern analysis, `spatstat` also contains many advanced tools, including results of recent research.

There are over 2000 user-level commands in the `spatstat` package, making it one of the largest contributed packages available for R. The package documentation is extensive: the help manual has more than 1300 pages, and there are several supporting documents. The package also includes 50 spatial datasets. The dedicated website `www.spatstat.org` gives information about the package.

Two decades of work by Adrian Baddeley and Rolf Turner, with support from the scientific community, have gone into the creation and development of `spatstat`. The package includes substantial contributions of code from Kasper Klitgaard Berthelsen, Abdollah Jalilian, Marie-Colette van Lieshout, Ege Rubak, Dominic Schuhmacher, and Rasmus Waagepetersen. In particular Ege Rubak has recently been playing a significant role and has now joined the `spatstat` developer team. The user community has generously contributed datasets, bug reports and fixes, and code for new purposes.

To obtain `spatstat` simply download it from `cran.r-project.org` as described in Subsection 2.2.2. The package is free open source software, under the GNU Public Licence (GPL) version 2 and above [291]. Several other packages (currently `deldir`, `abind`, `tensor`, `polyclip`, and `goftest`) are required by `spatstat`. The packages on which `spatstat` depends also have free open source licences so that there is no barrier to downloading, copying, installing, and using the package. The package is copyright, but can be reused under the conditions specified by the GPL.

2.3.2   Please acknowledge `spatstat`

If you use `spatstat` for research that leads to publications, it would be much appreciated if you could acknowledge `spatstat` in your publications, preferably citing this book. Citations help us to justify the expenditure of time and effort on maintaining and developing the package.

The help files for `spatstat` and the supporting documentation are *copyright*. Please do not copy substantial amounts of text from the `spatstat` help files into a scientific publication, without proper acknowledgement.

2.3.3   Overview of `spatstat`

As stated above, `spatstat` is designed to *support a complete statistical analysis* of spatial point pattern datasets. It includes implementations of a very wide range of techniques for spatial point pattern analysis, originating from many different fields of application. It includes some very recently developed techniques as well as classical procedures. It enables researchers to conduct a searching, critical, defensible analysis of their data. It also provides infrastructure for developing new kinds of analysis.

data types: The `spatstat` package provides specialised data types (classes) for spatial data. They include point patterns, spatial windows, pixel images, line patterns, tessellations, and linear networks.

data: A large and growing collection of example spatial datasets is installed with `spatstat`. They arise from a range of applications including cell biology, wildlife ecology, forestry, epidemiology, geology, and astronomy. These datasets have been generously contributed by the scientific community. They are used to demonstrate, test, and compare the performance of different methods.

data handling: Facilities in `spatstat` for handling spatial datasets include creation of datasets, basic data querying and modification (subsetting, splitting/merging, raster operations), printing, and summarising data. Many geometrical and mathematical calculations are supported, including distance measurement, distance transforms, nearest neighbours, mensuration (area and length measurement), numerical integration, filtering, and convolution. Available geometrical operations include shifts, rotations, rescaling, affine transformation, intersection, union, convex hull, and morphological operations (dilation, erosion, closing, opening). Tessellations such as the Dirichlet-Voronoï-Thiessen polygons and the Delaunay triangulation can be computed from point patterns.

data generation and sampling: Synthetic data can easily be generated in `spatstat` in a wide variety of ways, and for many purposes, including spatial sampling, Monte Carlo (simulation-based) statistical methods, experiments, and training. Point patterns can be generated according to random sampling designs (independent, stratified, systematic), point process models of many kinds (complete spatial randomness, cluster processes, Cox processes, Gibbs processes, sequential random processes), or by spatial resampling of existing data. For models which have been fitted to point pattern data, simulated realisations can be generated automatically. There is an interactive point-and-click interface enabling the user to manually create a point pattern dataset or manually digitise a scanned image. Random patterns of lines and other random sets can also be generated.

plotting: For each spatial data type, `spatstat` provides a `plot` method designed to produce an appropriate graphical display. Various alternative plotting procedures, adapted to different statistical purposes, are also provided. For instance a pixel image can be plotted as a colour image, a contour plot, a perspective view of a surface, a histogram, or a cumulative distribution function. Interactive plotting (zoom/pan/identify) allows data to be scrutinized closely. Different spatial datasets can be superimposed as 'layers' or plotted side by side or in an array. The `spatstat` package follows the R approach of providing sensible default behaviour while allowing detailed control over the plot. Auxiliary plotting information, for example a colour map, can be saved, examined, edited, and re-used. The plotting functions include algorithms for designing the layout automatically. The plot methods for summary functions such as the *K*-function are (usually) clever enough to avoid collisions between the curve and the legend, and to generate mathematical notation for the axis labels.

exploratory data analysis: The `spatstat` package supports a very wide range of techniques for exploratory data analysis and nonparametric analysis of spatial point patterns. Such techniques include classical methods such as spatial binning and quadrat counting, kernel estimation of intensity and relative risk, clustering indices (Clark-Evans index), scan statistics, Fry plots, and spatial summary functions (Ripley's K-function, the pair correlation function, the nearest-neighbour distance function, the empty-space hazard rate, the J-function, and the mark correlation function). The `spatstat` package includes extensions of these techniques (some of which are very recent) to inhomogeneous spatial patterns (e.g., the inhomogeneous K-function), to multitype point patterns (e.g., the bivariate K-function, mark connection function), to point patterns with *multivariate* marks (e.g., the reverse conditional moment function [601]) as well as local indicators of spatial association (e.g., the local K-function). Standard errors and confidence intervals for summary functions are supported by spatial bootstrap methods. Finally `spatstat` includes modern exploratory and nonparametric techniques such as data sharpening [151], cluster set estimation [11], nearest-neighbour cleaning [135], and kernel estimation of relative intensity [37].

model-fitting: The main distinguishing feature of `spatstat` is its extensive capability for fitting statistical models to point pattern data, and for performing statistical inference about them. These models are conceptualized as *point processes*, mechanisms which generate random point patterns. The `spatstat` package is able to fit Poisson, Gibbs, Cox, and Neyman-Scott point process models. Popular techniques such as maximum entropy and logistic regression are subsumed in the Poisson model. A very flexible syntax for specifying models is provided, using the powerful features of the R language. Terms in the model specification may represent spatial trend, the influence of spatial covariates, and interaction between points. By evaluating the statistical significance of these model terms the analyst can decide whether there is evidence for spatial trend or evidence for interpoint interaction (after allowing for covariate effects) and so on. The result of fitting a point process model to data is a fitted model object: there are facilities for printing, plotting, predicting, simulating from, and updating these objects.

statistical inference: For nonparametric estimation, `spatstat` computes standard errors and confidence intervals calculated using spatial bootstrap methods, and hypothesis tests based on Monte Carlo procedures and simulation envelopes. For testing whether a point process depends on a spatial covariate, `spatstat` supports nonparametric hypothesis tests (Berman, Lawson-Waller, Kolmogorov-Smirnov, Cramér-Von Mises, etc.). In the context of fitted point process models, it is possible to estimate model parameters with standard errors, compute predictions from the model with standard errors, perform significance tests for model terms (likelihood ratio test, score test), and to perform automatic stepwise model selection using likelihood ratios or Akaike's Information Criterion AIC. At the time of writing, `spatstat` does not support fully Bayesian inference, but does include variational Bayes methods (Section 9.11).

simulation, and simulation-based inference: Simulation-based techniques play an important role in the analysis of spatial point pattern data. To support them, `spatstat` has many commands for generating random point patterns. Points can be generated completely at random, or according to a specified probability model. For spatial sampling purposes, systematic and stratified random point patterns, and quasirandom point patterns, can be generated. Clustered point patterns and regular point patterns can be generated according to various point process models in a flexible way. Spatial dependence between points can be investigated in `spatstat` using Monte Carlo tests including tests based on simulation envelopes. For statistical inference it is possible to generate simulated realisations from a *fitted* point process model. Approximate maximum likelihood inference for point process models, using Monte Carlo simulation, is supported.

validation: An important part of statistical analysis is checking the validity and goodness-of-fit of a model fitted to the data. Such techniques, although well known in other areas of data analysis, have only recently been developed for spatial point patterns. The spatstat package is unique in providing a wide range of graphical diagnostic tools for validating models of spatial point patterns. Tools include residuals, leverage, influence, partial residual plots, added variable plots, compensators, and residual summary statistics.

The spatstat package is *organised around statistical concepts and terminology.* The name of a command in spatstat conforms to the standard statistical nomenclature used in R wherever possible. For example, the spatstat command that divides the points of a point pattern into several groups of points is called split because this is the name of the corresponding command in the base R system for dividing a dataset of numbers into several groups. This naming convention makes it easier for R users to find and remember command names.

The use of statistical terminology may be initially confusing for scientists from other fields. Unfortunately there is no agreement between scientific fields about terminology. In the book we try to mention equivalent terms for the same concept where relevant, and they are included in the index.

2.4 Getting started with spatstat

2.4.1 Installing and running spatstat

To start using spatstat:

1. Install R on your computer.

   Go to r-project.org and follow the installation instructions (or see Section 2.1.1).

2. Install the spatstat package in your R system. This, of course, only needs to be done once. (Well, you *will* need to re-install or update spatstat from time to time.)

   Start R and type install.packages("spatstat"). If that doesn't work, see Section 2.2.2 or go to cran.r-project.org to learn how to install Contributed Extension Packages.

3. Start R.

4. Type library(spatstat) to load the package.

5. Type beginner for information.

2.4.2 Demonstration example

Example code

Here is a short demonstration of data analysis using spatstat. We use the dataset swedishpines, installed in the package, which gives the locations of pine saplings in a survey plot in a Swedish forest. To inspect information about the dataset we simply type its name:

```
> swedishpines
Planar point pattern: 71 points
window: rectangle = [0, 96] x [0, 100] units (one unit = 0.1 metres)
```

Documentation about the dataset can be read by typing `help(swedishpines)` or `?swedishpines`. For neater results, we first convert the spatial coordinates from decimetres to metres:

```
> swp <- rescale(swedishpines)
```

To display the point pattern we type:

```
> plot(swp)
```

The result is shown in the left panel of Figure 2.2. To assess whether the density of trees is spatially varying, we apply kernel smoothing:

```
> DD <- density(swp)
> plot(DD)
```

The plot is shown in the middle panel of Figure 2.2. To assess whether the tree locations are completely random, or alternatively whether they are spatially clustered or spatially regular, we calculate Ripley's *K*-function:

```
> KK <- Kest(swp)
> plot(KK)
```

The plot is shown in the right panel of Figure 2.2.

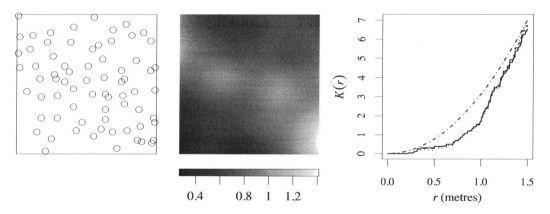

Figure 2.2. *Plots generated in the cookbook example.* Left: *Swedish Pines dataset.* Middle: *kernel estimate of forest density.* Right: *Ripley's K-function.*

On the basis of these plots we may decide that we need to take account of spatial inhomogeneity in the density of the forest as well as a tendency to spatial regularity (trees tend to maintain a certain distance apart). We fit such a model to the data:

```
> fitS <- ppm(swp ~ polynom(x,y,2), Strauss(0.9))
```

The model formula `swp ~ polynom(x,y,2)` describes a particular form of spatial inhomogeneity, while the model component `Strauss(0.9)` describes a tendency for trees to avoid coming closer than 0.9 metres apart. The result is a fitted point process model exhibiting both these features.

```
> fitS
```

```
Nonstationary Strauss process
Log trend:  ~x + y + I(x^2) + I(x * y) + I(y^2)
Fitted trend coefficients:
(Intercept)            x            y      I(x^2)     I(x * y)       I(y^2)
   0.223946     0.325417     0.401186    0.000459    -0.063262    -0.015963
Interaction distance:            0.9
Fitted interaction parameter gamma:         0.237
```

We may now conduct a formal statistical test for spatial regularity after accounting for spatial inhomogeneity:

```
> fitP <- ppm(swp ~ polynom(x,y,2), Poisson())
> anova(fitP, fitS, test="LR")
Analysis of Deviance Table
Model 1: ~x + y + I(x^2) + I(x * y) + I(y^2)          Poisson
Model 2: ~x + y + I(x^2) + I(x * y) + I(y^2)          Strauss
  Npar Df AdjDeviance Pr(>Chi)
1    6
2    7  1        98.5   <2e-16 ***
---
Signif. codes:  0 '***' 0.001 '**' 0.01 '*' 0.05 '.' 0.1 ' ' 1
```

The last result, using a modified Likelihood Ratio Test [56], is that there is significant evidence of spatial regularity. The fitted model fitS can be checked and validated in various ways. We can also generate simulations of the model (random point patterns obeying the probability distribution described by the model) by typing simulate(fitS).

Classes and methods in the example

The example code shows that spatstat commands are typically compact and easy to use. For the most effective use of spatstat, we need to understand what is happening behind these commands.

The spatstat package is designed around several *classes* of spatial data. The mechanism of classes and methods was explained in Section 2.1.7. Essentially every object handled by spatstat belongs to a class. In the examples above:

```
> class(swedishpines)
[1] "ppp"
```

The swedishpines dataset belongs to the class "ppp" standing for **p**lanar **p**oint **p**attern. In order to analyse your own point pattern data in spatstat, you will need to convert the data into this format, as explained in Chapter 3. When the data are plotted, the command plot(swedishpines) is dispatched to plot.ppp, the plot method for point patterns, producing the display in the left panel of Figure 2.2. To find out how to modify the plot, we would consult the help file for plot.ppp.

The results of calculations in spatstat are also stored as objects of an appropriate class:

```
> DD <- density(swp)
> class(DD)
[1] "im"
```

The object DD, an estimate of the spatially varying density of points, belongs to the class "im" of pixel images. A pixel image is an evenly spaced rectangular grid of spatial locations with values attached to each grid point. Some kinds of spatial data can be stored as pixel images. The command plot(DD) was dispatched to plot.im which plots the pixel values as colours or greyscale values as shown in the middle panel of Figure 2.2. The help for plot.im gives useful options for modifying this image plot, and pointers to alternative ways of plotting the same data as a contour map or a perspective view.

```
> KK <- Kest(swp)
> class(KK)
[1] "fv"           "data.frame"
```

The object KK, an estimate of Ripley's *K*-function for the Swedish Pines data, belongs to the class "fv" of summary functions. The command plot(KK) was dispatched to plot.fv to produce the right panel of Figure 2.2.

```
> fitS <- ppm(swp ~ polynom(x,y,2), Strauss(0.9))
> class(fitS)
[1] "ppm"
```

The object fitS is the result of fitting a point process model to the Swedish Pines data. It belongs to the class "ppm" for **p**oint **p**rocess **m**odel. The function which fits the model to the data is also called ppm. This is the standard convention for model-fitting functions in R, typified by the functions lm and glm for fitting linear models and generalised linear models, respectively. The model can be printed, plotted, simulated, predicted, and so on.

2.4.3   Information sources for spatstat

There is a Quick Reference guide for spatstat which summarises its capabilities and gives a comprehensive list of spatstat commands. To view the Quick Reference guide, load spatstat and type help(spatstat). Alternatively open the full spatstat package manual in a browser, and navigate to the first entry in the manual. You can also download a PDF file of the Quick Reference guide from the website www.spatstat.org.

To find a spatstat function that performs a desired task, we suggest looking through the Quick Reference guide, followed by the index of this book. If they are not helpful, try the keyword search function help.search narrowed to the spatstat package. To find a spatstat function that generates systematic random patterns,

```
> help.search("systematic", package="spatstat")
```

One can also type methods(class="*foo*") to find methods that apply to objects of class *foo*.

For information about a particular function in spatstat, either follow the links in the Quick Reference guide, or type help(*functionname*). Users of spatstat are strongly urged to **read the help files** for the package. Substantial effort has gone into writing clear, detailed, and comprehensive explanations of important concepts. The spatstat help files also include very practical guidance, warnings, and explanations of common error messages. For more general guidance about spatstat, the best source is this book, especially making use of the index.

The spatstat package is constantly evolving. Regular users of spatstat can stay up-to-date by typing latest.news to read about changes in the most recent version of spatstat, or news(package="spatstat") to read about changes in all previous versions. The package news file is also available at www.spatstat.org.

Several introductory 'vignettes' are installed with spatstat. They include a quickstart guide, Getting Started with Spatstat; a document about Handling shapefiles (a popular format for Geographical Information Systems); a summary of recent updates; and a summary of all installed datasets. To see a list of available vignettes from all installed packages, type vignette() in the R command line interface; type vignette(package="spatstat") for vignettes in spatstat. To read a vignette, type vignette("*name*") where *name* is the name of the document without the file extension, e.g., vignette("getstart"). Vignettes can also be accessed using the graphical help interface invoked by help.start.

The spatstat package includes about 50 example datasets, including all the examples used in this book. A useful overview of these datasets is vignette('datasets') and they are displayed in a slideshow by demo(data). Full documentation is given in the individual help files for each dataset.

Several demonstration scripts are also installed with spatstat. Type demo(spatstat) for a general overview. For a list of available demonstrations, type demo(package = spatstat).

The email forum R-sig-geo sometimes handles questions about spatstat. General questions about R can be addressed to the R-help forum. Join these and other forums by visiting r-project.org. The web forum stackexchange.com also handles questions and answers about spatstat.

If you have identified a bug in spatstat, please post a bug report to www.spatstat.org. You will need to supply a reproducible example of the bug.

2.4.4 The classes of objects available in spatstat

For the record, here is a list of all the classes defined in spatstat, with pointers to further information. This also gives an overall impression of what is possible in spatstat.

ppp — **planar point pattern:** A *planar point pattern* dataset is represented in spatstat by an object of class "ppp" (**p**lanar **p**oint **p**attern). See Figure 2.3. Point patterns in more than two dimensions are accommodated by other classes discussed below. The essential components of a point pattern object are the coordinates of the points in the pattern and (crucially!) the *observation window* (see below). In order to analyse your own point pattern data, you will need to convert your data into this format: a detailed explanation is given in Chapters 3 (particularly Section 3.3) and 4 (particularly Section 4.2).

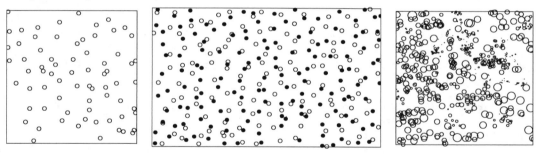

Figure 2.3. *Three objects of class* "ppp" *representing point patterns.* Left: *points of a single type;* Middle: *points of two types;* Right: *points with a continuous numerical mark value.*

owin — **observation window:** An *observation window* is an object of class "owin" which represents a specified region of space. It is often used to designate the region inside which a point pattern has been recorded, or the spatial region inside which the points are constrained to lie. It is an **essential** component of a point pattern object (see Section 3.1.2).

A window may be rectangular, polygonal, or a 'mask', this latter category being represented by an array of TRUE/FALSE values on a pixel grid. See Figure 2.4. Information on how to create an object of class "owin" is given in Section 3.5. Changing an "owin" object from one type to another is discussed in Section 4.2.2.

im — **pixel image:** A *pixel image* is an evenly spaced rectangular grid of spatial locations with values attached to them. Such an image is represented in spatstat by an object of class "im".

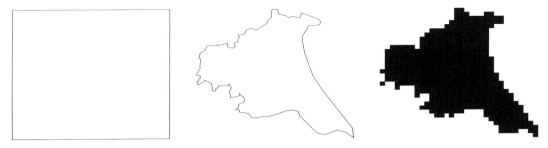

Figure 2.4. *Objects of class* "owin" *representing spatial windows.* Left: *rectangle;* Middle: *polygonal region;* Right: *binary pixel mask.*

The pixel values may be numeric (real or integer), logical, complex, character, or factor values. See Figure 2.5. Information on how to create an object of class "im" or to convert data stored in some other way into an object of this class is given in Section 3.6.

Figure 2.5. *Object of class* "im" *representing a pixel image.*

psp — **spatial pattern of line segments:** A *spatial pattern of line segments* is represented by an object of class "psp" (**p**lanar **s**egment **p**attern). See the left panel of Figure 2.6. Such objects can be used to represent patterns of geological faults, roads, fibres, and so on. Objects of class "psp" are discussed in Section 3.7.

tess — **tessellation:** A *tessellation* is a division of space into non-overlapping regions. Such divisions are represented in spatstat by objects of class "tess". See the right panel of Figure 2.6. Tessellations are discussed in Chapter 4, particularly in Section 4.5.

pp3 — **three-dimensional point pattern:** A *three-dimensional point pattern* is represented as an object of class "pp3" in spatstat. At present spatstat has rudimentary facilities for dealing with "pp3" objects. More facilities will be added in the future.

ppx — **multidimensional space-time point pattern:** A *multidimensional space-time point pattern* is represented as an object of class "ppx" in spatstat. This class accommodates patterns of points in space or space-time of arbitrary dimension. This is the completely general class for representing point patterns in spatstat. Objects of class "ppx" have a component coord.type which allows the user, amongst other things, to distinguish between a temporal coordinate, spatial coordinates, and local coordinates. This provides a potential for space-time modelling which is envisaged as a future enhancement to spatstat.

hyperframe — **array of highly structured data:** An *array of highly structured data* can be represented in spatstat as an object of class "hyperframe". This class generalizes the class "dataframe". An object of class "hyperframe" is 'like a data frame' except that the entries of its columns need not be scalars: they can be objects of any class. The entries of any given column must all be of the *same* class. For example the first column of such an object might consist of point patterns, objects of class `"ppp"` and the second column might consist of images (objects of class `"im"`) serving as covariate data. Hyperframes are often used to represent the results of an experiment involving spatial data. The class `"hyperframe"` is discussed in Chapter 16.

ppm — **point process model:** The result of fitting a Gibbs *point process model* by means of the function ppm is stored as an object of class `"ppm"`. Many tools are provided to handle them easily and efficiently. The ability to fit point process models in a straightforward and user-friendly manner is a distinguishing feature of spatstat. References to ppm are ubiquitous throughout this book. See Sections 9.3 and 13.5.

kppm — **cluster process model:** Some commonly used point process models cannot be fitted using ppm. They include *cluster process models* and *Cox models*. These models can be fitted using the function kppm and the result is an object of class `"kppm"`. Details are given in Chapter 12, particularly in Section 12.4.

mppm — **model fitted to multiple point patterns:** when the outcome of an experiment includes several different point patterns, we need a way to fit a model to these point patterns simultaneously. This is achieved using the function mppm which takes as input a hyperframe (see above) of data and returns an object of class `"mppm"`. There are many methods (see `methods(class="mppm")`) for handling such objects; they are analogous to methods for class `"ppm"`. See Chapter 16.

fv — **summary function:** Many point pattern analyses involve estimating a function rather than a numerical value. There are usually several different estimates of the same function available, typically based on different bias corrections. The class `"fv"` is provided to support the convenient plotting of *several estimates of a function*. In particular the 'summary functions' Kest (see Chapter 7) and Fest, Gest, and Jest (see Chapter 8) all return objects of class `"fv"`. An object of class `"fv"` is actually a data frame with a number of special 'attributes'. There are quite a few methods in addition to the plot method for handling `"fv"` objects. (See `methods(class="fv")`.) A noteworthy method is `as.function.fv` which turns an `"fv"` object into a real live function.

envelope — **simulation envelopes of a function:** The function envelope produces *simulation envelopes of a summary function*. For example, the grey shaded regions in Figure 1.16 on page 13 were computed by generating random point patterns, calculating the pair correlation function estimate for each simulated pattern, and finding the maximum and minimum values of these pair correlations for each distance r. We cannot emphasize strongly enough that **envelopes are not**

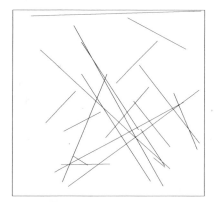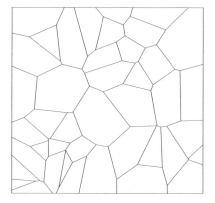

Figure 2.6. Left: *line segment pattern (object of class* `"psp"`*)*. Right: *tessellation (object of class* `"tess"`*)*.

confidence intervals — they are *critical bands* or *significance bands* associated with a test of statistical significance. The object returned by `envelope` is of class `"envelope"` and `"fv"`. It is an `"fv"` object containing the original function estimate for the data, the upper and lower envelopes, and auxiliary information. Envelopes are explained thoroughly in Sections 10.7 and 10.8.

`solist, anylist` — **lists of objects:** An object of class `"solist"` ('spatial object list') is a list of spatial objects, for example, a list of point patterns. This is often convenient when we want to handle a collection of related datasets, such as several point patterns. There is a specialised method for plotting a `"solist"` which plots the individual objects side by side.

An object of class `"anylist"` is a list of objects (not necessarily spatial objects). There is no conceptual difference between an ordinary list and an object of class `"anylist"` except for its class designation. However, the class designation allows us to have special methods for plotting and printing. It is often convenient to represent the results of calculation as an `"anylist"`. The classes `"solist"` and `"anylist"` are discussed in Section 3.8.1 and used heavily throughout this book.

`layered` — **successive layers of a plot:** To draw intricate plots in which several graphical elements are superimposed, an object of class `"layered"` can be used. A layered object is a list of spatial objects, each capable of being plotted, which are thought of as the successive layers of a complicated display. See Section 4.1.6.1. There are methods for manipulating `"layered"` objects which ensure that the layers are treated consistently, and allow the user to plot only selected layers.

`interact` — **interpoint interaction:** A very large and important collection of models for spatial point processes is that of Gibbs models. These are determined by the nature of the *interpoint interaction* that the points of these processes exhibit. The `spatstat` package provides a number of functions which create objects of class `"interact"` which describe and specify the interpoint interaction for a variety of models. These objects can then be passed to the function `ppm` which uses them in order to fit the appropriate model. All of these interaction functions are listed in the *Quick Reference Guide*; see Section 2.4.3. Fitting Gibbs models and dealing with their interaction components is dealt with in Chapter 13; see especially Section 13.5.

`quad` — **quadrature scheme:** Fitting point process models involves numerical approximation. Underlying this procedure is an object called a *quadrature scheme*: 'quadrature' is another term for 'numerical integration'. The quadrature scheme is an object of class `"quad"` and consists essentially of quadrature points with numerical weights. The quadrature points are of two kinds: data points (the points of the pattern to which the model is being fitted) and 'dummy' points. Increasing the number of dummy points improves the accuracy of the numerical approximation. The underlying ideas are explained in Chapter 9; see especially Section 9.3 and (if you are interested in the gory details) Section 9.7. Further discussion can be found in Chapter 13.

`slrm` — **spatial logistic regression model:** For methodological research purposes, `spatstat` provides an algorithm for *spatial logistic regression* via the function `slrm`. This procedure is mathematically equivalent to the procedure used in `ppm`, apart from differences in numerical approximation [34, 691]. See Section 9.9.

`msr` — **measure on a spatial domain:** An object of class `"msr"` represents a mathematical *measure on a spatial domain*. The measure may be either 'signed' (real valued) or vector valued and will in general have both a discrete (atomic) and a continuous ('absolutely continuous') component. Such objects are produced by the function `msr`, but this function is not usually called directly by the user. Such measures arise in *residual analysis* for point process models, and in particular the function `residuals.ppm` returns an object of class `"msr"`. Residual analysis and the associated measures are discussed in Chapter 11, particularly in Section 11.3, and in Chapter 13, particularly in Section 13.10.3.

linnet — **linear network:** A network of straight lines in two-dimensional space is represented in spatstat as an object of class "linnet". The straight lines might represent roads, rivers, wires, nerve fibres, pipes, mortar lines between bricks, and so on. The class "linnet" is discussed in Chapter 17.

lpp — **point pattern on a linear network:** In some cases, the spatial locations of events are constrained to lie on a network of lines. For example, road accidents can occur only on the road network. A point pattern on a linear network is represented in spatstat as an object of class "lpp", discussed in Chapter 17.

2.5 FAQs

• *What are the connections and differences between* spatstat *and other packages?*

See the Spatial Task View on CRAN for an overview of packages for analysing spatial data. Some packages listed there are intended to provide basic data analysis capabilities, file conversion, etc.

The spatstat package is focused on statistical analysis of spatial point pattern data. It is based on statistical principles, and follows the standard practice and terminology of spatial statistics. It is designed to contain all the standard tools that would be expected in a package for spatial point pattern analysis, as well as a wide selection of advanced methods.

The spatstat package is not derived from, or dependent on, or superseded by, any other package in spatial statistics. Several packages depend on spatstat; see the Reverse Depends: and Reverse Suggests: lists on the CRAN page for spatstat.

• *I want to test whether the point pattern is random. Can* spatstat *do that?*

Yes, and much more. The spatstat package provides facilities for formal inference (such as hypothesis tests) and informal inference (such as residual plots).

If you want to formally test the hypothesis of complete spatial randomness (CSR) you can do this using the χ^2 test based on quadrat counts (quadrat.test), the Kolmogorov-Smirnov test based on values of a covariate (cdf.test), graphical Monte Carlo tests based on simulation envelopes of the K-function (envelope), non-graphical tests like the Diggle-Cressie-Loosmore-Ford test (dclf.test) or the maximum absolute deviation test (mad.test), or the likelihood ratio test for parametric models (anova.ppm).

If you want to formally test the goodness-of-fit of another point process model, you can again use the χ^2 test (for Poisson models only), or graphical or non-graphical Monte Carlo tests (for Poisson, Gibbs, Cox, or cluster process models).

Formal tests are not the only procedure that should be used: informal validation of a point process model is equally important. By inspecting various kinds of diagnostic plots, we can assess whether the assumptions underlying our analysis appear to be valid. Tools include the point process residual plots (diagnose.ppm), nonparametric regression (rhohat), partial residual plots (parres), and added variable plots (addvar).

• *What is the practical limit on the number of points in a point pattern?*

This depends on the memory available to R, and on the version of spatstat. On a typical laptop with 2Gb of RAM, it should currently be possible to perform exploratory analysis (such as the

K-function) for 1 million points, and model-fitting for about 300,000 points. The hard limit on the number of points in a point pattern is currently the maximum length of a vector in R, namely 2^{31} or about 2 billion points.

3

Collecting and Handling Point Pattern Data

This Chapter provides guidance on collecting spatial data in surveys and experiments (Section 3.1), wrangling the data into files and reading it into R (Sections 3.2 and 3.9–3.10), handling the data in `spatstat` as a point pattern (Section 3.3), checking for data errors (Section 3.4), and creating a spatial window (Section 3.5), a pixel image (Section 3.6), a line segment pattern (Section 3.7), or a collection of spatial objects (Section 3.8).

3.1 Surveys and experiments

Every field of research has its own specialised methods for collecting data in surveys, observational studies, and experiments. We do not presume to tell researchers how to run their own studies.[1] However, statistical theory gives very useful guidance on how to avoid fundamental flaws in the study methodology, and how to get the maximum information from the resources available.[2] In this section we discuss some important aspects of data collection, draw attention to common errors, and suggest ways of ensuring that the collected data can serve their intended purpose.

3.1.1 Designing an experiment or survey

The most important advice about designing an experiment is to *plan the entire experiment, including the data analysis.* One should think about the entire sampling process that leads from the real things of interest (e.g., trees in a forest with no observers) to the data points on the computer screen which represent them. One should plan how the data are going to be analysed, and how this analysis will answer the research question. This exercise helps to clarify what needs to be done and what needs to be recorded in order to reach the scientific goal.

A *pilot experiment* is useful. It provides an opportunity to check and refine the experimental technique, to develop protocols for the experiment, and to detect unexpected problems. The data for the pilot experiment should undergo a *pilot data analysis* which checks that the experiment is capable of answering the research question, and provides estimates of variability, enabling optimal choice of sample size. Experience from the pilot experiment is used to refine the experimental protocol. For example, a pilot study of quadrat sampling of wildflowers might reveal that experimenters were unsure whether to count wildflowers lying on the border of a quadrat. The experimental protocol should clarify this.

Consideration of the entire sampling process, leading from the real world to a pattern of dots on a screen, also helps to identify *sources of bias* such as sampling bias and selection effects. In a galaxy survey in radioastronomy, the probability of observing a galaxy depends on its apparent magnitude at radio wavelengths, which in turn depends on its distance and absolute magnitude. Bias

---

[1] "Hiawatha, who at college/ Majored in applied statistics,/Consequently felt entitled/ To instruct his fellow man/ In any subject whatsoever" [374]

[2] "To consult the statistician after an experiment is finished is often merely to ask him to conduct a post mortem examination. He can perhaps say what the experiment died of." *R.A. Fisher*

of this known type can be handled during the analysis. In geological exploration surveys, uneven survey coverage or survey effort introduces a sampling bias which cannot easily be handled during analysis unless we have information about the survey effort.

Ideally, a survey should be designed so that the surveyed items can be *revisited* to cross-check the data or to collect additional data. For example, GPS locations or photographic evidence could be recorded.

3.1.2   What needs to be recorded

In addition to the coordinates of the points themselves, several other items of information need to be recorded. Foremost among these is the *observation window*, that is, the region in which the point pattern was mapped. Recording the boundaries of this window is important, since it is a crucial part of the experimental design: there is information in where the points were *not* observed, as well as the locations where they *were* observed.

Even something as simple as estimating the intensity of the point pattern (average number of points per unit area) depends on the window of observation. Attempting to infer the observation window from the data themselves (e.g., by computing the convex hull of the points) leads to an analysis which has substantially different properties from one based on the true observation window. An analogy may be drawn with the difference between sequential experiments and experiments in which the sample size is fixed *a priori*. Analysis using an estimated window could be seriously misleading.

Another vital component of information to be recorded consists of *spatial covariates*. A 'covariate' or explanatory variable is any quantity that may have an effect on the outcome of the experiment. A 'spatial covariate function' is a spatially varying quantity such as soil moisture content, soil acidity, terrain elevation, terrain gradient, distance to nearest road, ground cover type, region classification (urban, suburban, rural), or bedrock type. More generally, a 'spatial covariate' is any kind of spatial data, recruited as covariate information (Section 1.1.4). Examples include a spatial pattern of lines giving the locations of geological faults, or another spatial point pattern.

A covariate may be a quantity whose 'effect' or 'influence' is the primary focus of the study. For example in the Chorley-Ribble data (see Section 1.1.4, page 9) the key question is whether the risk of cancer of the larynx is affected by distance from the industrial incinerator. In order to detect a significant effect, we need a covariate that represents it.

A covariate may be a quantity that is not the primary focus of the study but which we need to 'adjust' or 'correct' for. In epidemiological studies, measures of exposure to risk are particularly important covariates. The density of the susceptible population is clearly important as the denominator used in calculating risk. A measure of sampling or censoring probability can also be important.

Theoretically, the value of a covariate should be observable at any spatial location. In reality, however, the values may only be known at a limited number of locations. For example the values may be supplied only on a coarse grid of locations, or measured only at irregularly scattered sample locations. Some data analysis procedures can handle this situation well, while others will require us to interpolate the covariate values onto a finer grid of locations.

The minimal requirement for covariate data is that, in addition to the covariate values at all points of the point pattern, the **covariate values must be available at some 'non-data' or 'background' locations**. This is an important methodological issue. **It is not sufficient to record the covariate values at the data points alone.**

For example, the finding that 95% of kookaburra nests are in eucalypt trees is useless until we know what proportion of trees *in the forest* are eucalypts. In a geological survey, suppose we wish to identify geochemical conditions that predict the occurrence of gold. It is not enough to record the geochemistry of the rocks which host each of the gold deposits; this will only determine geochemical conditions that are *consistent* with gold. To *predict* gold deposits, we need to find geochemical conditions that are more consistent with the presence of gold than with the absence of gold, and that requires information from places where gold is absent. (Bayesian methods make it possible to substitute other information, but the basic principle stands.)

There are various ways in which a covariate might be stored or presented to a `spatstat` function for analysis. Probably the most useful and effective format is a *pixel image* (Section 3.6).

It is good practice to record the *time* at which each observation was made, and to look for any apparent trends over time. An unexpected trend suggests the presence of a *lurking variable* — a quantity which was not measured but which affects the outcome. For instance, experimental measurements may depend on the temperature of the apparatus. If the temperature is changing over time, then plotting the data against time would reveal an unexpected trend in the experimental results, which would help us to recognise that temperature is a lurking variable.

3.1.3 Risks and good practices

The greatest risk when recording observations is that important information will be omitted or lost. Charles Darwin collected birds from different islands in the Galapagos archipelago, but failed to record which bird came from which island. The missing data subsequently became crucial for the theory of evolution. Luckily Darwin was able to cross-check with the ship's captain, who had collected his own specimens and had kept meticulous records.

What information will retrospectively turn out to be relevant to our analysis? This can be difficult to foresee. The best insurance against omitting important information is to **enable the observations to be revisited** by recording the context. For example when recording wildflowers inside a randomly positioned wooden quadrat in a field, we could easily use a smartphone to photograph the quadrat and its immediate environment, and record the quadrat location in the field. This will at least enable us to revisit the location later. Scientific instincts should be trusted: if you feel that something *might* be relevant, then it should be recorded.

In particular **don't** discard recorded data or events. Instead annotate such data to say they 'should' be discarded, and indicate why. The data analysis can easily cope with such annotations. This rule is important for the integrity of scientific research, as well as an important precaution for avoiding the loss of crucial information.

Astronomers sometimes delete observations *randomly* from a survey catalogue to compensate for bias. For example, the probability of detecting a galaxy in a survey of the distant universe depends on its apparent brightness, which depends on its distance from Earth. Nearby galaxies would be overrepresented in a catalogue of all galaxies detected in the survey. Common practice is to delete galaxies at random from the survey catalogue, in such a way that the probability of retaining (not deleting) a galaxy is inversely proportional to the sampling bias (the conjectured probability of observing the galaxy). In some studies the randomly thinned catalogue becomes 'the' standard catalogue of galaxies from the survey. We believe this is unwise, because information has been lost, and because this procedure is unnecessary: the statistical analysis can cope with the presence of sampling bias of a known type. In other studies, the random thinning is done repeatedly (starting each time from the original observations); this is valid and is an application of bootstrap principles [316].

A particular danger is that events may be effectively deleted from the record when their spatial location ceases to exist. For example, in road traffic accident research, the road network changes from year to year. If a four-way road intersection has been changed into a roundabout, should traffic accidents that occurred at the old intersection be deleted from the accident record? If we did (or if the database system effectively ignored such records), it would be impossible to assess whether the new roundabout is safer than the old intersection.

Where data are missing, record the 'missingness'. That is, if no value is available for a particular observation, then the observation should be recorded as 'NA'. Moreover when recording the 'missingness' be sure to use proper missing value notation — do not record missing values as supposedly implausible numerical values such as '99' or '−99'. Doing so can have disastrous results in the analysis. Likewise **do** record 'zeroes' — e.g., zero point counts for quadrats in which no points appear. Do not confuse these two ideas: 'missing' or 'unobservable' (NA) is a *completely different concept* from 'absent' (0).

Data should be recorded at the same time as the observation procedure; record as you go. If writing observations down on paper or tablet is not feasible, use a recording device such as a mobile phone. A photograph of the immediate environment can also be taken with a mobile phone.

In accordance with Murphy's Law, it is imperative to keep backups of the original data, preferably in text files. Data that are stored in a compressed or binary format, or in a proprietary format such as a word-processing document, may become unreadable if the format is changed or if the proprietary software is updated.

To ensure good practice and forestall dispute, conditions for accessing the data should be clarified. Who owns the data, who has permission to access the data, and under what conditions? Privacy and confidentiality restrictions should be clarified.

Data processing (including reorganising and cleaning data) should be documented as it happens. Record the sequence of operations applied to the data, explain the names of variables, state the units in which they are expressed, and so on. Data processing and cleaning can usually be automated, and is usually easy to implement by writing an R script. The script effectively documents what you did, and can be augmented and clarified by means of comments.

Data analysis should also be documented as it happens. We strongly recommend writing R scripts for all data analysis procedures: this is easier in an environment such as RStudio or ESS. The interactive features of R are very handy for exploring possibilities, and it does provide a basic mechanism for recording the history of actions taken. The disadvantage is that it can be difficult to 'back out' (to return to an earlier stage of analysis) and the analysis may depend on the state of the R workspace. Once you have figured out what to do, we strongly advise writing code (with copious comments!) that performs the relevant steps from the beginning. This makes the analysis reproducible and verifiable.

3.2   Data handling

3.2.1   Data file formats

If you obtain data files from another source, such as a website, it is of course important to understand the file format in which the data are stored and to have access to the software needed to extract the data from the files in question. It is also important to obtain all of the available information about the protocols under which the data were gathered, the range of possible values for each variable, the precision to which the variables were recorded, whether measurements were rounded and if so how, the taxonomic system or nomenclature used, and the treatment of missing values. See Chapter 4 for some advice on these matters.

If you have collected data yourself it is, as was mentioned above, good practice to save the original data in a text file, so that it is not dependent on any particular software. The text file should have a clearly defined format or structure. Data in a text file can easily be read into R.

For storing the point coordinates and associated mark values (see Section 3.3.2 for a discussion of marks) we recommend the following file formats.

table format: the data are arranged in rows and columns, with one row for each spatial point. There is a column for each of the *x*- and *y*-coordinates, and additional columns for mark variables. See Figure 3.1. The first line may be a *header* giving the names of the column variables.

Character strings must be enclosed in quotes unless they consist entirely of letters. Missing values should be entered as NA. The usual file name extension is .txt or .tab (the latter is understood by R to indicate that the file is in table format).

comma-separated values (csv): Spreadsheet software typically allows data to be exported to or imported from a comma-separated values file (extension .csv). This format is slightly more compressed than table format. Data values are separated by a comma (or other chosen character) rather than by white space. This format is convenient because it is widely accepted by other software, and is more memory-efficient than table format. However, errors are more difficult to detect visually than they are in table format.

shapefiles: A shapefile is a popular file format for sharing vector data between geographic information system (GIS) software packages. It was developed and is maintained by the commercial software provider ESRI™. Most of the specification is publicly accessible [254]. Storing data in a shapefile will result in a handful of files, with different extensions (at least .shp, .shx, and .dbf) which refer to different information types, e.g., the coordinates and the geographical projection that was used. Reading data from shapefiles is described in Section 3.10.

```
Easting       Northing      Diameter    Species
176.111       32.105        10.4        "E. regnans"
175.989       31.979        7.6         "E. camaldulensis"
....          ....          ....
```

Figure 3.1. *Example of text file in table format.*

You will also need to store auxiliary data such as the coordinates of the (corners of the) window boundary, covariate data such as a terrain elevation map, and metadata such as ownership, physical scale, and technical references. The window boundary and covariate data should also be stored in text files with well-defined formats: we discuss this in Section 3.5. Metadata can usually be typed into a plain text file with free format.

3.2.2 Reading data into R

Data in a text file in table format can be read into R using the command read.table. A comma-separated values file can be read into R using read.csv. Set the argument header=TRUE if the file has a header line (i.e. if the first line of the file gives the names of the columns of data).

The original data files for the vesicles dataset are installed in spatstat as a practice example. To copy these files to the current folder, type

```
> copyExampleFiles("vesicles")
```

The coordinates of the vesicles can then be read by either of the commands

```
> ves <- read.table("vesicles.txt", header=TRUE)
> ves <- read.csv("vesicles.csv")
```

The resulting object ves is a 'data frame' in R. You may need to set various options to get the desired result: type help(read.csv) or help(read.table) for information.

Use colnames(ves) to see the names of the columns in the data frame ves: these may have changed if the original column names contained strange characters or were duplicated. Note that if

the original data file had no header line, the columns of the data frame will have the default names
`V1, V2, ....` Use `head(ves)` to see the first few rows of data, and `summary(ves)` to see a
summary of the values in each column of the data frame. See Section 2.1.6 for more on data frames.

It is important to check that each column of data belongs to the intended class. Note that a column of character strings in the text file will be converted to a factor (categorical variable) by default.
Conversion to a factor would probably be appropriate for the `Species` column in Figure 3.1. However, character strings could also represent date-and-time values, or text annotations. In this case
`read.table` or `read.csv` should be called with `stringsAsFactors=FALSE` to prevent automatic
conversion to factors (or `options` should be used to change the default behaviour); then each column should be converted to the desired type. Factors are created using `factor` or `as.factor`. For
more details on factors see Section 2.1.9. Strings representing date-time values are converted using
`as.Date` or `as.POSIXct`. For more details on handling dates in R see the help entries for `ISOdate`
and `ISOdatetime`, or the online resources [578], `www.statmethods.net/input/dates.html`
or `en.wikibooks.org/wiki/R_Programming/Times_and_Dates`.

Note that if a column of numbers in the text file has been 'corrupted' with non-numeric characters — possibly due to typing errors — then this column will be read in as *character* data (and then
by default converted to a factor). Checking on the class of each column serves to detect when such
errors have occurred. A quick and easy way to find out the class of data in each column of your data
frame `df` is `sapply(df,class)`. If conversion errors are found, the text file should be corrected,
and read in again. Alternatively the data frame can be viewed and edited in a spreadsheet-style
interface using the R functions `View` and `edit`.

3.3   Entering point pattern data into `spatstat`

A spatial point pattern in two-dimensional space is stored in `spatstat` as an object of class `"ppp"`
(for 'planar point pattern'). In order to use the capabilities of `spatstat`, a spatial point pattern
dataset should be converted into an object of this class.

A point pattern object contains the spatial coordinates of the points, the marks attached to the
points (if any), the window in which the points were observed, and the name of the unit of length for
the spatial coordinates. Thus, a single object of class `"ppp"` contains *all* the information required to
perform standard calculations about a point pattern dataset.

This section describes some basic ways to create `"ppp"` objects from raw data, or from data
stored in a text file. For data stored in a recognised GIS file format, alternative methods are described
in Section 3.10. Section 3.9 explains how to create a point pattern interactively using a point-and-
click interface, which can be useful when the original dataset is a digital photograph or another form
of spatial data.

3.3.1   Creating a `"ppp"` object

To create an object of class `"ppp"` from raw data, use the function `ppp`. Suppose that the x, y
coordinates of the points of the pattern are contained in vectors `x` and `y` (which must, of course, be
of equal length). Then

```
X <- ppp(x, y, other.arguments)
```

will create the point pattern object `X`. The *other.arguments* must determine a window for the
pattern. Table 3.1 shows the different options for specifying a window.

If the observation window is a rectangle, it is sufficient to specify the ranges of the x and y
coordinates:

`ppp(x, y, xrange, yrange)`	point pattern in rectangle
`ppp(x, y, poly=p)`	point pattern in polygonal window
`ppp(x, y, mask=m)`	point pattern in binary mask window
`ppp(x, y, window=w)`	point pattern in specified window

Table 3.1. *Basic options for creating a point pattern using the creator function* `ppp`.

```
> df <- read.table("vesicles.txt", header=TRUE)
> x <- df$x
> y <- df$y
> X <- ppp(x, y, c(22,587), c(11,1031))
```

or more compactly

```
> X <- with(df, ppp(x, y, c(22,587), c(11,1031)))
```

If the argument `window` is given, then it must be a window object (of class `"owin"`) specifying the window for the point pattern. Otherwise, the additional arguments are passed to the function `owin` to create a window object. Section 3.5 gives a detailed explanation of these arguments.

Often the window of observation is a rectangle, so this requirement just means that we have to specify the *x* and *y* dimensions of the rectangle when we create the point pattern. Windows with a more complicated shape can easily be represented in `spatstat`, as described below.

The term 'window of observation' presumes that the points are scattered in two-dimensional space but that observations were confined to a known study region ('Window Sampling', page 143). This may not be appropriate in some applications. However, many statistical techniques still require some kind of bounding region for the point pattern. If the points are confined to a bounded region of space, like fish in a lake, the 'Small World' model (page 145) is more appropriate. If the bounding region is really unknown, `spatstat` provides the function `ripras` to compute the Ripley-Rasson [580] estimator of the bounding region, given only the point locations.

After creating a point pattern object X, it is advisable to type X to print the object, `is.ppp(X)` to check that it is indeed a point pattern, `summary(X)` to summarise its contents, and `plot(X)` to plot the pattern. More about these commands is explained in Chapter 4.

The generic functions `View` and `edit` also have methods for `"ppp"` objects, allowing the user to inspect and edit the spatial coordinates in a spreadsheet-like interface.

3.3.2   Marks

Chapter 1 introduced the idea of a 'mark', an additional attribute of each point in a point pattern. For example, in addition to recording the locations of trees in a forest, we could also record the species, diameter, and height of each tree, a chemical analysis of the leaves of each tree, and so on.

Suppose x and y are vectors containing the coordinates of the point locations, as before, and for simplicity assume that the observation window is a rectangle with extent given by `xrange` and `yrange`. If there are marks attached to the points, store the corresponding marks in a vector m with one entry for each point or in data frame m with one row for each point and one column for each mark variable. (It is also possible to use a matrix rather than a data frame to store multiple marks, but such a matrix is just converted to a data frame internally by `ppp` and in general a data frame is preferred.) Then create the marked point pattern by

```
ppp(x, y, xrange, yrange, marks=m)
```

For example, the following code reads raw data from a text file in table format, and creates a point pattern with a column of numeric marks containing the tree diameters:

```
> copyExampleFiles('finpines')
> fp <- read.table('finpines.txt', header=TRUE)
> X <- with(fp, ppp(x, y, c(-5,5), c(-8,2), marks=diameter))
```

An even slicker way to do this is to convert the data frame directly into a point pattern using the conversion operator as.ppp:

```
> fp <- read.table("finpines.txt", header=TRUE)
> X <- as.ppp(fp, owin(c(-5,5), c(-8,2)))
```

Notice this requires that the first two columns of fp contain the x and y coordinates (which they do in this case). The two steps of reading in data and creating an object of class "ppp" can be reduced to one step by using scanpp:

```
> X <- scanpp("finpines.txt", owin(c(-5,5), c(-8,2)))
```

The handling of marks in spatstat depends on their type. Mark values may belong to any of the atomic data types: numeric, integer, character, logical, or complex. Marks may also be categorical values (see below), calendar dates, or date/time values. Character-valued marks are rarely used; they should usually be converted to categorical or date/time values. To check that your data has the intended type, use class(m) if m is a vector and sapply(m, class) if m is a data frame.

For a marked point pattern, the functions View and edit allow the user to inspect and edit both the spatial coordinates and the marks.

3.3.2.1 Categorical marks

When the mark is a categorical variable, we have a *multitype point pattern* as described in Section 1.1.2 (some authors call it a 'multivariate' pattern; see Section 14.2.5). **The mark values must be stored as a 'factor' in** R. The possible 'types' are the different levels of the mark variable.

The installed dataset demopat is an artificial (simulated) point pattern that was created for demonstration purposes. It is a pattern with categorical marks:

```
> demopat
Marked planar point pattern: 112 points
Multitype, with levels = A, B
window: polygonal boundary
enclosing rectangle: [525, 10575] x [450, 7125] furlongs
```

The output (from the spatstat function print.ppp) indicates that this is a multitype point pattern. Here is the vector of marks:

```
> marks(demopat)
  [1] A B B A B B B A A A B A A B B A A A B B A A A A B B B A A B B B B B B A A B
 [38] A A B B A A B B B B A B B B B B B B A A A B A B A B B B B B B A B B B A A B B
 [75] B B B A B B B A A B A B B B A B A B B B B B A A B A B B B B B B A A A B A B B
[112] A
Levels: A B
```

This output indicates that marks(demopat) is a factor with levels A and B in that order. To stipulate a different ordering of the levels, do something like

```
> marks(demopat) <- factor(marks(demopat), levels=c("B", "A"))
```

or use the function relevel.

> Tip: Whenever you create a factor f, check that the factor levels are as you intended, using
> `levels(f)`. Check that the values have been correctly matched to the levels, by printing f or
> using `any(is.na(f))`.

Other ways of adding marks to a point pattern are described in Sections 4.2.4, 14.3, and 15.2.1.

3.3.2.2 Multivariate marks

A point pattern may have *several* mark variables attached to each point. For example, the `finpines`
dataset installed in `spatstat` gives the locations of 126 pine saplings in a Finnish forest, as well as
the diameter and height for each tree.

```
> finpines
Marked planar point pattern: 126 points
Mark variables: diameter, height
window: rectangle = [-5, 5] x [-8, 2] metres
```

Each point of the pattern is now associated with a *multivariate* mark value, and we say that the point
pattern has *multivariate marks*. (Note the potential for confusion with the term 'multivariate point
pattern' used by other authors in a different sense.)

To create a point pattern with multivariate marks, the mark data should be supplied as a data
frame, with one row for each data point and one column for each mark variable. For example,
`marks(finpines)` is a data frame with two columns named `diameter` and `height`. It is important
to check that each column of data has the intended type. Chapter 15 covers the analysis of point
patterns with multivariate marks.

3.3.3 Units

A point pattern X may include information about the units of length in which the x and y coordinates
are recorded. This information is optional; it merely enables the package to print better reports and
to annotate the axes in plots. It is good practice to keep track of the units.

If the x and y coordinates in the point pattern X were recorded in metres, type

```
> unitname(X) <- "m"
```

to use the standard abbreviation or supply both a singular and plural form if the full version is
desired:

```
> unitname(X) <- c("metre", "metres")
```

The measurement unit can also be given as a multiple of a standard unit. If, for example, one unit
for the coordinates equals 42 centimetres, type

```
> unitname(X) <- list("cm", "cm", 42)
```

The name of the unit of measurement can also involve accents or characters from non-Latin alpha-
bets: see page 80.

Note that the `unitname` applies only to the coordinates, and not to the marks, of a point pattern.
The units in which (numeric) marks are recorded are usually unrelated to the units in which the
spatial coordinates are recorded.

Altering the `unitname` in an existing dataset, while possible, is usually not sensible; it simply
alters the name of the unit, without changing the values of the coordinates. To convert the coordinates into a different unit of measurement (e.g., from metres to kilometres) use the command
`rescale` as described in Section 4.2.5.

If you really want to change the coordinates by a linear transformation, producing a dataset that
is *not equivalent to the original*, use `affine` or `scalardilate`.

3.4　Data errors and quirks

Experienced applied statisticians expect data to have problems that need fixing before a reliable analysis can be performed. Problems can arise in various ways, such as: transcription and recording errors; unclear definitions of variables or units of measurement; unexplained conventions (e.g., recording missing values as 99); errors or omissions in metadata; discretisation of data; bugs in software interfaces and file conversions; software version conflicts; failures of recording equipment; or exigencies of the experiment. Here we discuss various techniques for detecting such problems.

3.4.1　Definition of variables

For the variables recorded in a dataset, we need to know the range of possible values for each variable, the units in which the variables are recorded, and any conventions used for recording special values (such as 'infinite' or 'missing' values). An unambiguous definition of the variable is also important — for example, for angular coordinates we need to know whether the angle is measured clockwise or anticlockwise.

If the data are obtained from another source, it is important to obtain this information, usually from supplementary files or metadata. If the data are your own, it is highly recommended to write a separate plain text file containing this information, as discussed in Section 3.2.

Units of measurement are vital. Some important scientific errors (including the loss of a \$300 million spacecraft) have occurred because the units were given incorrectly or misinterpreted. Abbreviations for units can be misinterpreted — for example the symbol " is used to denote seconds of time, seconds of arc, and inches. In astronomy, Right Ascension is an angular coordinate like longitude, but measured in the opposite direction, and expressed in hours, minutes, and seconds of elapsed *time* in a 24-hour clock.

A good way to check for misinterpretation of variables in a dataset is to plot the data (see Section 4.1). Anomalies such as periodic patterns, impossibly dense clusters, and large gaps suggest misinterpretation of a variable. If possible, compare your plot with an original graphic of the data — perhaps a figure in the original publication, or an illustration on a website. Superimpose your own plot on the original figure for comparison.

3.4.2　Missing values

Some observations may be missing or unavailable. It is a very common (but *very bad*) practice to encode missing values as strange numbers like 99 or -1. Some people do not distinguish between 'missing' and 'zero', and thus record missing values as 0. Errors of this latter sort can be very hard to detect, especially if there are genuine zeroes in the data.

To find out if your data have been affected by this problem, the first and best option is to check the available documentation to determine how missing values were recorded.

Otherwise, there are many tricks for guessing such conventions. We recommend a histogram or a stem-and-leaf plot, generated by the R commands `hist` and `stem`. Look for frequently occurring values that seem strange.

In R, the symbol `NA` represents a missing value, and the entire system is built to handle missing values. Even when reading a stream of numbers from a text file, R will recognise the string `NA` as denoting a missing value. If you know the convention for representing missing values in your data, we highly recommend that these values be rewritten as `NA` to avoid confusion. If the value `-999` is used to represent missing values in a vector `x`, these can be changed to `NA` by

```
> x[x == -999] <- NA
```

3.4.3 Data entry checking

Initial exploration of data should include checks for errors in data entry. Typing and transcription errors tend to produce outliers, which will be revealed by graphical methods such as histograms and boxplots of the data.

One very basic and easy step in checking over a point pattern for data problems is to print out the coordinate values and marks using `as.data.frame(X)` or view them in a spreadsheet-like interface using `View(X)`. Use `head(as.data.frame(X))` to print only the top few lines, or `page(as.data.frame(X),method="print")` to print the data a page at a time. Visually scanning the data in this way can often reveal obvious errors in data entry. Errors can be corrected manually using the spreadsheet interface `edit(X)`.

Another crucial step is to plot the point pattern data (see Section 4.1.2). Look for unexpected 'structure' in the points such as the presence of bands or periodic patterns: this can be caused by errors in transforming the spatial coordinates, misunderstandings about the definitions of the spatial coordinates, or the use of an inappropriate window.

If points lie outside the window, then there is either something wrong with the window or something wrong with the points, or both! When a point pattern object has been created using ppp, points that lie outside the window will already have been detected by ppp:

```
> mybad <- ppp(x=c(-0.2, runif(10)),
              y=c( 0.3, runif(10)), window=square(1))
```

```
Warning message:
1 point was rejected as lying outside the specified window
```

These 'reject' points are not treated as legitimate points of the pattern, but are retained as an auxiliary 'attribute' of the pattern:

```
> mybad
Planar point pattern: 10 points
window: rectangle = [0, 1] x [0, 1] units
*** 1 illegal point stored in attr(,"rejects") ***
> attr(mybad, "rejects")
Planar point pattern: 1 point
window: polygonal boundary
enclosing rectangle: [-0.640543, 1.1982717] x [-0.2263007, 1.3539303] units
```

When the point pattern is plotted, the rejects are also plotted (with a warning). The rejects can be removed using `as.ppp`:

```
> as.ppp(mybad)
Planar point pattern: 10 points
window: rectangle = [0, 1] x [0, 1] units
```

However, it is not advisable to remove the offending points until you understand the reason for their offence.

If you have concerns or suspicions about an individual point of the pattern you can, after plotting the pattern, identify that point by typing `identify(X)` and clicking on the point in question; see Section 4.1.5. Alternatively the interactive plotting function `iplot` can be used.

The ppp method for the summary function may reveal quirks and anomalies in the data. You may need to determine the specifics of these anomalies by visually (re-) scanning the data as described above. Simply type `summary(X)` to apply the appropriate summary method to X.

3.4.4 Duplication

If two entries in a dataset are identical, this may or may not be the result of an error. Duplication of entire lines of a data file may occur because of recording errors or data-entry errors, in which case the duplicated lines will usually be adjacent. Duplication of point coordinates (i.e. having two records refer to the same (x, y) location) may happen for a variety of reasons and is surprisingly common. One of the possible reasons for such duplication is rounding, as discussed in Section 3.4.5, but there are others.

Duplication of points is important, because statistical methodology for spatial point processes (as used in this book) is based largely on assumption that processes are *simple*, i.e. that points of the process can never be coincident. When the data have coincident points, some statistical procedures designed for simple point processes will be severely affected. For example, the pair correlation function (Chapter 7) will have an infinite value at distance zero. It is strongly advisable to check for duplicated points and to decide on a strategy for dealing with them if they are present.

You can check for duplication of entries in a dataset using the generic function `duplicated`. If your data are stored as a matrix or a data frame, this will invoke `duplicated.data.frame` which compares rows of the array. The result is a logical vector, with one entry for each row of data, that is TRUE if the current row is identical to an *earlier* row.

If X is a point pattern, `duplicated(X)` will invoke the method `duplicated.ppp`. The result is a logical vector, with one entry for each point, that is TRUE if the current point is identical to an earlier point in the sequence. Note that, by default, `duplicated.ppp` and `duplicated.data.frame` use different rules for deciding whether values are identical. The rule for data frames is less strict, and thus more likely to declare values to be identical. See `help(duplicated.ppp)` for options to make the two methods consistent.

For a *marked* point pattern, two points are declared to be identical when their coordinates *and* their marks are identical. Two points at the same location but with different marks are not considered duplicates. To check for duplication of point coordinates only, use `duplicated(unmark(X))` or `duplicated(X, rule="unmark")`.

To discard duplicate points, type `Y <- unique(X)` or `Y <- X[!duplicated(X)]`. This retains a data point if it is not identical to any earlier points in the sequence. The function `unique` is generic; the method for point patterns takes account of the marks of the points as well as their spatial coordinates. To ignore the marks when deciding whether points are identical, type `Y <- unique(X, rule="unmark")`. Note that if several marked points share the same spatial location, this command extracts the first of these points in the sequence.

To count the number of coincident points, use `multiplicity(X)`. This returns a vector of integers, with one entry for each point of X, giving the number of points that are identical to the point in question (including itself). The function `multiplicity` is generic. The method for point patterns again takes account of the marks. To ignore marks when computing multiplicity, use `multiplicity(unmark(X))`.

A handy syntax to use when checking for duplication is `any(duplicated(X))` which will reveal if any duplication occurs. Applying `which(multiplicity(X) > 1)` will allow you to locate where the duplication has occurred and perhaps help you to determine how to account for it.

What to *do* about duplicated points is often unclear; it depends on the context and on the objectives of the analysis. An alternative to deleting duplicate points is to perturb the coordinates slightly using `rjitter`. Another alternative is to make the points of the pattern unique using `unique`, and to attach the multiplicities of the points to the pattern as marks. This can be done by something like:

```
dup <- duplicated(X)
marks(X) <- cbind(marx=marks(X), mul=multiplicity(X))
Y <- unique(X)
```

Data with multiplicities require different analysis techniques, depending on the objective.

There are also cases where a single point is erroneously recorded twice with *slightly different* coordinate values, for example when points are entered using a graphical interface. These will not be detected by the code above. One would typically use `nndist`, `pairdist` or `closepairs` to identify such cases: see Chapter 8.

3.4.5 Rounding

Spatial coordinates have usually been *rounded* or *discretised* to a certain number of significant digits. This may have occurred when the coordinates were recorded, or when they were stored in a text file, or when the data were rescaled.

The effects of rounding can substantially change the results of some statistical techniques, particularly those which deal with distances between neighbouring points. Rounding can also cause duplication of points, because rounding could map two distinct points in space to the same rounded location.

It is important to check whether the spatial coordinates of the point pattern have been rounded. If no background information is available, the function `rounding.ppp` will try to guess the number of digits used, but it is not always correct. A plot of the data, especially the *Fry plot* (Section 7.2.2), will often reveal the discretisation.

Note that, in an R session, numbers are printed to a limited number of significant digits, determined by `options("digits")`. This may give a false impression that the values have been rounded.

3.5 Windows in `spatstat`

Many data types in `spatstat` require us to specify the region of space inside which the data were observed. This is the *observation window* and it is represented by an object of class `"owin"`. Objects of this class are created from raw data by the function `owin`, or converted from other types of data by `as.owin`.

An `"owin"` object belongs to one of three types: rectangles, polygonal regions, and binary pixel masks. See Figure 3.2. Table 3.2 summarises the main options for creating each type of window, using `owin`.

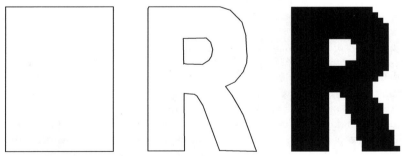

Figure 3.2. *Types of windows.* Left: *rectangle;* Middle: *polygonal;* Right: *binary mask.*

There are methods for printing and plotting windows, and there are numerous geometrical operations for manipulating window objects (described in Section 4.2). Here we describe how to create a window from raw data.

owin(xrange, yrange)	rectangle
owin(poly=p)	polygonal region
owin(mask=m)	binary pixel mask

Table 3.2. *Options for creating a window using the creator function* owin.

3.5.1   Rectangular window

A rectangular window in spatstat represents a rectangle with sides parallel to the coordinate axes. Rectangles can have zero width or zero height. To create a rectangular window, type owin(xrange, yrange) where xrange, yrange are vectors of length 2 giving the x and y dimensions, respectively, of the rectangle.

```
> owin(c(0,3), c(1,2))
window: rectangle = [0, 3] x [1, 2] units
```

Alternatives are as.owin and square:

```
> as.owin(c(0,3,1,2))
window: rectangle = [0, 3] x [1, 2] units
> square(5)
window: rectangle = [0, 5] x [0, 5] units
> square(c(1,3))
window: rectangle = [1, 3] x [1, 3] units
```

The function is.rectangle checks whether an object is a rectangular window.

3.5.2   Polygonal window

Any region drawn on a map (using vector graphics) can be represented as a *polygonal window.* Such windows are commonly used to represent national boundaries or administrative regions, such as the Chorley-South Ribble region (Figure 1.12 on page 9).

A polygonal window is defined as a region of space whose boundary is composed of straight line segments. The window may consist of several pieces which are not connected to each other. Each piece may have holes. The boundary of a polygonal window consists of several closed polygonal curves, which do not cross themselves or each other.

The spatstat package supports a full range of geometrical operations and analytic calculations on polygonal windows.

To create a polygonal window from raw data, type owin(poly=p, xrange, yrange) or just owin(poly=p). The argument poly=p indicates that the window is polygonal and its boundary is given by the dataset p. Note we must use the name=value syntax to give the argument poly. The arguments xrange and yrange are optional here; if they are absent, the x and y dimensions of the bounding rectangle will be computed from the polygon.

If the window boundary is a single polygon, then p should be a matrix or data frame with two columns, or a list with components x and y, giving the coordinates of the vertices of the window boundary, **traversed anticlockwise**[3] without repeating any vertex. For example, the triangle in the left panel of Figure 3.3 with corners $(0,0)$, $(10,0)$, and $(0,10)$ is created by

```
> Z <- owin(poly=list(x=c(0,10,0), y=c(0,0,10)))
```

Note that the first vertex in p should **not** be repeated as the last vertex. The same convention is used in the standard R plotting function polygon.

---

[3]To reverse the order of a numeric vector, use rev.

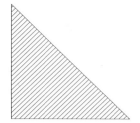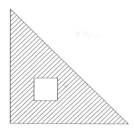

Figure 3.3. *Polygonal windows created in the text.* Left: *Triangle* Z. Right: *Triangle with a square hole* ZH. *Plotted with line shading (*hatch=TRUE*).*

If the window boundary consists of several separate polygons, then p should be a list, each of whose components p[[i]] is a matrix or data frame or a list with components x and y specifying one of the polygons. The vertices of each polygon should be traversed **anticlockwise for external boundaries** and **clockwise for internal boundaries (holes)**. For example, the following creates the triangle with a square hole displayed in the right panel of Figure 3.3.

```
> ZH <- owin(poly=list(list(x=c(0,10,0), y=c(0,0,10)),
                       list(x=c(2,2,4,4), y=c(2,4,4,2))))
```

Notice that the first boundary polygon is traversed anticlockwise and the second clockwise, because it is a hole.

The result of owin(poly=p) is a window object of class "owin" with type "polygonal". The function is.polygonal tests whether an object is a polygonal window.

It is usually practical to save the spatial coordinates of the polygonal boundary in a file and subsequently read them in to R. In manageable cases the data could be entered at the keyboard and saved in a text file. Moderately complicated boundaries could be traced roughly by hand, using a point-and-click or mouse-tracking interface to various software systems, and saved from the software into a text file. Very complicated boundaries, managed in a spatial database, can be exported to files to be read into R (see Section 3.10).

If a region boundary is a single polygon, with the vertices saved in a text file in table format with columns headed x and y like the file mitochondria.txt for the vesicles dataset, then the corresponding window can be created by

```
> bd <- read.table("mitochondria.txt", header=TRUE)
> W  <- owin(poly=bd)
```

If the region boundary consists of several polygons, one simple approach is to save the coordinates in a text file in table format with columns headed x, y and id, where id is an integer identifier specifying which of the polygons is being traced as exemplified in the file vesicleswindow.txt for the vesicles dataset. Then the window can be created by

```
> bd <- read.table("vesicleswindow.txt", header=TRUE)
> bds <- split(bd[,c("x","y")], bd$id)
> W  <- owin(poly=bds)
```

It is good practice to back up data as text files where possible. To save a window (that has been obtained by other means) as a text file, we recommend using the structure described above. A polygonal window can be converted back into this data frame format by as.data.frame.owin:

```
> as.data.frame(ZH)
```

```
    x  y id sign
1   0 10  1    1
2   0  0  1    1
3  10  0  1    1
4   2  2  2   -1
5   2  4  2   -1
6   4  4  2   -1
7   4  2  2   -1
```

The `spatstat` package also provides its own rudimentary point-and-click interface, `clickpoly`, which allows the user to create a window object directly. This was used to create the boundary of the `chorley` dataset by tracing a scanned image of a map. See Section 3.9.

Polygon data often contain small geometrical inconsistencies such as self-intersections and overlaps. These inconsistencies must be removed to prevent problems in other `spatstat` functions. By default, polygon data will be repaired automatically using polygon-clipping code, when `owin` or `as.owin` is called. The repair process may change the number of vertices in a polygon and the number of polygon components. For efficiency, the repair process can be disabled by setting `spatstat.options(fixpolygons=FALSE)`, but this should only be done if we are confident that the data are geometrically consistent.

3.5.3   Circular and elliptical windows

Circular (or disc-shaped) and elliptical windows are created by the `spatstat` functions `disc` and `ellipse`. In the current implementation these shapes are approximated by polygons. To make a circular window of radius 3 centered at the origin:

```
> W <- disc(radius=3, centre=c(0,0))
```

By default, a large number of polygon vertices is used to ensure a good approximation to the circle or ellipse.

One can use the same code to create a regular polygon with any desired small number of vertices. For example, to create a regular hexagon or equilateral triangle one can use `disc(npoly=6)` and `disc(npoly=3)`, respectively. The argument `radius` specifies the distance from the centre to each vertex of the regular, and equals the radius of the circumscribed circle.

3.5.4   Binary mask

A region of space may also be represented in discretised form using a finely spaced grid of test points. For each test point we record a logical value which is TRUE if the test point falls inside the window, and FALSE otherwise. The window is approximated by inferring that, if the value at a test point is TRUE, then the grid rectangle containing this test point lies entirely inside the window. See Figure 3.4. This is a 'pixel graphics' or *binary mask* representation of the window.

Spatial data files which specify the window as a binary mask are often obtained when the original data were a camera image or remotely sensed image, or when *some* of the original data were pixel-based and it was necessary to convert all of the data layers to a common pixel grid. Examples include objects of class `"SpatialGridDataFrame"` read in from a shapefile (see Section 3.10).

For some kinds of computation, it is much more efficient to represent the window by a binary mask than a polygonal window. Windows in the form of binary masks also arise from calculations with pixel-based data.

To create a binary mask directly from raw data, one can use the command

```
owin(mask=m, xrange, yrange)
```

where m is (or is interpreted as) a matrix with logical entries. Note carefully that the rows of the

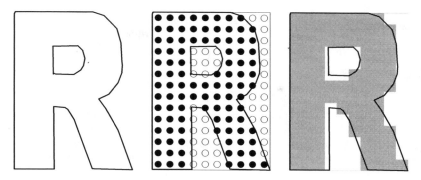

Figure 3.4. *Binary mask representation of a window.*

matrix are associated with the *y* coordinate, and the columns with the *x* coordinate. That is, the matrix entry m[i,j] is TRUE if the test point (xx[j],yy[i]) (sic) falls inside to the window, where xx, yy are vectors of coordinate values equally spaced over xrange and yrange, respectively. The length of xx is ncol(m) while the length of yy is nrow(m). The spatial indexing convention is explained further in Section 3.6.

Another possible syntax is owin(mask=m, xy=xy) where xy is a list of two vectors of coordinates, of the form list(x=xx, y=yy) where xx,yy are the vectors of *x*- and *y*-coordinates for the test points.

The resulting object is a window (object of class "owin") of type mask. The type can be determined using is.mask or print.owin or summary.owin.

The matrix m is usually large, and should be read in from a file which has been created by some other application. A safe strategy is to dump the data from the external application into a text file, and read the text file into R using scan. Next reformat the scanned-in data as a matrix, with the appropriate indexing convention, and finally use owin to create the window object.

When *saving* a mask window to a text file, it is simplest to save the binary pixel values in the order they are stored internally in R, so that they can later be read back into R in the same order. As mentioned in Section 2.1.6, a matrix is stored in 'column major' order in R, meaning that the the first column of an $m \times n$ matrix occupies the first *m* entries, the second column the next *m* entries, and so on. If W is a window of type mask, it can be stored as a text file in a manner something like write(as.matrix(W),file="W.txt"), which will automatically store the pixel values in column major order. The file can then be read back into R by M <- scan("W.txt",what=logical()) and Wnew <- owin(M,xrange=xr,yrange=yr) where xr and yr are the xrange and yrange of the original W. When storing a mask-type window as a text file, it is probably best to store xrange and yrange as 'metadata' in a separate file.

Rectangles and polygonal windows can be converted to binary masks using as.mask. For example the window in the right-hand panel in Figure 3.2 was created by as.mask(letterR, eps=0.1). See the help for as.mask for details about the eps argument. Several binary masks, based on different rectangular grids, can be converted to a common grid using harmonise.owin, a method for the generic function harmonise. The pixels of a binary mask can be extracted as a point pattern by pixelcentres.

Although a binary mask is very similar to a pixel image (Section 3.6) they are not equivalent: they have a different interpretation in some contexts, and their internal structures are slightly different.

3.6 Pixel images in `spatstat`

3.6.1 Pixel images and their uses

In a pixel image, the spatial domain is divided into a grid (of picture elements or 'pixels'), and a value is associated with each pixel. The pixel value could represent brightness (in a digital camera image or a remotely sensed image), terrain elevation (in a digital terrain model), soil pH or magnetic field strength (in a spatial survey), and other measurable quantities. Pixel values can be categorical values, representing a classification of space into different rock types, cell types, administrative regions, or land use types. Other types of spatial data can be converted into pixel images, so that the pixel value could represent (say) the distance from that pixel to the nearest geological fault. Many calculations in spatial statistics produce a pixel image as a result — for example, a kernel estimate of point process intensity.

A pixel image may be thought of as a spatial function $Z(u)$. The value of $Z(u)$ is the value associated with the pixel in which u lies. The value of $Z(u)$ is constant within each pixel (Z is a 'step function').

3.6.2 The class `"im"`

Pixel images are stored in `spatstat` as objects of class `"im"`. The pixel grid is rectangular and evenly spaced, and occupies a rectangular window in the spatial coordinate system. The pixel values are scalar: they can be real numbers, integers, complex numbers, single characters or strings, logical values, or categorical values. A pixel's value can also be `NA`, meaning that no value is defined at that location, and effectively that pixel is 'outside' the window. Photographic colour images (i.e., with red, green, and blue brightness channels) can be represented as character-valued images, using R's standard encoding of colours as character strings.

For basic information about an image Z, one can use `print(Z)` (or in interactive use simply type 'Z') or `summary(Z)`. There is a large number of tools for inspecting and manipulating pixel images, listed in Sections 4.3 and 4.3.2.

3.6.3 Spatial indexing of pixel images

Pixel images are handled by many different software packages. In virtually all of these, the pixel values are stored in a matrix, and are accessed ('addressed') using the row and column indices of the matrix. However, different pieces of software use different conventions for mapping the matrix indices (i, j) to the spatial coordinates (x, y). This is a frequent cause of head-scratching.

Three common conventions are sketched in Figure 3.5. In the *Cartesian* convention, the first matrix index i is associated with the first Cartesian coordinate x, and j is associated with y. This convention is used in the R base graphics function `image.default`. In the *European reading order* convention, a matrix is displayed in the spatial coordinate system as it would be printed in a page of text: i is effectively associated with the negative y coordinate, and j is associated with x. This convention is used in some image file formats. In the `spatstat` convention, i is associated with the y coordinate, and j is associated with x. This is also used in some image file formats.

To convert between these conventions, `spatstat` provides the function `transmat`. If a matrix m contains pixel image data that is correctly displayed by software that uses the Cartesian convention, and we wish to convert it to the European reading convention, we can type

```
> mm <- transmat(m, from="Cartesian", to="European")
```

The transformed matrix `mm` will then be correctly displayed by software that uses the European convention.

Cartesian

(1,4)	(2,4)	(3,4)
(1,3)	(2,3)	(3,3)
(1,2)	(2,2)	(3,2)
(1,1)	(2,1)	(3,1)

European

(1,1)	(1,2)	(1,3)	(1,4)
(2,1)	(2,2)	(2,3)	(2,4)
(3,1)	(3,2)	(3,3)	(3,4)

spatstat

(3,1)	(3,2)	(3,3)	(3,4)
(2,1)	(2,2)	(2,3)	(2,4)
(1,1)	(1,2)	(1,3)	(1,4)

Figure 3.5. *Spatial indexing conventions.*

Each of the arguments `from` and `to` can be one of the names `"Cartesian"`, `"European"`, or `"spatstat"` (partially matched) or it can be a list specifying the convention. For example `to=list(x="-i", y="-j")` specifies that rows of the output matrix are expected to be displayed as vertical columns in the plot, starting at the right side of the plot, as in the traditional Chinese, Japanese, and Korean writing order.

3.6.4   Creating pixel images from raw data

A pixel image can be created directly from raw data in `spatstat` by the function `im`; one form of the syntax is `A <- im(mat,xcol,yrow)`. (See `help(im)` for other forms.) Here `mat` is a matrix whose entries constitute the values associated with the appropriate pixels.

The reader may have noticed the somewhat idiosyncratic names of the last two arguments of `im`, namely `xcol` and `yrow`. They are given these names to remind the user of the convention for spatial indexing. The argument `xcol` is a vector of equally spaced *x*-coordinate values corresponding to the **columns** of `mat`, and `yrow` is a vector of equally spaced *y*-coordinate values corresponding to the **rows** of `mat`. These vectors determine the spatial position of the pixel grid. The length of `xcol` is `ncol(mat)` while the length of `yrow` is `nrow(mat)`. If `mat` is not a matrix, it will be converted into a matrix with `nrow(mat) = length(yrow)` and `ncol(mat) = length(xcol)`.

The value `mat[i,j]` is associated with the pixel whose centre is `(x[j],y[i])`. Note the switch in order of `i` and `j`.

3.6.5   Reading image files

Pixel images in standard image file formats, such as JPEG, can be read directly into the R session using contributed R packages that can be installed from CRAN. Available packages include `jpeg`, `tiff`, `png`, and `bmp`.

It is important to read the image metadata, especially to determine the pixel aspect ratio (height to width ratio of a single pixel). If the aspect ratio cannot be determined for a photographic image, the best guess is usually 2/3, whereas the `spatstat` default is 1.

The `spatstat` installation includes image files `vesiclesimage.tif` and `sandholes.jpg`. These files can be copied to the user's space by `copyExampleFiles`. Alternatively the location of the files can be found using `system.file`:

```
> fn  <- system.file("rawdata", "vesicles", "vesiclesimage.tif",
                  package="spatstat")
```

Here `rawdata` is a folder containing the subfolder `vesicles` which contains the TIFF image file `vesiclesimage.tif`. The advantage of the command above is that the system file separator is inserted automatically according to your system. However, R uses / on the major platforms (Windows®, OS X®, and Linux) and the command

```
> fn   <- system.file("rawdata/vesicles/vesiclesimage.tif",
                       package="spatstat")
```

would give the same result on these platforms. To read in the vesicles image:

```
> library(tiff)
> mat <- readTIFF(fn, as.is=TRUE, info=TRUE)
```

Now typing `str(mat)` would show the matrix dimensions and the auxiliary information from the image header, stating that the pixels are square, 72 pixels per inch, and are stored using the European indexing convention (`orientation` is given as `top.left`). To convert this to a `spatstat` pixel image we should change the indexing convention:

```
> smat <- transmat(mat, from="European", to="spatstat")
```

then convert using `im` or `as.im`. The scale of 72 pixels per inch is not the true physical scale of the microstructures: background information from the microscope determines that each pixel is 2.5 nanometres across, so the true physical scale is assigned by

```
> pixscale <- 2.5
> vim <- im(smat,
            xrange=c(0, ncol(smat) * pixscale),
            yrange=c(0, nrow(smat) * pixscale),
            unitname="nm")
```

It is then straightforward to plot the image using `plot(vim)`. The result is shown in Figure 3.6.

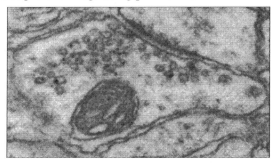

Figure 3.6. *The vesicles image, read in from a* `tiff` *file. Rotated 90 degrees anticlockwise. True physical size* 1019 × 563 *nanometres.*

The file `sandholes.jpg` is a colour image in `jpeg` format from a photograph by the first author.

```
> require(jpeg)
> fn   <- system.file("rawdata", "sandholes", "sandholes.jpg",
                       package="spatstat")
> arr <- readJPEG(fn)
> str(arr)
 num [1:600, 1:900, 1:3] 0.588 0.659 0.667 0.631 0.608 ...
```

The object `arr` produced by `readJPEG` is a three-dimensional array, in which the first two dimensions are spatial coordinates, and the third dimension contains the red, green, and blue channels.

Next we use the `rgb` command (from the standard `grDevices` package) to convert these numerical values to the colour values recognised by R, which are character strings like `"#96928F"`.

```
> mats <- rgb(arr[,,1], arr[,,2], arr[,,3])
> dim(mats) <- dim(arr)[1:2]
```

The matrix dimensions were lost, so they are reinstated using `dim<-`. Finally we convert the matrix of colour values to an image using `im`. To check the correct orientation and the pixel aspect ratio, we inspected the metadata for `sandholes.jpg` using the open source image editor `GIMP`.

```
> sand <- im(transmat(mats, "European", "spatstat"))
```

Since no other arguments are given to `im`, the pixels are squares of unit width. This is the correct aspect ratio according to the image metadata. We could alternatively have specified the arguments `xrange, yrange` to determine the image size and implicitly the aspect ratio. Another alternative is to use `rescale` or `affine` to rescale the pixel grid after it is created.

Figure 3.7. *The sandholes image, read in from a* `jpeg` *file.*

A plot of the image `sand` is shown in Figure 3.7. The true physical scale can be determined using the markings on the wooden ruler that is shown in the image. Using the command `clickdist` we click on two of the centimetre scale marks and read off the distance in pixel units. The full 30 centimetre length is about 609 pixel units, giving a physical scale of $30/609 = 0.049$ cm per pixel.

```
> unitname(sand) <- list("cm", "cm", 30/609)
> sand <- rescale(sand)
```

3.6.6 Factor-valued images

Making a factor-valued image is slightly tricky, because operations that create a factor in R usually discard information about array dimensions. To illustrate the problem, we read in categorical data, which are to be converted to an image, from a file.

The `spatstat` installation includes the file `vegetation.asc` which represents the vegetation covariate in the `gorillas` data (see Section 9.3.4.1). This is a text file, and the first few lines are:

```
ncols            181
nrows            149
xllcorner        580440.38505253
yllcorner        674156.51146465
cellsize         30.70932052048
NODATA_value     -9999
-9999 -9999 -9999 1 1 1 -9999 -9999 -9999
```

The file uses a simple format defined by the geospatial library GDAL. It could be read automatically using the function readGDAL from the package rgdal. If rgdal is not installed, we can simply read the body of the data using scan, skipping the first 6 lines of header information:

```
> fn  <- system.file("rawdata", "gorillas", "vegetation.asc",
                      package="spatstat")
> pixvals <- scan(fn, skip=6)
> pixvals[pixvals == -9999] <- NA
> mat <- matrix(pixvals, nrow=149, ncol=181, byrow=TRUE)
```

Note the use of byrow=TRUE because the rows of the data file are horizontal rows of pixels.

The entries in the matrix mat are the digits 1 to 6 corresponding to the following vegetation types:

```
> vtype <- c("Disturbed", "Colonising", "Grassland",
             "Primary", "Secondary", "Transition")
```

We convert mat to a factor:

```
> f <- factor(mat, labels=vtype)
> is.factor(f)
[1] TRUE
> is.matrix(f)
[1] FALSE
```

Although mat was a matrix, f is not. It is a factor, with no array dimensions. However, one can assign a dim attribute to a factor:

```
> dim(f) <- c(149,181)
```

It is then possible to convert the factor to a pixel image:

```
> factorim <- im(f)
```

By default the pixels have unit size. We would usually want to specify the correct spatial coordinates, given in the header above.

```
> x0 <- 580440.38505253 ; y0 <- 674156.51146465
> dx <- dy <- 30.70932052048
> factorim <- im(f, xrange=x0 + dx * c(0, 181),
                    yrange=y0 + dy * c(0, 149))
```

Alternatively we could have specified the arguments xcol, yrow giving the coordinates of each row and column of pixels. The image factorim is plotted in Figure 3.8.

A third alternative is to create an integer-valued matrix, and assign a levels attribute to it. This will be interpreted as a matrix with categorical values.

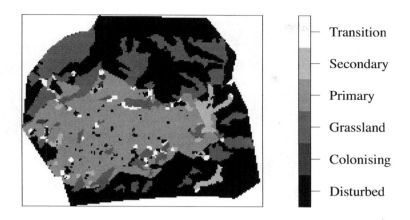

Figure 3.8. *The image* `factorim` *created in the text.*

3.6.7 Computed images

Many functions in `spatstat` return a pixel image. These include `pixellate` and `as.im` (which perform discretisation), `density.ppp`, `density.psp`, `blur`, `Smooth`, `relrisk` (kernel smoothing), `adaptive.density` (nonlinear smoothing), `Smooth`, `idw`, `nnmark` (interpolation), `distmap`, `nnmap` (distance functions), `predict.ppm`, `predict.kppm`, `intensity.ppm` (model prediction), and `rnoise` (which generates random pixel noise).

3.6.8 Images from functions

A mathematical function (described by an explicit formula) may be converted to a pixel image using `as.im`.

```
> f <- function(x,y){15*(cos(sqrt((x-3)^2+3*(y-3)^2)))^2}
> A <- as.im(f, W=square(6))
```

The image `A` is plotted in Figure 3.9. Note the mandatory observation window argument `W`; an image is always confined to a spatial region, which in this case must be given by the user, since it cannot be inferred from a function. Additional arguments to `as.im` control the pixel resolution.

3.6.9 Alternative to images: spatial function class `"funxy"`

Converting a function to a pixel image involves discretisation, which may be undesirable in some circumstances. An alternative to discretising the function `f` above would have been to register it as a 'spatial function':

```
> g <- funxy(f, W=square(6))
```

The result `g` is a copy of `f` with extra attributes, including the specified window `W`, and belongs to the special class `"funxy"`. This object can be used in many places where a pixel image is expected. It behaves like a pixel image in many ways, except that it is able to calculate the function value exactly at any spatial location.

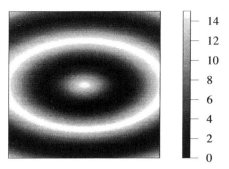

Figure 3.9. *A function converted to a pixel image.*

In a `"funxy"` object, computation of the function values is deferred until the last possible moment, that is, until `g(x,y)` is evaluated for coordinates `x,y`. In a pixel image the pixel values have already been computed when we create the image object.

3.7   Line segment patterns

Spatial data often include linear features such as roads, rivers, and geological faults. For example, the Queensland copper data shown in Figure 1.11 consist of a point pattern of known copper deposits and a spatial pattern of linear geological features, mostly faults, observed at the surface. A pattern of straight line segments can be stored in the `spatstat` package as an object of class `"psp"` (for **p**lanar **s**egment **p**attern).

Many functions are available for creating and manipulating `"psp"` objects. To create a `"psp"` object use `psp` or `as.psp`. The creator function `psp` requires vectors `x0`, `y0`, `x1`, and `y1` specifying the endpoints of the segments, and a window (object of class `"owin"`) in which the segments were observed. The conversion function `as.psp` allows the user to specify line segments in other ways, for example by specifying their midpoint, length, and orientation.

Random patterns of line segments may be created by randomly generating the vectors of endpoints (using `psp`), or randomly generating the midpoints, lengths, and orientations (using `as.psp`). A random line segment pattern from the *Poisson* line process may be generated using the function `rpoisline`. The infinite lines are clipped to the given window resulting in a pattern of segments.

The boundary edges of a window can be extracted as a `"psp"` object using the `edges` function.

Like planar point patterns, `"psp"` objects may be *marked* by a vector or data frame. The generic functions provided in `spatstat` for assigning, interrogating, and manipulating marks all have methods for the `"psp"` class.

See Section 4.4 for information on how to manipulate line segment patterns. Table 4.17 lists functions for extracting information from `"psp"` objects. Additionally Table 4.6 lists generic functions for performing geometric operations on spatial objects, including objects of class `"psp"`.

3.8 Collections of objects

3.8.1 The "solist" and "anylist" classes

In spatial statistics it is often necessary to handle a collection of several objects, such as a collection of point patterns. In the R language, a collection of things is usually organised as a list. The spatstat package supports two special classes of lists, "solist" and "anylist".

A solist object (spatial object list) is a list of 'spatial' objects in two dimensions. An object is recognised as 'spatial' if it occupies a definite region in two-dimensional space; examples include "owin", "ppp", "psp", "im", "ppm" and "layered" objects. In spatstat we use a solist object to store several different spatial objects of the same class, for example, several point patterns, or to store the results of several transformations applied to the same spatial dataset. The waterstriders dataset (Figure 1.2) is a "solist" object, essentially a list of three point patterns.

A "solist" object can be created explicitly by solist(entry1, entry2, ...) or by as.solist(xxx) where xxx is a list of (spatial) objects. For example:

```
> P <- solist(A=cells, B=japanesepines, C=redwood)
```

Various functions in spatstat produce objects of class "solist". There are numerous methods for the "solist" class, most notably a plot method (Section 4.1.6.2). The list P could be plotted immediately by plot(P) and this would display the three point patterns side by side.

An anylist object is a list of objects of a very general kind (not necessarily spatial objects) that we intend to treat in a similar way. One can, for example, use an anylist object to store the results of the same statistical technique applied to different spatial datasets, or the results of several different types of analysis applied to the same spatial dataset. For example the estimates of Ripley's *K*-function (Chapter 7) for each of the point patterns in the list P could be stored as

```
> KP <- anylist(A=Kest(cells), B=Kest(japanesepines), C=Kest(redwood))
```

or equivalently

```
> KP <- as.anylist(lapply(P, Kest))
```

There is also a plot method for "anylist" objects, which is only appropriate if each of the list entries can be plotted by its own plot method (see Section 4.1.6.2). The list KP could be plotted immediately by plot(KP) and would show the three *K*-functions side by side.

3.8.2 The "hyperframe" class

Another important class for storing collections of objects is the "hyperframe" class. A hyperframe is an array, 'like a data frame', but more general. Hyperframes allow the entries of columns to be objects of any class. The only constraint is that all the entries in a particular column must be of the same kind.

A hyperframe can be used to store the results of an experiment in which several point patterns were observed. One column of the hyperframe contains the observed point patterns, and other columns may contain covariate data. The point patterns may have been observed under identical conditions (*replicated* point patterns) or under different experimental conditions indicated by the covariates.

For example, the waterstriders dataset is a list of three point patterns obtained under identical conditions. It can be converted to a hyperframe with one column:

```
> ws <- hyperframe(Larvae=waterstriders)
```

Additional columns can be added in the same way as for a data frame.

Hyperframes are covered in Chapter 16, in particular Section 16.4. They appear again briefly in Section 3.10.3.6 below.

3.9 Interactive data entry in `spatstat`

Spatial data can also be entered interactively, using a graphical point-and-click interface. The facilities are listed in Table 3.3. They are rudimentary compared to other specialised graphics packages, but they have the advantage that the data will immediately be entered in a `spatstat` data format. The interface is robust and available on almost any computer platform, since it depends only on the base R graphics system.

FUNCTION	RESULT
clickppp	point pattern
clickbox	rectangle
clickpoly	polygonal window
clickdist	measured distance
clickjoin	adjacency matrix for linear network

Table 3.3. *Interactive data entry facilities in* `spatstat`.

These facilities are useful for rapid experimentation and exploration, and for annotating other kinds of spatial data. To 'annotate' spatial data, we display the original data, and use the graphical interface to superimpose new spatial information such as points, lines, or text. For these tasks it is recommended that RStudio users open the system's native R graphics device as explained in Section 2.1.11.

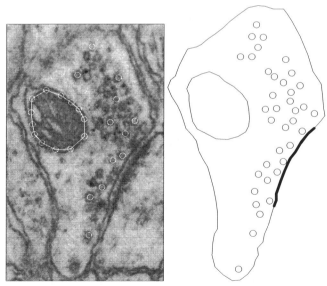

Figure 3.10.   Left: *Annotation of the vesicles image from Figure 3.6 using* `clickpoly` *and* `clickppp`.   Right: *Vesicles point pattern dataset* `vesicles` *and the active zone* `vesicles.extra$activezone` *(thick lines).*

Figure 3.10 shows the vesicles image from Figure 3.6 annotated by drawing the boundary of the mitochondrial region with `clickpoly` and marking the locations of some of the synaptic vesicles with `clickppp`:

```
> plot(vim)
> mito <- clickpoly(add=TRUE, col="white", win=Window(vim))
> vesi <- clickppp(add=TRUE, col="white", win=Window(vim))
```

Tracing the mitochondrial boundary is relatively easy because of the strong contrast. The vesicles have weaker contrast and fewer contextual cues, so they are more difficult to recognise without biological training and experience. In microscopy, each experimental protocol includes standardised criteria for recognising and counting the microstructures of interest. The `vesicles` point pattern installed in `spatstat` was annotated by a trained microscopist using such a protocol. It is an interesting exercise to compare your own guesses with the expert's annotation by typing

```
> plot(vesicles, add=TRUE, chars=3, col="green")
```

If that is too difficult, try annotating the sandholes image (Figure 3.7). Remember that `.Last.value` can also be used to capture the result of the last command.

For accurate annotation, it would be better to use specialised software from the field of application.

3.10 Reading GIS file formats

3.10.1 GIS file formats

Many different file formats are used to store spatial data for use in Geographical Information Systems (GIS) and other applications. Common formats include *shapefiles*, *NetCDF*, and *GRIB*.

Typically `spatstat` does not support these formats directly: this would not be good software design. Instead, we rely on specialised R packages which exist for handling different spatial data file formats. Table 3.4 lists some useful packages.

maptools	Tools for reading and handling spatial objects
shapefiles	Read and write ESRI™ shapefiles
RArcInfo	interface to ArcInfo system and data format
rgdal	interface to GDAL geographical data analysis system
GeoXp	interactive spatial exploratory data analysis
sp	spatial data classes and methods

Table 3.4. *Packages for handling GIS data files.*

For our purposes the most useful file-handling package is `maptools`. It recognises a large number of different file formats, and contains interface code for exchanging spatial objects between different R packages.

When a file is read by `maptools`, the data are represented in R using the data structures defined in the package sp. The sp package [111] supports a standard set of spatial data types in R. These standard data types can be handled by many other packages, so it is useful to convert your spatial data into one of the data types supported by sp.

The `maptools` package also contains code for converting sp data types to the data structures supported by `spatstat`. Our recommended strategy for converting spatial data from a standard GIS format into `spatstat` is: (1) using the facilities of `maptools`, read the data and store the data in one of the standard formats supported by sp; (2) convert the sp data type into one of the data types supported by spatstat, typically using maptools.

Using the sp data types as an intermediate stage is also useful if you plan to employ other R packages for spatial data analysis, which often use the sp data types.

3.10.2   Read shapefiles using maptools

A shapefile [254] represents a list of spatial objects — a list of points, a list of lines, or a list of polygonal regions — and each object in the list may have additional variables attached to it. A dataset stored in shapefile format is actually stored in a collection of text files, for example baltim.shp, baltim.prj, baltim.sbn, baltim.dbf, which all have the same base name baltim but different file extensions. To refer to this collection, always use the file name with the extension shp.

The maptools package contains facilities for reading and writing files in shapefile format. A spatial dataset is read in to R using x <- readShapeSpatial("*filename*.shp"). The class of the resulting object x may be "SpatialPoints" indicating a point pattern, "SpatialLines" indicating a list of polygonal lines, or "SpatialPolygons" indicating a list of polygons. It may also be "SpatialPointsDataFrame", "SpatialLinesDataFrame", or "SpatialPolygonsDataFrame" indicating that, in addition to the spatial objects, there is a data frame of additional variables. The classes "SpatialPixelsDataFrame" and "SpatialGridDataFrame" represent pixel image data.

Here are some examples, using the example shapefiles supplied in the maptools package itself.

```
> library(maptools)
> oldfolder <- getwd()
> setwd(system.file("shapes", package="maptools"))
> baltim   <- readShapeSpatial("baltim.shp")
> columbus <- readShapeSpatial("columbus.shp")
> fylk     <- readShapeSpatial("fylk-val.shp")
> setwd(oldfolder)
```

Then class(baltim) returns "SpatialPointsDataFrame", while class(columbus) returns "SpatialPolygonsDataFrame" and class(fylk) returns "SpatialLinesDataFrame".

3.10.3   Converting sp data to spatstat format

To convert a dataset in sp format to an object in the spatstat package, the subsequent procedure depends on the type of data, as explained below.

3.10.3.1   Objects of class "SpatialPoints"

An object x of class "SpatialPoints" represents a spatial point pattern. Use as(x, "ppp") or as.ppp(x) to convert it to a spatial point pattern in spatstat.

The window for the point pattern will initially be taken from the bounding box of the points. You will probably wish to change this window, usually by taking another dataset to provide the window information. Use [.ppp to change the window: if X is a point pattern object of class "ppp" and W is a window object of class "owin", type X <- X[W].

3.10.3.2   Objects of class "SpatialPointsDataFrame"

An object x of class "SpatialPointsDataFrame" represents a pattern of points with additional variables attached to each point. It includes an object of class "SpatialPoints" giving the point locations, and a data frame containing the additional variables attached to the points.

Use y <- as(x, "ppp") or y <- as.ppp(x) to convert a "SpatialPointsDataFrame" object x to a spatial point pattern y in spatstat. In this conversion, the data frame of additional variables in x will become the marks of the point pattern z. Before the conversion you can extract the data frame of auxiliary data by df <- x@data or df <- slot(x, "data"). After the conversion you can extract these data by df <- marks(y). For example:

```
> balt <- as(baltim, "ppp")
> bdata <- slot(baltim, "data")
```

3.10.3.3 Objects of class "SpatialLines"

A 'line segment' is the straight line between two points in the plane. In the spatstat package, an object of class "psp" ('planar segment pattern') represents a pattern of line segments, which may or may not be connected to each other (like matches which have fallen at random on the ground). In the sp package, an object of class "SpatialLines" represents a **list of lists** of **connected curves**, each curve consisting of a sequence of straight line segments that are joined together (like several pieces of a broken bicycle chain). These two data types do not correspond exactly: see Figure 3.11.

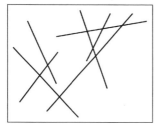 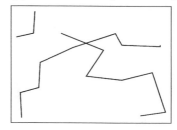

Figure 3.11. *Objects of class* "psp" *(Left) and* "SpatialLines" *(Right).*

The list-of-lists hierarchy in a "SpatialLines" object is useful when representing internal divisions in a country. For example, if USA is an object of class "SpatialLines" representing the borders of the United States of America, then USA@lines might be a list of length 51, with USA@lines[[i]] representing the borders of the i-th State. The borders of each State consist of several different curved lines. Thus USA@lines[[i]]@Lines[[j]] would represent the jth piece of the boundary of the i-th State.

If x is an object of class "SpatialLines", there are at least two different ways to convert x to a spatstat object. The first is to collect together all the line segments that make up all the connected curves and store them as a single object of class "psp". To do this, use as(x, "psp") or as.psp(x) to convert x to a spatial line segment pattern. The window for the line segment pattern can be specified as an argument window to as.psp.

The second way is to convert each connected curve to an object of class "psp", keeping different connected curves separate. To do this, type f <- function(z){ lapply(z@Lines, as.psp) } and out <- lapply(x@lines, f). The result will be a **list of lists** of objects of class "psp". Each one of these objects represents a connected curve, although the spatstat package does not know that. The list structure will reflect the list structure of the original "SpatialLines" object x. If that is not desired, then collapse the list-of-lists-of-"psp"'s into a list-of-"psp"'s using one of these two commands:

```
curvelist <- do.call("c", out)
curvegroup <- lapply(out, function(z) { do.call("superimposePSP", z)})
```

In the first case, curvelist[[i]] is a "psp" object representing the i-th connected curve. In the second case, curvegroup[[i]] is a "psp" object containing all the line segments in the i-th group of connected curves (for example the i-th State in the USA example).

3.10.3.4 Objects of class "SpatialLinesDataFrame"

An object x of class "SpatialLinesDataFrame" is a "SpatialLines" object with additional data. The additional data are stored as a data frame x@data with one row for each entry in x@lines, that is, one row for each group of connected curves.

In the spatstat package, an object of class "psp" may have a data frame of marks. Note that each individual line segment in a "psp" object may have different mark values.

If x is an object of class "SpatialLinesDataFrame", it can be converted to a single object of class "psp" using y <- as(x, "psp") or y <- as.psp(x). The mark variables attached to a particular *group of connected lines* in x will be duplicated and attached to each *line segment* in the resulting "psp" object y.

Alternatively x can be converted to a list of lists of "psp" objects as follows:

```
out <- lapply(x@lines, function(z) { lapply(z@Lines, as.psp) })
dat <- x@data
for(i in seq(nrow(dat)))
  out[[i]] <- lapply(out[[i]], "marks<-", value=dat[i, , drop=FALSE])
```

See the previous subsection for explanation on how to change this using c or superimposePSP.

3.10.3.5 Objects of class "SpatialPolygons"

First some terminology. A *polygon* is a closed curve that is composed of straight line segments. You can draw a polygon without lifting your pen from the paper. A *polygonal region* is a region in space whose boundary is composed of straight line segments. A polygonal region may consist of several unconnected pieces, and each piece may have holes. The boundary of a polygonal region consists of one or more polygons. To draw the boundary of a polygonal region, you may need to lift and drop the pen several times. See Figure 3.12.

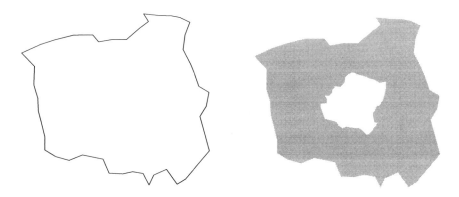

Figure 3.12. *Distinction between a polygon* (Left) *and a polygonal region* (Right).

An object of class "owin" in spatstat, if it is polygonal, represents a **single polygonal region**. It is a region of space that is delimited by boundaries made of lines. It may consist of several disconnected pieces, and may have holes.

An object x of class "SpatialPolygons" represents a **list of polygons**. For example, a single object of class "SpatialPolygons" could store information about every State in the United States of America (or the United States of Malaysia). Each State would be a separate polygonal region (and it might contain holes such as lakes).

There are two different ways to convert an object of class `"SpatialPolygons"`. The first is to combine all the polygonal regions together into a single polygonal region, and convert this to a single object of class `"owin"`. For example, we could combine all the States of the USA together and obtain a single object that represents the territory of the USA. To do this, use `as(x, "owin")` or `as.owin(x)`. The result is a single window (object of class `"owin"`) in the spatstat package.

The second way is to keep the different polygonal regions separate, and convert each one of the polygonal regions to an object of class `"owin"`. For example, we could keep the States of the USA separate, and convert each State to an object of class `"owin"`. To do this, type the following:

```
regions <- slot(x, "polygons")
regions <- lapply(regions,
          function(x) { SpatialPolygons(list(x)) })
windows <- solapply(regions, as.owin)
```

The result is a list of objects of class `"owin"`. Often it would make sense to convert this to a tessellation object, by typing `te <- tess(tiles=windows)`.

During the conversion process, the geometry of the polygons will be automatically 'repaired' if needed. Polygon data from shapefiles often contain geometrical inconsistencies such as self-intersecting boundaries and overlapping pieces. For example, these can arise from small errors in curve-tracing. Geometrical inconsistencies are tolerated in an object of class `"SpatialPolygons"` which is a list of lists of polygonal curves. However, they are not tolerated in an object of class `"owin"`, because an `"owin"` must specify a well-defined region of space. These data inconsistencies must be repaired to prevent technical problems. The spatstat package uses polygon-clipping code to automatically convert polygonal lines into valid polygon boundaries. The repair process changes the number of vertices in each polygon, and the number of polygons (if you chose option 1 above). To disable the repair process, set `spatstat.options(fixpolygons=FALSE)`.

3.10.3.6 Objects of class `"SpatialPolygonsDataFrame"`

An object x of class `"SpatialPolygonsDataFrame"` represents a list of polygonal regions, with additional variables attached to each region. It includes an object of class `"SpatialPolygons"` giving the spatial regions, and a data frame containing the additional variables attached to the regions. The regions are extracted by `y <- as(x, "SpatialPolygons")` and we then proceed as above to convert the curves to spatstat format.

The data frame of auxiliary data is extracted by `df <- x@data` or `df <- slot(x, "data")`. For example:

```
> cp <- as(columbus, "SpatialPolygons")
> cregions <- slot(cp, "polygons")
> cregions <- lapply(cregions, function(x){ SpatialPolygons(list(x)) })
> cwindows <- solapply(cregions, as.owin)
```

There is currently no facility in spatstat for attaching additional variables to an `"owin"` object directly. Marks can be attached to the tiles of a tessellation. Alternatively we can make use of the `"hyperframe"` class described in Section 3.8.2:

```
> ch <- hyperframe(window=cwindows)
> ch <- cbind.hyperframe(ch, columbus@data)
```

The resulting object ch is a hyperframe containing a column of `"owin"` objects followed by the columns of auxiliary data.

3.10.3.7   Objects of class `"SpatialGridDataFrame"` **and** `"SpatialPixelsDataFrame"`

An object x of class `"SpatialGridDataFrame"` represents a pixel image on a rectangular grid. It includes a `"SpatialGrid"` object `slot(x, "grid")` defining the full rectangular grid of pixels, and a data frame `slot(x, "data")` containing the pixel values (which may include `NA` values).

The command `as(x, "im")` converts x to a pixel image of class `"im"`, taking the pixel values from the *first column* of the data frame. If the data frame has multiple columns, these would currently have to be converted to separate pixel images in `spatstat`. For example

```
y <- as(x, "im")
ylist <- lapply(slot(x, "data"), function(z, y) { y[,] <- z; y }, y=y)
```

An object x of class `"SpatialPixelsDataFrame"` represents a *subset* of a pixel image. To convert this to a `spatstat` object, it should first be converted to a `"SpatialGridDataFrame"` by `as(x, "SpatialGridDataFrame")`, then handled as described above.

3.11   FAQs

- *Why doesn't* `spatstat` *use the same classes as* sp*?*

  Development of `spatstat` started long before sp. The data types in `spatstat` and sp are based on different abstractions and are not completely interchangeable, as explained in Section 3.10. The `spatstat` package uses S3 method dispatch while sp uses S4 classes and methods.

- *Can/should I record points lying just outside the sampling quadrat?*

  Yes, this can be done. Point pattern objects created with ppp can include such points (as 'rejects'). Of course this implies that you are not simply observing the point pattern through a 'window', and the sampling procedure is somewhat unclear (what rule exactly was applied to decide whether a point is recorded?). The presence of such points changes the treatment of edge effects. Many functions in `spatstat` can handle such data by setting the argument `domain` or `subset` to equal the sampling quadrat.

- *How can I include French or Scandinavian accents, Greek characters, or Māori language diacritical marks in the name of the unit of length, a factor level, an axis label, or the legend of a plot?*

  Find the Unicode number for the desired character in the data frame `tools::Adobe_glyphs`. Prefix this number by \u to include it in a character string. For example, to find the Greek letter μ:

```
> df <- tools::Adobe_glyphs
> ii <- match("mu", df$adobe)
> df[ii,]
        adobe unicode
2621      mu     00B5
```

  If X is a point pattern and we assign

```
> unitname(X) <- "\u00B5m"
```

  then the unit of length will be rendered as "µm" in `spatstat`'s printed output and graphics (provided the system recognises Unicode). Similarly Ångström (Å) is \u212B. To match all glyph names that include the string `macron`, use `grep("macron", df$adobe)`.

4

Inspecting and Exploring Data

This Chapter covers techniques for inspecting, manipulating and exploring spatial data.

4.1 Plotting

Graphics are essential in statistical science. In spatial statistics, plotting the original spatial data is useful for checking errors in the data record, for motivating the appropriate analysis, and for investigating anomalies. Plotting transformations of the data, and results of the analysis, is essential for exploratory and formal statistical analysis. Effective plots of spatial data are important for communication.

4.1.1 Overview of plot commands

The generic `plot` function in R is intended to 'generate a sensible graphical display' of the data, whatever that means for the particular dataset. In `spatstat`, every class of spatial data has a method for the generic `plot` function. Thus, even if you can't remember any other commands in `spatstat`, it is worth trying `plot(X)`, where X is the name of your dataset, to obtain some kind of reasonable plot. The display is generated by `plot.foo`, the `plot` method for class `"foo"`; for information on how to modify the display, consult the help for `plot.foo` (note that *foo* is not a real name, but just a placeholder name representing an arbitrary class or object). The `plot` methods for point patterns and other spatial objects in `spatstat` are discussed in this section, and in Section 4.3.

Additionally, R and its packages provide many other functions for generating graphical displays. Some of these functions are also generic. For example, `hist`, `contour`, `image` and `persp` are generic functions that display a histogram, contour plot, colour image, and surface perspective plot of the data. Only some classes of objects will have methods for these special plots. For pixel images, these and other plotting tools are presented in Section 4.3.

In `spatstat` there are some special classes of object whose main purpose is simply to indicate that the data should be plotted in a particular way. An object of class `"solist"` is a list of spatial objects of similar kind; `plot.solist` plots these objects side by side. An object of class `layered` is a list of spatial objects, possibly of different kinds, that occupy the same spatial region; `plot.layered` plots these objects on top of one another. Simple facilities are provided for selecting only some of the objects for plotting, and for modifying the plot in different ways.

Interactive display of point patterns is discussed in Section 4.1.5.

4.1.2 Plotting a point pattern

The `plot` method for point patterns, `plot.ppp`, displays the observation window for the pattern and plots the spatial locations of the points, using equal physical scales for the x- and y-coordinates. If the points carry marks, then the points are represented by different symbols according to their mark value, and the mapping from mark values to symbols is displayed in a legend. This mapping is also returned (invisibly) as the result of the `plot` command. It may sometimes be useful to generate an *empty* plot (to which various features may be added sequentially) using a command like `plot(X,type="n")` where X is the point pattern of interest.

Default behaviour

The precise behaviour of `plot.ppp` depends on the nature of the marks, and on arguments given to the `plot` function by the user. The *default* behaviour is illustrated in Figures 4.1 and 4.2.

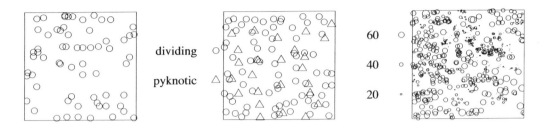

Figure 4.1. *Default behaviour of* `plot.ppp` *for different kinds of data.* Left: *unmarked point pattern;* Middle: *multitype point pattern;* Right: *point pattern with numeric marks.*

If the pattern is unmarked (has no marks), or if the argument `use.marks` is set equal to `FALSE`, then only the spatial locations are plotted. The plotting symbol is controlled by the argument `pch`. The default is an open circle (`pch=1`); popular choices are `pch=16` or `pch=3`. Use the character expansion factor `cex` to change the size of symbols. For very large patterns, use `pch="."` to represent each point as a tiny dot.

Notice that, in the help file for `plot.ppp`, the list of arguments does not include `pch` and `cex`. However it does include '...' which stands for 'other unrecognised arguments'. This mechanism allows such unrecognised arguments to be passed through the function `plot.ppp` to some other function which does recognise them. The help for `plot.ppp` states that such arguments are passed to the low-level R graphics function `points`. Looking up the help file for `points`, we find that `pch` and `cex` are recognised parameters. Possible values for `pch` are explained in `help(points)` and there is a beautiful demonstration of them all in `example(points)`.

If the marks of the pattern are categorical (of class `"factor"`) then each level of the factor is plotted as a different symbol. That is, a different value of `pch` is used to plot points corresponding to each factor level. A legend is plotted to the left of the main plot indicating which symbol corresponds to each mark. The default is to use the plotting characters $1, \ldots, k$ where k is the number of levels of the factor. A different choice may be specified using the `chars` argument which should be a vector of length equal to the number of levels. The entries of `chars` should be values which are acceptable as values of `pch`.

If the marks are real numbers, then points whose marks are positive numbers are plotted as circles of diameter proportional to the mark value. Points whose marks are negative numbers are plotted as squares of side length proportional to the (absolute value of the) mark. Again a legend is plotted to the left indicating the relationship between the mark value and the size of the plotted shape.

If the marks are dates, they are converted to seconds of time elapsed since a 'start' or 'origin' time, and plotted as circles, in the same manner as are positive real-valued marks.

If the marks are multivariate (i.e. if the `marks` component of the pattern is a data frame) then a separate plot is produced for each column of marks, as shown in Figure 4.2. The argument `which.marks` may be used to specify that plots are to be produced only for a subset of the columns of the `marks` data frame. For example, a single column of marks may be picked out for use.

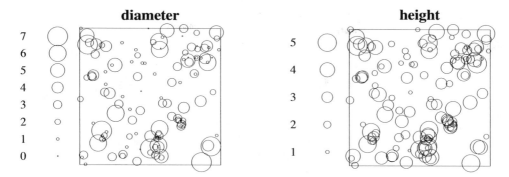

Figure 4.2. *Default behaviour of* `plot.ppp` *for a point pattern with several mark variables: a separate plot panel is displayed for each mark variable.*

The default behaviour of `plot.ppp` can be modified by giving additional graphics parameters, as we discuss below.

Symbol maps

The return value from `plot.ppp` is an object of class `"symbolmap"` representing the mapping from marks to symbols. This object can be examined, manipulated, and used for other plots.

```
> a <- plot(amacrine)
> a
Symbol map for discrete inputs:
[1] "off" "on"
chars: [1] 1 2
> (plot(longleaf))
Symbol map for real numbers in [2, 75.9]
shape: circles
size: function (x, scal = 0.172252245134423)
{
    scal * x/2
}
<environment: 0xc0b4e04>
```

A symbol map specifies the possible inputs (mark values that can be mapped) and the graphical parameter values to which these inputs will be mapped. The inputs may be a vector of categorical values or a range of real numbers. The graphical parameter names may be any of the graphics arguments `shape`, `pch`, `chars`, `size`, `cex`, `col`, `cols`, `fg`, `bg`, `lty`, `lwd`, `border`, `fill`, and `etch`. Each graphical parameter may be either a constant value, a vector of values corresponding to the inputs, or a function that will be applied to the input values.

The default symbol map used by `plot.ppp` was effectively described above. This default can be modified by specifying graphics parameters in the call to `plot.ppp`. For example, the default behaviour for a point pattern with positive numeric marks is to display circles of diameter proportional to the marks. If we wish to encode the mark values of the Longleaf Pines data (Figure 1.8 on page 7) as colours rather than sizes, one possibility would be

```
> A <- colourmap(heat.colors(128), range=range(marks(longleaf)))
> plot(longleaf, pch=21, bg=A, cex=1)
```

A greyscale version is shown in Figure 4.3. The effect of cex=1 is that all symbols will be the same size; plot symbol 21 is a filled circle; parameter bg is the background or interior colour for plot symbols 21 to 25; and the object A is a colour map, explained in Section 4.3.

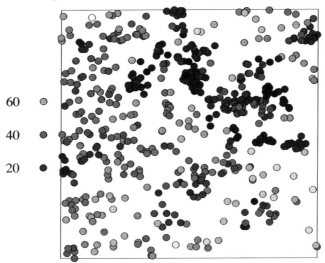

Figure 4.3. *Longleaf Pines data with marks encoded as greyscale values.*

If the argument symap is given in the call to plot.ppp, then this determines the symbol map used in the plot of the point pattern, overriding the default behaviour described above. A typical use of symap is to re-use a previous symbol map. For example, the following commands display the Longleaf Pines data, and next to it, the subset of trees with diameter less than 30 cm, *using the same scale* for the marks in both plots. See Figure 4.4.

```
> juveniles <- subset(longleaf, marks <= 30)
> a <- plot(longleaf, maxsize=15)
> plot(juveniles, symap=a)
```

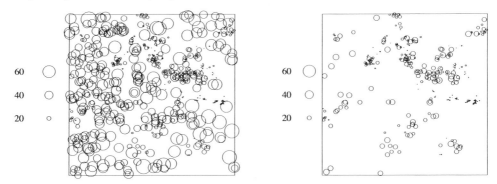

Figure 4.4. *Illustrating the re-use of a symbol map. Full dataset of Longleaf Pines data (*Left*) and subset consisting of juvenile trees (*Right*) plotted with the same symbol map.*

A symbol map can also be created directly using the function symbolmap.

```
> g1 <- symbolmap(inputs=letters[1:10], pch=11:20)
> g2 <- symbolmap(range=c(0,100), size=function(x) {x/50})
```

A symbol map can be modified using update.symbolmap, a method for the generic function update:

```
> g3 <- update(g2, col="red")
> g4 <- update(g3, col=function(x) ifelse(x < 30, "red", "black"))
> g4
Symbol map for real numbers in [0, 100]
col: function(x) ifelse(x < 30, "red", "black")
size: function(x) {x/50}
shape: circles
```

A symbolmap can be printed and plotted in its own right: the legend accompanying the plot of a point pattern is produced by plot.symbolmap.

4.1.3 Plotting a window

A window object (class "owin") often represents the observation window or survey region in which a spatial dataset was observed. A spatial covariate can take the form of a window object: for example, the dataset murchison includes an "owin" object murchison$greenstone specifying which parts of the survey region contained greenstone outcrop. Window objects can also arise from computations in spatstat. A window can be plotted using plot.owin:

```
> plot(murchison$greenstone, col="grey")
```

See Figure 4.5.

Figure 4.5. *Result of plotting the* "owin" *object* murchison$greenstone.

For rectangular and polygonal windows, the plotting is ultimately done by the R function polygon; arguments acceptable to this function can be given for finer control as explained in the help for plot.owin. For example, to paint the interior of the polygon in red, use the argument col="red". To draw the polygon edges in green, use border="green". To suppress the drawing of polygon edges, use border=NA. It is also possible to use transparent colours (the fourth argument to functions such as rgb and hsv), which can be convenient when plotting overlapping polygons:

```
> plot(square(c(-1,1)), main = "")
> plot(ellipse(1,0.5), col = rgb(0,0,0,.2), add = TRUE)
> plot(ellipse(0.5,1), col = rgb(0,0,0,.2), add = TRUE)
```

The resulting overlapping ellipses are shown in Figure 4.6. Note that transparency of colours is not supported by all graphics output devices. In particular it is *not* supported by postscript.

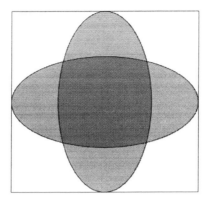

Figure 4.6. *Overlapping* `"owin"` *objects plotted with semi-transparent colours.*

4.1.4   Plotting an image

Several different views of the same data should be explored. Table 4.1 shows some of the most useful commands for displaying a pixel image (object of class `"im"`). Figure 4.7 shows two examples.

`plot(X)`	plot colour image
`image(X)`	plot colour image
`contour(X)`	draw contour map
`persp(X)`	display perspective view of surface
`iplot(X)`	launch interactive plot panel
`textureplot(X)`	plot values as textures
`plot(transect.im(X))`	plot pixel values along a line transect

Table 4.1. *Useful commands for plotting a pixel image. The standard functions* `plot`*,* `image`*,* `contour`*, and* `persp` *are generic, with methods for pixel images provided by* `spatstat`*. Additionally* `iplot` *is generic. Currently* `plot.im` *and* `image.im` *are identical.*

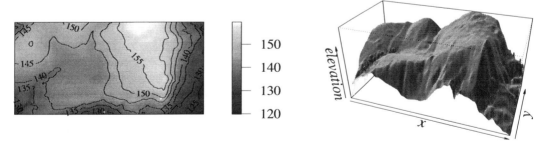

Figure 4.7. *Different displays of the same image data.* Left: *image plot with contour plot superimposed.* Right: *perspective plot.*

`plot.im` and colour values

In keeping with the design of R, the `plot` method for images, `plot.im`, gives the most 'natural' display of the data in most cases. Pixel values are rendered as colours or shades of grey, and displayed as a block of colour or greyscale on the plot region. Colours are rendered as greyscales in this book. The mapping from pixel values to colours or greyscales — the *colour map* — is displayed in a 'ribbon' to the right of the image. The return value from plot.im is an object of class "colourmap" representing the colour map used in the plot.

The observer's perception of details in a colour image depends greatly on the colour map used. In plot.im the colour map is controlled by the argument col, which can be either an explicit colour map object (of class "colourmap") or a vector of values interpretable as colours.

Values which specify colours in R can be strings containing names of colours (like "magenta") or strings of hexadecimal codes (like "#FF00FF") or integers in the range 1 to 8 indexing the standard palette of pen colours (see palette). There are many facilities available in R for generating smooth gradations of colour, for use in a colour map. Simple choices are the functions rainbow, heat.colors, terrain.colors, topo.colors, and cm.colors. For example rainbow(128, end=5/6) generates a vector of 128 colour values looking something like the visible rainbow. See help(colours) or the package RColorBrewer for further options.

Graphics experts caution against using rainbow colour maps for pixel image data [119] because these colour sequences are not 'perceptually uniform' and distort the information content of the image [392]. The spatstat dataset Kovesi provides several colour sequences that are perceptually uniform and are suited for different purposes. When images are plotted in spatstat, the default colour map is one of these perceptually uniform sequences.

In the command plot(X, col=V), if V is a vector of colours, then the range of pixel values in X will be mapped linearly to the colours in V. This is convenient for a single use, but if we plot two different images X and Y using plot(X, col=V) and plot(Y, col=V), direct comparison between the plots is not possible, because the same colour represents different pixel values in the two images. This can be overcome using a colour map as explained below.

Transformation of the pixel values themselves often improves interpretability of the image. One can try transforming to a uniform distribution (page 116), or plotting the image with log=TRUE which causes the colour map to be equally spaced on a logarithmic scale.

Colour maps

A colourmap object is a *fixed* correspondence between numerical pixel values and colours. If two images are plotted using the same colourmap object, we can be confident that a given colour represents the same pixel value in both plots. The user can make a colourmap object by hand using the function colourmap:

```
> g <- colourmap(rainbow(128), range=c(0,100))
> h <- colourmap(c("green", "yellow", "red"),
                 inputs=c("Low", "Medium", "High"))
```

Here g associates colours with numerical values, and h associates colours with categorical values. A colourmap object also behaves as a function, so that g(50) returns the colour associated with the pixel value 50:

```
> g(50)
[1] "#00FFFFFF"
> h("Medium")
[1] "yellow"
```

Colour map objects can be modified by the spatstat functions to.grey, to.transparent, to.opaque, interp.colourmap, tweak.colourmap, and complementarycolour.

To ensure that two images X and Y are plotted using a common colour map which embraces the range of values in both images, the usual idiom is

```
ra <- range(X, Y)
cm <- colourmap(rainbow(128), range=ra)
plot(X, col=cm)
plot(Y, col=cm)
```

One can also use plot.solist to plot several images side by side with the same colour map, as explained below.

Other plotting functions for images

In addition to plot.im, several other graphics commands for pixel image data are listed in Table 4.1.

The spatstat interactive plotting function iplot also applies to pixel images. It is effectively an interactive version of plot.im, and is useful for zooming in to view fine details.

The R generic functions contour and persp require the pixel values to be numerical, and treat them as the elevation (altitude) of a surface: contour produces a contour map of this surface, and persp produces a perspective view.

For a pixel image X, the command contour(X) is dispatched to contour.im which ultimately calls contour.default. Arguments to control the contour map can be found in the help files for contour.im and contour.default. Useful ones include nlevels for the number of contour lines, drawlabels to specify whether the contour lines should be labelled, and col,lty,lwd for the colour, type, and width of contour lines.

The method persp.im displays a perspective view of the surface, with equal scales for the *x*- and *y*-coordinates but a different scale for the *z*-coordinate. The ratio between the *z* and *x,y* scales is the argument expand. Plotting is performed by persp.default. This command is very flexible and can produce some beautiful pictures: see demo(persp).

The angle from which the surface is viewed can be quite influential; it is controlled by the arguments theta and phi. The default behaviour of persp is to draw the pixel edges in black, producing a wireframe representation of the surface. It is usually better to set border=NA to suppress the wireframe and shade=0.7 (for example) to fill the pixel surfaces with shades of grey calculated as if the sun were shining on the surface. For shades of red instead of grey, set col="red".

The spatstat method persp.im has some additional features allowing the surface colours to be controlled by spatial data. If the argument colin ('colour input') is given, it should be a pixel image, and the surface will be coloured according to the values of colin. Thus persp(X, colin=Y) looks as if we had printed plot(Y) on a sheet of paper and draped it over the wireframe surface in persp(X). The colour map is controlled by the argument colmap. If the argument colmap is given and colin is missing, then the surface colours are determined by the *surface elevation* according to the specified colour map: try colmap=terrain.colors(128). If the argument shade is given, the colours described above will be transformed to darker shades of the same colour according to the terrain lighting model. If apron=TRUE, the surface will be surrounded by vertical sides so that it looks more like a solid. For a realistic display of the *Beilschmiedia* terrain data, complete with 'snow-capped mountains', try

```
> persp(bei.extra$elev, expand=6, theta=-30, phi=20,
        colmap=terrain.colors(128), shade=0.2,
        apron=TRUE, main="", box=FALSE)
```

The return value of persp is a matrix representing the projection from 3D to 2D space. This can be used to add other graphics to the perspective plot, typically using the function trans3d from the graphics package, or the spatstat functions perspPoints, perspLines, and perspSegments. For example, Figure 4.8 shows the locations of the *Beilschmiedia* trees as dots on the terrain. It was produced by

```
> M <- persp(bei.extra$elev, theta=-45, phi=18, expand=7,
             border=NA, apron=TRUE, shade=0.3,
             box=FALSE, visible=TRUE)
> perspPoints(bei, Z=bei.extra$elev, M=M, pch=16)
```

The argument `visible=TRUE` causes `persp.im` to perform extra calculations which determine which parts of the surface are visible (rather than obscured) in the perspective view. This information is returned as an attribute of the matrix `M`. The function `perspPoints` then uses this information to determine which points are visible.

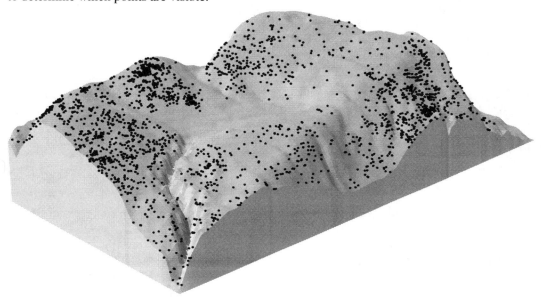

Figure 4.8. *Perspective view of rainforest terrain with visible* Beilschmiedia *tree locations super-imposed.*

The function `transect.im` extracts the pixel values of an image along a line in two-dimensional space — by default, the diagonal line from bottom left to top right of the image frame. The result is a function table (class `"fv"`) which can be plotted directly, to produce a 'profile' or 'vertical section' of the corresponding surface. Figure 4.9 shows the result of `with(bei.extra, plot(transect.im(elev)))`.

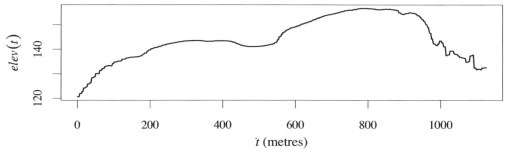

Figure 4.9. *Profile of terrain elevation along a line transect of the* Beilschmiedia *data. Transect runs from bottom left to top right of study area.*

The R generic function `cut` transforms numerical data into categorical values by dividing the range of numbers into bands, and categorising the numbers according to the band in which they fall. The method for images, `cut.im`, applies this to the pixel values of an image. This can be very useful for visualising pixel images which have continuous gradual changes in pixel value.

For images with categorical (factor) values, `plot.im` is usually the best option for plotting if colour display is available. Each level of the factor will be displayed as a different colour; careful choice of the colour map may be needed. The left panel of Figure 4.10 shows the result of `plot(U)` where

```
> U <- cut(bei.extra$elev, 4, labels=c("lo","mlo","mhi","hi"))
```

An alternative display for factor-valued images is `textureplot`, which finds the subregions associated with each level of the factor, and fills each subregion with a different geometric pattern. This only works well when the subregions do not contain small pieces or irregular boundaries. The right panel shows the result of `textureplot(U)`. The return value of `textureplot` is an object of class `"texturemap"` representing the mapping from factor levels to textures.

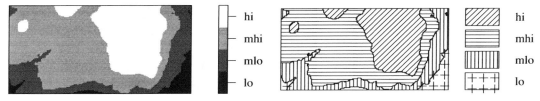

Figure 4.10. *Plotting a factor-valued image* U *using* plot *(Left) and* textureplot *(Right).*

Regions associated with levels of the factor can also be extracted with `tess` or `split` or `solutionset` (see below) and plotted separately. For example, with the factor-valued image U created above, `as.tess(U)` or `tess(image=U)` would be a tessellation whose tiles are the regions corresponding to each level of the factor, and `split(bei.extra$elev, U)` would be a list of pixel images obtained by restricting the terrain elevation image to each of these subregions.

Distribution of pixel values

The function `hist.im`, a method for the generic `hist`, displays a histogram of the pixel values in an image with numerical values, or a bar plot of the pixel values in an image with categorical values. The function `spatialcdf` computes the cumulative distribution function of the pixel values in an image with numerical values. The result is a function (of class `"stepfun"`) which can be plotted immediately.

```
> with(bei.extra, hist(grad, probability=TRUE))
> with(bei.extra, plot(spatialcdf(grad, normalise=TRUE)))
```

Inspection of such displays (shown in Figure 4.11) may suggest an appropriate transformation of the pixel values; see Section 4.3.2.

4.1.5 Interactive plotting

Interactive display, in which the user is able to point the mouse at the display and click to extract information or change the display, is also supported in a modest way. For these tasks it is recommended that RStudio users open the system's native R graphics device as explained in Section 2.1.11.

Mouse-click interaction

Virtually all installations of R support a simple but reliable form of interaction which responds only to mouse clicks (and not to mouse-down, mouse-up, and other signals) in the recommended package

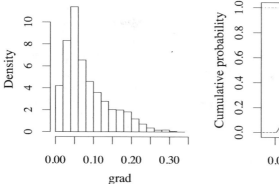
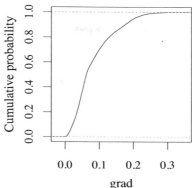

Figure 4.11. *Histogram (*Left*) and cumulative distribution function (*Right*) of terrain elevation in the* Beilschmiedia *data.*

graphics. The function locator records the spatial locations of the mouse when the left mouse button is clicked.

The R generic function identify allows the user to select data points in an existing plot, and read information about them. Methods for identify are provided in spatstat, including identify.ppp. After typing

```
> plot(amacrine)
> identify(amacrine)
```

the user can click on any point of the amacrine pattern. Information about the data point will be printed and returned, including its index number (in the sequence of points ordered as they appear in the "ppp" object) and its mark value.

The spatstat package also has interactive functions based on locator, including clickbox, clickpoly, and clickppp, which can be used to draw a rectangle, a polygonal window, or a point pattern, respectively. See Section 3.9. A common use of clickbox is to select a sub-region of a spatial dataset.

Elaborate interaction

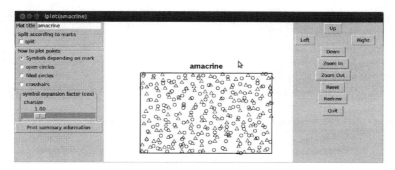

Figure 4.12. *Screenshot of the interactive graphics window started by* iplot *for a point pattern.*

Additionally spatstat has an experimental interactive plotting function iplot which depends on the rpanel package and is less stable. The function iplot is generic, with methods for objects of classes "ppp" and "layered" as well as a default method that works for classes "im", "psp", and "owin". When the user types library(rpanel) followed by iplot(X), a new pop-up window is launched. See Figure 4.12. The spatial dataset X is displayed in the middle of the window using the appropriate `plot` method. The left side of the window contains buttons and sliders allowing the user to change the plot parameters. The right side of the window contains navigation controls for zooming (changing magnification), panning (shifting the field of view relative to the data), redrawing and exiting. If the user clicks in the area where the point pattern is displayed, the field of view will be re-centred at the point that was clicked.

Zooming in to a spatial point pattern can be extremely useful for investigating point patterns with dense concentrations of points, and for checking whether the spatial coordinates were discretised.

4.1.6 Plotting several objects

4.1.6.1 Layered plots

Layering is a simple mechanism for controlling a high-level plot that is composed of several successive plots, for example, a background and a foreground plot. The layering mechanism makes it easier to issue the plot command, to switch on or off the plotting of each individual layer, to control the plotting arguments that are passed to each layer, and to zoom in.

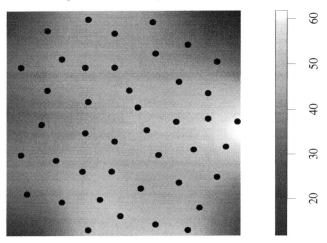

Figure 4.13. *Plot of a layered object.*

The command X `<- layered(...)` simply saves the objects ... as a list, of class `"layered"`. Each individual object should belong to some class of spatial objects that has a method for the generic functions `plot`, `shift`, and `"["`.

The layered object X can then be plotted by the method `plot.layered`. Thus, you only need to type a single `plot` command to produce the multilayered plot. Individual layers of the plot can be switched on or off, or manipulated, using arguments to `plot.layered`, or by selecting a subset using `"[.layered"`.

Default values of the plotting arguments for each layer can be set, by calling

```
X <- layered(..., plotargs=p)
```

or equivalently

```
X <- layered(...)
layerplotargs(X) <- p
```

where p should be a list, with one entry for each layer. Each entry of p should be a list of arguments in the form `name=value`, which are recognised by the `plot` method for the relevant layer.

For example, Figure 4.13 was generated by the following code:

```
> X <- layered(density(cells), cells)
> layerplotargs(X)[[2]] <- list(pch=16)
> plot(X, main="")
```

Here `cells` is one of the standard datasets installed in `spatstat`, giving the locations of the centres of biological cells observed in a histological section [572].

There is also an `iplot` method for layered objects, which allows the user to switch each layer on and off at the click of a mouse, to zoom in and out, and so on. Various other objects can be converted to layered objects by `as.layered`.

4.1.6.2 Plotting lists of objects

The classes `"solist"` and `"anylist"` were introduced in Sections 2.4.4 and 3.8.1. An object of class `"solist"` represents a list of spatial objects (such as point patterns, line segment patterns, pixel images). An object of class `"anylist"` is a list of objects of the same class, not necessarily spatial objects.

The `plot` methods for these classes, `plot.solist` and `plot.anylist`, will display the entries of the list side by side, or in a rectangular array. Using optional arguments to the `plot` methods, the user can specify that the objects should be plotted with equal scaling and compatible alignment.

In the following example, all the entries of the list are spatial objects.

```
> X   <- swedishpines
> QC <- quadratcount(X)
> QCI <- as.im(X,dimyx=5)
> DI <- density(X)
> L <- solist(X, QC, QCI, DI)
> names(L) <- c("Swedish Pines pattern", "Quadrat counts",
                "Quadrat count image", "Estimated intensity")
```

Figure 4.14 shows a plot of the resulting object L.

For a `solist` object, setting `equal.scales=TRUE` will ensure that all entries are plotted using the same physical scale for the coordinates. This might be undesirable if the objects have very different spatial sizes. Setting `valign=TRUE` will ensure that exactly the same *y*-coordinate system is used for all plots on a given row of the array of plots. Setting `halign=TRUE` will ensure that exactly the same *x*-coordinate system is used for all plots in a given column of the array. Alignment options are undesirable if the objects are not in the same spatial location. Try the examples

```
> P <- solist(A=cells, B=japanesepines, C=redwood)
> plot(P, equal.scales=TRUE)
> plot(P, equal.scales=TRUE, valign=TRUE)
```

In Figure 4.14 we needed to set `halign=TRUE`.

When all elements of the list are pixel images, `plot.solist` passes control to `image.imlist`. This has additional functionality including an argument `equal.ribbon` which forces all images to be plotted using the same colour map.

For a list of pixel images there is also the method `contour.imlist` which draws a contour plot of each image side by side.

4.1.6.3 Plotting several images

Table 4.2 lists commands which are useful for comparing several pixel images that are defined on the same spatial region.

Layered plots were discussed in Section 4.1.6.1. If X, Y, Z are pixel images defined on the same region, then `plot(layered(X,Y,Z))` would plot the images on top of one another, so that only

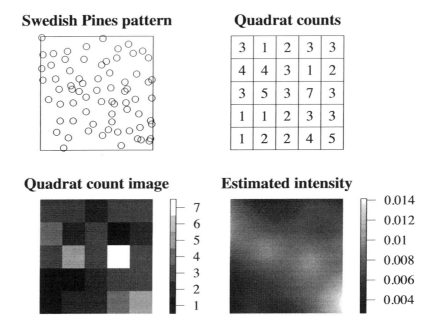

Figure 4.14. *Various plots associated with the* `swedishpines` *data, stored in the* `"solist"` *object* L *described in the text.*

Z would be visible. However, **iplot**(`layered(X,Y,Z)`) starts an interactive graphics panel which gives the option of switching off each of the layers X, Y, Z. This makes it possible to flip quickly between the different images.

`plot(solist(X,Y,...))`	plot images side by side
`contour(solist(X,Y,...))`	contour plots side by side
`iplot(layered(X,Y,...))`	interactively switch between images
`rgbim(X,Y,Z)`	RGB colour composite image
`hsvim(X,Y,Z)`	HSV colour composite image
`pairs(X,Y,...)`	scatterplot of corresponding pixel values

Table 4.2. *Useful commands for plotting or comparing several pixel images* X,Y,....

Several pixel images X,Y,Z,... can be assembled into an object of class `"solist"` by typing L <- `solist(X,Y,Z,...)` as discussed in Section 3.8.1. Then `plot(L)` invokes `plot.solist`, which displays the images side by side or in a rectangular array. Setting `equal.ribbon=TRUE` will ensure that the same colour scale is used for all images. The arguments `equal.scales=TRUE`, `halign=TRUE`, and `valign=TRUE` may or may not be needed, as explained in Section 4.1.6.2.

For a `"solist"` object whose entries are all pixel images, by default `plot.solist` calls the function `image.imlist` to display the images. (Note that `equal.ribbon` is an argument of `image.imlist`.) In other cases, `plot.solist` calls the generic `plot` function to display each entry in the list. This can be overridden by the argument `plotcommand`. For example, setting `plotcommand="hist"` would generate an array of histograms, and `plotcommand="contour"` an array of contour plots, of each entry in the list.

Given three images X,Y,Z with numerical pixel values, defined in the same spatial region, the function `rgbim` interprets the three images as the brightness values of the Red, Green, and Blue channels in an RGB colour space. The result is a pixel image whose pixel values are `colour` values (character strings of hexadecimal colour codes) which can be plotted directly by `plot.im`.

This is useful for detecting and locating differences between two very similar images, such as year-to-year changes in a remotely sensed image. For example `plot(rgbim(X,Y,0))` will show areas of disagreement between `X` and `Y` as patches of red or green, while the areas of agreement will be yellow. Similarly the function `hsvim` interprets three image inputs as the Hue, Saturation and Value channels of a colour. This is useful for combining image datasets that are fundamentally different.

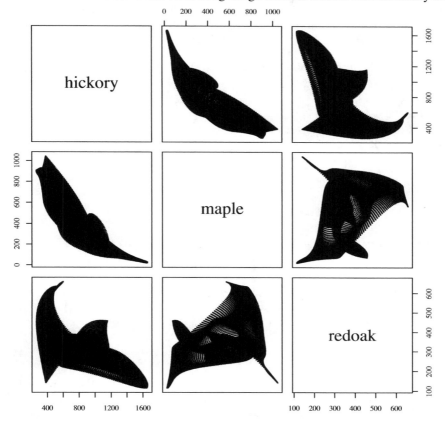

Figure 4.15. *Scatterplot pairs display for the densities of each tree species in Lansing Woods.*

To study relationships between variables in a dataset, the generic R function `pairs` can be useful. It displays an array of scatterplots for each pair of variables in the dataset. In `spatstat` we provide a method for images, `pairs.im`, which extracts the values of corresponding pixels in several images, and optionally generates the array of scatterplots using `pairs.default`. There is also a `pairs.solist` for handling a list of pixel images. Figure 4.15 shows the result of the command

```
> pairs(density(split(lansing)[c(2,3,5)]))
```

This divides the Lansing Woods data (see Section 14.3.3) into 6 sub-patterns of different tree species, extracts species 2, 3, and 5, computes the kernel smoothed intensity estimate for each species, and then displays scatterplots of the intensity estimates for each pair of species, plotted against each other.

The plot suggests that hickory and maple trees are strongly segregated from one another (since a high density of hickories is strongly associated with a low density of maples). There appears to be some segregation of hickory and red oak as well, although this association is not as strong. The relationship between red oak and maple exhibits no clear structure.

The scatterplots in each panel can be replaced with other kinds of plots, by specifying the argument `panel` (or `upper.panel` and `lower.panel`). The argument should be a function, as described in the help for `pairs.default`. In `spatstat` we provide panel functions `panel.contour`, `panel.image` which draw contour maps and image displays, respectively. It is also possible to fill the diagonal panels with another plot specified by the argument `diag.plot`: `spatstat` provides the panel function `panel.histogram` to draw a histogram of the pixel values.

The return value of `pairs.im` is a data frame containing the corresponding pixel values for each of the images. This is useful for calculating summary statistics. For example, to compute the sample correlation between the density estimates for the three dominant species in Lansing Woods,

```
> L <- density(split(lansing)[c(2,3,5)])
> df <- pairs(L, plot=FALSE)
> co <- cor(df)
> round(co, 2)
         hickory maple redoak
hickory     1.00 -0.85  -0.51
maple      -0.85  1.00   0.33
redoak     -0.51  0.33   1.00
```

4.2 Manipulating point patterns and windows

We now show how to manipulate point pattern data, for example, how to extract or change the spatial coordinates or the marks, extract a subset of the point pattern, split a pattern into several subsets, or combine several point patterns to make a single pattern.

4.2.1 Basic operations on a point pattern

A planar point pattern is represented in `spatstat` by an object of the class `"ppp"`. This object consists effectively of the coordinates of the points, the window or study region in which the points were observed, and optionally the mark values associated with each of the points.

Extracting data from a point pattern

Although it is possible to directly access the internal structure of a point pattern object, this is not advisable. *Modifying* the internal components is dangerous, because the data can become internally inconsistent. The internal format of objects in a package can change from one version of the package to another.

It is much safer and neater to use facilities defined in the package to extract and modify or manipulate the data in a point pattern object. Table 4.3 lists the most important functions for extracting data from a point pattern. These are all generic functions with methods for class `"ppp"`. For help on extracting the coordinates of a point pattern, see the help for `coords.ppp`.

The *Frame* of a spatial dataset, extracted by `Frame(X)`, is a rectangle which always contains the entire spatial extent of the dataset. The Frame is part of the data structure, and is used in many calculations which require a rectangle containing X, for example when allocating space for plotting the data, or when converting the data to a pixel image.

The Frame always contains the observation window `Window(X)` and is often the smallest possible rectangle containing the window, although it can be larger in some circumstances.

The *bounding box* of a spatial object in `spatstat`, computed by `boundingbox(X)`, is the smallest rectangle having sides parallel with the coordinate axes that encloses all of the 'contents' of the

`npoints(X)`	number of points in X
`marks(X)`	marks of X
`coords(X)`	coordinates of points in X
`as.data.frame(X)`	coordinates and marks of X
`Window(X)`	window of X
`Frame(X)`	containing rectangle of X

Table 4.3. *Important functions for extracting data from a point pattern.*

object. For a point pattern the bounding box is the smallest rectangle containing all the points of the pattern. For a window (object of class `"owin"`) the bounding box is the smallest rectangle containing the window. The bounding box of a point pattern is always contained in the bounding box of the window.

Note that the `ripras` function mentioned in Section 3.3, if called with `shape="rectangle"`, returns a slightly larger rectangle than does `boundingbox`. The `ripras` function attempts to accommodate the negative bias which is present when the window is 'estimated' by the bounding box [580].

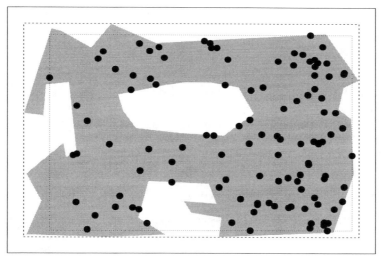

Figure 4.16. *Point pattern dataset showing the bounding box of the points (dotted lines), bounding box of the Window (dashed lines), the Window (grey shading), and the Frame (solid lines).*

Changing data in a point pattern

`marks(X) <- value`	change the marks of X
`coords(X) <- value`	change the coordinates of X
`Window(X) <- value`	change the window of X
`Frame(X) <- value`	change the containing rectangle of X

Table 4.4. *Important commands for altering the data in a point pattern.*

Table 4.4 lists the most important commands for altering data in a point pattern. For example,

```
> X <- redwood
> marks(X) <- nndist(X)
```

will attach a mark to each point in the `redwood` dataset, with mark value equal to the distance to the nearest other point. The side effect of `marks(X) <- m` is that the dataset X is changed.

It is useful to understand how these commands are executed. In R syntax, an assignment such as `marks(X) <- m` is translated into a call to the assignment function `"marks<-"`. This is a generic function, with a method for point patterns, `"marks<-.ppp"`. Consult the help for `"marks<-.ppp"` for information about assigning marks to a point pattern.

Note that R is clever enough to handle multiple layers of assignment in one expression. For example

```
> Y <- chorley
> levels(marks(Y)) <- c("case", "control")
```

is equivalent to

```
> m <- marks(Y)
> levels(m) <- c("case", "control")
> marks(Y) <- m
```

Similarly `marks(X)[3] <- 5` would set the mark value of the third data point in X to be equal to 5, leaving other mark values unchanged. It is syntactically expanded to something like

```
> m <- marks(X)
> m[3] <- 5
> marks(X) <- m
```

In commands like `Window(X) <- W`, if the new window W is smaller than the current `Window(X)`, points falling outside W will be discarded. Similarly for `Frame(X) <- B`, all data outside the box B will be removed, and the window will be intersected with B.

Note that it is quite possible to assign new marks or new values to any object, even a 'standard' or 'installed' dataset (i.e. one of the example datasets provided by `spatstat`.) In the examples above, rather than creating a new dataset X, we *could* have typed (for example):

```
> marks(redwood) <- nndist(redwood)
```

It is, however, rather unwise to do this because it is easy to forget that you have changed the data and assigning new data to a standard name like `redwood` can easily cause confusion. Note that assigning `marks(redwood) <- nndist(redwood)` would create a *new* pattern `redwood` in your workspace ('Global environment'). It would *not* change or overwrite the copy of `redwood` that exists in the `package:spatstat` position on the search path. The current workspace is searched before the libraries, so that the name `redwood` would be matched to the modified `redwood` object. The original `spatstat` package dataset `redwood` would be said to be *'masked'* by the new object `redwood` in the global environment. Warnings about 'masked' objects are sometimes given when a workspace or package is loaded.

4.2.2   Conversion of window types

As indicated above it is possible to change the observation window of a point pattern. One reason for doing this may be that another representation of the window is wanted. As described in Section 3.5 there are three different window formats in `spatstat`: rectangles, polygonal regions, and binary pixel masks. Table 4.5 lists some useful tools for converting between different window formats.

When applied to a window object of type `"rectangle"`, the function `as.polygonal` just changes the internal structure of that object to that of a window object of type `polygonal`. When applied to a window of type `"mask"` it creates a polygon all of whose edges are either horizontal or vertical, these edges being edges of the outer pixels comprising the mask-type window.

`as.polygonal`	convert a window to a polygonal window
`as.mask`	convert a window to a binary image mask window
`as.rectangle`	extract the Frame of a window
`as.matrix.owin`	convert a window to a logical matrix
`pixellate.owin`	convert window to pixel image
`simplify.owin`	approximate window by a polygon
`raster.xy`	raster coordinates of a pixel mask
`is.rectangle`	determine if a window is a rectangle
`is.polygonal`	determine if a window is polygonal
`is.mask`	determine if a window is a binary pixel mask

Table 4.5. *Tools for window format testing and conversion.*

The `simplify.owin` function converts a complicated polygon into one with fewer edges. We demonstrate using the `clmfires` dataset.

```
> W <- Window(clmfires)
> U <- simplify.owin(W,10) # Gives a polygon with about 200 edges.
```

The resulting windows are shown in Figure 4.17.

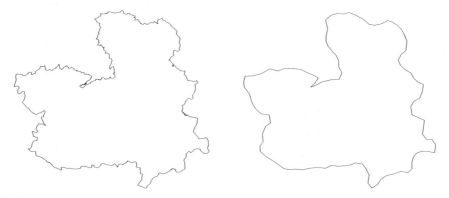

Figure 4.17. Left: *complicated polygonal window* W; Right: *simplified window* U *created in the text.*

4.2.3   Extracting subsets of a point pattern

In `spatstat` the tools for extracting subsets from a point pattern dataset are called `"[.ppp"` and `subset.ppp` and these are methods for the R generic commands `"["` and `subset`. This supports a versatile and easy-to-use system of subset extraction which is consistent with conventional R syntax.

Subset defined by an index

Subsets of a point pattern can be specified by indices using `"["`, in the same manner as for subsets of a vector. That is, if X is a point pattern and s is a vector of positive integers, of negative integers, or of logical values, then X[s] is also a point pattern, consisting of those points of X selected by s. For example we could take s to be the vector of positive integers 1:10.

```
> bei
Planar point pattern: 3604 points
window: rectangle = [0, 1000] x [0, 500] metres
```

```
> bei[1:10]
Planar point pattern: 10 points
window: rectangle = [0, 1000] x [0, 500] metres
```

This produces a point pattern of 10 points, namely the first ten points in the `bei` dataset. The window of the resulting pattern is the same as that of `bei`.

Note that, although a point pattern should in principle be treated as an unordered set, the coordinates are obviously stored in a particular order, and can be addressed using that order. If the points of `bei` had been written down in some other order, equally arbitrary but different, then `bei[1:10]` would consist of a different set of 10 points.

The index x can be a vector of negative integers:

```
> bei[-c(2,3,7)]
```

yields a point pattern consisting of all *except* the second, third, and seventh points of `bei`.

The index can be a vector of logical values with one entry for each point of the pattern, with value TRUE if the corresponding point is to be retained. For example,

```
> swedishpines[nndist(swedishpines) > 10]
```

extracts all points of the `swedishpines` pattern for which the nearest other point lies further than 10 units (1 metre) away; and

```
> longleaf[marks(longleaf) >= 42]
```

gives a (marked) point pattern consisting of all those points from `longleaf` where the mark value (the diameter at breast height of the tree) is at least 42 centimetres. The latter example could be done more elegantly using the `subset` function, explained below. A logical vector shorter than the required length will be 'recycled', so that

```
> longleaf[c(FALSE,TRUE)]
```

would extract *every second point* in the sequence.

> Tip: Put quotes around the subset operator when asking for help about it. The generic subset operator is `"["`; the right-hand square bracket that you type when using this operator is really just punctuation. The help file for this operator has to be summoned by typing `help("[")` or `?"["`, otherwise an error will result. Likewise the subset method for point patterns is called `"[.ppp"`; the help file for this method is summoned by typing `help("[.ppp")` or `?"[.ppp"`.

Subset defined by a window

In `spatstat` the indexing operator `"[.ppp"` can also extract a subset of a point pattern corresponding to *a spatial region*. If X is a point pattern and W is a spatial window (object of class `"owin"`) then X[W] is the point pattern consisting of all points of X that lie inside W.

```
> W <- owin(c(100,800), c(100,400))
> W
window: rectangle = [100, 800] x [100, 400] units
> bei[W]
Planar point pattern: 918 points
window: rectangle = [100, 800] x [100, 400] units
```

The window of the resulting pattern is W.

Note that there is also a `"["` method for the class `"owin"`. The expression A[B] is equivalent to `intersect.owin(A,B)`.

Subset defined by an expression

The method `subset.ppp` can be used to extract a subset of a point pattern defined by an expression involving the names of spatial coordinates x, y, the `marks`, and the names of individual columns of marks, if there are several columns. For example

```
> subset(cells, x > 0.5 & y < 0.4)
```

extracts the subset of the `cells` data consisting of all points with *x*-coordinate greater than 0.5 and *y*-coordinate less than 0.4. Note the single &, the vector-wise 'and' operator, so that the result of evaluating the expression is a logical vector of length equal to the number of points. Similarly

```
> subset(longleaf, marks >= 42)
```

extracts the trees from the Longleaf Pines data with mark value greater than or equal to 42. The `finpines` dataset has two columns of marks, named `diameter` and `height`, so

```
> subset(finpines, diameter > 2 & height < 4)
```

extracts the trees with diameter more than 2 centimetres and height less than 4 metres.

The argument `select` can be used to retain only some of the columns of marks. It is another expression, involving the name `marks` or the names of columns of marks. For example

```
> subset(finpines, diameter > 2, select=height)
```

retains only the `height` column of marks. In the expression `select`, the variable names are interpreted as column numbers in the data frame of marks, so

```
> subset(nbfires, year == 1999, select=cause:fnl.size)
```

is meaningful; it extracts the subset of the New Brunswick forestfires data where the year was 1999, and retains only the columns `cause`, `ign.src`, and `fnl.size`. We could also do

```
> subset(finpines, select = -height)
```

to remove the `height` column from the marks.

4.2.4 Manipulating marks

The commands `m <- marks(X)` for extracting marks, and `marks(X) <- m` for assigning marks, were discussed in Section 4.2.1 above.

To delete marks from a point pattern, assign the value NULL to the marks:

```
marks(X) <- NULL
```

For convenience, you can also perform these operations inside an expression, using the function `unmark` to remove marks and the binary operator `%mark%` to add marks. It is frequently convenient to make use of this facility when plotting. For example,

```
> plot(unmark(anemones))
> radii <- rexp(npoints(redwood), rate=10)
> plot(redwood %mark% radii)
```

The last line above plots the `redwood` data as a marked point pattern with random numeric marks.

A common task is to attach marks to a point pattern where the marks depend on another spatial dataset. If X is a point pattern and Z is a pixel image, then Z[X] gives the values of the image at the points of X, so X %mark% Z[X] attaches these values to the points.

```
> elev <- bei.extra$elev
> Y <- bei %mark% elev[bei]
```

To attach an *additional* column of marks to a dataset which already has marks, one would typically use cbind or data.frame.

```
> X <- amacrine
> marks(X) <- data.frame(type=marks(X), nn=nndist(amacrine))
> Y <- finpines
> vol <- with(marks(Y), (100 * pi/12) * height * diameter^2)
> marks(Y) <- cbind(marks(Y), volume=vol)
```

The generic function cut, briefly mentioned in Section 2.1.9, has a method for point patterns, cut.ppp. For a point pattern with real-valued marks, cut.ppp will divide the range of mark values into several discrete bands, yielding a point pattern with categorical marks:

```
> Y <- cut(longleaf, breaks=c(0, 5, 20, Inf))
> Y
Marked planar point pattern: 584 points
Multitype, with levels = (0,5], (5,20], (20,Inf]
window: rectangle = [0, 200] x [0, 200] metres
```

Here Inf is positive infinity, so the third category includes all values higher than 20. If breaks is a single number, it determines the number of bands:

```
> Y <- cut(longleaf, breaks=3)
> Y
Marked planar point pattern: 584 points
Multitype, with levels = (1.93,26.6], (26.6,51.3], (51.3,76]
window: rectangle = [0, 200] x [0, 200] metres
```

Used in this context, cut.ppp operates on the marks rather than the spatial coordinates.

If W is a window (class "owin"), then cut(X, W) will attach a logical-valued mark to each point, equal to TRUE if the point lies inside W, and FALSE otherwise. If A is a tessellation (class "tess", discussed in Section 4.5) then cut(X, A) will attach a factor-valued mark to each point, indicating which tile of the tessellation contains the point.

4.2.5 Converting to another unit of length

To convert the spatial coordinates to a different unit of length, use the function rescale:

```
Y <- rescale(X, s)
```

The spatial coordinates in the dataset X will be re-expressed in terms of a new unit of length that is s times the current unit of length given in X.

The spatstat generic function rescale is designed so that **the rescaled object is equivalent to the original**. For example if X is a dataset giving coordinates in metres, then rescale(X,1000) will divide the coordinate values by 1000 to obtain coordinates in kilometres, and the unit name will be changed from metres to 1000 metres. The resulting object is equivalent to the original; the values have just been converted to another unit of measurement.

If the argument unitname is given, it will be taken as the new name of the unit of length. It should be a valid name for the unit of length, as described in Section 3.3.3. In the example above, rescale(X, 1000, "km") will divide the original coordinate values (in metres) by 1000 to obtain coordinate values in kilometres, and the unit name will be changed to km (rather than to 1000 metres).

The unit name of a spatstat object may also be a multiple of a standard 'base' unit. For example, the Lansing Woods trees were originally observed in a square of physical side length 924 feet, then the coordinates were scaled to a unit square, so the lansing dataset in spatstat has unitname(lansing) = 924 feet. For such 'composite' units, the command rescale(X) with no further arguments will convert X to the base unit:

```
> rescale(lansing)
Marked planar point pattern: 2251 points
Multitype, with levels = blackoak, hickory, maple, misc, redoak, whiteoak
window: rectangle = [0, 924] x [0, 924] feet
```

Note that rescaling the point pattern has no effect on the marks. Conversion of the mark values to another unit would have to be done separately.

Sometimes spatial data are stored or presented as a list of related spatial objects. If rescaling is applied to one of the objects in such a list then the *same* rescaling almost certainly should be applied to *all* of the objects in the list so as to maintain consistency. For example, to convert the murchison coordinate data from metres to kilometres:

```
> murch2 <- solapply(murchison, rescale, s=1000, unitname="km")
```

Since murchison is a "solist", we have ensured that the transformed data also belong to this class by using solapply or equivalently as.solist(lapply(...)).

4.2.6 Geometrical transformations

Many geometrical transformations of spatial data are supported in spatstat. They are listed in Table 4.6.

shift(X)	translate (shift)
rotate(X)	rotate
reflect(X)	reflect about the origin
flipxy(X)	swap *x*- and *y*-coordinates
scalardilate(X)	expand or contract by a scale factor
affine(X)	general affine transformation
convexhull(X)	convex hull

Table 4.6. *Common geometrical transformations for point patterns, windows, and images supported by* spatstat *(for point patterns the transformation applies to both the points and the observation window). That is, the class of* X *can be* "ppp", "owin" *or* "im".

By default, rotation is performed about the origin in the spatial coordinate system. To rotate about another centre of rotation, specify the argument centre, which may be either a pair of coordinates, or a string indicating a special location. For example

```
> rotate(chorley, pi/2, centre="centroid")
```

applies a 90-degree anticlockwise rotation to the Chorley-Ribble data around the centroid of the window.

The command scalardilate(X, f) multiplies all the spatial coordinates of the object X by the factor f without changing any other information. It produces a dataset which is not equivalent to the original.

Table 4.7 lists some commonly used geometrical operations that only apply to windows (i.e. there is no method for "ppp" objects).

`intersect.owin`	intersection of two windows
`union.owin`	union of two windows
`setminus.owin`	set difference of two windows
`complement.owin`	swap inside and outside
`inside.owin`	determine whether a point is inside a window
`is.subset.owin`	determine whether one window contains another
`is.convex`	test whether a window is convex
`centroid.owin`	compute centroid (centre of mass) of window
`incircle`	find largest circle inside window
`trim.rectangle`	cut off side(s) of a rectangle

Table 4.7. *Geometrical operations on windows (objects of class* `"owin"`*).*

Further, `spatstat` supports the operations from mathematical morphology [611, 622] listed in Table 4.8. For a window W the morphological dilation $W_{\oplus r}$ by a disc of radius r is the set of points lying at most r units away from W. A point u belongs to $W_{\oplus r}$ if a disc of radius r centred at u intersects W. The morphological erosion $W_{\ominus r}$ is the set of points inside W lying at least r units away from the boundary of W. A point u belongs to $W_{\ominus r}$ if a disc of radius r centred at u is entirely contained in W. Erosion and dilation are often used in statistical calculations for spatial data.

`dilation.owin`	morphological dilation by disc
`erosion.owin`	morphological erosion by disc
`opening.owin`	morphological opening by disc
`closing.owin`	morphological closing by disc
`border`	create a border region around a window

Table 4.8. *Morphological operations on windows.*

The morphological opening is the result of erosion followed by dilation, $(W_{\ominus r})_{\oplus r}$. A point u belongs to the opening of W if there is some disc of radius r that contains u and lies entirely inside W. The morphological closing is the result of dilation then erosion, $(W_{\oplus r})_{\ominus r}$. A point u belongs to the closing of W if u does not lie inside any disc of radius r which avoids W entirely. Morphological closing is particularly useful for 'cleaning up' irregular windows and for filling tiny gaps.

In future versions of `spatstat`, morphological operations based on other structuring elements (rather than the disc) will be implemented.

4.2.7 Random perturbations of a point pattern

The `spatstat` package provides several functions for making random changes to a spatial point pattern.

The function `rjitter` displaces each point of the pattern by a small random distance in a random direction, independently of other points. If the original dataset contained some duplicate points, then jittering will separate these duplicates, making them visible in a plot of the data. If jittering the data causes a drastic change in the results of analysis, then *ipso facto* the results were highly sensitive to the precise locations.

The function `rshift` applies the *same* random shift to every point in a pattern, or to all points in a specified subset. In a multitype point pattern, all points *of the same type* will be subjected to the same shift, while the shifts applied to different types of points are independent. This can be used in a Monte Carlo test of the hypothesis that the different types of points are independent; see Chapter 10.

The function `rlabel` randomly assigns new mark values to the points in a pattern, by randomly permuting or resampling the original marks. This can be used in a Monte Carlo test of the hypothesis that the marks are independent of the points ('random labelling'). See Section 14.5.1.

The function `quadratresample` performs a block resampling procedure in which the window of the point pattern is divided into rectangles, and these rectangles are randomly resampled to generate a new point pattern.

The function `rthin` randomly deletes some of the points in a point pattern, according to specified probability rules.

4.2.8   Generating random point patterns

Apart from randomly modifying an existing point pattern as described above it can also be useful to randomly generate a point pattern from scratch. For example, random point patterns can be used for spatial sampling, and in simulation experiments. Table 4.9 contains a list of useful `spatstat` functions for generating random point patterns. They all (except `simulate.ppm`) start with a lower case 'r', which is the R convention for random generators.

`runifpoint`	generate n independent uniform random points
`rpoint`	generate n independent random points
`rmpoint`	generate n independent multitype random points
`rpoispp`	simulate the (in)homogeneous Poisson point process
`rmpoispp`	simulate the (in)homogeneous multitype Poisson point process
`runifdisc`	generate n independent uniform random points in disc
`rstrat`	stratified random sample of points
`rsyst`	systematic random sample of points
`rMaternI`	simulate the Matérn Model I inhibition process
`rMaternII`	simulate the Matérn Model II inhibition process
`rSSI`	simulate Simple Sequential Inhibition process
`rStrauss`	simulate Strauss process (perfect simulation)
`rHardcore`	simulate Hard Core process (perfect simulation)
`rDiggleGratton`	simulate Diggle-Gratton process (perfect simulation)
`rDGS`	simulate Diggle-Gates-Stibbard process (perfect simulation)
`rNeymanScott`	simulate a general Neyman-Scott process
`rPoissonCluster`	simulate a general Poisson cluster process
`rMatClust`	simulate the Matérn Cluster process
`rThomas`	simulate the Thomas process
`rGaussPoisson`	simulate the Gauss-Poisson cluster process
`rCauchy`	simulate Neyman-Scott Cauchy cluster process
`rVarGamma`	simulate Neyman-Scott Variance Gamma cluster process
`rcell`	simulate the Baddeley-Silverman cell process
`rmh`	simulate Gibbs point process using Metropolis-Hastings
`simulate.ppm`	simulate Gibbs point process using Metropolis-Hastings
`runifpointOnLines`	generate n random points along specified line segments
`rpoisppOnLines`	generate Poisson random points on specified line segments

Table 4.9. *Random point pattern generators in* `spatstat`.

4.2.9   Splitting and combining point patterns

It is sometimes useful to split a point pattern dataset into several sub-patterns, and perform calculations separately on each sub-pattern.

4.2.9.1   Splitting a point pattern into sub-patterns

The R generic function `split` is used to divide data into subsets: `split(x, f)` divides the dataset x into subsets determined by the grouping `f`.

In `spatstat` the method `split.ppp` enables the user to divide a point pattern x into sub-patterns using a variety of criteria.

If `f` is a factor, of length equal to the number of points in x, then `split(x,f)` separates the data points into groups according to the values of `f`. The result is a list of point patterns, each with the same window as x. The names of the list entries are the levels of `f`.

If `f` is a factor-valued image, then `split(x,f)` splits the *window* of x into sub-windows according to the pixel values of `f`. The result is a list of point patterns obtained by restricting x to each of these sub-windows.

If `f` is a tessellation, then `split(x,f)` causes the window of x to be sub-divided by the tiles of the tessellation. The result is a list of point patterns, each consisting of the points of x that fall in a given tile of the tessellation.

For a multitype point pattern x, the factor `f` defaults to `marks(x)`, so that `split(x)` separates the data into the patterns of points of each type. The resulting point patterns all have the same window as x.

If `marks(x)` is a data frame, then `f` may be the name of a column of marks. If `f` is omitted, it defaults to the first column of `marks(x)` containing factor values.

In all other cases, `f` must be specified explicitly and must be a factor.

The result of `split.ppp` also belongs to the classes `"solist"` and `"splitppp"` so that it can easily be printed and plotted. Figure 4.18 shows the result of typing `plot(split(amacrine))`.

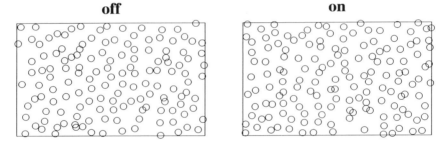

Figure 4.18. *The result of splitting the multitype point pattern* `amacrine` *into sub-patterns of points of each type, generated by* `split(amacrine)`.

To split the `nbfires` pattern into sub-patterns according to the cause of the fire:

```
> Y <- split(nbfires, "cause")
```

You can use the R command `lapply` to perform any desired operation on each element of the list. For example, to apply adaptive estimation of intensity (see Section 6.5.2) separately to each species of tree in the Lansing Woods data, do:

```
> V <- split(lansing)
> A <- solapply(V, adaptive.density)
```

A neater way to operate on sub-patterns is to use `by.ppp`, a method for the R function `by`. The call `by(X, INDICES=f, FUN=g)` is essentially equivalent to `lapply(split(X,f), g)`. The result above can be produced more briefly:

```
> A <- by(lansing, FUN=adaptive.density)
> plot(A)
```

> Tip: When writing R functions, avoid using argument names which consist entirely of capital letters. Such names tend to be used as arguments for R functions like `lapply` and `by`. Errors can result if your function has an argument with the same name.

The individual patterns which form the components of the list returned by `split.ppp` can be extracted and saved as individual patterns instead of list components. For example,

```
> S <- split(chorley)
> cases <- S$larynx
> controls <- S$lung
```

Note that if the marks of X consist of a single factor, then by default `split(X)` splits X into *unmarked* subpatterns, each corresponding to a level of these marks. In the foregoing example, `cases` and `controls` are unmarked point patterns. See `help(split.ppp)` for options to control the handling of marks.

4.2.9.2 Un-splitting a point pattern

There is an R generic function `split<-` which undoes the splitting operation, in the sense that it recombines the data, replacing each entry in its original position. A method `"split<-.ppp"` is provided in `spatstat`. This can be useful when we want to apply a transformation separately to each subset of the data. For example, in the `ants` data, suppose we want to apply jittering to the *Messor* nests only. This can be done simply by

```
> X <- ants
> u <- split(X)
> u$Messor <- rjitter(u$Messor)
> split(X) <- u
```

The resulting pattern X is identical to `ants` except that the points with type `Messor` have been slightly displaced. (Note we have avoided overwriting the standard dataset `ants`.) In this example, the last command `split(X) <- u` caused the displaced points to be put back into the dataset X in their original order. This only happens when the replacement value is compatible with the original data: that is, when each element of the list u contains the same number of points as the corresponding element of `split(ants)`. Otherwise, the entries of u will simply be concatenated.

4.2.9.3 Combining point patterns

The `spatstat` function `superimpose` combines any number of point patterns into a single pattern. Continuing the example above we could try to do the unsplitting by

```
> X <- superimpose(u$Messor, u$Cataglyphis)
```

However, the resulting pattern is unmarked. To keep track of which point came from which pattern, use argument names for the point patterns in the call to `superimpose`. The names will then become marks for the combined pattern (in addition to any existing marks).

```
> X <- superimpose(Cataglyphis=u$Cataglyphis, Messor=u$Messor)
```

In the combined pattern the points are stored in the order in which the arguments were given. This means that in the pattern X created above the Messor points are listed last while they are listed first in the original `ants` pattern. Thus, a bit of caution is needed.

The window for the combined pattern is taken to be the union of the windows of the patterns being superimposed, unless given explicitly by the argument W. Thus combining a pattern in the square $[0,2] \times [0,2]$ with a pattern in the square $[1,3] \times [1,3]$ gives a pattern in a polygonal region (the union of the two overlapping squares):

```
> X <- runifpoint(50, square(c(0,2)))
> Y <- runifpoint(50, square(c(1,3)))
> superimpose(X,Y)
Planar point pattern: 100 points
window: polygonal boundary
enclosing rectangle: [0, 3] x [0, 3] units
```

The window can be specified as the bounding square $[0,3] \times [0,3]$:

```
> superimpose(X, Y, W=square(3))
Planar point pattern: 100 points
window: rectangle = [0, 3] x [0, 3] units
```

Specifying the window W also reduces the amount of computation.

4.2.10 Basic summaries of point patterns and windows

Table 4.10 lists commands that are commonly used to obtain basic summary information about a point pattern. These are all generic functions so it is the method for class "ppp" that should be consulted for further information.

print(X)	Print basic information
X	Print basic information
summary(X)	Print detailed summary
npoints(X)	Number of points
coords(X)	Extract spatial coordinates
intensity(X)	Average density of points per unit area
density(X)	spatially varying density of points
istat(X)	Interactive data analysis
pairdist(X)	Distances between all pairs of points
nndist(X)	Distances to nearest neighbours
nnwhich(X)	Identify nearest neighbours
distmap(X)	Distance from each pixel to nearest data point

Table 4.10. *Commands for performing basic calculations on a point pattern* X.

The R generic function print can be invoked by typing print(X) or just X, the name of the object. For simple data types, this command causes all the data to be printed, but for datasets belonging to a class, the print method displays only minimal information, for the sake of efficiency. The method print.ppp displays the number of points, the type of marks (if any), and a brief description of the window (this window information is also printed by print.owin if one types Window(X)).

The R generic function summary yields a basic statistical summary of the data, typically including the mean and range of each variable. Typing summary(X), where X is an object of class "ppp", gives detailed information about the density of points, the distribution of the mark values, and the geometry of the window of X. Here the generic summary dispatches to the method summary.ppp. If we are only interested in the *window* of X and want to cut to the chase, we can just type summary(Window(X)) (which dispatches to summary.owin).

```
> summary(chorley)
```

```
Marked planar point pattern:   1036 points
Average intensity 3.287 points per square km

*Pattern contains duplicated points*

Coordinates are given to 1 decimal place
i.e. rounded to the nearest multiple of 0.1 km

Multitype:
       frequency proportion intensity
larynx        58    0.05598     0.184
lung         978    0.94400     3.103


Window: polygonal boundary
single connected closed polygon with 131 vertices
enclosing rectangle: [343.4, 366.4] x [410.4, 431.8] km
Window area = 315.155 square km
Unit of length: 1 km
```

A special trick used by many methods for summary (including summary.ppp and summary.owin) is that, instead of printing the summary information, the summary method only computes the summary information and returns it as an object. If the user types summary(X) then the print method for this special class is then invoked, and this generates the printed summary output. For a point pattern, the result of summary(X) is an object of the special class "summary.ppp" and the print method for this class is print.summary.ppp. Confused yet? The advantage is that if the user types A <- summary(X) then the summary information is stored in A for examination and re-use.

The command intensity(X) computes the average number of points per unit area, while density(X) computes a pixel image giving a spatially varying density of points per unit area; these are discussed in Chapter 6.

The command istat(X) (which requires library(rpanel)) starts an interactive interface in which the user can select different summary operations to be applied to the point pattern X. A screenshot of the interface is shown in Figure 4.19. This feature is experimental.

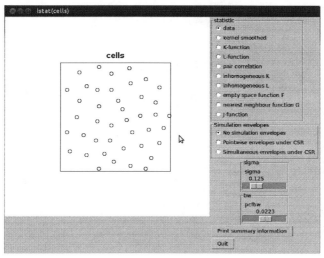

Figure 4.19. *Screenshot of the interactive graphics window started by* istat.

The command pairdist(X) returns a matrix containing the distances between all pairs of points in X. The command nndist(X) returns a numeric vector with one entry for each point of X, giving the distance from this point to the nearest neighbour (the nearest other point in X). Related to this, nnwhich(X) returns an integer vector giving for each point of X the index of the nearest neighbour. For example if m <- nnwhich(X) and m[3] = 5, then the nearest neighbour of the point X[3] is the point X[5]. A useful graphic related to nndist is the *Stienen diagram* (Section 8.2.3).

Table 4.11 lists some additional commands useful for summarising the marks in a point pattern. If the marks are categorical, then table(marks(X)) or barplot(table(marks(X))) are useful summaries of the distribution of marks. If the marks are numeric, then hist(marks(X)) or plot(ecdf(marks(X))) or plot(density(marks(X))) are useful summaries.

For numeric marks, Smooth(X) or markmean(X) will produce an image showing the spatially varying average (Nadaraya-Watson smoother) of the mark value, and markvar(X) produces an image of the spatially varying variance of the marks. These are discussed in Chapter 6.

The functions marktable, markstat, and applynbd perform calculations on the points and marks in a 'neighbourhood' of each data point, where the 'neighbourhood' of a point can be defined as the N nearest neighbours, or all data points within a distance R.

marks(X)	Extract marks
Smooth(X)	spatially varying average mark
markmean(X)	spatially varying average mark
markvar(X)	spatially varying variance of marks
marktable(X, ...)	Tabulate marks of neighbours
markstat(X, ...)	Summarise marks of neighbours
applynbd(X, ...)	Apply operation on each neighbourhood

Table 4.11. *Basic summaries of the marks of a point pattern.*

Table 4.12 lists the basic spatstat commands for calculating common summaries of a window such as the area or perimeter.

area	compute window's area
diameter	compute window's diameter
perimeter	compute window's perimeter length
deltametric	measure discrepancy between two windows
dilated.areas	compute areas of dilated windows
eroded.areas	compute areas of eroded windows
distmap.owin	distance transform image
distfun.owin	distance transform function

Table 4.12. *Basic summaries of a window.*

See help(spatstat) for an aide-mémoire list of operations that can be applied to or carried out on windows or point pattern objects.

4.2.11   Examples

We will now illustrate some of the ideas that were discussed in the previous material by applying them to an example dataset. We will use one of the point pattern datasets that is installed with the spatstat package, namely the NZ trees dataset nztrees. This represents the positions of 86 trees in a forest plot measuring 153 by 95 feet. The data are plotted in the left panel of Figure 4.20.

To get an impression of local spatial variations in intensity, we plot a kernel density estimate of intensity.

```
> contour(density(nztrees, 10), axes=FALSE)
```

We have chosen a value of 10 for the 'bandwidth' (smoothing parameter). The issue of the choice

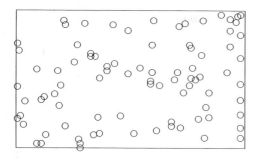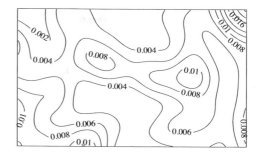

Figure 4.20. Left: *New Zealand trees data.* Right: *contour plot of kernel estimate of intensity.*

of smoothing parameter will be discussed elsewhere. (See Section 6.5.1.) We have also chosen to display the result as a contour plot. The result is shown in the right panel of Figure 4.20.

A striking feature of the intensity estimate is that it has a steep slope at the top right-hand corner of the study region. Looking at the plot of the point pattern itself, we can see a cluster of trees at the top right. You may also notice a line of trees at the right-hand edge of the study region. It looks as though the study region may have included some trees that were planted as a boundary or avenue. This feature of the data sticks out like a sore thumb if we plot a histogram of the x-coordinates of the trees, `hist(coords(nztrees)$x)`, shown in Figure 4.21.

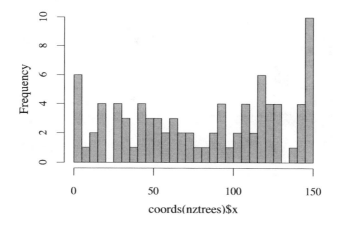

Figure 4.21. *Histogram of x-coordinates of New Zealand trees.*

Consequently we might want to exclude the right-hand boundary from the study region, so as to focus on the pattern of the remaining trees. Let's say we decide to trim a 5-foot margin off the right-hand side. First we create the new, trimmed study region:

```
> chopped <- owin(c(0,148),c(0,95))
```

or, more slickly,

```
> win <- Window(nztrees)
> chopped <- trim.rectangle(win, xmargin=c(0,5), ymargin=0)
> chopped
window: rectangle = [0, 148] x [0, 95] feet
```

Of course chopped is not a point pattern, but rather an owin object which is of type rectangle and hence simply a rectangle in the plane. We can, however, get the point pattern that we want by using the subset operator "[.ppp", to extract the subset of the original point pattern that lies inside the new window:

```
> nzchop <- nztrees[chopped]
```

We can now study the 'chopped' point pattern.

```
> summary(nzchop)
Planar point pattern:  78 points
Average intensity 0.005548 points per square foot

Coordinates are integers
i.e. rounded to the nearest foot

Window: rectangle = [0, 148] x [0, 95] feet
Window area = 14060 square feet
Unit of length: 1 foot
> plot(density(nzchop, 10))
> plot(nzchop, add=TRUE)
```

The results are shown in Figure 4.22. Removing the right margin seems to have produced a much more uniform pattern, although there is still a suggestion of variation in tree density.

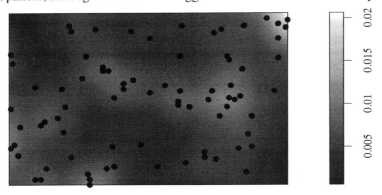

Figure 4.22. *New Zealand trees pattern with right-hand border removed, and kernel estimate of intensity.*

4.3   Exploring images

Pixel images were introduced in Section 3.6. This section explains how to plot pixel images in various ways, extract subsets of a pixel image, and manipulate image data.

The methods for print and summary provide useful basic information about an object of class "im". The array dimensions can be found from dim, nrow, and ncol. Summary statistics of the pixel values can be obtained from methods for range, max, min, mean, median, and quantile. The function integral.im calculates the sum of pixel values times pixel area for an "im" object, effectively producing the integral of the spatial function over the window.

4.3.1 Extracting pixel values and subimages

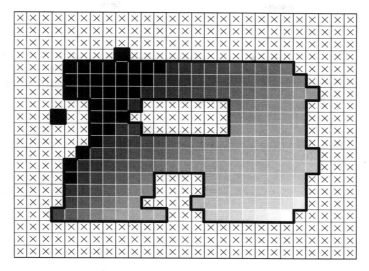

Figure 4.23. *Anatomy of a pixel image in* spatstat. *The* Frame *is the outer rectangle. Pixels marked* × *have the value* NA. *Shades of grey represent numerical pixel values. The* Window *is the set of pixels with a non-*NA *value, and is delineated by the thick black lines.*

Figure 4.23 shows the anatomy of a pixel image X in spatstat. Small squares represent pixels. The spatial coordinates of a pixel are the coordinates of the centre of the square. The outermost rectangle is Frame(X). Every pixel inside the frame has a pixel value. Pixels marked × have the value NA, indicating that the pixel lies outside the observation window. Thick black lines delineate Window(X).

When a pixel image is plotted, the pixels with NA values are not rendered on the graphics display. The corresponding part of the display will still show the background colour (or any previously displayed graphics if the plot was generated with add=TRUE). Confusion may arise if the background colour is also one of the colours in the colour map. This is most likely when plotting in a greyscale colour map when the background colour is white.

If you just want to get your hands on the pixel values of an image, use as.matrix.im, a method for the generic as.matrix. The resulting matrix contains the pixel values for all pixels in the frame, arranged according to the idiosyncratic convention of spatstat: columns correspond to increasing values of the *x*-coordinate, and rows to increasing values of the *y*-coordinate. Use transmat (page 66) to convert to another convention. If you need to refer to the spatial locations of the pixels, use the functions raster.x and raster.y, which make matrices containing the corresponding pixel coordinates. Note that the result of as.matrix.im may contain NA values, which come from pixels lying outside the window in which the image is defined.

Subsets of the pixel values can be extracted using the indexing operator [. Two common variants of the syntax are X[S] and X[S, drop=TRUE] where X is an image and the subset to be extracted is determined by the index argument S. If S is a point pattern, or a list(x,y), then X[S] extracts the values of the pixel image X at these points, and returns them as a vector. If S is a window (an object of class "owin"), then X[S] extracts the values of the image inside this window. The result is a pixel image if possible (see below), and a numeric vector otherwise. If S is a pixel image with logical values, it is interpreted as a window (with TRUE inside the window), and X[S] extracts the values of X inside this window.

The logical argument drop determines whether NA values (corresponding to pixels outside the observation window) should be omitted (drop=TRUE) or retained (drop=FALSE, the default). If P is a point pattern, X[P] is a vector with one entry for each point in P, but may contain NA values, while X[P,drop=TRUE] is generally a shorter vector. If S is a window or logical-valued image, then X[S] is an image obtained from X by setting all pixels outside S to have the value NA. That is, X[S] is the restriction of the image X to the region S. This changes the Window, but not the Frame or the pixel coordinates.

The argument tight determines whether the original Frame will be retained (tight=FALSE) or replaced by the smallest possible rectangle (tight=TRUE). The effect of X[S, tight=TRUE] can also be achieved by typing Y <- X[S] then Frame(Y) <- boundingbox(Y).

If S is a window or logical-valued image, X[S, drop=TRUE] consists only of pixel values inside the region S. If S is a non-rectangular window then the result is a vector containing the values for the selected pixels. However, if S is a rectangle, the result *could* be returned as a pixel image. The logical argument rescue determines whether this will happen: the default is rescue=TRUE. Thus, if R is a rectangle, X[R] and X[R, drop=TRUE] both return the image which is the restriction of X to the region R, but X[R, drop=TRUE, rescue=FALSE] yields the vector of pixel values inside this rectangle.

Like many methods for [, the method [.im can also be called without an index, in the forms X[,] and X[]. The form X[,] is equivalent to as.matrix(X): all the pixel values are returned, and the matrix structure is preserved. The form X[] generally means 'all relevant values'; for an image X the expression X[] represents all *defined* pixel values, that is, all pixel values of X which are not NA, returned as a vector.

The most common uses of [.im are summarised in Table 4.13. See help("[.im") for full details.

COMMAND	FORMAT	DESCRIPTION	NA's?
as.matrix(X)	matrix	all pixel values	Yes
X[,]	matrix	all pixel values	Yes
X[]	vector	all defined pixel values	No
X[P]	vector	values at all points of P	Yes
X[P, drop=TRUE]	vector	values defined at points of P	No
X[W]	image	image restricted to W	Yes
X[W, drop=TRUE]	vector	pixel values inside W	No
X[R]	image	image restricted to R	No
X[R, drop=TRUE, rescue=FALSE]	vector	pixel values inside R	No

Table 4.13. *Typical commands for extracting subsets of a pixel image* X *indexed by a point pattern* P, *a non-rectangular window* W, *or a rectangular window* R.

A handy trick is to use the subset index operator to look up the value of a pixel image at a single point:

```
> elev <- bei.extra$elev
> elev[list(x=142,y=356)]
[1] 147.1
```

As we have previously noted (page 91), this sort of subsetting can even be performed interactively, using the R function locator to click on a point in the window:

```
> plot(elev)
> elev[locator(1)]
```

For more frequent use, it would be easier to convert the image to a function using as.function.

To display a subregion of an image we could take the index argument to be a window object defining the subregion:

```
> S <- owin(c(200,300), c(100,200))
> plot(elev[S])
```

The subset can also be chosen interactively:

```
> plot(elev)
> S <- clickpoly(add=TRUE)
> plot(elev[S, drop=FALSE, tight=TRUE])
```

The result is shown in Figure 4.24.

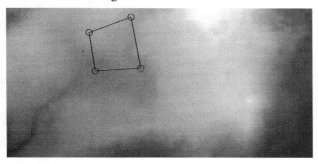 

Figure 4.24. *Interactive selection of a subset of an image.* Left: *Full image, with polygon selected using* clickpoly. Right: *Selected subset of image.*

4.3.2 Arithmetic on "im" objects

It is easy to perform mathematical calculations with the pixel values of a pixel image. Three options are supported: the pixel values can be extracted and manipulated as a numeric vector; the pixel values can be manipulated in place using eval.im, yielding another image; and most standard arithmetic operations can be applied to "im" objects directly, yielding another image.

4.3.2.1 Extracting pixel values

As we indicated above, Z[] simply returns all of the pixel values of Z which have a defined (non-NA) value. These can be manipulated as we choose, for example, by applying standard statistical summaries such as mean and median. A histogram of pixel values can be generated by hist(Z[]) and a kernel density estimate by plot(density(Z[])). One can also change pixel values by reassigning them using "[<-":

```
> Z <- distmap(cells)
> Z[] <- Z[] + 1
```

Using eval.im

The function eval.im performs pixel-by-pixel calculations on an image, or on several images, and returns a new image. For instance if Z is an image with numerical values, we can add 10 to each pixel value by typing

```
> Y <- eval.im(Z - 1.1)
```

The result is a new image Y, where the pixel value is equal to 10 plus the corresponding pixel value in Z. Note that eval is not generic, so we really have to type eval.im. If Z and Y are images, we can add their corresponding pixel values by

```
> X <- eval.im(Z + Y)
```

The argument of `eval.im` can be any R language expression in which the 'variables' are the names of pixel images. It may also involve the names of constants and functions, for example,

```
> eval.im(sqrt(Z))
> eval.im(sin(pi * Z))
> eval.im(Z > 3)
```

The expression can involve several images:

```
> eval.im(log(X) + Y - 3)
```

An important restriction is that the expression must operate 'pixel-by-pixel', in the sense that the output has as many pixel values as the input:

```
> W <- eval.im(max(0,Z))  # Throws an error.
> W <- eval.im(pmax(0,Z)) # Works.
```

The first expression throws an error because it gives a single value, whereas the second expression gives the same number of pixel values as are contained in Z.

```
> eval.im(if(X < 3) 3 else 1)   # Throws an error.
> eval.im(ifelse(X < 3, 3, 1))  # Works.
```

The expression must be able to handle `NA` pixel values. This is automatically true for most expressions in R, but is not automatic for some standard functions. If an expression produces infinite or NaN ('not a number') values, these will be converted to `NA`. For example we can safely take the logarithm

```
> logY <- eval.im(log(Y))
```

of any numeric-valued image Y. If any of the pixel values of Y are zero or negative, a warning is issued, and the relevant pixels of `logY` get the pixel value `NA`.

The pixel grids of the image objects are permitted to differ: grids will be resampled to a common resolution.

Another important restriction is that the variables in the expression passed to `eval.im` must be the *names* of pixel images. The following will fail, because `cells` is a point pattern, not an image:

```
> eval.im(density(cells) - 3)  # Throws an error
```

To get around this restriction you can do the calculations to create a pixel image before calling upon `eval.im`. For example:

```
> dcells <- density(cells)
> eval.im(dcells - 3) # Works
```

Another strategy is to use the second argument `envir` of `eval.im` to supply pixel images which need to be evaluated:

```
> eval.im(Z - 3, envir=list(Z=density(cells)))
```

It was suggested in Section 4.1.4 that after inspecting `hist(Z)` one might wish to consider appropriate transformations of the pixel values. These transformations can be performed using `eval.im`. We have already referred to the possibility of doing a logarithm transformation. Similarly a Box-Cox transformation can be carried out using `Y <- eval.im((Z^pwr-1)/pwr)` for some chosen value of `pwr`.

Another useful operation is the *transformation to uniform distribution* ('histogram equalisation' or the 'probability integral transformation')

```
> g <- ecdf(Z)  # Remember that g is a *function*.
> Y <- eval.im(g(Z))
```

Here g is the cumulative distribution function of the pixel values, so the image Y has pixel values between 0 and 1. The pixel values of Y effectively indicate the ranking of the corresponding pixel value of Z. A pixel value of 0.5 for Y means that the corresponding pixel value of Z was the median (0.5 or 50% quantile) of the pixel values of Z. Plots of the resulting image Y often reveal better detail than plots of the original.

Using unwrapped expressions

The spatstat package supports standard mathematical operations on images. Standard operations are those listed under the help for Math.im and include arithmetic operations (+, -, *, /) and mathematical functions (log, exp, sin, cos, etc.). One can, in many instances, treat "im" objects just as if they were numbers:

```
> Y <- Z+3
> X <- Z+Y
> W <- log(Z)
> U <- sin(pi*Z)
```

where Y and Z are objects of class "im". The results of the calculations will again be objects of class "im".

This scheme will also work with expressions like

```
> Y <- density(cells) - 3
```

because the intermediate result density(cells) is recognised as an image.

Operations which are not recognised or handled in this context include pmax and ifelse. For instance pmax(Z,42) and ifelse(Z < 42,0,Z) throw errors, while eval.im(pmax(Z,42)) or eval.im(ifelse(Z < 42,0,Z) work.

4.3.3 Manipulating images

Table 4.14 lists commands which change the spatial coordinates of the pixels but do not alter the pixel values. These are all generic functions defined in spatstat, with methods for class "im".

rescale(Z, s)	convert to different unit of length
scalardilate(Z, f)	multiply coordinates by f
shift(Z, v)	shift pixel grid
rotate(Z, a)	rotate pixel grid
reflect(Z, f)	reflect pixel grid about origin
flipxy(Z, f)	swap *x*- and *y*-coordinates
affine(Z, m, v)	affine transformation of coordinates

Table 4.14. *Commands which change the spatial coordinates of an image* Z *but do not affect the pixel values.*

Table 4.15 shows commands which operate on the pixel values of an image Z. In the expression Z[P], the argument P can be a point pattern or a list(x,y). The result is the vector of pixel values at the pixel centres *nearest to* the specified points. The alternative interp.im(Z,P) performs bilinear interpolation from the nearest pixels onto the specified points.

The use of methods for the generic function cut to discretize numerical values was discussed in Section 4.2.4. There is a "im" method for cut. For example, given an image Z with numeric values, plot(cut(Z, 3)) will divide the pixel values into 3 bands, and display the image with the

3 bands rendered in 3 different colours. This may serve to filter out extraneous detail that appears when the raw numeric values are plotted and thus reveal interesting features which would otherwise be obscured by the blur of the detail.

`Z[P]`	pixel value nearest to points
`interp.im(Z, P)`	interpolate to points
`scaletointerval(Z, lo, hi)`	scale pixel values to range
`cut(Z, ...)`	divide image into sub-images
`split(Z, f)`	divide image into sub-images
`by(Z, f, fun)`	apply `fun` to subsets
`levelset(Z, h)`	region where `Z <= h`
`solutionset(sin(Z) > 0)`	region where statement is true

Table 4.15. *Commands for manipulating pixel values of an image* Z.

The method `split.im` enables the user to divide a pixel image into sub-images using various criteria. The options are similar to those for `split.ppp` discussed in Section 4.2.9.1. The method `by.im` is more elegant for some purposes: the call `by(X, INDICES=f, FUN=g)` is essentially equivalent to `lapply(split(X,f), g)`.

The command `levelset(Z, h)` or `levelset(Z, h, "<=")` finds the region where the pixel value is less than or equal to `h`, and returns this region as a window. Suppose we wish to restrict the study region for the tropical rainforest data `bei` to altitudes higher than 145 metres. We could do this by forming a window, consisting of those pixels in the image whose values satisfy this constraint, as follows:

```
> elev <- bei.extra$elev
> W <- levelset(elev, 145, ">")
```

More complicated constraints can be accommodated by the `solutionset` function. This function uses `eval.im` to evaluate an expression passed to `solutionset` as its first argument. The result of applying `eval.im` is an image with logical values. The `solutionset` function returns the region where the value is TRUE as a window.

For an example we turn again to the `bei` data. The dataset `bei.extra$grad` accompanying the point pattern `bei` is a pixel image of the slope (gradient) of the terrain. We can use this image in conjunction with `bei.extra$elev` referred to above to restrict the study area to those locations where altitude is below 140 and slope exceeds 0.1. This is done by forming a window as follows:

```
> grad <- bei.extra$grad
> V <- solutionset(elev <= 140 & grad > 0.1)
```

The windows `W` and `V` produced by the two foregoing restriction operations are shown in Figure 4.25 and the results of restricting the `bei` pattern to these windows are shown in Figure 4.26.

Figure 4.25. *Binary masks* W *(Left) and* V *(Right) created in the text, with black colour representing* TRUE *(pixels inside window).*

In calculations involving several pixel images, we often strike the problem that the images are

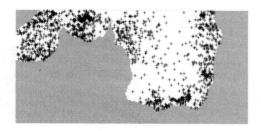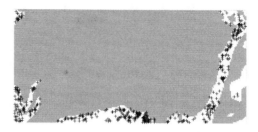

Figure 4.26. *The point pattern* bei *in the restricted study regions, with grey colour representing* FALSE *(pixels outside window).*

defined on different grids of pixels. Table 4.16 gives some useful tools for detecting and overcoming this problem. Many spatstat functions automatically convert their input images to a common grid.

compatible	Check whether images/windows have compatible pixel grids
commonGrid	Find a common spatial domain and pixel resolution
harmonise	Convert images/windows to a common pixel resolution

Table 4.16. *Tools for reconciling several pixel grids.*

4.4   Using line segment patterns

Line segment patterns (objects of class "psp") were introduced in Section 2.4.4; methods for creating them were described in Section 3.7. An object of class "psp" is effectively a list of line segments, together with a Window, a Frame, and an optional vector or data frame of marks. Each line segment must have both endpoints inside the observation window or on the boundary of the window.

Table 4.17 lists some of the more important functions for manipulating "psp" objects. The Window, Frame, and marks can be extracted and changed in the usual way. For the subset operator "[.psp" the subset index can be either an index vector (selecting some of the line segments by their position in the list of segments), or a spatial window of class "owin" (computing the intersection between the line segment pattern and the window).

Additionally Table 4.6 lists some generic geometrical operations that are defined for spatial objects including "psp" objects.

Many of the functions listed in Table 4.17 are methods for generic functions for the class "psp". However, many are not, despite the '.psp' suffix in their names. A complete list of all methods for class "psp" may be obtained by methods(class="psp"). To obtain the larger list of all functions with names ending in '.psp':

```
> ls(pos="package:spatstat",pattern="\\.psp$")
```

`"[.psp"`	subset of a line segment pattern
`angles.psp`	orientation angle of segment
`as.data.frame.psp`	extract endpoints and marks if any
`as.ppp.psp`	point pattern consisting of endpoints of segments
`closing.psp`	morphological closing of a point pattern
`crossdist.psp`	distances between all pairs of segments from two patterns
`crossing.psp`	find any crossing points of two patterns
`distfun.psp`	distance to nearest line segment (as a function)
`distmap.psp`	distance to nearest line segment (as a pixel image)
`endpoints.psp`	extract the endpoints of each segment
`identify.psp`	find the segment which is nearest to the mouse position
`lengths.psp`	compute the length of each segment in a pattern
`marks.psp`	extract marks for each segment
`"marks<-.psp"`	change marks for each segment
`midpoints.psp`	compute the midpoint of each segment
`nndist.psp`	distance from each segment to its nearest neighbour
`nnfun.psp`	nearest-neighbour index map (as a function)
`nnmap.psp`	nearest-neighbour index map (as a pixel image)
`nsegments.psp`	number of segments in pattern
`pairdist.psp`	matrix of distances between all pairs of segments
`periodify.psp`	make a periodic pattern from shifted copies
`selfcrossing.psp`	crossing points between the segments
`selfcut.psp`	cuts the segments into pieces at crossing-points

Table 4.17. *Functions for manipulating and obtaining information about a* `"psp"` *object.*

4.5 Tessellations

A *tessellation* is a division of a window into non-overlapping regions. These regions are called 'tiles' (Latin *tesserae*) although their shape is arbitrary, so they may not look anything like ceramic tiles. Tessellations are often used to represent administrative or political divisions, such as the subdivision of a country into states or provinces. They can be completely artificial, e.g., the rectangular quadrats which we use in quadrat counting. Tessellations can be observational data, such as the classification of a survey region into different rock types. A tessellation can be computed from data, e.g., the Dirichlet tessellation defined by a set of points.

Note that the tiles of a tessellation must be non-overlapping. The union of the tiles is the window associated with the tessellation: it is this window which is divided up by the tiles. Note that this window could be irregular or disconnected so that there could be apparent 'gaps' between the tiles, but these are not part of the window.

Tessellations have several uses in `spatstat`. A good way to conduct an initial analysis and exploration of a spatial object is to cut that object up into pieces, calculate summaries for each piece, and then examine in some way the relationship amongst these calculated quantities. Cutting an object up into pieces is effectively a 'tessellation'.

4.5.1 Creating a tessellation

In `spatstat` tessellations are represented as objects of class `"tess"`. Currently `spatstat` supports three kinds of tessellations:

- **rectangular tessellations** in which the tiles are rectangles with sides parallel to the coordinate axes;

- **tile lists**, tessellations consisting of a list of windows, usually polygonal windows;

- **pixellated tessellations**, in which space is divided into pixels and each tile occupies a subset of the pixel grid.

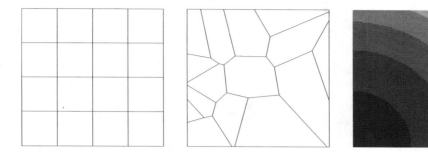

Figure 4.27. *Types of tessellations.* Left: *rectangular.* Middle: *tile list.* Right: *pixellated.*

All three types of tessellation can be created by the command `tess`. To create a rectangular tessellation:

```
tess(xgrid=xg, ygrid=yg)
```

where `xg` and `yg` are vectors of coordinates of vertical and horizontal lines determining a grid of rectangles. Alternatively, if you want to subdivide a window `W` (usually but not necessarily rectangular) by rectangles of equal size, you can use `quadrats`. Type:

```
quadrats(W, nx, ny)
```

where `nx,ny` are the numbers of rectangles in the x and y directions, respectively. A common use of this command is to create quadrats for a quadrat-counting method.

To create a tessellation from a list of windows,

```
tess(tiles=z)
```

where `z` is a list of objects of class `"owin"`. The windows should not be overlapping; currently `spatstat` does not check this. This command is commonly used when the study region is divided into administrative regions (states, départements, postcode regions, counties) and the boundaries of each sub-region are provided by GIS data files. The window that is tessellated by the list `z` is by default the union of the tiles, i.e. of the windows which constitute the components of `z`.

By default, the window of the tessellation will be computed as the union of the tiles. For efficiency, the `tess` function has an argument `window` which specifies the window. The `tess` function does not check whether this argument equals the union of the tiles; if it does not then the data will be internally inconsistent.

To create a tessellation from a pixel image:

```
tess(image=Z)
```

where `Z` is a pixel image with factor values. The pixels that have a particular value of the factor constitute a tile. For example, a pixel image indicating land cover type could be separated into tiles associated with each land cover type. These 'tiles' do not look like ceramic tiles and may even be disconnected.

The function `as.tess` can also be used to convert other types of data to a tessellation.

4.5.2   Computed tessellations

There are two commands which directly compute tessellations from point patterns.

The command `dirichlet(X)` computes the *Dirichlet* (or *Voronoï* or *Thiessen*) tessellation of the point pattern X. The tile associated with a given point of the pattern X is the region of space which is closer to that point than to any other point of X (see Section 8.2.3). The Dirichlet tiles are polygons. The command `dirichlet(X)` computes these polygons and intersects them with the window of X.

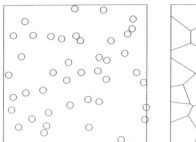 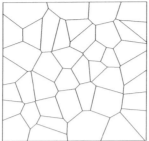 

Figure 4.28. *A point pattern* X, *its Dirichlet tessellation* `dirichlet(X)`, *and its Delaunay triangulation* `delaunay(X)`.

The *Delaunay triangulation* of a point pattern X is a network of straight lines connecting the points. Two points of X are joined if their Dirichlet tiles share a common edge. The resulting network defines a set of non-overlapping triangles, which cover the *convex hull* of X rather than the entire window containing X. The command `delaunay(X)` computes the Delaunay triangulation and returns it as a tessellation (object of class `"tess"`) which covers the convex hull of X.

4.5.3   Operations involving a tessellation

Tessellations can be plotted using the plot method `plot.tess`, which produced Figure 4.27. Alternatively the function `textureplot` displays each tile of the tessellation as a polygon filled with a different graphical pattern, as shown in the right panel of Figure 4.10.

There is a method for `"["` for subsetting tessellations. This can be used to select out a subset of the tiles which make up the tessellation in question. For example, one could create a tessellation B consisting of the first, third, and fifth tiles of a tessellation A by means of B <- A[c(1,3,5)]. Alternatively the subset index can be a window W: the result of A[W] is the restriction of A to W, that is, the tessellation of W formed by intersecting each tile of A with W.

Use the function `tiles` to extract a list of the tiles in a tessellation. The result is a list of windows (`"owin"` objects). This can be handy for computing some characteristic of the tiles in a tessellation, such as their areas or diameters:

```
> X <- runifpoint(10)
> V <- dirichlet(X)
> U <- tiles(V)
> sapply(U, diameter)
     1      2      3      4      5      6      7      8      9     10
0.4708 0.5066 0.4539 0.5076 0.6371 0.6763 0.6485 0.2855 0.2574 0.7334
```

Tile areas can be computed efficiently using `tile.areas`.

Tessellations are very handy for classifying the points of a pattern, so that the resulting classification can be used by `split.ppp`, `cut.ppp`, and `by.ppp`. If X is a point pattern and V is a tessellation (of the window of X) then

- `cut(X,V)` attaches marks to the points of X identifying which tile of V each point falls into (Figure 4.29);

- `split(X,V)` divides the point pattern into sub-patterns according to the tiles of V, and returns a list of the sub-patterns (Figure 4.30);

- `by(X,V,FUN)` divides the point pattern into sub-patterns according to the tiles of V, applies the function FUN to each sub-pattern, and returns the results as a list.

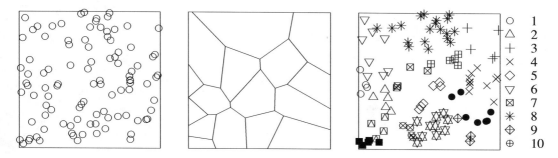

Figure 4.29. *A point pattern* X, *a tessellation* Z, *and the result of* cut(X,Z).

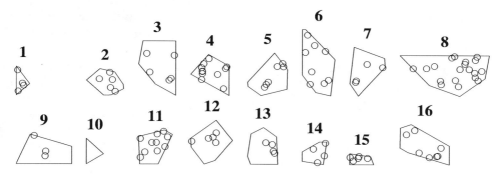

Figure 4.30. *The result of* split(Z,X).

If we superimpose plots of two tessellations on the same spatial domain (using `add=TRUE`) what we see is another tessellation. The *'intersection'* (or 'overlay' or 'common refinement') of two tessellations X and Y is the tessellation whose tiles are the intersections between tiles of X and tiles of Y. The command `intersect.tess` computes the intersection of two tessellations. The intersection of tessellations is illustrated in Figure 4.31.

Some other operations for tessellations are listed in Table 4.18. These operations are documented in the help files for `spatstat`.

4.6 FAQs

- *How do I remove the white space around the plot of a point pattern?*

Graphics code in `spatstat` uses the base R graphics system (the package `graphics`). General

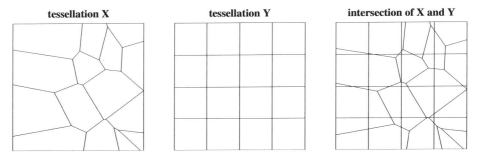

Figure 4.31. *Intersection of tessellations.*

`[.tess`	subset of tessellation
`tiles`	extract list of tiles of tessellation
`intersect.tess`	intersect two tessellations
`bdist.tiles`	compute distance from tile to window boundary
`chop.tess`	divide tessellation along a line
`rpoislinetess`	generate tessellation based on random lines

Table 4.18. *Operations on a tessellation.*

properties of a plot such as the text margin are controlled by the graphics parameters managed by the function `par`: see `help(par)`. To reduce the white space, change the parameter `mar`. Typically, `par(mar=rep(0.5, 4))` is adequate, if there are no annotations or titles outside the window.

In `spatstat`, spatial objects such as point patterns and pixel images are always plotted using equal scales on the *x* and *y* axes. White space is inevitable if the space available for plotting does not have the same *shape* (aspect ratio) as the spatial object to be plotted. If you are plotting on a computer screen, then you can change the white space by resizing the plot window using the mouse. (Re-issue the plot command to ensure the text scale is correctly updated.) If you are plotting to a file using `postscript` or `pdf`, try specifying the `width` and `height` parameters to have roughly the same ratio as the width and height of the spatial object to be plotted.

- *How do I add coordinate axes or labels to a plot produced by* `spatstat`*?*

You can use any base R graphics commands to annotate the plot, such as `axis` to add a coordinate axis (including tick marks if desired), `mtext` to add a name to the axis or other text in the margins, and `text` to add text inside the plot region.

It is strongly recommended to start each new plot with a `spatstat` command, before adding other graphics. This will ensure that the *x* and *y* scales are equal, and avoid many other problems.

- *Why doesn't* `spatstat` *show a scale bar on a plot of spatial data?*

Because scale markings are deprecated in some research fields, and mandatory in other fields. To plot a scale bar, see the help for `yardstick`.

- *My plot looks good on the computer screen, but when I send the same plot commands to a PDF or Postscript file, it looks different. For example the text is too large, or the plot symbols are too small.*

The layout of a plot is sensitive to the plot size parameters `width` and `height`, which typically have different default values in the graphics window and in the PDF or Postscript driver. To determine appropriate parameter values, draw the plot on the computer screen, adjust `par('mar')`, and resize the window using the mouse, to obtain a satisfactory plot. Repeat the plot command with these new settings to ensure the text size is correctly updated. Use a ruler to measure the physical size of the window in *inches* (one inch is 2.54 *cm*). Then pass these parameters as the `width` and `height` for `postscript` or `pdf`. Some graphics drivers also have an argument `pointsize` which can be used to control the size of text independently.

- *How can I see the internal structure of a* "ppp" *object, and extract some of the components?*

To extract components of a "ppp" object, use the functions `coords`, `marks`, `as.data.frame`, `Window`, or `npoints`.

We strongly advise against manipulating the internal structure of objects that belong to a class. This can cause the data to become internally inconsistent, leading to errors that are very hard to fix. You Have Been Warned.

The internal data in an object could also be quite voluminous, filling your screen with a mess. To see a compact description of the internal structure of an object x, type `str(x)`.

To strip away the class structure entirely, use `y <- unclass(x)`, which can then be printed to reveal the entire data structure. If this is too complicated, use `names(y)` to determine the names of the components of y, and inspect individual components `y[[ i ]]`.

- *Which kinds of spatial data processing should be conducted in* `spatstat`, *rather than a GIS or other software?*

The main strength of `spatstat` is statistical inference including exploratory statistics, model-fitting, model validation, nonparametric estimation, and hypothesis testing. For these purposes, `spatstat` typically has greater functionality than other systems, and the statistical procedures have been carefully checked.

The main strength of a typical GIS is its efficient handling of large spatially referenced datasets, including multiple data types, with efficient and flexible display and user interaction. Although many GIS systems include add-on modules for statistical analysis, the correctness of add-on modules is often difficult to verify.

Consequently, *large* spatial datasets should first be inspected and edited in specialised software appropriate to the application, such as a GIS. We recommend that statistical inference should usually be performed in `spatstat`.

- *Is it possible in* `spatstat` *to extract data from a 'stack' of multiple pixel images, in a manner analogous to appying* `extract()` *to an object of class* `RasterStack` *in the* `raster` *package?*

Currently `spatstat` has no objects directly analogous to `RasterStack` objects (although this will change in the future). Multiple pixel images are normally stored in a *list*-type object. Data can be extracted from such a list using a function of the `lapply` genre. Given a list L of images, one could extract the pixel values corresponding to the points of a pattern X via `LX<-lapply(L,"[",i=X)`. The result LX would be a list of vectors (or factors) of pixel values from the corresponding entries of L.

- *Do the Dirchlet/Voronoï tessellation facilities of* `spatstat` *provide for* weighted *tessellations?*

No.

- *Does* `spatstat` *make use of alternative graphical systems such as* `ggplot2` *or* `lattice`, *or provide facilities so that users can apply these systems?*

No. All of the graphical methods in spatstat are written using base R graphics. Users can, of course, extract any data they wish from `spatstat` objects and plot these data as they choose. All data are accessible. It is unlikely that `spatstat` will be converted to depend on `lattice` or `ggplot2` in the foreseeable future.

5

Point Process Methods

In order to develop statistical methods for analysing point patterns, and even in order to formulate clearly the scientific questions that can be answered by point pattern data, we need the concept of a **point process**. A point process is a random mechanism whose outcome is a point pattern.

This chapter develops terminology and basic concepts for point processes, and explains how these concepts can be applied to the analysis of point patterns.

5.1 Motivation

In analysing a spatial point pattern dataset, the researcher will pose questions such as: Are the points spread uniformly over the survey region? Does the density of points depend on an explanatory variable? Are the points randomly scattered? Is there evidence of clustering? Is the spatial pattern consistent with my scientific hypothesis? How accurate/reliable is my statistical analysis?

To give a sensible answer we must recognise that these questions are **not about the points themselves, but about the way the points were generated**.

If the researcher concludes that "the points are completely random" or "the points are uniformly spread", this cannot be a literal description of the point pattern dataset. Data are numbers that were observed and recorded: they are fixed, not random. A point in a point pattern is a discrete object, and cannot be 'spread uniformly' like butter. Instead, the conclusion is that the data points were *generated by* some mechanism which was completely random, or which spread the points uniformly, etc.

This is not to say that the individual data points are unimportant. In a spatial analysis of crimes or road accidents, each individual point has tragic significance. But in analysing the data we aim to learn about the spatial *pattern* of crime or the spatially varying *risk* of road accidents, to gain insight into the underlying causes. The risk of a certain type of accident is the *probability* that an accident will happen in the future. We study the past history of accidents in order to *infer* or *estimate* the probability of an accident in the future.

In the jargon of statistical science, a **random variable** is a numerical quantity that is observed, measured, or otherwise depends on the outcome of an experiment. If the experiment could be repeated under identical conditions, the observed value of the random variable would be different each time. The value is inherently variable, due to measurement error, sampling variability, or natural variability. For example, the number of road accidents that occur in one year on a particular stretch of road is a random variable.

The corresponding concept for spatial point patterns is a **(random) point process**. A point process is a random mechanism whose outcome is a point pattern. If the experiment could be repeated under identical conditions, the observed point pattern would be different each time. Figure 5.1 shows ten different possible outcomes or *realisations* of the same random point process. When we say 'point process' the reader can visualise something like Figure 5.1, but with infinitely many panels, each panel being a possible outcome of the point process. The different outcomes may have different probabilities assigned to them. This is what physicists would call an *ensemble* and statisticians call a *probability distribution* for the point process.

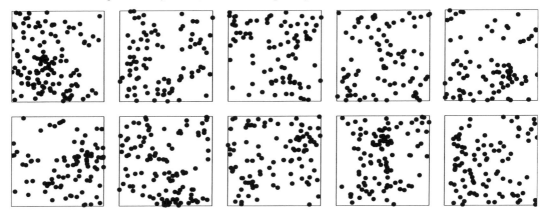

Figure 5.1. *Ten different possible outcomes ('realisations') of the same point process, generated by computer simulation.*

Any natural or synthetic phenomenon which results in a spatial point pattern can be viewed as a point process. This viewpoint does not involve any hidden assumptions: it is merely a way of thinking, a language for formulating scientific questions.

The concept of a point process is exactly what we need to formulate scientific questions about spatial pattern. In a forest ecology study, we observe a few dozen plants in a few selected plots, but we are interested in generalising to the entire forest, and ultimately we want to draw conclusions about the ecological processes which gave rise to the spatial arrangement of the plants. The focus is on the process which generated the observed arrangement of plants (and all the other unobserved plants in the forest), rather than the few observed plants themselves. Point process language helps us to formulate meaningful questions about this process, and puts them into a statistical framework.

5.2 Basic definitions*

Here we define some mathematical notation and give a few basic definitions. For full details see [198, 199, 484].

5.2.1 Basic notation

For most of this book we are working in the two-dimensional plane of euclidean geometry, denoted by $\mathbb{R}^2$. A point location in the plane is denoted by a lower case letter like u. Any location u can be specified by its Cartesian coordinates $u = (u_1, u_2)$; we shall usually not need to mention the coordinates explicitly. The euclidean distance between two points $u = (u_1, u_2)$ and $v = (v_1, v_2)$ is

$$\|u - v\| = \left[(u_1 - v_1)^2 + (u_2 - v_2)^2\right]^{1/2}.$$

A 'region' is a subset of the plane, denoted by a capital letter like A. The rectangle $A = [a, b] \times [c, d]$ is the set of all points $u = (u_1, u_2)$ with $a \leq u_1 \leq b$ and $c \leq u_2 \leq d$. For example $[0, 1] \times [0, 1] = [0, 1]^2$ is the unit square with bottom-left corner at the origin. The *disc* with centre point u and radius $r > 0$ is

$$b(u, r) = \{v : \|u - v\| \leq r\},$$

the set of points lying at most r units away from u. The *circle* with centre u and radius r is the boundary of the disc,

$$\partial b(u,r) = \{v : \|u - v\| = r\},$$

the set of points lying exactly r units away from u.

A point pattern is denoted by a bold lower case letter like $\mathbf{x}$. It is a set

$$\mathbf{x} = \{x_1, x_2, \ldots, x_n\}$$

of points x_i in two-dimensional space $\mathbb{R}^2$. The number $n = n(\mathbf{x})$ of points in the pattern is not fixed in advance, and may be any finite nonnegative number *including zero*. In practice, the data points are obviously recorded in some order $x_1, \ldots, x_n$; but this ordering is artificial, and we treat the pattern $\mathbf{x}$ as an unordered set of points. Duplicated points are allowed, that is, it is possible that $x_i = x_j$ for two different indices i and j. However, many methods in this book require that there should be no duplicated points.

If $\mathbf{x}$ is a point pattern and B is a region, we write $\mathbf{x} \cap B$ for the subset of $\mathbf{x}$ consisting of points that fall in B. The number of points of $\mathbf{x}$ falling in B is $n(\mathbf{x} \cap B)$. See Figure 5.2.

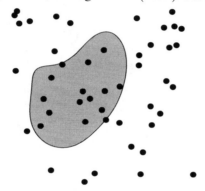

Figure 5.2. *A point pattern (dots) and a test set B (shaded).*

5.2.2 Point processes

A **point process** is a random mechanism whose outcomes are point patterns. Point processes are denoted by capital letters $\mathbf{X}, \mathbf{Y}$, and so on.

The rigorous mathematical definition of a point process is quite technical: see [197, 198, 199] for the general theory, and [484] for the theory that is particularly relevant to spatial statistics.

We can avoid many of these technicalities because, for statistical applications, we can usually assume that the number of points in the process is finite. A **finite point process** is a random mechanism for which (1) every possible outcome is a point pattern with a finite number of points; and (2) for any[1] region B, the number $n(\mathbf{X} \cap B)$ of points falling in B is a well-defined random variable.

These conditions are enough to support a full statistical theory for analysing spatial point patterns. They guarantee that all the statistics we might wish to calculate for a point process will be well-defined random variables.

It is sometimes necessary to consider a random pattern of infinitely many points scattered over the infinite two-dimensional plane. For this we need a slight extension of the previous definitions. A **locally finite point pattern** is a set $\mathbf{x} = \{x_1, x_2, \ldots\}$ of points in two-dimensional space $\mathbb{R}^2$ which has only a finite number of points in any bounded region B, that is, $n(\mathbf{x} \cap B)$ is finite. The total number of points in $\mathbf{x}$ may be infinite. A **locally finite point process** $\mathbf{X}$ in two-dimensional space

---

[1] the region B should be bounded, and topologically closed.

is defined to be any random mechanism for which (1) every possible outcome is a locally finite point pattern, and (2) for any bounded test region B, the number $n(\mathbf{X} \cap B)$ of points falling in B is a well-defined random variable.

5.2.3 Uniformly random points

A very simple point process is one that consists of a single random point. Suppose we need to pick one point at random: since a spatial location u is determined by its Cartesian coordinates (u_1, u_2), we only need to assign random values U_1, U_2 to the coordinates. This gives us a random point $U = (U_1, U_2)$.

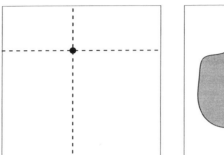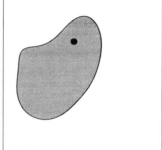

Figure 5.3. *Uniform random point.* Left: *a pair of random coordinates which are jointly uniformly distributed.* Right: *probability of falling in a test region B (shaded) is proportional to its area.*

To know the statistical properties of this random point, it is enough to know the joint probability distribution of the random coordinates U_1 and U_2. This could be given by a joint probability density $f(u_1, u_2)$. Then the probability that the random point $U = (U_1, U_2)$ falls in a test region B is

$$\mathbb{P}\{U \in B\} = \int_B f(u_1, u_2) \, \mathrm{d}u_1 \, \mathrm{d}u_2. \tag{5.1}$$

An important question in spatial statistics is whether points are 'uniformly spread' over the survey region. Say that a random point U is *uniformly distributed* in a spatial region W, if its Cartesian coordinates (U_1, U_2) have a joint probability density which is constant inside W and zero outside W. Since a probability density must integrate to 1, the constant value can be determined: it must be $1/|W|$, and the density is

$$f(u_1, u_2) = \begin{cases} 1/|W| & \text{if } (u_1, u_2) \text{ falls in } W \\ 0 & \text{if not} \end{cases}$$

where $|W|$ is the area of W. This is only meaningful if $|W|$ is non-zero and finite.

For example if W is a rectangle, say $W = [0, w] \times [0, h]$, then (u_1, u_2) falls in W if and only if $0 \le u_1 \le w$ and $0 \le u_2 \le h$. The joint probability density factorises into $f(u_1, u_2) = f_1(u_1) f_2(u_2)$ where

$$f_1(u_1) = \begin{cases} 1/w & \text{if } 0 \le u_1 \le w \\ 0 & \text{if not} \end{cases}$$

is the probability density of a random variable that is uniformly distributed on $[0, w]$, and similarly f_2 is the density of a random variable uniformly distributed on $[0, h]$. That is, to generate a uniformly distributed random point in a rectangle, we simply have to choose random coordinates U_1 and U_2 which are independent and uniformly distributed. A computer random number generator can provide such numbers. See Figure 5.3.

An important property of uniformly distributed random points is that, if B is a test region contained in W, the probability that U falls in B is

$$\mathbb{P}\{U \in B\} = \int_B f(u_1, u_2)\, du_1\, du_2 = \frac{1}{|W|} \int_B 1\, du_1\, du_2 = \frac{|B|}{|W|}, \tag{5.2}$$

the fraction of area occupied by B within W. This probability depends only on the *area* of the test set B, and not on its location. Indeed this is what would be expected for a random point that has no preference for particular locations.

5.2.4 Binomial point process

The next simplest example of a point process is one in which the number of points n is fixed, and only the locations of the points are random. This point process is a random set $\mathbf{X}$ containing exactly n random points $X_1, \ldots, X_n$.

To make the points 'uniformly spread' over a region W, let us assume that $X_1, \ldots, X_n$ are *independent* random locations, and that each X_i is *uniformly distributed* over the region W. This is enough information to generate simulated realisations of the process: see Figure 5.4.

If B is a test region, the number $n(\mathbf{X} \cap B)$ of random points falling in B is the number of indices i such that the random point X_i falls in B. It is clear that this is a well-defined random variable, as required, so that $\mathbf{X}$ satisfies the requirements of a finite point process.

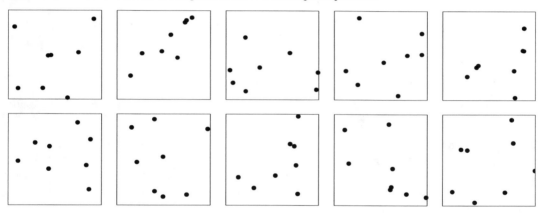

Figure 5.4. *Ten different realisations of the binomial point process with $n = 8$ points in a square.*

To determine the probability distribution of $n(\mathbf{X} \cap B)$, where B is a subset of W, notice that $n(\mathbf{X} \cap B)$ is the number of successes in n independent trials. That is, if we treat each random point X_i as a 'success' when it falls inside B and a 'failure' otherwise, these trials are independent, and each trial has success probability $p = |B|/|W|$. Consequently $n(\mathbf{X} \cap B)$ has a *binomial distribution*

$$\mathbb{P}\{n(\mathbf{X} \cap B) = k\} = \binom{n}{k} p^k (1-p)^{n-k}$$

for $k = 0, 1, \ldots, n$. For this reason the model is known as the *binomial point process*.

In spatstat the function runifpoint generates a random realisation of the binomial point process. (By convention, random number generators in R have names beginning with r.) The argument n gives the number of points, and win is the window in which to generate the points. Figure 5.4 was generated by runifpoint(8, square(1), nsim=10).

5.3   Complete spatial randomness

5.3.1   Introduction

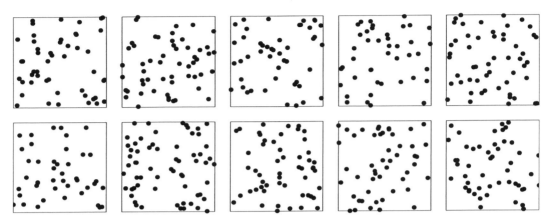

Figure 5.5. *Completely random point patterns: 10 simulated realisations of the Poisson point process with intensity 50 in the unit square.*

A point process which deserves to be called 'completely random' is illustrated in Figure 5.5. Each panel is a realisation of the *homogeneous Poisson point process*, also called *complete spatial randomness (CSR)*. The process is characterised by two key properties:

homogeneity: the points have no preference for any spatial location;

independence: information about the outcome in one region of space has no influence on the outcome in other regions.

CSR is important in many ways. It is a realistic model of some physical phenomena, such as radioactivity, rare events, and extreme events. It serves as a benchmark or standard reference model of completely random patterns, against which other patterns can be compared. In many statistical tests, CSR serves as the null hypothesis. Many other models are built starting from CSR, and many mathematical concepts are defined relative to CSR.

Unlike the binomial point process illustrated in Figure 5.4 which has a fixed total number of points, the completely random process shown in Figure 5.5 has a random number of points. This is emphasised by Figure 5.6 which shows realisations of CSR with the same average number of points as the patterns in Figure 5.4. Note the variation in the number of points in each panel of Figure 5.6.

Statisticians call this model the 'Poisson process' because the number of points falling in any region follows a Poisson distribution (as explained below). This can be confusing for some readers. Why should we assume that the number of points should follow a particular distribution? Actually we do not assume it. It turns out that the two properties of homogeneity and independence, described above, effectively *imply* that the distribution must be Poisson. We begin by explaining how this comes about.

5.3.2   Derivation from basic principles

Homogeneity (in this sense) means that the expected number of points falling in a region B should be proportional to its area $|B|$ on average, that is,

$$\mathbb{E}n(\mathbf{X} \cap B) = \lambda \, |B| \tag{5.3}$$

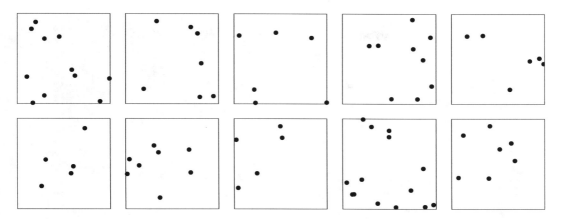

Figure 5.6. *Completely random point patterns: 10 simulated realisations of the Poisson point process with intensity 8 in the unit square. Note different numbers of points in each panel.*

where λ is constant. In effect λ is the average number of random points per unit area, and is known as the *intensity* of the point process.

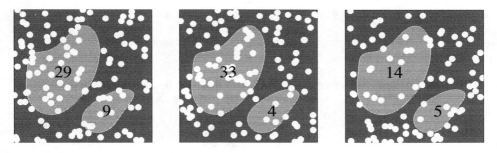

Figure 5.7. *Concept of independence between outcomes in disjoint regions (light grey shading).*

The concept of spatial independence is sketched in Figure 5.7. The three panels show three different possible realisations (outcomes) of the random point process. In each panel, the same two regions of space — say, A and B — are highlighted in light grey, and the numbers of random points falling in A and in B are displayed. A consequence of the independence assumption is that these counts $n(\mathbf{X} \cap A)$ and $n(\mathbf{X} \cap B)$ must be independent random variables. The value of $n(\mathbf{X} \cap A)$ carries no information affecting the probabilities of different possible values of $n(\mathbf{X} \cap B)$.

The binomial point process described in Section 5.2.4 does *not* have the property of spatial independence. Since there are known to be exactly n points altogether, the information that there are (say) 5 points in a region A implies that there are exactly $n - 5$ points in the complement of A, violating independence.

The independence assumption applies to any choice of disjoint regions A and B, and to any number of such regions. In particular, independence implies that quadrat counts are independent, for any size of quadrat. The left panel of Figure 5.8 shows a division of the rectangle into a 5×5 array of quadrats. The independence assumption implies that the numbers of points falling in each of the 25 quadrats must be independent random variables. The right panel of Figure 5.8 shows a finer division, and this too must have the same independence property.

Taking finer and finer divisions of space into squares, the same independence property must continue to hold in each case. When the squares are extremely small, most of the squares will not

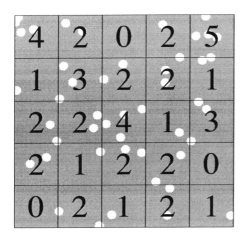

Figure 5.8. *The assumption of independence implies that quadrat counts are independent, for any size of quadrat.*

contain a random point, and a few squares will contain exactly one random point. Some squares could conceivably contain more than one point, so we need to impose a third assumption:

orderliness: there is negligible probability[2] that a region contains more than one point, when the region is small.

This implies that no two points can coincide, and also excludes fractal-like behaviour.

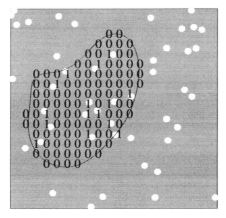

Figure 5.9. *Poisson limit. The number of points falling in a given region (indicated by the curved contour) is the result of a large number of trials (indicated by numerals), each having a small probability of success. If the trials are independent then the total number of points has a Poisson distribution.*

Figure 5.9 sketches a region of space A which has been subdivided into tiny squares. The number $n(\mathbf{X} \cap A)$ of random points falling in A is equal to the sum of the numbers falling in the squares inside A. Assuming, as above, that the outcomes in different squares are independent, and that there is negligible chance that some squares have more than one point, $n(\mathbf{X} \cap A)$ is the number

---

[2]To be precise, this probability, divided by the area of the region, must go to zero as the area goes to zero.

of successes in a large number of independent trials, each trial having a small probability of success. By a famous theorem in probability theory, this means that $n(\mathbf{X} \cap A)$ has a *Poisson distribution* [383].

The Poisson probability distribution is the classical law of the frequency of *rare* events. It is often used to model accidents, cases of rare diseases, and other rare events. Under the Poisson distribution, the probability of obtaining exactly k rare events is

$$\mathbb{P}\{N = k\} = e^{-\mu} \frac{\mu^k}{k!} \tag{5.4}$$

for any $k = 0, 1, 2, \ldots$. As usual in statistics, Greek letters represent parameters: here the Greek letter μ ('mu', cognate to 'm') is the mean of the Poisson distribution. For example, Figure 5.10 shows bar charts of the probabilities of the Poisson distribution with means $\mu = 0.6$ and $\mu = 1.7$, respectively.

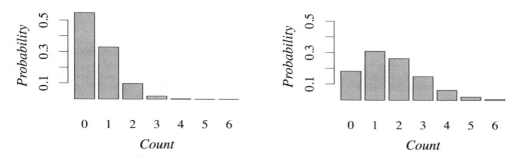

Figure 5.10. *Probabilities for the Poisson distributions with mean $\mu = 0.6$ (Left) and $\mu = 1.7$ (Right).*

Since the expected number of points falling in the test region A is $\mathbb{E} n(\mathbf{X} \cap A) = \lambda |A|$, we have just found that *the random number of points $n(\mathbf{X} \cap A)$ has a Poisson distribution with mean $\mu = \lambda |A|$.*

This finding is truly remarkable. Starting with two general assumptions (homogeneity and independence) we have come to a very specific conclusion: the number of random points falling in a test region follows a Poisson distribution. This may help to explain the importance of the Poisson distribution in point processes.

In summary, the *homogeneous Poisson point process* (or *'complete spatial randomness', CSR*) with intensity $\lambda > 0$ is a locally finite point process with the properties of

homogeneity: the number $n(\mathbf{X} \cap B)$ of random points falling in a test region B has mean value $\mathbb{E} n(\mathbf{X} \cap B) = \lambda |B|$;

independence: for test regions $B_1, B_2, \ldots, B_m$ which do not overlap, the counts $n(\mathbf{X} \cap B_1), \ldots, n(\mathbf{X} \cap B_m)$ are independent random variables;

Poisson distribution: the number $n(\mathbf{X} \cap B)$ of random points falling in a test region B has a Poisson distribution (5.4).

See Section 9.2.1 for more detail.

5.3.3 Useful properties of CSR

CSR has several important and useful properties which we mention briefly here.

For a test region B, suppose we are given the information that $n(\mathbf{X} \cap B) = n$, that is, exactly n points of the Poisson process fell in B. The **conditional property** of CSR is that these n points are independent and uniformly distributed in B. For example, each panel in Figure 5.6 containing exactly 8 points is statistically equivalent to one of the panels in Figure 5.4.

Thinning means deleting some of the points from a point pattern. Under '*completely random thinning*' each point of the point pattern is randomly deleted or retained, with probability p of retention, independently of the fate of other points. The **thinning property** of CSR is that if we start with a homogeneous Poisson process with intensity λ, and apply completely random thinning with retention probability p, the points retained after thinning constitute a homogeneous Poisson process with intensity $p\lambda$. See Figure 5.11.

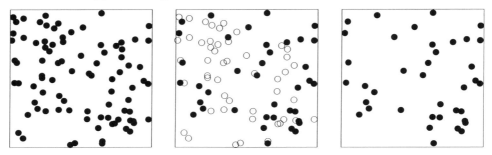

Figure 5.11. *Thinning property of Poisson process.* Left: *Poisson process* **X**. Middle: *points of* **X** *are randomly deleted or retained.* Right: *points that are retained constitute a Poisson process.*

Superimposing two point processes **X** and **Y** means that we combine the points from both processes into a new point process $\mathbf{Z} = \mathbf{X} \cup \mathbf{Y}$. The **superposition property** of the Poisson process is that if **X** and **Y** are homogeneous Poisson processes with intensities $\lambda_{\mathbf{X}}$ and $\lambda_{\mathbf{Y}}$, and if they are independent of each other, then their superposition **Z** is also a homogeneous Poisson process, with intensity $\lambda_{\mathbf{Z}} = \lambda_{\mathbf{X}} + \lambda_{\mathbf{Y}}$. See Figure 5.12.

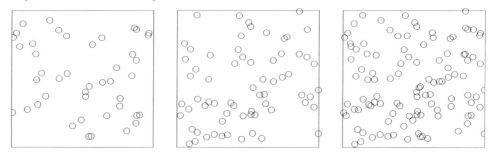

Figure 5.12. *Superposition property of Poisson process. When two independent Poisson point processes* **X** *and* **Y** *(*Left *and* Middle*) are superimposed, the result* **Z** *is also a Poisson process (*Right*).*

5.3.4   Simulation of CSR

It is easy to simulate the Poisson process directly, using the properties above. Given a region B where the realisation is to be generated, and an intensity value λ, we first determine the total number of points by generating a random number N according to a Poisson distribution (5.4) with mean $\mu = \lambda |B|$. These N points are then placed independently in B with a uniform distribution. In spatstat, use the command rpoispp. Each panel of Figure 5.5 was generated by the command

```
> plot(rpoispp(50))
```

To develop some intuition about completely random patterns, it is useful to repeat the command `plot(rpoispp(100))` several times (use the up-arrow key or `Ctrl-P` to recall the previous command line) so that you see several replicates of the Poisson process. Alternatively

```
> plot(rpoispp(100, nsim=9))
```

would produce a 3×3 array of different realisations of the same constant intensity Poisson process. In particular you will notice that the points in a homogeneous Poisson process are not 'uniformly spread': there are empty gaps and clusters of points. A typical realisation of the process does not show a uniform spread of points: after all, the points are independent of one another. A typical realisation will include some large gaps and some clusters of points.

There is even a small chance of having *no* random points in a region B: the probability that a Poisson variable with mean $\mu = \lambda |B|$ will yield the value 0 is $e^{-\lambda |B|}$ according to (5.4).

The command `rpoispp` has arguments `lambda` (the intensity) and `win` (the window in which to simulate). The default window is the unit square. Figure 5.13 shows a realisation of CSR inside the letter 'R', generated by the command

```
> plot(rpoispp(100, win=letterR))
```

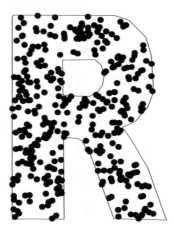

Figure 5.13. *Homogeneous Poisson process realisation inside the letter R.*

To simulate a Poisson process *conditionally* on a fixed number of points, use the command `runifpoint`, which generates a realisation of the binomial point process.

5.4 Inhomogeneous Poisson process

5.4.1 Definition

The most important model for many practical purposes is the *inhomogeneous* Poisson point process, a modification of complete spatial randomness in which the average density of points is spatially varying.

Suppose the average density of points is a function $\lambda(u)$ of spatial location u. Given a region of space B, divide this region into tiny pixels as described above. For a pixel at spatial location u, the expected number of points falling in the pixel is assumed to be $\lambda(u) \Delta u$ where Δu is the pixel area. The expected total number of points in B is the sum of these values $\lambda(u) \Delta u$, over all pixels with centres u inside B. In the limit as the pixel size becomes infinitesimal, this sum becomes the integral $\int_B \lambda(u) \, du$.

The *inhomogeneous Poisson point process* with intensity function $\lambda(u)$ is defined by the following properties:

intensity function: the expected number of points falling in a region B is the integral $\mu = \int_B \lambda(u) \, du$ of the intensity function $\lambda(u)$ over the region B;

independence: if space is divided into non-overlapping regions, the random patterns inside these regions are independent of each other;

Poisson-distributed counts: the random number of points falling in a given region has a *Poisson* probability distribution;

See Section 9.2.1 for more detail.

The all-important intensity function $\lambda(u)$ determines the overall abundance of points and their spatial distribution. This model is very general, since there are essentially no restrictions on the function $\lambda(u)$. Thus, a researcher who is studying 'the spatial distribution of the points' in some dataset will often find that an inhomogeneous Poisson process model is perfectly adequate. The main assumption is that the points are *independent* of each other.

The inhomogeneous Poisson process also enjoys the important properties listed in Section 5.3.3, with some modifications. The *conditional* property states that, given there are exactly n points falling in a region B, these points are independent, and each point has the same probability distribution over B, with probability density $f(u) = \lambda(u)/\mu$. The *random thinning* property states that, if an inhomogeneous Poisson process with intensity function $\lambda(u)$ is randomly thinned so that the probability of retaining a point at location u is $p(u)$, and the fate of each point is independent of the fates of other points, then the resulting process of retained points is also Poisson, with intensity function $p(u)\lambda(u)$. The *superposition* property states that if $\mathbf{X}$ and $\mathbf{Y}$ are independent point processes which are inhomogeneous Poisson processes with intensity functions $\lambda_{\mathbf{X}}(u)$ and $\lambda_{\mathbf{Y}}(u)$, respectively, then the superposition $\mathbf{X} \cup \mathbf{Y}$ is a Poisson process with intensity $\lambda_{\mathbf{X}}(u) + \lambda_{\mathbf{Y}}(u)$.

These properties mean that the inhomogeneous Poisson process is a plausible model for point patterns under several scenarios. This is discussed further in Section 9.2.2.

5.4.2 Simulation of inhomogeneous Poisson process

An inhomogeneous Poisson process with intensity function $\lambda(u)$ can be simulated by invoking the properties listed above.

The simplest strategy is the Lewis-Shedler thinning method [420, 421, 422] in which we first generate a homogeneous Poisson process, then apply random thinning with a retention probability $p(u)$ proportional to the desired intensity $\lambda(u)$.

Find an upper bound M such that $\lambda(u) \le M$ for all locations u in L. Generate a homogeneous Poisson process with intensity M as described above. For each random point x_i, evaluate $p_i = \lambda(x_i)/M$, and randomly delete or retain x_i with retention probability p_i, independently of the fate of other points. The result is a realisation of a Poisson process with intensity function $\lambda(u)$.

The Lewis-Shedler method is inefficient when the values of the intensity function $\lambda(u)$ vary substantially. In that case a large value of M will be required, so that a large number of points x_i will have to be generated; many of them will be discarded because the probabilities $p_i = \lambda(x_i)/M$ will often be small. An alternative strategy is to divide space into pixels, compute the probability that each pixel contains a point, and select pixels at random using these probabilities. This requires more initial computation than the Lewis-Shedler method, but requires less additional computation for each random point.

5.4.3 Simulation in `spatstat`

The Poisson process can be simulated using `rpoispp`. The intensity argument `lambda` can be a constant, a `function(x,y)` giving the values of the intensity function at coordinates x, y, or a pixel image containing the intensity values at a grid of locations.

```
> lambda <- function(x,y) { 100 * (x^2+y) }
> X <- rpoispp(lambda, win=square(1))
```

The result is shown in Figure 5.14.

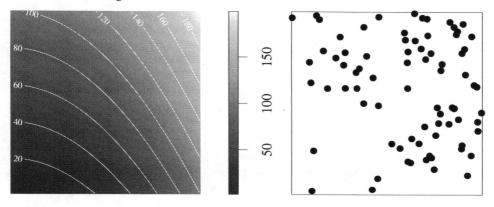

Figure 5.14. *Inhomogeneous Poisson process.* Left: *intensity function;* Right: *realisation of point process.*

5.5 A menagerie of models

A point process that is not Poisson can be said to exhibit 'interaction' or 'dependence' between the points. We now introduce some models for such processes. In this section we present some simple constructions of non-Poisson models, while a more thorough discussion of non-Poisson models is given later: Chapter 12 presents cluster processes and Cox processes; Chapter 13 presents Gibbs point processes.

5.5.1 Matérn cluster process

Consider a construction where first a process of parent points come from a homogeneous Poisson process with intensity κ, and second each parent has a Poisson (μ) number of offspring, independently and uniformly distributed in a disc of radius r centred around the parent. The process consisting of all the offspring (discarding the parents) is called the *Matérn cluster process*.

The Matérn cluster process can be generated in `spatstat` using the command `rMatClust`. [By convention, random data generators in R always have names beginning with `r`.]

```
> Xmc <- rMatClust(kappa=8, r=0.15, mu=5)
```

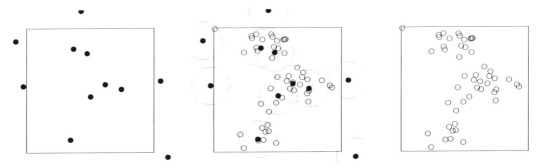

Figure 5.15. *Formation of Matérn cluster process.* Left: *parent points are generated according to a homogeneous Poisson process.* Middle: *for each parent, a Poisson random number of offspring (with mean μ offspring) are generated, and placed independently and uniformly in a circle of radius R around the parent.* Right: *the offspring constitute the Matérn cluster process.*

5.5.2 Dependent thinning

Under *'independent thinning'* the fate of each point is independent of other points. When independent thinning is applied to a Poisson process, the resulting process of retained points is Poisson. To get a non-Poisson process we need some kind of *dependent thinning* mechanism.

In *Matérn's Model I*, a homogeneous Poisson process **Y** is first generated. Then each point in **Y** that lies closer than a distance r from the nearest other point of **Y** is deleted. Thus, pairs of close neighbours annihilate each other. See Figure 5.16. The function rMaternI generates simulated realisations of this model:

```
> XmI <- rMaternI(30, 0.1)
```

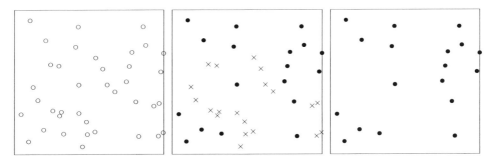

Figure 5.16. *Formation of Matérn's Model I.* Left: *points are generated by Poisson process.* Middle: *any point lying within distance R of another point is deleted* (×). Right: *resulting thinned process. Unit square window, interaction distance R = 0.1.*

In *Matérn's Model II*, illustrated in Figure 5.17, the points of the homogeneous Poisson process **Y** are marked by 'arrival times' t_i which are independent and uniformly distributed in $[0,1]$. Any point in **Y** that lies closer than distance r from another point with an earlier arrival time is deleted. Realisations are generated by rMaternII:

```
> XmII <- rMaternII(30, 0.1)
```

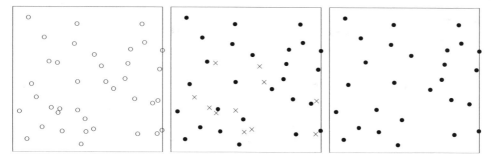

Figure 5.17. *Formation of Matérn's Model II.* Left: *points are generated by Poisson process and marked by independent arrival times.* Middle: *any point lying within distance R of an* earlier *point is deleted* (×). Right: *resulting thinned process. Unit square window, interaction distance R = 0.1.*

5.5.3 Sequential models

In a sequential model, we start with an empty window, and the points are placed into the window one at a time, according to some criterion.

In Simple Sequential Inhibition, each new point is generated uniformly in the window and independently of preceding points. If the new point lies closer than r units from an existing point, then it is rejected and another random point is generated. The process terminates when the desired number of points is achieved or no further points can be added (determined heuristically by giving up after a large number of failed attempts). Figure 5.18 shows a simulated realisation, generated by

```
> Xs <- rSSI(0.05, n = 200)
```

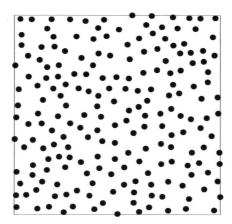

Figure 5.18. *Realisation of Simple Sequential Inhibition: 200 points in the unit square with inhibition radius 0.05 units.*

Sequential point processes are the easiest way to generate highly ordered patterns with high intensity. They are plausible models for real data in which the points have been added one at a time, such as the breakdown spots data (Section 16.6.2).

5.6   Fundamental issues

Before embarking on a statistical analysis of spatial point pattern data, it is important to ask two questions. Firstly, are point process methods appropriate to the scientific context? Secondly, are the standard assumptions for point process methods appropriate in the context?

5.6.1   Should I treat the data as a point process?

Treating a spatial point pattern as a realisation of a spatial point process effectively assumes that the pattern is random (in the general sense, i.e. the locations and number of points are not fixed) and that the point pattern is the *response* or observation of interest. The statistician should consider whether this is appropriate in the scientific context. Consider the following illustrative scenarios.

Scenario 5.1. *A silicon wafer is inspected for defects in the crystal surface, and the locations of all defects are recorded.*

This point pattern could be analysed as a point process in two dimensions, assuming the defects are point-like at the scale of interest. Questions for study would include frequency of defects, spatial trends in intensity, and spacing between defects.

Scenario 5.2. *Earthquake aftershocks in Japan are detected and the epicentre latitude and longitude and the time of occurrence are recorded.*

These data could be analysed as a point process in space-time, or as a marked point process in two-dimensional space, where the space is the two-dimensional plane or the Earth's surface. If the occurrence times are ignored, it may be analysed as a spatial point process. Spatiotemporal point processes are treated in [192, 226] and the CRAN packages `splancs` and `stpp`.

Scenario 5.3. *The locations of petty crimes that occurred in the past week are plotted on a street map of Chicago.*

This could be analysed as a two-dimensional spatial point process. Questions for study include the frequency of crimes, spatial variation in intensity, and evidence for clusters of crimes. One issue is whether the recorded crime locations can lie anywhere in two-dimensional space, or whether they are actually restricted to locations on the streets. In the latter case it would be more appropriate to treat the data as a point process on a network of one-dimensional lines. This is discussed in Chapter 17 and is supported in `spatstat`.

Scenario 5.4. *A tiger shark is captured, tagged with a satellite transmitter, and released. Over the next month its location is reported daily. These points are plotted on a map.*

It is probably *not* appropriate to analyse these data as a spatial point process. A realisation of a spatial point process is an unordered set of points, so the serial order in which the data were recorded is ignored by spatial point process methods.

At the very least, the date of each observation of the shark should be included in the analysis. The data could be treated as a space-time point process, except that this would be a strange process, consisting of exactly one point at each instant of time.

These data should properly be treated as a sparse sample of a continuous trajectory, and analysed using other methods: see [127, 593]. The CRAN packages `adehabitatLT`, `trip`, `crawl`, and `move` can be used to analyse such data.

Scenario 5.5. *A herd of deer is photographed from the air at noon each day for 10 days. Each photograph is processed to produce a point pattern of individual deer locations on a map.*

Each day produces a point pattern that could be analysed as a realisation of a point process. However, the observations on successive days are dependent (e.g., having constant herd size and systematic foraging behaviour). Assuming individual deer cannot be identified from day to day, this is effectively a 'repeated measures' dataset where each response is a point pattern. Methods for this problem are in their infancy. A pragmatic alternative may be to treat the data as a space-time point process.

Scenario 5.6. *In a designed controlled experiment, silicon wafers are produced under various conditions. Each wafer is inspected for defects in the crystal surface, and the locations of all defects are recorded as a point pattern.*

This is a designed experiment in which the response from each experimental unit is a point pattern. Methods for this problem are described in Chapter 16 and are supported in `spatstat`.

Scenario 5.7. *The points are not the original data, but were obtained after processing the data. For example,*

- *the original dataset is a pattern of small blobs, and the points are the blob centres;*

- *the original dataset is a collection of line segments, and the points are the endpoints, crossing points, midpoints, etc;*

- *the original dataset is a space-filling tessellation of biological cells, and the points are the centres of the cells.*

This is a grey area. Point process methodology can be applied, and may be more powerful or more flexible than existing methodology for the unprocessed data. However the origin of the point pattern may lead to artefacts (for example the centres of biological cells never lie very close together, because cells have nonzero size) which must be taken into account in the analysis.

This discussion is continued in Section 14.2 for marked point patterns. See [519, chap. 9].

5.6.2   The sampling context

Data cannot be analysed properly without knowledge of the scientific context and the sampling design.

The vast majority of statistical techniques for analysing spatial point patterns assume what we may call *'Window Sampling'*. The point process **X** of interest to us extends throughout 2-D space, but is observable only inside a bounded region W, the *'sampling window'*. See Figure 5.19. The data consist of a point pattern **x** in W, and knowledge of the window W. A typical example would be a dataset giving the locations of trees in a survey plot, within a much larger forest. It is the entire forest that is of interest, rather than just the trees that were surveyed.

WINDOW SAMPLING

- The point process **X** that is the focus of interest is a locally finite point process on the infinite two-dimensional plane;

- Points are only observed inside a region W, the 'sampling window' or 'study region', that is fixed and known.

The statistician's task is to draw inferences from the observed, *sampled* point pattern, about the entire point process. This will require some assumptions on the point process to ensure that the sample is 'representative' or more generally to ensure that inferences can be drawn from the sample. Historically it was often assumed that the point process is *stationary* (spatially homogeneous with respect to all statistical characteristics) as explained below.

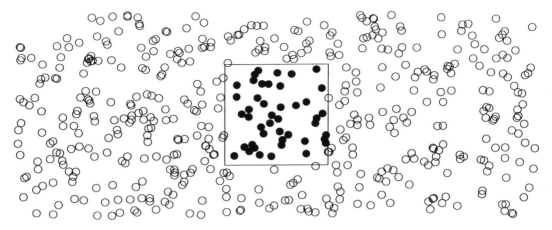

Figure 5.19. *Window Sampling. The point process lives in the infinite plane, but we only observe those points which fall in the observation window W. Lines show boundary of W. Filled dots are observed points. Open circles are unobserved points.*

For many statistical analyses it is important to *know* the sampling window W. This is a fundamental methodological issue that is unique to spatial point patterns. The data do not consist solely of the locations of the observed points. As well as knowing where points were recorded, we also need to know the limits of the sampling region where they could have been recorded. Even something as simple as estimating the density of points depends on the window. It would be wrong, or at least different, to analyze a point pattern dataset by 'guessing' the appropriate window.

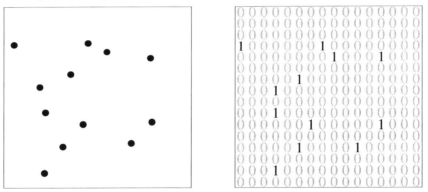

Figure 5.20. *Interpretation of a point process as a large number of trials associated with small pixels.*

When thinking about such methodological issues, it is often useful to think of the discretised counterpart of a point process on a grid of pixels. See Figure 5.20. The window W is divided into a large number of tiny pixels. Each pixel is assigned the value $I = 1$ if it contains a point of $\mathbf{X}$, and $I = 0$ otherwise. This array of 0's and 1's constitutes the data that must be modelled. Thus we clearly need to know where points did *not* occur, as well as where they *did* occur.

Difficulties arise when the sampling region W is not well defined, but we wish nevertheless to use a statistical technique that assumes Window Sampling and requires knowledge of W. It is a common error to take W to be the smallest rectangle containing the data points, or the convex hull of the data points. These are 'underestimates' of the true region W and typically yield overestimates of the density of points, the degree of clustering and so on. Some more defensible methods exist for estimating W if it is unknown [580, 486, 438, 72]. The ripras function in spatstat implements the methods described in [580].

One of the important consequences of Window Sampling is that information outside the window is lost. This leads to statistical difficulties, collectively called *edge effects*, which are discussed in Chapters 7 and 8.

Assuming that W is fixed, and known, implies that the sampling region does not depend on the data. This excludes some types of experiments in which we continue recording spatial points until a stopping criterion is satisfied. For example, a natural stand of trees surrounded by open grassland, or a cluster of fossil discoveries, would typically be mapped in its entirety. This does not fit the Window Sampling model; it may require different techniques, analogous to sequential analysis [187].

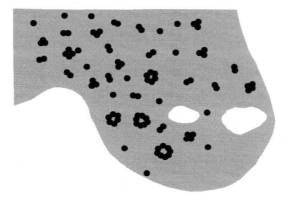

Figure 5.21. *Gordon Square data. Locations of 99 people (dots) sitting on grass (grey shading) in Gordon Square, London, UK on a sunny afternoon. Original data from [54], rotated.*

An alternative scenario is that the entire spatial point process 'lives' in a known spatial region. For example, Figure 5.21 shows the spatial locations of people sitting on the grass in a park on a sunny afternoon. There is no grass outside the park, so this pattern is confined within the boundary of the park.

SMALL WORLD MODEL

The point process **X** that is the focus of interest is a finite point process in a known, fixed, bounded region W.

In the Small World model, the point process is finite, and cannot be spatially homogeneous, which precludes or complicates the use of many classical techniques: for example the usual interpretation of the K-function (Chapter 7) assumes homogeneity.

In some cases the Small World model is appropriate but the region W is unknown. For example, a stand of trees may be confined to the spatial region where a particular type of soil is found. It would be valid to estimate the region W, using techniques such as the Ripley-Rasson estimator [580]. See [355, pp. 268–270].

In some situations, the points are not confined to lie inside a precise boundary, but are nevertheless spatially constrained by environmental or observational factors. If covariates are available, we may be able to use them to construct a reference intensity $\lambda_0(u)$ that represents the 'sampling design' or the 'null hypothesis'. For example, in spatial epidemiology, the covariates may include (a surrogate for) the spatially varying population density, which serves as the natural reference intensity for models of disease risk.

STANDARD ASSUMPTIONS FOR ANALYSIS

1. point locations are measured exactly;

2. no two points lie at exactly the same location;

3. points are mapped without omission, i.e. there are no errors in detecting the presence of points of the random process within the region W.

4. points *could* have been observed at any location in the region W.

Most current methods require Assumption 2. Duplicate points do commonly occur, through data entry errors, discretisation of the spatial coordinate system, reduction of resolution due to confidentiality or secrecy requirements, and other effects. See Section 3.4.4. Care should be taken in interpreting the output of statistical software if any duplicate points are present. A high prevalence of duplicate points precludes the use of many current techniques for spatial point processes.

Assumption 3 is usually implicit in our analysis. However, this does not preclude experiments where there is unreliable detection of the objects of interest. Examples include studies of the abundance of wild animal species, and galaxy surveys in radioastronomy. A galaxy catalogue is obtained by classifying each faint spot in an image as either a galaxy or noise. In such studies the analysis is consciously performed on the point process of *detected* points. Certain types of inference are then possible about the underlying process of *true* points, for example, estimation of the galaxy K-function, which is invariant under independent random thinning.

The points in a spatial point pattern dataset often represent the locations of physical objects. If the physical sizes of the objects cannot be neglected at the scale of interest, we may encounter methodological problems. Point process methods may still be applied to the locations, but may lead to artefacts. If the points are the centres of regions, such as cells observed in a microscope image, then a finding that the centres are regular or inhibited could simply be an artefact caused by the nonzero size of the cells. An extreme case occurs when the points are the centres of tiles in a space-filling tessellation. The strong geometric dependence in a tessellation causes striking artefacts in point process statistics; it is more appropriate to use methods specifically developed for tessellations.

Assumption 4 would be violated if, for example, the points were the locations of road traffic accidents, which can only occur along roads (see Chapter 17); or if the points were the nuclei of cells of a certain type, which can only occur inside certain biological tissues. In such cases there is a constraint on point locations, which may strongly affect the choice of statistical analysis.

5.6.3   Assumptions about the point process

In order to make progress in the analysis, researchers often need to make assumptions about the random point process itself. Some standard tools implicitly rely on these assumptions, so it is important to understand exactly what is being assumed.

A point process is called **stationary** if, when we view the process through a window W, its statistical properties do not depend on the location of the window in two-dimensional space. Imagine a sheet of cardboard with a hole cut in it. When we shift the cardboard around (without changing its directional orientation) and view the point process through the hole, the statistical properties of the observable point process are the same in each position.

To state this formally, if $\mathbf{x}$ is a point pattern, we write $\mathbf{x} + v$ for the result of shifting each point of the pattern by the same translation vector v, that is, $\mathbf{x} + v = \{x_i + v : i = 1, \ldots, n\}$. A point process $\mathbf{X}$ is called stationary (or 'strictly stationary') if the statistical properties of $\mathbf{X}$ and $\mathbf{X} + v$ are identical, for any choice of the translation vector v.

Stationarity implies, for example, that $\mathbf{X}$ must have homogeneous intensity, because the mean number of points falling in a set B is unaffected by shifting the set B. If B is any region, we have $n((\mathbf{X}+v)\cap B)=n(\mathbf{X}\cap(B-v))$ so that $n(\mathbf{X}\cap(B-v))$ has the same distribution as $n(\mathbf{X}\cap B)$, implying that $\mathbb{E}n(\mathbf{X}\cap B-v)=\mathbb{E}n(\mathbf{X}\cap B)$. It can be shown that this implies $\mathbb{E}n(\mathbf{X}\cap B)=\lambda|B|$ for some constant λ, so the point process has homogeneous intensity.

The assumption of stationarity is crucial for many of the classical tools of spatial statistics, such as Ripley's K-function (Chapter 7). If the point process is not stationary, the K-function is not even a well-defined concept.

A point process is called **stationary and isotropic** if its statistical properties are unaffected by *shifting or rotating* the point process. If we view such a process through a window W, the observable statistical properties do not depend on the *location or orientation* of the window in two-dimensional space. For example, CSR (the homogeneous Poisson process) is a stationary and isotropic point process.

A point process could be **isotropic** without being stationary. In this case the origin of the two-dimensional coordinate system is important. A point process is isotropic if its statistical properties are unchanged when it is rotated, through any given angle, around the origin. One application for which we might assume an isotropic but non-stationary point process is where we have recorded the spatial pattern of fungi growing around a tree (such as the dataset sporophores installed in spatstat). A simple model might assume that the statistical properties of the pattern depend only on distance from the tree, suggesting an isotropic point process. Very little work has been done on such models, however [136, 266].

5.6.4 Marks and covariates

In a dataset, the locations of points may be augmented by other numerical variables, which can usually be classified either as *marks* or as *covariates*. The main differences between marks and covariates are that marks are associated with data points, and marks are part of the 'response' (the point pattern) while covariates are 'explanatory'.

Marks

A mark variable may be interpreted as an additional coordinate for the point: for example a point process of earthquake epicentre locations (longitude, latitude), with marks giving the occurrence time of each earthquake, can alternatively be viewed as a point process in space-time with coordinates (longitude, latitude, time).

A marked point pattern is an unordered set

$$\mathbf{y}=\{(x_1,m_1),\ldots,(x_n,m_n)\},\quad x_i\in W,\quad m_i\in\mathcal{M}$$

where x_i are the locations and m_i are the corresponding marks. The analysis of marked point patterns is treated in detail in Chapters 14 and 15.

Covariates

Any kind of data may be recruited as an explanatory variable (covariate).

A 'spatial function', 'spatial covariate', or 'geostatistical covariate' is a function $Z(u)$ observable (potentially) at every spatial location $u\in W$. Values of $Z(u)$ may be available for a fine grid of locations u, for example, a terrain elevation map: see Figure 5.22 and Sections 1.1.4, 3.1.2, and 6.6.

Ideally, the values of a spatial function $Z(u)$ should be available at every spatial location u. In practice, $Z(u)$ may only be observable at some scattered sampling locations u. An example is the measurement of soil pH at a few sampling locations.

At a minimum, the covariate value must be available at every data point (i.e. the values $Z(x_i)$ must be available for every point x_i in the point pattern $\mathbf{x}$) and must also be available at some other 'non-data' or 'background' locations $u\in W$ with $u\notin\mathbf{x}$. It is not sufficient to observe the values

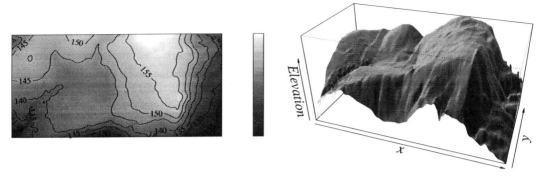

Figure 5.22. *A spatial covariate giving the terrain elevation, plotted as (Left) a pixel image with superimposed contours, and (Right) a perspective view of the terrain surface.*

of covariates at the data points only, if we wish to investigate the dependence of the point process on the covariate. Referring to the discretised point process in Figure 5.20, in order to investigate the dependence of the 0-or-1 response variable on the covariate Z, we need at least some instances where the response is 0. For further discussion see Section 14.2.2.

Instead of being a function, the covariate information may be a spatial pattern, such as a point pattern or a line segment pattern. The way in which this covariate information enters the analysis or statistical model depends very much on the context and the choice of model. Typically the covariate pattern would be used to define a surrogate spatial function Z, for example, $Z(u)$ may be the distance from u to the nearest line segment. For example, in the analysis of large-scale geological surveys, spatial covariate information may include a map of geological faults, effectively a pattern of line segments, as shown in the upper panel of Figure 5.23. Typically this would be converted to a spatial covariate by computing the *distance function*

$$d(u) = \text{distance from } u \text{ to nearest fault}$$

as shown in the lower panel of Figure 5.23.

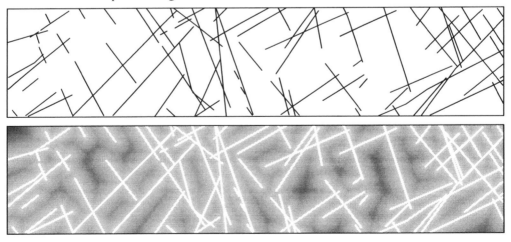

Figure 5.23. *Converting a spatial pattern to a spatial covariate function.* Top: *spatial pattern of geological faults.* Bottom: *distance function $d(u)$ for geological faults.*

5.7 Goals of analysis

Here we discuss strategies for analysing spatial point pattern data, foreshadowing the main themes of the book.[3]

The choice of strategy for modelling and analysing a spatial point pattern depends on the research goals. Our attention may be focussed primarily on the *intensity* of the point pattern, or primarily on the *interaction* between points, or equally on the intensity and interaction. There is a choice concerning the scope of statistical inference, that is, the 'population' to which we wish to generalise.

5.7.1 Intensity

The *intensity* is the (localised) expected density of points per unit area. It is typically interpreted as the rate of occurrence, abundance or incidence of the events recorded in the point pattern. When the prevention of these events is the primary concern (e.g., defects in crystal, petty crimes, cases of infectious disease), the intensity is usually the feature of primary interest. The main task for analysis may be to quantify the intensity, to decide whether intensity is constant or spatially varying, or to map the spatial variation in intensity. If covariates are present, then the main task may be to investigate whether the intensity depends on the covariate, for example, whether the abundance of trees depends on the acidity of soil.

The intensity is a first moment quantity (related to expectations of counts of points). Hence it is possible to study the intensity by formulating a *model for the intensity only*, for example, a parametric or semiparametric model for the intensity as a function of the Cartesian coordinates. See Chapter 6. In such analyses, stochastic dependence between points is a nuisance feature that complicates the methodology and inference.

Alternatively, we may formulate a complete stochastic model for the observed point pattern (i.e. a spatial point process model) in which the main focus is the description of the intensity. The model should exhibit the right type of stochastic dependence, and the intensity should be a tractable function of the model parameters. If points are independent, the correct model is a Poisson point process (Chapter 9). If there is positive association between points, useful models include cluster processes and Cox processes (Chapter 12). If there is negative association, Gibbs processes (Chapter 13) are appropriate, although the intensity is not a simple function of the model parameters.

5.7.2 Interaction

'Interpoint interaction' is the conventional term for stochastic dependence between points. This covers a wide range of behaviour, since the only point processes which do not exhibit stochastic dependence are the Poisson processes. The term 'interaction' can be rather prejudicial. One *possible* cause of stochastic dependence is a direct physical interaction between the objects recorded in the point pattern. For example, if the spatial pattern of pine seedlings in a natural forest is found to exhibit negative association at short distances, this might be interpreted as reflecting biological interaction between the seedlings, perhaps due to competition for space, light, or water.

The main task for analysis may be to decide whether there is stochastic dependence, to determine the type of dependence (e.g., positive or negative association), or to quantify its strength and spatial range.

Whereas intensity is a 'first moment' property, interpoint interaction is measured by second-order moment quantities such as the K-function (Chapter 7), or by higher-order quantities such as

---

[3] Some of the material in this section was previously published in [30, Sec. 20.2].

the distance functions G, F, and J (Chapter 8). Just as we must guard against spurious correlations in numerical data by carefully adjusting for changes in the mean, a rigorous analysis of interpoint interaction requires that we take into account any spatial variation in intensity.

A popular classical approach to spatial point pattern analysis was to assume that the point pattern is *stationary*. This implies that the intensity is constant. Analysis could then concentrate on investigating interpoint interaction. It was argued (e.g., [571]) that this approach was pragmatically justified when dealing with quite small datasets (containing only 30 to 100 points), or when the data were obtained by selecting a small subregion where the pattern appeared stationary, or when the assumption of stationarity is scientifically plausible.

Figure 5.24 shows the Swedish Pines dataset of Strand [642] presented by Ripley [575] as an example where the above-mentioned conditions for assuming stationarity were satisfied. There is nevertheless some suggestion of inhomogeneity. Contour lines represent the fitted intensity under a parametric model in which the logarithm of the intensity is a quadratic function of the Cartesian coordinates. Figure 5.25 shows the estimated K-function of the Swedish Pines assuming stationarity, and the *inhomogeneous* K-function which adjusts for the fitted log-quadratic intensity (see [46] and Section 7.10.2). The two K-functions convey a similar message, namely that there is inhibition between the saplings at distances less than one metre. They agree because gentle spatial variation in intensity over large spatial scales is irrelevant at shorter scales.

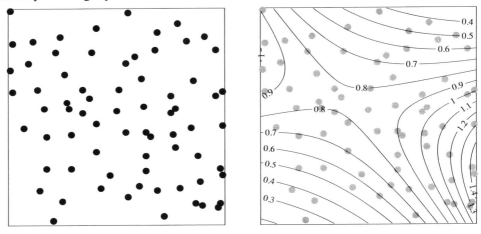

Figure 5.24. *Swedish Pines data* (Left) *with fitted log-quadratic intensity* (Right) *expressed in points per square metre.*

5.7.3   Intensity and interaction

In some applications, intensity and interaction are both of interest. For example, a cluster of new cases of a disease may be explicable either by a localised increase in intensity due to aetiology (such as a localised pathogen), sampling effects (a localised increase in vigilance, etc.), or by stochastic dependence between cases (due to person-to-person transmission, familial association, genetics, social dependence, etc.). The spatial arrangement of galaxies in a galaxy cluster invites complex space-time models, in which the history of the early universe is reflected in the overall intensity of galaxies, while the observed local arrangement of galaxies involves gravitational interactions in recent history.

When a point pattern exhibits both spatial inhomogeneity and interpoint interaction, several strategies are possible. In an *incremental* or *marginal* modelling strategy we try to estimate spatial trend, then 'subtract' or 'adjust' for spatial trend, possibly in several stages, before looking for evidence of interpoint interaction. In a *joint* modelling strategy we try to fit one stochastic model

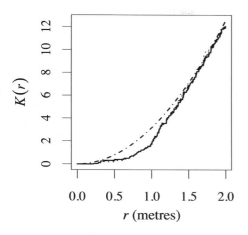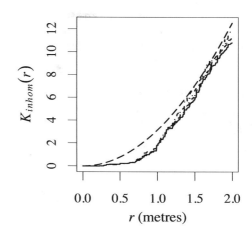

Figure 5.25. Left: *estimated K-function of Swedish Pines assuming stationarity.* Right: *estimated inhomogeneous K-function of Swedish Pines using fitted log-quadratic intensity.*

that captures all relevant features of the point process and, in particular, allows the statistician to account for spatial inhomogeneity during the analysis of interpoint interaction.

These choices are familiar from time series analysis. Incremental modelling is analogous to seasonal adjustment of time series, while joint modelling is analogous to fitting a time series model that embraces both seasonal trend and autocorrelation. Incremental modelling is less prone to the effects of model misspecification, while joint modelling is less susceptible to analogues of Simpson's Paradox. Joint modelling would normally be employed in the final and more formal stages of analysis, while incremental modelling would usually be preferred in the initial and more exploratory stages.

For example, in the analysis of the Swedish Pines data above, we first fitted a parametric intensity model, then computed the inhomogeneous K-function which 'adjusts' for this fitted intensity. This is an incremental modelling approach. A corresponding joint modelling approach is to fit a Gibbs point process (Chapter 13) with non-stationary spatial trend. Again we assume a log-quadratic trend. Figure 5.25 suggests fitting a Strauss process model (Section 13.3.7) with interaction radius r between 4 and 15 units. The model selected by maximum profile pseudolikelihood has $r = 9.5$ and a fitted interaction parameter of $\gamma = 0.27$, suggesting substantial inhibition between points.

5.7.4 Confounding between intensity and interaction

In analysing a point pattern, it may be impossible to distinguish between clustering and spatial inhomogeneity. Bartlett [83] showed that a single realisation of a point process model that is stationary and clustered (i.e. exhibits positive dependence between points) may be *identical* to a single realisation of a point process model that has spatially inhomogeneous intensity but is not clustered. Based on a single realisation, the two point process models are distributionally equivalent and hence unidentifiable. This represents a fundamental limitation on the scope of statistical inference from a spatial point pattern, assuming we do not have access to genuine replicate observations. The inability to separate trend and autocorrelation, within a single dataset, is also familiar in time series analysis.

This may be categorised as a form of *confounding*. In the theory of design and analysis of experiments, a linear model $Y = X\beta + \varepsilon$ is confounded (in the context of a particular experiment) if the columns of the design matrix X are not linearly independent, so that the parameter vector β is not identifiable. Bartlett's examples show that a point process model involving both spatial inhomogeneity and interpoint interaction may be confounded, that is, unidentifiable, given only a single realisation of the spatial point process.

The potential for confounding spatial inhomogeneity and interpoint interaction is important in the interpretation of summary statistics such as the K-function. In Figure 5.26 the left panel shows a realisation of a spatially inhomogeneous Poisson process, its intensity a linear function of the Cartesian coordinates. The right panel is a plot of $\hat{L}(r) - r$ against r, where $\hat{L}$ is the estimate of $L(r) = \sqrt{K(r)/\pi}$ assuming the point process is stationary. The right-hand plot invites the incorrect interpretation that the points are clustered.

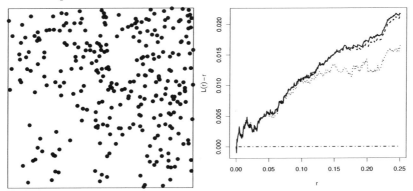

Figure 5.26. *Illusory clustering.* Left: *realisation of a nonstationary Poisson process.* Right: *plot of* $\hat{L}(r) - r$ *against r for the same point pattern, inviting the interpretation that the pattern is clustered.*

Formal tests such as the χ^2 test of CSR using quadrat counts (Section 6.4.2) are afflicted by a similar weakness (see Section 6.4.3).

5.7.5   Covariate effects

When a point pattern dataset is accompanied by covariate data (Sections 1.1.4 and 5.6.4), we typically want to investigate whether the intensity depends on the covariates and to quantify this dependence. It may be enough to conduct a formal statistical test of the hypothesis that the point pattern does not depend on the covariate.

For a numerical spatial covariate $Z(u)$, we could assume (in a simple case) that the intensity of the point process depends on Z through the relationship $\lambda(u) = \rho(Z(u))$ where ρ is a function to be estimated. For example, $\rho(z)$ could express the preference of a tree species for a particular habitat, or the likelihood of finding mineral deposits in a particular geochemical environment. The function ρ could be estimated using nonparametric methods such as kernel smoothing, or by fitting a parametric model.

When several covariates are present, we could focus on one covariate $Z(u)$ and assume that it has a multiplicative effect on the point process intensity, $\lambda(u) = \rho(Z(u))B(u)$ where $B(u)$ is a 'baseline' or 'reference' intensity that takes the other covariates into account. Again ρ could be estimated nonparametrically or parametrically. When many covariates are present, or when the effect of a covariate is not so simple, a parametric model is usually preferable.

One reason for investigating covariate effects is to adjust for them when studying interaction between points. For example, the standard analysis of correlation between points using Ripley's K-function is sensitive to spatial variation in the underlying intensity. The inhomogeneous K-function [46] adjusts for this spatial variation: it requires an accurate estimate of the intensity function.

5.7.6 Multitype point patterns

A *multitype* point pattern (Section 1.1.2) is a pattern of points of several different types. It is usually represented as a marked point pattern where the marks are categorical (factor) values.

Multitype point patterns introduce many new scientific questions. Under the heading of 'intensity' we may want to know whether the intensity functions of the points of each type are proportional to each other, implying that the relative proportions of each type of point are constant over the study region. If not, the most extreme alternative is that the different types of points are 'segregated', tending to occupy different parts of the study region.

Under the heading of 'interaction' we can investigate dependence between points of the *same* type or *different* types. Dependence between points of the same type has the usual interpretations of clustering, inhibition and so on. Independence between points of the same type implies that they form a Poisson process. However, dependence between points of *different* types has a completely different interpretation. Independence between points of types i and j means that the two point processes, consisting of points of type i and points of type j, respectively, are independent point processes, but does not imply anything about their spatial pattern.

Chapter 14 discusses the issues and techniques involved in analysing multitype point patterns.

5.7.7 Scope of inference

There is a choice concerning the scope of statistical inference, that is, the 'population' to which we wish to generalise from the data.

At the lowest level of generalisation, we are interested only in the region that was actually surveyed. In applying precision agriculture to a particular farm, we might use the observed spatial point pattern of tree seedlings, which germinated in a field sown with a uniform density of seed, as a means of estimating the unobservable, spatially varying, fertility of the soil in the same field. Statistical inference here is a form of interpolation or prediction. The modelling approach is influenced by the prediction goals: to predict soil fertility it may be sufficient to model the point process intensity only, and ignore interpoint interaction.

At the next level, the observed point pattern is treated as a 'typical' sample from a larger pattern which is the target of inference. To draw conclusions about an entire forest from observations in a small study region, we treat the forest as a spatial point process $\mathbf{X}$, effectively extending throughout the infinite two-dimensional plane. In order to draw inferences based only on a sample of $\mathbf{X}$ in a fixed bounded window W, we might assume that $\mathbf{X}$ is *stationary* and/or *isotropic*, meaning that statistical properties of the point process are unaffected by vector translations (shifts) and/or rotations, respectively. This implies that our dataset is a typical sample of the process, and supports nonparametric inference about distributional properties of $\mathbf{X}$ such as its intensity and K-function. It also supports parametric inference, for example about the interaction parameter γ of a Strauss process model for the spatial dependence between trees.

At a higher level, we seek to extract general 'laws' or 'relationships' from the data. This involves generalising from the observed point pattern to a hypothetical population of point patterns which are governed by the same 'laws' but which may be very different from the observed point pattern. One important example is modelling the dependence of the point pattern on a spatial covariate (such as terrain slope). This is a form of regression. We might assume that the intensity $\lambda(u)$ of the point process at a location u is a function $\lambda(u) = \rho(Z(u))$ of the spatial covariate $Z(u)$. The regression function ρ is the target of inference. The scope of inference is a population of experiments where the same variables are observed and the same regression relationship is assumed to hold. A model for ρ (parametric, non-, or semi-parametric) is formulated and fitted. More detailed inference requires either replication of the experiment, or an assumption such as joint stationarity of the covariates and the response, under which a large sample can be treated as containing sufficient replication.

At the highest level, we seek to capture all sources of variability that influence the spatial point pattern. Sources of variability may include 'fixed effects' such as regression on an observable spatial covariate, and also 'random effects' such as regression on an unobserved, random spatial covariate. For example, a Cox process (Section 5.5 and Chapter 12) is defined by starting with a random intensity function $\Lambda(u)$ and, conditional on the realisation of Λ, letting the point process be Poisson with intensity Λ. In forestry applications, Λ could represent the unobserved, spatially inhomogeneous fertility of soil, modelled as a random process. Thus Λ is a 'random effect'. Whether soil fertility should be modelled as a fixed effect or random effect depends on whether the main interest is in inferring the value of soil fertility in the study region (fixed effect) or in characterising the variability of soil fertility in general (random effect).

Part II

EXPLORATORY DATA ANALYSIS

6

Intensity

Part II covers methods for exploratory data analysis, starting in this chapter with the fundamental concept of the intensity of a point process.

6.1 Introduction

Average intensity

A wildlife ecologist who maps bird nests in a survey might choose to report the result as an average density of nests per hectare. Dividing the total number of points by the area of the survey region gives the average density of points per unit area. In different contexts, the average number of points per unit area could be a measure of abundance (for example, the abundance of virus particles on a cell surface), density (of light-sensitive cells in the retina), productivity (of crops), risk (of crimes), intensity (of a lightning storm), or prospectivity (of undiscovered mineral deposits). The standard generic term is *intensity*.

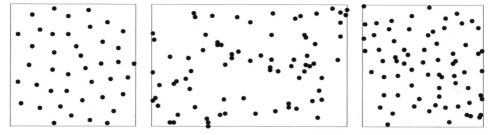

Figure 6.1. *Classic point pattern datasets with (roughly) homogeneous intensity.* Left: *biological cell centres in histological section [572, 575].* Middle: *trees in a New Zealand forest, excluding a five-foot border [441, 575].* Right: *Swedish Pine saplings [642, 572].*

Investigation of the intensity of a point pattern is one of the first and most important steps in data analysis. The intensity is a basic descriptive characteristic of a point process, an average ('expectation' or 'first moment') analogous to the average of a population of numbers. Compared to other properties of a point process, the intensity requires relatively few modelling assumptions.

This chapter deals with *exploratory* investigation of intensity using *nonparametric* tools (i.e. avoiding restrictive model assumptions). Parametric modelling of intensity is covered in Chapter 9.

In dividing the total number of points by the total area of the survey plot, the ecologist has effectively assumed that the spread of points is 'uniform' or 'homogeneous', as illustrated in Figure 6.1. Homogeneity may be a tentative, working assumption for some kinds of analysis. Homogeneity may be assumed if there is theoretical justification: for example, much of modern cosmology assumes that the universe is homogeneous on sufficiently large scales [445].

Spatially varying intensity

However, this assumption would be inappropriate for other point patterns, where the intensity is spatially varying ('nonuniform', 'inhomogeneous', or 'heterogeneous' intensity) as illustrated in Figure 6.2.

Sometimes the most important scientific question is whether the intensity is homogeneous or not. Inhomogeneity of the intensity can reflect spatial variation in abundance (of a bird population), fertility (of a natural forest), or risk (of tornadoes). It can reflect preference (of wild animals for certain types of habitat), avoidance, or segregation (between different species of tree). It can reflect dependence on spatially varying external factors: for example, the likelihood of finding a gold deposit often depends on proximity to geological faults. When the intensity is spatially varying, it is effectively a function of spatial location, and we can use statistical methods to estimate this function from data.

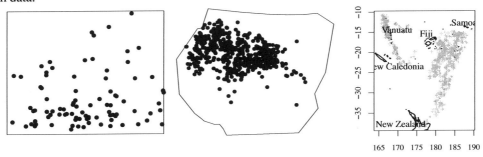

Figure 6.2. *Point pattern datasets with inhomogeneous intensity.* Left: *enterochromaffin-like cells in histological section of gastric mucosa (Dr Thomas Bendtsen; see [484, pp. 2, 169]), interior of stomach towards top of picture.* Middle: *gorilla nesting sites in Kagwene Gorilla Sanctuary, Cameroon (see Section 9.3.4.1).* Right: *epicentres of 1000 earthquakes of Richter magnitude 4.0 or greater in the South Pacific showing concentrations along two plate boundaries (*R *package* datasets; *Harvard PRIM-H project; Dr John Woodhouse).*

Relative intensity

Dividing the number of points by the area of the survey region also assumes that the area is the appropriate 'denominator'. This would not be appropriate in spatial epidemiology, for example, where the number of disease cases reported to a health authority would usually be expressed as a ratio relative to the *total population* in the same region, rather than the area. The ratio of cases to population is an estimate of the *disease risk per head of population*. That calculation effectively assumes that this risk is constant. This is equivalent to assuming that the spatially varying intensity of disease cases (cases per square kilometre) is proportional to the spatially varying population density (people per square kilometre).

To compare the *relative* spatial distribution of two different kinds of points, we can study the ratio of their intensities. For example a spatial case-control dataset gives the spatial locations of a set of disease cases, and of a separate set of controls (notionally a sample from the population at risk of the disease). If the disease risk per head of population does not depend on spatial location, then we would expect the spatially varying intensities of these two point processes to be proportional.

Weighted intensity

The points in a point pattern may carry different 'weights'. For example, astronomers are interested in the spatial distribution of *mass* in the universe, so the analysis of a spatial pattern of galaxies might take into account the estimated mass of each galaxy. The appropriate definition of intensity is then the average *total mass* of galaxies per unit volume of space. This reflects a different choice for the 'numerator' of the intensity. Similarly there are cases where each data point represents one or more events that occurred at the same spatial location, such as multiple disease cases at the same residence. It is then appropriate to weight each residential location by the number of cases, so that the intensity would be the average *total number of cases* per unit area, not the number of affected residences per unit area.

Adjusting for spatially varying intensity

Even when the intensity is not a major focus of statistical analysis, it remains an important part of the analysis, because spatial variation in intensity can easily be confused with other pattern characteristics such as clustering [83]. In order to establish convincingly that a point pattern is clustered, we need to eliminate alternative explanations, including spatial variation in the intensity. If the intensity is known to be spatially varying, some of the standard tools for investigating clustering can be *adjusted* for this effect, provided we have estimated the intensity faithfully. This is similar to the principle that, when analysing a time series, a test for the presence of autocorrelation should involve either adjustment for the trend in the mean, or confirmation that there is no trend.

6.2 Estimating homogeneous intensity

Analysis of a point pattern often starts by tentatively assuming homogeneity. In order to assess or validate this assumption, we must define exactly what is meant by homogeneous intensity. Figure 6.3 shows synthetic patterns generated at random by a mechanism that is spatially homogeneous. But a single point pattern such as those in Figure 6.3 cannot be perfectly homogeneous, since a point cannot be spread uniformly over the plane like butter. The only way to make rigorous sense of the assumption of 'homogeneity' is by thinking statistically. The outcome of the ecologist's survey of bird nests is one of many possible outcomes that could have been obtained, depending on random events in the history of the forest, its colonization by birds, and the selection of the survey plot. It is the *average over all possible outcomes* which is homogeneous.

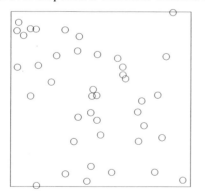 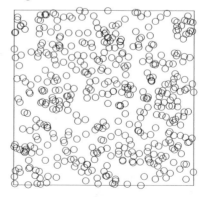

Figure 6.3. *Simulated point patterns with homogeneous intensity.* Left: *50 points per unit area.* Right: *500 points per unit area.*

Accordingly a point process **X** is defined to have *homogeneous intensity* if, for any sub-region B of two-dimensional space, the *expected* number of points of **X** falling in B is proportional to the area of B:

$$\mathbb{E}[n(\mathbf{X} \cap B)] = \lambda \, |B| \qquad (6.1)$$

where λ is a constant called the *intensity*.

The intensity λ is the expected number of points per unit area. The value of λ depends on the unit of measurement: it has dimensions $length^{-2}$. For example, 300 points per square kilometre is equivalent to 0.0003 points per square metre.

In basic statistics, the sample mean of a dataset is an unbiased estimate of the population mean. Similarly for point processes, the empirical density of points,

$$\overline{\lambda} = \frac{n(\mathbf{x})}{|W|}, \qquad\qquad (6.2)$$

is an unbiased estimate of the true intensity λ, assuming the point process has homogeneous intensity. Here $\mathbf{x}$ is the point pattern dataset, observed in a window W, and $n(\mathbf{x})$ is the number of points in $\mathbf{x}$.

The estimate $\overline{\lambda}$ could be computed by hand using the functions `npoints` and `area.owin`, or using the special function `intensity.ppp`, a method for the generic `intensity`. For example, the Swedish Pines pattern shown in the right panel of Figure 6.1 is available in `spatstat` as the dataset `swedishpines`. The intensity estimate $\overline{\lambda}$ could be computed by hand as

```
npoints(swedishpines)/area(Window(swedishpines))
```

or just by typing `intensity(swedishpines)`, which both return the value 0.0074, meaning that the estimated intensity is $\overline{\lambda} = 0.0074$ points *per square unit*. However, the units of length for the `swedishpines` dataset are unconventional: `unitname(swedishpines)` returns "0.1 metres", so the estimated intensity is 0.0074 trees per square *decimetre*, equivalent to 0.74 trees per square metre or 7400 trees per hectare. One could also type `summary(swedishpines)` which gives a longer summary of the data including both the intensity and the unit of length.

The function `rescale` can be used to convert the coordinates to a standard unit of length, before calculating the intensity. If the call to `rescale` does not specify the scaling factor, the default is to convert a composite unit like '0.1 metres' to the base unit 'metres'. For example, `intensity(rescale(swedishpines))` gives 0.74, the intensity in points per square metre.

To quote a standard error for our estimate of the intensity, we would need to make additional assumptions. If we provisionally assume that the point process is *Poisson* (Section 5.3) then the observed total number of points $n(\mathbf{x})$ is a realisation of a Poisson random variable with mean $\lambda|W|$. For the Poisson distribution, the variance is equal to the mean, so $\overline{\lambda}$ has variance $\text{var}\,\overline{\lambda} = \text{var}[n(\mathbf{X} \cap W)]/|W|^2 = \lambda/|W|$. An estimate of the standard error of $\overline{\lambda}$ is $\sqrt{\overline{\lambda}/|W|}$:

```
> X <- rescale(swedishpines)
> lam <- intensity(X)
> (sdX <- sqrt(lam/area(Window(X))))
[1] 0.08777
```

The estimated standard error is 0.0878 points per square metre. Standard errors and confidence intervals can also be obtained using quadrat counts (Section 6.4) or by fitting parametric models (Chapters 9, 12, and 13).

If the point pattern is multitype (with a vector of marks that is a `factor` classifying each point into one of several possible types, as explained in Sections 3.3.2.1, 4.2.4, and 14.3) then `intensity.ppp` will return a vector of intensities, one for each possible type.

```
> unitname(amacrine)
662 microns
> X <- rescale(amacrine, 1000/662, "mm")
> intensity(X)
   off    on
202.4 216.6
```

To estimate the intensity of the pattern of all points regardless of their type, use `sum(intensity(X))` or `intensity(unmark(X))`.

If the point pattern has weights, use the argument `weights` to `intensity.ppp`. For example, in the `finpines` dataset (page 57) each point is the location of a pine sapling, and has two marks, giving the tree's diameter in centimetres and its height in metres.

```
> finpines
Marked planar point pattern: 126 points
Mark variables: diameter, height
window: rectangle = [-5, 5] x [-8, 2] metres
```

A rough estimate of the sapling's volume is obtained by pretending that it is a cone (since a cone of height h with base diameter d has volume $\pi h d^2 / 12$):

```
> height <- marks(finpines)$height
> diameter <- marks(finpines)$diameter
> volume <- (pi/12) * height * (diameter/100)^2
```

or more elegantly

```
> volume <- with(marks(finpines), (pi/12) * height * (diameter/100)^2)
```

We can then use the estimated tree volumes as weights:

```
> intensity(finpines, weights=volume)
[1] 0.001274
```

There is an estimated average *standing volume* of 0.00127 cubic metres of wood per square metre of forest floor.

Other principles for estimating the intensity are discussed in Section 6.5.2.

6.3 Technical definition

We need a technical definition of 'intensity' which applies even when the point pattern is not homogeneous. Examples of inhomogeneous patterns are shown in Figure 6.4.

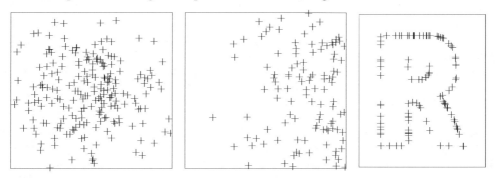

Figure 6.4. *Simulated point patterns with different intensities.* Left: *diffuse intensity, higher at centre of plot.* Middle: *diffuse intensity, higher at right-hand edge of plot.* Right: *singular intensity concentrated on edges of the letter R.*

The left and middle panels of Figure 6.4 can be described by saying that the intensity value λ is spatially varying. At a spatial location u, the intensity is $\lambda(u)$, where $\lambda(u)$ is some function of location. More precisely, in a small neighbourhood of the location u, with small area a, the expected number of points is approximately equal to $\lambda(u)a$. For any region B, we may imagine dividing B into pixels, calculating the expected number of points in each pixel, and adding up these expected numbers to obtain the expected total number of points in B. This sum is effectively the integral of the intensity function

$$\mathbb{E}[n(\mathbf{X} \cap B)] = \int_B \lambda(u)\,du \tag{6.3}$$

for any given region B. We can think of $\lambda(u)$ as an undulating surface whose height represents the intensity. From this standpoint, equation (6.3) states that the expected number of points falling in some region of interest is equal to the volume under the surface. If (6.3) holds, then $\lambda(u)$ is called the *intensity function* of the point process. The values of $\lambda(u)$ are intensity values (points per unit area) with dimensions *length*$^{-2}$.

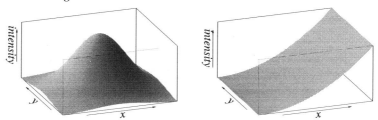

Figure 6.5. *Perspective views of the intensity functions $\lambda(u)$ of the point processes shown in the left and middle panels of Figure 6.4.*

Figure 6.5 shows the intensity functions of the point patterns in the left and middle panels of Figure 6.4 portrayed as surfaces. For example, the middle panel of Figure 6.4 is a realisation of a point process in the unit square with intensity function $\lambda(x,y) = 300x^2$ where x and y are the Cartesian coordinates.

However, the right panel of Figure 6.4 is a realisation of a (simulated) point process which *does not have an intensity function*. The points are concentrated along the edges of the letter R. Point processes of this kind are common in seismology (since many earthquakes occur exactly at a tectonic plate boundary). To handle such cases we need a more general language.

For any point process $\mathbf{X}$, the intensity characteristics are completely described if we know the expected number of points

$$\Lambda(B) = \mathbb{E}[n(\mathbf{X} \cap B)] \tag{6.4}$$

falling in any given set B. If the point process has homogeneous intensity, then $\Lambda(B) = \lambda|B|$ for all B, where the constant $\lambda \geq 0$ is the intensity value. A compact way to say this is that 'Λ is proportional to area'.

Notice that Λ is a set function — a function whose argument is a *set* or *region B*, and which returns a numerical value $\Lambda(B)$. In the same way, 'area' is a set function which, when applied to any region B, yields the area of the region. We call Λ the *mean measure* of the point process $\mathbf{X}$. This enables us to speak about the intensity of any[1] point process.

In some other texts, Λ is called the *intensity measure*. This can lead to confusion, because the values $\Lambda(B)$ are not intensities (numbers of points per unit area); they are mean numbers of points.

The mean measure Λ is called *diffuse* if it has an intensity function $\lambda(u)$ in the sense of (6.3). It is called *singular* if the points are confined to a set with zero area, such as the linear boundary in the right panel of Figure 6.4.

---

[1]The only requirement is that the expectation in (6.4) should be finite whenever the region B is bounded (contained in a rectangle of finite size).

6.4 Quadrat counting

If it is suspected that the intensity may be inhomogeneous, it can be estimated nonparametrically by techniques such as quadrat counting and kernel estimation.

6.4.1 Quadrat counts

A simple way to check for inhomogeneity is to check whether regions of equal area contain roughly equal numbers of points (as they must do if the point process is homogeneous).

In *quadrat counting*, the observation window W is divided into subregions $B_1, \ldots, B_m$ called quadrats.[2] For simplicity, suppose the regions have equal area. We count the numbers of points falling in each quadrat, $n_j = n(\mathbf{x} \cap B_j)$ for $j = 1, \ldots, m$. Since these counts are unbiased estimators of the corresponding expected values $\mathbb{E}[n(\mathbf{X} \cap B_j)]$, they should be equal 'on average' if the intensity is homogeneous. In particular, any apparent spatial trend in the counts n_j suggests that the intensity is inhomogeneous.

Quadrat counting is performed in spatstat by the function quadratcount.

```
> swp <- rescale(swedishpines)
> Q3 <- quadratcount(swp, nx=3, ny=3)
> Q3
            x
y            [0,3.2] (3.2,6.4] (6.4,9.6]
   (6.67,10]     8        6         7
   (3.33,6.67]   8       11         9
   [0,3.33]      5        6        11
```

The value returned by quadratcount is an object belonging to the special class "quadratcount". We used the print method for this class to obtain the text output above, and the plot method to get the display in the left panel of Figure 6.6. Note that the plot and print methods display the counts in the same spatial arrangement.

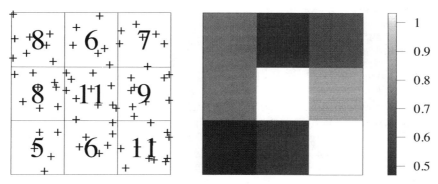

Figure 6.6. *Quadrat counting for Swedish Pines data.* Left: *quadrat counts.* Right: *intensity estimates (points per square metre).*

The arguments nx,ny to quadratcount specify that the quadrats should be an nx by ny grid

---

[2]A 'quadrat' (German for 'square') originally meant a rectangular wooden frame used for random sampling in the field. It now refers to a spatial sampling region of any shape. A quadrat is very different from a 'quadrant', which is one-quarter of the infinite Euclidean plane.

of rectangles. This ensures that the quadrats have equal area, provided the observation window is a rectangle. Other options are discussed below.

In choosing the size of quadrats, there is a tradeoff between bias and variability: choosing larger quadrats reduces the relative error (standard error divided by mean) of the counts n_j but also obliterates the spatial variation in intensity within each quadrat.

If the quadrat counts are divided by the areas of the corresponding quadrats, we obtain the average intensity in each quadrat, which is a simple estimate of the intensity function. The method `intensity.quadratcount` calculates these intensity estimates from a `"quadratcount"` object.

```
> intensity(Q3)
            x
y           [0,3.2] (3.2,6.4] (6.4,9.6]
  (6.67,10]  0.7500    0.5625    0.6562
  (3.33,6.67] 0.7500   1.0312    0.8437
  [0,3.33]   0.4687    0.5625    1.0312
```

Use `intensity( , image=TRUE)` to obtain a pixel image that is an estimate of the intensity function. Setting `L3 <- intensity(Q3, image=TRUE)` and calling `plot(L3)` gives the display in the right panel of Figure 6.6.

Quadrat counts, in quadrats of equal size and shape, can also be used to calculate a standard error for the overall estimate of intensity. If we are not willing to assume a Poisson process, but willing to assume that the counts in different quadrats are approximately independent variables with the same (unknown) distribution, we can apply the usual estimator of standard error for a mean:

```
> l3 <- as.numeric(intensity(Q3))
> sem <- sqrt(var(l3)/(length(l3)-1))
> sem
[1] 0.07118
```

For comparison, the estimated standard error assuming CSR was `sdX = 0.08777`.

In general, the quadrats could have unequal sizes and shapes. The alternative argument `tess` to `quadratcount` allows the quadrats to be any tessellation of the window. For example, hexagonal tiles can be created using `hextess`:

```
> H <- hextess(swp, 1)
> hQ <- quadratcount(swp, tess=H)
```

The counts are plotted in the left panel of Figure 6.7. Since the quadrat areas are not all equal, the counts of points in these quadrats should not be compared directly. Under the assumption of homogeneity, equation (6.1), the expected count in each quadrat is *proportional to the area* of the quadrat. The average intensity (6.2) in each quadrat is an unbiased estimator of the homogeneous intensity λ. For exploratory purposes, we can plot the average intensity in each quadrat. The right panel of Figure 6.7 shows a plot of `intensity(hQ, image=TRUE)`.

Note that `quadratcount` is not designed to handle very large numbers of quadrats. To count the number of points falling in each pixel of a fine grid of pixels, use `pixellate`.

6.4.2   Quadrat counting test of homogeneity

Historically the `swedishpines` dataset has been analysed under the assumption of homogeneous intensity [643, 575]. However, the quadrat counts in Figures 6.6 and 6.7 suggest the intensity may be slightly elevated along a diagonal swath from top left to bottom right of the plot.

One way to assess the evidence for inhomogeneity is to conduct a formal test of statistical

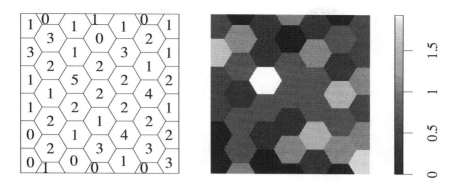

Figure 6.7. *Quadrat counting with hexagonal quadrats.* Left: *quadrat counts superimposed on point pattern.* Right: *intensity estimates (points per square metre). Swedish Pines data.*

significance. The principles of hypothesis tests are explained in Chapter 10: here we just run through the procedure for the test.

The null hypothesis is that the intensity is homogeneous, and the alternative hypothesis is that the intensity is inhomogeneous in some unspecified fashion. For practical purposes we will assume, provisionally, that the point process is Poisson. Then the null hypothesis is CSR and the alternative is an inhomogeneous Poisson process.

As before we divide the window W into quadrats $B_1, \ldots, B_m$ and count the numbers of points $n_1, \ldots, n_m$ of points in each quadrat. If the null hypothesis is true, the n_j are realisations of independent Poisson random variables with expected values $\mu_j = \lambda a_j$ where λ is the unknown intensity and a_j is the area of B_j. If the quadrats all have *equal* area $a_j = a$, then the counts n_j are independent Poisson random variables with *equal* mean λa.

The χ^2 (chi-squared) test could be used in two different ways here: to test goodness-of-fit to the Poisson distribution assuming homogeneity [113], or to test homogeneity assuming independence [296]. Our focus is on homogeneity, so we shall do the latter, applying the 'χ^2 test of uniformity'.

Given the total number of points $n = \sum_j n_j$, and the total window area $a = \sum_j a_j$, the estimated intensity is $\overline{\lambda} = n/a$, and the expected count in quadrat B_j is $e_j = \overline{\lambda} a_j = n a_j / a$. The test statistic is

$$X^2 = \sum_j \frac{(\text{observed} - \text{expected})^2}{\text{expected}} = \sum_j \frac{(n_j - e_j)^2}{e_j} = \sum_j \frac{(n_j - \overline{\lambda} a_j)^2}{\overline{\lambda} a_j}. \tag{6.5}$$

If the quadrats all have equal area, then the n_j are independent with *equal* expected value under the null hypothesis. The test statistic reduces to

$$X^2 = \sum_j \frac{(n_j - n/m)^2}{n/m}. \tag{6.6}$$

Under the null hypothesis, the distribution of the test statistic is approximately a χ^2 distribution with $m - 1$ degrees of freedom. The approximation is traditionally deemed to be acceptable when the expected counts e_j are greater than 5 for all quadrats.

The quadrat counting test is performed in `spatstat` by `quadrat.test`.

```
> tS <- quadrat.test(swp, 3,3)
> tS
```

```
      Chi-squared test of CSR using quadrat counts
      Pearson X2 statistic

data: swp
X2 = 4.7, df = 8, p-value = 0.4
alternative hypothesis: two.sided

Quadrats: 3 by 3 grid of tiles
```

The value returned by quadrat.test is an object of class "htest" (the standard R class for hypothesis tests). Printing the object (as shown above) gives comprehensible output about the outcome of the test. Inspecting the *p*-value, we see that the test does not reject the null hypothesis of CSR for the Swedish Pines data. The *p*-value can also be extracted by

```
> tS$p.value
[1] 0.4169
```

The return value of quadrat.test also belongs to the special class "quadrat.test". Plotting the object will display the quadrats, annotated by their observed and expected counts and the Pearson residuals. Figure 6.8 shows the result of plot(swp); plot(tS, add=TRUE). In each quadrat the observed counts n_j are displayed at top left; expected counts e_j at top right; and Pearson residuals $r_j = (n_j - e_j)/\sqrt{e_j}$ at bottom.

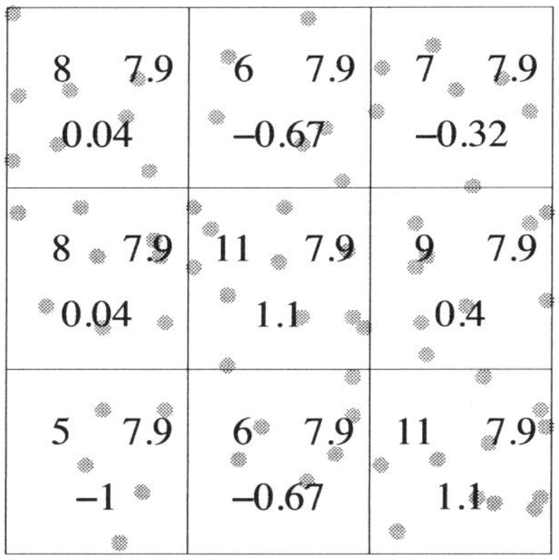

Figure 6.8. *Quadrat counting test of CSR for Swedish Pines data.*

Other arguments to quadrat.test make it possible to conduct a one-sided test, and to compute the *p*-value using Monte Carlo simulation instead of the χ^2 approximation.

```
> quadrat.test(swp, 5, alternative="regular", method="MonteCarlo")
      Conditional Monte Carlo test of CSR using quadrat counts
      Pearson X2 statistic

data: swp
X2 = 18, p-value = 0.2
alternative hypothesis: regular

Quadrats: 5 by 5 grid of tiles
```

The function quadrat.test is generic, with methods for "ppp" objects (which we have used above), but also for "splitppp" and "quadratcount" objects. It is possible to perform a χ^2 test using previously computed counts, for example the counts Q3 above:

```
> quadrat.test(Q3)

        Chi-squared test of CSR using quadrat counts
        Pearson X2 statistic

data:
X2 = 4.7, df = 8, p-value = 0.4
alternative hypothesis: two.sided

Quadrats: 3 by 3 grid of tiles
```

The results of several quadrat tests can also be pooled. For example, suppose an ecologist has recorded the spatial pattern of trees in three separate plots in the same forest. The data from each plot have been subjected to a quadrat counting test as described above. Then an overall test of uniform intensity is performed by applying pool.quadrattest to the three test results:

```
test1 <- quadrat.test(X1, 3)
test2 <- quadrat.test(X2, 3)
test3 <- quadrat.test(X3, 5)
pool(test1, test2, test3)
```

The quadrat test of homogeneity can be generalised to a test of any model for the intensity: see Section 10.4. More powerful tests become available if it is possible to specify a *covariate* upon which the intensity might depend (instead of being homogeneous). Specifying such a covariate makes the alternative hypothesis more precise, allowing the analyst to choose a more powerful test. Covariate-dependent quadrat tests are discussed in Sections 6.7.1 and 10.4. Other covariate-dependent tests are introduced in Sections 6.7 and 10.5.

The χ^2 test is simple to apply, but it is not necessarily the best test to apply to the quadrat counts. Other options are the likelihood ratio test and the Cressie-Read [566] divergence family, which can be selected in quadrat.test using the argument CR. The values CR=1, CR=0, and CR=-1/2 correspond to the χ^2 test, likelihood ratio test, and Freeman-Tukey test, respectively.

6.4.3 Critique

Since this technique is often used in the applied literature, a few comments are appropriate.

The main critique of the quadrat test described above is the lack of information. This is a goodness-of-fit test in which the alternative hypothesis H_1 is simply the negation of H_0, that is, the alternative is that 'the process is not a homogeneous Poisson process'. A point process may fail to be a homogeneous Poisson process either because it fails to have homogeneous intensity, or because it violates the property of independence between points. There are too many types of departure from H_0.

The usual justification for the classical χ^2 goodness-of-fit test is to assume that the counts are independent, and derive a test of the null hypothesis that all counts have the same expected value. Invoking it here is slightly naive, since the independence of counts is also open to question.

Indeed we can also turn things around and view the χ^2 test as a test of the independence property of the Poisson process, assuming that the intensity is homogeneous. The Pearson χ^2 test statistic (6.6) coincides, up to a constant factor, with the sample variance-to-mean ratio of the counts n_j, which is often interpreted as a measure of over-/underdispersion of the counts n_j assuming they have constant mean.

The power of the quadrat test depends on the size of quadrats, and is optimal when the quadrats are neither very large nor very small. The power also depends on the alternative hypothesis, in particular on the 'spatial scale' of any departures from the assumptions of constant intensity and independence of points. The choice of quadrat size is also an implicit choice of spatial scale, because the data points are aggregated at this scale. This is well understood in classic literature in ecology where it is recommended to compute the variance-to-mean ratio or χ^2 statistic for different sizes of quadrats, and to plot the statistic against quadrat size [296, 462]. We return to the topic of spatial scale in Chapter 7.

6.5　Smoothing estimation of intensity function

6.5.1　Kernel estimation

If the point process has an intensity function $\lambda(u)$, this function can be estimated nonparametrically by *kernel estimation*.

Our favorite analogy is to imagine placing one square of chocolate on each data point. Using a hair dryer we apply heat to the chocolate so that it melts slightly. The result is an undulating surface of chocolate; the height of the surface represents the estimated intensity function of the point process. The total mass of chocolate is unchanged.

6.5.1.1　Kernel estimators

The usual *kernel estimators* of the intensity function [222, 106] are:

$$\text{uncorrected:}\quad \widetilde{\lambda}^{(0)}(u) = \sum_{i=1}^{n} \kappa(u - x_i), \tag{6.7}$$

$$\text{uniformly corrected:}\quad \widetilde{\lambda}^{(U)}(u) = \frac{1}{e(u)} \sum_{i=1}^{n} \kappa(u - x_i), \tag{6.8}$$

$$\text{Diggle's [222] correction:}\quad \widetilde{\lambda}^{(D)}(u) = \sum_{i=1}^{n} \frac{1}{e(x_i)} \kappa(u - x_i), \tag{6.9}$$

for any spatial location u inside the window W, where $\kappa(u)$ is the *kernel function* (the shape of one melted square of chocolate) and

$$e(u) = \int_{W} \kappa(u - v)\, dv \tag{6.10}$$

is a correction for bias due to edge effects. Outside the window W, the estimated intensity is zero.

For a data point at location x_i, the function $f(u) = \kappa(u - x_i)$ represents the melted square of chocolate that was originally placed at x_i. The kernel κ must be a probability density, that is, $\kappa(u) \geq 0$ for all locations u, and $\int_{\mathbb{R}^2} \kappa(u)\, du = 1$. A common choice is the isotropic Gaussian (Normal distribution) probability density. The standard deviation of the kernel is the *smoothing bandwidth*: a larger bandwidth gives more smoothing. The choice of bandwidth involves a tradeoff between bias and variance: as bandwidth increases, typically the bias increases and variance decreases.

Figure 6.9 shows the different kernel estimates (6.7)–(6.9) for the Swedish Pines data using the same kernel. Note that the 'raw' or 'uncorrected' estimate (6.7) decreases close to the boundary of the observation window. This is an *edge effect*. Trees lying just outside the window were not observed, and did not contribute to the estimate (6.7), so that locations u closer to the boundary of the window receive fewer contributions in the sum (6.7). The raw estimate therefore has a strong negative bias at locations close to the boundary, due to edge effects. The raw estimate should be used only in those rare situations where there are no edge effects — for example, in mapping the density of an isolated stand of trees, where every tree in the stand is represented in the data.

The uniform correction (6.8) and Diggle's correction (6.9) are designed to compensate for the

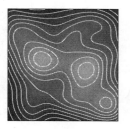 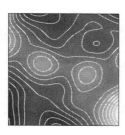 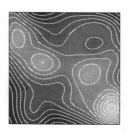 

Figure 6.9. *Kernel estimates of intensity for Swedish Pines using different edge corrections.* Left: *raw estimate.* Middle: *uniformly corrected estimate.* Right: *Diggle's correction estimate. Isotropic Gaussian kernel with standard deviation* 1 *metre. Intensity values are counts per square metre.*

edge effect arising when a point process is observed inside a window. The uniformly corrected estimator (6.8) is unbiased when the true intensity is homogeneous. Diggle's corrected estimator (6.9) has better performance overall (smaller mean square error) and is normalised so that the integral of $\widetilde{\lambda}^{(D)}(u)$ over the window is exactly equal to the observed number of points.

Kernel estimators of the intensity function are slightly biased in general, because they smooth out details in the intensity function. To understand their statistical properties we can use **Campbell's formula**. Suppose $f(u)$ is a function of spatial location u, and consider the random sum

$$T = \sum_i f(x_i)$$

of the values of f at each of the points x_i in a point process **X**. Campbell's formula states that the expected value $\mathbb{E}[T]$ of the random sum T is

$$\mathbb{E}[\sum_i f(x_i)] = \int_{\mathbb{R}^2} f(u)\lambda(u)\,du \tag{6.11}$$

where $\lambda(u)$ is the intensity function of **X**. This can be justified by dividing space into pixels and considering the contribution to T from each pixel.

To apply Campbell's formula to kernel estimation, suppose we fix a spatial location v, and let $f(u) = \kappa(v - u)$ if u is inside the window W, and $f(u) = 0$ if it is outside. Then

$$\sum_i f(x_i) = \sum_i \kappa(v - x_i) = \widetilde{\lambda}^{(0)}(v)$$

where the sum is over all points x_i in the window W. By Campbell's formula (6.11)

$$\mathbb{E}[\widetilde{\lambda}^{(0)}(v)] = \mathbb{E}[\sum_i f(x_i)] = \int f(u)\lambda(u)\,du = \int_W \kappa(v - u)\lambda(u)\,du.$$

The expected value of the estimate $\widetilde{\lambda}^{(0)}(v)$ is not equal to the true intensity value $\lambda(v)$. Even if the true intensity is constant, say $\lambda(v) \equiv \beta$ for all locations v, we get

$$\mathbb{E}[\widetilde{\lambda}^{(0)}(v)] = \int_W \kappa(v - u)\beta\,du = \beta \int_W \kappa(v - u)\,du = \beta\,e(v)$$

where $e(v)$ was defined in (6.10). This motivates the definition of the corrected estimator $\widetilde{\lambda}^{(U)}(u)$ which is then unbiased at least when the intensity is constant: if $\lambda(v) \equiv \lambda$, then $\mathbb{E}\widetilde{\lambda}^{(U)}(v) = \lambda$. However, in general, $\widetilde{\lambda}^{(U)}(v)$ is a biased estimator of $\lambda(u)$; the expected value of $\widetilde{\lambda}^{(U)}(v)$ is

$$\lambda^*(v) = \frac{1}{e(v)} \int_W \kappa(v - u)\lambda(u)\,du,$$

a smoothed version of the true intensity function $\lambda(v)$.

Kernel estimation is implemented in `spatstat` by the function `density.ppp`, a method for the generic command `density`.

```
> den <- density(swp, sigma=1)
```

The smoothing bandwidth is specified by the argument `sigma`. This may be a single numerical value (specifying the bandwidth of the kernel in the same units as the point pattern), or a pair of numerical values (specifying different standard deviations in the x and y directions), or a function which performs automatic bandwidth selection (see below). Currently the smoothing kernel is an isotropic Gaussian density; other options will be added soon.

By default, the uniformly corrected kernel estimator (6.8) is calculated. For Diggle's corrected estimator (6.9), specify `diggle=TRUE`: as discussed above, this has better statistical performance, but is slower to compute. For the uncorrected estimator (6.7), set `edge=FALSE`.

The value returned by `density.ppp` is a pixel image (object of class `"im"`). This class has methods for `print`, `summary`, `plot`, `contour` (contour plots), `persp` (perspective plots), and so on. The method for `plot` was used to produce Figure 6.9. The methods for `persp` and `contour` produced Figure 6.10.

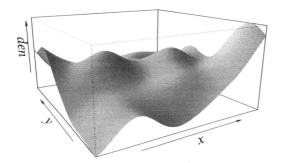

Figure 6.10. *Perspective plot* (Left) *and contour plot* (Right) *for a kernel estimate of intensity in the Swedish Pines data. Plots were generated using* `persp.im` *and* `contour.im`, *respectively.*

6.5.1.2   Bandwidth selection

The kernel bandwidth `sigma` controls the degree of smoothing (the amount of 'melting' in the chocolate analogy). As shown in Figure 6.11, a small value of `sigma` produces an irregular intensity surface, while a large value of `sigma` appears to oversmooth the intensity.

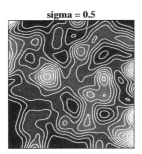

Figure 6.11. *Density estimates with different smoothing bandwidths.*

If the bandwidth `sigma` is not specified in the call to `density.ppp`, the default is to take `sigma` equal to one-eighth of the shortest side length of the enclosing rectangle. This is a very rough rule of thumb which may be unsatisfactory in many cases.

Several algorithms are available for automatically selecting the bandwidth `sigma` by minimising a measure of error. They include `bw.diggle` for Diggle and Berman's [222, 89] mean square error cross-validation method and `bw.ppl` for the likelihood cross-validation method [428, Sect. 5.3].

```
> b <- bw.ppl(swp)
> b
sigma
4.036
```

These commands return a numerical value, the optimised bandwidth, which also belongs to the special class `"bw.optim"`. The `plot` method for this class shows the objective function for the optimisation. Figure 6.12 shows the results of plotting the likelihood cross-validation value using `plot(b)` and zooming in using `plot(b, xlim=c(3,6))`. The first plot suggests that any smoothing bandwidth greater than about 2 metres would be adequate. This is what might be expected for a homogeneous point pattern.

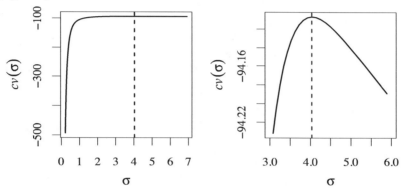

Figure 6.12. *Likelihood cross-validation criterion for smoothing bandwidth plotted against bandwidth σ in metres. Right panel is zoomed in to the range $3 \leq \sigma \leq 6$.*

Bandwidth selection may also be based on a fast rule of thumb. Examples include `bw.scott` for Scott's rule of thumb for bandwidth selection in multidimensional smoothing [608, p. 152], and `bw.frac` for a fast bandwidth selection rule based on the window geometry (explained in the `spatstat` help file).

Bandwidth selection commands can be invoked in either of the following ways:

```
> D <- density(swp, sigma=bw.diggle(swp))
> D <- density(swp, sigma=bw.diggle)
```

Different bandwidth selection methods can disagree substantially. For the Swedish Pines data, `bw.diggle` gives 0.571 metres, while `bw.ppl` gives 4.38 and `bw.scott` gives $(1.41, 1.32)$.

Any bandwidth selection rule gives unsatisfactory results in some cases, because it is based on assumptions about the dependence between points, which may be inappropriate. Likelihood cross-validation `bw.ppl` assumes an inhomogeneous Poisson process; `bw.diggle` assumes a Cox process, which is more clustered (positively correlated) than a Poisson process; both these assumptions are probably inappropriate for the Swedish Pines data which are somewhat more regular (negatively correlated) than a Poisson process. It is often convenient to be able to adjust the automatically selected bandwidth by specifying the argument `adjust`, a numeric value which multiplies the selected bandwidth:

```
> D <- density(swp, sigma=bw.diggle, adjust=2)
```

This is equivalent to selecting the bandwidth by `b <- bw.diggle(swp)` then computing `D <- density(swp, sigma=2*b)`.

6.5.1.3 Estimation of intensity at the data points

It is sometimes required to estimate the intensity values $\lambda(x_i)$ at the data points x_i themselves. For example, the estimated intensity values at the data points can be used as weights in some analysis procedures. However, the estimates $\widetilde{\lambda}^{(U)}(x_i)$ and $\widetilde{\lambda}^{(D)}(x_i)$ have a large positive bias, because of the term $\kappa(u - x_i) = \kappa(x_i - x_i) = \kappa(0)$ appearing in the sum in (6.8)–(6.9). To deal with this problem it is advisable to use a *leave-one-out estimator* in which the value of $\lambda(x_i)$ is estimated using all of the data points *except* x_i:

$$\widetilde{\lambda}_{-i}^{(U)}(x_i) = \frac{1}{e(x_i)} \sum_{j \neq i} \kappa(x_i - x_j) \tag{6.12}$$

$$\widetilde{\lambda}_{-i}^{(D)}(x_i) = \sum_{j \neq i} \frac{1}{e(x_j)} \kappa(x_i - x_j). \tag{6.13}$$

Typically the leave-one-out estimates have a slight negative bias.

To compute intensity estimates at the data points, invoke `density.ppp` with the argument `at="points"`. The result is a numeric vector of density values for each data point. The default is to compute the leave-one-out estimates; this can be suppressed by setting `leaveoneout=FALSE`.

```
> dX <- density(swp, sigma=1, at="points")
> dX[1:5]
[1] 0.3750 0.7880 0.6397 0.6144 0.3938
```

6.5.1.4 Computation

The `spatstat` package uses different algorithms to compute the intensity estimates at data points and on a pixel grid. The intensity estimates at the data points are computed to high precision using the formulae (6.8)–(6.9) or (6.12)–(6.13) in double precision arithmetic. For the intensity estimates on a pixel grid, exact calculation would be too slow, so the pixel values are computed by spatially discretising the point pattern and convolving using the Fast Fourier Transform [171]. Thus, the following are approximately but not exactly equal:

```
> den <- density(swp, sigma=1)
> denXpixel <- den[swp]
> denXpixel[1:5]
[1] 0.9177 1.1003 0.9126 1.0252 0.6044
> denXexact <- density(swp, sigma=1, at="points", leaveoneout=FALSE)
> denXexact[1:5]
[1] 0.9211 1.0836 0.9145 0.9051 0.6038
```

6.5.1.5 Standard errors

To compute standard errors and confidence intervals for the intensity function, additional assumptions are required. For example, assume a Poisson point process with intensity function $\lambda(u)$, and estimate the intensity by a kernel estimator of the general form

$$\widehat{\lambda}(u) = a(u) \sum_i b(x_i) \kappa(x_i - u) \tag{6.14}$$

where $a(u)$ and $b(x_i)$ are edge correction weights, embracing the three edge corrected estimators (6.7)–(6.9). Then the variance of $\widehat{\lambda}(u)$ is, for a Poisson process only [197, p. 188],

$$V(u) = \operatorname{var} \widehat{\lambda}(u) = a(u)^2 \int_W b(v)^2 \kappa(u-v)^2 \lambda(v)\,dv. \qquad (6.15)$$

An unbiased consistent estimator of $V(u)$ is

$$\widehat{V}(u) = a(u)^2 \sum_i b(x_i)^2 \kappa(u-x_i)^2. \qquad (6.16)$$

This takes the form of a weighted kernel estimate of intensity. If $\kappa(x) = \kappa_\sigma(x)$ is the isotropic Gaussian kernel with standard deviation σ, then a little algebra shows that $\kappa_\sigma(x)^2 = c\,\kappa_\tau(x)$ where $\tau = \sigma/\sqrt{2}$ and $c = 1/(8\pi\tau^2) = 1/(4\pi\sigma^2)$. That is, the variance $V(u)$ can effectively be estimated by smoothing the data with bandwidth $\tau = \sigma/\sqrt{2}$ and multiplying the result by c. Taking the square root gives the standard error for the intensity estimate. This calculation is performed by `density.ppp` when `se=TRUE`:

```
> dse <- density(swp, 1, se=TRUE)$SE
```

The result is shown in Figure 6.13: note the standard error increases near the boundary, because intensity estimates nearer the boundary are based on fewer data points. Similar calculations can be made when `at="points"`.

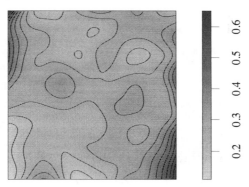

Figure 6.13. *Estimate of standard error for the kernel estimate of intensity for Swedish Pines. Uniform edge correction, bandwidth 1 metre.*

Be warned that, although the standard error provides an indication of accuracy, and is justified by asymptotic theory, confidence intervals based on the standard error are notoriously unreliable [315], essentially because the estimates $\widehat{\lambda}(u)$ and $\widehat{V}(u)$ are strongly correlated.

6.5.1.6 Weighted kernel estimators

If the data points x_i have numerical weights w_i, we can use weighted versions of the kernel estimators described above. The contribution from a point x_i to the estimator is simply multiplied by the weight w_i, so that (for example) the raw intensity estimator $\widehat{\lambda}^{(0)}(u) = \sum_i \kappa(u-x_i)$ becomes $\widehat{\lambda}^{(0,w)}(u) = \sum_i w_i \kappa(u-x_i)$. Using the chocolate analogy (page 168) the data point x_i is represented by w_i units of chocolate rather than one unit.

Weighted kernel estimators are natural if the weight of a point represents its multiplicity (e.g., number of disease cases at the same residence) or physical mass (e.g., mass of galaxy) or economic value (e.g., total endowment of a mineral deposit).

For example, in a forest inventory we could take the 'weight' of each tree to be its estimated

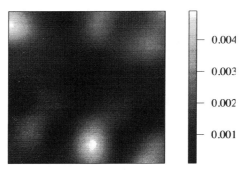

Figure 6.14. *Volume-weighted intensity for Finnish Pines data.*

volume. The volume-weighted intensity is the average *standing volume of wood* per unit area of forest. Figure 6.14 shows this quantity for the Finnish Pines data. The scale is in metres (cubic metres of wood per square metre of forest).

The argument `weights` to `density.ppp` specifies weights for the density calculation. Figure 6.14 was generated by

```
> vols <- with(marks(finpines),
            (pi/12) * height * (diameter/100)^2)
> Dvol <- density(finpines, weights=vols, sigma=bw.ppl)
```

The average standing volume of wood per square metre was calculated at the end of Section 6.2: it is

```
> intensity(finpines, weights=vols)
[1] 0.001274
```

6.5.2   Spatially adaptive smoothing

The kernel estimators described above are *fixed-bandwidth* smoothers: they use the same kernel and the same bandwith to compute estimates at different spatial locations. This approach has several weaknesses. A fixed smoothing bandwidth is unsatisfactory if the true intensity varies greatly across the spatial domain, because it is likely to cause oversmoothing in the high-intensity areas and undersmoothing in the low intensity areas. Kernel estimation is unsatisfactory when there is a sharp boundary between areas of high and low intensity, because this boundary will be smoothed out. These problems militate against the use of kernel estimation in seismology, for example.

Strategies for avoiding this problem include *variable-bandwidth smoothing* where the smoothing bandwidth is spatially varying and data-dependent [617, 203], [190, p. 654], and more generally *adaptive smoothing*. The contributed R package `sparr` [202] provides a suite of adaptive kernel spatial smoothing techniques and related tools.

Adaptive estimators of intensity can be based on Dirichlet-Voronoï tessellations (Section 8.2.3). The Dirichlet-Voronoï estimator [81] of intensity $\lambda(u)$ at a location u is $\widetilde{\lambda}(u) = 1/|C(u;\mathbf{x})|$, the reciprocal of the area of the tile $C(u;\mathbf{x})$ containing u in the Dirichlet-Voronoï tessellation defined by the data point pattern $\mathbf{x}$. Estimators of this type have been used in statistical seismology [509] and perform well when there is an abrupt change in intensity. The Dirichlet-Voronoï estimator is computed in `spatstat` by the function `adaptive.density` with argument `f=1`.

```
> vden <- adaptive.density(swp, f=1)
```

The value returned by `adaptive.density` is another pixel image (object of class `"im"`).

The algorithm in `adaptive.density` is more general. A specified fraction f of the points in the point pattern are selected at random, and used to construct a Dirichlet tessellation. A quadrat counting estimator of the intensity is based on this tessellation. This process is repeated `nrep` times and the results are averaged. The left panel of Figure 6.15 shows the result of

```
> aden <- adaptive.density(swp, f=0.1, nrep=30)
```

Another strategy is to measure the distance $R = d(u, \mathbf{x})$ from a fixed point u to the nearest data point x_i, and to compute the area $A = \pi R^2$ of the corresponding disc. For a homogeneous Poisson process with intensity λ, the random area A is negative exponential (λ) distributed, and the maximum likelihood estimate of λ based on R is $\hat{\lambda} = 1/(\pi R^2)$. Similarly we could use the distance R_k to the k-th nearest data point (for $k \geq 1$) and set $\hat{\lambda}_k = k/(\pi R_k^2)$. See [617, p. 96], [190, p. 654]. This intensity estimator can be calculated rapidly for all points u in a pixel grid: the spatstat function `nndensity` computes it. The right panel of Figure 6.15 shows the result of `nndensity(swp, k=10)`.

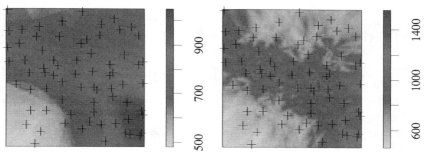

Figure 6.15. *Dirichlet-type adaptive density estimate* (Left) *and 10th nearest-neighbour density estimate* (Right) *for the Swedish Pines data. Density values multiplied by 1000.*

Intensity can also be estimated using nearest-neighbour distances [543, 216, 573, 575, 137] and similar principles [613]. These are often used as the plug-in estimates of λ in other statistics related to interpoint distances, such as Ripley's K-function (Chapter 7), or nearest-neighbour distances, and the G-function (Chapter 8). Bayesian estimation of the intensity is described in [332, 77].

6.5.3 Projections, transformations, change of coordinates

In Section 6.2 and 6.3 we saw that the intensity depends on the unit of length. In fact, changing the spatial coordinate system in any way — through a change of units, a change of scale, a geometric transformation, or a geographic projection — affects the intensity.

Intensity is the expected number of points per unit *area*, so any geometric transformation which changes the value of area also changes the value and the very *meaning* of the intensity.

The intensity function of a point process after a spatial transformation has been applied is related to the intensity function of the original point process, through the general principle of 'change of coordinates'.

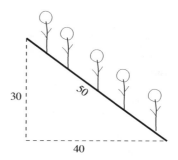

Figure 6.16. *Illustration of change of coordinates for intensity.*

We motivate this with an example. Imagine a hillside, covered in trees: see Figure 6.16. Walking straight up the hill involves a walking distance of 50 metres and an increase in altitude of 30 metres, so using trigonometry, the horizontal distance travelled is 40 metres. Suppose the hillside is 100 metres wide. The total surface area *of the hillside* is $50 \times 100 = 5000$ square metres or half a hectare. If there are 800 trees on the hillside, then the estimated intensity is $800/0.5 = 1600$ trees per hectare *of hillside surface*. However, if we photograph the hillside from directly above in a survey aircraft, or map the tree locations using a GPS device, the hillside is represented by a rectangle only $40 \times 100 = 4000$ square metres or 0.4 hectare in area on the map, and the estimated intensity is $800/0.4 = 2000$ trees per hectare *of map area*.

Which of these calculations is 'right'? Actually, both are correct. The 'forest density' could be defined either as a density per hectare of soil (which might be more useful for understanding soil ecology) or as a density per hectare of map area (perhaps more useful for understanding competition in the forest canopy).

The key is that the two measures of intensity are inter-related. Map area is equal to hillside surface area multiplied by the cosine of the slope angle, which is $40/50 = 0.8$ in the example above. Therefore the intensity per unit area of map is equal to the intensity per unit area of hillside, *divided* by the cosine of the slope angle.

The general principle is the following. Suppose that a point process $\mathbf{X}$ has intensity function $\lambda(u)$. We now apply a geographic projection, a geometrical transformation, or a change of the coordinate system, so that the points x_i are mapped to new coordinate positions $y_i = T(x_i)$. The transformed point process $\mathbf{Y} = T(\mathbf{X})$ has intensity function

$$\lambda_{\mathbf{Y}}(u) = J(u)\,\lambda_{\mathbf{X}}(T^{-1}(u)) \tag{6.17}$$

where T^{-1} is the inverse mapping (that is, $T^{-1}(u)$ is the point mapped onto u), and $J(u)$ is the *Jacobian* (determinant of the derivative matrix) of the inverse mapping.

Invoking this general principle, we can apply the same simple trigonometry to real terrain data where the slope is spatially varying. The tropical rainforest point pattern dataset `bei` comes with an extra set of covariate data `bei.extra`, which contains a pixel image of terrain elevation `bei.extra$elev` and a pixel image of terrain slope `bei.extra$grad` (for 'gradient'), at a coarse spatial resolution of 5 metres. The command `density(bei)` computes the estimated intensity of trees relative to map area. To convert this to an estimate of the intensity relative to terrain surface area, we need the cosine of the slope angle. The covariate `grad` is given as the number of metres of elevation increase for every metre on the map, so that a `grad` value of 1 corresponds to a 45 degree slope. That is, `grad` is the tangent of the slope angle. Recalling our high school trigonometry, $\cos^2(x) = 1/(1 + \tan^2(x))$, so the conversion is:

```
> grad <- bei.extra$grad
> dens.map <- density(bei, W=grad)
> dens.ter <- dens.map * sqrt(1+grad^2)
```

The two estimates are not very different in this case, because the maximum value of `grad` is only 0.33, so that the maximum inflation factor is only $\sqrt{(1 + 0.33^2)} = 1.05$.

Figure 6.17 shows a perspective view of the rainforest terrain, shaded according to the estimated density of *Beilschmiedia* trees, using

```
> persp(bei.extra$elev, colin=dens.ter)
```

An alternative to the calculation above is to introduce the Jacobian weight into the kernel smoother. That is, we smooth the point pattern in the projected space, but weight each data point by the Jacobian term at that point:

```
> dens.ter2 <- density(bei, weights=sqrt(1+grad[bei]^2))
```

This is justified by Campbell's formula (page 169). An advantage of this approach is that we only need to know the Jacobian values at the data points.

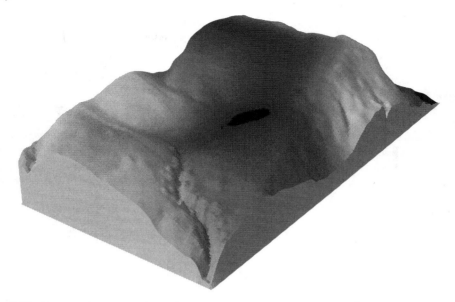

Figure 6.17. *Perspective view of rainforest terrain, shaded according to the estimated density of* Beilschmiedia *trees per unit area of soil. Lighter shades represent higher predicted densities. Scale on vertical axis is 6 times the scale on horizontal axes.*

6.6 Investigating dependence of intensity on a covariate

6.6.1 Spatial covariates

Often we want to know how the intensity of points depends on the values of a covariate. For example, for the *Beilschmiedia* data mentioned above, it is of interest to determine whether the trees prefer steep or flat terrain, and whether they prefer a particular altitude. Other applications include spatial epidemiology (e.g., disease risk as a function of environmental exposure [545]), spatial ecology (e.g., habitat preferences of organisms [439]), exploration geology (e.g., prospectivity of mineral deposits predicted from survey data [116]), and seismology.

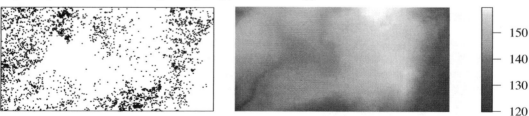

Figure 6.18. Beilschmiedia *data. Trees* (left) *and terrain elevation* (right) *in a 1000 by 500 metre research plot.*

6.6.2 Quadrats determined by a covariate

In quadrat counting methods, any choice of quadrats is permissible. From a theoretical viewpoint, the quadrats do not have to be rectangles of equal area, and could be regions of any shape.

Quadrat counting is more useful if we choose the quadrats in a meaningful way. One way to do this is to define the quadrats using covariate information.

For the tropical rainforest data `bei`, it might be useful to split the study region into several sub-regions according to the terrain elevation:

```
> elev <- bei.extra$elev
> b <- quantile(elev, probs=(0:4)/4, type=2)
> Zcut <- cut(elev, breaks=b, labels=1:4)
> V <- tess(image=Zcut)
```

The call to `quantile` gave us the quartiles of the elevation values, so the four tiles in the tessellation V have equal area (ignoring discretisation effects). In other words, we have divided the study region into four zones of equal area according to the terrain elevation. (The same calculation can be performed using the function `quantess`.) The resulting tessellation is shown in Figure 6.19 using `textureplot`.

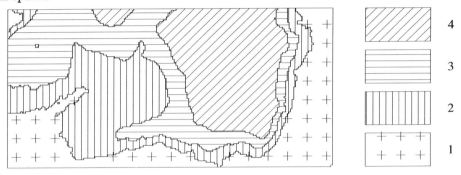

Figure 6.19. *Equal-area tessellation corresponding to the quartiles of terrain elevation in the* Beilschmiedia *dataset.*

We can now use this tessellation to study the point pattern `bei`. We could invoke the commands `split`, `cut`, or `by` to divide the points according to this tessellation and manipulate the sub-patterns. The command `quadratcount` also works with tessellations:

```
> qb <- quadratcount(bei, tess=V)
> qb
tile
   1    2    3    4
 714  883 1344  663
```

The output shows the number of trees in each region. Since the four regions have equal area, the counts should be approximately equal if there is a uniform density of trees. Obviously they are not equal; there appears to be a strong preference for higher elevations (dropping off for the highest elevations).

In the following calculation we divide the range of elevations into intervals of equal width, estimate the average intensity for each interval, and plot the result as a bar chart.

```
> b5 <- seq(0, 5 * ceiling(max(elev)/5), by=5)
> Zcut5 <- cut(elev, breaks=b5, include.lowest=TRUE)
> Q5 <- quadratcount(bei, tess=tess(image=Zcut5))
> lam5 <- intensity(Q5)
```

The result of `barplot(lam5)` is shown in Figure 6.20.

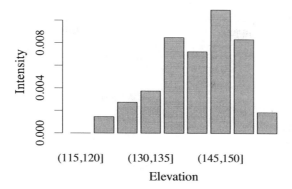

Figure 6.20. *Estimates of intensity of* Beilschmiedia *trees in each 5-metre band of terrain elevation.*

6.6.3 Relative distribution estimate

In the previous analysis we were effectively assuming that the intensity of the point process is a *function* of the covariate Z. At any spatial location u, let $\lambda(u)$ be the intensity of the point process, and $Z(u)$ the value of the covariate. Then we are assuming

$$\lambda(u) = \rho(Z(u)) \tag{6.18}$$

where ρ is a function that we want to investigate, telling us how the intensity of points depends on the value of the covariate.

In ecological applications where the points are the locations of individual organisms, ρ is a *resource selection function* [439] reflecting preference for particular environmental conditions Z. In geological applications where the points are the locations of valuable mineral deposits, ρ is an index of the *prospectivity* [116] or predicted frequency of undiscovered deposits as a function of geological and geochemical covariates Z.

Estimation of ρ

Nonparametric estimation of ρ is closely connected to estimation of a probability density from biased sample data [367, 251] and to the estimation of relative densities [318]. Under regularity conditions, ρ is proportional to the ratio of two probability densities, the numerator being the density of covariate values at the points of the point process, while the denominator is the density of covariate values at random locations in space. Kernel smoothing can be used to estimate the function ρ as a relative density [37, 301].

For a numerical covariate Z, the important tool is the *spatial distribution function*, the cumulative distribution function of the covariate value $Z(U)$ at a random point U uniformly distributed in W:

$$G(z) = \frac{1}{|W|} \int_W \mathbf{1}\{Z(u) \le z\} \, du. \tag{6.19}$$

Here we use the 'indicator' notation: $\mathbf{1}\{\dots\}$ equals 1 if the statement '...' is true, and 0 if the statement is false. Equivalently $G(z) = |W_z|/|W|$ where $W_z = \{u \in W : Z(u) \le z\}$ is the level set consisting of all locations in W where the covariate value is less than or equal to z. In practice $G(z)$ would often be estimated by evaluating the covariate at a fine grid of pixel locations, and forming the cumulative distribution function

$$G(z) = \frac{\#\{\text{pixels } u : Z(u) \le z\}}{\#\text{pixels}}. \tag{6.20}$$

For a numerical covariate Z, three estimators of ρ are

$$\text{ratio:} \quad \hat{\rho}_{R}(z) = \frac{1}{|W| G'(z)} \sum_{i} \kappa(Z(x_i) - z) \tag{6.21}$$

$$\text{reweighting:} \quad \hat{\rho}_{W}(z) = \sum_{i} \frac{1}{|W| G'(Z(x_i))} \kappa(Z(x_i) - z) \tag{6.22}$$

$$\text{transformation:} \quad \hat{\rho}_{T}(z) = \frac{1}{|W|} \sum_{i} \kappa(G(Z(x_i)) - G(z)) \tag{6.23}$$

where $x_1, \ldots, x_n$ are the data points, $Z(x_i)$ are the observed values of the covariate Z at the data points, $|W|$ is the area of the observation window W, and κ is a *one-dimensional* smoothing kernel — smoothing is conducted on the observed values $Z(x_i)$ rather than in the window W. The derivative $G'(z)$ is usually approximated by differentiating a smoothed estimate of G.

The estimators (6.21)–(6.23) were developed in [37] by adapting estimators from kernel smoothing [367, 251]. An estimator similar to (6.21) was proposed in [301].

Implementation in `spatstat`

In `spatstat` these estimators are computed by the command `rhohat`.

```
> rh <- rhohat(bei, elev)
```

Extra arguments to `rhohat` control the choice of estimator and the spatial resolution. As an alternative to kernel smoothing, local likelihood smoothing [429] can also be used.

The result of `rhohat` is an object belonging to the special class `"rhohat"` which represents the estimated function ρ together with additional information. The `print` method (not shown here) gives detailed information about the smoothing technique and the results that have been calculated.

The `plot` method generates a plot of the estimated function $\hat{\rho}(z)$ against covariate values z, together with 95% confidence bands assuming an inhomogeneous Poisson point process. It is shown in the left panel of Figure 6.21. The plot indicates that the *Beilschmiedia* trees are more likely to be found at elevations between 135 and 155 metres than would be expected if the intensity was constant.

There is also a method for the generic function `predict`. This computes the predicted intensity $\hat{\lambda}(u) = \hat{\rho}(Z(u))$ at each spatial location u, and returns it as a pixel image. The result of `predict(rh)` is plotted in the right panel of Figure 6.21.

An object of class `"rhohat"` can be converted to a function in the R language by `as.function`. To obtain the predicted intensity at terrain elevation 130 metres:

```
> rhf <- as.function(rh)
> rhf(130)
[1] 0.00325
```

The estimated function values can also be extracted as a data frame using `as.data.frame`.

Validation

This analysis assumes that the intensity at a location u depends *only* on the covariate value $Z(u)$. If this is not true, $\hat{\rho}(z)$ is still meaningful: it is effectively an estimate of the average intensity $\lambda(u)$ over all locations u where $Z(u) = z$.

To validate the assumption (6.18) we can compare the predicted intensity $\hat{\rho}(Z(u))$ assuming (6.18) with a (spatial) kernel estimate $\hat{\lambda}(u)$ which does not assume (6.18).

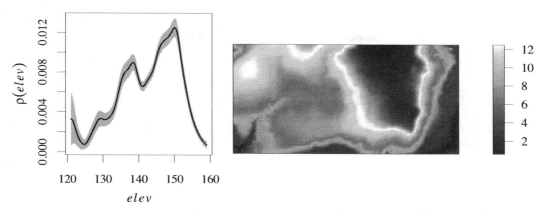

Figure 6.21. *Intensity as a function of terrain elevation for the* Beilschmiedia *data.* Left: *estimated function* $\widehat{\rho}(z)$ *giving forest density as a function of terrain elevation. Solid lines show function estimate; grey shading is pointwise 95% confidence band.* Right: *predicted forest density* $\widehat{\lambda}(u) = \widehat{\rho}(Z(u))$ *(scaled* $\times 1000$*) at each spatial location u.*

```
> pred <- predict(rh)
> kden <- density(bei, 50)
```

A scatterplot of the two estimates at corresponding pixels can be generated by `pairs(pred, kden)`; the scatterplot should concentrate around the diagonal under the assumption. The difference of estimates at corresponding pixels can be plotted as an image using `plot(eval.im(kden - pred))`; the difference should be roughly equal to 0 everwhere. For the *Beilschmiedia* data with terrain elevation as the covariate, these graphics suggest that (6.18) is a reasonable approximation. By contrast, a similar exercise performed for the terrain *slope* covariate suggests that forest density is not simply a function of terrain slope.

Baseline

In some circumstances there is a natural 'baseline' intensity function $B(u)$ such that the *relative intensity* $\lambda(u)/B(u)$ can be assumed to depend only on the covariate Z. That is, we assume

$$\lambda(u) = \rho(Z(u))B(u) \tag{6.24}$$

instead of (6.18). The interpretation of the function $\rho(z)$ is different in this case: values of $\rho(z)$ are dimensionless, and the value $\rho(z) = 1$ corresponds to the baseline intensity. The same estimators (6.21)–(6.23) can be used, provided we replace $|W|$ and $G(z)$ by their B-weighted counterparts

$$W_B = \int_W B(u)\,\mathrm{d}u \quad \text{and} \quad G_B(z) = \frac{1}{W_B}\int_W \mathbf{1}\{Z(u) \le z\}B(u)\,\mathrm{d}u,$$

respectively. Another strategy for correcting sampling bias, originating from sample surveys, is the *Horvitz-Thompson* estimator [341] in which each item in the sample is weighted by the reciprocal of its sampling probability. This approach can be applied in spatial statistics [58, 173]. In the Horvitz-Thompson estimator of ρ, the contribution from each data point x_i to the estimators (6.21)–(6.23) is weighted by a factor $1/B(x_i)$. Then the denominators $|W|$ and $G(z)$ are unchanged.

To compute the modified estimators in `spatstat`, the baseline $B(u)$ should be given as the argument `baseline` to `rhohat`. The baseline may be either a `function(x,y)` or a pixel image. Horvitz-Thompson weighting is selected by setting `horvitz=TRUE`.

Two covariates

If there are two numerical covariates $Z_1(u), Z_2(u)$ assumed to determine the intensity together, that is $\lambda(u) = \rho(Z_1(u), Z_2(u))$, then similar techniques can be applied [37] to estimate the function $\rho(z_1, z_2)$. These are implemented in the spatstat function rho2hat. For example

```
> with(bei.extra, rho2hat(bei, grad, elev))
```

would compute an estimate of the intensity of *Beilschmiedia* trees as a function jointly of the terrain slope and terrain elevation.

6.6.4   Distance map

One particularly important kind of spatial covariate is a distance function. The dataset copper is described in Section 1.1.4. Figure 6.22 shows the southern half of this dataset, rotated by 90 degrees, computed by the two commands

```
> X <- rotate(copper$SouthPoints, pi/2)
> L <- rotate(copper$SouthLines, pi/2)
```

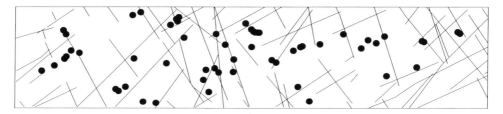

Figure 6.22. *Queensland copper data, southern half.*

To apply the methods described above, the covariate information contained in the map of geological faults L must be converted into a covariate that is a function $Z(u)$ of spatial location u. A natural choice is the *distance function* $Z(u) =$ distance from u to L. This can be computed by the command distmap, which returns a pixel image containing the values $Z(u)$ at a fine grid of pixels u. Figure 6.23 shows a contour plot of Z <- distmap(L). A pixel image of Z is shown in Figure 5.23 on page 148.

Figure 6.23. *Contour plot of distance map of geological lineaments in Queensland copper data, southern half.*

Having created this covariate image we can now apply the other techniques such as relative distributions. Figure 6.24 shows the nonparametric estimate rhohat(X, Z) of the intensity of copper deposits as a function of distance to the nearest lineament.

A slightly more sophisticated version of distmap is the command distfun. Whereas distmap returns a pixel image at a certain spatial resolution, distfun returns a function with arguments (x,y) that can be evaluated at any spatial location.

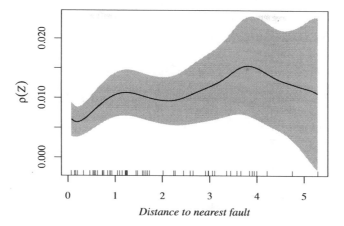

Figure 6.24. *Nonparametric estimate* (6.21) *of intensity of copper deposits as a function of distance to the nearest lineament.*

```
> f <- distfun(L)
> f
Distance function for line segment pattern
 planar line segment pattern: 90 line segments
window: rectangle = [-158.23, -0.19] x [-0.34, 35] km
> f(-42, 10)
[1] 2.387
```

In most commands in `spatstat` where a spatial covariate is required, a `"distfun"` can be used instead of a pixel image. This increases the precision of many calculations. It is usually advisable to call `distfun` rather than `distmap`, unless you really need a pixel image.

6.7 Formal tests of (non-)dependence on a covariate

It is often important to decide whether or not the intensity depends on a nominated spatial covariate Z. A formal hypothesis test may be useful, even in exploratory analysis. For example, Figure 6.24 seems to show very slight evidence that the intensity of copper deposits depends on distance to the nearest lineament. A formal test would enable us to assess the strength of this evidence.

For practical purposes we provisionally assume the process is Poisson. The null hypothesis is that the intensity does not depend on Z, while the alternative is that the intensity does depend on Z in some unspecified way.

6.7.1 Quadrats defined by a covariate

In the χ^2 quadrat counting test (Section 6.4.2) we can divide the survey region into quadrats of any desired shape and size. One possible choice is a partition *defined by ranges of values of the covariate*. The observed number of observed data points falling in each range of values of the covariate is compared with the number predicted by the model. This technique is common in logistic regression modelling [172, 343, 342].

For example, our exploration of the *Beilschmiedia* data in Section 6.6.2 above suggested that the density of trees depends on terrain elevation. We could divide the survey region into four irregular quadrats of equal area, defined by ranges of the elevation covariate, as we did on page 178, or equivalently by

```
> V <- quantess(Window(bei), elev, 4)
```

The command quadrat.test accepts a tessellation and uses the tiles of the tessellation as the quadrats:

```
> quadrat.test(bei, tess=V)

        Chi-squared test of CSR using quadrat counts
        Pearson X2 statistic

data:  bei
X2 = 350, df = 3, p-value <2e-16
alternative hypothesis: two.sided

Quadrats: 4 tiles (levels of a pixel image)
```

The test could also be performed using the previously computed quadrat counts qb by typing quadrat.test(qb). Because of the large counts in these regions, we can probably ignore concerns about independence, and conclude that the trees are not uniform in their intensity.

This test should be used in preference to the 'naive' χ^2 test described in Section 6.4.2 whenever we have good grounds for suspecting that the covariate Z has an influence on the intensity of the point process. Although the two tests have the same general form, the test based on ranges of values of the covariate is more sensitive (has greater statistical power) for detecting dependence on Z.

6.7.2   Tests based on exact values of a covariate

Instead of dividing the range of values of the covariate into discrete bands, it would be more informative to use the exact values of the covariate at each data point. Several tests using this approach are supported in spatstat. Details are given in Sections 10.5 and 10.3.5; here we simply present applications.

Cumulative distribution function (CDF) tests of CSR

One strategy is to compare the observed distribution of the values of the covariate at the data points with the values of that covariate at *all* spatial locations in the observation window. The principle is that, if the point process is completely random, then the data points are effectively a random sample of spatial locations in the window, so the values of the covariate at the data points, $z_i = Z(x_i)$, should be a random sample of the values of the covariate at all spatial locations in the window.

Tests based on the cumulative distribution function of a covariate are performed in spatstat by the generic function cdf.test. Here we focus on the method for point patterns, cdf.test.ppp, which performs a test of CSR. If X is the data point pattern, then

```
cdf.test(X, covariate, test="ks")
```

performs a test of CSR using the Kolmogorov-Smirnov test statistic, which is a measure of discrepancy between cumulative distribution functions (see equation (10.10) on page 381). By changing the argument test we can replace the Kolmogorov-Smirnov statistic by other measures of discrepancy such as the Cramér-Von Mises [183, 675, 194] (test="cvm") or Anderson-Darling [16, 17, 443] (test="ad") statistics.

The argument covariate is the spatial covariate that will be used. This may be a pixel image, a function(x,y) in the R language, or one of the strings "x" or "y" indicating one of the Cartesian coordinates.

Let us test the *Beilschmiedia* data for homogeneity against the alternative of dependence on the elevation covariate:

```
> cdf.test(bei, elev)

        Spatial Kolmogorov-Smirnov test of CSR in two dimensions

data:  covariate 'elev' evaluated at points of 'bei'
       and transformed to uniform distribution under CSR
D = 0.11, p-value <2e-16
alternative hypothesis: two-sided
```

The result of cdf.test is an object of class "htest" (the standard R class for hypothesis tests) and also of class "cdftest" so that it can be printed and plotted. The print method (demonstrated above) reports information about the hypothesis test such as the *p*-value. The plot method (illustrated in Figure 6.25) displays the observed and expected distribution functions. The Kolmogorov-Smirnov test statistic is the maximum vertical separation between these functions.

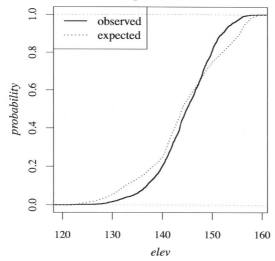

Figure 6.25. *Plot of* cdf.test(bei, elev), *illustrating the Kolmogorov-Smirnov test for dependence of the intensity of* Beilschmiedia *trees on terrain elevation.*

Figure 6.25 is the default plot, showing the two cumulative distribution functions $\hat{F}(z)$ and $F_0(z)$ plotted against z. Alternatively, selecting style="PP" in the plot command gives a P–P plot, showing $\hat{F}(z)$ plotted against $F_0(z)$; selecting style="QQ" gives a Q–Q plot, showing empirical quantiles $\hat{F}^{-1}(p)$ against theoretical quantiles $F_0^{-1}(p)$ for each probability $0 < p < 1$.

A warning message from cdf.test about tied values (which does not arise in this example) indicates that several values z_i are equal. Typically this occurs because the covariate values have been rounded, so that several data points give the same observed value of the covariate. A pragmatic solution is to 'jitter' or randomly perturb the covariate values, which is done by default in cdf.test. If the discretisation is severe, or if the covariate is an intrinsically discrete-valued variable or a factor, the Kolmogorov-Smirnov test is not supported by theory (and is ineffective because of tied values), and the χ^2 test based on quadrat counts would be preferable.

Choosing covariate="x" means that the covariate is the cartesian *x*-coordinate, $Z(x,y) = x$, so we are simply comparing the observed and expected distributions of the *x*-coordinate. This is a useful strategy for testing the null hypothesis of homogeneity against the alternative of a large-scale spatial trend, especially when there are no covariate data:

```
> cdf.test(swp, "x")
        Spatial Kolmogorov-Smirnov test of CSR in two dimensions

data:  covariate 'x' evaluated at points of 'swp'
    and transformed to uniform distribution under CSR
D = 0.088, p-value = 0.6
alternative hypothesis: two-sided
```

Warnings about tied values of the *x*-coordinate are common when the data are pixellated.

Berman's tests

Berman [88] proposed two tests (designated Z_1 and Z_2) for the dependence of a point process on a spatial covariate. These tests have better performance than the Kolmogorov-Smirnov test against certain alternatives. Full details are given in Section 10.5; briefly, the Z_1 test is based on the sum $T_1 = \sum_i Z(x_i)$ of the covariate values at all data points x_i, while the Z_2 test is based on $T_2 = \sum_i G(Z(x_i))$ where G is the cumulative distribution function of $Z(u)$ over all locations u in the window, defined in (6.19). Closely related tests were proposed independently by Waller *et al.* [685] and Lawson [413], so Berman's Z_1 test is often termed the 'Lawson-Waller' test in epidemiological literature. The Berman-Lawson-Waller test is a special case of the score test, as explained in Section 10.3.5.

Berman's tests are performed in spatstat by the command berman.test. The default is the Z_1 test.

```
> elev <- bei.extra$elev
> B <- berman.test(bei, elev)
> B
        Berman Z1 test of CSR in two dimensions

data:  covariate 'elev' evaluated at points of 'bei'
Z1 = -0.73, p-value = 0.5
alternative hypothesis: two-sided
```

Like cdf.test, the function berman.test returns an object of class "htest". The result also belongs to the special class "bermantest" for which there is a plot method. The plot is identical to a plot of the result of cdf.test except for the addition of two vertical lines which show the mean values of the empirical and null distributions.

In the example above, the *p*-value from the Berman Z_1 test is about 0.47 suggesting no evidence that *Beilschmiedia* density depends on terrain elevation, whereas the Kolmogorov-Smirnov test gave a *p*-value close to zero, suggesting strong evidence. The discrepancy between test outcomes is explained by Figure 6.25. The deviation between the two distribution curves is roughly symmetrical, so the two distributions have approximately equal means — that is, the average terrain elevation of a *Beilschmiedia* tree is roughly equal to the average terrain elevation in the study region. The Z_1 test statistic is a comparison between the two mean values, and is not sensitive to the type of deviation shown in Figure 6.25. The Z_2 test

```
> berman.test(bei, elev,"Z2")
```

gives a much smaller *p*-value (0.0142) although still not nearly as small as the *p*-value produced by cdf.test.

For a more searching investigation of the dependence of intensity on a covariate, the analyst would typically choose a class of parametric models for the intensity, and test hypotheses about the model parameters, as explained in Chapters 9 and 10.

When the covariate Z is the distance to a spatial pattern, another useful statistic is Foxall's *J*-function [269], explained in Section 8.10 and available using `Jfox`.

Tests for a point source of elevated risk

An important problem in spatial epidemiology is to determine whether a point source of radiation or pollution gave rise to elevated risk of disease. This is the special case of testing whether intensity depends on a covariate, where the covariate Z is distance to the point source. The tests described above can be applied, but a better option is a 'focused test' or 'point source test' designed to have good performance for this particular covariate [628, 108, 107], [545, pp. 56–63]. The key problem is the weakness of evidence: elevated risk will be strongest at short distances from the point source, but relatively few cases are expected to occur in this small area. The R package `DCluster` [287] provides implementations of many focused tests.

6.7.3 Strength of dependence on a covariate

Even if we find significant evidence that the point process intensity depends on the covariate Z, the effect of the covariate could still be quite weak. We need a measure of the strength of the effect of the covariate. Traditional tools for this purpose are the **Receiver Operating Characteristic (ROC)** curve, and the area under this curve, **AUC**. These are supported by the `spatstat` functions `roc` and `auc`.

If high densities of points are likely to be associated with low values of the covariate, the ROC plot is equivalent to a P–P plot of $\hat{F}(z)$ against $F_0(z)$ as described above. If high densities of points are associated with high values of the covariate, the ROC plot is a plot of $1 - \hat{F}(z)$ against $1 - F_0(z)$. Here $\hat{F}(z)$ is the cumulative distribution function of the values $Z(x_i)$ at the data points, while $F_0(z)$ is the c.d.f. of the values $Z(u)$ at all locations u.

The left panel of Figure 6.26 shows the ROC plot for the Queensland copper data against distance to the nearest lineament, assuming that copper deposits are likely to be found close to the lineaments. The plot was generated by

```
> coproc <- with(copper,
                 roc(SouthPoints, distfun(SouthLines), high=FALSE))
> plot(coproc)
```

To interpret this plot, suppose we consider the part of the study window that lies within a distance z of the nearest lineament. The fraction of window area occupied by this region is shown on the horizontal axis in Figure 6.26. The fraction of copper deposits which fall in this region is shown on the vertical axis. Varying the threshold z gives the curve shown in the figure.

In this example, the distance covariate has essentially no discriminatory power — thresholds of the distance covariate do not divide the window into regions of high and low density of copper deposits. For a covariate with strong discriminatory power, the ROC curve would lie substantially above the diagonal line.

The right panel of Figure 6.26 shows the ROC plot for another geological survey, the Murchison gold data (Section 9.3.2.2), computed in a similar way:

```
> murroc <- with(murchison, roc(gold, distfun(faults), high=FALSE))
```

In this case, the distance to the nearest fault has very strong discriminatory power. The plot shows that, by considering parcels of land lying closer than a specified distance away from the nearest geological fault, we can find 60% of all gold deposits in 10% of the land area, and 80% of gold deposits

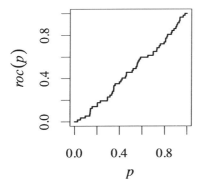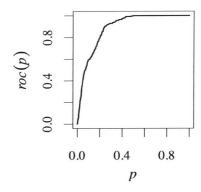

Figure 6.26. *ROC curves for* (Left) *the southern half of the Queensland copper deposits against distance to nearest lineament, and* (Right) *the Murchison gold deposits against distance to nearest geological fault.*

in 20% of land area. The ROC plot allows the distance threshold to be selected, for example, for mineral exploration planning.

A simple numerical index of discriminatory strength is the area under the curve, called AUC. The AUC is a number between 0 and 1, with values close to 1 or 0 indicating strong discrimination, and the value 0.5 indicating no discriminatory power. If high densities of points are expected at low values of the covariate, the AUC can be interpreted as the probability that a randomly selected data point will have a lower value of the covariate than a randomly selected spatial location in the window. The AUC is closely related to Berman's Z_2 test but is a measure of the magnitude of the covariate effect rather than the strength of evidence for an effect. For the two examples above:

```
> with(copper, auc(SouthPoints, distfun(SouthLines), high=FALSE))
[1] 0.4612
> with(murchison, auc(gold, distfun(faults), high=FALSE))
[1] 0.8873
```

The results suggest that the distance covariate has no discriminatory power in the Queensland copper data, and has strong discriminatory power in the Murchison gold data.

6.8 Hot spots, clusters, and local features

It is good practice to examine any spatial point pattern dataset for anomalies. Their presence can skew the results of statistical analysis if not handled appropriately. The presence of anomalies may be the main target of investigation, for example, in public health surveillance.

6.8.1 Hot spots

Often it is desired to detect anomalies in the intensity of a point process. A *hot spot* in a point process is a zone of elevated intensity. In archaeology, it often happens that ancient artefacts such as pottery fragments are much more abundant in one area than another. In seismology, earthquake epicentres are typically highly concentrated along a plate boundary. Old minefields can sometimes be recognised by their elevated density of metal fragments.

In spatial epidemiology one of the important goals is to detect spatial 'clusters' of disease cases. A cluster is a group of cases, lying close together, which are more numerous or more dense than expected, relative to the background pattern of such cases. A confirmed cluster (that is, one which is judged statistically significant) could be explained by a common source of infection or toxicity, by contagion in the local population, or by observer effects such as increased vigilance by the health services. A source of toxicity would lead to an *elevated intensity* of disease cases near the source, while contagion would cause *positive correlation* in disease status between neighbouring people. It can be difficult, contentious, and even impossible [83] to distinguish between these two effects. Thus, a hot spot of disease cases is only one possible kind of disease cluster — although it can be difficult to distinguish it from other kinds.

Detecting and locating hot spots is an ill-defined or open-ended problem. The choice of detection method will depend on the kind of anomaly we are looking for, and on prior information and subject knowledge.

Exploratory techniques for investigating localised features in a point pattern include scan statistics, model-based clustering, cluster set estimation, nearest-neighbour cleaning, and data sharpening.

The simplest place to start is with a kernel estimate of the intensity function. Zones of elevated intensity are often recognisable in an image plot of the kernel estimate.

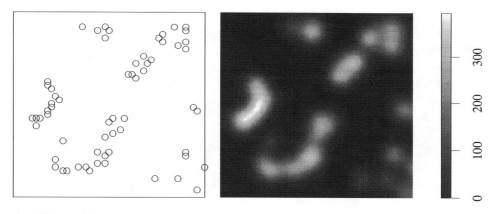

Figure 6.27. *California redwood seedlings and saplings (Ripley's subset).* Left: *point pattern scaled to unit square.* Right: *kernel estimate of density, with bandwidth selected by likelihood cross-validation.*

The left panel of Figure 6.27 shows the California redwood seedlings and saplings data (Ripley's [572] subset of a dataset of Strauss [643]). Clusters of points, clearly visible in the plot, are thought to have arisen by propagation from a parent tree which has since died [643]. Depending on the purpose of analysis, it might be appropriate to locate the positions of these clusters. The right panel of Figure 6.27 shows a kernel estimate of intensity:

```
> denRed <- density(redwood, bw.ppl, ns=16)
```

(The argument ns is passed to bw.ppl.) We selected the bandwidth by likelihood cross-validation (bw.ppl) as in our experience this tends to select more appropriate values than other methods when the pattern consists *predominantly* of tight clusters. However, to detect a single tight cluster in the midst of random noise, mean square cross-validation (bw.diggle) often seems to work best.

To determine whether such a zone is 'significant' we might use a Monte Carlo test:

```
> obsmax <- max(denRed)
> simmax <- numeric(99)
> lamRed <- intensity(redwood)
> winRed <- as.owin(redwood)
> for(i in 1:99) {
    Xsim <- rpoispp(lamRed, win=winRed)
    denXsim <- density(Xsim, bw.ppl, ns=16)
    simmax[i] <- max(denXsim)
  }
> (pval <- (1+sum(simmax > obsmax))/100)
[1] 0.01
```

A closely related strategy is to look for evidence of elevated intensity inside a circle, of fixed radius *R*, placed somewhere in the study region. The evidence can be assessed using the **scan test** [394, 395, 496, 497]. At each spatial location u, we draw a circle centred at u with a chosen radius r, denoted $b(u,r)$, and count the numbers of points inside and outside the circle, $n_{in} = n(\mathbf{x} \cap b(u,r))$ and $n_{out} = n(\mathbf{x} \cap W \setminus b(u,r))$. Assuming the point process is Poisson, we test the null hypothesis that the intensity is homogeneous, against the alternative that the intensity is different inside and outside the circle. The likelihood ratio test statistic is

$$\Gamma(u,r) = 2n_{in}\log(n_{in}/A_{in}) + 2n_{out}\log(n_{out}/A_{out}) - 2n(\mathbf{x})\log(n(\mathbf{x})/|W|) \tag{6.25}$$

where $A_{in} = |W \cap b(u,r)|$ and $A_{out} = |W \setminus b(u,r)|$ are the areas of the regions inside and outside the circle.

Figure 6.28 shows an image of $\Gamma(u,r)$ as a function of the centre location u, for a fixed value of r. This was computed by

```
> LR <- scanLRTS(redwood, r = 2 * bw.ppl(redwood))
```

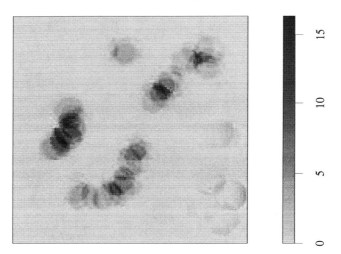

Figure 6.28. *Image of likelihood ratio test statistic* $\Gamma(u,r)$ *for fixed r, computed for the* redwood *data.*

The radius r was chosen so that the circle is comparable in size to the Gaussian kernel with bandwidth selected by likelihood cross-validation.[3] For fixed values of u and r, the null distribution of $\Gamma(u,r)$ is approximately χ^2 with 1 degree of freedom. The corresponding p-values can be computed by hand:

---

[3]The bandwidth of a kernel is defined as the standard deviation, or in higher dimensions the root-mean-square length

```
> pvals <- eval.im(pchisq(LR, df=1, lower.tail=FALSE))
```

Figure 6.29 shows the resulting *p*-values, and the set of locations where the *p*-value is less than 0.01.

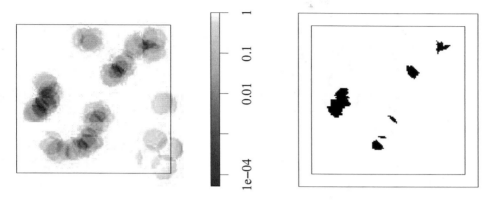

Figure 6.29. *Scan test with fixed circle radius.* Left: *p-values of likelihood ratio test statistic, with logarithmic colour scale.* Right: *locations where $p < 0.01$.*

For fixed u and r, the likelihood ratio test is theoretically optimal for detecting a difference in intensity inside and outside the circle $b(u,r)$.

If we consider all possible locations u for the centre of the circle, then there is a problem of multiple testing: even in a completely random pattern, it is likely that a circle somewhere in the study region will happen to contain a large number of points. The **scan statistic** Γ^* is the maximum value of $\Gamma(u,r)$ over all locations u in the window, and optionally over a range of different values of r. The scan test rejects the null hypothesis if Γ^* is large. The critical value of Γ^* is difficult to calculate, because the null distribution of Γ^* is much more complicated than in the case of fixed r [12, 429]. The stand-alone software package SaTScan™ (www.statscan.org) implements the scan test. The `spatstat` function `scan.test` implements a practical approximation to the scan test, using a Monte Carlo test (Chapter 10).

A procedure very similar to the scan test, the 'geographical analysis machine' (GAM), was proposed independently in [530]. The original GAM proposal fails to take account of the problem of multiple testing, as pointed out in [195, 659, 102, 444]. Alternative tests for 'spatial clustering' (meaning the presence of a hot spot at a location which is not specified in advance) are proposed in [102, 143, 195, 201, 210, 276, 415, 444, 607, 659, 704].

Test procedures applied to each local neighbourhood in a spatial domain are a special case of *local likelihood* methods [335, 428], also associated with the term *geographically weighted regression* [267]. See Sections 9.13, 12.5 and 13.11.

6.8.2 High and low intensity zones

Sometimes it is suspected that the spatial domain can be divided into two distinct regions of low and high intensity respectively. An artificial example is shown in the left panel of Figure 6.30. Specialised estimators of the high-intensity region are available for this situation. The nonparametric

---

(square root of the expected sum of squared coordinate values), of the distribution represented by the kernel. The expected sum of squares of an isotropic Gaussian vector with mean zero and standard deviation σ is $2\sigma^2$ so the bandwidth is $\sqrt{2}\,\sigma$. The expected squared length of a random vector uniformly distributed in a disc of radius r is $r^2/2$ so the bandwidth is $r/\sqrt{2}$. The bandwidths match if $r = 2\sigma$.

maximum likelihood estimator is the Allard-Fraley [11] cluster set. The Dirichlet tessellation induced by the point pattern is computed; the k smallest tiles (ranked by tile area) are identified, where k is determined by maximising a likelihood criterion analogous to the likelihood ratio test statistic described above; the union of these k smallest tiles is the cluster set.

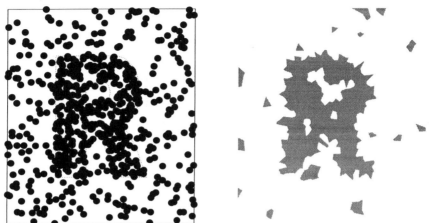

Figure 6.30. *Artificial example of high-intensity region.* Left: *simulated data from Poisson process with intensity* 100 *inside the letter R, and intensity* 20 *outside.* Right: *Nonparametric (Allard-Fraley) estimate of high-intensity region (grey shading).*

The right panel of Figure 6.30 shows the result of `clusterset(X, what="domain")` where X is the point pattern in the left panel of the Figure. The set estimate tends to be slightly smaller than the true set (because the Dirichlet tiles associated with the points closest to the boundary of the true set often have relatively large area) apart from including small isolated fragments outside the true set (because of random noise). These artefacts can be mitigated by constraining the set estimate to be a connected set; Allard and Fraley propose a more complex optimization algorithm [11] which is not yet implemented in `spatstat`.

If it is a requirement that the estimated set be connected, a fast alternative to constrained optimization of the likelihood, canvassed in the GIS literature under the heading of 'regionalisation' [306, 305], is to apply hierarchical clustering methods to the tiles, after constructing an appropriate measure of dissimilarity between tiles.

In large datasets, computation of the Dirichlet tessellation, the tile areas, and the union of tiles can be expensive. There is a `fast=TRUE` option to `clusterset` which uses discretisation to accelerate the calculations.

A quick and useful alternative is to compute the distance from each point to its nearest neighbour. *Nearest-neighbour cleaning* [135] groups the points into two classes — 'feature' and 'noise' — on the basis of their nearest-neighbour distances. In a homogeneous Poisson process of intensity λ, the distance R_k from a typical point to its kth-nearest neighbour is such that the disc area $A_k = \pi R_k^2$ has a gamma distribution with rate λ and shape k. If we suspect that there are two regions with different intensities, one strategy is to calculate A_k for each data point, and estimate the two intensities λ_1, λ_2 by fitting a two-component mixture model. The model states that each A_k value is drawn either from the Gamma(λ_1, k) distribution (with probability p) or from the Gamma(λ_2, k) distribution (with probability $1 - p$). Using the E–M algorithm, we estimate the unknown parameters λ_1, λ_2, p. From the fitted mixture model we can also estimate the probability that each observation A_k belongs to the first or second component.

The `spatstat` command `nnclean` performs nearest-neighbour cleaning [135]. Calling

```
> Z <- nnclean(X, k=10, plothist=TRUE)
```

fits the mixture model, prints a report, and generates the diagnostic plot in the left panel of Figure 6.31, in which the probability density of the fitted mixture model is superimposed on the histogram of observed values of the tenth-nearest-neighbour distance. The resulting point pattern Z is identical to X except for two columns of marks: class, shown in the middle panel of Figure 6.31, a factor which classifies the points into 'feature' (+) and 'noise' (·); and prob, shown in the right panel, giving the fitted probability that the point belongs to the cluster component of the mixture model. The choice of $k = 10$ was arbitrary; the optimal choice of k depends on characteristics of the pattern, so k is generally chosen by trial and error [135].

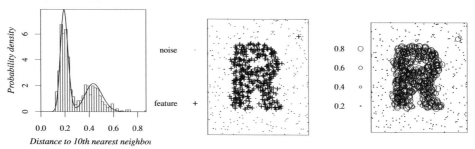

Figure 6.31. *Nearest-neighbour cleaning for artificial example of high-intensity region.* Left: *histogram of tenth-nearest-neighbour distances, and fitted density of mixture model.* Middle: *classification of data points into 'noise' (.) and 'feature' (+) classes.* Right: *data points marked by their fitted probability of belonging to the 'feature' class.*

An extreme example of high concentration of intensity is the South Pacific earthquakes data shown in the right panel of Figure 6.2 on page 158. This is a projection of the dataset quakes in the standard package datasets:

```
> require(datasets)
> qk <- ppp(quakes$long, quakes$lat, c(164, 190), c(-39,-10))
```

We used the mapdata and maps packages to plot the Pacific islands.

The different bandwidth selection procedures disagree widely: bw.diggle gives a bandwidth of 0.108 while bw.ppl gives 0.343 and bw.scott gives $(1.92, 1.59)$. This is a common feature of such highly concentrated patterns. We shall arbitrarily take a bandwidth of 0.5:

```
> dq.5 <- density(qk, 0.5)
```

To obtain comparable results from quadrat counting with hexagonal tiles, the tile size should be chosen to match the bandwidth of the kernel smoother. The mean squared length of a random point in a hexagon of side length s centred at the origin is $(5/12)s^2$. Setting $(5/12)s^2 = 2\sigma^2$ gives $s = \sqrt{24/5}\sigma \approx 2.19\sigma$. Taking $\sigma = 0.5$:

```
> sig <- 0.5
> (s <- sqrt(24/5) * sig)
[1] 1.095
> ht.5 <- hextess(as.owin(qk), s)
> hq.5 <- intensity(quadratcount(qk, tess=ht.5), image=TRUE)
```

The results are shown in Figure 6.32.

The left panel of Figure 6.33 shows the Allard-Fraley estimator computed by clusterset(qk, what="domain"). The right panel shows the classification of earthquakes into cluster and noise based on fifth-nearest-neighbour distances, computed by nnclean(qk, k=5, d=c(1,2)). All the clear outliers are classified as 'noise', but so are some earthquakes at the edge of the main clusters.

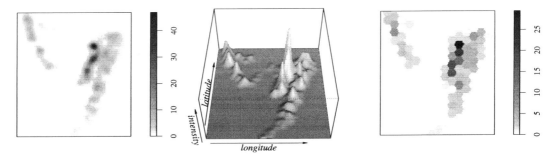

Figure 6.32. *Intensity estimates for South Pacific earthquakes.* Left and Middle: *kernel smoothing, isotropic Gaussian kernel with standard deviation 0.5 degrees.* Right: *intensity estimates on hexagonal quadrats of side length 1.09 degrees.*

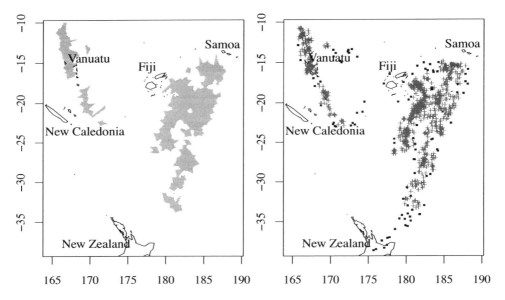

Figure 6.33. *Cluster analysis for South Pacific earthquakes.* Left: *Allard-Fraley cluster set estimate (grey shading).* Right: *classification into cluster (+) and noise (.) points based on distances to fifth-nearest neighbours.*

More formal procedures for detecting particular types of feature include *model-based clustering* [78] in which an explicit model of cluster shape is fitted to the data, allowing us to infer the exact locations of clusters and formally test for their presence; and *wombling* [75, 423] in which the curve of sharpest gradient in intensity is estimated, allowing us to detect irregular boundaries between two spatial textures.

6.8.3 Local features

The shapley dataset, shown in the left panel of Figure 6.34, is an example of a point pattern which is clearly not homogeneous. The data come from a radioastronomical survey of galaxies in the

Shapley Galaxy Concentration: each point is a galaxy in the distant universe. There are very dense concentrations of galaxies in some parts of the survey area.

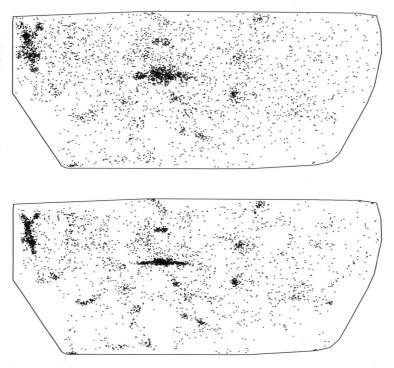

Figure 6.34. *Shapley Galaxy Concentration survey dataset.* Top: *original data.* Bottom: *result of Choi-Hall data sharpening.*

In Choi-Hall *data sharpening* [151] the points effectively exert a force of attraction on each other, and are allowed to move in the direction of the resultant force. This tends to enhance tight linear concentrations of points. The left panel of Figure 6.34 shows the result of applying the `spatstat` function `sharpen`:

```
> Y <- sharpen(unmark(shapley), sigma=0.5, edgecorrect=TRUE)
```

Another approach to detecting local features is LISA (Local Indicators of Spatial Association) methods, in which a summary statistic is separated into contributions from each of the data points. See Section 7.11.

6.9 Kernel smoothing of marks

It is often useful to apply spatial smoothing to the *mark values* attached to the points of a point pattern.

The left panel of Figure 6.35 shows the Longleaf Pines data (introduced on page 7). Circle diameters are proportional to each tree's diameter at breast height (*dbh*), a convenient surrogate measure of size and age. There appear to be some areas where the trees tend to be younger than in other areas. This is also suggested by Figures 4.3–4.4 on page 84. To investigate this we could compute a spatially varying average diameter of the trees in each neighbourhood.

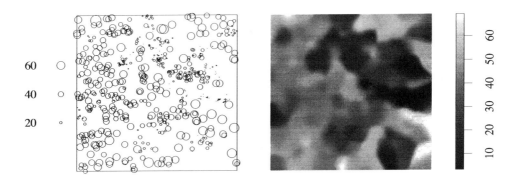

Figure 6.35. *Smoothing of mark values.* Left: *Longleaf Pines data: tree locations, marked by tree diameter at breast height, in a 200 metre square plot. Diameters are not to scale.* Right: *spatially varying average tree diameter (Nadaraya-Watson smoother of mark values) in centimetres.*

Suppose the data are points $x_1, \ldots, x_n$ with corresponding marks $m_1, \ldots, m_n$ which are real numbers. The *Nadaraya-Watson smoother* [494, 694, 495] is the spatial function

$$\widetilde{m}(u) = \frac{\sum_i m_i \kappa(u - x_i)}{\sum_i \kappa(u - x_i)} \tag{6.26}$$

for each spatial location u. The corresponding estimate using Diggle's correction is

$$\widetilde{m}^{(D)}(u) = \frac{\sum_i m_i \kappa(u - x_i)/e(x_i)}{\sum_i \kappa(u - x_i)/e(x_i)} \tag{6.27}$$

where $e(x_i)$ is the edge correction factor defined in (6.10). The function $\widetilde{m}(u)$ or $\widetilde{m}^{(D)}(u)$ can be taken as an estimate of the spatially varying average mark value.

The spatstat package has a generic function Smooth with a method Smooth.ppp for point patterns. The syntax of Smooth.ppp is similar to density.ppp and its return value is again a pixel image. Also available is the generic function Smoothfun which produces a function instead of a pixel image. See the FAQ for the distinction between Smooth and density.

The right panel of Figure 6.35 shows the result of Smooth.ppp applied to the Longleaf Pines data, with the smoothing bandwidth selected by least-squares cross-validation using the command bw.smoothppp. The result strengthens the conclusion that there is a swath of younger trees across the top right quarter of the plot.

For very large values of the smoothing bandwidth, the result of Smooth.ppp will become approximately constant and equal to the average mark value in the entire dataset. As bandwidth goes to zero, the expressions (6.26) and (6.27) converge to the mark value of the nearest data point (the data point x_i that is closest to the querying location u).

Many other smoothing techniques have been developed, but these are generally not supported in spatstat. See Chapters 14 and 15 for further techniques including inverse-distance weighted interpolation (idw). The contributed R packages akima, automap, locfit, and sm offer a wide range of tools for smoothing.

6.10 FAQs

- *What is the difference between 'intensity' and 'density'?*

  They are interchangeable to some extent. 'Intensity' is the technical term for the expected number of points per unit area. 'Density' is a synonym for intensity in some fields of research. However, in statistics, 'density' would often mean 'probability density', which we can think of as a normalised version of the intensity.

- *Why is the function for estimating point process intensity called* `density.ppp`*?*

  Many generic functions in R were originally developed by generalising from the simplest case. Their names are usually taken from the name of the simplest case. Analysis of variance (ANOVA) is a special case of analysis of deviance, but the generic function is named `anova`. Kernel estimation of a probability density is a special case of kernel estimation of intensity, but the generic function is named `density`.

- *What is the difference between* `density.ppp` *and* `Smooth.ppp`*?*

  Kernel estimation of point process intensity is performed by `density.ppp`. Kernel *smoothing* of the *mark values* of a point process is performed by `Smooth.ppp`. The result of `density.ppp` at a location *u* is a measure of how many points are found in the vicinity of *u*. The result of `Smooth.ppp` at a location *u* is a spatially weighted average of the marks attached to the data points in the vicinity of *u*.

- *When I apply* `density.ppp` *to my point pattern data, the resulting image has some pixel values which are zero or negative.*

  Although it is mathematically impossible, the intensity can take zero values or very small negative values in software because of numerical errors inherent in the Fast Fourier Transform algorithm. Such values can be removed, and replaced by a very small positive number, by setting `positive=TRUE` in `density.ppp`. This is not the default behaviour because it takes additional time.

- *When I type* `density(redwood, n=1000)` *the result is a* 128×128 *pixel image, no matter what value of n is used.*

  This is a misunderstanding about classes and methods. Typing `help(density)` shows the help for the generic `density` and its default method `density.default`. This method has an argument n. However `redwood` is a point pattern object of class `"ppp"`, so `density(redwood, ...)` is dispatched to `density.ppp`. Looking at the help for `density.ppp`, we see that this function does not have an argument n, and ignores it. To control the resolution of the result of `density(redwood)`, use the arguments `dimyx` or `eps`.

- *I have divided the window of a point pattern X into several quadrats. I would like to find out which points of X fall in each quadrat, and extract these points.*

  To label the points of X according to which quadrat they fall inside, use `Z <- cut(X,Q)` where Q is the tessellation of quadrats. The result Z is a point pattern, obtained from X by adding marks that indicate which quadrat contains each of the data points. This grouping factor can be extracted by `marks(Z)`.

  To divide the point pattern into sub-patterns according to which of the quadrats they fall inside, use `Y <- split(X, Q)` or `Y <- split(Z)`. The result is a list of point patterns, one for each quadrat.

- *I encounter technical problems when I use* quadratcount *to count the number of points falling in each pixel in a rectangular grid of pixels.*

  The quadratcount function is not efficient for this purpose; use pixellate instead.

7

Correlation

This chapter deals with measuring dependence between points in a point pattern, using the concept of correlation.

7.1 Introduction

Often the motivation for analysing point pattern data is to determine whether the points appear to have been placed independently of each other, or whether they exhibit some kind of interpoint dependence.

Figure 7.1 shows three archetypal point patterns representing 'regularity' (where points tend to avoid each other), 'independence' (complete spatial randomness), and 'clustering' (where points tend to be close together).

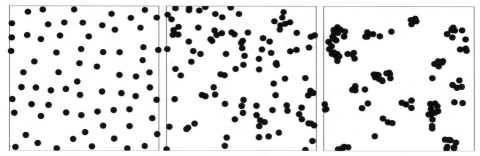

Figure 7.1. *Classical trichotomy between regular* (Left), *independent* (Middle), *and clustered* (Right) *point patterns. All three patterns are in the unit square.*

A standard statistical tool for measuring dependence is *correlation*, or more generally *covariance*. In this chapter we explain how to define and measure covariance in a point process, in such a way that (roughly speaking) the clustered point pattern in the right panel of Figure 7.1 has positive covariance, the completely random pattern in the middle panel has zero covariance, and the regular pattern in the left panel has negative covariance.

In statistical theory, correlation is classified as a *second moment* quantity. The 'first moment' of a random variable X is its mean value; the 'second moment' or 'mean square' is the mean of X^2. Together the first and second moments of random variables determine important quantities such as variance, standard deviation, covariance, and correlation.

Second moment quantities for point processes are intimately related to counting *pairs* of points, or adding up contributions from each pair of points in the process. In a point process $\mathbf{X}$, the squared point count $n(\mathbf{X} \cap B)^2$ can be interpreted as the number of *pairs of points* x_i, x_j in $\mathbf{X}$ which fall in the nominated set B, including identical pairs where $i = j$. The second moment of $n(\mathbf{X} \cap B)$ is the expected number of pairs of points falling in B.

Correlation has the great virtue of being easy to calculate and handle, and is a powerful tool for the data analyst. There are two important caveats. First, accurate measurement of correlation requires faithful estimation of the mean (first moment) if we are to avoid problems of spurious correlation and confounding. In the point process context, this means that we must have good knowledge of the intensity before we can trust the correlation. Second, the correlation is merely a summary index of statistical association, not a characterisation of dependence or causality: 'correlation is not causation'. Using only correlations, we cannot discriminate between different possible causes of spatial clustering.

7.2   Manual methods

We start by describing two simple, manual techniques which motivate and give insight into the more advanced methods for assessing corrlation in point pattern data.

7.2.1   Greig-Smith plot and Morisita index

If the observation window is a rectangle, a simple strategy for assessing spatial correlation is to subdivide the window into rectangular quadrats of equal size, and to count how often a pair of data points falls in the same quadrat. If there are n data points altogether, there are $n(n-1)$ ordered pairs of distinct points. If we use m quadrats and these are found to contain $n_1, \ldots, n_m$ points, respectively, then the jth quadrat contains $n_j(n_j-1)$ ordered pairs of distinct points. The total number of ordered pairs of distinct points which fall inside the same quadrat is thus $\sum_j n_j(n_j-1)$. The ratio

$$\frac{\sum_j n_j(n_j-1)}{n(n-1)}$$

is the fraction of all pairs of data points in which both points fall in the same quadrat. In a completely random (homogeneous Poisson) process, where points are independent of each other, two points fall in the same quadrat with probability $1/m$, where m is the number of quadrats, so the fraction above is expected to equal $1/m$. The ratio of the observed and expected fractions is the *Morisita index* [488]

$$M = m \frac{\sum_j n_j(n_j-1)}{n(n-1)}. \tag{7.1}$$

This index should be close to 1 if the points are independent, greater than 1 if they are clustered, and less than 1 if they are regular.

Repeating this calculation using different subdivisions of the window into quadrats, and plotting the Morisita index against the diameter of the quadrats, yields a *Morisita index plot*. Figure 7.2 shows that this has the ability to distinguish between the three archetypal point patterns in Figure 7.1. In spatstat the Morisita index plot of a point pattern X is generated by miplot(X).

The Morisita index is closely related to the *index of dispersion* for quadrat counts. Suppose we calculate the sample variance of the quadrat counts,

$$s^2 = \frac{1}{m-1} \sum_{j=1}^{m} (n_j - \overline{n})^2$$

where $\overline{n}$ is the average quadrat count, $\overline{n} = n/m$. If the point process is Poisson, then the counts $n_1, \ldots, n_m$ are observations of independent Poisson random variables with the same, unknown mean μ. The variance of a Poisson random variable is equal to its mean. A standard index of overdispersion or underdispersion for count variables is the sample variance divided by the sample mean: for

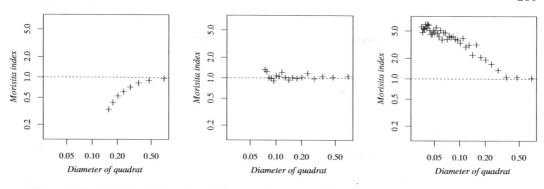

Figure 7.2. *Morisita index plots (on logarithmic scale) for the three patterns in Figure 7.1.*

a Poisson distribution this ratio should be approximately equal to 1. Dividing the sample variance s^2 by the sample mean $\bar{n}$ gives the *index of dispersion*

$$I = \frac{s^2}{\bar{n}} = \frac{m}{n(m-1)} \sum_{j=1}^{m} (n_j - \frac{n}{m})^2.$$

An earlier paper by Greig-Smith [296] had proposed plotting the index of dispersion against quadrat size. A little algebra shows that these are equivalent: the dispersion index I and Morisita index M are directly related by $I = (m/(m-1))[(n-1)M - (n/m-1)]$. As discussed on page 167, the index of dispersion is also closely related to the χ^2 test of CSR based on quadrat counts. Thus, the Morisita and Greig-Smith plots are effectively plots of the test statistic for a χ^2 quadrat counting test of CSR plotted against the size of the quadrats. See [462, 575, 225].

For the purposes of this chapter we will focus on the Morisita index M. Note especially that M *assumes the intensity is homogeneous.* If this is not the case, large values of M could arise simply from spatial inhomogeneity, rather than from some form of correlation between the points.

The Morisita index is not sensitive to subtle differences in spatial scale, because it is based on subdividing the observation window coarsely into quadrats. This suggests that we look for better ways to summarise the information about pairs of points.

There is also something unsatisfactory about the theoretical derivation of the Morisita index given above. That derivation referred only to the homogeneous Poisson point process, and calculated that the Morisita index should be about 1 for that process. But if the data were generated by another kind of point process, it is unclear (at least from the previous discussion) how to interpret the value of the Morisita index. Indeed the Morisita index might not even be a well-defined property of the point process: for example, it might depend on the size of the observation window. Although these questions can be resolved, the answers are not very simple.

A good statistical index should not only be accessible by simple direct calculation from the data, but it should have a simple, direct interpretation for the point process which generated the data. In the rest of the chapter we look for such indices.

7.2.2 Fry plot

More information about spacings in the point pattern can be obtained using a *Fry plot* or Patterson plot. Originally developed for crystallography by Patterson [534, 535] this technique was independently reinvented in geophysics by Fry [271, 321].

A Fry plot can be drawn by hand, as follows (see Figure 7.3). First print the point pattern on a sheet of paper. Take a transparency or sheet of tracing paper, and mark a cross in the middle. Place the transparency over the printout, so that the cross on the transparency lies on one of the data points.

Copy the positions of all the other data points onto the transparency in the same relative position. (Of course some data points may be too far away to be copied onto the transparency, depending on its size.) Now move the cross to another data point, and repeat the copying process. After every data point has been visited, the pattern on the transparency is the Fry plot.

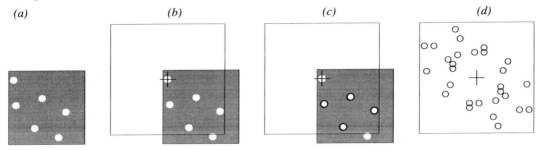

Figure 7.3. *Stages in forming the Fry plot.* (a): *data point pattern, printed on paper.* (b): *transparency superimposed on point pattern so that centre of transparency (+) lies above the first data point.* (c): *other data points are copied (○) onto the transparency.* (d): *final result.*

In mathematical terms the Fry plot is a scatterplot of the vector differences $x_j - x_i$ between all pairs of distinct points in the pattern.

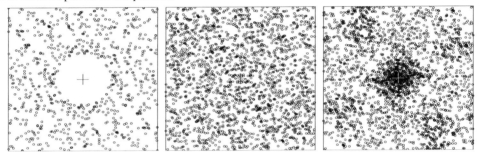

Figure 7.4. *Fry plots for the three patterns in Figure 7.1.*

In `spatstat` the Fry plot of a point pattern X is generated by the command `fryplot(X)` or `plot(frypoints(X))`. Additional arguments allow the plot to be restricted to a smaller area around the origin $(0,0)$, where the interesting detail is found, or restricted to certain groups of data points.

Figure 7.4 shows the Fry plots for the three archetypal point patterns in Figure 7.1, restricted to squares of width 0.6 units. The origin is in the middle of each panel of Figure 7.4 and is indicated by the '+' symbol, visible only in the left panel.

The origin in the Fry plot represents a typical point of the point pattern, and the dots in the Fry plot represent the positions of other nearby points, relative to the typical point. In the left panel of Figure 7.4 there is a clear absence of dots in the middle of the panel, indicating that data points never come closer to each other than a certain minimum distance. The middle panel of Figure 7.4 shows no obvious pattern, while the right panel shows a higher concentration of dots near the origin, indicating a clustered pattern. Thus, the Fry plot is easily able to distinguish the three basic kinds of dependence between points.

Fry plots are implicitly based on the assumption that the underlying process is stationary (Section 5.6.3). Under this assumption, the Fry plot contains essentially all information about correlations in the point process.

Fry plots can be very useful for spotting features of the point pattern which might not otherwise be obvious, such as discretisation of the coordinates. In geophysics, Fry plots have proven to be very useful for inferring mechanical strain in rocks, which is reflected in the shape of an elliptical 'hole' in the Fry plot. However, in many other applications, the interpretation of the Fry plot is too subjective. Fry plots often need to be simplified, reduced, or summarized in order to extract usable information.

7.3 The K-function

A very popular technique for analysing spatial correlation in point patterns is the K-function proposed[1] by Ripley [572].

7.3.1 The empirical K-function

Suppose that the primary research question concerns the distance or spacing between points in the point pattern. It would then be natural to measure the distances $d_{ij} = \|x_i - x_j\|$ between all ordered pairs of distinct points x_i and x_j in the point pattern **x** under consideration. These distances clearly capture a great deal of information about the spatial pattern. If the pattern is clustered, many of the pairwise distances will be small; if the pattern is regular, few of the distances will be small. This suggests that we might look at a statistical summary of the distances d_{ij}, such as the histogram.

We have argued that a good statistical summary of a point pattern should have a simple, direct interpretation in terms of the *point process* which generated the data. The histogram of observed pairwise distances d_{ij} is difficult to interpret in this way, because it depends on the shape and size of the observation window: the same point process, viewed through different windows, yields different histograms of pairwise distances.

Consider instead the empirical cumulative distribution function of the pairwise distances,

$$\widehat{H}(r) = \text{fraction of values } d_{ij} \text{ less than } r$$

$$= \frac{1}{n(n-1)} \sum_{i=1}^{n} \sum_{\substack{j=1 \\ j \neq i}}^{n} \mathbf{1}\left\{d_{ij} \leq r\right\} \tag{7.2}$$

defined for each distance value $r \geq 0$. Here we use the 'indicator' notation: $\mathbf{1}\{\ldots\}$ equals 1 if the statement '$\ldots$' is true, and 0 if the statement is false. The sum is taken over all ordered pairs i, j of indices which are not equal. The sum of these indicators is simply the number of times that the statement is true, that is, the number of values d_{ij} which are less than or equal to r. The denominator $n(n-1)$ is the total number of pairs of distinct points, so $\widehat{H}(r)$ is the fraction of pairs for which the distance is less than or equal to r.

Notice that $\widehat{H}(r)$ is analogous to the Morisita index: both quantities report the fraction of pairs of points which lie close together. The distance argument r in $\widehat{H}(r)$ defines 'closeness', and is analogous to the size of quadrats in the Morisita index.

We can also visualise the calculation of $\widehat{H}(r)$ using the Fry plot. The dots in the Fry plot are the vector differences $x_i - x_j$ between all pairs of distinct points in the point pattern dataset. The lengths of these vectors are the distances d_{ij}. To count the number of distances d_{ij} that are less than or equal to r, we simply draw a circle of radius r, centred at the origin of the Fry plot, and count the number of dots in the Fry plot which fall inside this circle. That is, $\widehat{H}(r)$ is the fraction of dots in the Fry plot which fall inside the circle of radius r.

---

[1] There are similar concepts in statistical physics [531] and astronomy (cf. [445]).

The contribution from each data point x_i to the sum in (7.2) is

$$t_i(r) = \sum_{j \neq i} \mathbf{1} \left\{ d_{ij} \leq r \right\},$$

the number of *other* data points x_j which lie closer than a distance r. We might call this the number of *r-neighbours* for the point x_i. Equivalently $t_i(r)$ is the number of data points which fall inside a circle of radius r centred at x_i, not counting x_i itself. Then

$$\widehat{H}(r) = \frac{1}{n(n-1)} \sum_{i=1}^{n} t_i(r) = \frac{1}{n-1} \bar{t}(r)$$

where $\bar{t}(r) = (1/n) \sum_i t_i(r)$ is the average number of r-neighbours per data point.

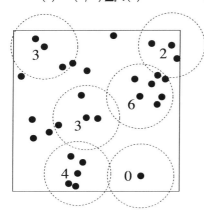

Figure 7.5. *Counting r-neighbours. In each circle, the numeral shows the value of $t_i(r)$ for the data point x_i at the centre of the circle.*

This is an important clue that the quantity we really want to estimate is *the average number of r-neighbours of a typical random point*. Figure 7.5 shows an intermediate stage in the calculation of $\bar{t}(r)$. The number of r-neighbours is shown for each of several data points. The average of these numbers for all data points is $\bar{t}(r)$.

The average number of r-neighbours of a data point will depend on the overall average density of points in the dataset. In a completely random pattern, we would expect $\bar{t}(r)$ to be about $\lambda \pi r^2$, since $\bar{t}(r)$ counts the number of points falling in a circle of radius r which has area πr^2. In order to be able to compare datasets with different numbers of points, it makes sense to standardise $\bar{t}(r)$ by dividing by λ. Since the maximum number of neighbours of any data point is $n - 1$, it may be more appropriate to divide by $\tilde{\lambda} = (n-1)/|W|$, where n is the number of points and $|W|$ is the area of the observation window. The result of this standardisation is $\bar{t}(r)/((n-1)/|W|) = |W|\widehat{H}(r)$.

The function $|W|\widehat{H}(r)$ is the standardised average number of r-neighbours of a typical data point. In order to be fully able to compare datasets observed in different windows, we also need to take account of *edge effects* as explained in Section 7.4. This leads to a slight modification of the function $|W|\widehat{H}(r)$, called the **empirical K-function**,

$$\widehat{K}(r) = \frac{|W|}{n(n-1)} \sum_{i=1}^{n} \sum_{\substack{j=1 \\ j \neq i}}^{n} \mathbf{1} \left\{ d_{ij} \leq r \right\} e_{ij}(r) \tag{7.3}$$

where $e_{ij}(r)$ is an *edge correction weight* described in Section 7.4.

In summary, the empirical K-function $\widehat{K}(r)$ is the cumulative average number of data points lying within a distance r of a typical data point, corrected for edge effects, and standardised by dividing by the intensity. The standardisation and edge correction make it possible to compare point patterns with different numbers of points, observed in different windows.

Figure 7.6 displays the empirical K-functions for the three archetypal point patterns of Figure 7.1, showing that the empirical K-function clearly has the ability to discriminate between the three basic kinds of interpoint dependence. The empirical K-function for the clustered pattern lies above the empirical K-function for a completely random pattern, which in turn lies above the empirical K-function for a regular pattern. This is equivalent to saying that, after adjusting for intensity, a typical point in the clustered pattern has more close neighbours than a typical point in the completely random pattern, which in turn has more close neighbours than a typical point in the regular pattern.

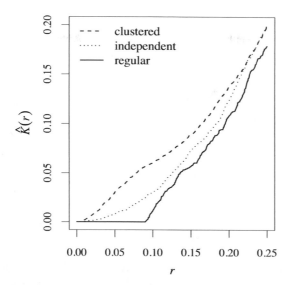

Figure 7.6. *Empirical K-functions for the three patterns in Figure 7.1. Solid line: regular pattern. Dotted line: independent pattern. Dashed line: clustered pattern.*

When interpreting graphs of the empirical K-function, it helps to remember that $\widehat{K}(r)$ is a standardised or relative quantity, rather than a direct physical quantity. The empirical K-function of a completely mapped forest gives us, for each r, the average number of neighbour trees lying within distance r of a typical tree, *divided by* the density of the forest in trees per unit area. The values of $\widehat{K}(r)$ are expressed in units of (number)/(number/area) = area. Standardisation makes it possible to compare the degree of regularity or clustering in forests with different average densities of trees. To recover the physically meaningful average number of neighbours $\bar{t}(r)$, we would need to know the forest density $\widehat{\lambda}$, and multiply $\widehat{K}(r)$ by $\widehat{\lambda}$ to obtain $\bar{t}(r)$.

> **Warning:** using the K-function implicitly assumes that the point process has homogeneous intensity. See Section 7.3.5.

7.3.2   The true K-function of a point process

The empirical function $\widehat{K}(r)$ is a summary of the pairwise distances in the point pattern dataset, normalised to enable us to compare different datasets. But the key question about any summary statistic for a point pattern is what it means for the *point process* which generated the pattern.

The K-function of a point process $\mathbf{X}$ will be defined as the expected number of r-neighbours of a typical point of $\mathbf{X}$, divided by the intensity λ. For this we need to assume[2] that $\mathbf{X}$ is **stationary** (see Section 5.6.3: the distribution of $\mathbf{X}$ is the same as the distribution of the shifted process $\mathbf{X} + v$, for any vector v). This implies that $\mathbf{X}$ has homogeneous intensity λ. We may then define

$$K(r) = \frac{1}{\lambda} \, \mathbb{E} \left[\text{number of } r\text{-neighbours of u} \mid \mathbf{X} \text{ has a point at location } u \right] \tag{7.4}$$

for any $r \geq 0$ and any location u. Since the process is stationary, this definition does not depend on the location u. On the right-hand side of (7.4), the symbol '|' indicates that this is a conditional expectation. Intuitively, we assume there is a random point of $\mathbf{X}$ at the location u; given this, we

---

[2]Slightly weaker assumptions are enough, and these will be stated below.

find the expected number of other points of $\mathbf{X}$ lying within a distance r; and finally we divide by the intensity λ, to obtain $K(r)$.

Extending the notation $t_i(r)$, let us define for any spatial location u

$$t(u,r,\mathbf{x}) = \sum_{j=1}^{n(\mathbf{x})} \mathbf{1}\left\{0 < \|u - x_j\| \le r\right\}, \tag{7.5}$$

the number of points in the point pattern $\mathbf{x}$ that lie within a distance r of the location u, but not at u itself.

Definition 7.1. *If* $\mathbf{X}$ *is a stationary point process, with intensity* $\lambda > 0$, *then for any* $r \ge 0$

$$K(r) = \frac{1}{\lambda}\mathbb{E}\left[t(u,r,\mathbf{X}) \mid u \in \mathbf{X}\right] \tag{7.6}$$

does not depend on the location u, *and is called the* K*-function of* $\mathbf{X}$.

Explicit formulae for the K-function have been derived for a few point process models. For the homogeneous Poisson point process (CSR), since the points are independent, intuitively speaking, the presence of a random point at the location u will have no bearing on the presence of points at other locations, so

$$\mathbb{E}\left[t(u,r,\mathbf{X}) \mid u \in \mathbf{X}\right] = \mathbb{E}\left[t(u,r,\mathbf{X})\right].$$

But $t(u,r,\mathbf{X})$ is the number of points of $\mathbf{X}$ falling in the disc $b(u,r)$ of radius r centred at u. The expected number of such points is $\lambda \times |b(u,r)| = \lambda \pi r^2$. Dividing by λ shows that, for a homogeneous Poisson process,

$$K_{pois}(r) = \pi r^2 \tag{7.7}$$

regardless of the intensity. This calculation is for two dimensions; for the case of three dimensions see Chapter 15.

7.3.3   Use of the empirical K-function

Visual inspection of empirical K-function

To study correlation in a point pattern dataset, assuming the intensity is homogeneous, we can plot the empirical K-function $\widehat{K}(r)$ calculated from the data, together with the theoretical K-function of the homogeneous Poisson process $K_{pois}(r) = \pi r^2$, which serves as the benchmark of 'no correlation'. Figure 7.7 shows this graphic for each of the three archetypal patterns in Figure 7.1.

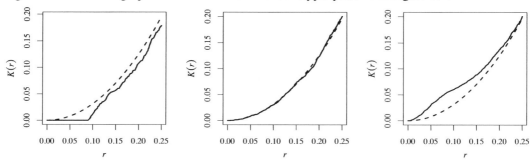

Figure 7.7. *Empirical K-function (solid lines) for each of the three patterns in Figure 7.1, and the theoretical K-function for a Poisson process (dashed lines).*

Figure 7.7 shows that this graphic easily detects the presence and type of correlation between points in a point pattern. In the left panel, the curve for the empirical K-function (solid lines) is lower than the theoretical curve for a completely random pattern (dashed lines), $\widehat{K}(r) < K_{pois}(r)$, indicating that a typical point in this pattern has fewer neighbours than would be expected if the pattern were completely random. This is consistent with a regular point process. Similarly in the right panel, the empirical curve is higher than the theoretical curve, $\widehat{K}(r) > K_{pois}(r)$, indicating that a typical point has more neighbours than would be expected if the pattern were completely random; this is consistent with clustering. The K-function is a powerful tool for investigating point patterns, but we need to remember that 'correlation is not causation'. If analysis shows that $\widehat{K}(r) > K_{pois}(r)$, the careful scientist will not say that this 'indicates' clustering, but that it is 'consistent with' clustering, or that it indicates 'positive association' between points. See Section 7.3.5 for more discussion.

Transformation of K

The *centred* version of the K-function is $K(r) - K_{pois}(r) = K(r) - \pi r^2$. This can be useful for classifying a point pattern as random, clustered, or regular, because the function is zero if the point pattern is completely random. It is less useful for other purposes.

A commonly used transformation of K proposed by Besag [103] is the **L-function**

$$L(r) = \sqrt{\frac{K(r)}{\pi}} \tag{7.8}$$

which transforms the theoretical Poisson K-function $K_{pois}(r) = \pi r^2$ to the straight line $L_{pois}(r) = r$, making visual assessment of the graph much easier. Figure 7.8 shows the empirical L-functions for the three archetypal point patterns. The square root transformation also approximately stabilises the variance of the estimator (that is, the variance of the empirical function $\widehat{L}(r)$ is roughly constant as a function of r), making it easier to assess deviations.

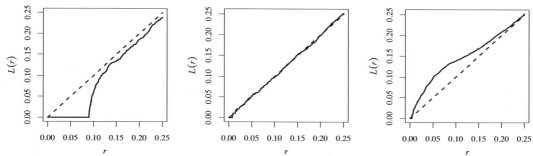

Figure 7.8. *Empirical L-function (solid lines) for each of the three patterns in Figure 7.1, and the theoretical L-function for a Poisson process (dashed lines).*

Besag's L-function has become such a popular transformation of Ripley's K-function that many writers in applied fields, and some software packages, apply the transformation without mentioning it. They plot $L(r)$ against r, or the centred version $L(r) - r$ against r, but still call it Ripley's K-function. This is not advisable because K and L are not equivalent in some contexts. When we say $K(r)$ we shall always mean $K(r)$.

Statistical inference

Options for formal statistical inference about the K-function include *confidence intervals* and *hypothesis tests* (the latter associated with *simulation envelopes*). These ideas are illustrated in Figure 7.9 using the clustered pattern in the right panel of Figure 7.1.

The left panel of Figure 7.9. shows a *confidence interval* for the true K-function of the point pattern. The right panel shows an *acceptance interval* (or 'non-rejection' interval) for testing whether the pattern is completely random.

It is very important to understand the difference between these two techniques. A **confidence interval** is designed to contain the true value of the target quantity with a specified degree of confidence. It is centred around an estimated value of the target quantity, and its width is an indication of the precision of the estimation. In the left panel of Figure 7.9, the shaded region is a pointwise 95% confidence interval for the true K-function of the clustered pattern; the shading is centred around the estimated K-function; the width of the shaded region reflects the precision of the estimate. Section 7.7 explains how to construct confidence intervals for the K-function.

On the other hand, an **acceptance interval** (or 'non-rejection interval') contains the *hypothesised* value of the target quantity. It is the range of values that are deemed to be not significantly different from the hypothesised value. The width of the acceptance interval reflects the inherent variability of the observations when the hypothesis is true. In the right panel of Figure 7.9, the shaded region is the acceptance interval for a formal test, with significance level 0.05, of the hypothesis that the pattern is completely random. The acceptance interval is centred around the K-function of a completely random point pattern, $K_{pois}(r) = \pi r^2$, and the width of the interval reflects the variability of $\widehat{K}(r)$. For a fixed value of r, the test rejects the null hypothesis of CSR if the value of $\widehat{K}(r)$ lies outside this acceptance interval. Section 7.8 explains how to construct acceptance intervals for testing CSR. More explanation and important caveats about hypothesis tests are given in Chapter 10.

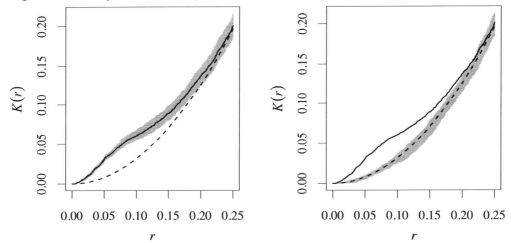

Figure 7.9. *Tools for formal statistical inference, demonstrated on the clustered pattern in the right panel of Figure 7.1. Left: 95% confidence intervals (shaded) for the true value of the K-function, obtained using Loh's bootstrap (function* lohboot, *Section 7.7.3). Right: acceptance region (shaded) for a hypothesis test of complete spatial randomness, with significance level 5%, using the envelopes of the K-functions of 39 simulated realisations of CSR. Generated using the function* envelope *(Sections 7.8 and 10.7).*

The K-function as a guide for building models

The K-function can serve as a guide for building a point process model of the phenomenon being studied. Chapter 12 presents one class of models which can be fitted directly to point pattern data using the K-function. More commonly the K-function is only a rough guide to the required behaviour of the model.

A point process model of a natural phenomenon could include several stages in which points are randomly added, combined, displaced, or removed. In this regard there are two important properties of the *K*-function.

Firstly **the *K*-function is invariant under random thinning**. Suppose **X** is a stationary point process, and **Y** is the point process obtained by independent random thinning — randomly deleting or retaining each point of **X**, with a constant probability p of retaining each point (Section 5.3.3). Then the *K*-function of **Y** is identical to the *K*-function of **X**.

Secondly there is the effect of **superposition**. Suppose **X** and **Y** are two independent, stationary point processes, with intensities λ_X, λ_Y and *K*-functions $K_X(r), K_Y(r)$. Superimposing these two processes gives a stationary point process with intensity $\lambda = \lambda_X + \lambda_Y$ and *K*-function

$$K(r) = p_X^2 K_X(r) + p_Y^2 K_Y(r) + 2p_X p_Y \pi r^2 \tag{7.9}$$

where $p_X = \lambda_X/\lambda$ and $p_Y = \lambda_Y/\lambda$ are the relative proportions of points coming from **X** and **Y**. The *centred K*-functions are more simply related:

$$(K(r) - \pi r^2) = p_X^2(K_X(r) - \pi r^2) + p_Y^2(K_Y(r) - \pi r^2). \tag{7.10}$$

7.3.4   Estimating the *K*- and *L*-functions in `spatstat`

The `spatstat` function `Kest` computes several estimates of $K(r)$ using different edge corrections. It also returns the 'benchmark' value πr^2 which is the theoretical value for $K(r)$ for a homogeneous Poisson process.

```
> K <- Kest(cells)
> Ki <- Kest(cells, correction="isotropic")
```

Here `cells` is Ripley's [572] biological cells dataset (shown on page 93) which we shall use frequently as an example.

The argument `correction` specifies which estimate or estimates will be computed by `Kest`. Options include `"isotropic"` for Ripley's isotropic correction, `"translation"` for the translation correction, `"rigid"` for the rigid motion correction, `"border"` for the border correction, and `"none"` for the uncorrected estimate. These are explained in Section 7.4. Any number of edge corrections may be selected. Specifying a single edge correction will reduce computation time, especially in simulation experiments.

Additionally `Kest`, like all standard summary functions in `spatstat`, recognises the options `"best"` (representing the estimate with the best statistical performance regardless of computational cost) and `"good"` (the estimate with the best statistical performance for reasonable computational cost).

If no `correction` argument is given, the default is to compute the isotropic, translation, and border corrections, unless the point pattern contains more than `nlarge` points, when only the border correction will be computed. The default threshold is `nlarge = 3000` points. Setting `nlarge=Inf` will suppress this behaviour.

The `spatstat` function `Lest` computes estimates of $L(r)$ directly from point pattern data.

```
> Lc <- Lest(cells)
```

The arguments of `Lest` are identical to those of `Kest`.

See Section 7.5 for a detailed explanation of how to plot and manipulate estimates of the *K*-function and other summary statistics. A previously computed *K*-function can be converted to the *L*-function using the syntax `L <- with(K, sqrt(./pi))`, explained on page 224, or plotted as an *L*-function using the syntax `plot(K, sqrt(./pi)~r)`, explained on page 221.

7.3.5   Caveats about the *K*-function

The use of the *K*-function for analysing point patterns has become established across wide areas of applied science, following Ripley's influential paper [572] and many subsequent textbooks [190, 221, 225, 661, 575, 576, 638]. Useful and powerful as this methodology is, there is an unfortunate tendency to apply it uncritically and to neglect other methods of analysis.

7.3.5.1   *K*-function assumes homogeneity

The *K*-function is defined and estimated under the *assumption that the point process is stationary.* If the process is not stationary, deviations between the empirical and theoretical functions (e.g., $\hat{K}$ and K_{pois}) are not necessarily evidence of interpoint interaction, since they may also be attributable to variations in intensity [83].

Figure 7.10 shows a realisation of an inhomogeneous Poisson process, and its empirical *K*-function. The plot of the *K*-function shows that, on average, a point in this pattern has more *r*-neighbours than would be expected for CSR. However, this happens because there is a higher density of points in one corner of the window. The points are positively associated, but not because they are clustered in the usual sense.

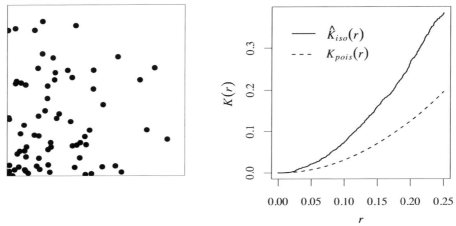

Figure 7.10. *The K-function is fooled by spatial inhomogeneity.* Left: *Inhomogeneous Poisson point pattern.* Right: *Empirical K-function.*

7.3.5.2   Correlation is not causation

There are many possible causes and mechanisms of correlation between points. True spatial regularity could be the result of ecological processes (territorial behaviour, competition for resources), physical forces (repulsion between objects), or human intervention. Apparent spatial regularity can occur as an artefact of the treatment of data, for example, if biological cells of appreciable size are treated as ideal points, or if spatial coordinates are discretised. Spatial clustering does not imply that the points are organised into identifiable 'clusters'; merely that they are closer together than would be expected for a completely random pattern. True spatial clustering could be the result of biological processes (reproduction, contagion), physical forces (electrostatic or magnetic attraction), or space-time history (clustering of meteorite impacts). Apparent clustering can occur if parts of a spatial pattern are obliterated (e.g., native forest partially destroyed by fire) or if objects of one type are physically *repelled* by another type of object. Spatial inhomogeneity is often mistaken for spatial clustering as we mentioned above.

7.3.5.3 Lack of correlation does not prove independence

Correlation is a summary measure of stochastic dependence, but absence of correlation does not necessarily indicate independence. Examples are well known in elementary statistics.

Similarly, there exist point processes whose K-functions are equal to πr^2 and yet the processes are *not* Poisson processes (so that there is dependence amongst the points). Therefore the K-function *does not completely characterise the point process.*

An example is the *cell process* [69]. Space is divided into equal tiles or cells; in each cell we place a random number N of points, according to a probability distribution whose variance is equal to its mean ($\mathbb{E}N = \mathrm{var}\,N$); the points are positioned independently and uniformly within the cell. The cell process has exactly the same theoretical K-function as the homogeneous Poisson process, but is manifestly different from a Poisson process. The left panel of Figure 7.11 shows a realisation of the cell process, generated by the `spatstat` function `rcell`:

```
> Xcell <- rcell(nx=15)
```

The right panel shows the empirical K-function, which falsely suggests that the process is Poisson. In fact the cell process would defeat *any* technique based on second moments, because all second-moment quantities for the cell process are identical to those for the Poisson process. Ripley [576, p. 23] commented: "It is a measure of the success of second-order methods that they have been mistakenly assumed to be all-powerful!"

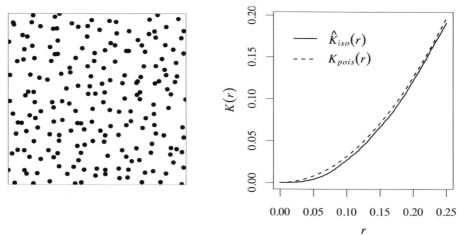

Figure 7.11. *Realisation of Baddeley-Silverman cell process* (Left) *and its empirical K-function* (Right).

7.3.5.4 Spatial scale of interaction

Summary functions like Ripley's K-function convey information across a range of spatial scales. This is an important motivation for using empirical functions, rather than simple numerical summary statistics, and for displaying them graphically.

Many researchers use a graph of the K-function to infer 'the' scale of spatial interaction in a point pattern. This scale is often estimated by reading off the position where the empirical function lies furthest away from the theoretical Poisson value, or furthest outside the region enclosed between the simulation envelopes.

This interpretation is not correct for several reasons. Firstly, 'the scale of interaction' is not a well-defined concept for point processes in general. It is only meaningful for certain point process models, such as Markov point processes (Chapter 13) and some Neyman-Scott cluster processes (Chapter 12).

Secondly, even for point processes which have a well-defined scale of interaction, it is not always true that the greatest deviation in the K-function occurs when r is equal to the scale of interaction. Examples are given in [40, p. 487]. The K-function reflects *correlation* between pairs of points, not direct dependence. Dependence between points at one scale can give rise to correlation between points at another scale.

Thirdly, the K-function is *cumulative*: $K(r)$ accumulates contributions from all distances *less than or equal to r*. If a pattern of emergent seedlings yields $\widehat{K}(r) > \pi r^2$ for the distance $r = 10$ metres, this does not necessarily indicate that seedlings are clustered **at** distances of 10 metres. A plausible explanation is that seedlings are organised in clusters at a much smaller spatial scale, and the cumulative effect is still evident at 10 metres. An alternative, non-cumulative summary statistic is the pair correlation function (Section 7.6). The K-function is optimal for detecting interpoint interaction that occurs equally at all distances up to a certain maximum distance r.

7.4   Edge corrections for the K-function[*]

This section gives more detail about the edge correction techniques used in estimating the K-function of a point process.

Most users of spatstat will not need to know the theory behind edge correction, the details of the techniques, or their relative merits. So long as some kind of edge correction is performed (which happens automatically in spatstat), the particular choice of edge correction technique is usually not critical.

However, there is a danger that the *implementations* of edge corrections in some software packages may be incorrect, which could cause substantial bias. There are certainly some misunderstandings about edge corrections that have crept into the applied literature. This section attempts to set the record straight.

For advanced users, understanding the assumptions behind edge correction is helpful in appreciating potential weaknesses of the analysis, and in resolving discrepancies between results. Edge corrections are reviewed in [576, chap. 3], [59, 355].

7.4.1   Estimation without edge effects

We recall that $d_{ij} = \|x_i - x_j\|$ is the distance between a pair of distinct data points x_i, x_j, and

$$t(u,r,\mathbf{x}) = \sum_{j=1}^{n(\mathbf{x})} \mathbf{1}\left\{0 < \|u - x_j\| \leq r\right\}$$

is the number of r-neighbours of the location u. An alternative expression for the K-function, that does not involve conditional expectation, is

$$K(r) = \frac{\mathbb{E}\sum_{x \in \mathbf{X} \cap B} t(x,r,\mathbf{X})}{\lambda\,\mathbb{E}n(\mathbf{X} \cap B)} \tag{7.11}$$

holding for a stationary point process $\mathbf{X}$, and for any bounded region B with area $|B| > 0$. Thus $\lambda K(r)$ is the expected number of r-close pairs of points in which the first point falls in B, divided by the expected number of points falling in B.

---

[*] Starred sections contain advanced material, and can be skipped by most readers.

A straightforward way to estimate the K-function is suggested by the representation in equation (7.11). Given a point pattern dataset $\mathbf{x}$, we simply replace the numerator and denominator of (7.11) by data-based estimates:

$$\widetilde{K}(r) = \frac{\sum_{x_i \in \mathbf{x} \cap B} t(x_i, r, \mathbf{x})}{\overline{\lambda}\, n(\mathbf{x} \cap B)}. \tag{7.12}$$

The numerator is the total number of r-neighbours of all random points falling in B (an arbitrary set with nonzero area), and in the denominator, $n(\mathbf{x} \cap B)$ is the number of points falling in B, while $\overline{\lambda} = n(\mathbf{x} \cap B)/|B|$ is an estimate of λ.

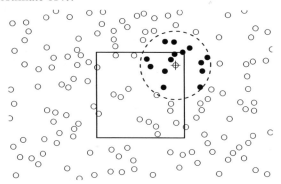

Figure 7.12. *Counting, without edge effects, the number of neighbours of each wildflower within a sampling frame (black rectangle) in a field of wildflowers.*

Figure 7.12 sketches how this estimator could be applied in practice. In a homogeneous field of wildflowers, we have laid out a sampling frame B (black square). Visiting one of the flowers inside the sampling frame, we push a thin stake into the soil near the flower (crosshairs), tied to a string exactly one metre long. Pulling the string taut, we describe a circle around the stake (dashed lines), and count how many other flowers are inside the circle (black dots) *regardless of whether they lie inside or outside the frame.* For the example in Figure 7.12 the count is 14. We have just counted the number of r-neighbours $t(x_i, r, \mathbf{x})$ for one flower x_i, for the distance $r = 1$ metre.

Repeating this process for each flower inside the study area would give us the neighbour counts $t(x_i, r, \mathbf{x})$ for flowers $x_1, \ldots, x_n$, say, where n is the number of flowers in the study area. The average observed neighbour count is $\bar{t}(r) = (1/n) \sum_i t(x_i, r, \mathbf{x})$. Dividing by the estimated intensity $\overline{\lambda} = n/|B|$, where $|B|$ is the area of the study frame B, gives us an estimate of $K(r)$ as in (7.12).

7.4.2 Edge effects

The situation sketched in Figure 7.12 is unusual, because we were able to look outside the sampling frame. In most research studies, this information is lost; points are recorded only if they fall inside the study region; the situation is more like Figure 7.13.

If points are observed only inside a window W, then the estimator (7.12) with $B = W$ is not feasible. The number of points inside a circle of radius r, centred on a point of the process inside W, is *not observable* if the circle extends outside W. This is an *edge effect* problem.

It might be tempting to ignore this problem and use the 'uncorrected' empirical function

$$\widehat{K}_{un}(r) = \frac{|W|}{n(n-1)} \sum_{i=1}^{n} \sum_{\substack{j=1 \\ j \neq i}}^{n} \mathbf{1}\{d_{ij} \leq r\} = |W|\widehat{H}(r), \tag{7.13}$$

the counterpart of (7.3) without the weighting terms $e_{ij}(r)$. However, a simple experiment shows that (7.13) is severely biased as an estimator of $K(r)$. We generate a completely random point

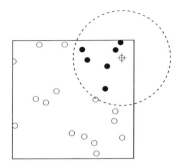

Figure 7.13. *Edge effect problem for estimation of the K-function. If we can only observe the points inside a window W (solid lines), then the number of points inside a circle (dashed lines) of radius r, centred on a point of the process inside W, is not known if the circle extends outside W. The number of points* observed *inside the circle (i.e. counting only points within the window) is typically less than the true number.*

pattern, according to a uniform Poisson process with intensity $\lambda = 100$ in the unit square. The uncorrected function $\widehat{K}_{un}(r)$ is plotted in Figure 7.14 together with the correct K-function $K(r) = \pi r^2$. The uncorrected function is clearly an underestimate of the correct K-function.

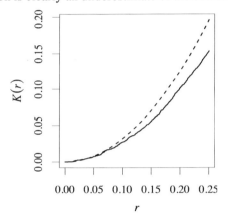

Figure 7.14. *Bias in estimating K(r) due to edge effects. Solid lines: uncorrected estimate $\widehat{K}_{un}(r)$ for a completely random point pattern. Dashed lines: correct K-function $K(r) = \pi r^2$.*

The edge effect bias can be predicted for a given window W using the function `distcdf`, which computes the true cumulative distribution function $H(r)$ of the distance between two independent random points in the window.

7.4.3   Border correction

An *edge correction* is a strategy for eliminating the edge effect bias. Edge corrections are surveyed in [576, chap. 3], [59], [355, sec. 4.3.3].

One simple strategy is the *border method*. When estimating $K(r)$ for a particular distance r, we restrict attention to cases where the circle of radius r lies entirely inside the window, so that the edge effect does not occur. See Figure 7.15.

In the numerator and denominator of (7.12) we restrict the summation to data points x_i for which $b(x_i, r)$ lies entirely inside W. For such points, $t(x_i, r, \mathbf{X} \cap W) = t(x_i, r, \mathbf{X})$, so the true number of r-

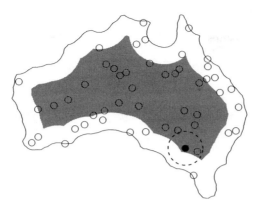

Figure 7.15. *Border method of edge correction for the K-function. When estimating $K(r)$, distances d_{ij} are measured only from points x_i lying at least r units away from the boundary. These are the points falling inside $W_{\ominus r}$ (shaded region).*

neighbours is observable. If $d(u, \partial W)$ denotes the shortest distance from a location u to the window boundary ∂W, then we are restricting attention to data points x_i which satisfy $d(x_i, \partial W) \geq r$. These are the data points which fall in the *eroded set*

$$W_{\ominus r} = W \ominus b(0, r) = \{u \in W : d(u, \partial W) \geq r\} \tag{7.14}$$

consisting of locations in W that are at least r units away from the boundary ∂W. The eroded set is shaded in Figure 7.15.

Thus we estimate $K(r)$ by the *border correction estimate*

$$\widehat{K}_{bord}(r) = \frac{\sum_{x_i \in \mathbf{x} \cap W_{\ominus r}} t(x_i, r, \mathbf{x})}{\overline{\lambda}\, n(\mathbf{x} \cap W_{\ominus r})} = \frac{\sum_{i=1}^{n} \mathbf{1}\{b_i \geq r\}\, t(x_i, r, \mathbf{x})}{\overline{\lambda} \sum_{i=1}^{n} \mathbf{1}\{b_i \geq r\}} \tag{7.15}$$

where $b_i = d(x_i, \partial W)$ is the distance from the data point x_i to the window boundary, and $\overline{\lambda}$ is an estimate of the intensity, usually $\overline{\lambda} = n(\mathbf{x} \cap W)/|W|$.

The estimate (7.15) is well defined so long as the denominator is non-zero, that is, so long as $r < \max_i b_i$. The maximum possible value of r for which (7.15) can ever be computed is[3] the *inradius* of W, the radius of the largest disc contained in W, since this is the maximum possible value of $d(\cdot, \partial W)$.

The border correction estimate (7.15) can be justified as a data-based estimate of the right-hand side of (7.11) taking $B = W_{\ominus r}$. This suggests that it will have reasonable statistical properties in sufficiently large datasets. The numerator of (7.15) is an unbiased estimator of the numerator of (7.11), while the denominator of (7.15) is the product of two terms which are unbiased estimators of the corresponding terms in the denominator of (7.11). In large datasets, the border-correction estimate is consistent and approximately unbiased, under reasonable assumptions.

Like many estimators in spatial statistics, the empirical K-function in (7.15) is a slightly biased estimator of the true K-function. It is composed of several terms, each of which is an unbiased estimator of a corresponding theoretical quantity. This implies that the overall bias of the empirical K-function will be small in large samples. Another way to write equation (7.11) is

$$\lambda^2 K(r) = \frac{1}{|B|} \mathbb{E} \sum_{x \in \mathbf{X} \cap B} t(x, r, \mathbf{X}) \tag{7.16}$$

---

[3] Here and throughout the discussion we assume W is a topologically regular set, meaning that it is the closure of its interior.

so that $\lambda^2 K(r)$ is the expected number of r-close pairs of points with the first point falling in B, divided by the area of B. In essence it is really $\lambda^2 K(r)$ that we need to estimate.

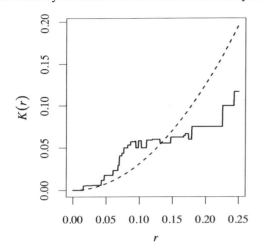

Figure 7.16. *Border correction estimate of the K-function (solid lines) for 20 uniformly random points in the unit square.*

The border correction estimate of the K-function is relatively simple to implement in software for any shape of window, and is fast to calculate. However, in small datasets, it can be inaccurate when compared to other methods, because it discards substantial amounts of data. For example, if W is the unit square and $r = 0.1$, the eroded window $W_{\ominus r}$ is a square of side 0.8 and area 0.64 so that almost 40% of the data is being discarded in order to estimate $K(0.1)$. In small datasets or in unusual situations, the graph of the estimated function $\widehat{K}_{bord}(r)$ can have an erratic trajectory, as shown in Figure 7.16. For large values of r, the denominator of (7.15) is small, magnifying the effect of discrete jumps in the numerator and denominator. Although the true K-function must be an increasing function of r, the border correction estimate need not be so.

When the number of data points is large, the border method is usually preferable to more computationally intensive methods, because the interesting distances r are small, and the loss of statistical efficiency in the border method is negligible.

The border method is a useful general-purpose remedy for edge effects in many spatial problems. It is related to the 'local knowledge principle' of mathematical morphology [622, 65].

7.4.4   Isotropic correction

An alternative approach to edge effects is to regard them as a form of *sampling bias*.

Visualise the entire point process **X** extending throughout two-dimensional space. Consider a pair of points x, x' from **X**, and assume that x falls in the observation window. Given the location of the first point x and the distance $d = \|x - x'\|$, the second point x' must lie somewhere on the circle $b(x, d)$ of radius d centred at x. In general, only part of this circle lies inside the window: see Figure 7.17.

If the point process is also *isotropic* (statistically invariant under rotation, Section 5.6.3), then roughly speaking, the second point x' is equally likely to lie anywhere on this circle. The probability that x' falls inside W is the fraction of length of the circle lying within W,

$$p(x, d) = \frac{\ell(W \cap \partial b(x, d))}{2\pi d}$$

where ℓ denotes length. As d increases, the probability $p(x, d)$ typically decreases. This gives rise to a sampling bias: larger distances are less likely to be observed.

A standard strategy for correcting sampling bias is the *Horvitz-Thompson* estimator [341, 58, 173] introduced on page 181. This strategy can be applied to estimators of the K-function. If each pair of points x_i, x_j is weighted by the reciprocal of the probability $p(x_i, d_{ij})$, we obtain Ripley's *isotropic correction* estimator [572, 569]

$$\widehat{K}_{iso}(r) = \frac{1}{\overline{\lambda} n} \sum_{i=1}^{n} \sum_{\substack{j=1 \\ j \neq i}}^{n} \mathbf{1}\{d_{ij} \leq r\} \frac{1}{p(x_i, d_{ij})}. \tag{7.17}$$

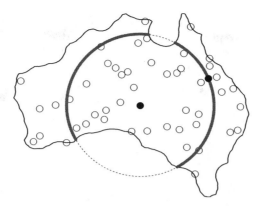

Figure 7.17. *Calculation of sampling bias for the isotropic correction. The contribution from a pair of data points x_i, x_j is determined by drawing a circle through x_j centred at x_i. The fraction p of the circle's perimeter which falls inside the observation window (grey line) is measured. The edge correction weight for this pair of points is $1/p$.*

Warning: Some authors state, incorrectly, that Ripley's isotropic correction uses the fraction of *area* rather than the fraction of *length* of the circle. This error may have found its way into computer code.

The isotropic-correction estimate is a non-decreasing function of r, unlike the border-correction estimate.

The Horvitz-Thompson principle is valid provided each item in the population has a non-zero probability of being sampled. In the isotropic correction, it is possible to have $p(x, d) = 0$ when d is large enough. Let $\widehat{R}_{iso}$ be the smallest distance for which there is some location u in W (not necessarily a data point) where $p(u, d) = 0$. Then $\overline{\lambda}^2 \widehat{K}_{iso}(r)$ is an unbiased estimator of $\lambda^2 K(r)$ for all $r < \widehat{R}_{iso}$. The distance $\widehat{R}_{iso}$ is the circumradius of W, the radius of the smallest circle containing W (or the minimum of the circumradii of the connected components of W, if W consists of several pieces). If W is a rectangle, $\widehat{R}_{iso}$ is equal to half the diagonal of W.

The isotropic estimator can be modified to work for longer distances. For a given distance r, let $W^*(r)$ be the subset of W consisting of locations u where $p(u, r) > 0$. Then the modified estimator of Ohser [518] is

$$\widehat{K}_{iso*}(r) = \frac{1}{\overline{\lambda}n} \sum_{i=1}^{n} \sum_{\substack{j=1 \\ j \neq i}}^{n} \mathbf{1}\left\{d_{ij} \leq r\right\} \frac{2\pi d_{ij}}{\ell(W \cap \partial b(x_i, d_{ij}))} \frac{|W|}{|W^*(d_{ij})|} \tag{7.18}$$

and $\overline{\lambda}^2 \widehat{K}_{iso*}(r)$ is an unbiased estimator of $\lambda^2 K(r)$ for all $r < \widehat{R}_{iso*}$. Here $\widehat{R}_{iso*}$ is the smallest distance R such that $|W^*(R)| = 0$, which is equal to the diameter of W if W is a connected set. The extra factor in (7.18) is the reciprocal of the fraction $|W^*(d_{ij})|/|W|$ of window area occupied by $W^*(d_{ij})$. Of course (7.18) agrees with (7.17) for $r < \widehat{R}_{iso}$ so this calculation is only required when $\widehat{R}_{iso} \leq r < \widehat{R}_{iso*}$.

For any $r < \widehat{R}_{iso}$, the double sum in (7.17), namely $\overline{\lambda}n\widehat{K}_{iso}(r)$, is an unbiased estimator of $\lambda^2 |W| K(r)$.

The Ohser modified isotropic correction provides estimates of $K(r)$ for much larger distances r than are possible with the border correction. However, at large distances, the variance of the estimator increases rapidly. This is a well-known problem with the Horvitz-Thompson estimator: if the probability of sampling an item is very small, it will be given a very large weight if it is sampled, so its contribution has large variance.

A pragmatic solution is to restrict the range of r values, and to truncate the edge correction weight in (7.18) so that it never exceeds a specified upper limit. This is the procedure used in `spatstat`.

Stein [625, 626] proposed a more satisfactory solution, in which contributions to the estimator are downweighted if they have large variance. Future versions of `spatstat` will include this estimator.

7.4.5   Translation correction

Another approach to edge correction can be followed if we are not willing to assume the point process is isotropic, or if the computations required for the isotropic correction are too intensive.

Visualise a stationary point process $\mathbf{X}$ (a 'homogeneous field of wildflowers'). Stationarity means that if we apply any translation (shift) vector s to the entire point process, the shifted process $\mathbf{X} + s$ is statistically equivalent to $\mathbf{X}$. Focus on a particular pair of random points x and x' in $\mathbf{X}$. If $\mathbf{X}$ is shifted to $\mathbf{X} + s$, the points x, x' are shifted to new positions $x + s$ and $x' + s$. They are in the same *relative* position: that is, the vector difference $v = x' - x$ from x to x' remains unchanged.

We may imagine that x and x' are joined by an arrow. When $\mathbf{X}$ is shifted to $\mathbf{X} + s$, the arrow's length and direction (given by $v = x' - x$) remain fixed, but the arrow's position is shifted by the vector s.

An arrow will be observed inside the window W only when both its endpoints x, x' fall in W. Intuitively it is clear that a shorter arrow is more likely to fall inside W than a longer arrow, and this gives rise to a sampling bias: the observed arrows are a biased sample of all arrows, with the bias favoring shorter arrows.

Consider the possible positions of the starting point x for which *both* endpoints x and $x' = x + v$ fall inside W. Notice that $x + v \in W$ if and only if x belongs to $W - v$, a copy of the window W shifted by the vector $-v$. Both endpoints of the arrow fall inside W if and only if $x \in W \cap (W - v)$. The region $W \cap (W - v)$ is shaded in Figure 7.18. It is the intersection of the window W with a shifted copy of itself. The probability that the starting point x falls in the shaded region $W \cap (W - v)$ is related to the *area* of this region.

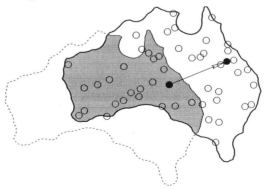

Figure 7.18. *Sampling bias calculation for the translation edge correction. An arrow joining a pair of random points will be observed only if both endpoints fall in the window W. The shaded region shows the possible locations for the start of the arrow which guarantee that both ends of the arrow will be observed.*

Applying the Horvitz-Thompson principle leads to the *translation correction* [518, 520] estimator of the *K*-function,

$$\widehat{K}_{trans}(r) = \frac{1}{\lambda n} \sum_{i=1}^{n} \sum_{\substack{j=1 \\ j \neq i}}^{n} \mathbf{1}\left\{d_{ij} \leq r\right\} \frac{|W|}{|W \cap (W - (x_j - x_i))|}. \tag{7.19}$$

In this estimator, each pair of data points (x_i, x_j) is weighted by the reciprocal of the fraction of window area in which the first data point x_i could be placed so that both points x_i, x_j would be observable (assuming their relative positions were held fixed). This effectively compensates for the sampling bias in observing such pairs.

The edge correction weight in (7.19) can be calculated using simple geometry if W is a rectangle. Otherwise, the most efficient algorithm computes the *set covariance function* $C_W(v) = |W \cap (W - v)|$ for all vectors v on a fine grid using the Fast Fourier Transform (because C is the convolution of two indicator functions) and then extracts the values $C_W(x_j - x_i)$ for each pair of points.

The Horvitz-Thompson principle gives an unbiased estimator provided that each item in the population has a non-zero probability of being sampled. Consequently $\overline{\lambda}^2 \widehat{K}_{trans}(r)$ is an unbiased estimator of $\lambda^2 K(r)$ for all $r \leq R$, where R is the length of the shortest vector v such that $C_W(v) = 0$. If W is a rectangle, then R is the length of the shortest side.

The translation correction provides estimates of $K(r)$ for larger distances r than are possible with the isotropic correction, but again suffers from inflated variance at these large distances. Again the pragmatic solution used in `spatstat` is to restrict the range of r values, and to truncate the edge correction weight in (7.19) so that it never exceeds a specified upper limit.

If the point process is assumed to be isotropic as well as stationary (Section 5.6.3), the translation correction can be modified to give the *rigid motion correction* [468, 402, 520]

$$\widehat{K}_{rigid}(r) = \frac{1}{\lambda n} \sum_{i=1}^{n} \sum_{\substack{j=1 \\ j \neq i}}^{n} \mathbf{1}\left\{d_{ij} \leq r\right\} \frac{|W|}{\overline{C}_W(d_{ij})} \tag{7.20}$$

where $\overline{C}_W$ is the rotational average of C_W, that is, $\overline{C}_W(r)$ is the average value of $C_W(v)$ over all vectors v of length $\|v\| = r$.

7.4.6 Discussion

Edge corrections are discussed and compared in [576, chap. 3], [355, sec. 4.3.3], [59, 640]. The choice of edge correction depends partly on the size of dataset. In small datasets (say, fewer than 100 points) statistical performance is very important, and the methods of choice are the Horvitz-Thompson style weighted edge corrections (translation, isotropic or rigid motion correction); the border method is too imprecise. In moderately large datasets (say 1000 to 10,000 points) the border method performs satisfactorily. In huge datasets no edge correction is necessary, and it is computationally efficient to avoid edge correction. The `spatstat` algorithm for `correction="none"` will handle datasets containing many millions of points.

The choice of edge correction does not, in most instances, seem to be very important, as long as *some* edge correction is applied. (Although it must be applied correctly!) Discrepancies between the results of different edge corrections often tend to indicate unusual features in the data, and usually suggest that the assumption of stationarity is violated.

The accuracy of estimators of the *K*-function is also affected by the choice of estimator for the squared intensity λ^2. Substituting the natural estimator of λ does not necessarily lead to the most efficient estimator of $K(r)$. Because $\widehat{K}(r)$ is the ratio of an estimator of $\lambda^2 K(r)$ divided by an estimator of λ^2, positive correlation between the numerator and denominator will generally reduce the variability of $\widehat{K}(r)$. Various options for the denominator are canvassed in [355, sec. 4.3.3]. One example favoured in astronomy [317, 401] is

$$\widehat{\lambda}_S(r) = \frac{1}{\overline{C}_W(r)} \sum_i p(x_i, r) \tag{7.21}$$

where, as above, $p(u, r)$ is the length fraction of the circle of radius r centred at location u that lies inside the window W, and $C_w(r)$ is the rotational average of the set covariance function.

7.5   Function objects in `spatstat`

7.5.1   Objects of class `"fv"`

An object of class `"fv"` ('function value table') is a convenient way of storing and plotting several different estimates of the same function.

The value returned by Kest is an object of class `"fv"`:

```
> KC <- Kest(cells)
> KC
Function value object (class 'fv')
for the function r -> K(r)
.............................................
        Description
r       distance argument r
theo    theoretical Poisson K(r)
border  border-corrected estimate of K(r)
trans   translation-corrected estimate of K(r)
iso     isotropic-corrected estimate of K(r)
.............................................
Default plot formula:  .~r
where "." stands for 'iso', 'trans', 'border', 'theo'
Recommended range of argument r: [0, 0.25]
Available range of argument r: [0, 0.25]
```

An `"fv"` object is a data frame (that is, it also belongs to the class `"data.frame"`) with attributes giving extra information such as the recommended way of plotting the function. One column of the data frame contains evenly spaced values of the distance argument r, while the other columns contain estimates of the value of the function, or the theoretical value of the function under CSR, corresponding to these distance values. The printout for KC above indicates that the columns in the data frame are named r, theo, border, trans, and iso, and explains their contents. For example, the column iso contains estimates of the K-function using the isotropic edge correction. The function argument in an `"fv"` object is usually, but not always, called r.

7.5.2   Plotting `"fv"` objects

If f is an object of class `"fv"`, then `plot(f)` is dispatched to the method `plot.fv`. The default behaviour of `plot(f)` is to generate a plot containing several curves, each representing a different edge-corrected estimate of the same target function, plotted against the distance argument r. For example, the command `plot(Kest(swedishpines))` generates the left panel of Figure 7.19, which shows three different estimates of the K-function for the swedishpines dataset, together with the 'theoretical' (expected) curve $K_{pois}(r) = \pi r^2$ for CSR, all plotted against the distance argument r. The legend indicates the meaning of each curve. The main title identifies the function object that was plotted.

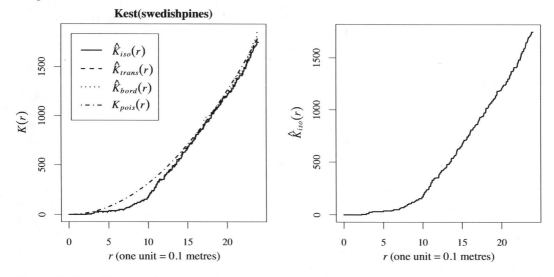

Figure 7.19. *Plots produced by* `plot.fv`. *Left:* *default plot in response to the command* `plot(Kest(swedishpines))`. *Right:* *plot of isotropic correction estimate* $\widehat{K}_{iso}(r)$ *against r.*

The return value from `plot.fv` is a data frame containing more detailed information about the meaning of the curves. For the left panel of Figure 7.19, the return value is

```
        lty col              label                           meaning
iso       1   1  italic(hat(K)[iso](r))    isotropic-corrected estimate of K(r)
trans     2   1 italic(hat(K)[trans](r)) translation-corrected estimate of K(r)
border    3   1 italic(hat(K)[bord](r))     border-corrected estimate of K(r)
theo      4   1     italic(K[pois](r))            theoretical Poisson K(r)
```

Here `lty` and `col` are the graphics parameters controlling the line type and line colour, and `label` is the mathematical notation for each edge-corrected estimate, in the syntax recognised by R graphics functions.

The default plot in the left panel of Figure 7.19 can easily be modified. Set `legend=FALSE` to suppress the legend, and `main=""` to suppress the main title. The legend position is automatically computed to avoid overlap with the plotted curves, but this can be overridden by `legendpos`. For further options see `help(plot.fv)`.

The printout of an `"fv"` object indicates the range of values of `r` in the table as the 'available range'. It also gives a 'recommended range' which is generally shorter than the available range. *The default plot of the object will only show the function values over the recommended range* and not over the full range of values available. This is done so that the interesting detail is clearly visible in the default plot. Values outside the recommended range may be unreliable due to increased variance or bias, depending on the edge correction. To change the range of `r` values, use the argument `xlim` in the plot command.

The variables plotted on the x and y axes are determined by the second argument to `plot.fv`, which should be a `formula`, involving the names of columns of the object. The left side of the formula represents what variables will be plotted on the y-axis, and the right side determines the x variable for the plot. For example, in the object `Kest(swedishpines)`, the column named `iso` contains the values of the isotropic correction estimate. The command

```
> Ks <- Kest(swedishpines)
> plot(Ks, iso ~ r)
```

plots the isotropic correction estimate against *r* as shown in the right panel of Figure 7.19.

The many uses of formulae for controlling plots were introduced in Section 2.1.10. In `plot.fv`, both sides of the plot formula are interpreted as *mathematical expressions*, so that operators like '+', '-', '*', '/' have their usual meaning in arithmetic. The right-hand side of the formula can be any expression that, when evaluated, yields a numeric vector, and the left-hand side is any expression that evaluates to a vector or matrix of compatible dimensions.

If evaluation of the left-hand side yields a matrix, then each column of that matrix is plotted against the specified *x* variable as a separate curve. In particular the left-hand side of the formula may invoke the function `cbind` to indicate that several different curves should be plotted. For example, to plot only the translation and isotropic correction estimators and the theoretical curve, an appropriate syntax is: `plot(Ks, cbind(iso, trans, theo) ~ r)` (results not shown).

The plot formula may also involve the names of constants like `pi`, standard functions like `sqrt`, and some special abbreviations listed in Table 7.1. The symbol `.x` represents the function argument, usually named `r`. One of the columns of function values will be designated as the 'best' estimate, for use by some other commands in `spatstat`. This column is identified by the symbol `.y`. Some or all of the columns of function values are designated as 'acceptable' estimates for the default plot, and these are identified by the symbol '`.`'. The default plotting formula is `. ~ .x` indicating that the acceptable estimates will be plotted against the function argument.

`.x`	argument of function
`.y`	best estimate of function
`.`	all estimates of function for default plot
`.a`	all columns of function values
`.s`	upper and lower limits of shading

Table 7.1. *Recognised abbreviations for columns of an* `"fv"` *object. To expand these abbreviations, use* `fvnames(f, ".x")` *and so on.*

A plot formula can be used to specify a transformation that should be applied to the function values before they are displayed. For example, to divide each of the function estimates by the theoretical Poisson value, one could use `plot(Ks, . / theo ~ r)`. The resulting plot is shown in the left panel of Figure 7.20. Alternatively one could plot the function estimates *against* the Poisson value, using `plot(Ks, . ~ theo)`. This is shown in the right panel of Figure 7.20.

The mathematical labels for the plot axes, and for the individual curves, are constructed automatically by `spatstat` from the plot formula. If the plot formula involves the names of external variables, these will be rendered in Greek where possible. For example, to plot the average number of trees surrounding a typical tree in the Swedish Pines data,

```
> lambda <- intensity(swedishpines)
> plot(Ks, lambda * . ~ r)
```

Here we use the name `lambda` so that it will be rendered as the Greek letter λ in the graphics: the *y*-axis will be labelled $\lambda K(r)$.

In the discussion of Ripley's paper, Cox [178] proposed that $\widehat{K}(r)$ should be plotted against r^2. This also linearises the plot of the *K*-function and has some theoretical support. This plot can easily be generated, using the plot formula `. ~ r^2`.

7.5.3 Manipulating `"fv"` objects

An `"fv"` object can be manipulated using the operations listed in Table 7.2.

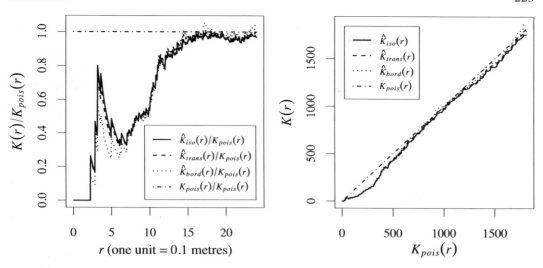

Figure 7.20. Left: *Estimates of $K(r)$ divided by πr^2, plotted against r.* Right: *Estimates of $K(r)$ plotted against πr^2.*

`f`	print a description
`print(f)`	print a description
`plot(f)`	plot the function estimates
`as.data.frame(f)`	strip extra information (returns a data frame)
`f$iso`	extract column named `iso` (returns a numeric vector)
`f[i,j]`	extract subset (returns an `"fv"` object)
`subset(f, ...)`	extract subset (returns an `"fv"` object)
`with(f, expr)`	perform calculations with columns of data frame
`eval.fv(expr)`	perform calculations with several `"fv"` objects
`cbind(f1, f2, ...)`	combine `"fv"` objects `f1`, `f2`, ...
`bind.fv(f, d)`	combine an `"fv"` object `f` and data frame `d`
`collapse.fv(f1, f2, ...)`	combine several redundant `"fv"` objects
`compatible(f1, f2, ...)`	check whether `"fv"` objects are compatible
`harmonise(f1, f2, ...)`	make `"fv"` objects compatible
`min(f),max(f),range(f)`	range of function values
`Smooth(f)`	apply smoothing to function values
`deriv(f)`	derivative of function
`stieltjes(g,f)`	compute Stieltjes integral with respect to `f`
`as.function(f)`	convert to a function

Table 7.2. *Operations for manipulating an `"fv"` object `f`.*

Values of the function

An `"fv"` object is essentially a table, containing the values of the desired function (such as $K(r)$) at a finely spaced grid of values of the function argument r. The function values can be extracted directly from an `"fv"` object since it is also a data frame. A single column of values can be extracted using the $ operator in the usual way: `Ks$iso` would extract a vector containing the isotropic correction estimates of $K(r)$.

The subset extraction operator '[' has a method for `"fv"` objects. This always returns another `"fv"` object, so it will refuse to remove the column containing values of the function argument r, for example. To override this refusal, convert the object to a data frame using `as.data.frame` and then use '`[`': the result will be a data frame or a vector.

Commands designed for data frames often work for `"fv"` objects as well. The functions `head` and `tail` extract the top (first few rows) and bottom (last few rows) of a data frame. They also work on `"fv"` objects: the result is a new `"fv"` object containing the function values for a short interval of r values at the beginning or end of the range. The function `subset` selects designated subsets of a data frame using an elegant syntax (see page 101) and this also works on `"fv"` objects. To restrict Ks to the range $r \leq 0.1$ and remove the border correction,

```
> Ko <- subset(Ks, r < 0.1, select= -border)
```

Convert to a true function

The table of function values can also be converted to a true function in the R language using `as.function`. This makes it easy to evaluate the function at any desired distance r.

```
> Ks <- Kest(swedishpines)
> K <- as.function(Ks)
> K(9)
[1] 129.096
```

By default, the result K is a function in R, with a single argument `r` (or whatever the original function argument was called). The new function accepts numeric values or numeric vectors of distance values, and returns the values of the 'best' estimate of the function, interpolated linearly between entries in the table. If one of the other function estimates is required, use the argument `value` to `as.function` to select it. Several estimates can be chosen:

```
> K <- as.function(Ks, value=".")
> K(9)
[1] 129.096
> K(9, "trans")
[1] 130.4998
```

Calculating with columns

To manipulate or combine one or more columns of data in an `"fv"` object, it is typically easiest to use `with.fv`, a method for the generic `with`. For example:

```
> Kr <- Kest(redwood)
> y <- with(Kr, iso - theo)
> x <- with(Kr, r)
```

The results x and y are numeric vectors, where x contains the values of the distance argument r, and y contains the difference between the columns `iso` (isotropic correction estimate) and `theo` (theoretical value for CSR) for the K-function estimate of the redwood seedlings data. For this to work, we have to know that Kr contains columns named `r`, `iso` and `theo`. Printing the object will reveal this information, as would typing `names(Kr)` or `colnames(Kr)`.

The general syntax is `with(X, expr)` where X is an `"fv"` object and `expr` can be any expression involving the names of columns of X. The expression can include functions, so long as they are capable of operating on numeric vectors. The expression can also involve the abbreviations listed in Table 7.1. Thus: `Kcen <- with(Kr, . - pi*r^2)` subtracts the 'theoretical' value from all the available edge correction estimates. In this case the result Kcen is an `"fv"` object. You can also get a result which is a vector or single number:

```
> with(Kr, max(abs(iso-theo)))
[1] 0.04945199
```

Calculating with several `"fv"` objects

To manipulate or combine several `"fv"` objects, `spatstat` provides the special function `eval.fv`. Its argument should be an expression involving *the names of* the `"fv"` objects which are to be combined. For example, to find the difference between the K-functions of the `redwood` and `cells` datasets,

```
> K1 <- Kest(redwood) ; K2 <- Kest(cells)
> DK <- eval.fv(K1-K2)
```

The result `DK` is another `"fv"` object. Note that something like `eval.fv(Kest(redwood) - Kest(cells))` would not work, because `redwood` and `cells` are not `"fv"` objects.

The objects in the expression should be 'compatible' in the sense that they have the same column names, and the same vector of r values. However, `eval.fv` will attempt to reconcile incompatible objects. (The `spatstat` generic function `compatible` determines whether two or more objects are compatible, and the generic function `harmonise` makes them compatible, if possible.)

Objects of class `"fv"` are data frames of function values, with extra attributes which explain the function values and determine how they are plotted. The function `eval.fv` performs arithmetic on the columns of function values, but also manipulates these extra attributes, so that when the result of `eval.fv` is plotted, the labels on the plot are constructed in a meaningful way.

7.6 The pair correlation function

7.6.1 The pair correlation function of a point process

A plot of the K-function can be difficult to interpret correctly in terms of the behaviour of the point process, because of its 'cumulative' nature (Section 7.3.5). The value of $K(r)$ contains contributions for all interpoint distances *less than or equal to r*.

An alternative tool is the **pair correlation function** $g(r)$ which contains contributions only from interpoint distances *equal to r*. In two dimensions, it can be defined by

$$g(r) = \frac{K'(r)}{2\pi r} \tag{7.22}$$

where $K'(r)$ is the derivative of the K-function with respect to r. See [355, pp. 218–223, 232–244]. The pair correlation function has a long history in astronomy [204, 317, 401, 552], mostly for three-dimensional patterns. It has also arisen independently in geographical analysis and other fields [285].

In geometric terms, $K(r)$ is defined by drawing a circle of radius r centred on a point of the point process, and counting the number of other points falling in the circle (see left panel of Figure 7.21). To define $g(r)$, draw two concentric circles of radius r and $r+h$, where h is a small increment of distance, and count only those points falling in the ring between the two circles (see right panel of Figure 7.21). This captures the interpoint distances d_{ij} that lie in the narrow range between r and $r+h$. The expected count of such distances is $\lambda K(r+h) - \lambda K(r)$. We standardise the expected count by dividing it by the expected value for complete spatial randomness, which is $\lambda \pi (r+h)^2 - \lambda \pi r^2$, yielding

$$g_h(r) = \frac{\lambda K(r+h) - \lambda K(r)}{\lambda \pi (r+h)^2 - \lambda \pi r^2} = \frac{K(r+h) - K(r)}{2\pi rh + \pi h^2}.$$

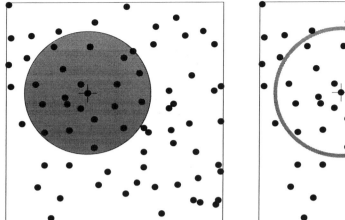

Figure 7.21. *Geometry of the K-function (Left) and the pair correlation function (Right).*

When h is very small, πh^2 becomes negligible, and $(K(r+h) - K(r))/h$ becomes the derivative of K, so that $g_h(r)$ becomes the pair correlation function (7.22).

The concept of the pair correlation function has also been reinvented in different guises. The *O-ring statistic* [700] is essentially $K(r+h) - K(r)$, where now the ring thickness h is an algorithm parameter that is held fixed and should not be too small. The O-ring statistic is approximately $hg(r)$. The K *density* [247] is defined as the derivative of the K-function, which is therefore exactly equal to a multiple of the pair correlation function, $K'(r) = 2\pi r g(r)$. Many writers seem to be unaware of the existence of the pair correlation function and its long history in astronomy and other fields.

Imagine the observation window is divided into a fine grid of pixels. The probability that a given pixel contains a random point is $p \doteq \lambda a$ where a is the pixel area. Here we use the symbol $\doteq$ to mean that the quantities are equal when they are very small.

Figure 7.22. *Interpretation of the pair correlation $g(r)$ as the (normalised) probability that two pixels, a distance r apart, both contain random points.*

Now consider two pixels with centres u and v, separated by a distance r. See Figure 7.22. Let $p_2(u,v)$ be the probability that *both* pixels contain random points. If the process is Poisson, then events in these pixels are independent, and $p_2(u,v) = p^2$. For any stationary and isotropic point process, it turns out that

$$g(r) \doteq \frac{p_2(u,v)}{p^2}. \qquad (7.23)$$

That is, $g(r)$ *is the probability of observing a pair of points of the process separated by a distance r, divided by the corresponding probability for a Poisson process.* In other words, if U and V are two 'infinitesimal' regions with areas du and dv, respectively, separated by a distance r then

$$\mathbb{P}\{\mathbf{X} \text{ has a point in } U \text{ and a point in } V\} \doteq \lambda^2 g(r) \, du \, dv \qquad (7.24)$$

where λ is the intensity of the process.

The value $g(r) = 1$ is consistent with complete spatial randomness, because of the way the function has been standardised. A value $g(r) < 1$ indicates that interpoint distances equal to r are less frequent than would be expected for a completely random process, so this suggests regularity. A value $g(r) > 1$ indicates that this interpoint distance is more frequent than expected for a completely random pattern, which suggests clustering. These interpretations are less ambiguous than the corresponding statements for the K-function, because they refer only to interpoint distances equal to r.

Notice that the value $g(r)$ is not a correlation in the sense used by statisticians. The correlation between two random variables is a number standardised to lie in the range between -1 and $+1$, with the value 0 indicating a lack of correlation. It turns out to be impractical to standardise the correlation structure of point processes in this way. Instead, $g(r)$ is the kind of correlation often used in physics: the possible values range from 0 to infinity, and the value 1 is associated with a lack of correlation.

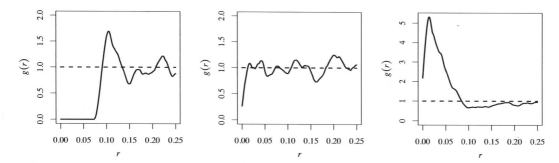

Figure 7.23. *Empirical pair correlation function for each of the three patterns in Figure 7.1.*

Figure 7.23 shows estimates of the pair correlation function for the three archetypal point patterns. Compare these with the empirical K-functions shown in Figure 7.7 on page 206.

The left panel of Figure 7.7 shows that the K-function for the regular pattern is below the Poisson curve for all distances r, suggesting (if we are not careful) that there is inhibition at *all* distances. The plot of the pair correlation has a different message: $g(r)$ is below the Poisson value, $g(r) < 1$, for distances r up to 0.07 units, then it climbs steeply to a value *greater than* the Poisson value, $g(r) > 1$ at about $r = 0.10$ units, before declining back to 1. In fact $g(r)$ is zero for $r \leq 0.07$ because this extremely regular pattern has no interpoint distances shorter than 0.07 units. The reason for the apparent discrepancy between g and K is clear: short interpoint distances are completely absent in this pattern, and this deficit affects the cumulative count of pairwise distances in the K-function, even at much larger distances r.

The K-function and the pair correlation function g are mathematically interrelated: g can be obtained from K through (7.22), and K can be obtained from g by the reverse relationship

$$K(r) = 2\pi \int_0^r sg(s)\,ds. \tag{7.25}$$

One may ask why anyone would use the K-function, if the pair correlation function g is easier to interpret. The main reason is that K seems to perform better as a basis for statistical inference (such as hypothesis tests).

Also, in theory g does not necessarily exist: it is essentially the derivative of the K-function, and this does not exist if the K-function has jumps. For example, for a regular grid of points (randomly translated to make it a stationary point process), with points spaced at multiples of s units, the K-function has a jump at $r = s$ and at $r = 2s$, $r = s\sqrt{2}$, and so on. This process has no pair correlation function. This example could be realistic for some crystalline structures.

The practical interpretation of the pair correlation function is discussed in [355, sec. 4.3.4].

7.6.2 Estimating the pair correlation function

A popular way to estimate the pair correlation function is by kernel smoothing [635, p. 126], [242, 257, 629, 637, 453, 531]. In very large samples, histogram-based methods could be used, and these are common in astronomical applications.

Taking any edge-correction estimator of the K-function of the general form (7.3), suppose we replace the indicator $\mathbf{1}\{d_{ij} \leq r\}$ by a kernel term $\kappa(d_{ij} - r)$ to obtain a smooth estimate of $K'(r)$, then plug into the formula (7.22) for the pair correlation function, to obtain the 'fixed-bandwidth' kernel estimator [638, eq. (15.15), p. 284]

$$\widehat{g}(r) = \frac{|W|}{2\pi rn(n-1)} \sum_{i=1}^{n} \sum_{\substack{j=1 \\ j \neq i}}^{n} \kappa_h(r - d_{ij}) e_{ij}(r) \tag{7.26}$$

where $d_{ij} = \|x_i - x_j\|$ is the interpoint distance between the ith and jth data points, $e_{ij}(r)$ is an edge correction weight, and κ_h is the smoothing kernel, with smoothing bandwidth $h > 0$.

The kernel κ_h is a rescaled version of a template kernel κ,

$$\kappa_h(x) = \frac{1}{h} \kappa \left(\frac{x}{h} \right) \tag{7.27}$$

where κ is any chosen function that is a probability density over the real line with mean 0. For example, κ could be the standard normal (Gaussian) density, so that κ_h would be the normal density with mean 0 and standard deviation h. The convention followed in R and in `spatstat` is that the template κ has standard deviation 1, so that κ_h has standard deviation h. The 'smoothing bandwidth'' is the standard deviation of the smoothing kernel. (Note that other writers may use a different convention.)

The usual choice of smoothing kernel for pair correlation functions is the *Epanechnikov kernel* with half-width w,

$$\varepsilon_w(x) = \frac{3}{4w} \left(1 - \frac{x^2}{w^2} \right)_+ \tag{7.28}$$

where $(x)_+ = \max(0, x)$. The kernel is a quadratic function truncated to the interval $[-w, w]$. The standard deviation of $\varepsilon_w(x)$ is $h = w/\sqrt{5}$, so the kernel of bandwidth h is $\kappa_h(x) = \varepsilon_{h\sqrt{5}}(x)$.

To compute (7.26) in practice we must choose a bandwidth value h, and as is common in data analysis, this choice involves a compromise between bias and variability. For excessively large bandwidth h, variability will be well controlled but the shape of the function $\widehat{g}$ will be oversmoothed, so that important detail in the pair correlation function may be missed. For excessively small bandwidth, the estimates $\widehat{g}(r)$ will have high variance, and the shape of $\widehat{g}$ will be erratic.

The standard rule of thumb [355, p. 236] is to take the half-width w of the Epanechnikov kernel (7.28) to be $w = c/\sqrt{\widehat{\lambda}}$ where c is a constant between 0.1 and 0.2. This corresponds to taking the bandwidth

$$h = \frac{c'}{\sqrt{\widehat{\lambda}}} \tag{7.29}$$

where $c' = c/\sqrt{5}$ is between 0.045 and 0.090. Alternative bandwidth selection and bias correction methods are discussed in [300], [355, p. 230 ff.]. The variance of $\widehat{g}(r)$ is discussed in [355, p. 232 ff.], [632]. The choice of denominator (related to estimation of the intensity) is also important to controlling the variance, as mentioned in Section 7.4.6.

To compute the estimated pair correlation function in `spatstat` use `pcf`.

```
> g <- pcf(cells)
> plot(g)
```

The function `pcf` is generic, with a method for point patterns. By default, `pcf.ppp` uses the estimator (7.26), with the Epanechnikov kernel, with bandwidth h selected by Stoyan's rule of thumb (7.29) with $c = 0.15$ so that $c' = 0.067$.

In analysing a point pattern dataset, attention is often focused on the behaviour of $g(r)$ for small r, which reflects the correlation between close pairs of points. Unfortunately, estimation of $g(r)$ at small distances is intrinsically difficult, because geometry dictates that there will typically be few observations at small distances. The estimator (7.26) has poor performance at small distances. Theoretically, as r approaches zero, the variance of $\widehat{g}(r)$ becomes infinite, for many point processes including CSR. The left panel of Figure 7.24 shows a real example where $\widehat{g}(r)$ has an infinite asymptote at $r = 0$. These problems are due to the factor $1/r$ in (7.26). An alternative estimator with usually better performance is

$$\widehat{g}(r) = \frac{1}{2\pi} \sum_{i=1}^{n} \sum_{\substack{j=1 \\ j \neq i}}^{n} \frac{\kappa_h(r - d_{ij})}{d_{ij}} e_{ij}(r) \tag{7.30}$$

in which the contribution to $\widehat{g}(r)$ from an observed interpoint distance d_{ij} involves a factor $1/d_{ij}$ rather than $1/r$. This estimator can be invoked by specifying `divisor="d"` in `pcf.ppp`. The right panel of Figure 7.24 shows this estimator computed for the same real example.

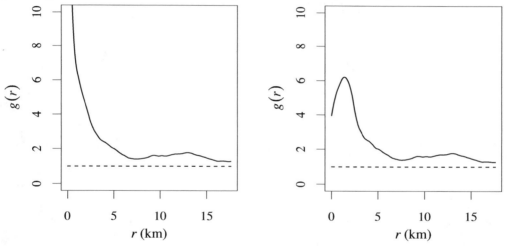

Figure 7.24. *Estimates of the pair correlation function of the Queensland copper deposits of Figure 1.11, using* (Left) *the divide-by-r estimator* (7.26) *and* (Right) *the divide-by-d estimator* (7.30).

Another useful trick applies when an estimate of $K(r)$ is already available: then we can calculate $g(r)$ from $K(r)$ through equation (7.22). This is particularly useful in large datasets, where direct estimation of $g(r)$ can be time-consuming. The conversion of $K(r)$ to $g(r)$ is done by the method `pcf.fv`, which computes the derivative by numerical differentiation using spline smoothing. Effectively the function $K(r)$ is first smoothed by fitting polynomial approximations to $K(r)$ over successive ranges of r values. The derivatives of the polynomials are then computed, using elementary calculus.

```
> K <- Kest(shapley)
> g <- pcf(K, spar=0.5)
```

Since K is an object of class `"fv"`, the last command is dispatched to `pcf.fv`. Additional arguments to `pcf.fv`, such as the argument `spar` above, are passed to the function `smooth.spline` in the `stats` package, which performs the spline smoothing. Using a little algebra, we may apply smoothing either to $K(r)$ or to the transformed functions $K(r)/(2\pi r)$, $K(r)/(\pi r^2)$, or $L(r) = \sqrt{K(r)/\pi}$, by specifying `method="a"`, `"b"`, `"c"` or `"d"`, respectively. Methods `"b"` to `"d"` offer improved performance at small distances r. The default is `method="c"` which seems to perform best overall. See `help(pcf.fv)` for more information.

7.7   Standard errors and confidence intervals

Statistical estimates should be accompanied by a measure of accuracy (such as a standard error) or uncertainty (such as a confidence interval). This is not straightforward for the K-function: there is no simple expression for the standard error or variance of $\widehat{K}(r)$, even in a Poisson process, because the contributions to $\widehat{K}(r)$ from different data points x_i are not independent.

7.7.1   Variance under CSR

Approximations to the standard error of $\widehat{K}(r)$ *assuming complete spatial randomness* are available, using the techniques of U-statistics [576, 518, 433, 225]. Ripley [576] gave an approximate variance estimate for the isotropic corrected K-function estimator,

$$\text{var}\,\widehat{K}_{iso}(r) \approx 2\left(\frac{A}{n-1}\right)^2 \left(\frac{\pi r^2}{A} + 0.96\frac{Pr^3}{A^2} + 0.13\frac{nPr^5}{A^3}\right) \qquad (7.31)$$

where n is the number of data points and A and P are the area and perimeter length of the window W. Lotwick and Silverman [433] gave a more intricate and accurate approximation for the case of a rectangular window. These approximations are chiefly useful for predicting the accuracy that can be expected from a survey region of given size and shape. They are computed by `Kest` if the argument `var.approx=TRUE` is given. For the L-function, the corresponding approximations for $\text{var}\,\widehat{L}_{iso}(r)$ are calculated by the delta method.

7.7.2   Block bootstrap

Bootstrap methods [250, 340] make it possible to estimate variance from the data, without assuming the Poisson model. A very simple version of this approach is *block variance estimation*. Assuming the window W is a rectangle, divide it into equal quadrats $B_1, \ldots, B_m$. Suppose our estimator of $K(r)$ is (7.3). For a particular quadrat B, let

$$\widehat{K}(r,B) = \frac{m|W|}{n(n-1)} \sum_{x_i \in B} \sum_{j \neq i} \mathbf{1}\{d_{ij} \leq r\}\, e_{ij}(r) \qquad (7.32)$$

be an estimate of $K(r)$ based only on the pairs (x_i, x_j) for which x_i falls in the quadrat B, while x_j may lie anywhere in W. The usual estimate $\widehat{K}(r)$ is the average of the estimates $\widehat{K}(r, B_k)$ over all $k = 1, \ldots, m$. We compute the pointwise sample mean, sample variance, and sample standard deviation of these estimates, yielding an elementary bootstrap estimate [225, eq. (4.21), p. 52] of the standard error for $\widehat{K}(r)$. Assuming $\widehat{K}(r)$ is approximately normally distributed, we can then calculate pointwise 95% confidence intervals for the true value of $K(r)$.

The block variance estimate is computed by the `spatstat` function `varblock`:

```
> swp <- rescale(swedishpines)
> Kvb <- varblock(swp, Kest, nx=3, ny=3)
> plot(Kvb)
```

The result is shown in the left panel of Figure 7.25. For a fixed value of r, the grey shading gives a 95% confidence interval for the true value of $K(r)$. Formally this means that, in 95% of outcomes of this procedure, the random interval computed here would embrace the true value of $K(r)$.

Notice that the confidence interval becomes wider as r increases, because the variance of $\widehat{K}(r)$ is roughly proportional to r^2. If we use $L(r)$ instead of $K(r)$, the variance will be stabilised: $\mathrm{var}\widehat{L}(r)$ is approximately constant as a function of r. The confidence intervals will then have roughly constant width.

An important caveat is that these are *pointwise* calculations (i.e. performed separately for each value of distance r) yielding *pointwise* confidence intervals (i.e. valid only for a single value of r at a time). Our confidence that the true K-function lies *entirely inside* the grey shaded region, over all values of r, is vastly less than 95%. A solution to this problem is explained below.

7.7.3 Confidence intervals from Loh's bootstrap

A more flexible bootstrap approach was developed by Loh [430]. First we decompose the empirical K-function (7.3) into the contributions from each individual data point x_i,

$$\widehat{K}(r,x_i) = \frac{|W|}{n-1} \sum_{j \neq i} \mathbf{1}\left\{d_{ij} \leq r\right\} e_{ij}(r), \quad \text{for } i = 1, \ldots, n \tag{7.33}$$

called the **local** K-functions. The usual estimate $\widehat{K}(r)$ is the average of the estimates $\widehat{K}(r,x_i)$ over all $i = 1, \ldots, n$. We could again calculate the pointwise sample mean and sample variance of these local K-functions. Instead, Loh's approach is to resample from the suite of local K-functions. Choose an independent random sample of size n, with replacement, from the numbers 1 to n. If the sample is $i_1, \ldots, i_n$, we calculate the average of the corresponding local K-functions,

$$K^*(r) = \frac{1}{n} \sum_{k=1}^{n} \widehat{K}(r,x_{i_k}). \tag{7.34}$$

This gives a resampled version of $\widehat{K}(r)$. Repeating this procedure a large number N of times, we obtain a bootstrap sample of the distribution of $\widehat{K}(r)$. Pointwise confidence intervals can then be computed directly from the observed quantiles of the bootstrap sample. This calculation is performed by the `spatstat` function `lohboot`:

```
> Kloh <- lohboot(swp, Kest)
> plot(Kloh)
```

The result is shown in the right panel of Figure 7.25. The default behaviour is to compute a pointwise 95% confidence interval.

The grey-shaded intervals in Figure 7.25 are pointwise confidence intervals. Ideally we would prefer a *global* confidence interval or confidence band, having a 95% chance of containing the entire graph of the true K-function. This is possible in Loh's approach. First we replace $K(r)$ by $L(r) = \sqrt{K(r)/\pi}$ to stabilise the variance. For each resampled L-function $L^*(r)$, we calculate the deviation

$$D^* = \max_{0 \leq r \leq r_{max}} |L^*(r) - \widehat{L}(r)|,$$

the maximum vertical separation between the graphs of $L^*(r)$ and $\widehat{L}(r)$ over the interval of r values up to a nominated maximum distance r_{max}. If there are N resampled L-functions, this gives N deviation values; we take the 95th percentile of these deviations, say $D_{0.95}$, and construct the **global confidence band** with boundaries

$$L_-(r) = \widehat{L}(r) - D_{0.95}$$
$$L_+(r) = \widehat{L}(r) + D_{0.95}. \tag{7.35}$$

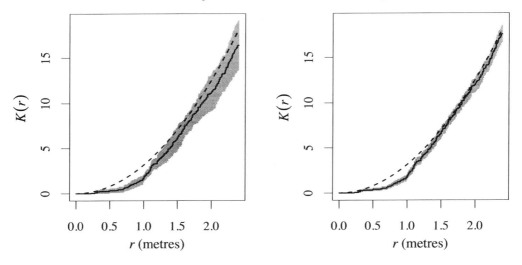

Figure 7.25. *Pointwise 95% confidence intervals for the true K-function of the Swedish Pines using (Left) the spatial block bootstrap* varblock, *and (Right) Loh's bootstrap* lohboot. *Solid lines: estimated K-function. Dashed lines: the theoretical K-function under CSR, i.e.,* $K(r) = \pi r^2$.

This has the desired interpretation that, in 95% of outcomes of this procedure, the true L-function will lie entirely between the two limits $L_-(r), L_+(r)$. This procedure is performed by lohboot with the argument global=TRUE:

```
> Lg <- lohboot(swp, Lest, global=TRUE)
```

A global confidence band for $K(r)$ can then be obtained by back-calculation:

```
> Kg <- eval.fv(pi * Lg^2)
```

The results are plotted in Figure 7.26.

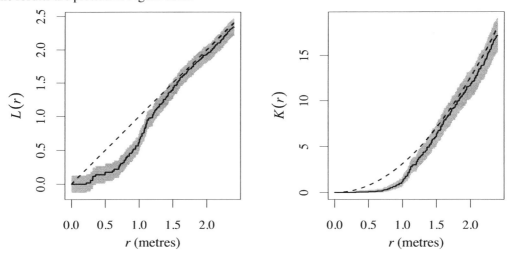

Figure 7.26. *Global 95% confidence interval for the true L-function (Left) and back-transformed confidence interval for the K-function (Right) of the Swedish Pines, using Loh's bootstrap.*

7.8 Testing whether a pattern is completely random

The K-function estimated from a point pattern dataset, $\widehat{K}(r)$, is often compared graphically with the theoretical K-function for the homogeneous Poisson process, $K_{pois}(r) = \pi r^2$, serving as the benchmark of a completely random pattern. In the examples above, large discrepancies between $\widehat{K}(r)$ and $K_{pois}(r)$ suggested that the Swedish Pines are not completely random. However, even with a completely random pattern, we will never obtain perfect agreement between $\widehat{K}$ and K_{pois}, because of random variability. We need a rule for deciding whether the difference between $\widehat{K}$ and K_{pois} is large enough (relative to the scale of variability) to convince us that the point process is not completely random: that is, whether the discrepancy between $\widehat{K}$ and K_{pois} is *statistically significant*. This rule is a *significance test* or *hypothesis test*. The principles of statistical tests are explained in Chapter 10; here we explain the basic idea and show how to conduct the test in `spatstat`.

7.8.1 Pointwise envelopes

In order to assess the statistical significance of deviations between the observed K-function and the theoretical K-function for CSR, essentially we need to know how much variability should be expected in the estimate $\widehat{K}(r)$ if the pattern is completely random.

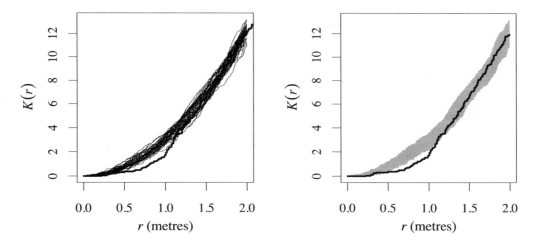

Figure 7.27. *Construction of envelopes of the K-function.* Left: *the K-function estimate of the Swedish Pines dataset (thick line), and of 39 simulated realisations of CSR with the same number of points (thin lines).* Right: *grey shading between the upper and lower* envelopes *(maximum and minimum values) of the K-functions of the simulated patterns.*

The left panel of Figure 7.27 shows the K-function estimated from the rescale `swedishpines` dataset (thick line), and also the K-functions of 39 simulated realisations of CSR with the same number of points (thin lines). Each thin line could have been generated by

```
plot(Kest(runifpoint(npoints(swp), Window(swp))))
```

The spread of the thin lines typifies the variability in $\widehat{K}(r)$ that we should expect if the pattern is completely random with the same number of points as the data. Clearly, the estimate of $K(r)$ for the Swedish Pines data lies outside this range, for some values of r.

In the right panel of Figure 7.27 we have shaded the region between the highest and lowest lines, that is, the minimum and maximum values of the K-functions of the simulated patterns, called the upper and lower *envelopes* of the simulated functions.

The envelope plot can be interpreted as a statistical significance test, as follows. Without peeking at the data, choose a particular distance r, say $r^* = 1$ metre. If the swedishpines data were completely random, then the data and simulated patterns would be statistically equivalent, since they would all be completely random patterns with the same number of points. Consequently the values of $\widehat{K}(r^*)$ for the data and simulated patterns would be statistically equivalent numbers. By symmetry, since there are 40 values altogether (one data value and 39 simulated values), there would be a chance of 1 in 40 that the value for the data is the smallest, i.e. that the data value is smaller than all of the simulated values. Similarly there would be 1 chance in 40 that the data value is larger than all the simulated values. This makes 2 chances in 40, or a probability of $1/20 = 0.05$ that the data value lies outside the range of the simulated values purely by chance. This has in fact occurred in Figure 7.27, so we could declare the result to be statistically significant with a p-value of 0.05.

This is an example of a *Monte Carlo test*, a test based on simulations from the null hypothesis, using a symmetry principle. A full explanation of Monte Carlo tests is given in Chapter 10.

The spatstat function envelope computes simulation envelopes. The right panel of Figure 7.27 was generated by the commands

```
> E <- envelope(swp, Kest, nsim=39, fix.n=TRUE)
> plot(E)
```

The result of envelope is an object of class "envelope" which also belongs to the class "fv". It can be plotted, printed, and manipulated using the tools for "fv" objects described in Section 7.5. See Table 7.2 on page 223. The print method gives a lot of detail:

```
> E
Pointwise critical envelopes for K(r)
and observed value for 'swp'
Edge correction: "iso"
Obtained from 39 simulations of CSR
Alternative: two.sided
Significance level of pointwise Monte Carlo test: 2/40 = 0.05
...........................................................
      Description
r     distance argument r
obs   observed value of K(r) for data pattern
theo  theoretical value of K(r) for CSR
lo    lower pointwise envelope of K(r) from simulations
hi    upper pointwise envelope of K(r) from simulations
...........................................................
Default plot formula:  .~r
where "." stands for 'obs', 'theo', 'hi', 'lo'
Columns 'lo' and 'hi' will be plotted as shading (by default)
Recommended range of argument r: [0, 2.4]
Available range of argument r: [0, 2.4]
Unit of length: 1 metre
```

Arguments to envelope allow us to specify the data pattern, the summary function to be used, the number of simulations, one-sided or two-sided tests, and so on. Instead of testing CSR, we can use envelope to test essentially any null hypothesis. Full details of the envelope function are given in Section 10.8. Table 10.1 on page 397 lists the most useful arguments to envelope.

Envelopes computed in the manner of Figure 7.27 are called *pointwise* envelopes because the maximum and minimum values are calculated separately for each distance r (i.e. for each 'point' on the horizontal axis). A very important caveat about pointwise envelopes is that they can only be interpreted as a statistical significance test if the distance value r^* was chosen in advance. It is not valid to first generate the Figure and then choose a favourable value of r; this would be 'data snooping'. If we allow ourselves this freedom then we are much more likely to find a value of r where the empirical $K(r)$ lies outside the envelopes, so the result is much less significant. The same caveat is emphasised in [572, 431, 40] and is discussed at length in Chapter 10.

7.8.2 Global envelopes

The problem of 'data snooping' can be avoided using *global envelopes*, shown in Figure 7.28. As distinct from the pointwise envelopes in Figure 7.27, the global envelopes in Figure 7.28 delimit a zone of *constant width*. The width is determined by finding the most extreme deviation from the theoretical K-function that is achieved by any of the simulated K-functions, at any distance r along the horizontal axis.

That is, for each simulated dataset, we compute the maximum vertical deviation D between the graphs of $\widehat{K}$ and K_{pois} over some range of distances. The maximum D_{max} of the deviations for all simulated datasets is taken. The global envelopes are

$$E_-(r) = K_{pois}(r) - D_{max}$$
$$E_+(r) = K_{pois}(r) + D_{max}. \tag{7.36}$$

If the graph of the estimated K-function for the data transgresses these limits, it is statistically significant with a p-value of $1/(m+1)$ where m is the number of simulated patterns. Taking $m = 19$ gives a test with significance level 0.05.

The left panel of Figure 7.28 shows global envelopes for the K-function computed by

```
> E <- envelope(swp, Kest, nsim=19, rank=1, global=TRUE)
```

A more powerful test is obtained if we (approximately) stabilise the variance, by using the L-function in place of K, as shown in the right panel of Figure 7.28. This clearly rejects the null hypothesis of CSR at the 0.05 significance level. The two panels used the *same* set of simulated point patterns.

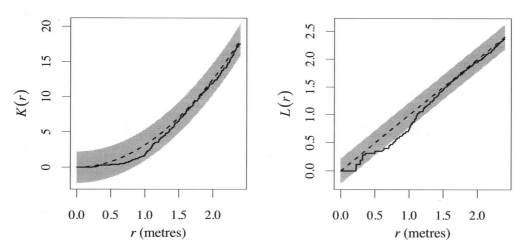

Figure 7.28. *Global envelopes relating to the Swedish Pines data.* Left: *global envelopes for the K-function.* Right: *global envelopes for the L-function. The same simulated patterns were used for both panels.*

Global envelopes were proposed by Ripley over 30 years ago [572, 574, 575], but they are still not widely used in applications, despite the fact that they provide a simple solution to the data snooping problem.

7.8.3 Non-graphical tests

Statistical tests can also be carried out without needing any graphical display. Two useful procedures are the MAD (maximum absolute deviation) test, and the DCLF (Diggle-Cressie-Loosmore-Ford) test, described in Sections 10.7.3 and 10.7.4, respectively. The MAD test [575] is equivalent to plotting the global envelopes described above and declaring a statistically significant outcome if the estimated K-function (etc.) wanders outside the global envelope anywhere:

```
> mad.test(swp, Lest, nsim=99, rmax=2, use.theo=TRUE)
        Maximum absolute deviation test of CSR
        Monte Carlo test based on 99 simulations
        Summary function: L(r)
        Reference function: theoretical
        Alternative: two.sided
        Interval of distance values: [0, 2] metres

data:  swp
mad = 0.29999, rank = 1, p-value = 0.01
```

The DCLF test [223, p. 122], [190, eq. (8.5.42), p. 667], [431] is based on the mean squared deviation between the empirical summary function and the theoretical function over a range of distance values. The DCLF test is more powerful than the MAD test provided the maximum distance is chosen carefully; the MAD test is more reliable when little information is available about the range of dependence [40].

```
> dclf.test(swp, Lest, nsim=99, rmax=2, use.theo=TRUE)$p.value
[1] 0.01
```

Both tests indicate there is strong evidence that the Swedish Pines data are not completely random.

7.9 Detecting anisotropy

A point process is 'isotropic' if all its statistical properties are unchanged when it is rotated (Section 5.6.3). If there is any statistical property of the process that does change when the process is rotated, then the process is 'anisotropic'. A point process could be stationary, but not isotropic. The microstructure of a mineral could be homogeneous, but with a preferential direction, perhaps reflecting the history of deformation [519, 355].

The following code generates a synthetic point pattern which is homogeneous and anisotropic. First we generate a Simple Sequential Inhibition process, and then deform space by squashing the y-axis.

```
> X <- rSSI(0.05, win=owin(c(0,1), c(0, 3)))
> Y <- affine(X, mat=diag(c(1, 1/3)))
```

Figure 7.29 shows the simulated pattern Y and its Fry plot, clearly indicating that the pattern is anisotropic.

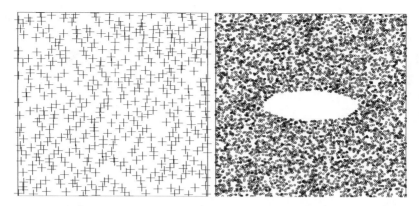

Figure 7.29. *Anisotropy.* Left: *synthetic point pattern exhibiting anisotropy.* Right: *Fry plot.*

7.9.1 Sector *K*-function

Various modifications of the *K*-function have been suggested [520, 636] for measuring anisotropy.

Instead of counting all data points falling inside a circle of radius *r* centred at a data point, we could replace the circle by another geometrical shape.

The *sector* shown in Figure 7.30 is that part of the disc of radius *r* lying between two lines at orientations α and β. Here *r* is the function argument, while the orientations α and β are fixed. The *sector K-function* of a stationary point process is $1/\lambda$ times the expected number of points lying within this sector, given that the vertex of the sector is a point of the process. The sector *K*-function can be estimated using the same edge-correction techniques as the original *K*-function.

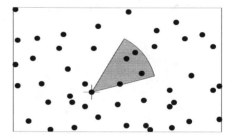

Figure 7.30. *Geometry of the sector K-function.*

The `spatstat` function `Ksector` computes the empirical sector *K*-function using the translation edge correction, together with the theoretical value for a Poisson process, $K_{sector,\,pois}(r) = (\beta - \alpha)r^2/2$ if α, β are expressed in radians.

For example, to compute the sector *K*-functions for two 30-degree angle sectors centred on the *x* and *y* axes, respectively:

```
> Khoriz <- Ksector(Y, begin = -15, end = 15, units="degrees")
> Kvert <- Ksector(Y,  begin = 90-15, end = 90+15, units="degrees")
```

Note that angles are measured anticlockwise from the *x*-axis. Anisotropy would be suggested if these two functions appeared to be unequal. We may then compare them by superimposing the plots, or by using `eval.fv` to compute the difference between the functions. The left panel of Figure 7.31 shows the result of

```
> plot(Khoriz, trans/theo ~ r, lty=2)
> plot(Kvert, trans/theo ~ r, add=TRUE)
```

where we have divided each estimate by the theoretical value.

The right panel of Figure 7.31 shows the difference `Khoriz-Kvert` and a 95% confidence interval computed by the following code:

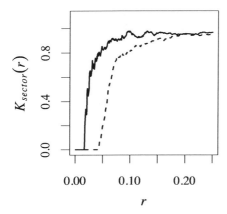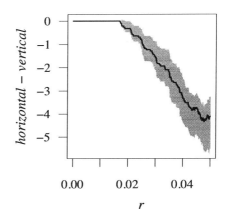

Figure 7.31. *Anisotropy analysis using* Ksector *for the synthetic pattern in Figure 7.29.* Left: *Superimposed plot of normalised* Ksector *functions for 30-degree sectors centred on the horizontal axis (dashed line) and vertical axis (solid line).* Right: *95% confidence interval for the difference ($\times 10^4$) between horizontal and vertical sector K-functions using block bootstrap.*

```
> dK <- function(X, ...) {
    K1 <- Ksector(X, ..., begin = -15, end = 15, units="degrees")
    K2 <- Ksector(X, ..., begin = 90-15, end = 90+15, units="degrees")
    eval.fv(K1-K2)
  }
> CIdK <- varblock(Y, dK, nx=5)
```

The sector K-function requires a choice of the limiting angles α and β, so is best used when there is some prior knowledge about the preferential directions in the pattern.

7.9.2   Pair orientation distribution

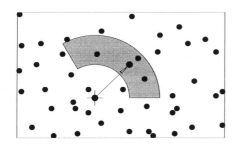

Figure 7.32. *Geometry for the point pair orientation distribution function.*

Anisotropy can also be measured by observing using the *directions* of arrows joining pairs of points.

Consider all pairs of points in a point pattern that lie more than r_1 and less than r_2 units apart, where $r_1 < r_2$ are fixed distances. For each such pair, we measure the direction of the arrow joining the points, as an angle in degrees anticlockwise from the x-axis. The probability distribution of these angles is the *point pair orientation distribution*.

Figure 7.32 shows the geometry for estimating the cumulative distribution function of the angles, called the 'point pair orientation distribution function' [638, (14.53), p. 271]

$$O_{r_1,r_2}(\phi) = \frac{\mathbb{E}\left[\text{number of points in } O(u,r_1,r_2,\phi) \,\middle|\, \mathbf{X} \text{ has a point at location } u\right]}{\mathbb{E}\left[\text{number of points in } O(u,r_1,r_2,360) \,\middle|\, \mathbf{X} \text{ has a point at location } u\right]}, \quad 0 \le \phi \le 360$$
(7.37)

where $O(u,r_1,r_2,\phi)$ is the region bounded by circles of radius r_1 and r_2 centred at u and lines at orientation 0 and ϕ through u, shown in Figure 7.32. As Figure 7.32 suggests, this summary function can be estimated in the same way as the K-function.

The `spatstat` function `pairorient` computes an estimate of either the cumulative distribution function or the probability density of angles. The default is to compute the probability density, using a kernel smoothing estimate:

```
> f <- pairorient(Y, r1=0.02, r2=0.05, sigma=5)
```

The result is an `"fv"` object. Although this could be plotted by `plot.fv`, the display is challenging to interpret. For probability distributions of angles, it is usually better to display a rose diagram or *rose of directions* plot, in which the function value is interpreted as radial distance from the centre of the plot.

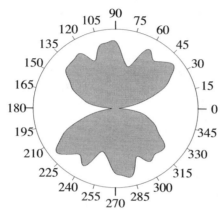

Figure 7.33. *Rose diagram of estimated probability density for the point pair orientation distribution of the synthetic pattern in Figure 7.29.*

Figure 7.33 shows a rose diagram for the estimated probability density `f`, generated by `rose(f)`. It strongly indicates a preferential direction around 90 and 270 degrees. That is, at distances between 0.02 and 0.05, pairs of data points tend to be either directly above or directly below each other, and are rarely found directly to one side of each other.

The point pair orientation distribution can be an effective tool for finding a preferential direction, but seems to be quite sensitive to the choice of the distances r_1 and r_2. An alternative tool is the nearest-neighbour orientation distribution (Section 8.8).

7.9.3 Anisotropic pair correlation function

These considerations lead us back to the Fry plot (Section 7.2.2). In a stationary point process, virtually all information about the correlation structure is contained in the Fry plot. For example, the K-function could be calculated from the Fry plot, by counting how many dots in the Fry plot fall in a circle of radius r centred at the origin, and standardising appropriately. Replacing the circle by another 'test set', such as a sector or a ring, gives the other summary functions described in this chapter.

By counting dots in the Fry plot we are effectively treating the Fry plot as a point process in its own right, and finding the *intensity* of this process. The intensity measure of the Fry plot, appropriately standardised, is called the *reduced second moment measure* of the original point process. It generalises Ripley's K-function from circles to test sets of general shape [674, 520, 638].

Definition 7.2. *For a stationary point process* **X** *with intensity* λ, *the* ***reduced second moment***

measure *is the measure defined for any test set A in two-dimensional space by*

$$\mathcal{K}(A) = \frac{1}{\lambda} \mathbb{E}\left[\text{number of points of } \mathbf{X} \text{ (except } u\text{) falling in } A + u \mid \mathbf{X} \text{ has a point at } u\right] \quad (7.38)$$

for an arbitrary location u.

For example, if we take the test set $A = b(0, r)$, the disc of radius r centred at the origin, then $A + u$ is the disc of radius r centred at u, and equation (7.38) reduces to (7.4), so that $\mathcal{K}(A) = K(r)$, the value of Ripley's K-function for distance r. See Section 7.13 for details.

The measure $\mathcal{K}$ can be regarded as the generalisation of the K-function, and also as a standardised version of the intensity measure of the Fry plot. The connection between (7.38) and the Fry plot is that a point x_j in $\mathbf{X}$ falls in $A + u$ if and only if the vector difference $x_j - u$ falls in A.

Be warned that some writers use the term *reduced second moment measure* when they mean the K-function. This has caused confusion. As originally defined [457, 197], the reduced second moment measure is a measure, while the K-function is a function obtained by evaluating this measure for discs of increasing radius.

In most applications we can assume that the Fry plot has an intensity function, so that $\mathcal{K}$ has an intensity or density function, the *anisotropic pair correlation function* $g(v)$ defined for vectors v. This can be estimated by kernel smoothing *of the Fry plot* with edge correction and standardisation. The left panel of Figure 7.34 shows an estimate of the anisotropic pair correlation function for the point pattern data in Figure 7.29.

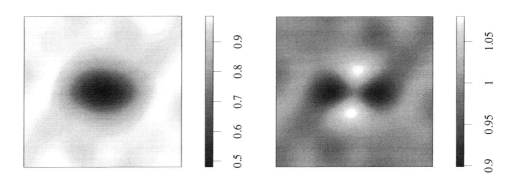

Figure 7.34. *Anisotropic pair correlation function.* Left: *image of estimated anisotropic pair correlation $g(u)$ for the synthetic pattern in Figure 7.29, as a function of vector difference u, with origin at centre of image. Detail of function inside the square of side length 0.2 centred at the origin.* Right: *anisotropy ratio $r(u) = g(u)/\overline{g}(\|u\|)$ where $\overline{g}$ is the rotational mean of g.*

The anisotropic pair correlation function $g(v)$ is a generalisation of the pair correlation function $g(r)$ to anisotropic, stationary point processes. It allows the correlation between pairs of points to be dependent on their relative position (the vector v) and not only on the distance r between them. Again, $g(v)$ is standardised so that it is equal to 1 when the process is completely random.

The anisotropic pair correlation function can be interpreted as a joint probability, as in Figure 7.22 on page 226. Imagine the observation window is divided into a fine grid of pixels, and consider two pixels u and v separated by the vector $v - u$. If $p_2(u, v)$ is the probability that *both* pixels contain random points, then

$$g(v - u) \doteq \frac{p_2(u, v)}{p^2} \quad (7.39)$$

where p^2 is the corresponding probability for a Poisson process.

The `spatstat` package provides a function `Kmeasure` which computes an estimate of the anisotropic pair correlation function (the density of the reduced second moment measure $\mathcal{K}$), effectively by smoothing the Fry plot. The algorithm approximates the point pattern and its window by pixel images, introduces a Gaussian smoothing kernel, and uses the Fast Fourier Transform to compute a density estimate of $\mathcal{K}$, effectively using the translation edge correction. The standard deviation σ of the kernel must be chosen by hand; there do not seem to be any established rules for bandwidth selection in this context.

The left panel of Figure 7.34 shows the estimated anisotropic pair correlation function $g(v)$ as a function of the two-dimensional vector v, with the origin in the centre of the picture. This is effectively a smoothed and renormalised version of the Fry plot. It was generated by the commands

```
> ganiso <- Kmeasure(Y, sigma=0.02, eps=0.001)
> detail <- square(c(-0.1,0.1))
> plot(ganiso[detail])
```

Choosing the bandwidth `sigma=0.02` and pixel resolution `eps=0.001` ensures that we do not oversmooth or pixellate interesting detail, which (according to the Fry plot) is at a scale of about 0.1 units.

Evidence for anisotropy can be judged by comparing the value of $g(v)$ for different vectors v of the same length but different orientations. A useful tool is to compute the *rotational mean*

$$\overline{g}(t) = \frac{1}{2\pi} \int_0^{2\pi} g((t\cos\theta, t\sin\theta)) \, \mathrm{d}\theta \tag{7.40}$$

which gives, for each distance t, the average value of $g(v)$ over all vectors v of length $\|v\| = t$. Then the *anisotropy ratio* $r(v) = g(v)/\overline{g}(\|v\|)$ is a comparison of the anisotropic pair correlation with its rotational mean.

The `spatstat` function `rotmean` computes the rotational mean of any image. The right panel of Figure 7.34 shows the result of plotting the anisotropy ratio $g(v)/\overline{g}(\|v\|)$, computed by

```
> giso <- rotmean(ganiso, result="im")
> grel <- ganiso/giso
> plot(grel[detail])
```

The right panel of Figure 7.34 gives a clear impression of anisotropy, with anisotropy ratios above 1 in the vertical direction and below 1 in the horizontal direction, at distances of about 0.04 units. The left panel shows that this corresponds to pair correlation values closer to 1 in the vertical direction, and further away from 1 in the horizontal direction, at this distance range. This is consistent with the appearance of the Fry plot.

For a stationary isotropic point process there is a close relationship between the K-function and the isotropic pair correlation function through equations (7.22) and (7.25). Equation (7.25) can be generalised to a stationary, *anisotropic* point process:

$$\mathcal{K}(A) = \int_A g(v) \, \mathrm{d}v. \tag{7.41}$$

In particular, for $A = b(0, r)$,

$$K(r) = \int_{b(0,r)} g(v) \, \mathrm{d}v. \tag{7.42}$$

Equation (7.41) can be used to compute an estimate of the second moment measure $\mathcal{K}(A)$ for any region A. We first use `Kmeasure` to compute an estimate of g as a pixel image, then compute the integral of the pixel image over the domain A using `integral.im`. For example, if A is the square window `detail` created above,

```
> integral(ganiso, detail)
[1] 0.03681638
```

The pair correlation function also has a direct relationship to the second moment of the number of points falling in a region B. If $N = n(\mathbf{X} \cap B)$ then $N(N-1)$ is the number of ordered pairs of distinct points falling in B; the expected number of such pairs is

$$\mathbb{E}[N(N-1)] = \lambda^2 \int_B \int_B g(v-u)\,\mathrm{d}u\,\mathrm{d}v \qquad (7.43)$$

as we can see intuitively by dividing B into a fine grid of pixels, and summing over all pairs of distinct pixels the joint probability that both pixels contain random points.

The result above is a special case of Campbell's formula for *second* moments. Consider the sum, over all pairs of distinct random points x_i, x_j where $i \neq j$, of some function $f(x_i, x_j)$. The second moment version of Campbell's formula states that

$$\mathbb{E}\left[\sum_{x_i \in \mathbf{X}} \sum_{\substack{x_j \in \mathbf{X} \\ x_j \neq x_i}} f(x_i, x_j)\right] = \lambda^2 \int \int f(u,v) g(v-u)\,\mathrm{d}u\,\mathrm{d}v \qquad (7.44)$$

provided the expectation is finite. This is the analogue of equation (6.11) on page 169, for second moments.

7.10 Adjusting for inhomogeneity

If a point pattern is known or suspected to be spatially inhomogeneous, then our statistical analysis of the pattern should take account of this inhomogeneity. For general discussion see [355, sec. 4.10], [230, 225].

7.10.1 The general pair correlation function

First we need to extend the concept of the pair correlation function to point processes which cannot be assumed to be stationary or isotropic.

Suppose the point process $\mathbf{X}$ has intensity function $\lambda(u)$, so that

$$\mathbb{E}n(\mathbf{X} \cap A) = \int_A \lambda(u)\,\mathrm{d}u$$

for any bounded region A. Additionally suppose that $\mathbf{X}$ has **second moment intensity function** $\lambda_2(u,v)$, also called the **product density**, defined as the function that satisfies

$$\mathbb{E}[n(\mathbf{X} \cap A)\, n(\mathbf{X} \cap B)] = \int_A \int_B \lambda_2(u,v)\,\mathrm{d}u\,\mathrm{d}v \qquad (7.45)$$

for any disjoint regions A and B. For a Poisson process with intensity function $\lambda(u)$, we have $\lambda_2(u,v) = \lambda(u)\lambda(v)$.

The second moment intensity function has a simple interpretation. Again imagine that the observation window is divided into a fine grid of pixels of area a. Let $p(u)$ be the probability that the pixel with centre u contains a random point. Then $p(u) \doteq \lambda(u)a$. Let $p_2(u,v)$ be the joint probability that the two pixels with centres u and v both contain random points. Then $p_2(u,v) \doteq \lambda_2(u,v)a^2$.

In this general setting we can define the general *pair correlation function*

$$g_2(u,v) = \frac{\lambda_2(u,v)}{\lambda(u)\lambda(v)} \tag{7.46}$$

for any two different spatial locations u and v. Using the approximations above, $g_2(u,v)$ is approximately $p_2(u,v)/(p(u)p(v))$, the probability that pixels centred at u and v both contain random points, divided by the corresponding probability for a Poisson process with the same intensity function $\lambda(u)$.

7.10.2 Inhomogeneous K and g functions

Estimation of the function $g_2(u,v)$ is not practically possible without some further assumption on the form of the function. A strategy proposed in [46] effectively assumes that the point process is *correlation-stationary*[4]

$$g_2(u,v) = g(v-u), \tag{7.47}$$

that is, the pair correlation between u and v depends only on their relative position. This would be true if the process is stationary, but is also true for an inhomogeneous Poisson process, and for many other processes [678].

Assuming (7.47), it is possible to define a counterpart of $K(r)$ called the *inhomogeneous K-function*. The idea is that each point x_i will be weighted by $w_i = 1/\lambda(x_i)$, the reciprocal of the intensity at x_i, and each pair of points x_i, x_j will be weighted by $w_{ij} = w_i w_j = 1/(\lambda(x_i)\lambda(x_j))$.

The *inhomogeneous K-function* is defined as

$$K_{inhom}(r) = \mathbb{E}\left[\sum_{x_j \in \mathbf{X}} \frac{1}{\lambda(x_j)} \mathbf{1}\left\{0 < \|u - x_j\| \le r\right\} \,\middle|\, u \in \mathbf{X}\right] \tag{7.48}$$

assuming this does not depend on location u. If (7.47) holds, then (7.48) does not depend on u.

Thus, $K_{inhom}(r)$ is the expected total 'weight' of all random points within a distance r of the point u, where the 'weight' of a point x_i is $1/\lambda(x_i)$. If the process is stationary, then $\lambda(u)$ is constant and $K_{inhom}(r)$ reduces to the usual K-function (7.6).

For an inhomogeneous Poisson process with intensity function $\lambda(u)$, the inhomogeneous K-function is

$$K_{inhom,\,pois}(r) = \pi r^2 \tag{7.49}$$

exactly as for the homogeneous case.

The standard estimators of K can be extended to the inhomogeneous K-function:

$$\widehat{K}_{inhom}(r) = \frac{1}{D^p |W|} \sum_i \sum_{j \ne i} \frac{\mathbf{1}\{\|x_i - x_j\| \le r\}}{\widehat{\lambda}(x_i)\widehat{\lambda}(x_j)} e(x_i, x_j; r) \tag{7.50}$$

where $e(u,v,r)$ is an edge correction weight as before, and $\widehat{\lambda}(u)$ is an estimate of the intensity function $\lambda(u)$. The constant D^p in (7.50) is the pth power of

$$D = \frac{1}{|W|} \sum_i \frac{1}{\widehat{\lambda}(x_i)} \tag{7.51}$$

which has expected value 1 if the intensity is estimated without error. Choosing the power p equal to 1 or 2 improves statistical performance. Theoretical results in [46] show that, if the intensity function is *known*, or can be estimated with very high accuracy, then $\widehat{K}_{inhom}(r)$ is an unbiased estimator of $K_{inhom}(r)$ when $p = 0$, and is approximately unbiased when $p = 1$ or 2.

---

[4]Called 'second order intensity-reweighted stationary' (soirs) in [46].

In practice, the intensity function will need to be estimated from data, so the estimator (7.50) could be biased. The intensity function is usually estimated from the *same* point pattern data, which can lead to substantial bias in the estimate of $K_{inhom}(r)$. This justifies the data-dependent denominator D. Nonparametric estimates of intensity, being more responsive to local fluctuations in the data, tend to produce more biased estimates of $K_{inhom}(r)$ than parametric estimates of intensity, provided the intensity model is correctly specified [46, 225, 274].

The estimator (7.50) requires the estimated intensities $\widehat{\lambda}(x_i)$ at the data points x_i. To reduce bias it is advisable to use a *leave-one-out* intensity estimator as described in Section 6.5.1.3.

The empirical inhomogeneous K-function is computed in `spatstat` by the command `Kinhom(X, lambda)` where `X` is the point pattern and `lambda` is the estimated intensity function. Here `lambda` may be a pixel image, a `function(x,y)` in the R language, a fitted point process model, a numeric vector giving the values $\widehat{\lambda}(x_i)$ at the data points x_i only, or it may be omitted (and will then be estimated from `X`). By default, the data-dependent denominator D^1 is used: the power p is controlled by the argument `normpower`.

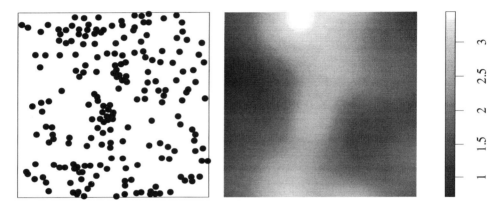

Figure 7.35. *Full dataset of Japanese Black Pines of Numata and Ogata (*Left*) and kernel-smoothed intensity estimate (*Right*). Smoothing bandwidth selected by likelihood cross-validation.*

Figure 7.35 shows the full dataset of Japanese Black Pine seedlings and saplings in a 10 metre square study region recorded by Numata [503] and extensively studied by Ogata [514, 516]. A kernel-smoothed intensity estimate is shown, with smoothing bandwidth selected by likelihood cross-validation using `bw.ppl` (see Section 6.5.1.2 on page 170):

```
> numata <- residualspaper$Fig1
> lambda <- density(numata, bw.ppl)
```

The chosen bandwidth seems appropriate, and the intensity estimate gives a strong impression of inhomogeneity. We now compute the inhomogeneous K-function:

```
> numataK <- Kinhom(numata, lambda)
```

The result is shown in Figure 7.36. The edge-corrected estimates are close to the Poisson theoretical value, suggesting that the data are consistent with an inhomogeneous Poisson process with the intensity shown in Figure 7.35.

If the intensity function is not given in the call to `Kinhom` then it will be estimated by kernel smoothing using the leave-one-out estimator. We could have obtained the same result by

```
> numataK <- Kinhom(numata, sigma=bw.ppl)
```

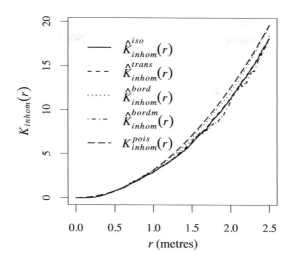

Figure 7.36. *Inhomogeneous K-function for full Japanese Pines data, using the estimated intensity function from the right panel of Figure 7.35.*

The argument `sigma=bw.ppl` is passed to `density.ppp`. In practice the results obtained from this procedure can be quite sensitive to the choice of bandwidth.

The inhomogeneous analogue of the L-function is

$$L_{inhom}(r) = \sqrt{K_{inhom}(r)/\pi}. \tag{7.52}$$

For an inhomogeneous Poisson process, $L_{inhom,\,pois}(r) \equiv r$. The corresponding empirical L-function is again justified by the fact that it approximately stabilises the variance. The inhomogeneous L-function can be computed in `spatstat` using `Linhom`, which has the same arguments as `Kinhom`.

The relationship (7.42) between the K-function and pair correlation function of a stationary point process extends to the inhomogeneous K-function:

$$K_{inhom}(r) = \int_{b(0,r)} g_2(0,v)\,dv. \tag{7.53}$$

In general $g_2(0,v)$ depends on the orientation of v as well as the distance $\|v\|$. The rotational mean of the pair correlation

$$\overline{g}(r) = \frac{1}{2\pi} \int_0^{2\pi} g_2(0,(r\cos\theta, r\sin\theta))\,d\theta$$

is the 'inhomogeneous pair correlation function' $g_{inhom}(r)$, and satisfies

$$g_{inhom}(r) = \frac{K'_{inhom}(r)}{2\pi r}.$$

The inhomogeneous pair correlation function is computed by `pcfinhom`:

```
> g <- pcfinhom(bei)
```

The previously mentioned method `pcf.fv`, which converts a K-function estimate into a pair correlation function estimate by numerical differentiation, also works for estimates of the inhomogeneous K-function. Thus the following is an alternative to the foregoing estimation procedure:

```
> g <- pcf(Kinhom(bei))
```

To construct confidence intervals for the true value of $K_{inhom}(r)$ or $g_{inhom}(r)$ one can use `varblock` or `lohboot` as described in Section 7.7.

To test the hypothesis that the point process is an inhomogeneous Poisson process, use `envelope`, `dclf.test`, or `mad.test`. Some care is required: as explained in Chapter 10, it is important to ensure that the simulated patterns are treated in exactly the same way as the original data pattern was treated. Figure 7.37 shows the global envelopes of the centred inhomogeneous L function for the full Japanese Pines data of Figure 7.35, generated by

```
> lam <- density(numata, bw.ppl)
> E <- envelope(numata, Linhom, sigma=bw.ppl,
              simulate=expression(rpoispp(lam)),
              use.theory=TRUE, nsim=19, global=TRUE)
> plot(E, . - r ~ r)
```

The plot indicates significant evidence against the inhomogeneous Poisson model, with a suggestion that there is inhibition at small distances.

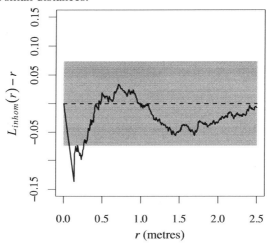

Figure 7.37. *Centred inhomogeneous L-function $L_{inhom}(r) - r$ for the Numata-Ogata Japanese pines data (black line) and global envelopes from 19 simulated realisations of an inhomogeneous Poisson process with intensity estimated by a leave-one-out kernel smoother.*

We emphasise that — despite its name — the inhomogeneous K-function does not apply to every spatially inhomogeneous point process; it applies only when the point process is correlation-stationary, equation (7.47). Substantial differences between the estimates of the inhomogeneous K-function obtained from different subsets of the data, or obtained using different edge corrections, would suggest that this assumption is false. A formal hypothesis test for correlation-stationarity is described in Section 16.8.5.

7.10.3   Local scaling

The inhomogeneous K-function effectively assumes that the *spatial scale* of interaction remains constant, while the intensity is spatially varying. This is not an appropriate assumption for the bronze filter data in Figure 1.9 on page 8, an inhomogeneous point pattern in which the spacing between points increases gradually from left to right.

An alternative approach to inhomogeneity [312, 556, 309, 311] is to assume that the point process is equivalent, in small regions, to a rescaled version of a 'template' process, where the template process is stationary and isotropic, and the rescaling factor can vary from place to place. This could be an appropriate model for Figure 1.9.

We would then be assuming that, for two locations u and v sufficiently close together,

$$g(u,v) = g_1\left(\frac{\|u-v\|}{s}\right) \tag{7.54}$$

where g_1 is the pair correlation function of the 'template' process, and s is the local scaling factor applicable to both locations u and v. Rescaling the spatial coordinates by a factor $1/s$ rescales the intensity by s^2, so the appropriate value is $s = 1/\sqrt{\lambda}$ where λ is the local intensity in the neighbourhood of u and v.

In practice, we would first estimate the intensity function of the original data by $\widehat{\lambda}(u)$, then for each pair of data points x_i, x_j define the rescaled distance

$$d_{ij}^* = \frac{\|x_i - x_j\|}{s(x_i, x_j)}$$

where the rescaling factor is

$$s(x_i, x_j) = \frac{1}{2}\left(\frac{1}{\sqrt{\widehat{\lambda}(x_i)}} + \frac{1}{\sqrt{\widehat{\lambda}(x_j)}}\right).$$

An edge-corrected estimator of Ripley's original K-function is then applied to the distances d_{ij}^* to give the *locally scaled K-function* [312, 556, 311].

The `spatstat` package provides the commands `Kscaled` and `Lscaled` which compute the locally scaled K- and L-functions. Their syntax is similar to `Kinhom`.

To estimate a locally scaled K-function for the bronze filter data (Figure 1.9), we first estimated the intensity, assuming it is an exponential function of the x-coordinate, by fitting a point process model (see Chapter 9):

```
> X <- unmark(bronzefilter)
> fit <- ppm(X ~ x)
> lam <- predict(fit)
```

The locally scaled K-function was then estimated by

```
> Kbro <- Kscaled(X, lam)
```

The result is plotted in Figure 7.38.

Again we emphasise that the locally scaled K-function is applicable only when the point process is locally scaled. A formal hypothesis test for local scaling is described in Section 16.8.5.

7.11 Local indicators of spatial association

Another tool for handling inhomogeneous patterns is a LISA (Local Indicator of Spatial Association), obtained by decomposing a summary statistic into contributions from each of the data points.

For example, the empirical K-function (7.3) can be broken into contributions $\widehat{K}(r, x_i)$ from each individual data point x_i, shown in (7.33), known as the *local K-functions*. The usual estimate $\widehat{K}(r)$ is the average of the estimates $\widehat{K}(r, x_i)$ over all $i = 1, \ldots, n$.

If it is suspected that the pattern may be a patchwork of different textures in different places, or

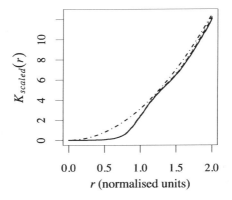

Figure 7.38. *Locally scaled K-function of the bronze filter data.*

that the pattern contains some anomalous features, then it can be useful to classify the n local K-functions $\widehat{K}(r, x_1), \ldots, \widehat{K}(r, x_n)$ into several groups of functions, perhaps using hierarhical clustering techniques. This approach was proposed independently in [281, 20] and developed in [186, 185]. Repeated measures analysis or functional principal component analysis [562] could be used to identify the main differences between the local functions.

The `spatstat` functions `localK`, `localL`, `localpcf` compute local versions of the K-function, L-function, and pair correlation function, respectively. Of course the L-function is not a *sum* of contributions from individual data points: the local L-function is arbitrarily defined as $\widehat{L}(r, x_i) = (\widehat{K}(r, x_i)/\pi)^{1/2}$. There are also *inhomogeneous* counterparts `localKinhom`, `localLinhom`, and `localpcfinhom` which are the contributions to the inhomogeneous K-function, etc, from each data point.

Local K-functions of the Swedish Pines data are computed by

```
> lK <- localK(swedishpines)
```

The result is an `"fv"` object with $n + 2 = 73$ columns, where $n = 71$ is the number of data points. The first n columns give the local K-functions $\widehat{K}(r, x_i)$ for each data point, in order. The final two columns contain the distance values r and the K-function for a Poisson process. Extracting only the function values by

```
> locK <- as.data.frame(lK)[, fvnames(lK, ".a")]
> rr   <- with(lK, r)
```

gives the data in a format ready for cluster analysis or functional data analysis. Normally `locK` would be transposed to `t(locK)` so that each row of the matrix contains one local K-function. Hierarchical clustering of the local K-functions, based on the mean square distance between functions, is implemented by

```
> locH <- hclust(dist(t(locK)))
```

A plot of `locH` would show a dendrogram of the local K-functions. See [186, 185] for further discussion.

The average of the local K-functions for the data points falling in a region A can also be computed by `Kest` using the argument `domain`.

7.12 Third- and higher-order summary statistics

We saw on page 211 that two different point processes can have identical first-order (intensity) and second-order (correlation) properties. In order to distinguish between such processes we need another approach. One possible strategy is to estimate the *third-order* moments, which would involve counting *triples* rather than *pairs* of data points.

For a stationary point process $\mathbf{X}$ with intensity λ, the triangle-counting function $T(r)$ is defined as the normalised expected number of triangles formed by points of $\mathbf{X}$, with all side lengths less than or equal to r, with one corner at a specified point of the process:

$$T(r) = \frac{1}{\lambda^3} \mathbb{E} \left[\sum_{i=1}^{n} \sum_{\substack{j=1 \\ j \neq i}}^{n} m(x_i, x_j, u) \mid u \in \mathbf{X} \right] \tag{7.55}$$

where $m(a,b,c)$ is the maximum side length of the triangle with vertices a,b,c,

$$m(a,b,c) = \max\{\|a-b\|, \|a-c\|, \|b-c\|\}.$$

For a homogeneous Poisson process with intensity λ we have

$$T_{pois}(r) = \frac{\pi}{2}(\pi - \frac{3\sqrt{3}}{4})r^4. \tag{7.56}$$

Formal theory and edge corrections are given in [600].

Estimates of the triangle-counting function $T(r)$ are computed in spatstat by Tstat. For the point pattern Xcell shown in Figure 7.11, the empirical estimate $\widehat{T}(r)$ computed by Tstat(Xcell) is shown in Figure 7.39. This shows clear deviation from the values expected for a Poisson process. Simulation envelopes can be computed in the usual way.

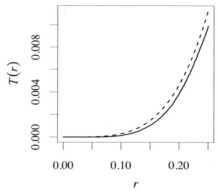

Figure 7.39. *Triangle-counting function $\widehat{T}(r)$ for the simulated realisation of the cell process in Figure 7.11. Solid line: translation correction estimate. Dashed line: theoretical value for Poisson process.*

Three-point correlation functions are frequently used in astronomy [537] and even n-point correlation functions are used [378, sec. 3.2], [445], [355, pp. 4.4.2].

7.13   Theory*

This section covers some basic theory for second moments of point processes. For more detail, see [484, Appendix C], [355, sec. 1.5], [27, 32, 569], [576, chap. 3]. For a full account, see [198, 199].

7.13.1   Second moment measures

Let $\mathbf{X}$ be a point process. We are interested in the variance of the count $n(\mathbf{X} \cap B)$,

$$\operatorname{var} n(\mathbf{X} \cap B) = \mathbb{E}[n(\mathbf{X} \cap B)^2] - [\mathbb{E}n(\mathbf{X} \cap B)]^2$$

and the covariance of two such counts,

$$\operatorname{cov}[n(\mathbf{X} \cap B_1), n(\mathbf{X} \cap B_2)] = \mathbb{E}[n(\mathbf{X} \cap B_1)n(\mathbf{X} \cap B_2)] - [\mathbb{E}n(\mathbf{X} \cap B_1)]\,[\mathbb{E}n(\mathbf{X} \cap B_2)].$$

A key observation is that $n(\mathbf{X} \cap B_1)n(\mathbf{X} \cap B_2)$ is equal to the number of ordered pairs (x, x') of points in the process $\mathbf{X}$ such that $x \in B_1$ and $x' \in B_2$.

Definition 7.3. *Let $\mathbf{X}$ be a point process in $\mathbb{R}^2$. Then $\mathbf{X} \times \mathbf{X}$ is a point process on $\mathbb{R}^2 \times \mathbb{R}^2$ consisting of all ordered pairs (x, x') of points $x, x' \in \mathbf{X}$. The intensity measure ν_2 of $\mathbf{X} \times \mathbf{X}$ is a measure on $\mathbb{R}^2 \times \mathbb{R}^2$ satisfying*

$$\nu_2(A \times B) = \mathbb{E}[N_{\mathbf{X}}(A)N_{\mathbf{X}}(B)].$$

This measure ν_2 is called the second moment measure *of $\mathbf{X}$.*

The second moment measure contains all information about the variances and covariances of the variables $N_{\mathbf{X}}(A)$. Campbell's formula (6.11) applied to $\mathbf{X} \times \mathbf{X}$ becomes

$$\mathbb{E}\Big[\sum_{x_i \in \mathbf{X}}\sum_{x_j \in \mathbf{X}} f(x_i, y_j)\Big] = \int_{\mathbb{R}^2}\int_{\mathbb{R}^2} f(u, v)\,\mathrm{d}\nu_2(u, v)$$

for a measurable function $f : \mathbb{R}^2 \times \mathbb{R}^2 \to$ R.

For example, for the uniform Poisson point process of intensity $\lambda > 0$ in R^d, the second moment measure satisfies $\nu_2(A \times B) = \lambda^2|A|\,|B| + \lambda|A \cap B|$. To simplify the calculation of certain moments, we introduce the *second factorial moment measure*

$$\nu_{[2]}(A \times B) = \mathbb{E}[n(\mathbf{X} \cap A)n(\mathbf{X} \cap B)] - \mathbb{E}[n(\mathbf{X} \cap A \cap B)].$$

This is the intensity measure of the process $\mathbf{X} * \mathbf{X}$ of all ordered pairs of *distinct* points of $\mathbf{X}$. It satisfies

$$\mathbb{E}\Big[\sum_{x_i \in \mathbf{X}}\sum_{x_j \in \mathbf{X},\ j \neq i} f(x_i, x_j)\Big] = \int_{\mathbb{R}^2}\int_{\mathbb{R}^2} f(u, v)\,\mathrm{d}\nu_{[2]}(u, v).$$

The name 'factorial' is derived from

$$\nu_{[2]}(A \times A) = \mathbb{E}[n(\mathbf{X} \cap A)^2] - \mathbb{E}[n(\mathbf{X} \cap A)] = \mathbb{E}[n(\mathbf{X} \cap A)[n(\mathbf{X} \cap A) - 1]].$$

For example, for the uniform Poisson process of intensity λ, the second factorial moment measure is $\nu_{[2]}(A \times B) = \lambda^2|A||B|$.

∗ Starred sections contain advanced material, and can be skipped by most readers.

Definition 7.4. *A point process* $\mathbf{X}$ *on* $\mathbb{R}^2$ *is said to have* second moment intensity λ_2 *if*

$$\nu_{[2]}(C) = \int_C \lambda_2(u,v)\,\mathrm{d}u\,\mathrm{d}v \qquad (7.57)$$

for any compact $C \subset \mathbb{R}^2 \times \mathbb{R}^2$.

Informally, $\lambda_2(u,v)$ gives the joint probability that there will be points of $\mathbf{X}$ in infinitesimal regions around two specified locations u and v:

$$\mathbb{P}\{N(\mathrm{d}u) > 0,\, N(\mathrm{d}v) > 0\} \doteq \lambda_2(u,v)\,\mathrm{d}u\,\mathrm{d}v.$$

For example, the uniform Poisson process has second moment density $\lambda_2(x,y) = \lambda^2$. The binomial process of n points in W has $\lambda_2(u,v) = n(n-1)/|W|^2$ if $x,y \in W$, and zero otherwise.

Definition 7.5. *Suppose* $\mathbf{X}$ *is a point process on* $\mathbb{R}^2$ *which has an intensity function* $\lambda(x)$ *and a second moment intensity* $\lambda_2(u,v)$. *Then we define the* pair correlation function *of* $\mathbf{X}$ *by*

$$g_2(u,v) = \frac{\lambda_2(u,v)}{\lambda(u)\lambda(v)}.$$

For a uniform Poisson process of intensity λ, we have $\lambda(u) \equiv \lambda$ and $\lambda_2(u,v) \equiv \lambda^2$, so that $g_2(u,v) \equiv 1$. For a binomial process of n points in a region W, we have $g_2(u,v) \equiv 1 - 1/n$.

7.13.2 Second moments for stationary processes

For a *stationary* point process in $\mathbb{R}^2$, there is a 'disintegration' of the second moment measure.

Theorem 7.1. *Let* $\mathbf{X}$ *be a stationary point process on* $\mathbb{R}^2$ *with intensity* λ. *Then there is a measure* $\mathcal{K}$ *on* $\mathbb{R}^2$ *such that, for a general integrand* f,

$$\mathbb{E}\left[\sum_{x_i \in \mathbf{X}}\sum_{x_j \in \mathbf{X},\, j \neq i} f(x_i, x_j)\right] = \lambda \int\int f(u, u+v)\,\mathrm{d}\mathcal{K}(v)\,\mathrm{d}u. \qquad (7.58)$$

$\mathcal{K}$ *is called the* reduced second moment measure *of* $\mathbf{X}$.

To understand the measure $\mathcal{K}$, we notice that for $A, B \subset \mathbb{R}^2$

$$\lambda|A|\mathcal{K}(B) = \int\int 1_A(u) 1_B(v-u)\,\mathrm{d}\nu_{[2]}(u,v) = \mathbb{E}\left[\sum_{x_i \in \mathbf{X}}\sum_{x_j \in \mathbf{X},\, j \neq i} 1_A(u) 1_B(v-u)\right].$$

This may also be obtained directly from (7.58) by taking $f(u,v) = 1_A(u) 1_B(v-u)$. Since $\lambda|A| = \mathbb{E}n(\mathbf{X} \cap A)$, we have

$$\mathcal{K}(B) = \frac{\mathbb{E}\sum_{x_i \in \mathbf{X} \cap A} n(\mathbf{X} \cap (B + x_i) \setminus x_i)}{\mathbb{E}n(\mathbf{X} \cap A)}. \qquad (7.59)$$

The right-hand side of (7.59) may be interpreted as the average, over all points x_i of the process, of the number of other points x_j of the process such that $x_j - x_i \in B$.

For the uniform Poisson process, $\mathcal{K}(B) = \lambda|B|$.

Suppose $\mathbf{X}$ is a stationary process on $\mathbb{R}^2$ which has a second moment density function λ_2. Then by comparing (7.57) with (7.58) we can see that $\lambda_2(u,v)$ depends only on $v - u$, say $\lambda_2(u,v) = \lambda^2 g(v-u)$ for some function g, and we can write

$$\mathcal{K}(B) = \lambda \int_B g(u)\,\mathrm{d}u.$$

This is the fundamental relationship between the pair correlation and the second moment measure.

7.14 FAQs

- *Should I use Ripley's K-function, the inhomogeneous K-function, or the scaled K-function?*

  This depends on the assumptions that are appropriate for your data. Using the K-function assumes the point process has homogeneous intensity and is correlation-stationary. Using the inhomogeneous K-function assumes the process is correlation-stationary. Hypothesis tests for checking these assumptions are described in Section 16.8.5.

- *When I plot the estimated K-function of my data using the command* `plot(Kest(X))`, *what is the quantity on the horizontal axis, and what units is it expressed in?*

  The horizontal axis is distance between points. If the plot annotation does not indicate the units in which the distance is measured, then the point pattern dataset X did not include a specification of the units of length. This should be repaired using the function `unitname`.

- *When I plot the estimated K-function of my data using the command* `plot(Kest(X))`, *the horizontal axis is the distance in metres, but what is the quantity on the vertical axis, and what units is it expressed in?*

  The quantity on the vertical axis is the average number of neighbours of a typical point, *divided by* the intensity (average number of points per unit area) so that different patterns can be compared. It is measured in units of area (number of points divided by points-per-unit-area). The reference benchmark is that for a completely random process the value of $K(r)$ is the area of the disc of radius r.

- *When I plot the estimated K-function of my data using the command* `plot(Kest(X))`, *the scale marks on the y-axis are huge numbers like* `1e9`. *Is this wrong? I don't see anything like this in your book.*

  The values of $K(r)$ are areas, and should be of the same order as the area of the observation window, expressed in the units you are using for the spatial coordinates. If your window is 30 kilometres across and the coordinates are recorded in metres, the window area is about $30000^2 \approx 10^9$ square metres. To avoid numerical overflow, it would be wise to rescale the spatial coordinates, for example converting metres to kilometres.

- *When I plot the estimated K-function of my data using the command* `plot(Kest(X))`, *I don't understand the meaning of the different coloured curves.*

  These curves are different estimates of the K-function, using different 'edge correction' techniques. Usually one of the curves is the theoretical value of the K-function, $K(r) = \pi r^2$, corresponding to a completely random pattern.

  The accompanying legend (plotted unless `legend=FALSE`) gives a mathematical expression for each of the edge corrections, and shows the corresponding line colour and line type used to plot the estimate.

  The return value of `plot.fv` is a table giving the line colour, line type, keyword, mathematical expression, and long text description of each curve that was plotted. For further information about the different edge corrections, see `help(Kest)`.

  The estimates of $K(r)$ by different edge correction techniques should be roughly equal. If the curves for the isotropic correction (`iso`), translation correction (`trans`), and border correction (`border`) estimates are wildly different, this suggests that estimation of K is difficult for these data (e.g., because there are too few data points, or the window is very narrow, or there are data points very close to the edge of the window).

For information on how to modify the plot of the *K*-functions, see help(plot.fv) or the examples in help(Kest).

- *I have computed the inhomogeneous K-function using* Kinhom. *The different edge-corrected estimates have very different values. Which correction should I use?*

Discrepancies between the edge-corrected estimates of the inhomogeneous *K*-function computed by Kinhom(X, lambda) usually occur when the intensity estimate lambda is inaccurate, or when the data are not correlation-stationary. First check that the estimated intensity appears to be reasonably accurate. If so, try changing the arguments normalise and normpower. If large discrepancies still remain, check for data points very close to the window boundary, which are treated differently by the different edge corrections. If there are no such points, the tentative conclusion is that the assumption of a correlation-stationary process is violated. Consider performing a test of this assumption (Section 16.8.5).

- *I can't seem to control the range of r values in* plot(Kest(X)). *How can I control it? How is the default plotting range determined?*

Use the argument xlim to control the range of the *x*-axis, as documented in help(plot.fv). For example, plot(Kest(X), xlim=c(0, 7)).

An object of class "fv" contains function values for a certain range of *r* values (the '*available range*', which is usually the maximum possible range). However, the default behaviour of plot.fv is to plot the function values for a narrower range of *r* values (the '*recommended range*') which usually contains the important detail in the function. Both the available range and recommended range are printed when the "fv" object is printed.

Using the argument xlim when plotting the *K*-function will override the recommended range. However, obviously we cannot choose xlim to be wider than the *available* range of *r* values. To extend the available range, you would need to re-compute the *K*-function, specifying the argument r. This should be a vector of closely spaced values of *r* ranging from 0 to the desired maximum.

- *When plotting a summary function using* plot.fv, *how can I control the legend position, text size, and other parameters?*

See the help file for plot.fv. To set the legend text to half normal size, for example, set legendargs=list(cex=0.5).

To move the legend *outside* the plot area, call plot.fv with legend=FALSE, save the result, and use the columns in this data frame in a subsequent call to the legend function.

- *I have plotted a summary function using a formula such as* sqrt(./pi) ~ r. *Is it possible to save the plotted curves as another summary function?*

Yes, if the right-hand side of the formula is the function argument. Use with.fv, for example with(Kest(cells), sqrt(./pi))

- *I used the* envelope *command to compute envelopes of the L-function. The printout for the envelope object says that it contains 'pointwise envelopes'. What value of r was used for these?*

'Pointwise' means that for *each* chosen value of the distance *r*, the range of simulated values of $\widehat{L}(r)$ has been calculated, and a test based on these values would be valid. Plotting the pointwise envelopes shows what the result of the test would have been if we had chosen any particular value of *r*. For further explanation, see Chapter 10.

- *Is it possible to combine the empirical K-functions from several point patterns into a common K-function?*

Yes; see Section 16.8.1.

- *I have mapped the locations of bird nests on a small island. Do I need to use an edge correction for K(r)?*

  This is an example of the 'Small World Model' (page 145). Strictly speaking the K-function is not defined because the point process is not stationary. A sensible strategy is to use the empirical K-function *without* edge correction (Kest with correction="none") and to base inference on simulations from the appropriate null hypothesis.

8

Spacing

This chapter discusses summary statistics based on the spacing between points in a point pattern.

8.1 Introduction

Statistics such as Ripley's K-function, which measure the spatial correlation in a point process, are the most popular tools for assessing dependence. However, correlation is only a simple summary of dependence, and is blind to some aspects of dependence, as we saw in Chapter 7.

Additional information about a point pattern can often be revealed by measuring *spacings* or *shortest distances* in the pattern, such as the 'nearest-neighbour' distance from each data point to the nearest other data point. In a forestry survey, it would make good practical sense to measure the distance from each sampled tree to the nearest other tree. Obviously this carries important information about the spatial arrangement of trees. This chapter introduces analytical tools based on shortest distances.

There is a kind of duality between measuring shortest distances between points and counting points in a given area. At a bus stop we can either say that (on average) there are 4 buses per hour, or that there is $1/4$ hour between buses. This duality runs deep in the mathematical theory, telling us that the shortest distances in a point pattern provide information which is complementary to the correlation structure. A good statistical analysis of spatial point patterns should therefore study both the correlation and the shortest distances.

8.2 Basic methods

8.2.1 Measuring distances in point patterns

There are several different kinds of distances that could be measured in a point pattern:

- the **pairwise distances** $d_{ij} = ||x_i - x_j||$ between all distinct pairs of points x_i and x_j ($i \neq j$) in the pattern;

- the **nearest-neighbour distances** $d_i = \min_{j \neq i} d_{ij}$, the distance from each point x_i to its nearest neighbour (left panel of Figure 8.1);

- the **empty-space distance** $d(u) = \min_j ||u - x_j||$, the distance from a fixed reference location u in the window to the nearest data point (right panel of Figure 8.1).

Table 8.1 shows the `spatstat` functions for computing these distances in a point pattern dataset. If X is a point pattern object, `pairdist(X)` returns the matrix of pairwise distances:

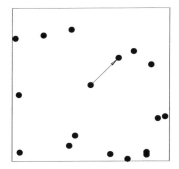 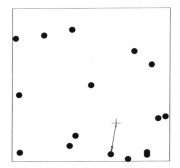

Figure 8.1. Left: *the nearest-neighbour distance is the distance from a data point (•) to the nearest other data point.* Right: *the empty-space distance is the distance from a fixed location (+) to the nearest data point.*

```
> M <- pairdist(redwood)
> M[1:3, 1:5]
        [,1]  [,2]  [,3]  [,4]  [,5]
[1,] 0.000 0.082 0.120 0.134 0.141
[2,] 0.082 0.000 0.045 0.057 0.060
[3,] 0.120 0.045 0.000 0.060 0.028
```

while nndist(X) returns the vector of nearest-neighbour distances:

```
> v <- nndist(redwood)
> v[1:3]
[1] 0.082 0.045 0.028
```

The command distmap(X), introduced in Section 6.6.4, returns a pixel image whose pixel values are the empty-space distances to the pattern X measured from every pixel:

```
> Z <- distmap(redwood)
```

The result can be plotted in various ways; a contour plot produced by contour(Z) is shown in the left panel of Figure 8.2.

DISTANCES	
pairdist(X)	matrix of pairwise distances in X
crossdist(X,Y)	matrix of pairwise distances from X to Y
nndist(X)	vector of nearest-neighbour distances in X
distmap(X)	pixel image of empty-space distances to X
distfun(X)	function that computes empty-space distance to X
nncross(X,Y, what="dist")	vector of distances from X to nearest point in Y
INDICES	
nnwhich(X)	vector of indices of nearest neighbours in X
nncross(X,Y, what="which")	vector of indices from X to nearest point in Y
nnmap(X)	pixel image of index of nearest point of X
nnfun(X)	function that computes index of nearest point of X

Table 8.1. *Functions for measuring distances in a point pattern* X *and for identifying the nearest point.*

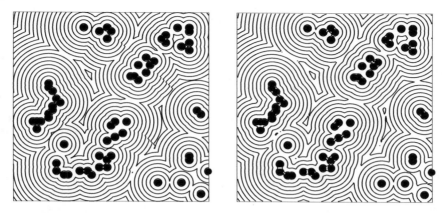

Figure 8.2. *Contours of the distance function $d(u)$ for the* redwood *dataset.* Left: *computed approximately by* distmap. Right: *computed exactly by* distfun. *Note slight artefacts in the left panel.*

The function distmap uses a very fast image processing algorithm for calculating approximate distances [586, 587, 117, 118], which we have modified[1] to compute 'almost exact' distances $d(u)$ from the centre of each pixel. This will be sufficient for most purposes.

However, if exact values of empty-space distance $d(u)$ are required, or if only a few values of $d(u)$ need to be computed, nncross or distfun should be used. If U and X are point patterns, the command nncross(U,X,what="dist") computes the exact distance from each point in U to the nearest point in X, and returns a numeric vector with one entry for each point of U. The command f <- distfun(X) returns a function f in the R language that can be evaluated at any spatial location to give the exact value of empty-space distance.

```
> U <- runifpoint(3, Window(redwood))
> Z <- distmap(redwood)
> Z[U]
[1] 0.052 0.073 0.043
> nncross(U, redwood, what="dist")
[1] 0.052 0.073 0.040
> f <- distfun(redwood)
> f(U)
[1] 0.052 0.073 0.040
```

The result of distfun also belongs to class "funxy" as described in Section 3.6.9. This is the recommended way to build a spatial covariate function for use in point process modelling.

A pixel image containing the *exact* distance values at the pixel centres can be computed by as.im(distfun(X), ...) where the additional arguments control the pixel resolution. This idiom was used to make the right panel of Figure 8.2. Note slight discrepancies between the left and right panels.

Instead of the nearest neighbour we may also consider the second-nearest, third-nearest, and so on. The functions nndist, nncross, and distfun have an argument k specifying the order of neighbour: k=2 would specify the second-nearest neighbour.

---

[1]The modified algorithm runs the classical distance transform algorithm to determine, for each pixel centre u, which data point x_i is the closest. It then computes the exact distance from u to x_i. In marginal cases, the wrong index i may be selected, and the result will be slightly larger than it should be.

```
> nncross(U, redwood, k=2, what="dist")
[1] 0.084 0.105 0.053
> f2 <- distfun(redwood, k=2)
> f2(U)
[1] 0.084 0.105 0.053
```

To determine *which* point is the nearest neighbour of a given point, use one of the functions nnwhich, nncross, nnmap, or nnfun. For a point pattern X, the result of nnwhich(X) is a vector v containing the index of the nearest neighbour for each point of X. That is, v[i]=j if the nearest neighbour of X[i] is X[j]. For two point patterns U and X, the result of nncross(U, X, what="which") is a vector v such that v[i]=j if, starting from the point U[i], the nearest point in X is X[j]. The result of nnmap(X) is a pixel image with integer values giving for each pixel the index of the nearest point of X. The result of nnfun(X) is a function in the R language (of class "funxy") which can be evaluated at any spatial location *u* to give the index of the nearest point of X. These functions all accept the argument k specifying the order of neighbour.

The nearest neighbour of a specified kind — for example, the nearest point of a specified type in a multitype point pattern — can be found using nndist and nnwhich with the argument by, as described in Section 14.6.1.

The functions minnndist and maxnndist are faster algorithms for computing the minimum and maximum of the nearest-neighbour distances in a point pattern.

8.2.2   Tests of CSR based on shortest distances

Early literature in statistical ecology includes several methods for deciding whether a point pattern is completely random, using nearest-neighbour and empty-space distances that could be measured in a field study. We present two of them here.

The nearest-neighbour distances d_i, d_j of two nearby plants x_i, x_j are statistically dependent. To avoid problems arising from this dependence, the classical techniques measure the nearest-neighbour distances only from a *random sample* of data points. Similarly the empty-space distances for two nearby locations are statistically dependent, so a *random sample* of spatial locations will be used to measure empty-space distances.

Clark and Evans [155] proposed taking the average of the nearest-neighbour distances d_i for *m* randomly sampled points in a point pattern (or for all data points), and dividing this by the expected value $\mathbb{E}[D]$ for a completely random process with the same intensity, to obtain an index of spatial regularity. In a Poisson process of intensity λ the expected distance to the nearest neighbour is $\mathbb{E}[D] = 1/(2\sqrt{\lambda})$. The *Clark-Evans index* is

$$R = \frac{\overline{d}}{\mathbb{E}[D]} = \frac{2\sqrt{\overline{\lambda}}}{m} \sum_{i=1}^{m} d_i \qquad (8.1)$$

where *m* is the number of sampled points, and $\overline{\lambda} = n/|W|$ is the estimated intensity for the entire point pattern consisting of *n* data points in a window *W*. The index *R* is dimensionless; the value $R = 1$ is consistent with a completely random pattern; $R > 1$ suggests regularity, while $R < 1$ suggests clustering.

The *Clark-Evans test* of CSR is performed by approximating the distribution of *R* under CSR by a normal distribution with mean 1 and variance $s^2 = (1/\pi - 1/4)/(m\lambda)$. For a two-sided test, we reject the null hypothesis of complete spatial randomness at the 5% significance level if the standardised statistic $(R-1)/s$ lies outside the critical boundaries ± 1.96.

Without correction for edge effects, the value of *R* will be positively biased [244]. Edge effects arise because, for a data point close to the edge of the window, the true nearest neighbour may actually lie outside the window. Hence observed nearest-neighbour distances tend to be larger than the true nearest-neighbour distances.

The spatstat functions clarkevans and clarkevans.test perform these calculations. The argument correction specifies an edge correction to be applied. In some cases the test will be performed using Monte Carlo simulation.

```
> clarkevans(redwood)
   naive Donnelly      cdf
  0.6187   0.5850   0.5836
> clarkevans.test(redwood, correction="donnelly",
                alternative="clustered")
          Clark-Evans test
          Donnelly correction
          Monte Carlo test based on 999 simulations of CSR with fixed n

data:  redwood
R = 0.58, p-value = 0.001
alternative hypothesis: clustered (R < 1)
```

An important weakness of the Clark-Evans test is that it *assumes that the point process is stationary* (as defined in Section 5.6.3). An inhomogeneous point pattern will typically give $R < 1$, and can produce spurious significance.

Hopkins and Skellam [339, 618] proposed taking the nearest-neighbour distances d_i for m randomly sampled data points, and the empty-space distances $e_j = d(u_j)$ for an equal number m of randomly sampled spatial locations. If the point pattern is completely random, points are independent of each other, so the distance from a data point to the nearest other point should have the same probability distribution as the distance from a fixed spatial location to the nearest data point. That is, the values d_i and e_j should have the same distribution. The *Hopkins-Skellam index* is

$$A = \frac{\sum_{i=1}^{m} d_i^2}{\sum_{j=1}^{m} e_j^2}. \tag{8.2}$$

Again this is a dimensionless index; $A = 1$ is consistent with a completely random pattern; $A < 1$ is consistent with clustering, and $A > 1$ with regularity. The *Hopkins-Skellam test* compares the value of A to the F distribution with parameters $(2m, 2m)$. In spatstat the functions hopskel and hopskel.test perform these calculations.

```
> hopskel(redwood)
[1] 0.142
> hopskel.test(redwood, alternative="clustered")
        Hopkins-Skellam test of CSR
        using F distribution

data:  redwood
A = 0.17, p-value <2e-16
alternative hypothesis: clustered (A < 1)
```

Interestingly the Hopkins-Skellam index is much less sensitive than the Clark-Evans index to problems such as edge effect bias and spatial inhomogeneity. These phenomena have roughly equal effect on the nearest-neighbour distances and the empty-space distances, so the ratio (8.2) is insensitive to them. We return to this principle in Section 8.6.

Summary statistics like the average nearest-neighbour distance, Clark-Evans index, and Hopkins-Skellam index are simple and practical to apply in the field. They may be useful when a simple numerical index of spatial pattern is needed, for example, when we need to compare a large number of different point patterns, or to monitor changes in spatial clustering over time. Their main weakness is that they compress all the spatial information into a single number, conflating information from different spatial scales and different spatial locations.

8.2.3 Exploratory graphics

Nearest-neighbour and empty-space distances can also be represented spatially and graphically. This can be useful for exploratory investigation of spatial point pattern data.

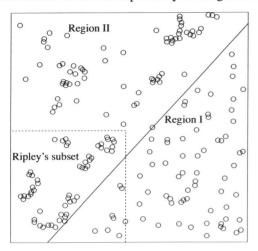

Figure 8.3. *California redwood seedlings and saplings, full dataset of Strauss [643].*

Figure 8.3 shows the locations of 195 seedlings and saplings of California giant redwood trees in a square sampling region, described and analysed by Strauss [643], and available in `spatstat` as the dataset `redwoodfull`. Strauss [643] divided the sampling region into two subregions I and II demarcated by a diagonal line shown in the Figure. The spatial pattern appears to be slightly regular in region I and strongly clustered in region II. Most writers have analysed a subset of the data extracted by Ripley [572], which is available in `spatstat` as `redwood` and is also frequently used in this book; see Section 6.8. Its position in the full dataset is shown in Figure 8.3.

The *Stienen diagram* of a point pattern is obtained by drawing a circle around each data point, of diameter equal to its nearest-neighbour distance. Imagine circular balloons, centred at each point of the pattern, growing at a constant rate. Each balloon stops growing when it touches another balloon. The result is the Stienen diagram, generated in `spatstat` by the function `stienen`, or by `plot(X %mark% nndist(X), markscale=1)` where X is the point pattern. The left panel of Figure 8.4 shows the Stienen diagram for the full California redwoods data, generated by

```
> stienen(redwoodfull)
```

Circles are shaded in grey if they are observed without edge effects — that is, if the observed nearest-neighbour distance is shorter than the distance to the window boundary.

The Stienen diagram can be useful for revealing aspects of spatial pattern, such as a trend in spatial scale. A pair of circles touching each other represents a pair of points which are mutual nearest neighbours. The fraction of area covered by the Stienen circles is related to the mean square nearest-neighbour distance.

As mentioned in Chapter 4, the *Dirichlet tile* associated with a particular data point x_i is the region of space that is closer to x_i than to any other point in the pattern $\mathbf{x}$:

$$C(x_i \mid \mathbf{x}) = \{u \in \mathbb{R}^2 : \|u - x_i\| = \min_j \|u - x_j\|\}. \tag{8.3}$$

The Dirichlet tiles are convex polygons (also known as Voronoï polygons, Thiessen polygons, fundamental polygons, or domains of influence) which divide two-dimensional space into disjoint regions, forming the *Dirichlet tessellation* or Voronoï diagram. The right panel of Figure 8.4 shows the Dirichlet tessellation of the full California redwoods data, generated by

```
> plot(dirichlet(redwoodfull))
```

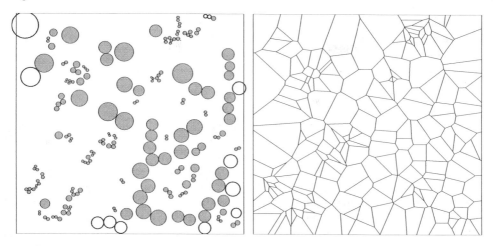

Figure 8.4. *Stienen diagram (Left) and Dirichlet tessellation (Right) for the full California redwoods data of Figure 8.3.*

The function `dirichlet` computes the Dirichlet tiles themselves, as polygonal windows, returning a tessellation object. The functions `nncross`, `nnmap`, `nnfun`, or `tile.index` can be used to identify *which* Dirichlet tile contains a given spatial location u. The tile containing u is $C(x_i \mid \mathbf{x})$ where x_i is the point of $\mathbf{x}$ nearest to u. An image plot of `nnmap(X)` or `nnfun(X)` is essentially a discretised image of `dirichlet(X)`.

The Dirichlet tessellation finds many other applications [6, 11, 149, 213, 521]. Pioneering epidemiologist John Snow [621] used a Dirichlet tessellation to show that the majority of people who died in the Soho cholera epidemic of 1854 lived closer to the infected Broad Street pump than to any other water pump.

8.3 Nearest-neighbour function G and empty-space function F

Valuable information about the spatial arrangement of points is conveyed by the nearest-neighbour distances. Much of this information is lost if we simply take the average of the nearest-neighbour distances. A better summary of the information is the cumulative distribution function of the nearest-neighbour distances, $G(r)$, called the *nearest-neighbour distance distribution function.*

Correspondingly, instead of taking the mean or mean square of the empty-space distances, a better summary is their cumulative distribution function $F(r)$, called the *empty-space function.*

The nearest-neighbour distance distribution function G and empty-space function F are important properties of the point process, analogous to the K-function in some ways, but based on a completely different construction.

8.3.1 Definitions of F and G for a stationary point process

Empty-space function F of a point process

If $\mathbf{X}$ is a spatial point process, the distance

$$d(u, \mathbf{X}) = \min\{\|u - x_i\| : x_i \in \mathbf{X}\} \tag{8.4}$$

from a fixed location $u \in \mathbb{R}^2$ to the nearest point of the process is called the 'empty-space distance' (or 'spherical contact distance' or 'void distance'). See the right panel of Figure 8.1.

For a stationary point process (defined in Section 5.6.3), the cumulative distribution function of the empty-space distance is the *empty-space function*

$$F(r) = \mathbb{P}\{d(u,\mathbf{X}) \le r\}, \tag{8.5}$$

defined for all distances $r \ge 0$, where u is an arbitrary reference location. Since the process is stationary, this definition does not depend on u. Alternative names for F are the 'spherical first contact distribution' and the 'point-to-nearest-event distribution'.

An estimate of F can be interpreted as a summary of spatial pattern [575], [611, pp. 488,491], [221, 576], [188, §4], [635, pp. 43,75,80–92,126–127], [190, chaps. 8–9], [484], [355, §4.2].

The values of $F(r)$ are probabilities (between 0 and 1) giving, for any fixed reference location u, the chance that there will be a point of $\mathbf{X}$ lying within distance r of this location. The value of $F(r)$ increases as a function of r, starting from $F(0) = 0$. For a stationary process, F is always differentiable, so that it has a probability density function [64, 322].

Nearest-neighbour distance distribution function G of a point process

If x_i is one of the points in a point pattern $\mathbf{x}$, the nearest-neighbour distance $d_i = \min_{j \ne i} \|x_j - x_i\|$ can also be written as

$$d_i = d(x_i, \mathbf{x} \setminus x_i),$$

the shortest distance from x_i to the pattern $\mathbf{x} \setminus x_i$ consisting of all points of $\mathbf{x}$ *except* x_i. We use this notation frequently below.

For a stationary point process $\mathbf{X}$ the *nearest-neighbour distance distribution function $G(r)$* is defined by

$$G(r) = \mathbb{P}\left\{d(u,\mathbf{X} \setminus u) \le r \,\middle|\, \mathbf{X} \text{ has a point at } u\right\} \tag{8.6}$$

for any $r \ge 0$ and any location u. That is, $G(r)$ is the cumulative distribution function of the nearest-neighbour distance $d(u,\mathbf{X} \setminus u)$ at a typical point of $\mathbf{X}$. Since the process is stationary, this definition does not depend on the location u.

The values of $G(r)$ are probabilities and are non-decreasing as a function of distance r, starting from $G(0) = 0$.

In general, the nearest-neighbour function G might *not* be differentiable, so that G may not have a probability density. For example, a randomly translated regular grid of points (as described at the end of Section 7.6.1) is a stationary point process; the distance from each point to its nearest neighbour is a fixed value, so $G(r)$ is a step function, and does not have a derivative.

8.3.2 Values of F and G for complete randomness

Empty-space function F for Poisson process

Figure 8.5 sketches the fundamental duality between distances and counts. The distance to the nearest point of $\mathbf{X}$ is greater than r, if and only if the disc of radius r contains no points of $\mathbf{X}$. Formally

$$d(u,\mathbf{X}) > r \quad \text{if and only if} \quad n(\mathbf{X} \cap b(u,r)) = 0 \tag{8.7}$$

where $b(u,r)$ is the disc of radius r centred at u.

This fact makes it possible to calculate $F(r)$ for a completely random pattern. If $\mathbf{X}$ is a uniform Poisson process in $\mathbb{R}^2$ of intensity λ, the number of points falling in a set B has a Poisson distribution with mean $\lambda|B|$. The probability that no points fall in B is the probability of zero

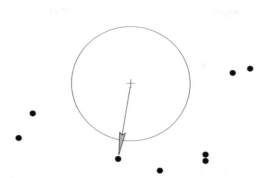

Figure 8.5. *Duality between distances and counts. The distance $d(u, \mathbf{X})$ from a fixed location u (+) to the nearest random point ($\bullet$) satisfies $d(u, \mathbf{X}) > r$ if and only if there are no random points in the disc of radius r centred at u.*

counts in the Poisson distribution, $\exp(-\lambda |B|)$. If B is a disc of radius r, then $|B| = \pi r^2$, so we get $\mathbb{P}\{d(u, X) > r\} = \mathbb{P}\{n(\mathbf{X} \cap b(u, r)) = 0\} = \exp(-\lambda \pi r^2)$, which gives

$$F_{pois}(r) = 1 - \exp(-\lambda \pi r^2). \tag{8.8}$$

This is the theoretical empty-space function for complete spatial randomness. Notice that (8.8) depends on the intensity λ, unlike the theoretical K-function which does not depend on intensity. The function is plotted in Figure 8.6.

Under complete spatial randomness, the empty-space distance has mean value $\mathbb{E}[d(u, \mathbf{X})] = 1/(2\sqrt{\lambda})$ and variance $\mathrm{var}[d(u, \mathbf{X})] = (1/\pi - 1/4)/\lambda$. The mean squared empty-space distance is $\mathbb{E}[d(u, \mathbf{X})^2] = 1/(\pi \lambda)$.

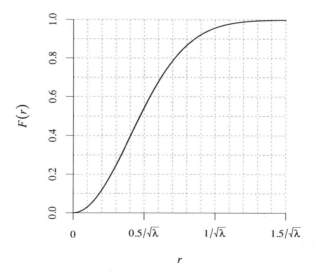

Figure 8.6. *Theoretical empty-space function for a homogeneous Poisson process with intensity λ. The mean distance is $0.5/\sqrt{\lambda}$ and the median is about $0.47/\sqrt{\lambda}$.*

Another way to describe the empty-space function is using the 'contact disc', the disc of radius $d(u, \mathbf{X})$ centred at u. The area of the contact disc, $A = \pi d(u, X)^2$, has a negative exponential distribution with rate λ under CSR:

$$\mathbb{P}\{A \leq a\} = 1 - \exp(-\lambda a), \qquad a \geq 0.$$

Equivalently $F_{pois}(r)$ is the distribution function of the square root of an exponential random variable with rate $\pi\lambda$.

Nearest-neighbour distance distribution function G for Poisson process

The nearest-neighbour distance distribution function for complete spatial randomness is now easy to calculate. Suppose $\mathbf{X}$ is a uniform Poisson process in R^2 of intensity λ. Since the points of $\mathbf{X}$ are independent of each other, the information that there is a point of $\mathbf{X}$ at the location u does not affect the probability distribution of the rest of the process. Therefore for CSR

$$
\begin{aligned}
G(r) &= \mathbb{P}\left\{d(u, \mathbf{X} \setminus u) \leq r \,\middle|\, \mathbf{X} \text{ has a point at } u\right\} \\
&= \mathbb{P}\left\{d(u, \mathbf{X} \setminus u) \leq r\right\}.
\end{aligned}
$$

But this is the empty-space function of the Poisson process, which has already been calculated in (8.8). That is, for a homogeneous Poisson process, the nearest-neighbour distance distribution function is

$$G_{pois}(r) = 1 - \exp(-\lambda \pi r^2) \tag{8.9}$$

identical to the empty-space function for the same process, $G_{pois} \equiv F_{pois}$. For a completely random pattern, the distribution of the nearest-neighbour distance is the same as that of the empty space distance.

For a general point process, of course, F and G will be different functions.

8.3.3 Estimation from data

Estimators of the empty-space function $F(r)$ and the nearest-neighbour function $G(r)$ from a point pattern dataset are described in Section 8.11. These estimators *assume that the point process is stationary*. They are edge-corrected versions of the empirical cumulative distribution functions of the nearest-neighbour distances at all data points (for G) and the empty-space distances at a grid of test locations (for F).

Experience suggests that the choice of edge correction is not very important so long as some edge correction is performed. However, some software implementations of these edge corrections may be incorrect, which could introduce substantial bias. Trustworthy R packages include spatial, splancs, and spatstat.

In spatstat, estimates of $F(r)$ and $G(r)$ are computed from a point pattern X by the commands Fest(X) and Gest(X), respectively. The syntax for these commands is very similar to that of Kest.

```
> Fs <- Fest(swedishpines)
> Gs <- Gest(swedishpines)
```

Tip: Avoid using the single letter F as the name of an object. The letter F is a recognised variable name in the R base environment, initially set to the logical value FALSE. Other single-letter names that should be avoided are C, c, D, I, T, t, and q.

The result of Fest or Gest, like Kest, is an object of class "fv", containing several estimates of the function using different edge corrections, together with the theoretical value for a homogeneous Poisson process with the same average intensity.

```
> Fs
Function value object (class 'fv')
for the function r -> F(r)
.............................................................
          Math.label        Description
r         r                 distance argument r
theo      F[pois](r)        theoretical Poisson F(r)
cs        hat(F)[cs](r)     Chiu-Stoyan estimate of F(r)
rs        hat(F)[bord](r)   border corrected estimate of F(r)
km        hat(F)[km](r)     Kaplan-Meier estimate of F(r)
hazard    hat(h)[km](r)     Kaplan-Meier estimate of hazard function h(r)
theohaz   h[pois](r)        theoretical Poisson hazard h(r)
.............................................................
Default plot formula:  .~r
where "." stands for 'km', 'rs', 'cs', 'theo'
Recommended range of argument r: [0, 7.6875]
Available range of argument r: [0, 22.312]
Unit of length: 0.1 metres
```

Like the *K*-function, the functions *F* and *G* are typically inspected by plotting the empirical function calculated from the data, together with the theoretical empty-space function of the homogeneous Poisson process with the same average intensity, plotted against distance *r*. This is the default behaviour in spatstat. Figure 8.7 shows the plots generated by the commands

```
> swp <- rescale(swedishpines)
> plot(Fest(swp))
> plot(Gest(swp))
```

Each plot has four curves, as indicated in the legend: three very similar estimates of the function using different edge corrections, and the theoretical function for complete spatial randomness.

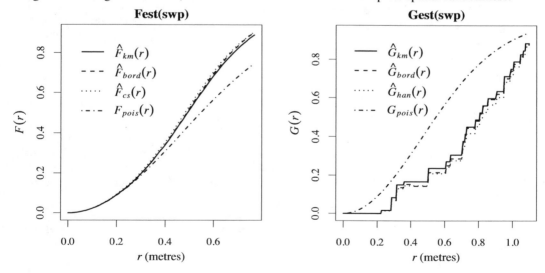

Figure 8.7. *Distance analysis of Swedish Pines data: empty-space function (Left) and nearest-neighbour distance distribution function (Right).*

Reading off the right panel in Figure 8.7, the value $\widehat{G}(r) = 0.5$ is achieved at about $r = 0.8$ metres, so the median of the distribution of nearest-neighbour distance in the Swedish Pines is about 80 centimetres. For a completely random process of the same intensity, the median nearest-neighbour distance would be about 55 centimetres according to the same plot.

8.3.4   Interpretation of the empty-space function F

Figure 8.8 shows the estimated empty-space function $\widehat{F}(r)$ for each of the three archetypal patterns in Figure 7.1, using only one edge correction for simplicity.

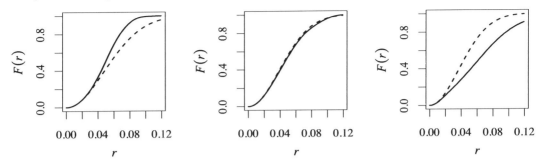

Figure 8.8. *Empirical empty-space function $\widehat{F}$ (solid lines) for each of the three patterns in Figure 7.1, and the theoretical function for a Poisson process (dashed lines). Generated using the plot formula* cbind(km, theo) ~ r.

The interpretation of deviations in F is the opposite of that for the K-function. In the left panel of Figure 8.8, the curve for the empirical empty-space function (solid lines) is above the theoretical curve for a completely random pattern (dashed lines), $\widehat{F}(r) > F_{pois}(r)$. That is, for a given distance r, the estimated probability that $d(u, \mathbf{X}) \leq r$ is greater than it would be in a completely random pattern. Consequently the empty-space distances are *shorter* than expected if the pattern were completely random. This is consistent with a regular pattern.

Similarly in the right panel, the empirical curve is below the theoretical curve, $\widehat{F}(r) < F_{pois}(r)$, indicating that empty-space distances are larger than would be expected if the corresponding pattern were completely random; this is consistent with clustering.

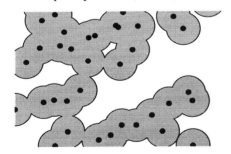

It may be helpful to interpret the value of $F(r)$ as the average *fraction of area* occupied by the level set of the distance function

$$\mathbf{X}_{\oplus r} = \{u \in \mathbb{R}^2 : d(u, \mathbf{X}) \leq r\}, \qquad (8.10)$$

the set of all spatial locations which lie at most r units away from the point process $\mathbf{X}$. This region is sketched in Figure 8.9: it is the union of discs of radius r centred at each point of $\mathbf{X}$. It is variously called a 'buffer region', 'Steiner set', or 'morphological dilation' of $\mathbf{X}$ by distance r. As r increases, the shaded region in Figure 8.9 grows and eventually fills the space. The space is filled faster if the data points are regularly spaced, and is filled more slowly if the data points are clustered together.

Figure 8.9. *Dilation set $\mathbf{X}_{\oplus r}$ (grey shading) of a point process $\mathbf{X}$ (points shown as •). The empty-space probability $F(r)$ is the average fraction of area occupied by $\mathbf{X}_{\oplus r}$.*

8.3.5   Interpretation of the nearest-neighbour distance distribution function G

Figure 8.10 shows the estimated nearest-neighbour distance distribution function $\widehat{G}(r)$ for each of the three archetypal patterns in Figure 7.1, using only one edge correction.

The interpretation of deviations in $G(r)$ is similar to that for the K-function, and the opposite of that for $F(r)$. In the left panel of Figure 8.10, the curve for the empirical nearest-neighbour distance distribution function (solid lines) is below the theoretical curve for a completely random pattern

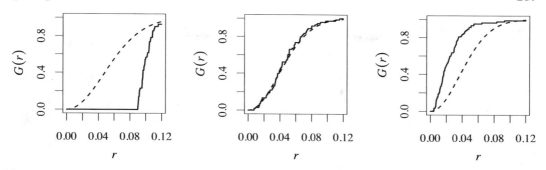

Figure 8.10. *Empirical nearest-neighbour distance distribution function* $\widehat{G}$ *(solid lines) for each of the three patterns in Figure 7.1, and the theoretical function for a Poisson process (dashed lines). Generated using the plot formula* cbind(km, theo) ~ r.

(dashed lines), $\widehat{G}(r) < G_{pois}(r)$. The nearest-neighbour distances in the data are *longer* than would be expected for a completely random pattern with the same average intensity. This is consistent with a regular pattern. Indeed, $\widehat{G}(r)$ is zero for distances less than about 0.09, which indicates that there are no nearest-neighbour distances shorter than 0.09. In the right panel of Figure 8.10, the empirical curve lies above the theoretical curve for a completely random pattern, indicating that the nearest-neighbour distances in the data are *shorter* than expected for a completely random pattern. This is consistent with clustering.

8.3.6 Implications for modelling

If the purpose of estimating the functions F and G is to guide the selection of appropriate point process models, then the functions $1 - F(r)$ and $1 - G(r)$ may be more useful.

In a modelling scenario where random points come from several different sources, $1 - F(r)$ is obtained by multiplying contributions from each of the sources. If **X** and **Y** are two *independent* stationary point processes with empty-space functions $F_{\mathbf{X}}(r)$ and $F_{\mathbf{Y}}(r)$, then their superposition $\mathbf{X} \cup \mathbf{Y}$ has empty-space function $F_{\mathbf{X} \cup \mathbf{Y}}(r)$ where

$$1 - F_{\mathbf{X} \cup \mathbf{Y}}(r) = (1 - F_{\mathbf{X}}(r))\,(1 - F_{\mathbf{Y}}(r)) \tag{8.11}$$

as we can see easily by remembering that $1 - F(r)$ is the probability that a disc of radius r contains no points.

The corresponding relation for G is more complicated:

$$1 - G_{\mathbf{X} \cup \mathbf{Y}}(r) = \frac{\lambda_{\mathbf{X}}}{\lambda_{\mathbf{X}} + \lambda_{\mathbf{Y}}}(1 - G_{\mathbf{X}}(r))\,(1 - F_{\mathbf{Y}}(r)) + \frac{\lambda_{\mathbf{Y}}}{\lambda_{\mathbf{X}} + \lambda_{\mathbf{Y}}}(1 - F_{\mathbf{X}}(r))\,(1 - G_{\mathbf{Y}}(r)) \tag{8.12}$$

where $\lambda_{\mathbf{X}}, \lambda_{\mathbf{Y}}$ are the intensities of **X** and **Y**, respectively. This arises because a typical point $\mathbf{X} \cup \mathbf{Y}$ originally came from **X** with probability $p_{\mathbf{X}}$ and from **Y** with probability $1 - p_{\mathbf{X}}$, where $p_{\mathbf{X}} = \lambda_{\mathbf{X}}/(\lambda_{\mathbf{X}} + \lambda_{\mathbf{Y}})$. For a typical point of **X**, say, the probability that there are no other points of **X** within a distance r is of course $1 - G_{\mathbf{X}}(r)$, and because of independence, the probability that there are no points of **Y** within distance r is $1 - F_{\mathbf{Y}}(r)$. The complexity of (8.12) is best handled using the J-function, discussed in Section 8.6.

8.4   Confidence intervals and simulation envelopes

Simulation envelopes for the empty-space function and nearest-neighbour function can be obtained using the methods discussed in Section 7.8 and spelt out in full in Sections 10.7 and 10.8.2. Confidence intervals are available using the block bootstrap procedure `varblock` described in Section 7.7. (Loh's bootstrap would not be useful in this context, as we discuss below.) Since $F(r)$ and $G(r)$ are cumulative distribution functions, some additional techniques are available, which we now present.

Figure 8.11 shows pointwise 95% confidence intervals for the true values of $F(r)$ and $G(r)$ for the Swedish Pines data, obtained by the spatial block bootstrap with a 5×5 grid of blocks, using

```
> swp <- rescale(swedishpines)
> Fci <- varblock(swp, Fest, nx=5, correction="best")
> Gci <- varblock(swp, Gest, nx=5, correction="best")
```

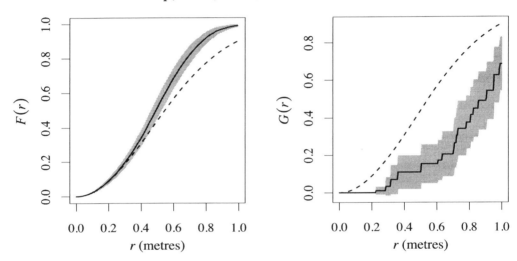

Figure 8.11. *Pointwise 95% confidence intervals for the true value of the empty-space function* (Left) *and the nearest-neighbour function* (Right) *for the rescaled Swedish Pines data.*

Figure 8.12 shows pointwise 5% significance envelopes for the null hypothesis of complete spatial randomness, generated by

```
> Fenv <- envelope(swp, Fest, nsim=39, fix.n=TRUE)
> Genv <- envelope(swp, Gest, nsim=39, fix.n=TRUE)
```

The shaded regions in Figures 8.11 and 8.12 have a 'spindle' shape which can be revealed more clearly by

```
> plot(Fenv, hi - lo ~ r, shade=NULL, xlim=c(0, 2))
```

This shape is to be expected, because $F(r)$ and $G(r)$ are probabilities, and the estimate of a population proportion is usually governed by the variance properties of the binomial distribution. This holds in our context, where limit theorems [57, 366] imply that $\text{var}[\widehat{F}(r)]$ should be approximately proportional to $F(r)(1 - F(r))$, and $\text{var}[\widehat{G}(r)]$ should be approximately proportional to $G(r)(1 - G(r))$, except for values of r close to 0 or 1. This can be confirmed by plotting the same objects using the formula `hi - lo ~ sqrt(theo * (1-theo))`. If Loh's bootstrap were implemented for the G-function, it would also effectively compute the same variance estimate.

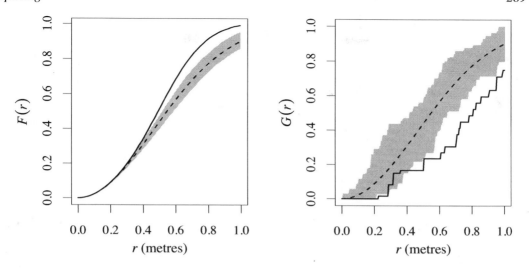

Figure 8.12. *Pointwise 5% significance envelopes for the empty-space function (Left) and the nearest-neighbour function (Right) for testing the hypothesis of complete spatial randomness on the rescaled Swedish Pines data.*

The variance-stabilising transformation for the binomial distribution is *Fisher's arcsine transformation*

$$\Phi(p) = \arcsin \sqrt{p} \tag{8.13}$$

so $\Phi(\widehat{F}(r)) = \arcsin \sqrt{\widehat{F}(r)}$ should have approximately constant variance as a function of r.

Variance stabilisation is particularly useful for improving the performance of deviation tests, such as the Diggle-Cressie-Loosmore-Ford test (Section 10.7.4), and of global simulation envelopes (Section 10.7.3). Figure 8.13 shows global envelopes for $\Phi(F(r))$ based on simulations from complete spatial randomness, generated by

```
> Phi <- function(x) asin(sqrt(x))
> Fglob <- envelope(swp, Fest, nsim=19, fix.n=TRUE,
                    global=TRUE, ginterval=c(0,1),
                    transform=expression(Phi(.)))
```

and plotted in the default style. Note that, although we could have given any name to the R function which implements the arcsine transformation (8.13), by choosing the name Phi we ensured that it would be rendered as a Greek letter in the plot labels.

Another feature in Figure 8.12 is the 'elephant's foot' at the bottom left corner of the right-hand panel. For any point pattern **x**, the graph of the empirical nearest-neighbour function $\widehat{G}(r)$ has a flat segment at zero height at the bottom left, because $\widehat{G}(r) = 0$ for all r less than the *minimum nearest-neighbour distance* $d_{min} = \min_i d_i$. See Section 8.2.1 for the notation.

At such small values of r, the asymptotic normal approximation to $\widehat{G}(r)$ breaks down completely, and Φ does not stabilise the variance. Small values of r should therefore be avoided when constructing global envelopes or a global confidence interval for $G(r)$. How small? For a homogeneous Poisson process observed in a window W, the *minimum* nearest-neighbour distance has very roughly[2] the distribution (8.8) with the parameter λ replaced by $\lambda^2 |W|/2$. The expected minimum nearest-neighbour distance is about $r_0 = 1/(\lambda \sqrt{2|W|})$ which we could estimate by $\hat{r}_0 = \sqrt{|W|}/(n\sqrt{2})$

---

[2] This is based on the heuristic that the n points are $n/2$ pairs of mutual nearest neighbours, with nearest-neighbour distances independent between each pair. Since πd_i^2 is exponential with rate λ, the minimum of $n/2$ such values, namely πd_{min}^2, is exponential with rate $n\lambda/2$.

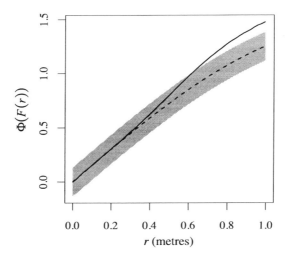

Figure 8.13. *Global envelopes of the variance-stabilised empty-space function. Solid line: transformed empirical estimate* $\Phi(F_{km}(r))$ *for rescaled Swedish Pines data. Dashed line: transformed Poisson function* $\Phi(F_{pois}(r))$. *Grey shading: region between global envelopes of* $\Phi(F_{km}(r))$ *over* $0 \le r \le 1$ *for 19 simulations of CSR with fixed number of points.*

where n is the number of data points. Figure 8.6 shows the cumulative distribution function (8.8) with mean 0.5, indicating that the value rarely exceeds 1.5, three times the mean. Consequently the minimum nearest-neighbour distance in our case is very unlikely to exceed $3r_0$. Thus, we should avoid distances smaller than about $3\hat{r}_0$.

P–P and Q–Q plots [701] are useful statistical graphics for comparing observed and predicted distributions. The P–P plot for the empty-space function (say) is a plot of the empirical probability $\widehat{F}(r)$ against the theoretical probability $F_{pois}(r)$ for every r. This can be generated using a plot formula in `plot.fv` (Section 7.5.2, page 220):

```
> plot(Fest(swp), . ~ theo)
```

Again the symbol '.' stands for 'all acceptable estimates of the function'. The result is shown in the left panel of Figure 8.14. Theoretical probabilities are plotted on the x-axis, and empirical probabilities on the y-axis. The diagonal line $y = x$ corresponds to perfect agreement between the observed data and a Poisson process. Reading from the plot, we find that the median empty-space distance in a Poisson process ($x = 0.5$ in the plot) is equal to the 60th percentile of empty-space distances in the data ($y = 0.6$ in the plot).

The *Q–Q plot* for the nearest-neighbour distance distribution function (say) is the plot of empirical quantiles $\widehat{G}^{-1}(p)$ against theoretical quantiles $G_{pois}^{-1}(p)$ for every p between 0 and 1, where $\widehat{G}^{-1}$ and G_{pois}^{-1} are the inverse functions. This can be generated by

```
> plot(QQversion(Gest(swp)))
```

The result is shown in the right panel of Figure 8.14. Empirical quantiles are plotted on the x-axis, and the corresponding theoretical quantiles on the y-axis. For example the observed proportion of nearest-neighbour distances less than 80 centimetres ($x = 0.8$) is equal to the theoretically expected proportion of nearest-neighbour distances less than 55 centimetres ($y = 0.55$). The diagonal line $y = x$ corresponds to perfect agreement between the observed data and a Poisson process. Another straight line would correspond to a linear relationship between the observed and theoretical nearest-

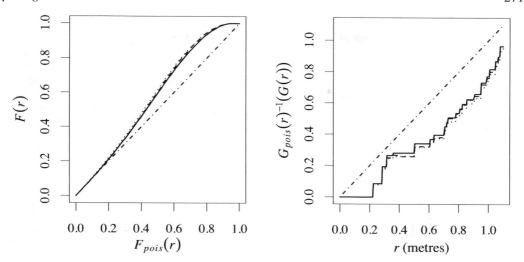

Figure 8.14. *P–P and Q–Q plots.* Left: *P–P plot of empty-space function;* Right: *Q–Q plot of nearest-neighbour distance distribution function. Rescaled Swedish Pines data. Diagonal line is theoretical curve for CSR; other lines are empirical estimates.*

neighbour distances: a straight line through the origin would correspond to a Poisson process with a different intensity.

Aitkin and Clayton [8] suggested applying Fisher's variance-stabilising transformation (8.13) to *both* axes in a P–P plot to preserve the linear relationship. For the previously computed pointwise confidence intervals and envelopes `Fci`, `Gci`, `Fenv`, `Genv` this could be achieved by plotting the object using the formula `Phi(.) ~ Phi(theo)`.

The global envelopes contained in `Fglob` were computed using the argument `transform = expression(Phi(.))`, so the appropriate plot formula is `. ~ theo`. The result is shown in Figure 8.15. To make global envelopes for the variance-stabilised nearest-neighbour distance distribution function $\Phi(G(r))$, we should avoid small distances as discussed above:

```
> r0 <- sqrt(area(swp)/2)/npoints(swp)
> Gglob <- envelope(swp, Gest, nsim=19, fix.n=TRUE,
                    global=TRUE, ginterval=c(3*r0,1),
                    transform=expression(Phi(.)))
```

then plot this with the formula `. ~ theo` to produce the P–P plot. The result is shown in the right panel of Figure 8.15.

8.5 Empty-space hazard

In Chapter 7 we saw that the cumulative nature of the K-function has some disadvantages. Similar comments apply to the cumulative distribution functions $F(r)$ and $G(r)$. While they are very useful for assessing statistical significance, their interpretation is complicated by the fact that the value of $F(r)$ or $G(r)$ contains information from all distances s less than or equal to r.

There is a need for alternative summary functions, derived from F and G, which would only contain contributions from distances *equal to* r. For technical reasons we will concentrate mostly on the empty-space function F.

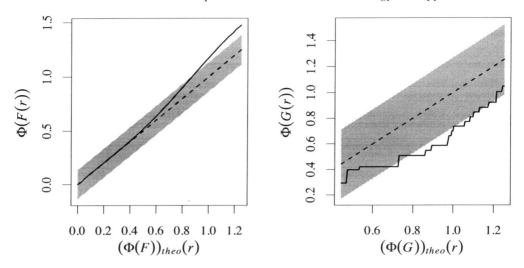

Figure 8.15. *Global envelopes for variance-stabilised summary functions.* Left: *empty-space function.* Right: *nearest-neighbour distance distribution function, avoiding short distances.*

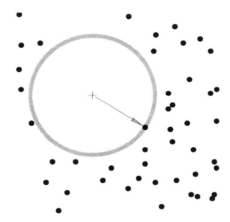

Figure 8.16. *Geometry for derivative and hazard rate of empty-space function.*

The most obvious choice for a non-cumulative alternative to F is the *derivative* $f(r) = F'(r)$, which is the probability density function of the empty-space distance $d(u, \mathbf{X})$ from an arbitrary fixed location u to the nearest point in $\mathbf{X}$. Kernel estimators of the density of F have been proposed in [243, 257]. Its geometrical interpretation is sketched in Figure 8.16. Consider two circles, of radius r and $r + \Delta r$, centred at the reference location u. The distance $d(u, \mathbf{X})$ lies between r and $r + \Delta r$ if and only if there are no points of $\mathbf{X}$ in the inner circle of radius r, *and* there is a point of $\mathbf{X}$ in the outer circle of radius $r + \Delta r$. This has probability $F(r + \Delta r) - F(r)$. Dividing by Δr and taking the limit as $\Delta r \to 0$ we obtain the derivative $f(r)$.

For a homogeneous Poisson process, the density $f(r)$ of the empty-space function can be calculated from (8.8):

$$f_{pois}(r) = F'_{pois}(r) = 2\pi r \lambda \exp(-\lambda \pi r^2). \tag{8.14}$$

This is not very convenient as a benchmark for practical interpretation.

Refer again to the geometry of Figure 8.16. The probability of the event depicted in Figure 8.16

is approximately $f(r)\Delta r$, or

$$[\lambda 2\pi r\Delta r]\,\exp(-\lambda\pi r^2).$$

The term $\exp(-\lambda\pi r^2)$ is the probability of the event that there are no points of **X** in the disc of radius r. The term $\lambda 2\pi r\Delta r$ is (for small Δr) the probability of the event that at least one point of **X** falls in the ring between the two circles of radius r and $r+\Delta r$. For a Poisson process, these two events are independent, so the probability that they both occur is the product of their probabilities. This intuitively explains the form of (8.14).

A better alternative to $f(r)$ is the *hazard rate*

$$h(r) = \frac{f(r)}{1 - F(r)}. \tag{8.15}$$

The hazard rate has the following simple interpretation: *given* that the empty-space distance $d(u,\mathbf{X})$ is greater than r, the probability that it falls between r and $r+\Delta r$ is equal to $h(r)\Delta r$. In Figure 8.16, $h(r)\Delta r$ is the conditional probability that a point falls in the shaded ring, *given* that no point falls in the inner circle.

For complete spatial randomness, the hazard rate for the empty-space distance is

$$h_{pois}(r) = 2\pi\lambda r, \tag{8.16}$$

a linear function of r, making it very useful as a benchmark. The hazard rate of a distribution is a useful tool in biostatistics and risk analysis for studying the distribution of lifetimes, survival times, time-to-first-failure, and similar quantities. For human lifetimes, the hazard rate answers the question: given that I have reached age n years, what is the chance that I will die before my $(n+1)$th birthday? Human mortality, machine reliability, and so on are much easier to interpret using the hazard rate than the probability density.

The `spatstat` command `Fest`, in addition to calculating estimates of the empty-space function $F(r)$, also computes an estimate $\widehat{h}(r)$ of the hazard rate $h(r)$, and the corresponding theoretical hazard $h_{pois}(r)$ for a Poisson process with the same intensity. The hazard rate estimate is not plotted in the default plot, because $h(r)$ is on a completely different scale from $F(r)$. However, the estimated and theoretical hazard functions are shown as columns of the result of `Fest` if the result is printed (see page 8.3.3 for an example). These columns can be plotted by specifying them in the plot formula, for example,

```
> plot(Fest(cells), cbind(hazard, theohaz) ~ r)
```

Figure 8.17 shows the estimated empty-space hazard $h(r)$ for the three archetypal point patterns in Figure 7.1.

The `spatstat` function `Fhazard` is a wrapper for `Fest` which extracts the hazard estimates. It should be used when computing confidence intervals and simulation envelopes. Figure 8.18 shows an example computed by

```
> hazenv <- envelope(swp, Fhazard, nsim=39, fix.n=TRUE,
                 transform=expression(./(2*pi*r)))
```

Dividing the hazard rate $h(r)$ by $2\pi r$ means that if the point process is CSR, the ratio $h(r)/(2\pi r)$ will be constant and equal to λ. This transformation does not completely stabilise variance but is preferable to the un-transformed hazard rate for use in simulation envelopes.

Estimates of the hazard rate are usually computed directly from data using specialised estimators described in Section 8.11. If an estimate of the function $F(r)$ is already available, then $h(r)$ can be computed by numerical differentiation, using either (8.15) or the identity

$$h(r) = \frac{\mathrm{d}}{\mathrm{d}r}\left[-\log(1 - F(r))\right]. \tag{8.17}$$

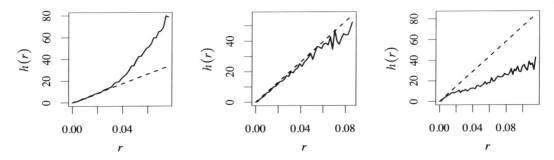

Figure 8.17. *Estimated empty-space hazard* $\widehat{h}$ *(solid lines) for each of the three patterns in Figure 7.1, and the theoretical hazard function for a Poisson process (dashed lines). Generated using the plot formula* cbind(hazard, theohaz) ~ r.

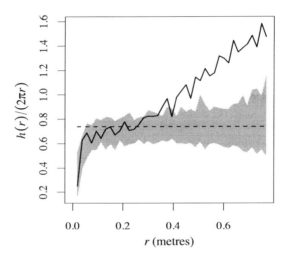

Figure 8.18. *The standardised hazard function* $h(r)/(2\pi r)$ *for the rescaled Swedish Pines data, and the pointwise envelopes of 39 simulations of complete spatial randomness.*

The function $-log(1 - F(r))$ is called the *integrated hazard* or *cumulative hazard* since it satisfies

$$-\log(1 - F(r)) = \int_0^r h(s)\, ds. \tag{8.18}$$

The cumulative hazard of a lifetime or survival time is important in survival analysis. The cumulative hazard of empty-space distance has been much used in connection with parameter estimation for the Boolean model [190, §9.6], [314, pp. 291–294], [611, pp. 495–502], [635, pp. 84–92], [576, chap. 6]. It has fundamental importance because it is a special case of the *capacity functional* in stochastic geometry, connected with characterisations of infinite divisibility, convexity, and other properties [456, chap. 3, thm. 5.3.1, 5.5.1].

The interpretation of empty-space distance and nearest-neighbour distance as waiting times or failure times was proposed in [62, 284]. Many distance and size variables arising in stochastic geometry are generalisations of one-dimensional waiting times [466, 469, 485, 713].

The transformation from F to h is a convenient simplification, at least in the Poisson case. For Poisson cluster processes, the hazard rate of F is discussed in [197, pp. 242–245].

8.6 *J*-function

8.6.1 Comparing nearest-neighbour and empty-space distances

Nearest-neighbour distances and empty-space distances have the same probability distribution if the pattern is completely random. Under various departures from complete spatial randomness, the nearest-neighbour and empty-space distances tend to respond in opposite directions — one becoming larger while the other becomes smaller. This suggests that a comparison of these two types of distance could be useful in assessing departure from CSR.

The Hopkins-Skellam test (Section 8.2.2) is an example of such a technique. The test statistic (8.2) is the ratio of the mean squares of the nearest-neighbour and empty-space distances.

Diggle [218, eq. (5.7)] proposed the diagnostic $D_{max} = \max_r |\widehat{G}(r) - \widehat{F}(r)|$ as a measure of deviation from the Poisson process. A Monte Carlo test based on D_{max} is equivalent to a global envelope test (Section 10.7.3) based on the summary function $D(r) = G(r) - F(r)$. This function is identically equal to zero for a homogeneous Poisson process. Values $D(r) > 0$ are consistent with clustering, while $D(r) < 0$ is consistent with regularity. The function can be calculated in spatstat using eval.fv:

```
> FS <- Fest(swp)
> GS <- Gest(swp)
> DD <- eval.fv(GS-FS)
```

In order to construct a confidence interval or simulation envelopes, this should be coded as a single function:

```
> DigDif <- function(X, ..., r=NULL) {
    FX <- Fest(X, ..., r=r)
    GX <- Gest(X, ..., r=FX$r)
    eval.fv(GX-FX)
  }
```

The argument r is not necessary here, but ensures that the two function objects have exactly the same vector of *r* values. Then simulation envelopes for testing CSR (for example) could be constructed by

```
> DE <- envelope(swp, DigDif, nsim=19, fix.n=TRUE, global=TRUE)
```

Weaknesses of the Hopkins-Skellam index and the diagnostic D_{max} include the loss of information about spatial scale, and the lack of a clear interpretation for their numerical values, other than with reference to CSR. The function $D(r)$ does have an interpretation as a score residual for testing the null hypothesis of CSR against a particular alternative hypothesis; see Section 11.6.4.

8.6.2 The *J*-function

A useful combination of F and G, suggested by fundamental theory, is the *J-function* [663] of a stationary point process,

$$J(r) = \frac{1 - G(r)}{1 - F(r)}, \tag{8.19}$$

defined for all $r \geq 0$ such that $F(r) < 1$. For a homogeneous Poisson process, $F_{pois} \equiv G_{pois}$, so that

$$J_{pois}(r) \equiv 1. \tag{8.20}$$

Values $J(r) > 1$ are consistent with a regular pattern, and $J(r) < 1$ is consistent with clustering, at scales less than or equal to r.

The value of $J(r)$ is the dimensionless ratio of two probabilities: $1 - G(r)$ is the probability that the nearest-neighbour distance will be *greater than r*, and $1 - F(r)$ is the corresponding probability for the empty-space distances. A value $J(10) = 3$, for example, would mean that the distance threshold of 10 units is exceeded 3 times more often by the nearest-neighbour distances than by the empty-space distances.

Figure 8.19 shows the estimated J-functions for the three archetypal point patterns in Figure 7.1, calculated by substituting estimates of F and G into the definition (8.19). The J-function clearly discriminates between clustering and regularity.

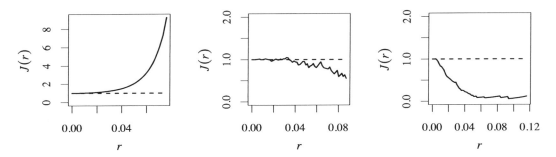

Figure 8.19. *Empirical J-function (solid lines) for each of the three patterns in Figure 7.1, and the theoretical function for a Poisson process (dashed lines). Generated using the plot formula* `cbind(km, theo)` ~ `r`. *Note different axis scale for left panel.*

In a modelling scenario where random points come from several different sources, the J-function is a simple mixture of the J-functions of the sources. The superposition $\mathbf{X} \cup \mathbf{Y}$ of two *independent* point processes $\mathbf{X}$ and $\mathbf{Y}$ has J-function

$$J_{\mathbf{X} \cup \mathbf{Y}}(r) = \frac{\lambda_{\mathbf{X}}}{\lambda_{\mathbf{X}} + \lambda_{\mathbf{Y}}} J_{\mathbf{X}}(r) + \frac{\lambda_{\mathbf{Y}}}{\lambda_{\mathbf{X}} + \lambda_{\mathbf{Y}}} J_{\mathbf{Y}}(r) \qquad (8.21)$$

where $J_{\mathbf{X}}$ and $J_{\mathbf{Y}}$ are the J-functions of $\mathbf{X}$ and $\mathbf{Y}$, respectively, and $\lambda_{\mathbf{X}}, \lambda_{\mathbf{Y}}$ are the intensities of the two processes. This follows from equations (8.11) and (8.12).

The J-function can be evaluated for many point processes even when the F and G functions cannot [663]. For example, for Neyman-Scott cluster processes there is an exact formula for the J-function; for Gibbs processes there is a general expression (Section 13.12.4) which can be used at least to derive qualitative properties.

The `spatstat` function `Jest` computes edge-corrected estimates of the J-function from a point pattern dataset, assuming the point process is stationary. Its syntax is identical to that of `Gest` and `Fest`.

The empirical J-function is insensitive to edge effects, because the nearest-neighbour and empty-space distances are subject to similar edge effects which 'cancel out' in the ratio $(1 - G)/(1 - F)$ to a first approximation [44]. In many applications the uncorrected estimate of the J-function can be used, saving computation time and maximising the use of data.

The J-function has found numerous applications in astronomy [377, 380, 379], soil science [624, 263], and forestry [536].

Like the K-function, the functions F, G, and J do not completely characterise the point process. In particular, if data analysis suggests that $J(r) \equiv 1$, this does not guarantee that the point process is completely random. A counterexample in one-dimensional space was found in [86].

A rough approximation to $J(r)$, obtained by ignoring all moments higher than the second moment, is [662, p. 186]

$$J(r) - 1 \approx -\lambda (K(r) - \pi r^2) \tag{8.22}$$

so that we can think of the K-function as a second-order approximation to the J-function. It will therefore not be too surprising if the empirical J- and K-functions of a point pattern dataset support the same conclusion about the data.

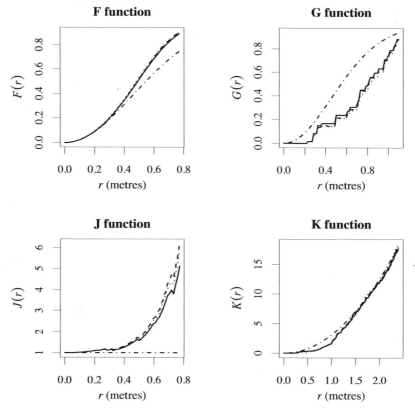

Figure 8.20. *Result of* `allstats` *for the Swedish Pines data.*

The function `allstats` efficiently computes all of the F-, G-, J- and K-functions for a dataset. The result can be plotted directly. Figure 8.20 was generated by `plot(allstats(swp))`. The plots of F and J in the left column are expected to respond to interpoint interaction in the opposite sense from the plots of G and K in the right column.

8.7 Inhomogeneous F-, G- and J-functions

The summary functions F, G, and J assume that the point process is stationary (as defined in Section 5.6.3). Van Lieshout [662] introduced 'inhomogeneous' versions of F, G, and J which can be used when the point process is not stationary, subject to special conditions.

The definition of the inhomogeneous K-function (Section 7.10.2) required the assumption that the point process is correlation-stationary, equation (7.47). Similarly the inhomogeneous versions of F, G, and J assume that the 'k-point correlation functions' for all $k \geq 2$ are invariant under translation. For definitions and details, see [662]. If $\lambda(u)$ is the true intensity function of the point process **X**, it is additionally assumed that $\lambda(u) \geq \lambda_{min} > 0$ for all locations u. Then the inhomogeneous empty-space function F_{inhom} can be defined by

$$F_{inhom}(r) = 1 - \mathbb{E}\left[\prod_{x_i \in \mathbf{X} \cap b(u,r)} \left(1 - \frac{\lambda_{min}}{\lambda(x_i)}\right)\right] \tag{8.23}$$

and the inhomogeneous nearest-neighbour function G_{inhom} by

$$G_{inhom}(r) = 1 - \mathbb{E}\left[\prod_{x_i \in \mathbf{X} \cap b(u,r)} \left(1 - \frac{\lambda_{min}}{\lambda(x_i)}\right) \;\middle|\; \mathbf{X} \text{ has a point at } u\right] \tag{8.24}$$

where u is an arbitrary location. Under the assumptions stated above, these definitions do not depend on the choice of u. The inhomogeneous J-function is then defined as

$$J_{inhom}(r) = \frac{1 - G_{inhom}(r)}{1 - F_{inhom}(r)}. \tag{8.25}$$

The spatstat functions Finhom, Ginhom, and Jinhom compute estimates of F_{inhom}, G_{inhom}, and J_{inhom}, respectively, using the border method of edge correction. They require an estimate of the intensity function $\lambda(u)$, which may be provided as a vector of intensity values at the data points, a pixel image, a function, or a fitted point process model. If no intensity values are provided by the user, then an intensity function will be estimated from the data by a leave-one-out kernel smoother, in the same way as for Kinhom (Section 7.10.2).

Limited practical experience is available with this new technique but it seems likely to be useful in cases where the intensity is *gradually* varying across the sampling window. For example, the Swedish Pines data show evidence of slight inhomogeneity, discussed in Section 6.5.1.2. Figure 8.21 shows the homogeneous J-function for Swedish Pines and the inhomogeneous J-function using the leave-one-out kernel smoother with bandwidth selected by Scott's rule. The discrepancy between the two J-functions suggests that some of the evidence for interpoint interaction is better explained by inhomogeneity. To confirm this we should check visually whether the selected bandwidth seems appropriate, since the technique is sensitive to the choice of bandwidth. See also [355, sec. 4.10].

8.8 Anisotropy and the nearest-neighbour orientation

Figure 7.29 showed a synthetic point pattern exhibiting anisotropy. Second-order methods for analysing the directional preference include the sector K-function (Section 7.9.1), point pair orientation distribution (Section 7.9.2), and the anisotropic pair correlation function (Section 7.9.3).

An alternative approach is to measure the direction of the vector joining each data point to its nearest neighbour. The probability distribution of these directions is the *nearest-neighbour orientation distribution*. An estimate of this distribution is computed by the function nnorient using edge correction as described in [355, eq. (4.5.3.), p. 253].

Figure 8.22 shows a kernel-smoothed estimate of the probability density of the nearest-neighbour orientation distribution for the synthetic pattern Y in Figure 7.29, computed and plotted by rose(nnorient(Y)). Results are often similar to those obtained from pairorient, but nnorient has the advantage that it does not require a choice of distance range, and can be computed quickly.

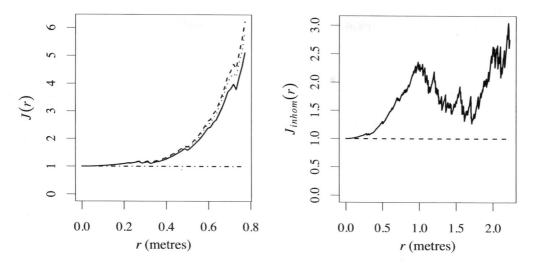

Figure 8.21. *Estimated J-function assuming stationarity (Left) and inhomogeneous J-function assuming correlation stationarity of all orders (Right) for the Swedish Pines data. Note that panels have different axis scales. Horizontal line is theoretical value for CSR; other lines are empirical estimates.*

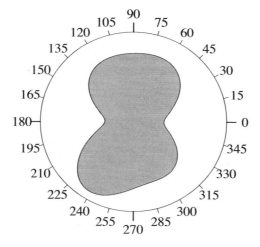

Figure 8.22. *Rose diagram of estimated probability density for the nearest-neighbour orientation distribution of the synthetic pattern in Figure 7.29.*

8.9 Empty-space distance for a spatial pattern

The concept of empty-space distance is not restricted to point patterns. Any spatial pattern of geometrical objects — such as lines, circles, or irregular shapes — can also be analysed by measuring the shortest distance from a reference point to the spatial pattern.

In particular this provides one easy way in which *some* statistical methods for point patterns can be extended to apply to patterns of objects of appreciable size [519, chap. 9].

The empty-space distance can be defined for any subset Y of two-dimensional space. We assume

Y is non-empty and closed. Then for any location u, the empty-space distance is the minimum distance from u to any point in Y,

$$d(u,Y) = \min\{\|u - z\| : z \in Y\}. \qquad (8.26)$$

See Figure 8.23. If u falls inside Y, then $d(u,Y) = 0$.

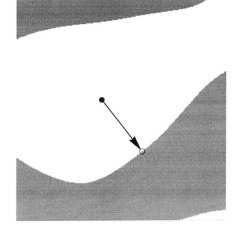

Figure 8.23. *Distance from a location u to the nearest point in a set Y (grey shading).*

Now consider a *random* set $\mathbf{Y}$, so that $d(u, \mathbf{Y})$ becomes a random variable. Assume that $\mathbf{Y}$ is a *stationary* random set (for details of these concepts, see [473, 636].) The unconditional *empty-space function* or *spherical contact distribution function* of $\mathbf{Y}$ is

$$H^*(r) = \mathbb{P}\{d(u, \mathbf{Y}) \le r\}, \quad r \ge 0 \qquad (8.27)$$

where u is an arbitrary fixed location. The definition of H^* does not depend on the choice of location u. The function H^* is the generalisation to random sets of the empty-space function F for point processes.

If the random set $\mathbf{Y}$ happens to cover the location u, then $d(u, \mathbf{Y}) = 0$. Typically there is a non-zero probability p that this will occur. This means that the function H has a jump at the origin: $H^*(0) = p > 0$ is a positive number, equal to the probability that $\mathbf{Y}$ covers u. This is equal to the average fraction of area covered by $\mathbf{Y}$. It is often desirable to remove this jump, by defining the *conditional* empty-space or spherical contact function

$$H(r) = \mathbb{P}\{d(u,\mathbf{Y}) \le r \mid d(u,\mathbf{Y}) > 0\} = \frac{H^*(r) - H^*(0)}{1 - H^*(0)}, \quad r \ge 0. \qquad (8.28)$$

This function satisfies $H(0) = 0$ and gives the cumulative distribution function of the empty-space distance from an arbitrary location u, given that this location falls outside $\mathbf{Y}$.

The `spatstat` function `Hest` computes edge-corrected estimates of $H^*(r)$, $H(r)$ and the hazard rate of $H(r)$ using the methods of [323]. Its first argument X may be a point pattern (object of class `"ppp"`), a line segment pattern (object of class `"psp"`), or a window (object of class `"owin"`). It is assumed to be a realisation of a stationary random set. The argument `conditional` determines whether H^* or H is computed.

Figure 8.24 shows the spatial locations and diameters of sea anemones (beadlet anemone *Actinia equina*) in a sample plot on a boulder between low and high tide levels at Quiberon, Bretagne, France [389, 390], [660, pp. 64–67]. The data are given in the `spatstat` dataset `anemones`, a marked point pattern with marks representing diameters, expressed in the same units as the spatial coordinates. The diameters were plotted to scale in Figure 8.24 by `plot(anemones, markscale=1)`. Although these data are superficially similar to the Longleaf Pines data (Figure 1.8 on page 7), they require different treatment, because the physical sizes of the anemones must be taken into account. One possible exploratory tool is the spherical contact function H.

The cited literature gives conflicting definitions for the unit of length in this dataset: see the help entry for `anemones`. We shall take the frame to be exactly 145 centimetres wide:

```
> Anemones <- anemones
> width <- sidelengths(Window(Anemones))[1]
> unitname(Anemones) <- list("cm", "cm", 145/width)
```

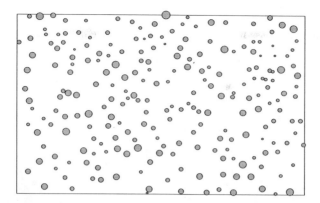

Figure 8.24. *Sea anemones on a boulder near the waterline. About 145 cm across. Anemone sizes are to scale.*

This retains the original data in the original units, and simply declares one unit to be equal to `145/width = 145/280 = 0.518` centimetres. To convert all the data values to centimetres, we then type

```
> Anemones <- rescale(Anemones)
> marks(Anemones) <- marks(Anemones) * 145/width
```

where the last line is needed because `rescale` only changes the numerical values of the spatial coordinates, and not the mark values.

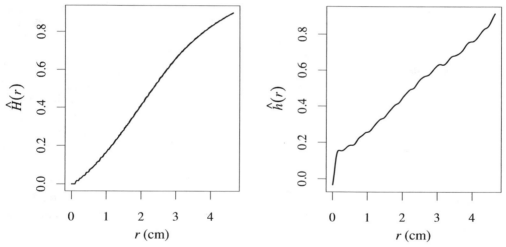

Figure 8.25. *empty-space statistics for anemones data.* Left: *conditional empty-space function* $\widehat{H}(r)$. Right: *smoothed hazard rate* $\widehat{h}(r)$.

To compute the spherical contact function, we need to convert these location-and-size data to a spatial region, that is, the region of space covered by the anemones. The `spatstat` command `discs` is built for this purpose. By default it assumes that the mark values are diameters, which is correct in this case, so we simply type

```
> AnemoneSet <- discs(Anemones, eps=0.1, mask=TRUE)
```

where `eps` is the desired pixel resolution in spatial coordinate units (centimetres). The result

AnemoneSet is a binary mask window, since we chose mask=TRUE. A plot of AnemoneSet (not shown) looks very similar to Figure 8.24, as required. Then we apply Hest to this window:

```
> Hanem <- Hest(AnemoneSet, conditional=TRUE)
```

The resulting estimate of $H(r)$ is shown in the left panel of Figure 8.25. The right panel shows the estimate of the hazard rate of $H(r)$, after smoothing by

```
> Hsmooth <- Smooth(Hanem, which="hazard")
```

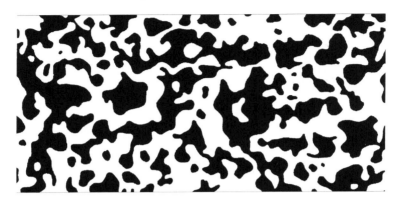

Figure 8.26. *Diggle's heather data, high resolution. Land covered by* Calluna vulgaris *(black) in a* 10×20 *metre sampling plot. Image rotated by 90 degrees using* flipxy.

Figure 8.26 shows the spatial mosaic of vegetation of the heather plant *Calluna vulgaris* in a 10 by 20 metre sampling plot in a meadow near Jädraås, Sweden [542]. They were recorded and first analysed by Diggle [220] and subsequently studied in [576, pp. 121–122, 131–135], [313], [314, pp. 301–318], [190, pp. 763–770]. It is of interest to characterize the 'spatial pattern' of heather [297]. Diggle [220] fitted a 'Boolean model' in which the heather component is the union of discs of independent random radius centred at the points of a Poisson point process.

Several versions of this dataset are provided in spatstat in the object heather. Figure 8.26 shows the highest-resolution version, heather$fine, which was scanned from Diggle's original hand-drawn map and processed by Chris Jonker.

Figure 8.27 shows the estimates of $H(r)$ and its smoothed hazard rate, computed by

```
> HH <- with(heather, Hest(fine))
> SH <- Smooth(HH)
```

The estimated hazard rates in these two analyses are roughly linear except at small distances. The linear part is broadly consistent with a Boolean model of discs (essentially the union of discs centred at the points of a Poisson process). Formal interpretation is more complicated than for point processes, and we do not elaborate here. Theoretical results were developed in [63, 323, 64, 322, 351, 350, 408, 407, 409].

8.10   Distance from a point pattern to another spatial pattern

Often a spatial point pattern dataset is accompanied by covariate data in the form of another spatial pattern. For example, geological surveys of mineral deposits, such as the Queensland copper deposits and the Murchison gold data, usually include a map of the relevant geological faults and other

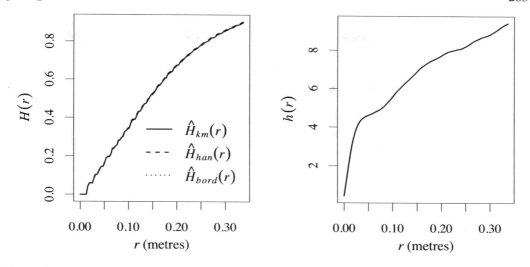

Figure 8.27. *Conditional empty-space function H(r) (Left) and empty-space hazard h(r) (Right) for Diggle's heather data.*

structural features. In these surveys, the distance from each mineral deposit to the nearest geological fault is an important quantity. Accordingly we need techniques for analysing the distance *from* a point pattern *to* another spatial pattern.

For example, the Queensland copper data (Figure 1.11 on page 9) consist of a point pattern **x** giving the locations of copper deposits in a survey region, together with a line segment pattern **y** giving the locations of geological lineaments (mostly geological faults). The aim is to decide whether the lineament pattern, which can easily be observed and mapped in an aerial survey, carries any predictive information about the location of copper deposits. It is conjectured that copper is more likely to be deposited close to a fault.

Techniques suited to this purpose include the estimator of intensity as a function $\rho(Z(u))$ of a covariate Z (Section 6.6.3) and formal tests of dependence on a covariate Z (Section 6.7) where in both cases the obvious covariate would be the distance to the nearest lineament, $Z(u) = d(u, \mathbf{y})$. These analyses are *conditional* on the pattern **y**, effectively taking **y** to be fixed, so that $Z(u) = d(u, \mathbf{y})$ is a known function.

An alternative approach is to view **x** and **y** as the results of random processes. We can then define an analogue of the J-function which compares 'nearest-neighbour' distances from **x** to **y** to the 'empty-space' distances from any spatial location to **y**. Formally we need to assume that **X** and **Y** are 'jointly stationary', in the sense that the pair $(\mathbf{X}, \mathbf{Y})$ has the same joint probability distribution as the pair $(\mathbf{X} + v, \mathbf{Y} + v)$ for any shift vector v. This implies that **X** is a stationary point process and **Y** is a stationary random set. *Foxall's G-function* [269] is

$$G^*_{\mathbf{X},\mathbf{Y}}(r) = \mathbb{P}\{d(u, \mathbf{Y}) \le r \mid \mathbf{X} \text{ has a point at } u\}, \quad r \ge 0 \tag{8.29}$$

for any location u. Assuming **X** and **Y** are jointly stationarity, the definition does not depend on u. *Foxall's J-function* [269] is

$$J_{\mathbf{X},\mathbf{Y}}(r) = \frac{1 - G^*_{\mathbf{X},\mathbf{Y}}(r)}{1 - H^*_{\mathbf{Y}}(r)} \tag{8.30}$$

for any $r \ge 0$ such that $H^*_{\mathbf{Y}}(r) < 1$, where $H^*_{\mathbf{Y}}(r)$ is the unconditional spherical contact distribution function for **Y**. The spatstat functions Gfox and Jfox compute estimates of $G_{\mathbf{X},\mathbf{Y}}(r)$ and $J_{\mathbf{X},\mathbf{Y}}(r)$, respectively.

The interpretation of $J_{\mathbf{X},\mathbf{Y}}$ is different from that of the J-function of a point process: it carries no information about the spatial interaction between points of **X**, only about the dependence of **X** on **Y**. If **X** and **Y** are *independent* random patterns, then $J_{\mathbf{X},\mathbf{Y}}(r) \equiv 1$, regardless of whether **X** is a Poisson process or another stationary point process. Values $J_{\mathbf{X},\mathbf{Y}}(r) > 1$ mean that **X** 'avoids' **Y**: the points of **X** are more likely than random points to lie far away from **Y**. Conversely $J_{\mathbf{X},\mathbf{Y}}(r) < 1$ means that the points of **X** tend to be closer to **Y** than expected for random points. For further discussion see [269].

Since this analysis assumes stationarity, we shall restrict calculations to the southern half of the Queensland copper survey, shown in Figure 6.22 on page 182, which appears to be homogeneous. The Foxall functions are calculated by

```
> GC <- with(copper, Gfox(SouthPoints, SouthLines))
> JC <- with(copper, Jfox(SouthPoints, SouthLines, eps=0.1))
```

The results are shown in Figure 8.28. The Foxall *J*-function is close to 1 for distances less than 2 *km*, and climbs slightly higher for larger distances. Simulation envelopes (not shown) suggest there is no evidence of dependence between deposits and lineaments.

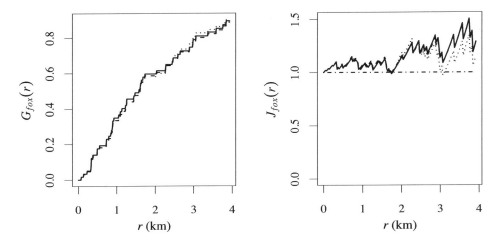

Figure 8.28. *Foxall G-function* (Left) *and Foxall J-function* (Right) *from copper deposits to lineaments, in the southern half of the Queensland copper data of Figure 1.11.*

This technique has found applications in geological prospectivity analysis [134, 269]. Another useful formulation of Foxall's *J*-function for the copper deposits is that $\widehat{J}_{\mathbf{X},\mathbf{Y}}(r)$ is the estimated intensity of copper deposits in the region lying more than r units away from the nearest lineament, divided by the average intensity of copper deposits over the entire survey. A decreasing Foxall *J*-function would indicate concentration of the copper deposits around the lineaments.

Foxall's *J*-function is closely related to the analysis using relative distributions (Section 6.6.3) and to Berman's tests (Sections 6.7.2 and 10.5.2) as explained in [269]. An advantage of $J_{\mathbf{X},\mathbf{Y}}$ over these other techniques is that it is correctly adjusted for edge effects. There is also a connection with the ROC curve (Section 6.7.3), which in this context is a plot of $1 - G^*_{\mathbf{X},\mathbf{Y}}(r)$ against $1 - H^*_{\mathbf{Y}}(r)$.

8.11 Theory for edge corrections*

The estimation of summary functions is hampered by edge effects, arising when we cannot see beyond the boundary of our study region. If no correction is made for edge effects, then estimates of the functions F, G, and K are negatively biased, with the magnitude of bias increasing with r.

Edge corrections are routinely used to avoid this problem, and are strongly recommended, unless the data are huge (when the interesting values of r will be small, and the edge effect bias also small).

This section gives essential details about edge corrections for the distance distributions F and G. Readers are warned that the implementations of edge corrections in some software may be incorrect. Trustworthy R packages include `spatial`, `spatstat`, and `splancs`.

For discussion of edge corrections, see [59, 355, 640], [576, chap. 3].

8.11.1 Estimating F and G if edge effects are absent

To establish the basic theory, we first study a different scenario in which edge effects do not arise. As sketched in Figure 7.12 in Section 7.4.1, visualise a sampling frame B placed in a homogeneous field of wildflowers. The spatial pattern of wildflowers over the entire field will be denoted by $\mathbf{x}$, while the wildflowers falling inside the sampling frame B are the point pattern $\mathbf{x} \cap B = \mathbf{x} \cap B$.

8.11.1.1 G-function

To estimate the nearest-neighbour distance distribution function $G(r)$, we simply visit each wild-flower inside the sampling frame, measure its true nearest-neighbour distance, and form the cumulative distribution function of these distances. That is, we consider each point x_i in $\mathbf{x} \cap B$, measure the true nearest-neighbour distance $d_i = d(x_i, \mathbf{x} \setminus x_i)$, and form the empirical cumulative distribution function

$$\widehat{G}(r) = \frac{1}{n(\mathbf{x} \cap B)} \sum_i \mathbf{1}\{d_i \leq r\}. \tag{8.31}$$

For a stationary point process, $\widehat{G}(r)$ is not exactly unbiased as an estimate of $G(r)$; it is approximately unbiased and consistent. In Section 8.12.2 it is shown that

$$\mathbb{E}[\sum_i \mathbf{1}\{d_i \leq r\}] = \lambda |B| G(r) \tag{8.32}$$

so that $n(\mathbf{x} \cap B)\widehat{G}(r) = \sum_i \mathbf{1}\{d_i \leq r\}$ is an (exactly) unbiased estimate of $\lambda |B| G(r)$, and the denominator $n(\mathbf{x} \cap B)$ is of course an unbiased estimate of $\lambda |B|$. That is, $\widehat{G}(r)$ is the ratio of two unbiased estimates, $\widehat{G}(r) = A/B$ where $\mathbb{E}(A)/\mathbb{E}(B) = G(r)$. This is the common structure of many estimators in spatial statistics; they are often called '*ratio-unbiased*' [58]. They are not unbiased, because $\mathbb{E}(A/B) \neq \mathbb{E}(A)/\mathbb{E}(B)$. However, under reasonable conditions, they are approximately unbiased and consistent [59, 640].

8.11.1.2 F-function

Taking any spatial location u inside the sampling frame, we measure the true distance $d(u, \mathbf{x})$ to the nearest wildflower, *whether it lies inside or outside the sampling frame*. We repeat this process for several spatial locations $u_1, \ldots, u_m$. The sampling locations u_j could be laid out in a grid inside B,

---

* Starred sections contain advanced material, and can be skipped by most readers.

or could be randomly generated inside B, in any manner that is independent of the underlying point process $\mathbf{X}$. We compile the empirical cumulative distribution function

$$\widehat{F}(r) = \frac{1}{m} \sum_{j=1}^{m} \mathbf{1}\left\{d(u_j, \mathbf{x}) \le r\right\} \tag{8.33}$$

as a function of distance $r \ge 0$. The expectation of a sum is the sum of the expectations, so $\widehat{F}(r)$ is an unbiased estimate of the true empty-space function $F(r)$, for each distance r, regardless of the layout of the sampling locations u_j.

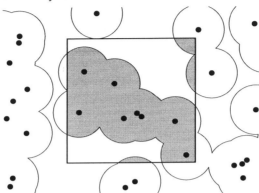

Figure 8.29. *Estimation of the empty-space function as the estimation of the area fraction of* $\mathbf{X}_{\oplus r}$.

In geometrical terms, we saw that $F(r)$ is the average fraction of area covered by the dilation set $\mathbf{x}_{\oplus r}$, defined in equation (8.10) and sketched in Figure 8.9 on page 266. This fraction is estimated in (8.33) by visiting several (usually *many*) test locations u_j and counting the fraction of them which are covered by $\mathbf{x}_{\oplus r}$. Taking the limit of (8.33) as the grid becomes infinitely fine, we get

$$\widehat{F}(r) = \frac{1}{|B|} \int_B \mathbf{1}\left\{d(u, \mathbf{x}) \le r\right\} \mathrm{d}u = \frac{|B \cap \mathbf{x}_{\oplus r}|}{|B|}, \tag{8.34}$$

the *fraction of area* of B occupied by the set $\mathbf{x}_{\oplus r}$. See Figure 8.29.

Some software packages such as `splancs` compute an estimate of $F(r)$ using relatively coarse grids of sampling locations u_j following [233]. Paradoxically, this may be more efficient than taking a finer grid of sampling locations. A finer grid does not necessarily increase the accuracy of the estimator (8.33), because there is strong positive correlation between the empty-space distances [61, 599], [575, p. 151]. However, the optimal choice of grid depends on the particular point process [599] and on the distance r, so it is difficult to choose a coarse grid in an optimal way. In `spatstat` we evaluate the empty-space distances on a fine grid of pixels: this makes it easier to estimate related quantities such as the probability density or hazard rate of F. The algorithms in `spatstat` also *require* a fine grid spacing.

8.11.2   Edge effects for F and G

In most applications, the spatial locations of points are recorded only inside a study region or 'window' W. This is 'window sampling' as defined in Section 5.6.2. The observable data are the points inside the window, $\mathbf{x} \cap W = \mathbf{x} \cap W$, rather than the entire point pattern $\mathbf{x}$ extending throughout space.

Confining observations to a window W creates the problem (sketched in Figure 8.30) that observed nearest-neighbour distances are larger, in general, than the true nearest-neighbour distances. For a location u close to the boundary of the window W, there may be points lying within distance r of u, but falling outside W, so that they are not observed. Hence, the observed empty-space distance

$d(u, \mathbf{x} \cap W)$, may be *larger* than the true empty-space distance $d(u, \mathbf{x})$. Indeed it is always true that $d(u, \mathbf{x} \cap W) \geq d(u, \mathbf{x})$.

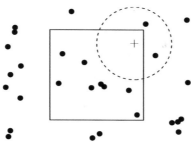

Figure 8.30. *Edge effect problem for estimation of the empty-space function F. Dots represent a stationary point process, which we observe only inside a window (black square). For a spatial location inside the window (+), the nearest neighbour may lie outside the window.*

If no correction is made for edge effects, then estimates of the functions F and G are negatively biased. The *uncorrected* or *raw* estimates of F and G based on $\mathbf{x}$ are

$$\widehat{F}_{un}(r) = \frac{1}{m} \sum_{j=1}^{m} \mathbf{1}\{d(u_j, \mathbf{x} \cap W) \leq r\} \tag{8.35}$$

$$\widehat{G}_{un}(r) = \frac{1}{n(\mathbf{x})} \sum_{i} \mathbf{1}\{d(x_i, \mathbf{x} \cap W \setminus x_i) \leq r\}. \tag{8.36}$$

The observed shortest distance is greater than or equal to the true shortest distance, so

$$\mathbf{1}\{d(u, \mathbf{x} \cap W) \leq r\} \leq \mathbf{1}\{d(u, \mathbf{x}) \leq r\}$$

so that $\widehat{F}_{un}(r)$ is negatively biased as an estimator of $F(r)$. Similarly $\widehat{G}_{un}(r)$ is negatively biased for $G(r)$. The result *edge effect bias* can be substantial and has the potential to be misleading.

8.11.3   Border correction

The border method of edge correction (Section 7.4.3) can be applied to the empty-space function and nearest-neighbour function.

8.11.3.1   Border correction for F

In the border method, when estimating the value of $F(r)$ for a particular distance r, we simply restrict attention to those test locations u for which the disc $b(u, r)$ lies entirely inside W. For such locations, the value of $d(u, \mathbf{x})$ is observed correctly, that is, $d(u, \mathbf{x} \cap W) = d(u, \mathbf{x})$. These are the locations u falling in the eroded window $W_{\ominus r}$ defined in (7.14) and sketched in Figure 7.15. In set theoretic terms

$$\mathbf{x}_{\oplus r} \cap W_{\ominus r} = (\mathbf{x} \cap W)_{\oplus r} \cap W_{\ominus r} \tag{8.37}$$

and the right-hand side is computable from the data $\mathbf{x} \cap W$.

We can obtain unbiased estimators of $F(r)$ by taking $B = W_{\ominus r}$ in (8.33) or (8.34). In the first case, suppose $u_1, \ldots, u_m$ is a grid (or other arrangement) of test points in W. Let $b_j = d(u_j, W^c)$ be the distance from u_j to the boundary of the window. Then u_j belongs to $W_{\ominus r}$ if $b_j > r$. Taking $B = W_{\ominus r}$ in (8.33), the border corrected estimate of $F(r)$ based on test points u_j is

$$\widehat{F}_{bord}(r) = \frac{\sum_j \mathbf{1}\{d(u_j, \mathbf{x} \cap W) \leq r\} \mathbf{1}\{b_j > r\}}{\sum_j \mathbf{1}\{b_j > r\}}, \tag{8.38}$$

the fraction of test points, amongst those falling in $W_{\ominus r}$, which have an empty-space distance less than or equal to r.

Similarly, taking $B = W_{\ominus r}$ in (8.34), the border corrected estimate of $F(r)$ is

$$\widehat{F}_{bord}(r) = \frac{|\mathbf{x}_{\oplus r} \cap W_{\ominus r}|}{|W_{\ominus r}|}, \tag{8.39}$$

the fraction of *area* of the eroded set $W_{\ominus r}$ covered by the dilated set $\mathbf{x}_{\oplus r}$. See the left panel of Figure 8.31.

The estimator (8.38)–(8.39) is variously known as the 'minus sampling' [467], 'border correction' [576, chap. 3], or 'reduced sample' [62] estimator, and equation (8.37) is sometimes said to be an instance of the 'local knowledge principle' [611, pp. 1149,62], [59]. The empty-space distance can be computed for all locations u on a fine grid, using the `spatstat` function `distmap`. The border correction can then be calculated rapidly using (8.38). However, it is statistically inefficient, because it discards a substantial amount of data. Although $\widehat{F}_{bord}(r)$ is an unbiased estimator of $F(r)$ for each r, the function $\widehat{F}_{bord}$ may not be a distribution function: it may not be an increasing function of r.

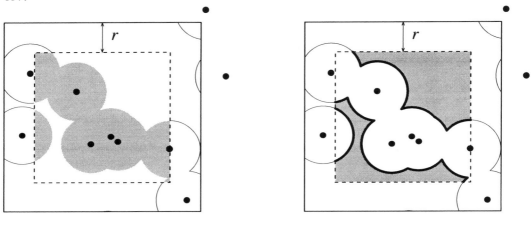

Figure 8.31. *Geometry for edge-corrected estimators of $F(r)$.* Left: *border correction.* Right: *Kaplan-Meier correction.*

8.11.3.2 Border correction for G

Similarly when estimating the value of $G(r)$ for a particular distance r, for the border correction we restrict attention to those data points x_i for which the disc $b(x_i, r)$ lies entirely inside W. For such points, the nearest-neighbour distance is observed correctly, that is, $d_i = d(x_i, \mathbf{x} \cap W \setminus x_i) = d(x_i, \mathbf{x} \setminus x_i)$.

The selected data points x_i are those falling in the eroded window $W_{\ominus r}$ defined in (7.14) and sketched in Figure 7.15, or equivalently, the points for which the boundary distance $b_i = d(x_i, W^c)$ is greater than or equal to r. The border correction estimate of $G(r)$ is [572]

$$\widehat{G}_{bord}(r) = \frac{\sum_i \mathbf{1}\{b_i \geq r \text{ and } d_i \leq r\}}{\sum_i \mathbf{1}\{b_i \geq r\}}, \tag{8.40}$$

the fraction of data points, amongst those falling in $W_{\ominus r}$, for which the nearest neighbour is closer than r.

Taking $B = W_{\ominus r}$ in (8.31) we find that $\widehat{G}_{bord}(r)$ is an approximately unbiased estimator of $G(r)$.

8.11.4 Kaplan-Meier style correction

A completely different approach to edge correction for distance distributions was developed in [63, 323, 64, 322].

The basic problem of edge effects, it could be said, is that some relevant information is not observed. There are many powerful statistical methodologies for dealing with missing information, including survey sampling, incomplete data likelihoods, data augmentation, and survival analysis. Most of the existing methods for edge correction in spatial statistics are related to survey sampling theory [59] and are designed to correct the edge-effect bias, rather than improve overall accuracy. Incomplete data likelihoods and data augmentation methods have also been developed [282] but require parametric model assumptions.

Survival analysis provides an alternative nonparametric approach. It has been known for many years that distances in spatial processes are analogous to lifetimes or waiting times, and have many similar properties [466]. For example, the distance from a fixed location u to the nearest point of a point pattern $\mathbf{x}$ can be viewed as the lifetime of a balloon centred at u, which grows at a uniform rate until it meets a point of $\mathbf{x}$, when it pops. The possibility of using methods of survival analysis to study the distribution of these 'lifetimes' was first raised in [406, 405].

If our observations are confined to the window W, then the 'true' lifetime $d(u,\mathbf{x})$ of the balloon might not be observed, because the nearest point of $\mathbf{x}$ might lie outside the window. If the balloon has not popped by the time it reaches the *boundary* of the window, then all we know is that the true lifetime is longer than this. The true lifetime or empty-space distance $d(u,\mathbf{x})$ is then said to be *censored* by the boundary distance $b(u) = d(u, W^c)$. The observed empty-space distance is $d(u) = d(u, \mathbf{x} \cap W)$. If $d(u) \le b(u)$, then we know that the true lifetime has been observed, and is equal to $d(u)$. If $d(u) > b(u)$, then we know only that the true lifetime is greater than $b(u)$.

For censored lifetime data, the theoretically optimal estimator of the true lifetime distribution is the *Kaplan-Meier estimator* [369]. A counterpart of the Kaplan-Meier estimator for the distribution of shortest distances in spatial processes was developed in [63, 323, 64, 322].

8.11.4.1 Kaplan-Meier style estimator for G

It is easiest to begin with the estimation of G. Suppose that $\mathbf{X}$ is a stationary point process, which we observe through a window W. For each point x_i in $\mathbf{X}$, the distance to the nearest neighbour, $T_i = d(x_i, \mathbf{X} \setminus x_i)$, can be regarded as a 'survival time'. However, we do not observe T_i directly; since we see only those points inside W, we are only able to measure $t_i = d_i = d(x_i, \mathbf{X} \cap W \setminus x_i)$, the nearest neighbour distance in the pattern $\mathbf{X} \cap W$. The analogue of the 'censoring time' is the distance from x_i to the window boundary, $c_i = d(x_i, \partial W)$. If $t_i < c_i$ then we know that the true lifetime T_i has been observed; otherwise, we know only that the true lifetime T_i is greater than c_i.

Define the 'observed failure distance'

$$\widetilde{t}_i = \min(t_i, c_i) = \min(T_i, c_i)$$

and the censoring indicator

$$d_i = \mathbf{1}\{t_i \le c_i\} = \mathbf{1}\{T_i \le c_i\}.$$

The *Kaplan-Meier estimator* of $G(r)$ is

$$G_{km}(r) = 1 - \prod_{s \le r}\left(1 - \frac{\#\{i : \widetilde{t}_i = s,\, d_i = 1\}}{\#\{i : \widetilde{t}_i \ge s\}}\right). \tag{8.41}$$

The symbol $\prod$ on the right-hand side denotes the product rather than the sum $\sum$. The product is taken over all distance values s less than or equal to r. Although there are infinitely many such values, the term in parentheses will be equal to 1 (meaning that it does not contribute to the result) unless $\widetilde{t}_i = s$ for some i. The product can be computed by sorting the values $\widetilde{t}_1, \ldots, \widetilde{t}_n$ and removing duplicated values to obtain the sequence $s_1 < s_2 < \cdots < s_k$ of unique distances, then taking the product over these values s_j.

To understand the rationale for (8.41), notice that the denominator $\#\{i : \tilde{t}_i \geq s\}$ is the number of points x_i which are 'still alive at time s', while the numerator $\#\{i : \tilde{t}_i = s,\ d_i = 1\}$ is the number of points which are '*observed* to die at time s'. The ratio of these numbers is an estimate of the conditional probability of dying exactly at time s given survival at least to time s. The term in parentheses is an estimate of the probability of surviving longer than time s, given survival at least to time s. Multiplying these conditional probabilities together for all $s \leq r$ gives an estimate of the probability of surviving longer than time r, which is $1 - G(r)$.

For any point pattern $\mathbf{x}$, the Kaplan-Meier estimate of $G(r)$ in (8.41) is a (possibly defective) distribution function, that is, it is a non-decreasing, right continuous function bounded by 0 and 1.

8.11.4.2   Kaplan-Meier correction for F

It was proved in [322, Theorem 1] that for a stationary point process $\mathbf{X}$ (in fact for any stationary random set) the empty-space function F is absolutely continuous on $r > 0$, with density

$$f(r) = \frac{\mathbb{E}\,\ell(W \cap \partial(\mathbf{X}_{\oplus r}))}{|W|} \tag{8.42}$$

for almost all $r > 0$ and any sufficiently regular domain $W \subset \mathbb{R}^2$. Here $\ell(S)$ denotes the total length of a set S consisting of one-dimensional curves. Thus $f(r)$ is the length density (expected length per unit area) of the boundary of $\mathbf{X}_{\oplus r}$. The hazard rate of F equals

$$h(r) = \frac{f(r)}{1 - F(r)} = \frac{\mathbb{E}\,\ell(W \cap \partial(\mathbf{X}_{\oplus r}))}{\mathbb{E}\,|W \setminus \mathbf{X}_{\oplus r}|}$$

for *almost every* $r > 0$. The Kaplan-Meier estimator F_{km} of the empty-space function F of $\mathbf{X}$, based on data $\mathbf{X} \cap W$, is

$$F_{km}(r) = 1 - \frac{|W \setminus \mathbf{X}|}{|W|} \exp\left(-\int_0^r \frac{\ell(\partial(\mathbf{X}_{\oplus s}) \cap W_{\ominus s})}{|W_{\ominus s} \setminus \mathbf{X}_{\oplus s}|}\,ds\right). \tag{8.43}$$

Note that (8.43) is computable from the data $\mathbf{X} \cap W$ since $W \setminus \mathbf{X} = W \setminus (\mathbf{X} \cap W)$ and, by (8.37), we can replace $\mathbf{X}$ by $\mathbf{X} \cap W$ in the numerator and denominator of the integrand. See the right panel of Figure 8.31 for the geometry of the estimator.

To see that (8.43) deserves the epithet Kaplan-Meier, define for each $u \in W$

$$t(u) = d(u, \mathbf{X} \cap W) \quad \text{and} \quad c(u) = d(u, \partial W),$$

respectively the distance to 'failure' and the censoring distance. Define the 'observed failure distance'

$$\tilde{t}(u) = t(u) \wedge c(u) = d(u, \mathbf{X} \cap W) \wedge d(u, \partial W) = d(u, \mathbf{X}) \wedge d(u, \partial W)$$

where $a \wedge b = \min(a, b)$, and the censoring indicator

$$d(u) = \mathbf{1}\{t(u) \leq c(u)\} = \mathbf{1}\{d(u, \mathbf{X} \cap W) \leq d(u, \partial W)\} = \mathbf{1}\{d(u, \mathbf{X}) \leq d(u, \partial W)\}$$

where in each case the last line follows by (8.37). Then the integrand of (8.43) has denominator

$$|W_{\ominus s} \setminus X_{\oplus s}| = |\{u \in W : \tilde{t}(u) \geq s\}|,$$

the measure of the set of points 'at risk' at distance s, and numerator

$$\ell(\partial(X_{\oplus s}) \cap W_{\ominus s}) = \ell(\{u \in W : \tilde{t}(u) = s,\ d(u) = 1\}),$$

the measure of the set of points observed to 'fail' at distance s. Thus (8.43) is at least intuitively the analogue of the usual Kaplan-Meier estimator for the continuum of data $\{(\tilde{t}(u), c(u)) : u \in W\}$.

For an arbitrary closed set $\mathbf{X}$ and regular compact set W, the statistic F_{km} is a distribution function (possibly defective), i.e. it is nondecreasing, right-continuous, and bounded by 0 and 1. Indeed it is continuous, and absolutely continuous for $r > 0$, with hazard rate

$$h_{km}(r) = \frac{\ell(\partial(\mathbf{X}_{\oplus r}) \cap W_{\ominus r})}{|W_{\ominus r} \setminus \mathbf{X}_{\oplus r}|}. \tag{8.44}$$

The estimator $h_{km}(r)$ of $h(r)$ is 'ratio-unbiased' in the sense that $h_{km}(r) = U/V$ where U, V are such that $h(r) = \mathbb{E}U/\mathbb{E}V$.

In practice one would not actually compute the surface areas and volumes required in (8.43). Rather, the sampling window W would be discretised on a regular lattice (see, e.g., [233]). A natural possibility is to calculate for each lattice point u_j the censored distance $d(u_j, \mathbf{X}) \wedge d(u_j, \partial W)$ and the indicator $\mathbf{1}\{d(u_j, \mathbf{X}) \leq d(u_j, \partial W)\}$. Then one would calculate the ordinary Kaplan-Meier estimator based on this finite dataset.

Let $t_i = d(u_j, \mathbf{X} \cap W)$, $c_i = d(u_j, \partial W)$ and $\tilde{t}_i = t_i \wedge c_i$, $d_i = \mathbf{1}\{t_i \leq c_i\}$ be the observations at the points of $W \cap L$, where $L = \varepsilon M + b$ is a rescaled, translated copy of a fixed regular lattice M. Calculate the discrete Kaplan-Meier estimator $F_{km}^L(r)$ using the analogue of (8.41). Then [64] as the lattice mesh ε converges to zero, $F_{km}^L(r) \to F_{km}(r)$ for any $r < R$, where

$$R = \inf\{r \geq 0 : W_{\ominus r} \cap \Phi_{\oplus r} = \emptyset\}.$$

8.11.5 Hanisch-Chiu-Stoyan correction

8.11.5.1 Hanisch correction for G

Hanisch [320, 319] proposed the following estimators for the nearest-neighbour distance distribution function G. For each data point x_i in the pattern $\mathbf{x}$, write $d_i = d(x_i, \mathbf{x} \setminus x_i)$ for the nearest-neighbour distance and $b_i = d(x_i, \partial W)$ for the distance to the window boundary. Then Hanisch's pointwise unbiased estimator for $A(r) = \lambda G(r)$ is

$$A_H(r) = \sum_i \frac{\mathbf{1}\{d_i \leq b_i\}\mathbf{1}\{d_i \leq r\}}{|W_{\ominus d_i}|} \tag{8.45}$$

and a consistent estimator of $G(r)$ is

$$G_H(r) = \frac{\sum_i \mathbf{1}\{d_i \leq b_i\}\mathbf{1}\{d_i \leq r\}}{\sum_i \mathbf{1}\{d_i \leq b_i\}}. \tag{8.46}$$

See also [191, p. 638], [150, 631].

8.11.5.2 Chiu-Stoyan correction for F

Using unpublished results by Baddeley and Gill [63], Chiu and Stoyan [150] developed a counterpart of the Hanisch estimator for the empty-space function F:

$$F_{CS}(r) = \int_0^r \frac{\ell(W_{\ominus s} \cap \partial(\mathbf{x}_{\oplus s}))}{|W_{\ominus s}|} \, \mathrm{d}s. \tag{8.47}$$

By Theorem 1 of [63, 64], the integrand is a pointwise unbiased estimator of the empty-space function's probability density $f(r) = F'(r)$.

8.11.6   Comparison of methods

For the estimators described above, a comparison of performance is complicated, because they are not all designed for the same objective. The classical edge corrected estimates are designed to provide approximately unbiased estimates of $F(r)$ and $G(r)$, while the Kaplan-Meier style estimator is designed to provide efficient unbiased estimates of the hazard rate and the integrated hazard $-\log(1-F(r))$ or $-\log(1-G(r))$. Some writers have reported experiments suggesting that the Kaplan-Meier style estimators of F and G have higher mean squared error than other estimators, but at least one of these experiments was based on an incorrect implementation of the Kaplan-Meier estimator and has been withdrawn. Perhaps the most reliable study so far is [631] which shows only very small differences in performance between the estimators other than the border correction.

Many alternative methods for estimation of F and G have been proposed. Kernel estimators for the density of G were proposed in [241, 257], [464, pp. 137–138]. Further discussion of edge corrections can be found in [636, 59]. Proposals for edge correction using data augmentation [282, 656] and other promising techniques [265] have not been widely used.

8.12   Palm distribution∗

The nearest-neighbour distance distribution function $G(r)$ was defined in (8.6) as the *conditional* probability of finding at least one point within a distance r, given that the point process $\mathbf{X}$ has a point at the specified location u. The K-function was defined in (7.4) as the *conditional* expectation of the number of r-neighbours, given that $\mathbf{X}$ has a point at u.

These conditional probabilities and conditional expectations cannot be defined and handled using elementary probability theory, because the event that a point process $\mathbf{X}$ has a point at a specified location u is typically an event of probability zero. Ignoring this technical problem can cause trouble [559].

In this section we explain how to make rigorous sense of these conditional probabilities and expectations using the *Palm distribution* of the point process. Engineer and statistician Conny Palm [533] analysed probability models of telephone traffic by studying the conditional probability of any event A given that a new telephone call started at time t. These conditional probabilities are denoted $\mathbb{P}^t(A)$. Together they form the Palm distribution of the telephone traffic at time t, denoted $\mathbb{P}^t$. Similarly for a spatial point process $\mathbf{X}$, the Palm distribution $\mathbb{P}^u$ is the conditional distribution of $\mathbf{X}$ given that $\mathbf{X}$ has a point at the location u. See [198, pp. 13,14] for a historical review.

The Palm distribution gives us powerful new tools for analysing point patterns, especially the *Campbell-Mecke formula*. These tools allow us to define new characteristics of a point process, construct new estimators, and check the validity of estimators.

8.12.1   Basic formulae

Here we present a very simplified explanation of the Palm distribution theory, omitting many technical conditions. See [27, 368] for more details.

Suppose $\mathbf{X}$ is a point process in two dimensions, with intensity function $\lambda(u)$. Suppose there is a function $Q(u,A)$ such that, for any random event A and any spatial region B,

$$\mathbb{E}[n(\mathbf{X}\cap B)1_A] = \int_B Q(u,A)\lambda(u)\,\mathrm{d}u$$

∗ Starred sections contain advanced material, and can be skipped by most readers.

where 1_A is the indicator random variable which equals 1 if the event A occurs, and equals 0 otherwise. Then we say that $Q(u,A) = \mathbb{P}^u(A)$ is the *Palm probability* of the event A given a point at u, and write

$$\mathbb{E}[n(\mathbf{X} \cap B)1_A] = \int_B \mathbb{P}^u(A)\lambda(u)\,du. \tag{8.48}$$

In principle the formula (8.48) can be used to calculate Palm probabilities from basic principles. We do this for the Poisson process below.

Under reasonable conditions, for any fixed u, the values $\mathbb{P}^u(A)$ for different events A constitute a probability distribution $\mathbb{P}^u$, the *Palm distribution* of the process conditioned on there being a point at u. We can then define the expected value $\mathbb{E}^u[Y]$ of any random variable Y with respect to the Palm distribution, which satisfies

$$\mathbb{E}[n(\mathbf{X} \cap B)Y] = \int_B \mathbb{E}^u[Y]\lambda(u)\,du. \tag{8.49}$$

Then by extension, for any function $f(u,\mathbf{X})$ we have the *Campbell-Mecke formula* [463]

$$\mathbb{E}[\sum_{x \in \mathbf{X} \cap B} f(x,\mathbf{X})] = \int_B \mathbb{E}^u[f(u,\mathbf{X})]\lambda(u)\,du. \tag{8.50}$$

This formula is the theoretical basis for estimators of summary functions like $G(r)$ and $K(r)$.

8.12.2 Summary functions for stationary processes

Suppose $\mathbf{X}$ is a *stationary* point process, with intensity λ. Apply the Campbell-Mecke formula (8.50) to the function $f(u,\mathbf{X}) = g(\mathbf{X} - u)$, where g is any function, and $\mathbf{X} - u$ is the point process $\mathbf{X}$ shifted by the vector $-u$, that is, shifted so that the location u is moved to the origin. Stationarity implies that $\mathbb{E}^u[g(\mathbf{X} - u)]$ is constant for all locations u, and

$$\mathbb{E}[\sum_{x \in \mathbf{X} \cap B} g(\mathbf{X} - x)] = \lambda|B|\mathbb{E}^0[g(\mathbf{X})]. \tag{8.51}$$

This is the basis for the formal definition of summary functions such as $K(r)$ and $G(r)$, and also for constructing estimators.

For the K-function, define

$$g(\mathbf{X}) = n(\mathbf{X} \cap b(0,r) \setminus 0);$$

then

$$g(\mathbf{X} - u) = n((\mathbf{X} - u) \cap b(0,r) \setminus 0) = n(\mathbf{X} \cap b(u,r) \setminus u).$$

If we define $K(r) = (1/\lambda)\mathbb{E}^u[g(\mathbf{X} - u)]$ then this is equivalent to the heuristic definition (7.4); we know immediately that the definition does not depend on u; and from (8.51) we immediately get the alternative formula (7.11).

For the G-function, define

$$g(\mathbf{X}) = \mathbf{1}\{d(0,\mathbf{X} \setminus 0) \leq r\};$$

then

$$g(\mathbf{X} - u) = \mathbf{1}\{d(u,\mathbf{X} \setminus u) \leq r\}.$$

If we define $G(r) = \mathbb{E}^u[g(\mathbf{X} - u)]$ then this is equivalent to the heuristic definition (8.6), the definition does not depend on the choice of u, and from (8.51) we immediately get the alternative formula (8.32).

8.12.3   Palm distribution for Poisson process

Suppose $\mathbf{X}$ is a Poisson point process with constant intensity λ. Consider the event A that there are no points of $\mathbf{X}$ in a given set S,

$$A = \{\mathbf{X} \cap S = \emptyset\}.$$

We want to determine $\mathbb{P}^u(A)$ for every location u.

In equation (8.48), if the set B is disjoint from S, then $n(\mathbf{X} \cap B)$ and 1_A are independent events because of the independence properties of the Poisson process, so

$$\mathbb{E}[n(\mathbf{X} \cap B)1_A] = \mathbb{E}[n(\mathbf{X} \cap B)]\,\mathbb{E}[1_A] = \mathbb{E}[n(\mathbf{X} \cap B)]\,\mathbb{P}(A) = \lambda\,|B|\exp(-\lambda\,|S|).$$

If B overlaps S, we split it into $B \cap S$ and $B \setminus S$. Then

$$\mathbb{E}[n(\mathbf{X} \cap B)1_A] = \mathbb{E}[n(\mathbf{X} \cap B \cap S)1_A] + \mathbb{E}[n(\mathbf{X} \cap B \setminus S)1_A].$$

If $n(\mathbf{X} \cap B \cap S) > 0$ then there is a random point in S, so the event A does not occur, and $1_A = 0$. Therefore $n(\mathbf{X} \cap B \cap S)1_A = 0$ always, so its expectation is zero, and we get

$$\mathbb{E}[n(\mathbf{X} \cap B)1_A] = \lambda\,|B \setminus S|\exp(-\lambda\,|S|).$$

Thus we have calculated that, for any set B,

$$\lambda \int_B \mathbb{P}^u(A)\,\mathrm{d}u = \lambda\,|B \setminus S|\exp(-\lambda\,|S|).$$

It follows that

$$\mathbb{P}^u(A) = \mathbb{P}^u\{\mathbf{X} \cap S = \emptyset\} = \begin{cases} \exp(-\lambda\,|S|) & \text{if } u \text{ is not in } S, \\ 0 & \text{if } u \text{ is in } S, \end{cases} \qquad (8.52)$$

This is what we would have expected from the intuition that the points of a Poisson process are independent. If u is not in S, then the knowledge that there is a random point at u has no bearing on the number of random points in S, so the probability of the event A is unchanged. If u is in S, then knowing that there is a random point at u tells us of course there is at least one random point in S, and the event A does not occur.

The right-hand side of (8.52) is the same as the probability of the event A for the point process $\mathbf{Y} = \mathbf{X} \cup \{u\}$, the superposition of the Poisson process $\mathbf{X}$ and a fixed point at location u. That is, *under the Palm distribution at location u, a Poisson process $\mathbf{X}$ behaves like the superposition of $\mathbf{X}$ with a fixed point at u.*

It follows immediately that, for a homogeneous Poisson process, $K(r) = \pi r^2$ and $G(r) = F(r) = 1 - \exp(-\lambda \pi r^2)$.

8.12.4   Reduced Palm distribution

Many calculations for point processes can be simplified by using the *reduced* Palm distribution. The reduced Palm distribution $\mathbb{P}^{!u}$ of a point process $\mathbf{X}$ at a location u is the distribution of $\mathbf{X} \setminus u$ under $\mathbb{P}^u$. For any integrable function $f(u, \mathbf{x})$,

$$\mathbb{E}^{!u} f(u, \mathbf{X}) = \mathbb{E}^u f(u, \mathbf{X} \setminus u).$$

The Campbell-Mecke formula (8.50) has a 'reduced' version

$$\mathbb{E}\Big[\sum_{x \in \mathbf{X} \cap B} f(x, \mathbf{X} \setminus u)\Big] = \int_B \mathbb{E}^{!u}[f(u, \mathbf{X})]\lambda(u)\,\mathrm{d}u \qquad (8.53)$$

which is more convenient for studying summary functions. For example, the K-function and G-function of a stationary point process are

$$K(r) = \mathbb{E}^{!u} n(\mathbf{X} \cap ()b(u,r))$$
$$G(r) = \mathbb{P}^{!u}\{d(u,\mathbf{X}) \le r\}.$$

The statement at the end of the last section implies that *for a Poisson point process, the reduced Palm distribution is identical to the usual distribution of the process.*

8.13 FAQs

- *A plot of the G-function for my data suggests the pattern is regular, but the plot of the K-function suggests clustering. Which is right?*

The results from G and K are not contradictory, because a summary function does not completely characterise the point process. The G- and K-functions are sensitive to different aspects of the point process. The G-function summarises information at a shorter scale than the K-function. If you are confident that the intensity is constant, the most likely explanation is that there is both short-scale regularity and larger-scale clustering.

- *Which of the functions $K(r)$, $g(r)$, $F(r)$, $G(r)$, and $J(r)$ do you recommend?*

This depends on the type of dependence between points. Second order methods ($K(r)$ and $g(r)$) have the best performance overall, but $G(r)$ is more sensitive to the presence of a hard core, and $J(r)$ is the most robust against edge effects. In a detailed analysis these functions are often used in a complementary way: see [263, 378]. For a comparative analysis of the same data using many different summary functions, demonstrating their relative merits, look through the examples in this book, or in [355].

- *I can't seem to control the range of r values in `plot(Gest(X))`. How can I control it? How is the default plotting range determined?*

To control the range of r values in the plot, use the argument `xlim` documented in the help for `plot.fv`. For example, `plot(Gest(X), xlim=c(0, 7))`.

The default range of r values that is plotted depends on the 'default plotting range' of the object (of class `"fv"`) returned by `Gest`. See below.

- *How are the r values determined in `Gest(X)`?*

The maximum distance to be considered, `rmax`, is computed by the function `rmax.rule`. It depends on the window geometry and on the estimated intensity λ for the point pattern.

The values of r in the vector `r` must be finely spaced and equally spaced. The default spacing between successive values of `r` is `rmax/512`, or `eps/4` if the window is a binary mask with pixel width `eps`.

- *What is the 'recommended range' of r values, and how is it determined in `Fest(X)` or `Gest(X)`?*

Every `"fv"` object has an attribute `alim` containing the 'recommended range' of r. This is the range of r values that will be displayed in the default plot. It is typically shorter than the 'available range' of all r values for which the function was computed. The available range and recommended range are printed when the object is printed.

For the F and G functions and their relatives, the recommended range is determined after the function has been computed, so that the plot will show the most informative detail. For $F(r)$ and $G(r)$, the recommended range is from 0 to the 90th percentile. For example the recommended range for $F(r)$ is the range of r for which $\widehat{F}(r)$ is less than or equal to 0.9.

- *Gest(X) gives me a table of values of $\widehat{G}(r)$ for a sequence of r values. But how do I get the value of $\widehat{G}(r)$ at other values of r?*

The object G <- Gest(X) can be converted to a function by Gfun <- as.function(G). This function can then be evaluated at any distance value: Gfun(r). This interpolates between the values in the table.

For higher precision, call Gest with the argument r giving a more finely spaced, and evenly spaced, sequence of values of r starting from 0.

- *Your package gives the wrong answer for the border-corrected estimate of $G(r)$ for the distance $r = 0.1$ in the standard dataset xxx. I have calculated the correct answer by hand.*

First, check that you have used Gest correctly. If you used the argument r, please be aware that this is not the standard usage of Gest. The argument r cannot be a single value such as 0.1, or a handful of values. It must be a finely spaced sequence of values in order that the algorithm be valid.

Second, be aware that computers perform finite-precision binary arithmetic. The number 0.1 has an infinite binary expansion and must be approximated by a different binary number. Try the following (and variations):

```
> (1 - 0.9) == 0.1
[1] FALSE
> 1 - 10 * (1 - 0.9)
[1] 2.22e-16
```

The answers depend on the hardware. This is sometimes known as 'GCC bug 323' although computer scientists do not regard it as a bug; it is a numerical rounding effect.[3] Hence, if the spatial coordinates are multiples of 0.1, the estimate of $G(r)$ for r=0.1 may be incorrect. This has no perceptible effect on the plotted function estimates, since the calculations will be correct for values of r very close to 0.1.

Bugs are possible, of course. If you are convinced that you have found a bug, please find the *simplest reproducible example* of the bug, and post a bug report to www.spatstat.org.

- *I have divided the window of my point pattern into several polygons. I want to compute, for each pair of polygons i and j, the minimum distance from a data point in polygon i to a data point in polygon j.*

If X is the point pattern and P is the tessellation of polygons, first label each point of X by the polygon which contains it: Y <- cut(X, P). Next call D <- nndist(X, by=marks(X)). The result is a matrix with one row for each data point and one column for each polygon. The entry in row i, column j is the distance from the ith data point to the nearest point in polygon j. Finally M <- aggregate(D, by=list(from=marks(X)), min) yields a matrix with one row and column for each polygon, giving the desired result.

---

[3]Portrayed in the famous play, *"Much Ado About 2.2e-16"*.

Part III

STATISTICAL INFERENCE

9

Poisson Models

Part III deals with statistical modelling and inference for point pattern data, starting in this chapter with Poisson point process models.

9.1 Introduction

Poisson point processes were introduced briefly in Chapter 5. In this chapter we show how to use them as statistical models for point pattern data. We show how to formulate a Poisson model appropriate to the data, fit the model to the data, and interpret the results. Validation of a fitted model is discussed in chapters 10 and 11.

The key property of a Poisson process is that the random points are *independent* of each other. This property greatly simplifies the statistical analysis, and gives access to a wide variety of powerful statistical techniques.

A Poisson process is completely described by its intensity function $\lambda(u)$. Statistical models of Poisson point processes are easy to build: they are simply models of the intensity, that is, a model is just an equation stating the form of the function $\lambda(u)$. This also means that there is a close relationship between techniques for estimating an intensity function (Chapter 6) and methods for fitting a Poisson point process model.

Why Poisson?

The huge literature on spatial analysis contains many other techniques for analysing spatial point pattern data. Many of these techniques appear at first sight to have nothing to do with Poisson point processes — and even seem to have nothing to do with point processes at all.

In fact, several popular techniques are *equivalent* to fitting a Poisson point process model with a specific form of the intensity — namely, where the intensity is a loglinear function of the covariates. Such techniques include logistic regression [4, 658, 604, 326, 397, 116] and maximum entropy [248, 246, 252, 546].

Logistic regression is commonly used in GIS to predict the probability of occurrence of mineral deposits [4, 152], archaeological finds [604, 397], landslides [153, 288], and other events which can be treated as points at the scale of interest. The study region is divided into pixels; in each pixel the presence or absence of any data points is recorded; then logistic regression [458, 240, 344] is used to predict the probability of the presence of a point as a function of predictor variables.

Logistic regression implicitly assumes that the presence/absence values for different pixels are independent [458, 240, 344]. This implies that the underlying point process is Poisson (as explained in Chapter 5). Furthermore, the *logistic* relationship implies that the intensity of the Poisson process is a *loglinear* function of the covariates and the parameters. Thus, logistic regression is essentially equivalent to fitting a Poisson point process model with *loglinear* intensity [34, 691].

The principle of maximum entropy [248] is often used in ecology, for example, to study the influence of habitat variables on the spatial distribution of animals or plants [246, 252, 546]. Con-

ceptually we consider all possible spatial distributions, and find the spatial distribution which maximises a quantity called entropy, subject to constraints implied by the observed data. Surprisingly,
the solution is a probability density which is a *loglinear* function of the covariates. An analogy
could be drawn with the stretching of a string: a string may take on any shape, but if we demand
that the string be stretched as tight as possible, it will take up a straight line. Thus, again, this analysis principle is equivalent to fitting a Poisson point process model with loglinear intensity [568].

Advantages of statistical modelling

Statistical modelling is a much more powerful way to analyse data than simply computing summary
statistics. Formulating a statistical model, and fitting it to the data, allows us to adjust for effects
that might otherwise distort the analysis (such as uneven distribution of survey effort, and spatially
varying population density) by including terms that represent these effects. By fitting different
models that include or omit a particular term, and comparing the models, we can decide which
variables have a statistically significant influence on the intensity.

A great advantage of statistical modelling is that the assumptions of the analysis are openly
stated, rather than implicitly imposed. All model assumptions are 'on the table' and are amenable
to criticism. Chapter 11 provides tools for criticising each individual model assumption of a Poisson
point process model, and for identifying weaknesses of the data analysis.

9.2   Poisson point process models

Poisson processes were introduced briefly in Sections 5.3 and 5.4 of Chapter 5. Here we discuss
using Poisson processes as statistical models for data analysis of spatial point patterns.

9.2.1   Characteristic properties of the Poisson process

We start with a detailed statement of the characteristic properties of the Poisson point process. Apart
from clarifying our understanding of the model, this is a useful checklist of properties that need to
be verified before the analyst can be satisfied that a real point pattern dataset is well described by a
Poisson point process.

Homogeneous Poisson process (CSR)

The *homogeneous Poisson process* with intensity $\lambda > 0$ (also called *'complete spatial randomness'*,
CSR) has the properties that

(PP1) **Poisson counts:** the number $n(\mathbf{X} \cap B)$ of points falling in any region B has a Poisson distribution (see page 135);

(PP2) **homogeneous intensity:** the expected number of points falling in B is $\mathbb{E}[n(\mathbf{X} \cap B)] = \lambda \cdot |B|$;

(PP3) **independence:** if $B_1, B_2, \ldots$ are disjoint regions of space then $n(\mathbf{X} \cap B_1)$, $n(\mathbf{X} \cap B_2)$, $\ldots$ are
independent random variables;

(PP4) **conditional property:** given that $n(\mathbf{X} \cap B) = n$, the n points are independent and uniformly
distributed in B.

We are justified in calling λ the 'intensity' because (PP2) implies that this point process has
homogeneous intensity λ in the sense of Section 6.2. The intensity parameter λ has dimensions

length^{-2} and determines the average number of points per unit area. Scaling properties of λ (for example, the effect of changing the unit of length) are discussed in Section 6.2. Any value can be specified for λ, as long as it is non-negative and finite. If $\lambda = 0$ then all realisations of the point process are empty.

The properties (PP1)–(PP4) provide the basis for many statistical techniques. For example, in quadrat counting, (PP3) implies that the counts in different quadrats are independent random variables; (PP1) implies that these counts have Poisson distributions; and (PP2) implies that the mean values of the counts are proportional to the areas of the quadrats. If the quadrats have equal area, then the quadrat counts are Poisson random variables with equal mean values, so they are identically distributed. This justifies the use of the χ^2 test.

When we need to check whether a point process is CSR, the most important properties to check are independence (PP3) and homogeneity (PP2). As explained in Section 5.3, the other properties follow logically from these, under reasonable conditions.[1] However, if any one of these properties does not hold, then the process is not CSR. Statistical methods for checking CSR can focus, for example, on detecting a departure from the Poisson distribution (PP1) or from conditional independence and uniformity (PP4).

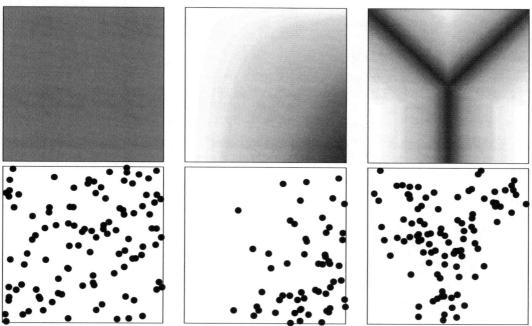

Figure 9.1. *Examples of intensity functions (top) and simulated realisations (bottom) for Poisson processes.*

Inhomogeneous Poisson process

The *inhomogeneous* Poisson process with intensity function $\lambda(u)$, $u \in \mathbb{R}^2$, is a modification of the homogeneous Poisson process, in which properties (PP2) and (PP4) above are replaced by

(PP2′): the number $n(\mathbf{X} \cap B)$ of points falling in a region B has expected value

$$\mathbb{E}[n(\mathbf{X} \cap B)] = \int_B \lambda(u)\,\mathrm{d}u. \tag{9.1}$$

---

[1] The reasonable condition is that the probability of *more than one* random point falling in a region B, divided by the area of B, falls to zero as the area of B goes to zero. This excludes coincident points and fractal-like behaviour.

(PP4′): given that $n(\mathbf{X} \cap B) = n$, the n points are independent and identically distributed, with common probability density

$$f(u) = \frac{\lambda(u)}{I} \qquad (9.2)$$

where $I = \int_B \lambda(u)\,\mathrm{d}u$.

Again we are justified in calling $\lambda(u)$ the 'intensity function' because, by virtue of (PP2′), it is the intensity function in the sense of Section 6.3. Its values have dimension $length^{-2}$ and determine the spatially varying intensity (average number of points per unit area).

The intensity function encapsulates both the abundance of points (by equation (9.1)) and the spatial distribution of individual point locations (by equation (9.2)).

Any function $\lambda(u)$ can be used, as long as its values are non-negative and *integrable* (the right-hand side of (9.1) must be finite).

9.2.2 Modelling scenarios giving rise to a Poisson process

There are many realistic models of physical processes which give rise to a Poisson point process, because the Poisson limit law is so widely applicable. We sketch some common examples below.

Random trials on a very fine grid

Imagine a radioactive material, consisting of N atoms arranged in a regular grid, where N is very large, and the spacing between atoms is tiny. In a specified time interval, each atom has a very small probability p of undergoing random decay. Decay events of different atoms are independent. The total number of decay events in the specified period is the number of successes in N trials with success probability p. Since N is large and p is small, the number of decay events has a Poisson distribution (to a very good approximation) and the locations of the decay events approximately constitute a Poisson point process.

Independent random thinning

Suppose that there is some initial population represented by a point process $\mathbf{X}$ (not assumed to be a Poisson process) and that each point of $\mathbf{X}$ is either deleted or retained, independently of other points. Suppose $\mathbf{X}$ has intensity function $\beta(u)$ and the probability of retaining a point at the location u is $p(u)$. The resulting process $\mathbf{Y}$ of retained points has intensity $\lambda(u) = p(u)\beta(u)$. If the original population density $\beta(u)$ is large and the retention probabilities $p(u)$ are small, then the resulting process $\mathbf{Y}$ is approximately an inhomogeneous Poisson process.

A model of plant propagation could assume that seeds are randomly dispersed according to some point process, and seeds randomly germinate or do not germinate, independently of each other, with a germination probability that depends on the local soil conditions. The resulting pattern of plants is well described by an inhomogeneous Poisson process.

In accident research, some models describe a population of 'near-accidents' or 'potential accidents', each of which has a small probability of progressing to a real accident. The thinning principle says that the real accidents are well described by an inhomogeneous Poisson process. Similarly in a galaxy survey of the distant universe, we may fail to detect some individual galaxies. The detected locations of very faint galaxies could be described by an inhomogeneous Poisson process.

In spatial epidemiology, $\mathbf{X}$ could represent the human population at risk of a rare disease, and $\mathbf{Y}$ would be the disease cases. The key assumption is that a person's disease status is statistically independent of other people.

Random strewing

Suppose a large number N of points is scattered randomly in a large region D according to a binomial point process. That is, each point is uniformly distributed over D, and different points are independent. If this random pattern is observed within a subregion W, where D is much larger than W, then the observed pattern is approximately a Poisson point process inside W.

Random displacement

Again we imagine an initial population represented by any point process **X**. Suppose that each point in **X** is subjected to a random displacement, independently of other points. A point at the original location x_i is moved to the new location $y_i = x_i + V_i$ where V_i is a random vector. Assume that $V_1, V_2, \ldots$ are independent random vectors. Then under reasonable conditions, the resulting point process **Y** is approximately an inhomogeneous Poisson process.

Spatial birth-and-death process

Consider a forest which evolves over time. In each small interval of time Δt, each existing tree has probability $m \Delta t$ of dying, independently of other trees, where m is the 'mortality rate' per tree per unit time. In the same time interval, in any small region of area Δa, a new tree germinates with probability $g \Delta a \Delta t$, independently of any existing trees, where g is the 'germination rate' per unit area per unit time. No matter what the initial state of the forest, over long time scales this 'spatial birth-and-death process' reaches an equilibrium in which any snapshot of the forest (the pattern of trees in existence at a fixed time) is a realisation of a Poisson process with intensity g/m.

9.2.3 Common parametric models of intensity

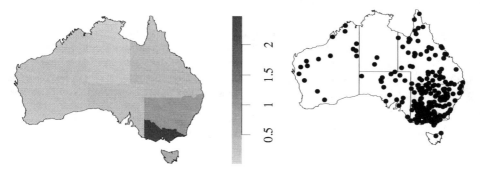

Figure 9.2. *Intensity (Left) and simulated realisation (Right) of the Poisson process with a different, homogeneous intensity in each Australian state and territory. Intensity is proportional to average population density within the state or territory.*

The intensity function $\lambda(u)$ of the Poisson process can be assumed to take any form that is appropriate to the context, so long as the intensity values are non-negative and the integral of the intensity over the observation window is finite. Some common models of the intensity function, and their interpretation, are discussed below.

homogeneous intensity: the Poisson process with constant intensity $\lambda(u) \equiv \lambda$ is the simplest model, called the homogeneous Poisson process or 'complete spatial randomness' (CSR).

homogeneous in different regions: Space is partitioned into non-overlapping regions $B_1, \ldots, B_m$. Within region B_j, the process is Poisson with intensity β_j, where $\beta_1, \ldots, \beta_m$ are parameters of the model. Then the overall process is an inhomogeneous Poisson process where the intensity function $\lambda(u)$ takes the value β_j when u is in region B_j. See Figure 9.2.

intensity proportional to baseline: A simple model is

$$\lambda(u) = \theta b(u) \tag{9.3}$$

where $b(u)$ is a known function ('baseline') and θ is an unknown parameter that must be estimated. If the baseline $b(u)$ represents the spatially varying density of a population, and we assume that each member of the population has equal chance θ of contracting a rare disease, then the cases of the disease will form a Poisson process with intensity (9.3), the 'constant risk' model. Modelling scenarios which involve random thinning often lead to a model of this form.

exponential function of covariate: A very important model is

$$\lambda(u) = \kappa e^{\beta Z(u)} = \exp(\alpha + \beta Z(u)) \tag{9.4}$$

where $Z(u)$ is a spatial covariate, and α, β, κ are parameters to be estimated. This is the model that is fitted in basic applications of logistic regression and maximum entropy. An important example is in prospectivity analysis, where $Z(u)$ is the distance from location u to the nearest geological fault. The parameter $\kappa = e^\alpha$ gives the intensity at locations where $Z(u) \approx 0$ (i.e. locations close to a geological fault), while the intensity changes by a factor e^β for every 1 unit of distance away from the faults. The Berman-Lawson-Waller test (Section 10.3.5) for a covariate effect is equivalent to testing whether $\beta = 0$ in this model.

raised incidence model: A combination of the last two models is

$$\lambda(u) = b(u)e^{\alpha + \beta Z(u)} = b(u)\exp(\alpha + \beta Z(u)) \tag{9.5}$$

where $b(u)$ is a known baseline function and $Z(u)$ is a spatial covariate. This model expresses the covariate effect relative to the baseline $b(u)$. For example in spatial epidemiology, $b(u)$ could be the spatially varying population density, and the term $\exp(\alpha + \beta Z(u))$ would then be the disease risk per person, depending on the value of Z. The exponential term could be replaced by something more complicated [224]. Modelling scenarios that involve independent random thinning often lead to models of this kind.

loglinear model: Embracing all of the models listed above is the general loglinear model

$$\lambda_\theta(u) = \exp(B(u) + \theta^\top \mathbf{Z}(u)) = \exp(B(u) + \theta_1 Z_1(u) + \theta_2 Z_2(u) + \ldots + \theta_p Z_p(u)) \tag{9.6}$$

where $B(u)$ and $Z_1(u), \ldots, Z_p(u)$ are known functions, $\theta_1, \ldots, \theta_p$ are parameters to be estimated, and we write $\mathbf{Z}(u) = (Z_1(u), \ldots, Z_p(u))$ for the vector-valued function. This was termed a *modulated* Poisson process by Cox [177].

As this sequence of models shows, we often build models for the intensity by multiplying together several terms which represent different effects. The logarithm of intensity is a sum of terms representing different effects. We can even include a term in the model that accounts for the probability of observing a point (e.g., related to the survey effort).

There are plausible modelling assumptions that give rise to Poisson point process models of either *additive* or *multiplicative* form. A model of the additive form has an intensity function

$$\lambda(u) = \gamma_1(u) + \gamma_2(u) + \ldots$$

where the terms $\gamma_k(u)$ are interpretable as the intensities of Poisson processes of points associated with different 'origins' or 'causes', that were superimposed to yield the observed point process. A model of multiplicative form has

$$\lambda(u) = \exp(A(u) + B_1(u) + B_2(u) + \ldots)$$

where the first term $A(u)$ is a baseline intensity (on a log scale) and the subsequent terms $B_1(u)$, $B_2(u)$, … can be interpreted as 'thinning' or 'modulation' effects.

The multiplicative form is more convenient for inference, and corresponds directly to the canonical choice of the log link in a Poisson Generalised Linear Model. Inference for models of the additive form is more difficult as it effectively leads to a missing value problem, the missing information being the attribution of each observed point to one of the component Poisson processes.

9.3 Fitting Poisson models in `spatstat`

Fitting a statistical model to data means choosing values for the parameters governing the model, so that the model fits the data as well as possible. For example, to fit a line $y = ax + b$ through a scatter of data points, we find the values of the parameters a (slope) and b (intercept) that give the best-fitting line.

To fit a Poisson point process model to a point pattern dataset, we would specify the form of the intensity function — perhaps using one of the common models described above — and find the values of the parameters which best fit the point pattern dataset. The definition of 'best fit' and the technical details of model-fitting are described in Sections 9.7–9.10. This section explains how to use `spatstat` to fit Poisson models to point pattern data in practice.

9.3.1 The `ppm` function

Syntax

Fitting a Poisson point process model to a point pattern dataset is performed in `spatstat` by the function ppm (for **p**oint **p**rocess **m**odel). This is closely analogous to the standard model fitting functions in R such as `lm` (for fitting linear models) and `glm` (for fitting generalized linear models). The essential syntax for specifying a Poisson point process model to be fitted by `ppm` is

```
ppm(X ~ trend, ...)
```

where the first argument 'X ~ trend' is a `formula` in the R language (Section 2.1.10).

The left-hand side of the formula gives the name of the point pattern dataset X. Alternatively the left-hand side may be an expression which can be evaluated to yield a point pattern.

The right-hand side of the model formula specifies the form of the *logarithm* of the intensity function for the model. Thus, any *loglinear* intensity model (9.6) can be fitted.[2]

Simplest example

The simplest possible model for a Poisson planar point process is the constant intensity or homogeneous model, equivalent to complete spatial randomness. A call to ppm to fit this model employs the simplest possible formula, X ~ 1. We illustrate this using the *Beilschmiedia* forest data (Figure 1.10) available in `spatstat` as the "ppp" object `bei`.

---

[2]Fitting Poisson models which are *not* loglinear is discussed in Section 9.12.

```
> fit <- ppm(bei ~ 1)
> fit
Stationary Poisson process
Intensity: 0.007208
              Estimate    S.E. CI95.lo CI95.hi Ztest   Zval
log(lambda)    -4.933 0.01666  -4.965    -4.9   *** -296.1
```

The result of ppm is a fitted point process model, an object belonging to the class "ppm". There are many facilities for handling such objects. For instance there is a print method print.ppm, which produced the output above. This output tells us that the fitted intensity was $\overline{\lambda} = 0.007208$ trees per square metre. For the *logarithm* of the intensity, the output gives an estimate, a standard error, and a 95% confidence interval. The fitted intensity can be recovered from this line of output by exp(-4.933) = 0.007208.

Variables in a formula

A 'variable name' in a formula is a name that plays the role of a variable rather than a function. For example, in the formula Y ~ f(X), the symbols X and Y are recognised by R as playing the role of 'variables', while the symbol f is recognised as playing the role of a function.

In a call to ppm of the form

        ppm(X ~ trend)

all the variable names appearing in the formula should be the names of existing objects in the R session. An alternative is to use the form

        ppm(X ~ trend, ..., data)

where the argument data is a list containing data needed to fit the model. In this case, for each variable name in the formula, ppm will first try to find a matching entry in the list data; if it is not found, then ppm will look for existing objects in the R session.

The left-hand side of the model formula should be either the name of a point pattern, or an expression which can be evaluated to yield a point pattern. On the left side of the formula, mathematical operators such as +, -, *, /, and ^ have their usual mathematical meaning.

The right-hand side of the model formula is an expression involving the names of spatial covariates which affect the intensity of the point process. On the right-hand side of a model formula, the operators +, -, *, /, and ^ have a special interpretation: they are model operators, which are used to combine terms in a model, as explained below.

The variable names x and y are reserved names which represent the Cartesian coordinates. Any other variable name in the formula should match the name of an existing object in the R session, or the name of an entry in the argument data if it is given. Each object should be either *(a)* a pixel image (object of class "im") giving the values of a spatial covariate at a fine grid of locations; *(b)* a function which can be evaluated at any location (x, y) to obtain the value of the spatial covariate. It should be a function(x, y, ...) in the R language; *(c)* a window (object of class "owin") which will be interpreted as a logical variable which is TRUE inside the window and FALSE outside it; *(d)* a tessellation (object of class "tess") which will be interpreted as a factor covariate. For each spatial location, the factor value indicates which tile of the tessellation contains the location in question; or *(e)* a single number, indicating a covariate that is constant in this dataset.

In what follows we explain how to use the ppm function and elaborate on the use of formulae introduced in Section 2.1.10. We illustrate the ideas by examples using point pattern datasets supplied with the spatstat package.

9.3.2 Models with a single numerical covariate

Recall that the interpretation of a formula depends on the types of variables involved: for example, numerical variables and categorical variables are interpreted differently. We start by dealing with models involving a single numerical covariate and proceed from there to more complicated models. Given a numerical covariate (predictor) Z, say, we could use a formula such as X ~ Z to specify the model. This formula indicates that the intensity is a loglinear function of Z as given by (9.4).

9.3.2.1 Analysis of *Beilschmedia* data

The data set `bei` installed in `spatstat` is a point pattern giving the locations of *Beilschmiedia* trees. Covariate data are supplied in a separate object `bei.extra`, which is a list with two entries, `elev` and `grad`, which are pixel images of values of the terrain elevation and terrain slope, respectively. See Section 6.5.3, page 176 and Section 1.1.4.

```
> bei.extra
List of pixel images
elev:
real-valued pixel image
101 x 201 pixel array (ny, nx)
enclosing rectangle: [-2.5, 1002] x [-2.5, 502.5] metres
grad:
real-valued pixel image
101 x 201 pixel array (ny, nx)
enclosing rectangle: [-2.5, 1002] x [-2.5, 502.5] metres
```

The following command fits a Poisson point process model in which the intensity of *Beilschmiedia* trees is a *loglinear* function of terrain slope:

```
> fit <- ppm(bei ~ grad, data=bei.extra)
> fit
Nonstationary Poisson process
Log intensity:   ~grad
Fitted trend coefficients:
(Intercept)        grad
     -5.391       5.022
```

The model fitted by the commands above is a Poisson point process with intensity

$$\lambda(u) = \exp(\beta_0 + \beta_1 S(u)) \tag{9.7}$$

where $S(u)$ is the terrain slope at location u. The argument to the exp function (in this case $\beta_0 + \beta_1 S(u)$) is referred to as the *'linear predictor'*. The component $\beta_1 S(u)$ is called the *'effect'* of the covariate S.

The printout above[3] includes the fitted coefficients marked by the labels "`(Intercept)`" and "`grad`". These are the maximum likelihood estimates of β_0 and β_1, respectively, the coefficients of the linear predictor; so the fitted model is

$$\lambda(u) = \exp(-5.391 + 5.022 S(u)). \tag{9.8}$$

These results tell us that the estimated intensity of *Beilschmiedia* trees on a level surface (slope $S = 0$) is about $\exp(-5.391) = 0.004559$ trees per square metre, or 45.59 trees per hectare, and

---

[3]To save space on the printed page, we have set `options(digits=4)` so that numbers are printed to four significant digits for the rest of this chapter. We have also set `spatstat.options(terse=2)` to shorten the output.

would increase by a *factor* of exp(5.022) = 151.7 if the slope increased to 1.0. The largest slope value in the data is about 0.3, at which stage the predicted intensity has risen by a factor of exp(0.3 × 5.022) = 4.511, that is, more than quadrupled from its value on a level surface.

Although it is possible to read off the fitted model from the printout and calculate intensity levels for various values of the predictors, this can be tedious, and becomes intricate when the model is more complicated. The generic R command `predict` (explained in Section 9.4.3) has a method for point process models (class `"ppm"`), which can be used to calculate fitted intensities and other properties of the model. In this simple model, the fitted intensity can also be plotted as a function of a covariate (in the current example, the covariate `grad`) using the `spatstat` utility `effectfun`, as shown in the left panel of Figure 9.3.

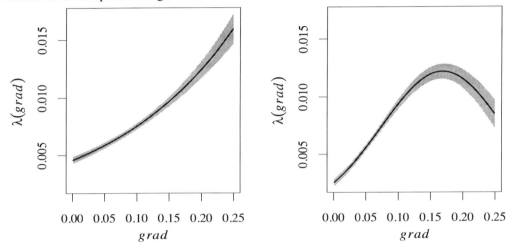

Figure 9.3. *Fitted intensity of* Beilschmiedia *trees as a function of slope. Solid line: maximum likelihood estimate. Shading: pointwise 95% confidence interval. Generated by* `plot(effectfun(fit, "grad", se.fit=TRUE))`. *Left:* `fit` *is a* "ppm" *object containing the loglinear model (9.7). Right:* `code` *is a* "ppm" *object containing the log-quadratic model (9.9).*

It should be clearly understood that the formula `bei ~ grad` does not represent a flexible model in which the abundance of trees depends, in some unspecified way, on the terrain slope. On the contrary, the dependence is very tightly specified: the formula `bei ~ grad` corresponds to the *loglinear* relationship (9.7), that is, it corresponds to assuming intensity is exponentially increasing or decreasing as a function of terrain slope.

If a loglinear relationship is *not* an appropriate assumption, but the intensity is believed to depend on terrain slope somehow, several strategies are available.

One strategy is to transform the slope values and fit a loglinear relationship to the transformed values. Transformations of the original covariates can be specified directly in the model formula. For example, the covariate `grad` gives the gradient as the *tangent* of the angle of inclination (vertical elevation divided by horizontal displacement) so that a gradient of 1.0 corresponds to a 45-degree incline. If we think it is more appropriate to use the *angle* of inclination as the covariate in the loglinear model, we can convert the gradient values to angles using the arctangent, simply by typing

```
> ppm(bei ~ atan(grad), data=bei.extra)
```

Any mathematical function can appear in a model formula. However, care is required with expressions that involve the symbols +, -, *, /, and ^, which have special syntactic meanings in a formula, except inside a function call.

For example, `atan` calculates angles in radians; to convert them to degrees, we would need to multiply by $180/\pi$, so the covariate should be `atan(grad) * 180/pi`. To prevent the mathematical symbols `*` and `/` from being interpreted as model formula operators, this expression should be enclosed inside the function `I` (for 'identity'):

```
> ppm(bei ~ I(atan(grad) * 180/pi), data=bei.extra)
```

The function `I` returns its argument unchanged, and the rules for interpreting a formula ensure that the expression inside the parentheses is evaluated using the usual mathematical operations. The command above fits a loglinear model depending on the angle of inclination, expressed in degrees. It may be neater to write a separate R function which can then appear in the model formula:

```
> degrees <- function(x) { x * 180/pi }
> ppm(bei ~ degrees(atan(grad)), data=bei.extra)
```

Another strategy is to replace the first-order function in (9.7) by a quadratic function,

$$\lambda(u) = \exp(\beta_0 + \beta_1 S(u) + \beta_2 S(u)^2), \tag{9.9}$$

which results in a model which is *log-quadratic* as a function of slope. Here there are three parameters, β_0, β_1, and β_2, to be estimated. This model can be fitted by specifying the `slope` and `slope^2` terms in the model formula:

```
> ppm(bei ~ grad + I(grad^2), data=bei.extra)
Nonstationary Poisson process
Log intensity:  ~grad + I(grad^2)
Fitted trend coefficients:
(Intercept)        grad    I(grad^2)
     -5.987      18.745      -55.602
```

An advantage of this approach is that there is a formal mechanism for deciding (in statistical terms) whether the `grad^2` term is needed, namely a test of the null hypothesis $H_0 : \beta_2 = 0$. We discuss this in Section 9.4. Higher-order polynomials can be fitted in the same way.

The command `effectfun` can be used to visualise the fitted effect of a covariate: in this context `effectfun(fit, "grad")` would show the combined effect (linear and quadratic terms together) of the slope variable as shown in the right panel of Figure 9.3.

From a theoretical viewpoint, all the models fitted above are 'loglinear' models, of the general form (9.6). It is useful to distinguish between *'original covariates'* (e.g., `grad` in the current example) and *'canonical covariates'* (the variables Z_j which appear in the loglinear model (9.6)). Canonical covariates may be transformations and combinations of the original covariates which were provided in the dataset.

Table 9.1 lists some of the common terms appearing in model formulae.

9.3.2.2   Murchison gold data

We continue to illustrate models involving a single numerical covariate.

The Murchison geological survey data shown in Figure 9.4 record the spatial locations of gold deposits and associated geological features in the Murchison area of Western Australia. They are extracted from a regional survey (scale 1:500,000) of the Murchison area made by the Geological Survey of Western Australia [693]. The features recorded are the known locations of gold deposits (point pattern `gold`); the known or inferred locations of geological faults (line segment pattern `faults`); and the region of greenstone outcrop (window `greenstone`). The study region is contained in a 330×400 kilometre rectangle. At this scale, gold deposits are point-like, i.e. their spatial extent is negligible. Gold deposits are strongly associated with greenstone bedrock and with faults,

TERM	MEANING
`(age > 40)`	logical covariate: TRUE if age exceeds 40
`cut(age, 3)`	factor covariate: age classified into 3 bands
`I(age^2)`	numerical covariate: age squared
`polynom(age, 3)`	cubic polynomial function of age
`bs(age, 5)`	smooth function of age
`1`	constant term (intercept)
`-1`	remove intercept term
`+0`	remove intercept term

Table 9.1. *Common terms in model formulae, assuming* `age` *is a numerical covariate. (Note:* `bs` *requires the* `splines` *package.)*

but the geology is of course three-dimensional, and the survey data are a two-dimensional projection. The survey may not have detected all existing faults, because they are usually not observed directly; they are observed in magnetic field surveys or geologically inferred from discontinuities in the rock sequence.

These data were analysed in [269, 134]; see also [298, 386]. The main aim is to predict the intensity of the point pattern of gold deposits from the more easily observable fault pattern. To be precise, the aim is to use the observed pattern of faults to specify zones of high 'prospectivity' to be explored for gold. See [116, 139] for general background.

First we rescale the data to kilometres:

```
> mur <- solapply(murchison, rescale, s=1000, unitname="km")
```

The full dataset is shown in Figure 9.4, generated using `plot(as.layered(mur))`, and a detail in the upper right corner is shown in Figure 9.5.

A common model for such data is a Poisson process with intensity which is a loglinear function of distance to the nearest fault,

$$\lambda(u) = \exp(\beta_0 + \beta_1 d(u)), \tag{9.10}$$

where β_0, β_1 are parameters and $d(u)$ is the distance from location u to the nearest geological fault. This is equivalent to the model customarily fitted in GIS software by logistic regression of presence-absence indicators on distance to nearest fault.

We compute the distance covariate $d(u)$ using `distfun` (Section 6.6.4) which has the advantage that distances will be computed exactly by analytic geometry. A less accurate alternative would be to compute the distance values at a grid of pixel locations using `distmap`.

```
> dfault <- with(mur,distfun(faults))
```

Next we fit the loglinear model (9.10):

```
> fit <- ppm(gold ~ dfault, data=mur)
> fit
Nonstationary Poisson process
Log intensity:  ~dfault
Fitted trend coefficients:
(Intercept)      dfault
    -4.3413     -0.2608
```

The fitted model is

$$\lambda(u) = \exp(-4.341 - 0.2608 d(u)). \tag{9.11}$$

That is, the estimated intensity of gold deposits in the immediate vicinity of a geological fault

Figure 9.4. *Murchison geological survey data. Gold deposits (+), geological faults (—), and greenstone outcrop (grey shading) in a survey region about 200 by 300 km across.*

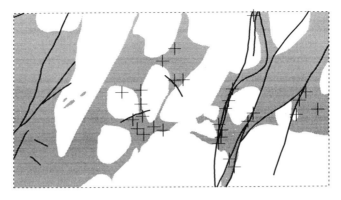

Figure 9.5. *Detail of Murchison geological survey data in a 70 by 40 km region near the abandoned town of Reedy. Gold deposits (+), faults (solid lines), and greenstone outcrop (grey shading).*

is about $\exp(-4.341) = 0.01302$ deposits per square kilometre or 1.302 deposits per 100 square kilometres, and decreases by a *factor* of $\exp(-0.2608) = 0.7704$ for every additional kilometre away from a fault. At a distance of 10 kilometres, the intensity has fallen by a factor of $\exp(10 \times (-0.2608)) = 0.07368$ to $\exp(-4.341 + 10 \times (-0.2608)) = 0.0009596$ deposits per square kilometre or 0.09596 deposits per 100 square kilometres. Figure 9.6 shows the effect of the distance covariate on the intensity function.

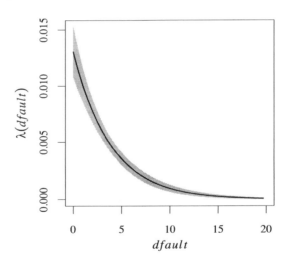

Figure 9.6. *Fitted intensity of Murchison gold deposits as a function of distance to the nearest fault, assuming it is a loglinear function of distance (equivalent to logistic regression). Solid line: maximum likelihood estimate. Shading: pointwise 95% confidence interval. Generated by* `plot(effectfun(fit, "dfault", se.fit=TRUE))`.

We caution that, by fitting the model with formula `gold ~ dfault` using `ppm` or equivalently using logistic regression, we have not fitted a highly flexible model in which the abundance of deposits depends, in some unspecified way, on the distance to the nearest fault. Rather, the very specific relationship (9.10) has been fitted. To fit more flexible models, we need to modify the model formula, as discussed below.

In prospectivity analysis it may or may not be desirable to fit any explicit relationship between deposit abundance and covariates such as distance to the nearest fault. Often the objective is simply to select a distance threshold, so as to delimit the area which is considered highly prospective (high predicted intensity) for the mineral. The ROC curve (Section 6.7.3) is more relevant to this exercise. However, if credible models can be fitted, they contain much more valuable predictive information.

9.3.3 Models with a logical covariate

A logical covariate takes the values TRUE and FALSE. Logical covariates can arise in many ways, as we shall see in this chapter.

The Murchison data include the polygonal boundary of greenstone outcrop, given as a window called greenstone. Setting aside the information about geological faults for a moment, we could fit a simple Poisson model in which the intensity of gold deposits is constant inside the greenstone, and constant outside the greenstone, but with different intensity values inside and outside.

```
> ppm(gold ~ greenstone, data=mur)
Nonstationary Poisson process

Log intensity:    ~greenstone

Fitted trend coefficients:
    (Intercept) greenstoneTRUE
         -8.103          3.980

                  Estimate   S.E. CI95.lo CI95.hi Ztest    Zval
(Intercept)         -8.103 0.1667  -8.430  -7.777   ***  -48.62
greenstoneTRUE       3.980 0.1798   3.628   4.333   ***   22.13
```

The function ppm recognises that greenstone is a spatial window, and interprets it as a covariate with *logical* values, equal to TRUE inside the greenstone outcrop and FALSE outside.

Since the model formula does not explicitly exclude a constant term ('intercept'), this term is assumed to be present. The model fitted is

$$\lambda(u) = \exp(\alpha + \beta G(u)) \tag{9.12}$$

where G is the indicator covariate for the greenstone, taking the value $G(u) = 1$ if the location u is inside the greenstone outcrop, and $G(u) = 0$ if it is outside. The printed output shows the estimates for the parameters α (labelled Intercept) and β (labelled greenstoneTRUE) as -8.103 and 3.98 respectively. The estimated intensities are $\lambda_{out} = \exp(-8.103) = 0.0003026$ outside the greenstone outcrop and $\lambda_{in} = \exp(-8.103 + 3.98) = 0.0162$ inside. The last few lines of output give the estimates of α and β together with estimates of standard error, 95% confidence intervals, and a report that the coefficient β is significantly different from 0.

Slightly different output is obtained with the command

```
> ppm(gold ~ greenstone - 1, data=mur)
Nonstationary Poisson process

Log intensity:    ~greenstone - 1

Fitted trend coefficients:
greenstoneFALSE  greenstoneTRUE
         -8.103          -4.123

                   Estimate    S.E. CI95.lo CI95.hi Ztest    Zval
greenstoneFALSE      -8.103 0.16667  -8.430  -7.777   ***  -48.62
greenstoneTRUE       -4.123 0.06757  -4.255  -3.990   ***  -61.01
```

The model formula now includes -1 forbidding the use of a constant term ('intercept'). The model fitted here is

$$\lambda(u) = \exp(\gamma G_{\text{out}}(u) + \delta G_{\text{in}}(u)) \tag{9.13}$$

where $G_{\text{in}}(u) = G(u)$ is the indicator of the greenstone as before, and $G_{\text{out}}(u) = 1 - G(u)$ is the indicator of the complement of the greenstone. The estimates of γ and δ are -8.103 and -4.123 respectively. The estimates of the intensities are $\lambda_{\text{out}} = \exp(-8.103) = 0.0003026$ outside the greenstone outcrop and $\lambda_{\text{in}} = \exp(-4.123) = 0.0162$ inside. The two fitted models agree, but are parametrised differently. The last few lines of output give standard errors and confidence intervals for the parameters γ and δ, and indicate that these parameters are significantly different *from zero*. The first parameterisation (i.e. with an intercept included) is more useful for deciding whether the greenstone has an effect on the intensity of gold deposits, while the second parameterisation is more useful for directly reading off the estimated intensities.

The foregoing discussion depends upon the default 'treatment contrasts' being used, as explained below.

9.3.4 Models with a factor covariate

A spatial covariate with *categorical* values could arise from a map which divides the spatial domain into regions of different type, according to criteria such as rock type, vegetation cover, land use, administrative region, socioeconomic level, government building zone type, or anatomical subdivision of tissue. The map itself is a tessellation of the spatial domain. The associated spatial covariate $J(u)$ tells us which type or category applies to each given location u, so that J is a spatial covariate with categorical values.

In some cases the original data will be in 'vector graphics' form, giving the spatial coordinates of the boundaries of the regions of each type. To preserve accuracy, vector data should not be discretised. The boundary polygons should be converted to a tessellation (object of class `"tess"`) which can then be passed directly as a covariate to ppm.

If the categorical data are provided as a pixel image, it is important to ensure that the pixel values are recognised as categorical values. Printing or summarising the image in question should indicate that it is `factor`-valued.

A numerical covariate Z can also be converted into a factor by dividing the range of values into a few bands, and treating each band of values as a category. This is often a useful exploratory tool, as we saw in Section 6.6.2.

9.3.4.1 Gorilla nest data

The `gorillas` dataset comes from a study [272] in the Kagwene Gorilla Sanctuary, Cameroon, by the Wildlife Conservation Society Takamanda-Mone Landscape Project (WCS-TMLP). A detailed description and analysis of the data is reported in [273]. The data were kindly contributed to `spatstat` by Dr Funwi-Gabga Neba. The collaboration of Prof Jorge Mateu is gratefully acknowledged.

The data include the marked point pattern of gorilla nest sites `gorillas`, and auxiliary data in a list `gorillas.extra` containing seven pixel images of spatial covariates. We rescale the data to kilometres:

```
> gor <- rescale(gorillas, 1000, unitname="km")
> gor <- unmark(gor)
> gex <- lapply(gorillas.extra, rescale,
               s=1000, unitname="km")
```

Figure 9.7 shows the spatial locations of the (unmarked) gorilla nests, and the covariate `vegetation` which reports the vegetation or cover type.

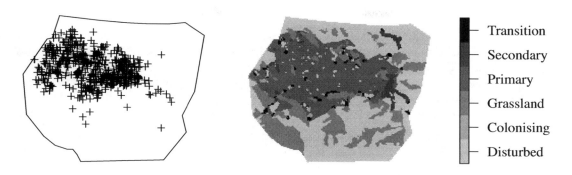

Figure 9.7. *Gorillas data.* Left: *gorilla nest locations.* Right: *vegetation type.*

To save space on the printed page, we will abbreviate the names of the covariates, and the names of the levels of factors. We could use `substr` to extract the first few characters of each name, or `abbreviate` to convert the names to shorthand.

```
> names(gex)
[1] "aspect"     "elevation"  "heat"       "slopeangle" "slopetype"
[6] "vegetation" "waterdist"
> shorten <- function(x) substr(x, 1, 4)
> names(gex) <- shorten(names(gex))
> names(gex)
[1] "aspe" "elev" "heat" "slop" "slop" "vege" "wate"
> names(gex)[4:5] <- c("sang", "styp")
> names(gex)
[1] "aspe" "elev" "heat" "sang" "styp" "vege" "wate"
> isfactor <- !sapply(lapply(gex, levels), is.null)
> for(i in which(isfactor))
      levels(gex[[i]]) <- shorten(levels(gex[[i]]))
> levels(gex$vege)
[1] "Dist" "Colo" "Gras" "Prim" "Seco" "Tran"
```

The following command fits a simple Poisson model in which the intensity is constant in each vegetation type, but may depend on vegetation type:

```
> ppm(gor ~ vege, data=gex)
Nonstationary Poisson process

Log intensity:  ~vege

Fitted trend coefficients:
(Intercept)     vegeColo     vegeGras     vegePrim     vegeSeco     vegeTran
     2.3238       2.0817      -0.7721       2.1148       1.1341       1.6151


              Estimate   S.E. CI95.lo CI95.hi Ztest    Zval
(Intercept)     2.3238 0.1060  2.1161  2.5316   *** 21.923
vegeColo        2.0817 0.5870  0.9312  3.2322   ***  3.546
vegeGras       -0.7721 0.2475 -1.2571 -0.2871    ** -3.120
vegePrim        2.1148 0.1151  1.8892  2.3403   *** 18.373
```

```
vegeSeco      1.1341 0.2426  0.6586  1.6096    ***  4.675
vegeTran      1.6151 0.2647  1.0964  2.1339    ***  6.102
```

For models which involve a factor covariate, the interpretation of the fitted coefficients depends on which *contrasts* are in force, as specified by getOption("contrasts"). We explain more about this below. By default the 'treatment contrasts' are assumed. This means that — if an intercept term is present — the coefficient associated with the first level of the factor is taken to be zero, and the coefficients associated with the other levels are effectively differences relative to the first level. In the model above, the fitted log intensities for each vegetation type are calculated as follows:

TYPE	RELEVANT COEFFICIENTS	LOG INTENSITY
Disturbed	(Intercept)	2.324
Colonising	(Intercept) + vegeColo	$2.324 + 2.082$
Grassland	(Intercept) + vegeGras	$2.324 - 0.7721$
Primary	(Intercept) + vegePrim	$2.324 + 2.115$
Secondary	(Intercept) + vegeSeco	$2.324 + 1.134$
Transition	(Intercept) + vegeTran	$2.324 + 1.615$

Rather than relying on such interpretations, it would be prudent to use the command predict to compute predicted values of the model, as explained in Section 9.4.1 below.

Another way to fit the same model is to remove the intercept, so that a single coefficient will be associated with each level of the factor.

```
> fitveg <- ppm(gor ~ vege - 1, data=gex)
> fitveg
Nonstationary Poisson process

Log intensity:  ~vege - 1

Fitted trend coefficients:
vegeDist vegeColo vegeGras vegePrim vegeSeco vegeTran
   2.324    4.406    1.552    4.439    3.458    3.939

         Estimate    S.E. CI95.lo CI95.hi Ztest     Zval
vegeDist    2.324 0.10600   2.116   2.532   *** 21.923
vegeColo    4.406 0.57735   3.274   5.537   ***  7.631
vegeGras    1.552 0.22361   1.114   1.990   ***  6.940
vegePrim    4.439 0.04486   4.351   4.527   *** 98.952
vegeSeco    3.458 0.21822   3.030   3.886   *** 15.846
vegeTran    3.939 0.24254   3.464   4.414   *** 16.241
> exp(coef(fitveg))
vegeDist vegeColo vegeGras vegePrim vegeSeco vegeTran
   10.21    81.90     4.72    84.66    31.75    51.37
```

For this simple model, where intensity is constant inside each region defined by vegetation type, an equivalent way to fit the same model is to estimate the intensities using quadrat counts. To check this, we convert the pixel image to a tessellation and apply quadrat counting:

```
> vt <- tess(image=gex$vege)
> intensity(quadratcount(gor, tess=vt))
tile
  Dist   Colo   Gras   Prim   Seco   Tran
10.201 69.155  4.781 84.012 32.651 50.922
```

The discrepancies are due to discretisation effects: refitting the model by calling ppm with nd=512 gives results consistent with the quadrat-counting estimates.

9.3.4.2 Factor effects and contrasts

To understand fully the results obtained for the `gorillas` data above, we need to know more about the handling of factors by the model-fitting code in R. If `f` is a factor then `X ~ f` specifies a point process model in which the intensity depends on the level of `f`. The model is

$$\lambda(u) = \exp(\mu + \alpha_{J(u)}) \tag{9.14}$$

where $J(u)$ is the level of the factor `f` at the location u. The parameters of this model are the *intercept* μ and the *effects* $\alpha_1, \ldots, \alpha_L$ of the different factor levels. Here we are assuming the factor has L different levels numbered 1 to L. For the ith level of the factor, the corresponding value of the intensity is $e^{\mu + \alpha_i}$.

While the model (9.14) is conceptually useful, it has the practical drawback that it is *overparameterised*. If there are L different levels of the factor, then there are $L+1$ parameters to be estimated from data, but only L different values of the intensity on the right-hand side of (9.14). In order to fit the model we need to reduce the number of parameters by 1.

One option is to remove the intercept parameter μ. The model is then

$$\lambda(u) = \exp(\alpha_{J(u)}) \tag{9.15}$$

so that the intensity value for the ith level of the factor is e^{α_i}. There are now only L parameters, the effects $\alpha_1, \ldots, \alpha_L$ of the factor levels. The model `ppm(gor ~ vege - 1)` has this form, so the fitted coefficients are the fitted estimates of the log intensity for each type of vegetation.

The default behaviour in R is to retain the intercept and instead to constrain the effect of the first level to be zero, $\alpha_1 = 0$. This convention is called the *'treatment contrast'*; it makes sense when the first level of the factor is a baseline or control, while the other levels are different possible 'treatments' that could be applied to a subject in an experiment. The model (9.14) then states that the intensity value for the first level is e^{μ}, while the intensity value for another level i is $e^{\mu + \alpha_i}$, so the coefficient α_i is the *difference* in effect between level i and the first level, on a log scale. The table on page 316 shows this calculation. An advantage of the treatment contrast is that we are often interested in whether there is any difference between the intensities associated with different factor levels: this can be assessed by testing the hypothesis $\alpha_i = 0$ for each $i > 1$.

In its linear modelling procedures, the commercial statistics package SAS® constrains the last coefficient α_L to be 0. Another commonly used constraint is $\alpha_1 + \ldots + \alpha_L = 0$. In R one can choose the constraint that is used by specifying the *contrast* or linear expression that will be set to zero. This is done using the `options` function, with a call of the form:

```
options(contrasts=c(arg1,arg2))
```

where `arg1`, `arg2` are strings which specify the contrasts to be used with 'unordered' and 'ordered' factors, respectively. All the factors discussed in this book are 'unordered'. Note that if you do set the contrasts, you have to specify both `arg1` and `arg2` even though you only care about `arg1`.

The possible values for either argument are the character strings `"contr.sum"`, `"contr.poly"`, `"contr.helmert"`, `"contr.treatment"` and `"contr.SAS"`. The default for `arg1` is the 'treatment contrasts' `"contr.treatment"` which impose the constraint $\alpha_1 = 0$ in the context of a single categorical predictor. We use this default throughout the book. If you are unsure what contrasts are currently in force, type `getOption("contrasts")`.

It is important to remember that models obtained by employing different constraints are *precisely equivalent*, although the interpretation of the parameter estimates is different.

The overparameterised form (9.14) of the model allows one to conveniently specify models that are of genuine interest and that would be very hard to specify otherwise. We elaborate on this when we come to models with two (or more) categorical predictors.

The model-fitting software in R converts logical variables to categorical variables, so that the variable `greenstone` in the models above is converted to a `factor` with levels `TRUE` and `FALSE`. The intercept term in the model is associated with the first level of the factor, which is `FALSE`, since this comes alphabetically before `TRUE`. Happily this convention gives a sensible result here.

9.3.5 Additive models

We now start looking at models which involve more than one 'original' covariate. The *Beilschmiedia* data (see Sections 9.3.1 and 9.3.2.1) include two (numerical) covariates, the terrain elevation and terrain slope. The simplest loglinear model depending on both covariates is the additive model

$$\lambda(u) = \exp(\beta_0 + \beta_1 E(u) + \beta_2 S(u)) \tag{9.16}$$

where $\beta_0, \beta_1, \beta_2$ are parameters to be estimated, $E(u)$ is the terrain elevation in metres at location u, and $S(u)$ is the terrain slope (a dimensionless gradient). Additive models are specified by joining the relevant model terms together using +:

```
> fitadd <- ppm(bei ~ elev + grad, data=bei.extra)
> fitadd
Nonstationary Poisson process
Log intensity:  ~elev + grad
Fitted trend coefficients:
(Intercept)          elev          grad
   -8.55862       0.02141       5.84104
```

The interpretation of the fitted model is straightforward, since both covariates are continuous numerical quantities. On a level parcel of terrain (slope zero) at sea level (elevation zero) the intensity of *Beilschmiedia* trees would be $e^{-8.559} = 0.0001918$ trees per square metre. For each additional metre of terrain elevation, the intensity of increases by a factor $e^{1 \times 0.02141} = 1.022$ or about 2.2 percent. For each additional increase of (say) 0.1 units in slope, the intensity increases by a factor $e^{0.1 \times 5.841} = 1.793$. These two effects are *additive* on the scale of the linear predictor: elevation increases by 1 metre and slope increases by 0.1 units, then the *log* intensity increases by the elevation effect 1×0.02141 **plus** the slope effect 0.1×5.841.

To assess the relative 'importance' of the slope and elevation variables in the model, we need to account for the range of values of each variable. The terrain slope varies from zero to `max(grad)` = 0.3285 so the effect of slope varies by a factor of $\exp(5.841 \times 0.3285) = 6.813$ between flattest and steepest slopes. Terrain elevation ranges between 119.8 and 159.5 metres, so the effect of elevation varies by a factor of $\exp(0.02141(159.5 - 119.8)) = 2.34$ between lowest and highest elevations. Hence the slope effect is more 'important' in magnitude.

For the Murchison data we have two covariates, namely `dfault` (a continuous numerical variable) and `greenstone` (a logical covariate). The additive model with these two covariates is

$$\lambda(u) = \exp(\beta_0 + \beta_1 d(u) + \beta_2 G(u)) \tag{9.17}$$

where $\beta_0, \beta_1, \beta_2$ are parameters to be estimated, $d(u)$ is the distance to nearest fault, and $G(u)$ is the indicator which equals 1 inside the greenstone outcrop.

```
> ppm(gold ~ dfault + greenstone,data=mur)
Nonstationary Poisson process
Log intensity:  ~dfault + greenstone
Fitted trend coefficients:
   (Intercept)          dfault greenstoneTRUE
       -6.6171         -0.1038         2.7540
```

The fitted model states that the intensity of gold deposits declines by a factor $e^{-0.1038} = 0.9014$ for every additional kilometre of distance from the nearest fault, and for a given distance, the intensity is $e^{2.754} = 15.71$ times higher inside the greenstone outcrop than it is outside the greenstone.

Similarly we could build an additive model with two factor covariates, such as

```
> ppm(gor ~ vege + styp, data=gex)
```

which states that the log intensity of gorilla nesting sites is the sum of an effect due to vegetation/cover type and an effect due to slope type. Equivalently the *intensity* is a vegetation effect *multiplied by* a slope effect.

Note that the model `bei ~ grad + I(grad^2)` discussed in Section 9.3.2 is also an additive model. It adds the effects of the two covariates `grad` and `I(grad^2)`. Although the two covariates are closely related, nevertheless this is formally an additive model.

Finally, note that adding the same covariate twice has no effect in a model formula: `grad + grad` is equivalent to `grad`.

9.3.6 Modelling spatial trend using Cartesian coordinates

The Cartesian coordinates x and y can also serve as spatial covariates. They are particularly useful for investigating spatial trend when there are no relevant covariate data, or perhaps when it is suspected that the intensity is not only dependent on the available covariates. A wide class of models can be built up by combining the Cartesian coordinates in convenient ways.

We illustrate this idea using the full Japanese Pines data of Ogata and Numata,

```
> jpines <- residualspaper[["Fig1"]]
```

A plot of these data (Figure 1.3 on page 4) suggests that they exhibit spatial inhomogeneity. No auxiliary covariate data are available, so we resort to the Cartesian coordinates. For brevity we shall write the intensity function for such models in the form $\lambda((x,y))$ rather than the more long-winded form '$\lambda(u)$ where u is the point with coordinates (x,y)'. For instance an inhomogeneous Poisson model with an intensity that is first order loglinear in the Cartesian coordinates will be written as $\lambda_\theta((x,y)) = \exp(\theta_0 + \theta_1 x + \theta_2 y)$. To fit such a model to the Japanese Pines data simply type

```
> ppm(jpines ~ x + y)
Nonstationary Poisson process
Log intensity:   ~x + y
Fitted trend coefficients:
(Intercept)            x            y
   0.591840     0.014329     0.009644
```

Here the symbols x and y are reserved variable names that always indicate the Cartesian coordinates. The fitted intensity function is

$$\lambda_\theta((x,y)) = \exp(0.5918 + 0.01433x + 0.009644y).$$

To fit an inhomogeneous Poisson model with an intensity that is log-quadratic in the Cartesian coordinates, i.e. such that $\log \lambda_\theta((x,y))$ is a quadratic in x and y:

```
> ppm(jpines ~ polynom(x,y,2))
Nonstationary Poisson process
Log intensity:   ~x + y + I(x^2) + I(x * y) + I(y^2)
Fitted trend coefficients:
(Intercept)           x           y        I(x^2)      I(x * y)       I(y^2)
   0.130330    0.461497   -0.200945     -0.044738      0.001058     0.020399
```

Notice that the expression `polynom(x,y,2)` has been syntactically expanded into

$$x + y + I(x^2) + I(x*y) + I(y^2)$$

This is a special `spatstat` trick and applies only to the model formulae passed to `ppm` or `kppm`.

In the same vein we could fit a log-cubic polynomial `polynom(x,y,3)` and so on. An alternative to the full polynomial of order 3, which has 10 coefficients, is the *harmonic* polynomial `harmonic(x,y,3)` which has only 7 coefficients.

Any transformation of the Cartesian coordinates can also be used as a covariate. For example, to fit a model with constant but unequal intensities on each side of the vertical line $x = 0.5$, we simply use the expression `x < 0.5` to make a logical covariate:

```
> ppm(jpines ~ (x < 0.5))
Nonstationary Poisson process
Log intensity:  ~(x < 0.5)
Fitted trend coefficients:
(Intercept)  x < 0.5TRUE
     0.7668       -1.8931
```

9.3.7   Models with offsets

An *offset* is a term in the linear predictor which does not involve any parameters of the model. Offsets are useful when the effect of one variable is already known, or when we want to fit the model relative to a known baseline.

For instance, in some cases it is appropriate to fit an inhomogeneous Poisson model with intensity that is *proportional* to the covariate,

$$\lambda(u) = \kappa Z(u) \tag{9.18}$$

where Z is the covariate and κ is the parameter to be estimated. We called this a 'baseline' model in Section 9.2.3. Taking logarithms, this model is equivalent to

$$\log \lambda(u) = \log \kappa + \log Z(u) = \theta + \log Z(u) \tag{9.19}$$

where $\theta = \log \kappa$ is the only parameter of the model. Note that there is no coefficient in front of the term $\log Z(u)$ in (9.19), so $\log Z(u)$ is an *offset*.

An important example of a baseline covariate $Z(u)$ is the spatially varying density of human population. The spatial point pattern of cases of a rare disease could reasonably be expected to follow a Poisson point process with intensity (9.18), where κ is the (constant) disease risk per head of population.

The Chorley-Ribble data (Figure 1.12 on page 9) give the locations of cases of the rare cancer of the larynx, and a sample of cases of the much more common lung cancer. The smoothed intensity of lung cancer cases can serve as a surrogate for the spatially varying density of the susceptible population.

```
>   lung <- split(chorley)$lung
>   larynx <- split(chorley)$larynx
>   smo <- density(lung, sigma=0.15, eps=0.1, positive=TRUE)
```

Here we have adopted the smoothing bandwidth $\sigma = 0.15km$ chosen by Diggle [224]; the different rules for automatic bandwidth selection give a wide range of results between 0.07 and 2 km. The pixels are `eps=0.1`km wide. The argument `positive=TRUE` ensures that negative or zero pixel values (due to numerical error) are replaced by a small positive number. To avoid numerical instability we shall raise the threshold slightly:

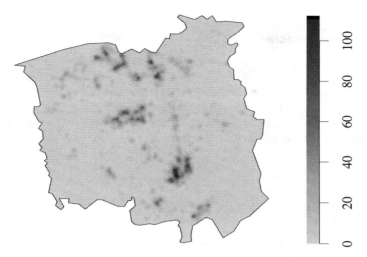

Figure 9.8. *Smooth intensity estimate for lung cancer cases in the Chorley-Ribble data.*

```
> smo <- eval.im(pmax(smo, 1e-10))
```

The resulting pixel image smo, serving as the baseline for our model, is shown in Figure 9.8.

In a model formula, we indicate an offset term by enclosing it in the function offset. Accordingly we fit the constant risk model (9.18) by

```
> ppm(larynx ~ offset(log(smo)))
Nonstationary Poisson process
Log intensity:   ~offset(log(smo))
Fitted trend coefficient:   (Intercept) = -2.939
```

The fitted coefficient (Intercept) is the constant $\log \kappa$ appearing in (9.19), so converting back to the form (9.18), the fitted model is

$$\lambda(u) = e^{-2.939} Z(u) = 0.05292 \, Z(u)$$

where in this case $Z(u)$ is the smoothed intensity of lung cancer cases.

In this example, note that the fitted parameter κ is not the estimated risk of laryngeal cancer per head of population, because the lung data are a subsample from the cancer registry, and lung cancer cases are a subset of the susceptible population. The best way to estimate the risk of laryngeal cancer assuming constant risk is to divide the total number of laryngeal cancer cases by an estimate of the total susceptible population, since the constant risk model does not involve spatial information. The model fitted above is useful mainly for comparison against alternative models where the risk of laryngeal cancer is spatially varying.

Spatial transformations, such as geographic projections, also introduce offset terms. Suppose that a point process **X** has intensity function $\lambda_X(u) = \exp(\beta^T Z(u))$. We change the coordinate system so that the points x_i are mapped to new coordinate positions $y_i = f(x_i)$. Changes of coordinates were discussed in Section 6.5.3: the transformed point process $\mathbf{Y} = f(\mathbf{X})$ has intensity function given by equation (6.17). The new model is

$$\lambda_Y(u) = \exp(\log J(u) + \beta^T Z(f^{-1}(u))) \qquad (9.20)$$

which includes the offset term $\log J(u)$, the log of the Jacobian of the change of coordinates.

In a study of the spatial pattern of tree deaths in a water catchment [684, 145] the spatial pattern of *live* trees, observed at a certain date, was smoothed to give a covariate image $A(u)$ representing the spatially varying density of the forest. Other covariates $Z_1(u), \ldots, Z_p(u)$ were available, including terrain elevation, depth to water table, and seasonal recharge flow of groundwater. Models of the form

$$\lambda(u) = A(u) \exp(\theta_0 + \theta_1 Z_1(u) + \ldots + \theta_p Z_p(u)) \tag{9.21}$$

were fitted, using a model formula of the type `~ offset(log(A)) + Z1 + ... + Zp`. The term in the exponential represents the spatially varying tree death risk.

When fitting a baseline model it is absolutely crucial to express the baseline term in the form `offset(log(baseline))`. If the `offset` is omitted, we get a completely different model, as discussed below.

9.3.8   Power law models

It is sometimes appropriate to fit a *power law* relationship

$$\lambda(u) = \alpha Z(u)^k \tag{9.22}$$

where the exponent k is not known, and must be estimated from data, along with the coefficient α. Taking logarithms, the power law (9.22) is equivalent to

$$\log \lambda(u) = \beta_0 + \beta_1 \log Z(u) \tag{9.23}$$

where $\beta_0 = \log \alpha$ and $\beta_1 = k$. This is a loglinear model with covariate $\log Z$.

In the *Beilschmiedia* data, suppose we believe that the forest density obeys a power law as a function of terrain slope. This can be fitted easily:

```
> ppm(bei ~ log(grad), data=bei.extra)
Nonstationary Poisson process
Log intensity:   ~log(grad)
Fitted trend coefficients:
(Intercept)    log(grad)
    -3.4797       0.5549
```

Note that there is no `offset` here. The printed output describes a model in which the intensity is proportional to the 0.55th power of the terrain slope, or roughly the *square root* of the slope.

Care needs to be taken with power models if the covariate data may include zero values. If the exponent $k = \beta_1$ is positive, then the power law (9.22) implies that, in places where the covariate value $Z(u)$ is zero, the intensity $\lambda(u)$ is also zero, so there is zero probability of observing a random point at such places. If a data point *is* observed at a place where the covariate is zero, this model is invalidated (it has likelihood zero). In the examples above, if there had been a *Beilschmiedia* tree growing on a perfectly level patch of land (i.e. where the terrain slope is zero) then the data would have been inconsistent with the model (9.18) and (9.22). In some cases this may be detected by the ppm function, but in many cases it will not be detected until the low-level code throws a rather undignified error. We can reproduce this scenario by artificially assigning a zero value to a pixel where there is a data point:

```
> G <- bei.extra[["grad"]]
> G[bei[42]] <- 0
> ppm(bei ~ log(G))
Error in glm.fit(x = structure(c(1, 1, 1, 1, 1, 1, 1, 1, 1, 1, 1, 1, 1,  :
NA/NaN/Inf in 'x'
```

If the fitted exponent $k = \beta_1$ is negative, then the power law implies that, in places where $Z(u) = 0$, the intensity is *infinite*. This means it is possible that the fitted model may be **improper**. In

Section 9.2.1 we noted that, in order for a Poisson process to be well defined, the integral of the intensity function over the window must be finite. Intensity functions of the form (9.22) may not be integrable if the exponent β_1 is negative.

For example with the Murchison data, the covariate $d(u)$, distance to nearest fault, takes zero values along the faults themselves. If we fit a power model,

```
> ppm(gold ~ log(dfault), data=mur)
Nonstationary Poisson process
Log intensity:  ~log(dfault)
Fitted trend coefficients:
(Intercept) log(dfault)
    -5.0576     -0.7188
```

the fitted model says that the intensity is the negative 0.72th power of $d(u)$. This function is not integrable so the model is improper, that is, it is not well defined.

Improper models can occur in other contexts where some values of the covariate are infinite (in this case $Z = 0$ gives $\log Z = -\infty$). An improper model will lead to difficulties with simulation code and various other algorithms.

9.3.9 Models with interaction between covariates

A model which is not additive is said to have 'interaction'. A statistical model involving two covariates A and B is additive if the linear predictor is the sum of a term depending only on A plus another term depending only on B. Otherwise, the model is said to exhibit *interaction* between the covariates A and B. Interaction means that the effect of a change in A depends on the current value of B, and *vice versa*.

Be warned that there is another use of the word 'interaction' in spatial statistics. A point process which is not Poisson is said to exhibit 'interaction' between the *points of the process*. This is a distinct and completely unrelated concept for which the same word is used. We will call this 'interpoint interaction' when the distinction needs to be emphasised.

In the present context we are talking about *interaction between covariates* in a model for the intensity of the point process. In a model formula, the expression A:B represents a particular model term called 'the' interaction between the covariates A and B. This is effectively a new covariate depending on both A and B. Its definition depends, as usual, on the type of variables involved: we explain further below.

Normally, a model that includes an interaction A:B should also include the 'main effects' A and B. Usually it is *possible* to write a model which includes only an interaction term but this makes the results difficult to interpret and reduces their usefulness. How to interpret them, and whether they actually make any sense at all, depends on circumstances.

To include interaction between A and B in a model we would thus generally write the formula in the form X ~ A + B + A:B. The expression A + B + A:B can be abbreviated to A * B which is read as 'A cross B' (and the operator * is referred to as the crossing operator). Examples are discussed below.

Table 9.2 lists operators that can be applied to terms in a model formula.

9.3.9.1 Interaction between two numerical covariates

If A and B are both numerical covariates, the expression A:B is interpreted as simply their numerical product, A times B. This makes sense: there is 'interaction' between A and B if the rate of change of the response with respect to A depends upon the value of B (and vice versa). This will be the case if the response depends on the product of A and B.

Thus, if A and B are numerical covariates, A:B is equivalent to I(A * B) and A * B is equivalent to X ~ A + B + I(A*B). Some analysts prefer to use the expanded forms to prevent confusion.

OPERATOR	EXAMPLE	MEANING
+	+A	include this term
–	–A	remove this term
:	A:B	include interaction between A and B
*	A*B	include these terms and their interaction
^	(A+B)^3	include these terms and interactions up to order 3
/	A/B	nesting: B nested within A
%in%	A %in% B	interaction between A and B
|	A|B	conditioning: A given B

Table 9.2. *Operators in model formulae. (The operator '|' is only recognised by some functions.)*

For the *Beilschmiedia* data, we have already looked at an additive model involving the terrain elevation elev and terrain slope grad. Fitting a model with interaction between these covariates is just as easy:

```
> fit <- ppm(bei ~ elev * grad, data=bei.extra)
> fit
Nonstationary Poisson process
Log intensity:  ~elev * grad
Fitted trend coefficients:
(Intercept)          elev          grad     elev:grad
  -4.389734     -0.007115    -36.608966      0.293532
```

9.3.9.2 Interaction between two factors

If A and B are both factors, the formula X ~ A * B represents the (overparameterised) mathematical model

$$\lambda(u) = \exp(\mu + \alpha_{A(u)} + \beta_{B(u)} + \gamma_{A(u),B(u)}) \tag{9.24}$$

where $A(u)$ is the level of factor A at the location u, and $B(u)$ is the level of B at u. This model has parameters μ (the intercept), $\alpha_1, \ldots, \alpha_L$ (the main effects of the levels of A), $\beta_1, \ldots, \beta_M$ (the main effects of the levels of B), and $\gamma_{1,1}, \ldots, \gamma_{L,M}$ (the interaction effects for each combination of levels of A and B). At a location u where $A(u) = i$ and $B(u) = j$, the predicted intensity is $\exp(\mu + \alpha_i + \beta_j + \gamma_{i,j})$. Compared with the additive model

$$\lambda(u) = \exp(\mu + \alpha_{A(u)} + \beta_{B(u)}), \tag{9.25}$$

the interaction model (9.24) has an additional 'synergy' or 'catalysis' term between the individual covariates (which could be negative; a 'counter-synergy') of the form $\gamma_{A(u),B(u)}$ which corresponds to the interaction term A:B.

The gorillas dataset (page 314) is supplied with factor covariates vegetation and heat. A model involving interaction between these two factors is:

```
> ppm(gor ~ vege * heat, data=gex)
Nonstationary Poisson process
Log intensity:   ~vege * heat
Fitted trend coefficients:
      (Intercept)           vegeColo           vegeGras           vegePrim
           2.6687           -13.9713            -1.1268             1.7189
         vegeSeco           vegeTran           heatMode           heatCool
           0.6746             0.8753            -0.7432            -0.8605
  vegeColo:heatMode  vegeGras:heatMode  vegePrim:heatMode  vegeSeco:heatMode
```

16.5848	0.7797	0.8326	0.9814
vegeTran:heatMode	vegeColo:heatCool	vegeGras:heatCool	vegePrim:heatCool
1.3636	17.9549	-11.9840	0.9214
vegeSeco:heatCool	vegeTran:heatCool		
-13.7854	NA		

```
*** Model is not valid ***
*** Some coefficients are NA or Inf ***
```

The result gives a fitted log intensity for every possible combination of vegetation type and heat load index. Note that one of the fitted coefficients is NA, because the corresponding combination of vegetation and heat does not occur in the data — at least, not at any of the quadrature points (sample locations used to fit the model).

9.3.9.3 Interaction between factor and numerical covariate

If A is a factor and Z is numerical, the additive model X ~ A + Z is

$$\lambda(u) = \exp(\mu + \alpha_{A(u)} + \beta Z(u)) \tag{9.26}$$

while the full interaction model X ~ A * Z is

$$\lambda(u) = \exp(\mu + \alpha_{A(u)} + \beta Z(u) + \gamma_{A(u)} Z(u)). \tag{9.27}$$

In the additive model (9.26) the parameters $\alpha_1, \ldots, \alpha_L$ are the effects of the different levels of the factor A, while the 'slope' parameter β is the effect of a unit increase in the numerical covariate Z. If we were to plot the linear predictor (the log intensity) as a function of Z for different levels of the covariate, the graph would consist of parallel lines with the same slope β but with different intercepts $\mu + \alpha_i$.

In the interaction model (9.27) there is an extra term $\gamma_{A(u)} Z(u)$ which causes the effect of a unit increase in Z to depend on the level of the factor A. The plot of the linear predictor described above would show lines with *different* slopes $\beta + \gamma_{A_i}$.

The Murchison data (Section 9.3.2.2) include two spatial objects serving as covariate data: a window greenstone and a line segment pattern faults. The greenstone window is automatically interpreted as a logical-valued covariate function, equal to TRUE inside the window. From the faults pattern we have constructed a numerical spatial covariate function dfault. The model with interaction is:

```
> ppm(gold ~ dfault * greenstone, data=mur)
Nonstationary Poisson process
Log intensity:  ~dfault * greenstone
Fitted trend coefficients:
         (Intercept)              dfault        greenstoneTRUE
             -6.0184             -0.2047                2.0013
   dfault:greenstoneTRUE
              0.1674
```

The estimates of the coefficients in (9.27) are $\mu = 315.4$, $\beta = 316.3$, $\alpha_{\text{FALSE}} = 0$, $\alpha_{\text{TRUE}} = 316.5$, $\gamma_{\text{FALSE}} = 0$, and $\gamma_{\text{TRUE}} = 317.6$.

9.3.9.4 Nested interaction

It is sometimes appropriate to fit a model with the formula X ~ A + A:B. That formula can also be rendered as y ~ A/B. The operator / stands for a *nested interaction* and we read A/B as 'B nested within A'.

For example, in a taxonomic classification of organisms, suppose G and S are the Genus and

Species names, treated as factors. The Species name only makes sense when the Genus is specified. We may wish to compare models which depend only on Genus, of the form y ~ G, with models which depend on Species, of the form y ~ G/S.

Nesting a numerical covariate inside a factor is effectively the same as crossing the two covariates. For the Murchison data, we may try nesting the numerical covariate dfault inside the factor greenstone:

```
> ppm(gold ~ greenstone/dfault, data=mur)
Nonstationary Poisson process
Log intensity:  ~greenstone/dfault
Fitted trend coefficients:
              (Intercept)          greenstoneTRUE greenstoneFALSE:dfault
                 -6.01845                 2.00131               -0.20466
  greenstoneTRUE:dfault
                 -0.03728
> ppm(gold ~ greenstone/dfault - 1, data=mur)
Nonstationary Poisson process
Log intensity:  ~greenstone/dfault - 1
Fitted trend coefficients:
          greenstoneFALSE          greenstoneTRUE greenstoneFALSE:dfault
                 -6.01845                -4.01714               -0.20466
  greenstoneTRUE:dfault
                 -0.03728
```

These two models are exactly equivalent, except for the way in which they are parametrised. They are also equivalent to the interaction model with formula gold ~ greenstone * dfault. Different choices of parameterisation make it easier or harder to extract particular kinds of information.

Nesting a factor A inside a numerical covariate Z gives a model where the effect of a unit change in Z depends on the level of A, but the factor A has no effect on the intercept. In the plot of the linear predictor against Z described above, the lines are not parallel, but they all have the same intercept. For the Murchison data, the model gold ~ dfault/greenstone stipulates that the intensity of gold deposits very close to a geological fault (dfault ≈ 0) must be the same inside and outside the greenstone, but as we increase distance from the faults, the intensity may decrease at different rates inside and outside the greenstone:

```
> ppm(gold ~ dfault/greenstone, data=mur)
Nonstationary Poisson process
Log intensity:  ~dfault/greenstone
Fitted trend coefficients:
              (Intercept)            dfault dfault:greenstoneTRUE
                  -4.3472           -0.4972                0.5020
```

Nested interactions also arise in connection with random effects, as discussed in Chapter 16.

In nested designs the operator %in% is sometimes useful. It is equivalent to an interaction: A %in% B is equivalent to A:B.

9.3.10 Formulae involving many variables

A model can involve any number of covariates. In an additive model, the variable names are simply joined by the '+' operator:

```
ppm(Y ~ X1 + X2 + X3 + ... + Xn)
```

This expression could be long and tedious to type out. A useful shortcut is the symbol '.' representing *all* the available covariates. This only works when the covariates are supplied in the argument data (see page 306). The additive model involving all covariates can be fitted using the formula ' Y ~ . ' as in

```
> ppm(gor ~ . , data=gex)
```

To fit the additive model involving all covariates *except* the heat covariate,

```
> ppm(gor ~ . - heat, data=gex)
```

Interactions between more than two predictors ('higher order interactions') are easy to define. For three factors A, B, and C the second-order interactions are A:B, A:C, and B:C. The third-order interaction is A:B:C. The mathematical expressions defining higher-order interactions get increasingly cumbersome to write down, but the ideas are basically the same as for second-order interactions. Higher-order interactions are often difficult to interpret in practical terms.

In the model formula context the + and : operators are what the pure mathematicians (poor dears!) call 'idempotent' operations. That is, A+A is just equal to A, and likewise A:A is equal to A.

The crossing operator * obeys the distributive and associative laws, so that (A + B + C) * (A + B + C) expands to A + B + C + A:B + A:C + B:C, and A * B * C expands to A + B + C + A:B + A:C + B:C + A:B:C.

We have now dealt with the model operators +, -, :, *, and /. The last operator to be mentioned is the power operator ^. If A is any model term, A^2 is equivalent to A * A, while A^3 is equivalent to A * A * A and so on. This is particularly useful when we want to specify all main effects and interactions up to a certain order. For example (A + B + C)^2 is equivalent to (A + B + C) * (A + B + C) which expands to A + B + A:B + A:C + B:C, containing all main effects and second order interactions between the covariates A, B, and C. One could also use the '.' symbol, for example

```
> ppm(gor ~ .^2, data=gex)
```

would fit all main effects and all second-order interaction terms involving all the covariates in the list gex.

> Tip: To check the expansion of a complicated formula f, try update(f, . ~ .) This uses the symbol '.' in another sense, explained in Section 9.4.4.

9.4 Statistical inference for Poisson models

9.4.1 Fitted models

The value returned by the model-fitting function ppm is an object of class ""ppm"" that represents the fitted model. This is analogous to the fitting of linear models (lm), generalised linear models (glm), and so on. There are many standard operations on fitted models in R which can be applied to point process models: these are listed in Table 9.3. That is, these generic operations have methods for the class "ppm". For information on these methods, consult the help for print.ppm, summary.ppm, plot.ppm, etc. The methods are described in the present chapter, except for residuals, influence, and dfbetas which are described in Chapter 11.

Additionally Table 9.4 lists some non-generic functions in the base R system which work on

print	print basic information
summary	print detailed summary information
plot	plot the fitted intensity
predict	compute the fitted intensity
simulate	generate simulated realisations of model
update	re-fit the model
coef	extract the fitted coefficient vector $\hat{\theta}$
vcov	variance-covariance matrix of $\hat{\theta}$
anova	analysis of deviance
logLik	loglikelihood value
formula	extract the model formula
terms	extract the terms in the model
fitted	compute the fitted intensity at data points
residuals	compute residuals
influence	compute likelihood influence diagnostics
dfbetas	compute parameter influence diagnostics
model.matrix	compute the design matrix

Table 9.3. *Standard* R *generic operations which have methods for point process models (class* "ppm"*).*

"ppm" objects. For information on these functions, consult the help for the function itself. Finally Table 9.5 on page 338 lists some generic functions defined in spatstat which apply to "ppm" objects and are useful for Poisson models.

confint	confidence intervals for parameters
step	stepwise model selection
stepAIC	(package "MASS") stepwise model selection
drop1	one step model deletion
add1	one step model augmentation
AIC	Akaike Information Criterion

Table 9.4. *Functions in the base* R *system which work on* "ppm" *objects.*

The print method gives important information about the model structure and the fitted model parameters. For Poisson models it also gives a table of parameter estimates with standard errors, confidence intervals, and the result of a test that the parameter is significantly different from zero. For example with the *Beilschmiedia* data we may do the following.

```
> beikm <- rescale(bei, 1000, unitname="km")
> bei.extrakm <- lapply(bei.extra, rescale, s=1000, unitname="km")
> fitkm <- ppm(beikm ~ x + y)
> fitkm
Nonstationary Poisson process

Log intensity:  ~x + y

Fitted trend coefficients:
(Intercept)          x          y
    9.0910    -0.8031     0.6496

            Estimate    S.E. CI95.lo CI95.hi Ztest     Zval
(Intercept)   9.0910 0.04306  9.0066  9.1754   *** 211.128
```

```
x                   -0.8031 0.05863 -0.9180 -0.6882   *** -13.698
y                    0.6496 0.11571  0.4228  0.8764   ***   5.614
```

This is the fitted model with intensity function

$$\lambda_{\theta}((x,y)) = \exp\left(\theta_0 + \theta_1 x + \theta_2 y\right) \tag{9.28}$$

where spatial coordinates x and y are measured in kilometres and, for example, the estimate of θ_1 is $\hat{\theta}_1 = -0.8031$ with standard error $\mathrm{se}(\hat{\theta}_1) = 0.05863$ and 95% confidence interval $[-0.918, -0.6882]$. The amount of information displayed, and the layout, depend on `spatstat.options('terse')` and `spatstat.options('print.ppm.SE')`.

The fitted coefficients of the model can be extracted by `coef.ppm`, a method for the generic function `coef`:

```
> coef(fitkm)
(Intercept)          x           y
    9.0910      -0.8031      0.6496
```

The `plot` method is useful for initial inspection of a fitted model, because it offers a wide range of displays. Figure 9.9 shows the result of the command

```
> plot(fitkm, how="image", se=FALSE)
```

with modifications to the graphics parameters.

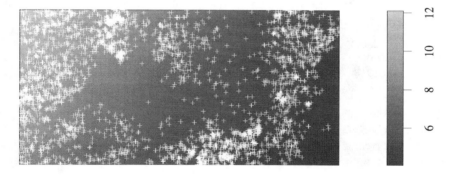

Figure 9.9. *Result of plotting a fitted Poisson model.*

Very detailed information about the model can be printed by typing `summary(fitkm)`. The result of `summary.ppm` is an object of class `summary.ppm` which can be printed and manipulated in other ways. For example the table of parameter estimates, standard errors, and confidence intervals is obtained by

```
> coef(summary(fitkm))
            Estimate    S.E. CI95.lo CI95.hi Ztest    Zval
(Intercept)   9.0910 0.04306  9.0066  9.1754   *** 211.128
x            -0.8031 0.05863 -0.9180 -0.6882   *** -13.698
y             0.6496 0.11571  0.4228  0.8764   ***   5.614
```

9.4.2   Standard errors and confidence intervals for parameters

The accuracy of the fitted model coefficients can be analysed by standard asymptotic theory as explained in Section 9.7 and in [396, 565]. For sufficiently large samples from a loglinear Poisson model (9.6), the estimated parameters $\hat{\theta}$ have approximately a multivariate normal distribution with mean equal to the true parameters θ and variance-covariance matrix $\mathrm{var}\,\hat{\theta} = I_\theta^{-1}$, where I_θ is the Fisher information matrix

$$I_\theta = \int_W Z(u)\,Z(u)^\mathsf{T}\lambda_\theta(u)\,\mathrm{d}u. \qquad (9.29)$$

The estimated variance-covariance matrix $\hat{\mathrm{var}}\,\hat{\theta} = I_{\hat\theta}^{-1}$ is computed by vcov.ppm, a method for the generic function vcov.

```
> vcov(fitkm)
            (Intercept)          x          y
(Intercept)    0.001854 -1.491e-03 -3.528e-03
x             -0.001491  3.438e-03  1.208e-08
y             -0.003528  1.208e-08  1.339e-02
```

The diagonal of this matrix contains the estimated variances of the individual parameter estimates $\theta_0, \theta_1, \theta_2$, so the standard errors are:

```
> sqrt(diag(vcov(fitkm)))
(Intercept)          x          y
    0.04306    0.05863    0.11571
```

Confidence intervals for the parameters θ can be obtained from the standard function confint:

```
> confint(fitkm, level=0.95)
                2.5 %  97.5 %
(Intercept)    9.0066  9.1754
x             -0.9180 -0.6882
y              0.4228  0.8764
```

and these could also be constructed manually from the parameter estimate coef(fitkm) and the standard error sqrt(diag(vcov(fitkm))).

The estimated correlation between individual parameter estimates can be useful in detecting collinearity and confounding. Correlations can be computed using vcov.ppm:

```
> co <- vcov(fitkm, what="corr")
> round(co, 2)
            (Intercept)     x     y
(Intercept)        1.00 -0.59 -0.71
x                 -0.59  1.00  0.00
y                 -0.71  0.00  1.00
```

This suggests fairly strong negative correlation between the intercept parameter estimate $\hat{\theta}_0$ and the estimates of the coefficients of x and y. However, the correlation between an intercept estimate and a slope estimate depends on the origin for the covariate. We get a different answer if we place the coordinate origin in the centre of the study region:

```
> fitch <- update(fitkm, . ~ I(x-0.5) + I(y-0.25))
> co <- vcov(fitch, what="corr")
> round(co, 2)
```

	(Intercept)	I(x - 0.5)	I(y - 0.25)
(Intercept)	1.00	0.23	-0.09
I(x - 0.5)	0.23	1.00	0.00
I(y - 0.25)	-0.09	0.00	1.00

Standard errors and confidence intervals for these parameters also depend on the origin for the covariate.

9.4.3 Prediction

For any Poisson model with intensity $\lambda(u) = \lambda_\theta(u)$ where θ is a parameter vector, the *fitted intensity* or *predicted intensity* is the function $\lambda_{\hat\theta}(u)$ obtained by substituting the fitted parameter estimates into the intensity formula.

The fitted intensity is computed by `predict.ppm`, a method for the generic function `predict`. By default it computes the fitted intensity at a regular grid of locations yielding a pixel image:

```
> fit <- ppm(bei ~ polynom(grad, elev, 2), data=bei.extra)
> lamhat <- predict(fit)
```

The result is shown in the left panel of Figure 9.10. Alternatively `predict.ppm` can be used to obtain predicted values at any locations, including the original data points:

```
> lamB <- predict(fit, locations=bei)
```

Figure 9.10. *Contour plots of the fitted intensity* `predict(fit)` (Left) *and standard error* `predict(fit, se=TRUE)$se` (Right).

There are many useful ways to plot the fitted intensity function, using the tools described in Section 4.1.4 of Chapter 4. The front cover illustration of this book (shown in greyscale in Figure 9.11) is a perspective view of the rainforest terrain, coloured according to the value of the fitted intensity, with the *Beilschmiedia* tree locations added. It was produced by the following code:

```
> M <- persp(bei.extra$elev, colin=lamhat, colmap=topo.colors,
        shade=0.4, theta=-55, phi=25, expand=6,
        box=FALSE, apron=TRUE, visible=TRUE)
> perspPoints(bei, Z=bei.extra$elev, M=M, pch=20, cex=0.1)
```

More detail about the use of `persp.im` and `perspPoints` can be found on page 88.

For a loglinear model (9.6), the asymptotic variance of the predicted intensity $\lambda_{\hat\theta}(u)$ at a given location u is

$$\mathrm{var}\,\hat\lambda(u) \sim \mathbf{Z}(u)\mathbf{I}_\theta^{-1}\mathbf{Z}(u)^\top \lambda_\theta(u). \tag{9.30}$$

The right panel of Figure 9.10 shows the standard error, i.e. the square root of this variance, which is computed by `predict(fit, se=TRUE)$se`.

A confidence interval for the true value of $\lambda(u)$ at each location u can be computed by

Figure 9.11. *Perspective view of rainforest terrain shaded according to intensity of fitted model, with original point pattern superimposed. Colour version on front cover of book.*

`predict(fit, interval="confidence")`. This yields a list of two images giving the lower and upper limits of the confidence interval.

It is also possible to plot the 'effect' of a single covariate in the model. The command `effectfun` computes the intensity of the fitted model as a function of one of its covariates. An example was shown in Figure 9.3.

We may also be interested in predicting the number of points in a spatial region B. For a Poisson process, the expected number of points in B is the integral of the intensity over B, equation (9.1) on page 301. The fitted mean (expected) number of points in B is

$$\hat{\mu}(B) = \int_B \hat{\lambda}(u)\,du.$$

This can be computed using `predict.ppm` by setting `type="count"` and specifying the region B as `window`. To find the expected number of trees at elevations below 130 metres:

```
> B <- levelset(bei.extra$elev, 130)
> predict(fit, type="count", window=B)
[1] 37.4
```

The asymptotic variance of $\hat{\mu}(B)$ is

$$\text{var}\,\hat{\mu}(B) \sim M(B)\,\boldsymbol{I_\theta}^{-1}M(B)^\top \tag{9.31}$$

where

$$M(B) = \int_B \mathbf{Z}(u)\lambda_{\boldsymbol{\theta}}(u)\,du \tag{9.32}$$

is the expected sum of the covariate values at the data points, by Campbell's formula (6.11). This allows us to compute the standard error for the estimate of the mean number of trees:

```
> predict(fit, B, type="count", se=TRUE)
$estimate
[1] 37.4

$se
[1] 3.631
```

or a confidence interval for the true expected number (9.1):

```
> predict(fit, B, type="count", interval="confidence")
 2.5% 97.5%
30.28 44.51
```

In this case it is also meaningful to compute a *prediction interval* for the random number of trees in the specified region:

```
> predict(fit, B, type="count", interval="prediction")
 2.5% 97.5%
   23    51
```

It is not clear exactly what this means for the *Beilschmiedia* data. A 95% prediction interval for the number of points in region B is designed so that, if the experiment were repeated, there is a 95% probability that the random value of $n(\mathbf{X} \cap B)$ would lie in the interval. The concept of 'repeating the experiment' makes more sense if there is a time dimension — for example if the point pattern is a record of accidents reported in a specific year. Assuming independence between successive years of observations, there is a 95% probability that next year's count of accidents in the same region will lie between the calculated limits. The calculation assumes that the model is true, but takes into account the uncertainty in the model parameters due to estimation from the data.

9.4.4 Updating a model

In data analysis we typically fit several different candidate models to the same data [181, 180]. The generic function `update` makes it easy to do this. It modifies a fitted model, for example by changing the model formula and re-fitting the model.

The method `update.ppm` is provided in `spatstat` for updating a fitted point process model. The syntax is

```
update(object, ...)
```

where `object` is a fitted point process model (class `"ppm"`) and the subsequent arguments '...', if any, determine how the model should be changed. The result is another fitted model of the same kind.

The syntax `update(object)` makes sense, and causes the `object` to be re-fitted. The result is different from the original `object` if any of the data used to fit the original model have changed.

The second argument of `update` may be a formula, specifying the new model formula to be used in re-fitting the model.

```
> fitcsr <- ppm(bei ~ 1, data=bei.extra)
> update(fitcsr, bei ~ grad)
Nonstationary Poisson process
Log intensity:  ~grad
            Estimate    S.E. CI95.lo CI95.hi Ztest     Zval
(Intercept)   -5.391 0.03002  -5.449  -5.332   *** -179.58
grad           5.022 0.24540   4.541   5.503   ***   20.46
```

Notice that we needed to provide the `data` argument to the first model `fitcsr`, even though that model does not depend on any covariates, in order for the covariate `grad` to be available for the updated model. This would have been unnecessary if `grad` had been an existing object in the R session.

The new formula may include the symbol '.' representing *'what was there before'*. To keep the same left-hand side of the formula, use '.' on the left-hand side:

```
> fitgrad <- update(fitcsr, . ~ grad)
```

To modify the right-hand side of the formula, use '.' on the right-hand side, modifying it with the model operators:

```
> fitall <- update(fitgrad, . ~ . + elev)
> fitall
Nonstationary Poisson process
Log intensity:  ~grad + elev
            Estimate    S.E.  CI95.lo  CI95.hi Ztest     Zval
(Intercept) -8.55862 0.341101 -9.22717 -7.89008   *** -25.091
grad         5.84104 0.255861  5.33956  6.34252   ***  22.829
elev         0.02141 0.002288  0.01693  0.02589   ***   9.358
```

To remove the intercept, use -1 or +0:

```
> fitp <- update(fitall, . ~ . - 1)
```

The symbolic manipulation of formulae involving the symbol '.' is handled by the update method for formulae, `update.formula`. This is quite useful in its own right as a quick way to see the effect of a change in a formula:

```
> update(gor ~ (heat + vege)^2,  . ~ . - heat:vege)
gor ~ heat + vege
```

and simply to expand a complicated formula:

```
> update(gor ~ (heat + vege + sang)^3, . ~ .)
gor ~ heat + vege + sang + heat:vege + heat:sang + vege:sang +
    heat:vege:sang
```

> *Warning:* the operator '-' *does not remove offset terms* in model formulae. A command like
> `update(fit, . ~ . - offset(A))` does not delete the term `offset(A)`.

9.4.5 Testing significance of a term in the model

Often we want to decide whether a particular covariate Z has an effect on the point pattern (see Sections 6.6–6.7). In a modelling context this can be studied by comparing two different models, which are identical except that one of the models includes a term depending on the covariate Z. For example, to decide whether the *Beilschmiedia* forest density depends on terrain slope, the two models would be:

```
> fit0 <- ppm(bei ~ 1)
> fit1 <- ppm(bei ~ grad, data=bei.extra)
```

We can assess the strength of evidence for the covariate effect by performing a *significance test* (Chapter 10) comparing the two models. The null hypothesis is the smaller model `fit0` which does not include the covariate effect; the alternative hypothesis is the larger model `fit1` including the covariate effect.

The theoretically optimal technique is the Likelihood Ratio Test (Section 10.3.2) which can be carried out using **analysis of deviance**, the generalisation of analysis of variance. It is performed in `spatstat` using `anova.ppm`, a method for the generic function `anova`:

```
> anova(fit0, fit1, test="LR")
```

```
Analysis of Deviance Table
Model 1: ~1              Poisson
Model 2: ~grad           Poisson
  Npar Df Deviance Pr(>Chi)
1   1
2   2  1      382    <2e-16 ***
---
Signif. codes:  0 '***' 0.001 '**' 0.01 '*' 0.05 '.' 0.1 ' ' 1
```

This example shows the likelihood ratio test of the null hypothesis of CSR against the alternative of an inhomogeneous Poisson process with intensity that is a loglinear function of the slope covariate (9.7). The *p*-value, shown under the heading `Pr(>Chi)`, is extremely small, indicating rejection of CSR in favour of the alternative. Note that 2e-16 or 2×10^{-16} is the smallest detectable difference between 'real numbers'[4] on the 32-bit computer which produced this output, so the output says that the *p*-value is effectively zero.

An advantage of this approach is that we can easily allow for the effects of other covariates, by including these 'other covariates' in both models. For example, we could test the significance of a terrain slope effect after accounting for the effect of terrain elevation:

```
> fit2e <- ppm(bei ~ polynom(elev, 2), data=bei.extra)
> fit2e1g <- update(fit2e, . ~ . + grad)
> anova(fit2e, fit2e1g, test="LR")
Analysis of Deviance Table
Model 1: ~elev + I(elev^2)           Poisson
Model 2: ~elev + I(elev^2) + grad         Poisson
  Npar Df Deviance Pr(>Chi)
1   3
2   4  1      418    <2e-16 ***
---
Signif. codes:  0 '***' 0.001 '**' 0.01 '*' 0.05 '.' 0.1 ' ' 1
```

Similarly, for the Chorley-Ribble data (page 9) the key question is whether distance to the incinerator has an effect on the intensity of laryngeal cancer, after allowing for the fact that the intensity of laryngeal cancer cases will depend on the spatially varying density of the population at risk. A likelihood ratio test can be used to compare two models, both involving a term related to the population density, but only one model including a term related to distance from the incinerator.

9.4.6 Model selection using AIC

Model selection [154] is the task of choosing the 'best fitting' statistical model from amongst several competing models for the same dataset.

Hypothesis testing, as described above, can be regarded as a special case of model selection. Alternatives to hypothesis testing are sometimes required. For example, the likelihood ratio test requires the null hypothesis to be a sub-model of the alternative. Unfortunately the model (9.18), in which forest intensity is proportional to terrain slope, does not include homogeneous intensity as a special case, so we cannot use analysis of deviance to test the null hypothesis of CSR against the alternative of an inhomogeneous Poisson process with intensity (9.18).

Models which are not nested can nevertheless be compared using the *Akaike Information Criterion* [9]

$$\text{AIC} = -2\log L_{max} + 2p \tag{9.33}$$

---

[4].`Machine$double.eps` gives the smallest double-precision floating-point number x that satisfies `1 + x != 1`.

where $L_{max} = L(\widehat{\theta})$ is the maximised likelihood for the model in question, and p the number of parameters for this model. The model with the *lowest* value of AIC is preferred. This strikes a balance between the degree of fit and the complexity of the model.

The R function AIC computes the AIC of any model, including a "ppm" object:

```
> fitprop <- ppm(bei ~ offset(log(grad)),data=bei.extra)
> fitnull <- ppm(bei ~1)
> AIC(fitprop)
[1] 42497
> AIC(fitnull)
[1] 42764
```

In this case the preferred model is fitprop, the model (9.18) with intensity proportional to slope.

Model selection techniques are especially valuable when there are many covariates, or many possible terms which could be included in the model. One very simple approach is 'forward stepwise selection', in which we start with a minimal acceptable model (containing few terms), and add new terms to the model, one by one. In 'backward stepwise selection' we start with a maximal model (containing many terms) which is believed to be adequate, and delete terms from the model, one by one.

We could use the Likelihood Ratio Test as the criterion for deciding whether to add a term (or to delete a term) at each step. However, this often results in models which are smaller than they should be: the customary significance level $\alpha = 0.05$ is an excessively stringent standard of evidence for including a proposed term in the model. A better approach is to compare models using the AIC.

In stepwise model selection, choosing the model with the smaller AIC is equivalent to applying the likelihood ratio test, but taking the critical value to be $2d$ where d is the number of degrees of freedom, i.e. the number of added or deleted parameters. The significance level of this test is $\alpha = 0.157, 0.135, 0.112$ when $d = 1, 2, 3$, respectively, dropping below 0.05 when $d = 8$.

The stats package (included in a standard installation of R) provides functions add1, drop1, and step to perform 'automatic' stepwise model selection. The MASS package provides stepAIC, an improved version of step. These functions can be applied to many kinds of models, including fitted point process models. The function drop1 compares a model with all the sub-models obtained by deleting a single term, and evaluates the AIC for each sub-model. For example, to explore the dependence of the intensity of the Swedish Pines data on the Cartesian coodinates, we could do:

```
> fitxy <- ppm(swedishpines ~ x + y)
> drop1(fitxy)
Single term deletions

Model:
~x + y
        Df AIC
<none>     844
x        1 843
y        1 842
```

The output indicates that the lowest AIC would be achieved by deleting the y term. Similarly add1 compares a model with all the 'super-models' obtained by adding a single term:

```
> fitcsr <- ppm(swedishpines ~ 1)
> add1(fitcsr, ~x+y)
Single term additions

Model:
```

```
~1
          Df AIC
<none>       841
x          1 842
y          1 843
```

The output indicates that CSR has lower AIC than the models in which we add the term x or add the term y.

The function `step` performs stepwise model selection using AIC. By default it performs backward stepwise selection. Starting from a 'maximal' model, the procedure considers each term in the model, and decides whether the term should be deleted. The deletion yielding the biggest reduction in AIC is carried out. This is applied recursively until no more terms can be deleted (i.e. until all further deletions would lead to an *increase* in AIC).

Continuing the example:

```
> fitxy <- ppm(swedishpines ~ x + y)
> fitopt <- step(fitxy)
Start:  AIC=843.6
~x + y

          Df AIC
- y        1 842
- x        1 843
<none>       844

Step:  AIC=841.6
~x

          Df AIC
- x        1 841
<none>       842

Step:  AIC=840.8
~1
> fitopt
Stationary Poisson process
Intensity: 0.007396
              Estimate    S.E. CI95.lo CI95.hi Ztest    Zval
log(lambda)     -4.907  0.1187  -5.139  -4.674   *** -41.35
```

The output here is different from the output of `drop1` or `add1`. At each step in the procedure, the possible deletions are ranked in ascending order of AIC: the deletion giving the greatest reduction in AIC is at the top of the table. The output shows that the term y was first deleted from the model, and then the term x was deleted, leaving only the formula ~1 corresponding to CSR.

In what follows, we set `trace=0` to suppress the progress reports, so the output shows only the final result. Larger values of `trace` give more information about the sequence of models considered. The default is `trace=1`. For brevity one can use `formula.ppm` to extract the model formula:

```
> bigfit <- ppm(swedishpines ~ polynom(x,y,3))
> formula(bigfit)
~x + y + I(x^2) + I(x * y) + I(y^2) + I(x^3) + I(x^2 * y) + I(x * y^2) + I(y^3)
> goodfit <- step(bigfit, trace=0)
> formula(goodfit)
```

```
~x + y + I(x * y) + I(y^2)
> AIC(goodfit)
[1] 841.5
```

The selected model `goodfit` has slightly larger AIC than the constant intensity model in `fitopt`. This illustrates the fact that dropping terms one at a time is not guaranteed to produce the submodel with minimal AIC.

Many other methods for model selection exist [154]. We also warn that there are some perils in automatic model selection techniques: see [470]. See Section 13.5.11 for a discussion of model selection using AIC for *Gibbs* point process models.

9.4.7 Simulating the fitted model

A fitted Poisson model can be simulated automatically using the function `simulate.ppm`. The result of the commands X `<- simulate(fitprop); plot(X[[1]])` is shown in Figure 9.12.

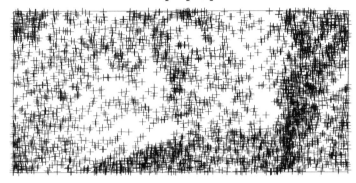

Figure 9.12. *Simulated realisation of a Poisson model fitted to the* Beilschmiedia *data.*

It is also possible to perform conditional simulation (conditional on the number of points, on the configuration of points in a particular subregion, or on the presence of certain points). See Section 13.9.1.4 or the help for `rmhcontrol`.

9.4.8 Additional capabilities

Table 9.5 lists some additional generic functions defined in `spatstat` which have methods for fitted models. For the full list, see Table 13.1 on page 508 and Table 13.5 on page 536, or type `methods(class="ppm")`.

envelope	Simulation envelopes of summary function
berman.test	Berman's tests
cdf.test	Spatial CDF tests
quadrat.test	χ^2 test based on quadrat counts
leverage	Leverage diagnostic
model.images	Pixel images of the canonical covariates
relrisk	Relative risk predicted by model
rhohat	Relative intensity as a function of covariate

Table 9.5. *Some of the generics defined in* `spatstat` *with methods for class* "ppm" *which are useful for Poisson models.*

The R generic function `model.matrix` computes the 'design matrix' of a model. For a Poisson point process model, `model.matrix.ppm` returns a matrix with one row for each quadrature point, and one column for each of the *canonical* covariates appearing in (9.6). Note that, if the model formula involves transformations of the original covariates, then `model.matrix(fit)` gives values of these transformed covariates.

The `spatstat` package also defines a new generic function `model.images` with a method for "ppm" objects. This produces a list of *pixel images* of the canonical covariates.

```
> fit2 <- ppm(bei ~ sqrt(grad) + x, data=bei.extra)
> mo <- model.images(fit2)
> names(mo)
[1] "(Intercept)" "sqrt(grad)"  "x"
```

Additionally `spatstat` has some functions that can be applied to any fitted model (of class "lm", "glm", "ppm", etc.) to understand the structure of the model: `model.covariates` determines which of the original covariates in `data` are used in the model; `model.depends` identifies which of the covariates is involved in each term of the model; `model.is.additive` determines whether the model is additive; and `has.offset` determines whether the model has an offset.

9.5 Alternative fitting methods

Here we review alternative options for fitting the loglinear Poisson model in `spatstat` and related packages.

Faster fitting

If `ppm` is called with the option `method="logi"`, it fits the point process model using conditional logistic regression, as described in Section 9.10. This method is typically faster than the default technique, and often more reliable, but involves randomisation.

More robust fitting

The *lasso* [655] is an improvement to classical regression which increases its robustness against anomalies or extremes in the data. The lasso version of linear regression, for example, finds the parameter value θ that maximises $RSS(\theta) - \alpha\|\theta\|$, where $RSS(\theta)$ is the usual sum of squared errors, $\|\theta\| = \sum_j |\theta_j|$ is the sum of absolute values of all parameters, and α is a tradeoff coefficient. A lasso version of ppm has been implemented in the R package `ppmlasso` [568, 567]. In a nutshell, this approach will fit *the same loglinear model* as is fitted by ppm, but with less sensitivity to anomalous observations. Variable selection for point process models using the lasso has recently been studied [654, 709]. In future there may also be a Stein estimator [156].

9.6 More flexible models

Until this point in the chapter, we have been concerned exclusively with fitting *loglinear* Poisson point process models. These models have a specially convenient structure in which the log intensity is a linear function of the *parameters*. Note that the *covariates* can be quite general functions; so, this is a very wide class of models.

However, there are many applications where we need to fit a model which does not have this loglinear form. An important example is in spatial epidemiology where the effect of a pollution source may not be easily expressible in loglinear form.

Technique for building and fitting such models are reviewed below.

Generalised additive models

Modern regression tools, including machine learning algorithms [328], smooth function estimators with roughness penalties, and semi-parametric models, allow more flexible modelling than classical regression. Some of these tools are available for point process models in `spatstat` and related packages.

For more flexible models of covariate effects, one standard tool is a generalised additive model [327] in which the covariate effect is modelled as a polynomial or spline function and the model is fitted by maximising $L(\boldsymbol{\theta}) - \alpha Q(\boldsymbol{\theta})$, where $Q(\boldsymbol{\theta})$ is a penalty for the roughness of the function, and again α is a tradeoff coefficient. These models can be fitted immediately in `ppm` by including appropriate terms in the model formula, and setting `use.gam=TRUE` to ensure that the fitting is performed by `gam` rather than `glm`. For example,

```
> ppm(gold ~ bs(dfault, 5), data=mur, use.gam=TRUE)
```

fits a model in which the log intensity of Murchison gold deposits is a smooth function of distance to the nearest fault. Here `bs` is part of the standard `splines` package and provides a B-spline basis for smooth functions.

General non-loglinear model

Often a model has a simple mathematical form, with some parameters φ that appear in loglinear form, but other parameters ψ that do not. In the model

$$\lambda(u) = \exp\left(\varphi Z(u) + \frac{1}{1 + \psi Y(u)}\right)$$

where $Y(u), Z(u)$ are known covariate functions, the parameter φ appears in loglinear form, but the parameter ψ does not. The principle of maximum likelihood can still be applied to estimate both parameters (φ, ψ), but the algorithm in `ppm` cannot be used to compute the maximum likelihood estimate.

Instead, such models are usually fitted by *maximum profile likelihood*, as described in Section 9.12. Two available algorithms for maximum profile likelihood are brute force maximisation, implemented in `spatstat` by the function `profilepl` described in Section 9.12.2, and Newton's method, implemented in `spatstat` by `ippm` as described in Section 9.12.3. Specialised algorithms also exist for particular models [236].

Local ('geographically weighted') models

In some studies, we may be willing to assume a simple loglinear model, but not willing to assume that the model parameters are constant across the entire study region. For example in the Queensland copper data (Figure 1.11) we may be willing to assume a loglinear relationship between the intensity of copper deposits and distance to the nearest lineament, $\lambda(u) = \exp(\alpha + \beta d(u))$, but we might believe that the coefficients α and β are spatially varying because of the geological history.

Models of this kind can be fitted using the methods of *local likelihood* [335, 428], also known in this context as *geographically weighted regression* [267]. Code for this purpose will be added to `spatstat` shortly. For details, see Section 9.13.

9.7 Theory*

This section covers the theory which underpins the methods for fitting a Poisson point process model to a point pattern dataset. Since there has been some confusion in the literature, we shall go to some lengths to explain maximum likelihood estimation for Poisson point processes. See also [484, chapter 3], [396].

9.7.1 Introduction to maximum likelihood

In mainstream statistical methodology, a standard way to fit models to data is by maximising the *likelihood* of the model for the data. For any choice of parameter values, the likelihood value is defined as the probability (or probability density) that the given observations would have been obtained, from the model with these parameter values. For different possible values of the model parameters, the likelihood gives a measure of relative plausibility of these values. The parameter values which maximise the likelihood are the 'most plausible' values, and are called the *maximum likelihood estimates (MLE)*.

Maximum likelihood is theoretically the *optimal* method for parameter estimation, provided (a) the model is true, (b) the model satisfies regularity conditions, and (c) there are many observations and few parameters. Under these conditions, the MLE is more precise than other estimators. The MLE is always logically consistent with the data. However, maximum likelihood can perform poorly if the model is wrong (even if it is only slightly wrong) or if the data contain anomalies.

As an example, suppose that we have counted the number n of insects caught in a trap, and we believe this number follows a Poisson distribution with some mean μ. Our goal is to infer ('estimate') the value of μ from the data. The likelihood is

$$L(\mu) = e^{-\mu} \frac{\mu^n}{n!}$$

from (5.4), where n is the observed number of insects. When we want to stress that this is the likelihood for μ given the data n, we may write $L(\mu) = L(\mu; n)$. We need to find the value of μ which gives the maximum value of $L(\mu)$. It is easier to work with the natural logarithm of the likelihood,

$$\log L(\mu) = -\mu + n \log \mu - \log(n!)$$

and maximising $\log L$ is equivalent to maximising L. Take the derivative of $\log L(\mu)$ with respect to μ (known as the score function or simply the *score*):

$$U(\mu) = U(\mu; n) = \frac{d}{d\mu} \log L(\mu) = -1 + \frac{n}{\mu}.$$

The likelihood is maximised when $U(\mu) = 0$, which gives $\mu = n$. That is, the *maximum likelihood estimate* of μ is $\hat{\mu} = n$, the observed number of insects.

Note that the term $\log(n!)$ disappeared in the score. This always happens to additive terms in the loglikelihood (or equivalently multiplicative factors in the likelihood) that only depend on data and not the parameters. Since such terms do not affect the location of the maximum of the (log)likelihood it is common practice to leave them out and still call the function the (log)likelihood.

∗ Starred sections contain advanced material, and can be skipped by most readers.

9.7.2 Maximum likelihood for CSR

For simplicity we start with the homogeneous Poisson point process (CSR). This model has a single parameter λ, the intensity of the process. Assume we have observed a point pattern $\mathbf{x}$ inside a spatial window W. The likelihood function is

$$L(\lambda) = L(\lambda; \mathbf{x}) = \lambda^{n(\mathbf{x})} e^{(1-\lambda)|W|} \tag{9.34}$$

where $n(\mathbf{x})$ is the number of points of $\mathbf{x}$. The right-hand side is the probability density of observing the pattern $\mathbf{x}$ if the true intensity is λ. Notice that this probability density depends only on the number of points, and not on their location, because all spatial locations are equally likely under CSR. The point locations are 'ancillary' for λ.

More details about point process likelihoods are given in Section 13.12.1. For the moment, it is useful to appreciate why the likelihood (9.34) consists of two terms, one associated with the data $\mathbf{x}$ and the other with the window W. Imagine that space is divided into pixels of area a, as illustrated in Figure 5.20 on page 144. If a is small, there is negligible chance that any pixel will contain more than one point, so that each pixel contains either 1 point (with probability λa) or 0 points (with probability $1 - \lambda a$). Pixels are independent of each other, so the probability of a particular configuration of zeroes and ones is obtained by multiplying together the probabilities of the outcomes for each separate pixel. If there are n pixels which contain random points, the presence probability λa will appear n times, giving a factor of $\lambda^n a^n$. For the remaining pixels, which do not contain random points, the contribution to the probability is $(1 - \lambda a)^m$ where m is the number of these pixels. Since m is large and a is small, the second term is close to $e^{-\lambda|W|}$. Thus, ignoring some rescaling,[5] the first term in the likelihood (9.34) is the probability of observing the data points in $\mathbf{x}$, and the second term is the probability of **not** observing any other points in the window W.

The likelihood (9.34) is conventionally used in theoretical work, but as mentioned at the end of Section 9.7.1 the constant factor $e^{|W|}$ can be omitted. Thus the likelihood we use in practice is

$$L(\lambda) = \lambda^{n(\mathbf{x})} e^{-\lambda|W|}. \tag{9.35}$$

The maximum likelihood estimate (MLE) of the intensity λ is the value $\hat{\lambda}$ which maximises $L(\lambda)$ defined in (9.35). As before it is easier to work with the logarithm of the likelihood, which in this case is

$$\log L(\lambda) = n(\mathbf{x}) \log \lambda - \lambda|W|. \tag{9.36}$$

The score (derivative of the loglikelihood) is

$$U(\lambda) = U(\lambda; \mathbf{x}) = \frac{\mathrm{d}}{\mathrm{d}\lambda} \log L(\lambda) = \frac{n(\mathbf{x})}{\lambda} - |W|. \tag{9.37}$$

The maximum of the likelihood is attained when this derivative is zero, so the MLE is

$$\hat{\lambda} = \frac{n(\mathbf{x})}{|W|}. \tag{9.38}$$

That is, the maximum likelihood estimate $\hat{\lambda}$ is the average intensity of $\mathbf{x}$, a good 'commonsense' estimate of the intensity λ.

Another way to estimate the intensity λ would have been to use the *'method of moments'*. We would equate the observed number of points, $n(\mathbf{x})$, to the theoretically expected number of points, $\mathbb{E}n(\mathbf{X} \cap W) = \lambda|W|$, and solve the equation $\lambda|W| = n(\mathbf{x})$ for λ, yielding $\widehat{\lambda} = n(\mathbf{x})/|W|$. In this case, the method-of-moments estimate agrees with the MLE.

---

[5]The point process likelihood is conventionally measured relative to the probability for a Poisson process of rate 1. To do this we divide the value obtained above, $\lambda^n a^n e^{-\lambda|W|}$, by the corresponding value for a Poisson process of intensity $\lambda = 1$, namely $a^n e^{-|W|}$ (which is a constant only depending on data, not the parameter), finally yielding $\lambda^n e^{(1-\lambda)|W|}$.

9.7.3 Maximum likelihood for general Poisson process

Maximum likelihood estimation for Poisson point processes is treated extensively in [565, 396].

9.7.3.1 General form of likelihood

Now consider a general, inhomogeneous Poisson process model, governed by a parameter θ. This model states that the intensity is $\lambda_\theta(u)$ where the value of θ is to be estimated. The likelihood for θ is

$$L(\theta) = L(\theta; \mathbf{x}) = \lambda_\theta(x_1)\lambda_\theta(x_2)\ldots\lambda_\theta(x_n)\exp\left(\int_W (1 - \lambda_\theta(u))\,\mathrm{d}u\right) \tag{9.39}$$

where $x_1, \ldots, x_n$ are the points of $\mathbf{x}$. This can be derived by pixel approximation as described above. This probability density *does* depend on the locations of the data points x_i, because the intensity function $\lambda_\theta(u)$ makes some locations more likely than others. Therefore the data on spatial locations x_i, and not just the total number of points, are needed for model-fitting.

The loglikelihood for θ is

$$\log L(\theta) = \sum_{i=1}^n \log \lambda_\theta(x_i) - \int_W \lambda_\theta(u)\,\mathrm{d}u \tag{9.40}$$

where we have omitted the constant term $\int_W 1\,\mathrm{d}u = |W|$.

The MLE $\widehat{\theta}$ is usually not a simple function of the data, and must be computed by maximising the likelihood numerically. In general, with no assumptions on the way $\lambda_\theta(u)$ depends on θ, the likelihood function might behave poorly, and might not have a unique maximum. Further analysis depends on assuming more about the intensity model.

9.7.3.2 Maximum likelihood for baseline model

One Poisson model that is easy to analyse is that in which the intensity is an unknown multiple of a known baseline (9.3). The loglikelihood (9.40) is

$$\log L(\theta) = \sum_{i=1}^n \log(\theta\,b(x_i)) - \int_W \theta\,b(u)\,\mathrm{d}u = n\log\theta + \sum_{i=1}^n \log b(x_i) - \theta\int_W b(u)\,\mathrm{d}u. \tag{9.41}$$

The loglikelihood is differentiable with respect to θ for fixed $\mathbf{x}$, even if $b(u)$ is not a continuous function. The score is

$$U(\theta) = \frac{\mathrm{d}}{\mathrm{d}\theta}\log L = \frac{n(\mathbf{x})}{\theta} - \int_W b(u)\,\mathrm{d}u. \tag{9.42}$$

If there are no constraints on θ, the maximum likelihood estimate (MLE) of θ is the solution of the score equation $U(\theta) = 0$, which is

$$\widehat{\theta} = \frac{n(\mathbf{x})}{\int_W b(u)\,\mathrm{d}u}. \tag{9.43}$$

This is also the method-of-moments estimate, because under the baseline model, the expected total number of points is

$$\mathbb{E}_\theta[n(\mathbf{X})] = \int_W \lambda_\theta(u)\,\mathrm{d}u = \theta\int_W b(u)\,\mathrm{d}u$$

so that $\widehat{\theta}$ is the solution of $n(\mathbf{x}) = \mathbb{E}_\theta[n(\mathbf{X})]$.

9.7.4 Maximum likelihood for loglinear Poisson models

9.7.4.1 Loglinear models

For the vast majority of Poisson models treated in this book, the intensity is a **loglinear** function of the parameters:

$$\lambda_\theta(u) = \exp(B(u) + \theta^\top \mathbf{Z}(u)) \tag{9.44}$$

where $B(u)$ is a known function (the 'offset' or 'log baseline'), $\theta = (\theta_1, \ldots, \theta_p)$ is the vector of parameters, $\mathbf{Z}(u) = (Z_1(u), \ldots, Z_p(u))$ is a vector of covariate functions, and

$$\theta^\top \mathbf{Z}(u) = \theta_1 Z_1(u) + \cdots + \theta_p Z_p(u).$$

Note that the model implies that the *logarithm* of the intensity is a linear function *of the parameters*:

$$\log \lambda_\theta(u) = B(u) + \theta^\top \mathbf{Z}(u). \tag{9.45}$$

The functions B and $Z_1, \ldots, Z_p$ could be spatially varying in any fashion, so this is a very wide and flexible class of models.

The loglinear intensity model has several advantages. The intensity of a point process must be greater than or equal to zero, and this is always satisfied by the loglinear model, regardless of the value of θ and the values of the functions B and $Z_1, \ldots, Z_p$, because of the exponent in (9.44). In statistical theory the logarithm is the 'canonical' transformation of the mean for a Poisson model, and this confers many advantages in theory and practice.

9.7.4.2 Likelihood for loglinear model

For the loglinear intensity the loglikelihood takes the form

$$\log L(\theta) = \sum_{i=1}^n B(x_i) + \theta^\top \sum_{i=1}^n \mathbf{Z}(x_i) - \int_W \exp(B(u) + \theta^\top \mathbf{Z}(u)) \, du. \tag{9.46}$$

This model is a canonically parametrised exponential family [80, p. 113], [416, pp. 23–24]. The loglikelihood (9.46) is a concave function of the parameter θ, and is differentiable with respect to θ, even if the functions B and Z_j are not continuous. If the matrix

$$M = \int_W \mathbf{Z}(u) \mathbf{Z}(u)^\top \, du$$

is positive definite, then the model is identifiable. If the data are such that $\sum_i Z_j(x_i) \neq 0$ for all j, the MLE exists and is unique. Unless there are further constraints on θ, the MLE is the solution of the *score equations* $\mathbf{U}(\theta) = 0$, where the score function is

$$\mathbf{U}(\theta) = \mathbf{U}(\theta; \mathbf{x}) = \sum_{i=1}^{n(\mathbf{x})} \mathbf{Z}(x_i) - \int_W \mathbf{Z}(u) \lambda_\theta(u) \, du. \tag{9.47}$$

The score is a vector $\mathbf{U}(\theta; \mathbf{x}) = (U_1(\theta; \mathbf{x}), \ldots, U_p(\theta; \mathbf{x}))$ with components

$$U_j(\theta; \mathbf{x}) = \sum_{i=1}^{n(\mathbf{x})} Z_j(x_i) - \int_W Z_j(u) \lambda_\theta(u) \, du$$

for $j = 1, \ldots, p$. The integral in (9.47) is the Laplace transform of $\mathbf{Z}$, which is generally intractable, so that in general the score equations cannot be solved analytically.

9.7.4.3 Accuracy of maximum likelihood estimate

In many statistical models, as the sample size increases, the maximum likelihood estimator $\widehat{\theta}$ follows a standard pattern of behaviour. It is a consistent estimator (it converges in probability to the true answer θ), and it asymptotically follows a normal distribution, with mean equal to the true parameter vector θ, and variance I_θ^{-1}, where I_θ is the *Fisher information* matrix. The Fisher information is defined as the variance-covariance matrix of the score:

$$I_\theta = \text{var}_\theta \left[\mathbf{U}(\theta; \mathbf{X}) \right] = \mathbb{E}_\theta \left[\mathbf{U}(\theta; \mathbf{X}) \mathbf{U}(\theta; \mathbf{X})^\top \right] = \mathbb{E}_\theta \left[-\frac{\partial}{\partial \theta} \mathbf{U}(\theta; \mathbf{X}) \right] \tag{9.48}$$

where $\mathbb{E}_\theta$ and var_θ denote the mean and variance when the true parameter value is θ. The negative Hessian matrix

$$H(\theta; \mathbf{x}) = -\frac{\partial}{\partial \theta} \mathbf{U}(\theta; \mathbf{x}) \tag{9.49}$$

when evaluated at $\theta = \widehat{\theta}$ is called the *observed information*, while the Fisher information is sometimes called the *expected information*.

For the loglinear Poisson point process model, asymptotic distribution theory is available [565, 396, 34] and the results conform to the pattern described above. Under suitable conditions,[6] the MLE $\widehat{\theta}$ is consistent, asymptotically normal, and asymptotically efficient in a 'large domain' limiting regime [565, Theorem 11, pp. 135–136], [396, Thm. 2.4, p. 51], [393]. The Hessian is

$$H(\theta; \mathbf{x}) = \int_W \mathbf{Z}(u) \mathbf{Z}(u)^\top \lambda_\theta(u) \, du. \tag{9.50}$$

Since the Hessian does not depend on the data $\mathbf{x}$, it is equal to the Fisher information I_θ. The Fisher information is a matrix with entries (on row i, column j)

$$(I_\theta)_{ij} = \int_W Z_i(u) Z_j(u) \lambda_\theta(u) \, du.$$

This is the basis of calculations of standard error and confidence intervals for Poisson models in `spatstat`. Asymptotic theory is also available [565, 396] to support the likelihood ratio test.

The integral in the loglinear Poisson process likelihood (9.46) is the Laplace transform of the covariate function $\mathbf{Z}$, which is not usually available in closed form. Consequently, it is not usually possible to find an exact analytic solution for the maximum likelihood estimate. Some form of numerical approximation is required. Strategies are presented in the next sections.

9.8 Coarse quadrature approximation∗

The likelihood function (9.40) of a Poisson point process involves an integral over the spatial window. Except in special cases, this means that the likelihood cannot be computed exactly, but must be approximated numerically.

A good strategy is to set up the approximation so that that the approximate likelihood function is equivalent to the likelihood of another, simpler, statistical model which we know how to handle. Then using statistical techniques for the simpler model, we can compute approximate parameter

---

[6]The 'suitable conditions' are essentially that every entry in the Fisher information matrix should be large, and that regularity conditions hold.

∗ Starred sections contain advanced material, and can be skipped by most readers.

estimates for the original, complex model. This strategy has been used in statistical science since earliest times.

For a point process model, approximation of the likelihood converts the point process into a *regression* model. Lewis [419] and Brillinger [124, 126, 125] showed that the likelihood of a general point process in one-dimensional time, or a Poisson point process in higher dimensions, can be usefully approximated by the likelihood of logistic regression for the discretised process. Asymptotic equivalence was established in [101]. This makes it practicable to fit spatial Poisson point process models of general form to point pattern data [90, 158, 50, 51] by enlisting efficient and reliable software already developed for generalized linear models. The approximation has effectively reduced the problem to a standard statistical model-fitting task. Approximation of a stochastic process by a generalized linear model is now commonplace in applied statistics [424, 425, 427, 426].

9.8.1 Berman-Turner device

Numerical quadrature is a simple and efficient computational strategy for numerical integration, in which the integral $\int_W f(u)\,du$ of some function f is approximated by a weighted sum $\sum_j w_j f(u_j)$ of values of the function at a finite list of 'quadrature points' u_j which have 'quadrature weights' w_j.

9.8.1.1 Berman-Turner quadrature

Berman and Turner [90] developed a numerical quadrature method for approximate maximum likelihood estimation for an inhomogeneous Poisson point process. Suppose we approximate the integral in (9.40) by a finite sum using any quadrature rule,

$$\int_W \lambda_\theta(u)\,du \approx \sum_{j=1}^m \lambda_\theta(u_j)\,w_j \tag{9.51}$$

where u_j, $j = 1, \ldots, m$ are points in W and $w_j > 0$ are quadrature weights summing to $|W|$. This yields an approximation to the loglikelihood,

$$\log L(\theta) \approx \sum_{i=1}^{n(\mathbf{x})} \log \lambda_\theta(x_i) - \sum_{j=1}^m \lambda_\theta(u_j)\,w_j. \tag{9.52}$$

Berman and Turner observed that if the list of points $\{u_j, j = 1, \ldots, m\}$ *includes all the data points* $\{x_i, i = 1, \ldots, n\}$, then we can rewrite (9.52) as a sum over quadrature points:

$$\log L(\theta) \approx \sum_{j=1}^m (I_j \log \lambda_j - w_j \lambda_j) \tag{9.53}$$

where $\lambda_j = \lambda_\theta(u_j)$ and

$$I_j = \begin{cases} 1 & \text{if } u_j \text{ is a data point} \\ 0 & \text{if } u_j \text{ is a dummy point.} \end{cases} \tag{9.54}$$

Berman and Turner [90] re-expressed this as

$$\log L(\theta) \approx \sum_{j=1}^m (y_j \log \lambda_j - \lambda_j)\,w_j \tag{9.55}$$

where $y_j = I_j / w_j$, and noted that the right side of (9.55), for fixed $\mathbf{x}$, is formally equivalent to the weighted loglikelihood of independent Poisson variables $Y_j \sim \text{Poisson}(\lambda_j)$ taken with weights w_j. The expression (9.55) can therefore be maximised using standard software for fitting generalised linear models [458].

Later it was pointed out [459] that (9.53) is the *unweighted* loglikelihood of independent Poisson variables $I_j \sim \text{Poisson}(w_j \lambda_j)$.

The main attraction of the Berman-Turner device is that the use of standard statistical packages rather than *ad hoc* software confers great advantages in applications. Modern statistical packages have a convenient notation for statistical models [7, 142, 669] which makes it very easy to specify and fit a wide variety of models. Algorithms in the package may allow one to fit very flexible model terms such as the smooth functions in a generalised additive model [327]. Interactive software allows great freedom to reanalyse the data. The fitting algorithms are typically more reliable and stable than in homegrown software. This approach is the basis of spatstat.

In summary, the procedure is as follows: (1) generate a set of dummy points, and combine it with the data points x_i to form the set of quadrature points u_j; (2) compute the quadrature weights w_j; (3) form the indicators I_j as in (9.54) and calculate $y_j = I_j/w_j$; (4) compute the (possibly vector) values $\mathbf{z}_j = \mathbf{Z}(u_j)$ of the covariates at each quadrature point; (5) invoke standard model-fitting software, specifying that the model is a loglinear Poisson regression $\log \lambda_j = \boldsymbol{\theta}^\mathsf{T} \mathbf{z}_j$, to be fitted to the responses y_j and covariate values $\mathbf{z}_j$, with weights w_j; (6) the coefficient estimates returned by the software give the (approximate) MLE $\widehat{\boldsymbol{\theta}}$ of $\boldsymbol{\theta}$.

In the spatstat implementation, step (5) is performed by the standard R model-fitting functions glm or gam. The estimates of standard errors returned by glm are also valid for the Poisson point process, because both the weighted loglinear Poisson regression and the loglinear Poisson point process model are canonically parametrised: they have the property that the Fisher information is equal to the negative Hessian of the loglikelihood, The negative Hessians of the two models are approximately equal by (9.55). Additionally glm returns the deviance D of the fitted model; this is related to the loglikelihood of the fitted model by

$$-\log L(\widehat{\boldsymbol{\theta}}; \mathbf{x}) = \frac{D}{2} + \sum_{j=1}^{m} I_j \log w_j + n(\mathbf{x}) \tag{9.56}$$

where the sum is effectively over data points only.

Conveniently, the null model $\lambda_j \equiv \lambda$ in the weighted loglinear Poisson regression corresponds to the uniform Poisson point process with intensity λ. The MLE is $\widehat{\lambda} = n(\mathbf{x})/\sum_j w_j = n(\mathbf{x})/|W|$ with corresponding loglikelihood $\log L(\widehat{\lambda}) = n(\mathbf{x})[\log n(\mathbf{x}) - \log|W| - 1]$.

Note that this formulation assumes $\lambda(u)$ is positive everywhere. Zero values are also permissible, provided the set of zeroes does not depend on $\boldsymbol{\theta}$. Thus we formally allow negative infinite values for $\mathbf{Z}(u)$. In the approximation (9.52) all points u_j with $\lambda(u_j) = 0$ will be dummy points. Their contribution is zero and so they should be omitted in all contexts.

9.8.1.2 Design of quadrature schemes

Berman and Turner [90] used the Dirichlet tessellation or Voronoï diagram [521] to generate quadrature weights. The data points are augmented by a list of dummy points; then the Dirichlet tessellation of the combined set of points is computed as sketched in the left panel of Figure 9.13. The quadrature weight w_j associated with a (data or dummy) point u_j is the area of the corresponding Dirichlet tile.

It is computationally cheaper to use the *counting weights* proposed in [50]. In the simplest form we assign equal weight to each quadrature point, namely $w_j = |W|/m$. In general we divide the window into 'tiles', and all quadrature points within a given tile T receive the same weight $w_j = |T|/k$ where k is the number of quadrature points in T. The right panel of Figure 9.13 shows an example where W is partitioned into equal rectangular tiles.

Other choices of quadrature scheme have been explored in [50, 412].

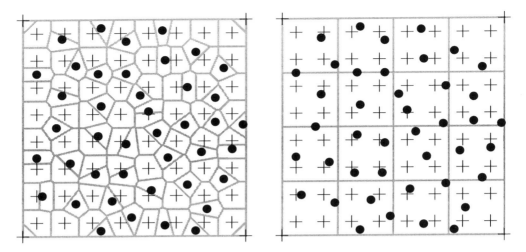

Figure 9.13. *Quadrature schemes for a point pattern dataset. Data points (•), dummy points (+), and tile boundaries (grey lines).* Left: *Dirichlet tiles and weights.* Right: *rectangular tiles, 'counting weights'. Coarsely spaced example for illustration only.*

9.8.1.3 Quadrature schemes in spatstat

The spatstat function ppm fits a point process model to an observed point pattern. By default, it uses Berman-Turner quadrature approximation.

In normal use, the quadrature scheme is generated automatically from the data point pattern, so the user does not need to think about it. However, the default quadrature scheme may give unsatisfactory results in some cases. The default rules for building a quadrature scheme are designed so that ppm will execute quickly, rather than guaranteeing a highly accurate result.

Users who wish to override the defaults may modify the rules for building a quadrature scheme, or may even provide their own quadrature scheme.

Quadrature schemes are created by the function quadscheme:

```
quadscheme(data, dummy, ..., method="grid")
```

The arguments data and dummy specify the data and dummy points, respectively.

quadscheme(X)	default for data pattern X
quadscheme(X, nd=128)	dummy points in 128×128 grid
quadscheme(X, eps=0.1)	dummy point spacing 0.1 units
quadscheme(X, random=TRUE)	stratified random dummy points
quadscheme(X, quasi=TRUE)	quasirandom dummy points
quadscheme(X, D)	data X, dummy D

Table 9.6. *Typical options for controlling the dummy points in a quadrature scheme (options are passed to* default.dummy).

For the dummy points, there is a sensible default, provided by default.dummy(X). Table 9.6 shows some of the important options for modifying the default dummy pattern. By default, the dummy points are arranged in a rectangular grid; recognised arguments include nd (the number of grid points in the horizontal and vertical directions) and eps (the spacing between dummy points). If random=TRUE, a systematic random (also called stratified random) pattern of dummy points is generated instead. If quasi=TRUE, a quasirandom pattern of dummy points is generated.

Alternatively the dummy point pattern may be specified arbitrarily and given in any format recognised by as.ppp. There are functions for creating special dummy patterns including corners, gridcentres, stratrand, and spokes. It is also possible to generate dummy points according to a quasirandom scheme, using the functions Halton and Hammersley. Quasirandom patterns fill the space uniformly but without any obvious regularities: they may be preferable to regular grids.

quadscheme(X, ..., method="d")	specify Dirichlet weights
quadscheme(X, ..., ntile=8)	8×8 array of counting tiles

Table 9.7. *Typical options for controlling the quadrature weights.*

The **quadrature weights** are determined by further arguments to quadscheme. If method = "grid" (the default) the window is divided into an ntile[1] by ntile[2] grid of rectangular tiles, and the 'counting weights' are applied: the weight for each quadrature point is the area of a tile divided by the number of quadrature points in that tile. By default the values ntile and nd are the same so all tiles (except the corners) contain exactly one dummy point as illustrated in the right panel of Figure 9.13. If method="dirichlet", the quadrature points (both data and dummy) are used to construct the Dirichlet tessellation. The quadrature weight of each point is the area of its Dirichlet tile inside the quadrature region.

A quadrature scheme (consisting of the original data point pattern, an additional pattern of dummy points, and a vector of quadrature weights for all these points) is represented by an object of class "quad". In principle, the user could create one of these objects from scratch, using the creator function quad. More details of the software implementation of ppm are given in [50, 51, 52] and particularly [54].

9.8.1.4 Gorilla nests example

In Section 9.3.4.1 we fitted a simple model to the gorillas data in which the intensity is constant inside each region defined by vegetation type:

```
> ppm(gor ~ vege, data=gex)
Nonstationary Poisson process
Log intensity:  ~vege
             Estimate   S.E. CI95.lo CI95.hi Ztest    Zval
(Intercept)    2.3238 0.1060  2.1161  2.5316   *** 21.923
vegeColo       2.0817 0.5870  0.9312  3.2322   ***  3.546
vegeGras      -0.7721 0.2475 -1.2571 -0.2871    ** -3.120
vegePrim       2.1148 0.1151  1.8892  2.3403   *** 18.373
vegeSeco       1.1341 0.2426  0.6586  1.6096   ***  4.675
vegeTran       1.6151 0.2647  1.0964  2.1339   ***  6.102
```

Fitting this model by maximum likelihood is equivalent to estimating the intensities using quadrat counts. To check this, we convert the pixel image to a tessellation and apply quadrat counting:

```
> vt <- tess(image=gex$vege)
> intensity(quadratcount(gor, tess=vt))
tile
  Dist   Colo   Gras   Prim   Seco   Tran
10.201 69.155  4.781 84.012 32.651 50.922
```

Again there are discrepancies due to discretisation of the integral in ppm in the Berman-Turner technique. Better agreement can be obtained by increasing the density of dummy points, for example, using the parameter nd:

```
> fitveg2 <- ppm(gor ~ vege - 1, data=gex, nd=256)
> exp(coef(fitveg2))
```

```
vegeDist vegeColo vegeGras vegePrim vegeSeco vegeTran
 10.188   69.040    4.786   83.807   32.329   51.199
```

Exact agreement (up to numerical rounding error) can be obtained by using a quadrature scheme with one point at each pixel of the covariate image. This is constructed by the command `pixelquad`:

```
> Q <- pixelquad(gor, gex$vege)
> fitveg3 <- ppm(Q ~ vege - 1, data=gex)
> exp(coef(fitveg3))
vegeDist vegeColo vegeGras vegePrim vegeSeco vegeTran
 10.201   69.155    4.781   84.012   32.651   50.922
```

9.8.1.5 Covariate values known only at some locations

In practice, covariate values may only be available at a small set of spatial sample locations. For example, measuring a covariate at a given location may require digging a bore hole at that location, or physically extracting material, or performing a complicated chemical assay. It may require the presence of a sensor at that location, such as a weather station or a ship. Some astronomical measurements require the fortuitous presence of a light source behind the region of interest.

One strategy for dealing with this situation is to estimate the covariate values at other locations by interpolating between the observed values. One could use the Nadaraya-Watson smoother (Section 6.9), or kriging interpolation, or many other methods.

An alternative strategy is to build a coarse quadrature scheme using only the available sample locations. This is more faithful to the observations, and avoids some potential artefacts of the smoothing. If X is the observed point pattern and D the pattern of locations where covariate values have been observed, then `quadscheme(X, D)` will build a quadrature scheme using these points. Additional arguments determine the weighting scheme: we recommend the Dirichlet weights. The corresponding covariate values should be organised in a data frame, with one column for each covariate, and one row for each quadrature point (the points of X are listed first, followed by the points of D). This data frame can then be passed to ppm as the argument `data`.

Just to demonstrate that this is possible, we take the gorilla nests data analysed above, and pretend that the land condition was only recorded at certain locations, namely the gorilla nest sites gor and some arbitrarily chosen sample locations `samp`:

```
> dfdata <- as.data.frame(lapply(gex, "[", i=gor))
> samp <- rSSI(0.5, 42, Window(gor))
> dfsamp <- as.data.frame(lapply(gex, "[", i=samp))
```

The previous model can be fitted to these data by building a quadrature scheme based on the available locations:

```
> G <- quadscheme(gor, samp, method="d")
> df <- rbind(dfdata, dfsamp)
> fitdf <- ppm(G ~ vege - 1, data=df)
> exp(coef(fitdf))
vegeDist vegeColo vegeGras vegePrim vegeSeco vegeTran
  9.765   24.175    5.468   93.612   20.337   26.731
```

We emphasise that covariate values must be available at *all data points* and at *some non-data points*. If covariate values have only been observed at the data points, it is effectively impossible to fit a point process model in which the presence or absence of points depends on the covariate value.

9.9 Fine pixel approximation∗

9.9.1 Pixel counts

Another quadrature strategy for approximating the Poisson process likelihood is to divide the window W into small pixels of equal area a. The integral over the window W is then approximated by a sum over pixels:

$$\int_W \lambda_\theta(u)\,du \approx \sum_j \lambda_\theta(u_j)a \tag{9.57}$$

where u_j is the centre of the jth pixel. We also discard the exact locations of the data points, and effectively move each data point to the centre of the pixel which contains it. Thus we approximate the sum over data points by a sum over pixels,

$$\sum_i \log \lambda_\theta(x_i) \approx \sum_j n_j \log \lambda_\theta(u_j) \tag{9.58}$$

where n_j is the number of data points falling in the jth pixel. Collecting (9.57) and (9.58), we approximate the true loglikelihood (9.40) by

$$\log L(\theta) \approx \sum_j [n_j \log \lambda_\theta(u_j) - \lambda_\theta(u_j)a] = \sum_j (n_j \log \lambda_j - \lambda_j a) \tag{9.59}$$

where $\lambda_j = \lambda_\theta(u_j)$. The adequacy of this approximation is discussed in Section 9.9.4.

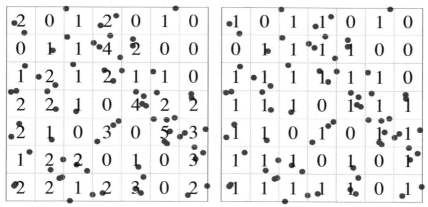

Figure 9.14. *Pixel counts* (Left) *and presence-absence indicators* (Right) *for the same point pattern.*

The right-hand side of (9.59) has the same form as the loglikelihood of independent Poisson random variables N_j with means $a\lambda_j$. This was to be expected, because the pixel counts N_j *are* independent Poisson random variables, and $a\lambda_j$ is an approximation to the true mean of N_j.

For a loglinear Poisson point process model (9.44), we have

$$\lambda_j = \lambda_\theta(u_j) = \exp(B(u_j) + \theta^\top \mathbf{Z}(u_j)) = \exp(b_j + \theta^\top \mathbf{z}_j) \tag{9.60}$$

where $b_j = B(u_j)$ and $\mathbf{z}_j = \mathbf{Z}(u_j)$. So the right-hand side of (9.59) is the loglikelihood of independent Poisson random variables N_j with means

$$\mu_j = a\lambda_j = \exp(\log a + b_j + \theta^\top \mathbf{z}_j) = \exp(\log o_j + \theta^\top \mathbf{z}_j) \tag{9.61}$$

∗ Starred sections contain advanced material, and can be skipped by most readers.

where $o_j = \log a + b_j$. This is *loglinear Poisson regression* with regression covariates $\mathbf{z}_j$ and offset $o_j = b_j + \log a$.

A practical strategy for fitting a loglinear Poisson point process model is therefore: (1) divide the window W into a fine grid of pixels of area a; (2) count the number n_j of data points falling in each pixel j; (3) evaluate the offset term $o_j = \log a + B(u_j)$ and the covariate vector $\mathbf{z}_j = \mathbf{Z}(u_j)$ at the centre u_j of each pixel j; (4) use standard statistical software (such as the glm function in R) to fit (by maximum likelihood) a loglinear Poisson regression model with responses n_j, regression covariates $\mathbf{z}_j$, and offsets o_j; (5) the fitted coefficients $\hat{\boldsymbol{\theta}}$ for the loglinear Poisson regression are the approximate maximum likelihood estimates $\hat{\boldsymbol{\theta}}$ for the loglinear Poisson point process model (9.44).

9.9.2 Pixel presence-absence indicators

A further simplification is to replace the counts n_j by *presence-absence indicators* $I_j = \mathbf{1}\{n_j > 0\}$ which tell us whether data points were present ($I_j = 1$) or absent ($I_j = 0$) in each pixel j. See the right panel of Figure 9.14. The presence-absence indicators are independent random variables, so their loglikelihood is the binomial

$$\log L = \sum_j (I_j \log p_j + (1 - I_j) \log(1 - p_j)) \tag{9.62}$$

where p_j denotes the probability that there are any data points in pixel j. Using the same approximation (9.57) we have $p_j = \mathbb{P}\{I_j = 1\} = \mathbb{P}\{N_j \geq 1\} = 1 - e^{-\mu_j} = e^{-a\lambda_j}$. For a loglinear Poisson point process model (9.44), substituting (9.61) we get $p_j = 1 - \exp(-\exp(o_j + \boldsymbol{\theta}^\top \mathbf{z}_j))$, and inverting this relationship gives

$$\log(-\log(1 - p_j)) = o_j + \boldsymbol{\theta}^\top \mathbf{z}_j \tag{9.63}$$

where the offset o_j is

$$o_j = b_j + \log a. \tag{9.64}$$

The transformation $\log(-\log(1 - p))$ is called the *complementary log-log* link. Thus, the presence-absence indicators follow a binary *complementary log-log regression* with offset o_j and regression covariates $\mathbf{z}_j$ as before.

Thus, the correct way to analyse the presence-absence indicator data from a loglinear Poisson point process is to fit a complementary log-log regression, using the offset (9.64) to account for pixel size. This ensures that estimates of $\boldsymbol{\theta}$ obtained using pixel grids of different size are exactly compatible. The parameter estimates $\hat{\boldsymbol{\theta}}$ obtained from this procedure are estimates of the parameters of the loglinear Poisson point process model (9.44).

9.9.3 Logistic regression in a fine pixel grid

There is a further simplification when the pixels are so small that they have negligible chance of containing more than one data point. Given that $N_j \leq 1$ for all pixels j, the conditional probability that $N_j = 1$ is

$$p_j^* = \mathbb{P}\{N_j = 1 \mid N_j \leq 1\} = \frac{\mathbb{P}\{N_j = 1\}}{\mathbb{P}\{N_j \leq 1\}} = \frac{\mathbb{P}\{N_j = 1\}}{\mathbb{P}\{N_j = 0\} + \mathbb{P}\{N_j = 1\}}.$$

Using the formula for the Poisson probabilities (5.4) this is

$$p_j = \frac{\mu_j \exp(-\mu_j)}{\exp(-\mu_j) + \mu_j \exp(-\mu_j)} = \frac{\mu_j}{1 + \mu_j}$$

and similarly the conditional probability that $N_j = 0$ is $1 - p_j^* = 1/(1 + \mu_j)$. We find that $p_j^*/(1 - p_j^*) = \mu_j$ so that the *odds* of presence (that is, the ratio of the presence probability divided by the absence probability) is equal to μ_j. For a loglinear Poisson process, substituting (9.61) and taking the logarithm gives

$$\log\left(\frac{p_j^*}{1-p_j^*}\right) = o_j + \boldsymbol{\theta}^\top \mathbf{z}_j \tag{9.65}$$

saying that the *log odds* of presence is equal to a linear function of the parameters. The transformation $\log(p/(1-p))$ is called the *logistic* link, and (9.65) states that the presence-absence indicators follow a *logistic regression* with offset o_j as in (9.64) and covariates $\mathbf{z}_j$ as before. Thus, *if the probability of getting more than one point in a pixel is negligible, the presence-absence indicators I_j approximately satisfy a logistic regression with the same linear predictor.*

For computation and for statistical inference, logistic regression has some practical advantages over complementary log-log regression, because the logistic link is the 'canonical link' for binary regression [458]. In logistic regression the estimated variance-covariance matrix of $\widehat{\boldsymbol{\theta}}$ depends only on $\widehat{\boldsymbol{\theta}}$, whereas for complementary log-log regression it depends on $\mathbf{x}$ as well.

Pixel-based logistic regression for point events was pioneered in geology by F.P. Agterberg [4] on the suggestion of J.W. Tukey [658]. It was later independently rediscovered in archaeology [604, 326, 397, 398] and is now a standard technique in GIS applications [116].

Despite its popularity, pixel logistic regression does not seem to be universally well understood. Some writers describe it as a 'nonparametric' technique [399, p. 24]. The interpretation of the fitted parameters is widely held to be obscure [698, p. 175] and is typically based only on the *sign* of the slope parameters, that is, parameters other than the intercept [288, pp. 405–407].

Contrary to these statements, there is a clear physical meaning for the model parameters of (pixel-based) logistic regression when pixels are sufficiently small. The model is effectively a Poisson point process with loglinear intensity (9.6): see [34, 691]. This is a relatively simple parametric model. The fitted parameters of the logistic regression have a direct interpretation as the parameters of the Poisson point process (provided the model is fitted using an offset that includes the logarithm of pixel area). The known properties of the Poisson model (Section 9.2.1) enable us to make numerous predictions about quantities of interest, such as the expected number of points in a target region, the probability of exactly k points in a target region, and the distribution of distance from a fixed starting location to the nearest random point. See Section 9.4.3.

9.9.4 The effect of pixel size

The need for small pixels

By converting a spatial point pattern into an array of pixel counts or pixel presence-absence indicators, we have effectively *aggregated* the individual points into groups. Each group consists of the points that fall in a given pixel. Normally the pixels would be so small that at most one point would fall in each pixel (unless of course there are duplicated points). The theory above relies on this approximation (9.58) and the corresponding approximation for the *expected* number of points in the pixel (9.57).

But how small is 'small enough'? This was investigated in [34]. The adequacy of the approximation depends on the spatial regularity of the function $\mathbf{Z}$, spelt out in [34, Theorem 10]. In the most optimistic case, $\mathbf{Z}$ is *constant within each pixel*,

$$\mathbf{Z}(u) = \mathbf{z}_j \text{ for } u \in S_j, \tag{9.66}$$

so that (9.57) and (9.58) are exact equations rather than approximations, and the loglikelihoods (9.59) and (9.62)–(9.64) are the exact loglikelihoods. Covariates of this kind include factor-valued classifications relating to ownership and management, and classifications of terrain into large geological units. In the next best case, $\mathbf{Z}$ is a smooth function.[7] Then (9.57), (9.58), (9.59), and (9.62)–

---

[7] Here $\mathbf{Z}$ should be a Lipschitz function, i.e. $\|\mathbf{Z}(u) - \mathbf{Z}(v)\| \leq C\|u - v\|$ where $C < \infty$ is constant.

(9.64) are good approximations when the pixels are small. Examples include distance transforms (e.g., $\mathbf{Z}(u)$ is the distance from u to the nearest geological fault), spatial coordinates, and kernel-smoothed geochemical assay values. In the least optimistic case, $\mathbf{Z}$ is a discontinuous function, such as the indicator of a very irregular spatial domain such as a rock outcrop. The approximations (9.59) and (9.62)–(9.64) can then give rise to considerable bias, even for quite small pixel sizes [34].

Replacing pixel counts by presence-absence indicators causes a loss of statistical efficiency (i.e. variance of the estimator increases). A rule-of-thumb [34] is that the relative efficiency (ratio of variances, assuming the same size of data) is $\bar{\mu}/(e^{\bar{\mu}} - 1) \approx 1 - \bar{\mu}/2$ where $\bar{\mu} = \frac{1}{N} \sum \mu_j$ is the average expected number of points per pixel, estimated by $n(\mathbf{X})/N$, the observed average number of data points per pixel. To ensure the loss of efficiency is less than 10%, we would require $\bar{\mu} < 0.2$, which would normally be satisfied in practice. Logistic regression involves a further approximation in which we say there is a negligible probability of getting more than one point in the same pixel, *in any pixel*. This again is not a very onerous condition, as shown in [34].

In summary, the accuracy of the pixel approximation depends mainly on the spatial irregularity of the covariates. Very small pixels may be necessary if the covariates are highly irregular.

Problems with small pixels

If very small pixels are used, practical problems can arise. There will be a very large number of pixels, and hence a large amount of data to handle. Software for logistic regression may not be designed for large amounts of data; for example, computing the likelihood simply by summing all the individual terms in (9.62) could lead to numerical overflow. In a fine pixellation, the overwhelming majority of pixels do not contain a data point, so the overwhelming majority of indicator values I_j are equal to zero. This causes numerical instability (leading to error messages about singular matrices, for example). It also causes some of the associated statistical tools to fail or behave unexpectedly, due to the *Hauck-Donner effect* [330], essentially the breakdown of the delta-method approximation.

One valid strategy for avoiding these problems is to take only a random sample of the zero-pixels (the pixels with $I_j = 0$), as discussed in Section 9.10.

Problems with large pixels

Some researchers choose to use quite large pixels, and apply logistic regression to the presence-absence indicators. This leads to difficult problems related to the aggregation of points into geographical areas. These problems have been studied extensively in epidemiology and in ecological and environmental statistics [253, 686, 682, 681]. Important questions include the equivalence of models fitted using different pixel grids (the 'modifiable area unit problem' [529] or 'change-of-support' [289, 76, 184]), the relation between discrete and continuous spatial models ('ecological fallacy' [583]), and bias due to aggregation over pixels ('ecological bias' [681, 682] or aggregation bias [207, 13]).

An unexpected finding in [34] is that it may be *impossible* to reconcile two spatial logistic regression models that were fitted to the same spatial point pattern data using different pixel grids. Two such models are logically incompatible except in simple cases. Perhaps the easiest way to understand this is to remember that the pixels are artificial. If we fit a logistic regression using pixels of a particular size, we are making an assumption about the underlying random process of points in continuous space. If we now choose pixels of a different size, this corresponds to a different assumption, and in general the two sets of assumptions are logically incompatible. There is no point process in continuous space which satisfies a logistic regression model when it is discretised on *any* pixel grid. The implication is that two research teams who apply spatial logistic regression to the same data, but using different pixel sizes, may obtain results that cannot be reconciled.

9.9.5 Implementation in `spatstat`

The `spatstat` package provides facilities for performing pixel logistic regression, mainly for teaching and for methodological research. Models are fitted using the command `slrm` (for spatial logistic regression model, pronounced '*SLURM*') with a syntax very similar to ppm.

```
> fit <- slrm(bei ~ grad, data=bei.extra)
> fit
Fitted spatial logistic regression model
Formula:        bei ~ grad
<environment: 0xbfa7098>
Fitted coefficients:
(Intercept)          grad
    -5.727         6.516
```

The result is a 'fitted spatial logistic regression model' object of class `"slrm"`. Methods for this class include `print`, `plot`, `predict`, `coef`, `fitted`, `update`, `terms`, `formula`, `anova`, and `logLik`. Additionally `step` and `AIC` can be used for model selection.

We reiterate that `slrm` is designed for methodological research rather than day-to-day analysis. For example, it helps to demonstrate that spatial logistic regression is equivalent to fitting a loglinear Poisson point process [34].

9.10 Conditional logistic regression∗

This section describes an alternative strategy for fitting point process models to point pattern data. Instead of using a dense grid of pixels or dummy points, we generate a smaller number of dummy points at *random* locations. Given only the data at these (data plus dummy) locations, a loglinear Poisson point process model becomes a logistic regression model, which we can fit using standard software.

Although there is a loss of statistical efficiency (increase in variance of parameter estimates) due to the sub-sampling, this method avoids some of the biases inherent in quadrature approximations, so it may have better mean square error overall.

9.10.1 Connection between loglinear Poisson and logistic models

Statisticians will be familiar with the connection between loglinear Poisson regression models and logistic models, for example in contingency tables. The basic principle is as follows.

Suppose X apples and Y oranges have fallen from the trees in an orchard, where X and Y are independent, Poisson random variables with means μ and ν. Then the special properties of the Poisson distribution imply various relationships. The total number of fruit $X + Y$ is a Poisson variable with mean $\mu + \nu$. Given that there are $X + Y = n$ fruit in total, each fruit is either an apple with probability p or an orange with probability $1 - p$, independently of other fruit, where $p = \mu/(\mu + \nu)$.

Suppose the expected number of apples μ depends on a covariate z through a *loglinear* relationship, $\log \mu = a + bz$, or equivalently, $\mu = e^{a+bz}$. Given there are n fruit in total, the probability that each fruit is an apple is $p = \mu/(\mu + \nu) = e^{a+bz}/(e^{a+bz} + \nu)$, and for an orange, $1 - p = \nu/(\mu + \nu) = \nu/(e^{a+bz} + \nu)$. The odds of an apple against an orange are $p/(1 - p) = e^{a+bz}/\nu$ so that the logarithm

---

∗ Starred sections contain advanced material, and can be skipped by most readers.

of the odds is

$$\log \frac{p}{1-p} = a + bz - \log v.$$

That is, the probability p of an apple is related to the covariate z through a *logistic regression* relationship. Thus, *loglinear Poisson regression becomes logistic regression when we condition on the total count.*

9.10.2   Random dummy points and logistic regression

Assume, as before, that the data points come from a Poisson process $\mathbf{X}$ with intensity function $\lambda_{\boldsymbol{\theta}}(u)$ where $\boldsymbol{\theta}$ is to be estimated. Suppose that we generate dummy points at random, according to a Poisson process $\mathbf{D}$ with known intensity function $\delta(u) > 0$, independently of the data points. Combining these two types of points, the superposition $\mathbf{Y} = \mathbf{X} \cup \mathbf{D}$ is again a Poisson point process (by the superposition property, Section 5.3.3) with intensity function $\kappa(u) = \lambda_{\boldsymbol{\theta}}(u) + \delta(u)$. For each point $y_i \in \mathbf{Y}$, let $I_i = \mathbf{1}\{y_i \in \mathbf{X}\}$ be the indicator that equals 1 if y_i was a data point, and 0 if it was a dummy point.

Let us now *condition on* $\mathbf{Y}$, so that we regard the points y_i as fixed locations. The data now consist only of the data/dummy indicators $I_1, \ldots, I_m$ where $m = n(\mathbf{X}) + n(\mathbf{D})$ is the total number of (data and dummy) points. Conditional on location, the indicators I_i are independent random variables, with probabilities

$$\mathbb{P}\{I_i = 1\} = p_i = \frac{\lambda_{\boldsymbol{\theta}}(y_i)}{\lambda_{\boldsymbol{\theta}}(y_i) + \delta(y_i)}, \qquad \mathbb{P}\{I_i = 0\} = 1 - p_i = \frac{\delta(y_i)}{\lambda_{\boldsymbol{\theta}}(y_i) + \delta(y_i)} .$$

so that the odds are

$$\frac{\mathbb{P}\{I_i = 1\}}{\mathbb{P}\{I_i = 0\}} = \frac{p_i}{1 - p_i} = \frac{\lambda_{\boldsymbol{\theta}}(y_i)}{\delta(y_i)}.$$

If the intensity of $\mathbf{X}$ is a loglinear function of the covariate, $\lambda_{\boldsymbol{\theta}}(u) = \exp(B(u) + \boldsymbol{\theta}^{\top} \mathbf{Z}(u))$, then the log odds are

$$\log \frac{\mathbb{P}\{I_i = 1\}}{\mathbb{P}\{I_i = 0\}} = B(y_i) + \boldsymbol{\theta}^{\top} \mathbf{Z}(y_i) - \log \delta(y_i). \tag{9.67}$$

The indicators I_i therefore satisfy a *logistic regression*. The loglikelihood given the data $\mathbf{x}$ and a realisation $\mathbf{d}$ of the dummy process is

$$\log L(\boldsymbol{\theta}; \mathbf{x}, \mathbf{d}) = \sum_i I_i \log(p_i) + (1 - I_i) \log(1 - p_i)$$

$$= \sum_{u \in \mathbf{x}} \log \frac{\lambda_{\boldsymbol{\theta}}(u)}{\lambda_{\boldsymbol{\theta}}(u) + \delta(u)} + \sum_{u \in \mathbf{d}} \log \frac{\delta(u)}{\lambda_{\boldsymbol{\theta}}(u) + \delta(u)}. \tag{9.68}$$

That is, if we treat the pattern of data and dummy points $\mathbf{y} = \mathbf{x} \cup \mathbf{d}$ as fixed locations, then the maximum likelihood estimator of the parameters $\boldsymbol{\theta}$ is obtained by fitting a logistic regression (9.67) to the data/dummy indicators I_i.

This approach has many advantages. Estimation can be implemented straightforwardly using standard software for generalised linear models. The loglikelihood (9.68) is a concave function of $\boldsymbol{\theta}$, and conditions for existence and uniqueness of the maximum are well known [616], [38].

9.10.3   Pixel logistic regression with subsampled zero pixels

Another motivation for the same technique comes from pixel logistic regression (Section 9.9). As noted in Section 9.9.4, a fine grid of pixels is needed in order to minimise bias, but this results in a large volume of data, in which most of the presence-absence indicators are zero, causing computational and statistical problems. One strategy is to take a random sub-sample of the 'zero pixels'

(pixels with presence-absence indicator equal to zero). Suppose each zero-pixel is sampled with probability s independently of other zero-pixels. Each pixel with the presence value 1 is retained. If the data on the full pixel grid satisfy a logistic regression

$$\log \frac{p_j}{1 - p_j} = \boldsymbol{\theta}^\top \mathbf{z}_j$$

where p_j is the presence probability for pixel j, then the subsampled data satisfy

$$\log \frac{p_j^*}{1 - p_j^*} = \boldsymbol{\theta}^\top \mathbf{z}_j - \log s \qquad (9.69)$$

where p_j^* is the presence probability for pixel j given that it has been sampled. Since s is a probability, $\log s$ is negative, so the effect of subtracting $\log s$ is that the log odds of presence are increased in the subsample, as we would expect.

In summary, if we discretise the point pattern data on a fine grid, set up the pixel presence-absence indicators I_j, then *randomly subsample* the zero pixels with retention probability s, a logistic regression is appropriate, with adjustment for the sub-sampling probability as shown in (9.69).

9.10.4   Conditional logistic regression as a point process method

A pivotal question about conditional logistic regression is whether it is statistically reliable. Quadrature methods use a fine grid of pixels or dummy points in an effort to achieve accurate approximation of the likelihood of the Poisson process. However, random sampling can produce unbiased estimates of the score using much fewer function evaluations, at the expense of increased variability due to the randomisation.

Suppose $\mathbf{x}$ is a point pattern dataset, and $\mathbf{d}$ is a random pattern of dummy points which we have generated with intensity function $\delta(u)$. Assume as usual that $\mathbf{x}$ is a realisation of a Poisson process with loglinear intensity. The score is

$$\mathbf{U}(\boldsymbol{\theta}; \mathbf{x}, \mathbf{d}) = \sum_{u \in \mathbf{x}} \frac{\delta(u) \mathbf{Z}(u)}{\lambda_{\boldsymbol{\theta}}(u) + \delta(u)} - \sum_{u \in \mathbf{d}} \frac{\lambda_{\boldsymbol{\theta}}(u) \mathbf{Z}(u)}{\lambda_{\boldsymbol{\theta}}(u) + \delta(u)}. \qquad (9.70)$$

By Campbell's formula (6.11) the expectation of the first sum in (9.70) over all outcomes of $\mathbf{X}$ is

$$\mathbb{E} \sum_{u \in \mathbf{X}} \frac{\delta(u) \mathbf{Z}(u)}{\lambda_{\boldsymbol{\theta}}(u) + \delta(u)} = \mathbb{E} \int_W \frac{\lambda_{\boldsymbol{\theta}}(u) \delta(u) \mathbf{Z}(u)}{\lambda_{\boldsymbol{\theta}}(u) + \delta(u)} \, \mathrm{d}u \qquad (9.71)$$

and for the second sum in (9.70) the expectation over all outcomes of $\mathbf{D}$ is

$$\mathbb{E} \sum_{u \in \mathbf{D}} \frac{\lambda_{\boldsymbol{\theta}}(u, X) \mathbf{Z}(u)}{\lambda_{\boldsymbol{\theta}}(u, X) + \delta(u)} = \mathbb{E} \int_W \frac{\lambda_{\boldsymbol{\theta}}(u) \delta(u) \mathbf{Z}(u)}{\lambda_{\boldsymbol{\theta}}(u) + \delta(u)} \, \mathrm{d}u. \qquad (9.72)$$

It follows that $\mathbb{E}_{\boldsymbol{\theta}} \mathbf{U}(\boldsymbol{\theta}; \mathbf{X}, \mathbf{D}) = 0$ where the expectation is taken over both $\mathbf{X}$ and $\mathbf{D}$. The logistic score (9.70) is an unbiased estimating function. This implies that, under reasonable conditions, the estimate of $\boldsymbol{\theta}$ obtained from logistic regression is consistent and asymptotically normal [38].

If we rearrange (9.70) as

$$\mathbf{U}(\boldsymbol{\theta}; \mathbf{x}, \mathbf{d}) = \sum_{u \in \mathbf{x}} \mathbf{Z}(u) - \sum_{u \in \mathbf{x} \cup \mathbf{d}} \frac{\lambda_{\boldsymbol{\theta}}(u)}{\lambda_{\boldsymbol{\theta}}(u) + \delta(u)} \mathbf{Z}(u) \qquad (9.73)$$

and apply Campbell's formula to the last term in (9.73), we obtain

$$\mathbb{E} \sum_{u \in \mathbf{X} \cup \mathbf{D}} \frac{\lambda_{\boldsymbol{\theta}}(u)}{\lambda_{\boldsymbol{\theta}}(u) + \delta(u)} \mathbf{Z}(u) = \int_W \mathbf{Z}(u) \lambda_{\boldsymbol{\theta}}(u) \, \mathrm{d}u. \qquad (9.74)$$

Thus, if the last term in (9.73) is replaced by its expectation, the score of the full point process likelihood is obtained. Hence the score of the conditional logistic regression may be viewed as a Monte Carlo approximation of the full maximum likelihood score to which it converges (in mean square) when $\inf_{u \in W} \delta(u) \to \infty$.

It is also possible to use dummy points generated by a point process other than a Poisson process, with appropriate adjustments to the likelihood [38].

9.10.5 Implementation in `spatstat`

If ppm is called with the option `method="logi"`, it will fit the point process model using conditional logistic regression.

```
> fitM <- ppm(bei ~ grad, data=bei.extra)
> fitL <- ppm(bei ~ grad, data=bei.extra, method="logi")
> coef(fitM)
(Intercept)        grad
     -5.391       5.022
> coef(fitL)
(Intercept)        grad
     -5.415       5.294
```

Dummy points will be generated at random, by default, but can be generated in several ways, including a deterministic grid. The default is a stratified random pattern, in which the window is divided into equal tiles and each tile contains a fixed number of independent random points.

In the usual case where the left-hand side of the model formula is a point pattern X, the function ppm will call `quadscheme.logi(X)` to assemble a 'quadrature scheme' object in which all data and dummy points have weight equal to 1.

Alternatively the left-hand side of the model formula may be such a quadrature scheme which the user has constructed by calling `quadscheme.logi(X, ...)`. The user may provide the dummy points by calling `quadscheme.logi(X, D)`, but this would be unusual: normally they are generated randomly in `quadscheme.logi` (ultimately by calling `dummy.logi`) taking account of the optional arguments `dummytype`, `nd`, and `mark.repeat`.

9.10.6 Logistic regression in case-control studies

In a spatial case-control study of disease we have two types of points: cases (people diagnosed with the disease of interest) and controls (people who are similar to the cases except that they do not have the disease).

Figure 1.12 on page 9 shows the Chorley-Ribble cancer data of Diggle [224] giving the residential locations of new cases of cancer of the larynx (58 cases) and cancer of the lung (978 cases) in the Chorley and South Ribble Health Authority of Lancashire, England, between 1974 and 1983. The location of a disused industrial incinerator is also given. The aim is to assess evidence for an increase in the incidence of laryngeal cancer close to the incinerator. The lung cancer cases can arguably be treated as 'controls', that is, as a sample from the susceptible population. Data analysis in [224, 236, 55] concluded there is significant evidence of an incinerator effect.

Assume that the susceptible population has some unknown spatial density $s(u)$ (people per square kilometre). As explained in Section 9.3.7, the null model of constant risk assumes that the cases are a Poisson process with intensity $\lambda(u) = \rho s(u)$ (cases per square kilometre) where ρ is the disease risk per head of population. Common alternative models assume that the cases are Poisson with intensity $\lambda(u) = r(u, \theta)s(u)$ where $r(u, \theta)$ is a spatially varying risk function.

Suppose we treat the locations y_i of cases and controls as fixed, and consider only the disease status indicators, $I_i = 1$ if y_i is a case and $I_i = 0$ if it is a control. Then, by the principle explained in Section 9.10.2, the disease status indicators satisfy a binary regression

$$\log \frac{\mathbb{P}\{I_i = 1\}}{\mathbb{P}\{I_i = 0\}} = \log \frac{p_i}{1 - p_i} = \log r(y_i, \boldsymbol{\theta}).$$

If the risk function $r(u, \boldsymbol{\theta})$ is loglinear in $\boldsymbol{\theta}$, the relationship is a logistic regression. This is a well-known principle in epidemiology. Diggle and Rowlingson [236] argue the advantages of this approach in a spatial context, which include not having to estimate the population density.

For the Chorley-Ribble dataset we can carry out such an analysis as follows:

```
> X <- split(chorley)$larynx
> D <- split(chorley)$lung
> incin <- as.ppp(chorley.extra$incin, W = Window(chorley))
> dincin <- distfun(incin)
> Q <- quadscheme.logi(X,D)
> fit <- ppm(Q~dincin)
```

For similar applications in forestry see [684, 145].

9.11 Approximate Bayesian inference

Recently Rajala [561] suggested using the conditional logistic regression likelihood as the basis for approximate Bayesian inference for point process models. In this setup (9.68) is treated as the true loglikelihood of the model and a multivariate Gaussian distribution is used as the prior for $\boldsymbol{\theta}$. The corresponding posterior density is not easy to derive, but applying a tangential lower bound to the log terms in (9.68) leads to a multivariate Gaussian distribution as a lower bound to the posterior. The mean of this lower bound distribution is used as the parameter estimate. However, the lower bound depends on so-called variational parameters and the estimate depends on how these are chosen. The best variational parameters are the ones that maximise the lower bound since then we are as close as possible to the true posterior distribution. In [561] the EM algorithm is used to find this optimal choice of the variational parameters. Rajala's code has been incorporated in ppm as the method "VBlogi":

```
> fitVB <- ppm(bei ~ grad, data=bei.extra, method="VBlogi")
> coef(fitVB)
(Intercept)        grad
    -5.417        5.305
```

The parameter estimates are very close to those obtained when method="logi" in Section 9.10.5. However, the results depend on the chosen Gaussian prior distribution, which by default is chosen to be somewhat 'uninformative'. It is recommended to set the prior mean and variance-covariance matrix explicitly, so that the dependence on the prior is more apparent. Suppose for example that our prior belief is that the terrain slope (grad) should have very little effect on the abundance of trees, while we have no strong prior knowledge about the intercept term (underlying abundance of trees). It would be appropriate to take the prior mean of both parameters to be zero, with large prior variance for the intercept parameter, and small prior variance for the coefficient of slope:

```
> fitVBp <- ppm(bei ~ grad, data=bei.extra,
                prior.mean = c(0,0), prior.var = diag(c(10000,0.01)))
```

```
> coef(fitVBp)
(Intercept)        grad
    -4.9794      0.5438
```

In this example our strong belief in the absence of a slope effect has substantially changed the estimate for the slope coefficient.

At the time of writing it is not possible to calculate the posterior variance-covariance matrix in `spatstat`, so there is currently no way to assess the uncertainty of this approximate Bayesian estimate.

9.12 Non-loglinear models

This section covers methods for fitting a *general* (not loglinear) Poisson point process model. It describes the model-fitting functions `ippm` (implementing Newton's method for differentiable models) and `profilepl` (applying brute force estimation for general models, such as those involving a threshold).

9.12.1 Profile likelihood

In many intensity models, *some* of the parameters appear in loglinear form, while other parameters do not. Such a model is of the form

$$\lambda_\theta(u) = \exp(\varphi^\top \mathbf{Z}(u, \psi)) \tag{9.75}$$

where $\theta = (\varphi, \psi)$ is a partition of the entries of the parameter vector θ into *regular parameters* φ which appear in loglinear form in (9.75) and *irregular parameters* ψ which do not appear in loglinear form.

If we *fix* the values of the irregular parameters ψ, then equation (9.75) is loglinear in the remaining parameters φ. Considered as a model with parameters φ only, this is a loglinear Poisson point process model, which can be fitted using the methods described in previous sections.

Thus, for any chosen value of ψ, the likelihood $L(\theta) = L(\varphi, \psi)$ can easily be maximised over all possible values of φ, giving us the *profile maximum likelihood estimate*

$$\widehat{\varphi}(\psi) = \operatorname{argmax}_\varphi L(\varphi, \psi). \tag{9.76}$$

The achieved maximum value of the likelihood for a given value of ψ is called the *profile likelihood*:

$$\mathrm{pL}(\psi) = \max_\varphi L(\varphi, \psi). \tag{9.77}$$

The maximum likelihood estimate of θ can be obtained by maximising the *profile* likelihood over ψ. That is, if $\widehat{\psi} = \operatorname{argmax}_\psi \mathrm{pL}(\psi)$ is the value of the irregular parameters that maximises the profile likelihood, then $\widehat{\theta} = (\widehat{\varphi}(\widehat{\psi}), \widehat{\psi})$ is the maximum likelihood estimate of θ.

9.12.2 Maximising profile likelihood by brute force

A simple strategy for maximising the profile likelihood is to evaluate $\mathrm{pL}(\psi)$ at a grid of test values of ψ and simply choose the value of ψ which yields the maximum.

The function `profilepl` performs this 'brute force' maximisation. Its syntax is

```
profilepl(s, f, ...)
```

The argument s is a data frame containing values of the irregular parameters over which the profile likelihood will be computed. The argument f should be set to Poisson for fitting Poisson processes. Additional arguments ... are passed to ppm to fit the model.

For example, the nztrees dataset appears to have a line of trees (perhaps a planted avenue) at right boundary. We fit a threshold model where the intensity is different on either side of the line $x = a$. The threshold value a can be found using profilepl:

```
> thresh <- function(x,y,a) { x < a }
> df <- data.frame(a=1:152)
> nzfit <- profilepl(df, Poisson, nztrees ~ thresh, eps=0.5)
```

This calculation may take considerable time. The result, nzfit, is an object of class "profilepl". Printing the object gives information about the fitting procedure, the fitted value of the irregular parameter, and the fitted values of the regular parameters.

```
> nzfit
Profile log pseudolikelihood
for model:  ppm(nztrees ~ thresh,  eps = 0.5,  interaction = Poisson)
fitted with rbord = 0
Interaction: Poisson process
Irregular parameter: a in [1, 152]
Optimum value of irregular parameter:  a = 150
```

The plot method generates a plot of the profile likelihood, as shown in Figure 9.15. The plot shows a clear preference for a threshold near the right-hand edge of the field. The optimal fitted model can be extracted as as.ppm(nzfit).

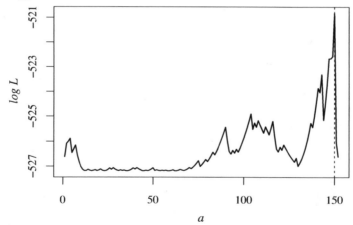

Figure 9.15. *Profile likelihood for changepoint model of NZ Trees data.*

Figure 9.15 shows the typical behaviour of profile likelihood in many applications. The profile likelihood is not continuous as a function of the threshold parameter a: it has discrete jumps at the values a_i which coincide with the horizontal coordinates of data points. Between these jumps, the profile likelihood is continuous and differentiable.

This is a simple instance of a *changepoint* model; the 'significance' of the threshold should be assessed using specialised methods for changepoints [147]. The R package changepoint may help.

9.12.3   Maximising profile likelihood by Newton's method

9.12.3.1   Newton's method

If the intensity function is differentiable with respect to the irregular parameters, then a more efficient technique for maximisation is Newton's method of root-finding, applied to the score. If $\lambda_\theta(u)$ is differentiable with respect to (all components of) θ, the score is

$$\mathbf{U}(\theta; \mathbf{x}) = \sum_{i=1}^{n} z_\theta(x_i) - \int_W z_\theta(u) \lambda_\theta(u) \, \mathrm{d}u \tag{9.78}$$

where $z_\theta(u) = (\partial/\partial\theta) \log \lambda_\theta(u)$, and the *observed* information is

$$H(\widehat{\theta}; \mathbf{x}) = -\sum_{i=1}^{n} \kappa_{\widehat{\theta}}(x_i) + \int_W \kappa_{\widehat{\theta}}(u) \lambda_{\widehat{\theta}}(u) \, \mathrm{d}u + \int_W z_{\widehat{\theta}}(u) z_{\widehat{\theta}}(u)^\top \lambda_{\widehat{\theta}}(u) \, \mathrm{d}u \tag{9.79}$$

where $\kappa_\theta(u) = (\partial/\partial\theta) z_\theta(u) = (\partial^2/\partial\theta^2) \log \lambda_\theta(u)$. In the Newton-Raphson method we repeatedly update our current estimate of θ by

$$\theta_{m+1} = \theta_m - H(\theta_m; \mathbf{x})^{-1} \mathbf{U}(\theta_m; \mathbf{x}). \tag{9.80}$$

In this case there is some asymptotic theory available for the statistical performance of the estimator [205, sec. 4.5.2].

9.12.3.2   Implementation in `spatstat`

The `spatstat` function `ippm` performs the iterative maximisation (9.80). For example, consider a Poisson point process with intensity function $\lambda(u)$ at u such that

$$\lambda(u) = \exp(\alpha + \beta Z(u)) f(u, \gamma)$$

where α, β, γ are parameters to be estimated, $Z(u)$ is a spatial covariate function, and f is some known function. Then the parameters α, β are called *regular* because they appear in a loglinear form; the parameter γ is called *irregular*.

To fit this model using `ippm`, we specify the model formula in the usual way for `ppm`. Recall that the right-hand side of the model formula is a representation of the *log* of the intensity. In the above example the log intensity is

$$\log \lambda(u) = \alpha + \beta Z(u) + \log f(u, \gamma)$$

so the model above would be encoded with the trend formula `~Z + offset(log(f))`. Note that the irregular part of the model is an *offset* term, which means that it is included in the log trend as it is, without being multiplied by another regular parameter.

The optimisation runs faster if we specify the derivative of $\log f(u, \gamma)$ with respect to γ. We call this the *irregular score*. To specify this, the user must write an R function that computes the irregular score for any value of γ at any location `(x,y)`.

Thus, to code such a problem, (1) the model formula should define the log intensity, with the irregular part as an `offset`; (2) the argument `start` should be a list containing initial values of each of the irregular parameters; and (3) the argument `iScore`, if provided, must be a list (with one entry for each entry of `start`) of functions with arguments `x,y,...`, that evaluate the partial derivatives of $\log f(u, \gamma)$ with respect to each irregular parameter. An example is given below.

For a general concave loglikelihood, with arbitrary data, the existence and uniqueness of the maximum likelihood estimate (MLE) are not guaranteed, and depend on the absence of 'directions of recession' [584, 697, 616, 10, 282].

9.12.3.3 Incinerator effect in Chorley-Ribble data

The Chorley-Ribble data (Figure 1.12 on page 9) were discussed in Sections 9.3.7 and 9.10.6.

```
>    lung <- split(chorley)$lung
>    larynx <- split(chorley)$larynx
>    Q <- quadscheme(larynx, eps=0.1)
```

Here we follow Diggle [224] in treating the cases of laryngeal cancer as the response, and taking the lung cancers as a covariate. We apply kernel smoothing [222] to the lung cancer locations to obtain an unnormalised estimate ρ of the spatially varying population density of susceptibles, shown in Figure 9.8 on page 321. This approach is open to critique [224, 236, 55] but is shown here for demonstration purposes.

```
> smo <- density(lung, sigma=0.15, eps=0.1)
> smo <- eval.im(pmax(smo, 1e-10))
```

The null model, of 'constant relative risk', postulates that the larynx cases are a Poisson process with intensity $\lambda(u) = \kappa \rho(u)$ at location u. The parameter κ adjusts for the relative abundance of the two types of data points; it is the baseline relative risk of larynx and lung cancer, multiplied by the ratio of sampling fractions used when these data were sampled from the cancer registry. This *could* be used to infer the absolute risk of laryngeal cancer if these sampling fractions were known. We fit the null model:

```
> ppm(Q ~ offset(log(smo)))
```

The fitted coefficient (`Intercept`)`=-2.824` is the estimate of $\log \kappa$.

In Section 10.5.2 we advocated using the Kolmogorov-Smirnov test or similar tests (Anderson-Darling, Berman-Lawson-Waller) to decide whether a point process depends on a covariate. However in this context, where the covariate is the distance from a fixed point and the alternative is that very small distance values are overrepresented, these tests have weak power, because the null and alternative differ only for a small range of distance values. A parametric modelling approach is more powerful in this context.

Diggle [224] considered alternative 'raised incidence' models that include an effect due to proximity to the incinerator. Assume the intensity of laryngeal cancer cases at a location u is

$$\lambda_\theta(u) = \kappa \, \rho(u) \, b_{\alpha,\beta}(d(u)) \tag{9.81}$$

where $d(u)$ is the distance in kilometres to the incinerator from location u. Here $\theta = (\kappa, \alpha, \beta)$ is the parameter vector and

$$b_{\alpha,\beta}(d) = 1 + \alpha \exp(-\beta d^2) \tag{9.82}$$

is the raised risk ratio at a distance d from the incinerator [224, eq. (6)]. The parameters α, β control the magnitude and dropoff rate, respectively, of the incinerator effect. The log intensity is

$$\log \lambda_\theta(u) = \log \kappa + \log \rho(u) + \log(1 + \alpha \exp(-\beta d^2)); \tag{9.83}$$

this is not a linear function of the parameters α and β, so these parameters are irregular, and we need to use another algorithm such as `ippm` or `profilepl` to fit the model. Since (9.82) is differentiable with respect to α and β, we can use `ippm`.

We start by defining the squared distance to the incinerator, as this function will be re-used frequently.

```
> d2incin <- function(x, y, xincin=354.5, yincin=413.6) {
    (x - xincin)^2 + (y - yincin)^2
  }
```

The arguments `xincin,yincin` are the coordinates of the incinerator. By specifying default values for them, we do not have to handle these values again. Next we define the raised incidence term $b_\theta(u)$:

```
> raisin <- function(x,y, alpha, beta) {
    1 + alpha * exp( - beta * d2incin(x,y))
  }
```

We then fit the model using `ippm` (this can be slow):

```
> chorleyDfit <- ippm(Q ~offset(log(smo) + log(raisin)),
                      start=list(alpha=5, beta=1))
```

The argument `start` lists the irregular parameters over which the likelihood is to be maximised, and gives their starting values for the iterative algorithm. The parameter names should match the names of arguments of the function `raisin`. As intended, the arguments `xincin, yincin` are held fixed, because they are not listed in `start`.

```
> chorleyDfit
Nonstationary Poisson process
Log intensity:   ~offset(log(smo) + log(raisin))
            Estimate   S.E.  CI95.lo CI95.hi Ztest    Zval
(Intercept)   -2.896 0.1313  -3.153  -2.638   ***  -22.05
```

The MLE's are $\hat\alpha = 22.2$ (dimensionless) and $\hat\beta = 0.89\,\mathrm{km}^{-1}$. The `spatstat` generic function `parameters` can be used to extract all the model parameters, including both regular and irregular trend parameters:

```
> unlist(parameters(chorleyDfit))
  trend.(Intercept)             alpha                beta
           -2.8958           22.2410              0.8892
```

Note that the iterative maximisation procedure can fail to converge, because the parameters do not appear in loglinear form and the log-likelihood may not be a convex function. In principle there could even be several different maxima of the likelihood. Tweaking the arguments passed to the optimisation function `nlm` usually sidesteps these problems.

9.13 Local likelihood

In local likelihood inference [335, 428] or geographically weighted regression [267], the model parameters are allowed to be spatially varying. More precisely, we start with a homogeneous model; at each spatial location u we fit the same model to data in a neighbourhood of u, obtaining a parameter estimate $\widehat\theta(u)$ that will vary between locations.

Suppose we define the neighbourhood of location u as the disc $b(u,R)$ centred at u, with some fixed radius R. At each location u we would fit a model to the data inside $b(u,R)$, obtaining a parameter estimate $\widehat\theta(u)$. As we move the disc over the study region, the fitted parameter values will change. This is reminiscent of the scan statistic (page 190).

Instead of including or excluding data according to whether they fall inside or outside a disc, we could weight the data using a smoothing kernel κ. For a Poisson point process model with intensity $\lambda_\theta(u)$, the *local log likelihood* or *geographically weighted likelihood* at location u is

$$\log L_u(\theta) = \sum_{i=1}^{n} \kappa(u - x_i) \log \lambda_\theta(x_i) - \int_W \kappa(u - v)\,\lambda_\theta(v)\,\mathrm{d}v. \qquad (9.84)$$

For each location u, the local log likelihood can be maximised using the same techniques as for the original likelihood, yielding a parameter estimate $\widehat{\theta}(u)$.

Code for local likelihood [33] will be added to `spatstat` soon. The fitting function is `locppm`. For the Queensland copper data:

```
> X <- rotate(copper$Points, pi/2)
> L <- rotate(copper$Lines, pi/2)
> D <- distfun(L)
> copfit <- locppm(X ~ D, eps=1.5, sigma=bw.locppm)
```

Here the smoothing bandwith σ was selected by cross-validation.

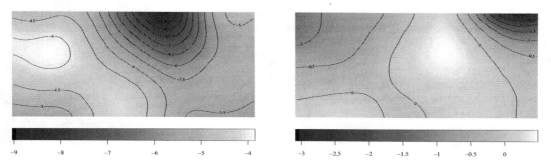

Figure 9.16. *Spatially varying estimates of intercept (*Left*) and slope coefficient (*Right*) from the local likelihood fit of the loglinear model to the Queensland copper data.*

Figure 9.16 is the result of `plot(copfit)`, showing the spatially varying fitted estimates $\widehat{\alpha}(u)$, $\widehat{\beta}(u)$ of the intercept parameter α and slope parameter β in the loglinear model $\lambda(u) = \exp(\alpha + \beta d(u))$ where $d(u)$ is the distance function of the geological lineaments. The estimates of β are close to zero, suggesting a weak relationship, across most of the survey area.

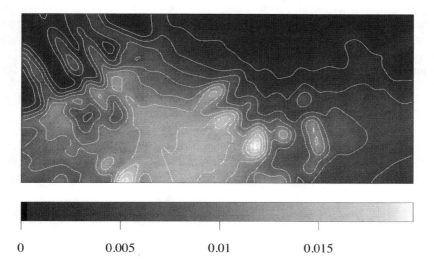

Figure 9.17. *Fitted intensity of copper deposits according to the local likelihood fit of the loglinear model.*

Figure 9.17 shows the fitted intensity $\widehat{\lambda}(u) = \exp(\widehat{\alpha}(u) + \widehat{\beta}(u)d(u))$ of each local model. This strongly suggests there is a higher intensity of copper deposits in the middle of the survey region.

Figure 9.17 is very similar to the picture that would be obtained simply by kernel smoothing the copper deposits without reference to the lineament pattern. However, the top left of Figure 9.17 shows an area where the predicted intensity under the local model is higher than the kernel-smoothed intensity, because of an apparent association between copper and lineaments in that area.

9.14   FAQs

- *How accurate are the parameter estimates computed by* ppm *?*

  This depends on the algorithm settings such as the number of quadrature points, the quadrature weighting method, and the choice of fitting algorithm. See [53, 38] for detailed measurements of accuracy and bias.

  The default settings are chosen to achieve speed rather than high accuracy, and are suitable for initial analysis. (High speed is a requirement of the R package checker.) For accurate or definitive analysis, a larger number of quadrature points should be used, by setting the argument nd or eps in ppm, or by setting spatstat.options("ndummy.min").

- *I get different parameter estimates from different versions of* spatstat.

  Later versions of spatstat typically give more accurate results. Progressive increases in computational efficiency allow us to change the default settings of ppm to increase accuracy.

  To ensure that calculations are exactly reproducible, the algorithm parameters would have to be fixed explicitly. In ppm this means the arguments method and correction, and a specification of the quadrature scheme, such as the arguments nd and nt.

- *Does* ppm *take account of correlation and confounding between the covariates?*

  Like most regression techniques, ppm treats the covariates as non-random quantities that are measured without error. The covariates do not have 'correlation' in the statistical sense. Any empirical correlation between the covariates is taken into account when calculating the variance-covariance matrix (vcov.ppm), when performing the likelihood ratio test (anova.ppm), and so on. If the covariates are confounded in the sense that they are approximately linearly related, the Fisher information matrix will be singular or ill-conditioned, and an error or warning will occur.

- *When I perform pixel logistic regression using* slrm*, the fitted presence probabilities in each pixel are tiny numbers like* 10^{-8}*. Is this physically meaningful?*

  Yes. The pixels are tiny, so the probability that a point falls in a given pixel is tiny. In any spatial region of appreciable size, the expected number of points is the sum of the pixel presence probabilities, a sum of a very large number of very small values, yielding a physically meaningful probability. The chance that I win the lottery is tiny; the chance that someone wins the lottery is high.

- *Is it true that a logistic regression model does not assume independence between pixels?*

  The simplest answer is that standard application of logistic regression gives the same result as if we had assumed independence between pixels.

  The logistic regression relationship $\log(p/(1-p)) = \beta_0 + \beta_1 x$ itself does not assume independence between pixels. However, in order to fit a regression model, we must also describe the probability distribution of the observations, including the dependence between them. The standard 'logistic regression model' does assume that observations are independent. More importantly, the standard algorithm for *fitting* a logistic regression assumes independence between pixels (since the correlation structure determines the Fisher information matrix which is used in the Iteratively Reweighted Least Squares algorithm).

A logistic regression relationship can still be fitted if we assume some kind of dependence between pixels. However, the fitting algorithm must be adjusted to incorporate this dependence. See Chapter 12.

- *I want to fit a model where the intensity depends on a covariate Z. I don't want to make any restrictive assumptions about the relationship between intensity and Z. As I understand it, a loglinear Poisson point process model is a very specific parametric model, whereas logistic regression and Maxent are flexible non-parametric models. Can you confirm this?*

Fitting a loglinear Poisson process model using ppm is **equivalent** to fitting a spatial (pixel) logistic regression model [34, 691], and also equivalent to fitting a Maxent model [568]. If mathematical proofs don't convince you, try comparing the results of ppm and slrm, or look at the examples in [34, 691, 568, 567].

Fitting loglinear Poisson models (or equivalently fitting spatial logistic regressions or fitting Maxent models) is quite a flexible approach, because we are allowed to build models using any transformation of the original covariates. For example, we can easily fit a model in which the log intensity is a cubic function of the covariate Z. The limitation is that the user must specify the transformed covariates.

If you believe that $\lambda(u)$ depends *only* on the covariate $Z(u)$, then the best place to start is a truly non-parametric estimator based on smoothing, such as rhohat (Section 6.6.3). Plotting the estimated relationship may suggest the form of a model which can then be fitted using ppm.

Very flexible regression relationships can also be fitted using generalised additive models as explained in Section 9.6. This also works when there are several covariates.

- *How do I obtain standard errors for the estimates of irregular parameters fitted by* ippm *or* profilepl?

At the time of writing, the only generally-applicable technique is a bootstrap approach in which the model is re-fitted to simulated realisations from the model, and the sample standard deviation of the parameter estimates is computed. A code snippet is available at www.spatstat.org.

10

Hypothesis Tests and Simulation Envelopes

This chapter explains how to conduct a variety of tests of statistical significance, and how to use simulation envelopes for hypothesis testing. It also attempts to correct some of the common misconceptions about these methods.

10.1 Introduction

Often the main purpose of statistical analysis is to decide whether a claim is true or false. For example, the Chorley-Ribble data (page 9) were studied in Section 9.12.3.3 to investigate whether an industrial incinerator did, or did not, increase the risk of laryngeal cancer for residents nearby. A standard tool for this purpose is *hypothesis testing* or *significance testing*, which is covered in this Chapter.

Hypothesis tests and their associated *p*-values are often used in scientific publications to reach a *formal conclusion* about the scientific question that is under study. However, they also have several other roles in statistical analysis.

Tests can be used to *guide the choice of strategy* when we start analysing a dataset. In the analysis of spatial point patterns, a hypothesis test can be used to decide whether we can provisionally treat the pattern as homogeneous (and then use the *K*-function, etc.) or whether spatial variation in intensity must be investigated.

Hypothesis testing is also one of the tools available for *model selection*, that is, for choosing the best among several competing models for the same data. This is useful at intermediate stages of data analysis, where we need to adopt a provisional model in order to perform detailed analysis. Model selection is also part of the formal process of assessing evidence for an incinerator effect in the Chorley-Ribble data, for example. Hypothesis testing is useful for selecting amongst a *small* number of competing models.

In the analysis of spatial point patterns, statistical tests are often based on *simulation envelopes*, such as the envelopes of simulations of Ripley's *K*-function (Chapter 7).

10.2 Concepts and terminology

Following is a brief summary of concepts and terminology for hypothesis tests. See [417] for details.

In a nutshell: since observations are affected by random variability, our data may show a trend or change or difference which is not 'real', but is simply due to this random variability. An apparent trend in the observed data (say, numbers of accidents) does not necessarily signify that there is a trend in the underlying variable (say, accident risk): it is not necessarily **significant** in the statistical sense. The purpose of statistical tests is to decide whether findings can or cannot be attributed to chance.

In most scientific literature, the convention for reporting a statistical test is actually a muddled compromise between two different procedures, R.A. Fisher's *significance testing* and Neyman-Pearson *hypothesis testing*. We sketch them both, to clarify some important misunderstandings.

Significance testing

In Fisher's significance testing approach, the researcher formulates a **null hypothesis** H_0 which represents the statement that 'nothing is happening' — the treatment has no effect, or there is no difference between groups, or there is no relationship between the variables measured.

The test involves a numerical quantity called the **test statistic** T. This quantity is selected because it takes smaller (or 'less extreme') values when the null hypothesis is true, and larger (or 'more extreme') values when it is false. Using the observed data, the researcher calculates the 'observed value' t_{obs} of the test statistic.

Finally, the researcher calculates the *p*-**value** defined as *the probability, assuming H_0 is true, that the test statistic would have given a value at least as extreme as the observed value t_{obs}*. In the left panel of Figure 10.1, the *p*-value is the area under the probability density curve to the right of the observed value t_{obs}.

According to Fisher, a small *p*-value suggests that *"either the null hypothesis is false, or a very unusual event has just occurred"* [260]. A very small *p*-value strongly suggests the observations are statistically significant, although the researcher always has the option to believe the contrary.

Notice that the *p*-value is calculated under the assumption that the null hypothesis is true. It is *not* 'the probability that H_0 is true'. A *p*-value of 5% does *not* give us '95% confidence' in the result. The *p*-value only has a rigorous interpretation under a hypothetical scenario, the null hypothesis, which is probably false in this context. The *p*-value is best interpreted using the quote from Fisher above.

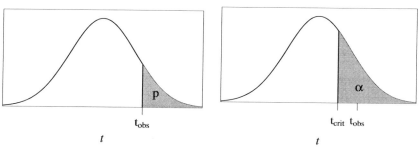

Figure 10.1. *Fisher (Left) and Neyman-Pearson (Right) versions of statistical testing. The curve is the probability density of the test statistic T assuming H_0 is true.*

Hypothesis testing

In the Neyman-Pearson [498] approach, the researcher formulates two hypotheses, the null hypothesis H_0 and the **alternative hypothesis** H_1. The null hypothesis is the statement that 'nothing is happening'. The alternative hypothesis effectively specifies what kinds of departures from the null hypothesis we wish to be able to detect efficiently. For example, in a comparison between a drug and a placebo, the null hypothesis is that there is no difference in effect between the drug and a placebo, while the alternative is usually that the drug has a *more beneficial* effect than the placebo.

Unlike Fisher's approach, a Neyman-Pearson test is a mechanical procedure which results in one of two possible outcomes: either 'H_0 is rejected' or 'H_0 is not rejected.' Roughly speaking, the procedure will decide either that there is sufficient evidence against H_0, or that there is not.

The test procedure can make two kinds of mistakes: a **Type I error** (false positive) occurs if we reject H_0 when H_0 is really true; a **Type II error** (false negative) occurs if we fail to reject

H_0 when H_0 is really false. Performance is measured by the probability α of Type I error and the probability β of Type II error. If we compare a Neyman-Pearson test to a criminal trial, then H_0 is the presumption of innocence, α is the probability that an innocent person will be found guilty, and β is the probability that a guilty person will be acquitted.

For the validity of the Neyman-Pearson approach, it is very important that the test procedure is specified *before* we conduct the experiment and look at the results. The researcher must also specify, in advance, what value of α will be tolerated — called the **significance level**. Typical choices for the significance level α are 0.05 and 0.01. The smaller the value of α, the more stringent the test.

The researcher calculates the 'observed value' t of the test statistic for the observed data. The test is performed by checking whether the observed value t is more extreme than the 'critical value' t_{crit}, defined as the value that would be exceeded with probability α if the null hypothesis is true. If t is more extreme than t_{crit}, then H_0 is rejected; otherwise it is not.

The right panel of Figure 10.1 shows a situation where H_0 is rejected. The critical value t_{crit} is determined by the requirement that the shaded area under the density curve to the right of t_{crit} should equal α. The observed value t_{obs} is more extreme than t_{crit}, so we reject H_0.

Performing a Neyman-Pearson test with significance level α is usually equivalent to calculating Fisher's p-value and then rejecting H_0 if the p-value is less than or equal to α.

The opposite of 'reject' is 'accept', so it is common to say that the null hypothesis was 'accepted'. Strictly speaking this is unsound (we can only 'fail to reject' the null hypothesis) but we use it sometimes to avoid awkward sentences. Note however that the *alternative* hypothesis can never be 'accepted'. The alternative hypothesis only specifies the *direction* of departures from the null hypothesis. Evidence *for* the alternative is not directly assessed by a Neyman-Pearson test.

The **power** of a test is the probability $1 - \beta$ of *not* making a Type II error. That is, power is the probability of making the correct decision when H_1 is true. The power depends on the alternative hypothesis H_1.

10.3 Testing for a covariate effect in a parametric model

10.3.1 General problem

A common task for analysis is to decide whether a point pattern **x** depends on a covariate Z, after adjusting for variables that are known to influence **x**. For example, in the Chorley-Ribble data (page 9) we wish to decide whether the rate of laryngeal cancer depends on distance from the incinerator, after allowing for spatial variation in the population density. In the Murchison gold data, we might wish to determine whether the abundance of gold deposits depends on distance from the nearest geological fault, after adjusting for rock type.

This is effectively a choice between two models, which differ only in that one of the models includes a term for the effect of the covariate Z in question, and the other does not. In order to 'adjust' for other variables that are known to influence **x**, these other variables are included in both the models. For example in the Murchison gold survey,

```
> mur <- lapply(murchison, rescale, s=1000, unitname="km")
> mur$dfault <- with(mur, distfun(faults))
```

we might compare the two Poisson point process models

```
> mfit0 <- ppm(gold ~ greenstone, data=mur)
> mfit1 <- ppm(gold ~ greenstone + dfault, data=mur)
```

or equivalently

```
> mfit1 <- update(mfit0, . ~ . + dfault)
```

If ψ is the vector of parameters for the first model `mfit0`, and φ is the vector of parameters corresponding to the extra covariate term Z, then the augmented model `mfit1` has parameters $\theta = (\psi, \varphi)$. We can regard `mfit0` as a special case of `mfit1` in which the extra parameters φ are equal to zero.

Effectively we wish to decide whether $\varphi = 0$ or not. A standard tool for this purpose is a test of the null hypothesis $H_0 : \varphi = 0$ against the alternative $H_1 : \varphi \neq 0$. Note that the null hypothesis corresponds to the *absence* of an effect due to the covariate Z.

Another example is a formal test of homogeneous intensity against the alternative that the intensity depends on a covariate Z. For example, for the Queensland copper data (Figure 1.11)

```
> Dist <- with(copper, distfun(SouthLines))
> cfit0 <- ppm(SouthPoints ~ 1, data=copper)
> cfit1 <- ppm(SouthPoints ~ Dist, data=copper)
```

Again the task is to compare the two models `cfit0` and `cfit1` where the null hypothesis, fitted by the smaller model `cfit0`, corresponds to the *absence* of an effect due to the covariate.

10.3.2 Likelihood ratio test

The **likelihood ratio test** of $H_0 : \varphi = 0$ against $H_1 : \varphi \neq 0$ in a statistical model with parameters $\theta = (\psi, \varphi)$ is based on the test statistic[1]

$$\Gamma = 2\log\frac{L_1}{L_0} = 2(\log L_1 - \log L_0) \tag{10.1}$$

where L_0 and L_1 are the maximum values of the likelihood achieved under H_0 and H_1, respectively. For loglinear Poisson point process models, under regularity conditions, the large-sample asymptotic distribution of Γ under H_0 is χ^2 with d degrees of freedom, where d is the dimension of φ [565, 396]. The likelihood ratio test is theoretically optimal (uniformly most powerful) in some contexts.

The likelihood ratio test for Poisson models can be carried out by `anova.ppm`, a method for the generic function `anova`. For the Murchison gold data:

```
> anova(mfit0, mfit1, test="LRT")
Analysis of Deviance Table
Model 1: ~greenstone          Poisson
Model 2: ~greenstone + dfault          Poisson
  Npar Df Deviance Pr(>Chi)
1    2
2    3  1     67.8   <2e-16 ***
---
Signif. codes:  0 '***' 0.001 '**' 0.01 '*' 0.05 '.' 0.1 ' ' 1
```

Although the name of the generic function is `anova`, the extension of Analysis of Variance to more general models is Analysis of Deviance, as indicated by the printed output. The argument `test="LRT"` or `test="Chi"` specifies the likelihood ratio test. The value of Γ is reported under the heading `Deviance`, and the number of degrees of freedom is `Df`. The p-value given under the heading `Pr(>Chi)` is obtained by referring Γ to the χ^2 distribution with 1 degree of freedom. The symbol `***` refers to the significance codes printed at the bottom of the output, and indicates that $p < 0.001$. The null hypothesis is emphatically rejected, indicating strong evidence that the

---

[1]The likelihood ratio test statistic is usually written as Λ in statistical literature, but in this book Λ and λ always refer to point process intensity.

abundance of gold deposits depends on distance to the nearest fault. However, see the caveats in Section 10.3.6 below.

```
> anova(cfit0, cfit1, test="LRT")
Analysis of Deviance Table
Model 1: ~1          Poisson
Model 2: ~Dist          Poisson
  Npar Df Deviance Pr(>Chi)
1   1
2   2 1    0.34    0.56
```

In this case the null hypothesis is 'accepted' (i.e. not rejected), indicating a lack of evidence that the abundance of copper deposits depends on distance to the nearest lineament.

```
>    lung <- split(chorley)$lung
>    larynx <- split(chorley)$larynx
>    Q <- quadscheme(larynx, eps=0.1)
```

The Chorley-Ribble data were introduced on page 9. Diggle's [224, 236] raised incidence model was fitted on page 364:

```
> chorleyDfit
Nonstationary Poisson process
Log intensity:  ~offset(log(smo) + log(raisin))
              Estimate  S.E. CI95.lo CI95.hi Ztest  Zval
(Intercept)      -2.9 0.131   -3.15   -2.64   ***  -22.1
```

To test whether the raised incidence term `raisin` is significant, we fit the model without this term:

```
> chorley0fit <- update(chorleyDfit, . ~ offset(log(smo)))
```

Warning: The operator '−' is *not supported for offsets* in model formulae. The command
          `update(chorleyDfit, . ~ . - offset(log(raisin)))`
would **not** delete the offset term.

Then we invoke `anova.ppm`:

```
> anova(chorley0fit, chorleyDfit, test="LRT")
Analysis of Deviance Table
Model 1: ~offset(log(smo))             Poisson
Model 2: ~offset(log(smo) + log(raisin))          Poisson
  Npar Df Deviance Pr(>Chi)
1   1
2   3 2    9.31   0.0095 **
---
Signif. codes:  0 '***' 0.001 '**' 0.01 '*' 0.05 '.' 0.1 ' ' 1
```

The result indicates strong evidence against H_0 in favour of the raised incidence alternative.

    Note that some special tricks are happening in the last call to `anova.ppm` above. At first glance, it appears to compare two ppm objects defined by formulae that contain only offset terms. This would not be possible if the models were both fitted by ppm, because two such models are not nested: one 'pure offset' model is not a special case of another. However, in this case, the alternative model `chorleyDfit` was fitted by `ippm` and includes two irregular parameters `alpha` and `beta`, the parameters of the function `raisin` appearing in the second offset term. The models *are* nested, because if we set `alpha=0` then the `log(raisin)` term is identically equal to zero. The likelihood ratio test is therefore valid, and `anova.ppm` correctly counts the degrees of freedom as $d = 2$. Note that `anova.ppm` is not smart enough to verify that the models are nested: it relies on the user to check this.

10.3.3 Wald test for single parameter

The Wald test of $H_0 : \varphi = 0$ for a *one-dimensional* parameter φ is based on the statistic

$$V = \frac{\widehat{\varphi}}{\text{se}(\widehat{\varphi})} \tag{10.2}$$

where $\widehat{\varphi}$ is the maximum likelihood estimate of φ under the alternative hypothesis H_1, and $\text{se}(\widehat{\varphi})$ is the plug-in estimate of standard error of $\widehat{\varphi}$. The asymptotic null distribution of V is standard normal. The likelihood ratio test is asymptotically equivalent to the Wald test, under regularity conditions.

For a loglinear Poisson point process model fitted by ppm, the Wald test for each coefficient is performed as part of the print method. It can be recovered using coef(summary(..)).

```
> coef(summary(mfit1))
              Estimate  S.E. CI95.lo CI95.hi Ztest    Zval
(Intercept)    -6.617 0.2171  -7.043 -6.1916  ***  -30.48
greenstoneTRUE  2.754 0.2066   2.349  3.1588  ***   13.33
dfault         -0.104 0.0179  -0.139 -0.0686  ***   -5.78
```

The column headed Ztest reports the outcomes of the Wald tests for each of the model coefficients, using the customary asterisk convention. In particular the covariate effect for dfault, distance to the nearest fault, is significant at the 0.001 level.

The output includes the value of the Wald statistic, under the heading Zval. This makes it possible to perform a *one-sided* test of $H_0 : \varphi = 0$ against $H_2 : \varphi < 0$, which is the alternative of interest for the Murchison data, corresponding to a concentration of gold deposits *near* the geological faults rather than *away* from the faults:

```
> V <- coef(summary(mfit1))["dfault", "Zval"]
> pnorm(V, lower.tail=TRUE)
[1] 3.69e-09
```

The use of character indices helps to ensure we are extracting the correct entry from the data frame. The call to pnorm computes the p-value corresponding to the left-sided test, because we requested the lower tail probability. The result gives strong evidence against H_0 in favour of H_2.

The Wald test for *irregular* parameters is not yet supported in spatstat. For example it is not yet straightforward to apply the Wald test to the parameters alpha and beta in the raised-incidence model for the Chorley-Ribble data.

10.3.4 Score test

The likelihood ratio test and Wald test require us to estimate the parameters under the alternative hypothesis — that is, to fit the full model including covariate effects — which can be difficult and computationally costly. It may be convenient to apply the Rao [563] *score test* (also known as the Lagrange multiplier test) which only requires us to fit the *null* model, that is, the model without the covariate effect. In the simplest case where we test $H_0 : \theta = \theta_0$ against $H_1 : \theta \neq \theta_0$ in a model with parameter θ, the score test statistic is

$$S = U(\theta_0) I_\theta^{-1} U(\theta_0) \tag{10.3}$$

where $U(\theta)$ and I_θ are the score function and Fisher information. If we split the parameter vector into $\theta = (\psi, \varphi)$ and wish to test $H_0 : \varphi = 0$ against $H_1 : \varphi \neq 0$, we first compute the maximum likelihood estimate of θ under H_0, say $\dot{\theta} = (\dot{\psi}, 0)$, then compute

$$S = U(\dot{\theta})_\varphi^\top \left(I_\theta^{-1} \right)_{\varphi\varphi} U(\dot{\theta})_\varphi \tag{10.4}$$

where the subscripts φ denote the components of the score vector and inverse Fisher information matrix (of the full model) that correspond to the parameter φ. Under H_0 the test statistic has a χ_d^2 distribution where d is the dimension of φ. The score test is less powerful than the likelihood ratio test or Wald test in small samples, but asymptotically equivalent to them in large samples, under regularity conditions.

In anova.ppm the score test is specified by test="Rao" or test="score":

```
> anova(mfit0, mfit1, test="Rao")
Analysis of Deviance Table
Model 1: ~greenstone           Poisson
Model 2: ~greenstone + dfault          Poisson
  Npar Df Deviance  Rao Pr(>Chi)
1    2
2    3  1    67.8 36.6  1.4e-09 ***
---
Signif. codes:  0 '***' 0.001 '**' 0.01 '*' 0.05 '.' 0.1 ' ' 1
```

10.3.5 Berman's tests

An important example of the score test in spatial statistics is the Berman-Lawson-Waller test [88, 685, 413] introduced in Section 6.7.2.

Suppose we want to test the null hypothesis that the point process intensity is constant (or more generally, proportional to a known baseline) against the alternative that the intensity depends on a specified covariate Z. Suppose $b(u)$ is a known baseline function, $Z(u)$ is a specified covariate function, and assume a Poisson process with intensity

$$\lambda_\theta(u) = b(u)\exp(\psi + \varphi Z(u)) \tag{10.5}$$

where $\theta = (\psi, \varphi)$ are parameters. The null hypothesis $H_0 : \varphi = 0$ states that the intensity is proportional to the baseline $b(u)$, and does not depend on Z. The alternative $H_1 : \varphi \neq 0$ means that the intensity is *loglinear* in $Z(u)$.

We could fit the model (10.5) using ppm and perform the test using anova.ppm. Either the likelihood ratio test, Wald test or score test could be used. There is, however, no analytically exact formula for the likelihood ratio test statistic Γ or the Wald test statistic Z in this model. On the other hand, the score test is easy to describe: straightforward calculation gives the score test statistic

$$S = \frac{(T - nM_1)^2}{n(M_2 - M_1^2)}$$

where $n = n(\mathbf{x})$ is the number of data points, $T = \sum_i Z(x_i)$ is the sum of covariate values over all data points, and

$$M_k = \frac{\int_W Z(u)^k b(u)\,du}{\int_W b(u)\,du}, \quad k = 1, 2$$

is the mean value of $Z(U)^k$ when U is a random point in the window W with probability density proportional to $b(u)$. To perform the score test, we refer the observed value of S to the χ^2 distribution with 1 degree of freedom. Equivalently, $V = (T - nM_1)/\sqrt{n(M_2 - M_1^2)}$ is referred to the standard normal distribution. The score test can be performed using anova.ppm:

```
> anova(cfit0, cfit1, test="score")
```

```
Analysis of Deviance Table
Model 1: ~1          Poisson
Model 2: ~Dist          Poisson
  Npar Df Deviance    Rao Pr(>Chi)
1   1
2   2  1     0.34 0.349      0.55
```

Alternatively if we condition on the number of points $n(\mathbf{x}) = n$, then the score test is based on

$$S' = \frac{(T - nM_1)^2}{nM_2} \quad \text{or} \quad V' = \frac{T - nM_1}{\sqrt{nM_2}}.$$

This is the Z_1 test of Berman [88], who showed that the one-sided version of the test is theoretically optimal (uniformly most powerful) for testing $H_0 : \phi = 0$ against $H_1 : \phi > 0$. It is closely related to the tests of Waller *et al.* [685] and Lawson [413] for discrete counts.

In Berman's Z_2 test, the observed values z_i are first transformed to $u_i = F_0(z_i)$ where

$$F_0(z) = \frac{1}{|W|} \int_W \mathbf{1}\{Z(u) \le z\}\, \mathrm{d}z$$

is the spatial cdf of the covariate Z. Under the null hypothesis, these transformed values would be uniformly distributed on $[0, 1]$. The test statistic is the standardised sum of these transformed values,

$$Z_2 = \sqrt{\frac{12}{n}}\left(\sum_i u_i - \frac{1}{2}\right).$$

Again Z_2 is asymptotically normal under the null hypothesis. The `spatstat` function `berman.test` described in Section 6.7.2 performs these tests. The Berman tests can also be applied to a *fitted* Poisson point process model, using the method for `"ppm"`:

```
> dcop <- distfun(copper$SouthLines)
> berman.test(cfit0, dcop)

        Berman Z1 test of CSR in two dimensions

data:  covariate 'dcop' evaluated at points of 'SouthPoints'
Z1 = 0.5, p-value = 0.6
alternative hypothesis: two-sided
```

10.3.6   Caveats

Scientific articles sometimes misstate the conclusion from a hypothesis test, exaggerate the strength of the conclusion, or overlook potential weaknesses of a test. Here we list some of the most important caveats.

The conclusion from a hypothesis test

Hypothesis testing focuses on the *truth or falsity of the null hypothesis* H_0. The possible conclusions from the test are either to 'accept H_0' or 'reject H_0'. If the test accepts the null hypothesis, we are effectively reporting there is insufficient evidence against it. If the test rejects H_0, we are reporting that there is sufficient evidence against H_0. Concluding that H_0 is false does not necessarily imply that H_1 is true.

The *p*-value, as discussed in Section 10.1, is a measure of the strength of evidence *against* the null hypothesis, with smaller values indicating stronger evidence. It is defined as the probability of obtaining data at least as extreme as the data that were observed, *assuming the null hypothesis is true*. 'Extreme' is defined below. The logic of significance testing is that, if the p-value is small, then "either an exceptionally rare chance has occurred, or the theory [null hypothesis] is not true" [262, p. 39]. The p-value measures how rare this 'rare chance' is.

A common fallacy is to interpret a p-value of 0.03 as meaning we are 97% confident that the alternative hypothesis is true (or equally bad, that there is a 0.97 probability that the null hypothesis is false). A hypothesis test never accepts the alternative; and the p-value is not a measure of confidence about the alternative. The p-value is a probability calculated under the assumption that the *null* hypothesis is true. *A hypothesis test does not evaluate the strength of evidence for the alternative hypothesis.*

Role of the alternative hypothesis

The alternative hypothesis serves mainly to define the term 'extreme' used in calculating the p-value. 'Extreme' observations are those which are more consistent with the alternative hypothesis than the null hypothesis. For example, in the Wald test for the Murchison gold example (Section 10.3.3) the null hypothesis is $H_0 : \varphi = 0$. Specifying the alternative $H_1 : \varphi \neq 0$ implies that extreme values of the test statistic V are those which are large and positive, or large and negative. Specifying $H_1 : \varphi < 0$ implies that only large *negative* values of V are considered extreme. Essentially the alternative hypothesis determines the *direction* of departures from the null hypothesis that are relevant to the test.

Changing the alternative hypothesis can change the outcome of the test. Making the alternative hypothesis more restrictive will reduce the p-value, strengthening the apparent evidence against the null hypothesis.

Tests depend on assumptions

A hypothesis test involves assumptions, and the conclusions may be highly sensitive to violations of these assumptions.

In all the examples in Sections 10.3.2–10.3.4 above, we assumed that the point process was a Poisson process, and entertained various possibilities for the intensity function. The calculated p-values therefore assume that the point process is Poisson, among other assumptions.

For the Murchison gold data, the null model is a Poisson process with two different, constant intensities inside and outside the greenstone. The p-value is calculated under these assumptions. The p-value close to zero indicates that, assuming this null model to be true, it would be virtually impossible to obtain a value of Γ as large as the value obtained. As discussed above, this does not imply the truth of the alternative model, in which the intensity of gold deposits is an exponential function of distance to the nearest fault. Essentially we have not evaluated the strength of evidence for an exponential relationship. Even if we are confident that the gold deposits are a Poisson process (which seems implausible from a geological standpoint) the best we can conclude from the Likelihood Ratio Test is that the intensity is not constant inside the greenstone *and/or* is not constant outside the greenstone.

The null hypothesis

Hypotheses are often written in shorthand — for example, '$H_0 : \mu_1 = \mu_2$' or 'no difference between groups'. However, this is dangerous if the user does not appreciate the full meaning of the shorthand. Does 'no difference between groups' signify that the two groups have the same population means, or the same population distributions, or something else appropriate to the scientific context?

Strong [644] gave an important example related to the study of species diversity on islands. The scientific null hypothesis is that small and large island communities are governed by the same ecological processes. Researchers have tested this by measuring species diversity on large and small islands, and testing whether the diversity index values are equal. However, in a smaller population chosen at random from a larger population, less species diversity is to be expected. A test of 'no difference in species diversity' based on the diversity index values would be erroneous.

The null hypothesis represents what we expect to see if 'nothing is happening'. It can be very important to spell out exactly what we expect to see under the null hypothesis.

Asymptotic distributions

A technical point about the tests described in Sections 10.3.2–10.3.4 above is that they use the *asymptotic* (large sample limit) distributions of the test statistics, namely the χ^2 and normal distributions. In small samples, these are only approximations to the true distribution of the test statistic, and therefore the p-value is an approximation. More accurate p-values can be obtained by simulation from the null hypothesis, with some care.

10.4 Quadrat counting tests

Quadrat counts were introduced in Chapter 6 for testing whether a point process has homogeneous intensity. Quadrat counts can be used for several other purposes, including tests of goodness-of-fit of a fitted model, and tests of independence between points.

10.4.1 Quadrat counting tests of fitted models

The χ^2 test of homogeneity based on quadrat counts (Sections 6.4.2 and 6.7.1) can be applied more generally to a fitted inhomogeneous Poisson model. Suppose we have fitted an inhomogeneous Poisson process with intensity function $\lambda_\theta(u)$ to a point pattern dataset **x**. *Assuming* the true point process is Poisson, we will test the null hypothesis H_0 that the intensity is of the form $\lambda_\theta(u)$ for some value of the parameter θ, against the alternative H_1 that the intensity is of some other form. Under the null hypothesis, the quadrat counts n_j in quadrats B_j are realisations of independent Poisson variables with different mean values μ_j. The means μ_j have to be *estimated* using the intensity of the fitted model:

$$\hat{\mu}_j = \int_{B_j} \hat{\lambda}(u)\,du$$

where $\hat{\lambda}(u) = \lambda_{\hat{\theta}}(u)$ is the fitted intensity obtained by substituting the parameter estimate $\widehat{\theta}$ into the formula for the intensity in the model. The χ^2 test statistic is

$$X^2 = \sum_j \frac{(\text{observed} - \text{expected})^2}{\text{expected}} = \sum_j \frac{(n_j - \hat{\mu}_j)^2}{\hat{\mu}_j}. \tag{10.6}$$

The test is conducted by referring the observed value of X^2 to the χ^2 distribution with $m - p$ degrees of freedom, where m is the number of quadrats and p is the number of model parameters that were estimated. Subtracting p can be regarded as an adjustment for the fact that the parameters were estimated. This calculation is performed in spatstat by the method quadrat.test.ppm.

```
> fit2e <- ppm(bei ~ polynom(elev,2), data=bei.extra)
> M <- quadrat.test(fit2e, nx=4, ny=2)
> M

        Chi-squared test of fitted Poisson model 'fit2e' using quadrat counts
        Pearson X2 statistic
```

```
data:   data from fit2e
X2 = 500, df = 5, p-value <2e-16
alternative hypothesis: two.sided
```

Notice that the degrees of freedom for X^2 are $m - p = 4 \times 2 - 3 = 5$.

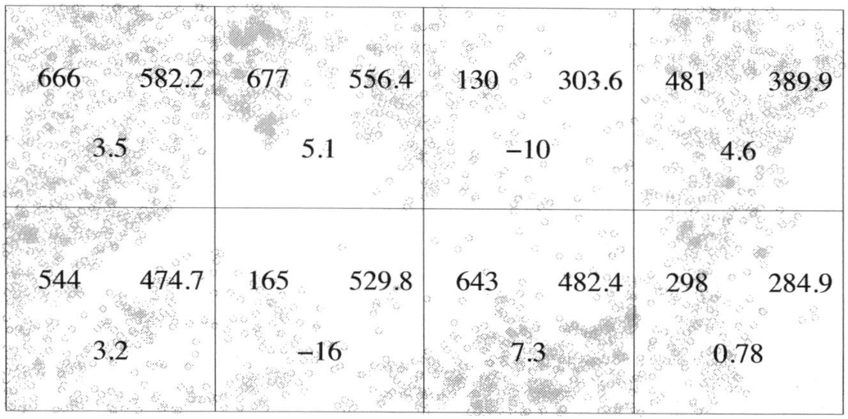

666	582.2	677	556.4	130	303.6	481	389.9
3.5		5.1		−10		4.6	
544	474.7	165	529.8	643	482.4	298	284.9
3.2		−16		7.3		0.78	

Figure 10.2. *Plot of quadrat counting test of fitted log-quadratic model for* Beilschmiedia *data, superimposed on the original point pattern.*

If (as in this case) the formal goodness-of-fit test rejects the fitted model, we would then like to understand in what way the data do not conform to the predictions of the model. Figure 10.2 shows the result of plot(M) which displays, for each quadrat, the observed number of points (top left), the expected number of points under the model (top right), and the Pearson residual (bottom) defined by

$$\text{Pearson residual} = \frac{(\text{observed}) - (\text{expected})}{\sqrt{\text{expected}}} = \frac{n_j - \hat{\mu}_j}{\sqrt{\hat{\mu}_j}}. \tag{10.7}$$

If the original data were Poisson, this transformation approximately standardises the residuals, so that they have mean zero and variance 1 (approximately) when the model is true. A Pearson residual of plus or minus 2 is unusual; values of -16 or 7 indicate gross departures from the fitted model.

The χ^2 test of homogeneity (Sections 6.4.2 and 6.7.1) is a special case of the test described above, because CSR is a special case of a Poisson process model. In principle we could conduct a quadrat test of CSR using quadrat.test.ppm rather than quadrat.test.ppp. The results are not exactly identical because of the numerical approximations involved.

Section 10.3 described three tests which can be applied in many contexts: the likelihood ratio test, Wald test, and score test. In the present context of an inhomogeneous Poisson model, the score test is identical to the χ^2 test described above [620]. The theoretically optimal test is the likelihood ratio test, based on

$$\Gamma = 2 \sum_j \left(n_j \log \frac{n_j}{\hat{\mu}_j} + \hat{\mu}_j - n_j \right) \tag{10.8}$$

instead of (10.6). This test can be selected by setting the argument CR=0 in quadrat.test.ppm or quadrat.test.ppp as explained in Section 6.4.2:

```
> quadrat.test(fit2e, nx=4, ny=2, CR=0)
```

The value CR=1 gives the χ^2 test, and other values of CR give other tests from the power-divergence family [566].

When a spatial covariate Z is available, then as suggested in [172, 343, 342], we could divide the range of Z values into several bands, divide the data correspondingly, and apply a χ^2

test to this grouping of the data. This was done in Section 6.6.2 when the null hypothesis is CSR. Similar code works when the null hypothesis is a fitted inhomogeneous Poisson model, using quadrat.test.ppm:

```
> V <- quantess(Window(bei), bei.extra$grad, 4)
> quadrat.test(fit2e, tess=V)
        Chi-squared test of fitted Poisson model 'fit2e' using quadrat counts
        Pearson X2 statistic

data:  data from fit2e
X2 = 700, df = 1, p-value <2e-16
alternative hypothesis: two.sided
```

Notice that there is only one degree of freedom remaining, as there were 4 quadrats and 3 parameters.

The null hypothesis is rejected, which might suggest including grad in the model, but there is no assurance that this would help. Note that it would make perfect sense to form the quadrats on the basis of elev, rather than of grad in the foregoing test, even though elev was used to create the model fit2e. Rejecting the null hypothesis in this case might suggest that a different structure of model is called for, but again there is no assurance. The results of goodness-of-fit tests are always somewhat uninformative: see Section 10.5.4.

For a continuous spatial covariate it is probably better to avoid discretisation, and to use one of the tests based on the spatial distribution of the covariate, as discussed in Section 10.5.2.

10.4.2 Quadrat counting tests of Poisson distribution

In a Poisson process, the number of points falling in any quadrat follows a Poisson distribution. This fact can be used to test whether a point process is Poisson, *assuming* it has homogeneous intensity. We must divide the window into quadrats of *equal area*. Then under the null hypothesis of CSR, the counts in each quadrat are Poisson random variables with the same mean.

```
> X <- bdspots[[2]][square(c(-200,200))]
> qX <- quadratcount(X, 20, 20)
```

We wish to test whether these 400 numbers could have come from a Poisson distribution. We construct a table giving, for each possible count $k = 0, 1, 2, \ldots$, the *number of quadrats* where this count was obtained:

```
> tX <- table(factor(as.numeric(qX), levels=0:max(qX)))
> tX

  0   1   2   3   4   5   6   7
 61 133 108  63  20  11   3   1
```

We now apply the χ^2 goodness-of-fit test for the Poisson distribution. The mean quadrat count is npoints(X)/400 = mean(qX) = 1.745. The predicted frequencies are:

```
> eX <- 400 * dpois(0:max(qX), mean(qX))
> eX
[1]  69.86 121.90 106.36  61.87  26.99   9.42   2.74   0.68
```

To avoid small expected counts, we merge the categories 5, 6, and 7 into a single category:

```
> fX <- factor(as.numeric(qX), levels=0:max(qX))
> fX <- mergeLevels(fX, ">=5"=5:7)
> tX <- table(fX)
> tX
fX
   0   1   2   3   4 >=5
  61 133 108  63  20  15
> p04 <- dpois(0:4, mean(qX))
> eX <- 400 * c(p04, 1 - sum(p04))
> eX
[1]  70 122 106  62  27  13
```

Finally the χ^2 test of goodness-of-fit is performed by:

```
> X2 <- sum((tX - eX)^2/eX)
> (pval <- pchisq(X2, df=length(tX)-1, lower.tail=FALSE))
[1] 0.509
```

This suggests no evidence against the Poisson distribution.

10.5 Tests based on the cumulative distribution function

Section 6.7.2 introduced several tests of the null hypothesis of homogeneous intensity against the alternative that the intensity depends on a spatial covariate Z. Here we explain the basis for these tests, which are applications of classical goodness-of-fit tests for cumulative distribution functions.

10.5.1 Goodness-of-fit of a probability distribution

A 'goodness-of-fit' test of a statistical model is a test of the null hypothesis that the model is true, against the very general alternative hypothesis that the model is not true [196, 566].

Suppose we wish to test whether some numerical data $z_1, \ldots, z_n$ follow a specified probability distribution (such as the standard normal distribution) with cumulative distribution function (cdf) $F_0(z)$. This is a goodness-of-fit problem for which standard, generic tools are the Kolmogorov-Smirnov [388, 619], Cramér-Von Mises [183, 675] and Anderson-Darling [16, 17] tests.

Assume the observations are independent, and come from a common distribution with cdf F. The null hypothesis is $H_0 : F_0 \equiv F$ and the alternative is $H_1 : F_0 \not\equiv F$. First form the *empirical* cdf of the data,

$$\widehat{F}(z) = \frac{1}{n} \sum_{i=1}^{n} \mathbf{1}\{z_i \leq z\}, \tag{10.9}$$

so that $\widehat{F}(z)$ is the fraction of observations less than or equal to z.

A test can be based on the discrepancy between the functions $\widehat{F}$ and F_0. The Kolmogorov-Smirnov test statistic is the maximum vertical separation between the graphs of $\hat{F}(z)$ and $F_0(z)$:

$$D = \max_z |\hat{F}(z) - F_0(z)|. \tag{10.10}$$

The Cramér-Von Mises test statistic is the average squared separation:

$$\omega^2 = n \int_{\infty}^{\infty} [\hat{F}(z) - F_0(z)]^2 \, dF_0(z) \tag{10.11}$$

and the Anderson-Darling test statistic is a weighted average:

$$A = n \int_{\infty}^{\infty} \frac{(\hat{F}(z) - F_0(z))^2}{F_0(z)(1 - F_0(z))} \, \mathrm{d}F_0(z). \tag{10.12}$$

Assuming F_0 is differentiable (which importantly implies that tied values $z_i = z_j$ cannot occur), the null distributions of the test statistics D, ω^2 and A are known exactly, so that a test can be conducted. The Anderson-Darling test is typically the most powerful of the three tests, but also the most computationally intensive.

The Kolmogorov-Smirnov test is implemented in the function `ks.test` in the standard `stats` package. Efficient methods for performing the Cramér-Von Mises and Anderson-Darling tests were developed recently [194, 443] and are implemented as `cvm.test` and `ad.test` in the contributed package `goftest`.

10.5.2   CDF tests of homogeneous intensity

The classical goodness-of-fit tests described above can be adapted to point patterns to test for homogeneous intensity. As explained in Section 6.7.2, the test is performed by comparing the observed distribution of the values of a spatial covariate at the data points with the values of that covariate at *all* spatial locations in the observation window.

Any spatial function Z can be used, including observed data and artificially constructed functions. Different choices of Z change the sensitivity of the test to different types of departure from the null hypothesis. The covariate Z is evaluated at each of the data points, yielding $z_i = Z(x_i)$, and the cumulative distribution function $\hat{F}(z)$ of these values is compiled. Then Z is evaluated at 'every' spatial location u in the window (in practice evaluated at the centre of every pixel in a grid) and the cumulative distribution function $F_0(z)$ of these values is formed:

$$F_0(z) = \frac{|\{u \in W : Z(u) \le z\}|}{|W|} \approx \frac{\#\{\text{pixels } u : Z(u) \le z\}}{\#\text{pixels}}.$$

The discrepancy between $\hat{F}(z)$ and $F_0(z)$ is then measured using the Kolmogorov-Smirnov statistic (10.10), Cramér-Von Mises statistic (10.11), or Anderson-Darling statistic (10.12). The null distribution of the test statistic depends only on the number of data points, assuming that the point process is Poisson and that F_0 is a continuous function. These tests are performed in `spatstat` by the function `cdf.test` as described in Section 6.7.2.

This technique is implicit in the work of Kolmogorov, but for spatial point processes it was first mentioned by Berman [88] to the best of our knowledge. The technique is closely related to nonparametric estimation of the intensity as a function of a covariate (Section 6.6.3).

10.5.3   CDF tests of a fitted intensity model

The cumulative distribution function of covariate values can also be used to test the goodness-of-fit of an inhomogeneous Poisson process model. The null hypothesis is a Poisson process with intensity $\lambda_0(u)$ on the window W. We start by extracting the covariate values $z_i = Z(x_i)$ at the data points, and form the empirical cdf $\hat{F}(z)$ of the observed values z_i. If the null hypothesis is true, then $z_1, z_2, \ldots$ constitute a Poisson process on the real line. The individual values z_i are independent and identically distributed, with c.d.f.

$$F_0(z) = \frac{\int_W \mathbf{1}\{Z(u) \le z\} \lambda_0(u) \, \mathrm{d}u}{\int_W \lambda_0(u) \, \mathrm{d}u}. \tag{10.13}$$

We test the goodness-of-fit of the theoretical c.d.f. $F_0(z)$ to the empirical c.d.f. $\hat{F}(z)$ using the Kolmogorov-Smirnov, Cramér-Von Mises, or Anderson-Darling test. If the hypothesised intensity function is constant, $\lambda_0(u) \equiv \lambda$, then this reduces to the cdf test of homogeneity described in Section 10.5.2.

The `spatstat` function `cdf.test.ppm` performs this test. For example we could test the model `fit2e` (see page 379) for goodness of fit as follows:

```
> grad <- bei.extra$grad
> cdf.test(fit2e, grad, test="ks")

        Spatial Kolmogorov-Smirnov test of inhomogeneous Poisson process in two
        dimensions

data:   covariate 'grad' evaluated at points of 'bei'
        and transformed to uniform distribution under 'fit2e'
D = 0.2, p-value <2e-16
alternative hypothesis: two-sided
```

By choosing the covariate `grad` for the test, we ensure that the test is particularly sensitive to departures from the null hypothesis in which the intensity depends on `grad`. A significant lack of fit in this test may prompt us to try including the slope covariate in the model, but does not guarantee that this will remedy the lack of fit.

If additional covariates had not been available, we could have made use of the x-coordinate by typing `cdf.test(fit2e,"x")`, or of some user-defined function of x and y if this seemed to be appropriate.

The test is not adjusted for the fact that the parameters were estimated, which can cause the test to be slightly conservative.

10.5.4 Caveats

Goodness-of-fit tests have weak power (low sensitivity) for detecting some kinds of departure from the fitted model, so a non-significant result is not always reassuring. On the other hand, if there is found to be significant evidence of a lack-of-fit, then we can only conclude that at least one of the model assumptions is false, without knowing which assumption it is.

The major underlying assumption of the foregoing goodness-of-fit tests is that the point process is *Poisson*. If the Poisson assumption holds true, then the behaviour of the test is well understood, because it reduces to a classical statistical test of goodness-of-fit to a probability distribution. However, if the true point process is not Poisson, the behaviour of the test can be difficult to predict. In respect of the tests based on cumulative distribution functions, clustering increases the expected size of deviations between F_0 and $\widehat{F}$, thus increasing the probability of type I error, while inhibition has the opposite effect. Consequently if we are not sure that the point process is Poisson, a non-significant result from the goodness-of-fit test could be explained by the presence of inhibition between points; a significant result could be an artefact of clustering between points.

The test *statistics* employed in these tests can be used to test goodness-of-fit of point process models other than the Poisson model. The null distribution of the test statistic is not the same as it is in the Poisson setting, and is essentially analytically intractable. However this difficulty may be circumvented by simulation from the fitted model. Doing so is justified by the Monte Carlo test principle (see Section 10.6). Such Monte Carlo tests are supported by all the test functions: `quadrat.test.ppm`, `cdf.test.ppm`, and `berman.test.ppm`. If the null model involves parameters which must be estimated, then (10.13) involves estimated parameters, and the theory is not strictly applicable. By failing to account for the effect of estimating the parameters, the test typically becomes slightly *conservative*, meaning that the computed p-value is slightly too high.

10.6   Monte Carlo tests

Monte Carlo methods use random simulation to replace complicated calculations in algebra and calculus. The use of Monte Carlo methods typically involves large numbers of simulations, in order to achieve high accuracy.

However, a *Monte Carlo* *<u>test</u>* is different: it uses a relatively small number of simulations from the null hypothesis, and appeals to a symmetry principle instead of the law of large numbers.

Monte Carlo tests were developed independently by Barnard [79] and Dwass [249], and further elaborated by Hope [338]. They were first applied in spatial statistics by Ripley [572, 575] and Besag and Diggle [99, 104]. Monte Carlo tests are related to the *randomisation tests* which are commonly used in nonparametric statistics.

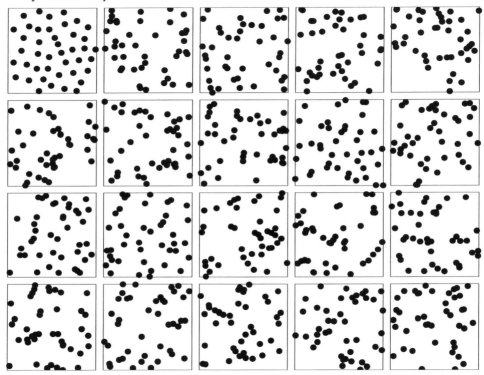

Figure 10.3. *The* cells *data (*Top Left*) and 19 simulated realisations of complete spatial randomness with the same number of points.*

10.6.1   Basic principle

The principle of a Monte Carlo test can be explained by a simple example. Suppose we wish to test whether the cells point pattern (see Section 4.1.6.1) was generated by complete spatial randomness. Figure 10.3 shows the cells data plotted side by side with 19 simulated point patterns, each containing the same number of points, generated according to CSR. If the null hypothesis of CSR were true, then these 20 point patterns should be statistically equivalent, and we should not be able to guess which of the 20 patterns was the original data.

In order to make a formal test, we need to reduce each point pattern to a single number, and compare the different numbers obtained. For argument's sake we use the Clark-Evans [155] index. The value for the `cells` data is 1.672 while the (sorted) values for the simulated patterns are

```
[1]  0.834 0.977 0.983 0.992 1.006 1.012 1.034 1.038 1.039 1.053 1.058 1.065
[13] 1.066 1.085 1.154 1.168 1.169 1.200 1.233
```

We find that the Clark-Evans index for the data is greater than all of the simulated values. This suggests that the `cells` data is not a realisation of CSR.

The formal rationale for the Monte Carlo test is as follows. If the null hypothesis were true, then these 20 point patterns (the `cells` data and the 19 simulated patterns) would be statistically equivalent, because they would be independent random outcomes of the *same* point process. The 20 values of the Clark-Evans index would be statistically equivalent, because they would be independent random numbers with the same probability distribution. By symmetry, there would be 1 chance in 20 that the index value for the `cells` data is the largest of the 20 numbers. Therefore, the result we obtained is statistically significant at the level $1/20 = 0.05$ and the p-value is equal to 0.05.

The main advantage of a Monte Carlo test is that it is cheap — only a small number of simulations are needed — and the p-value is exactly correct, rather than an approximation. The main disadvantages are that the test involves randomisation, so the outcome may change if the test is repeated; there is some loss of power (reduced sensitivity to alternatives); and very strong evidence may be understated, because the smallest possible p-value is $1/(1+m)$ where m is the number of simulations.

10.6.2 Details of Monte Carlo test

10.6.2.1 Simple form

In its simplest form, a Monte Carlo test for spatial point patterns involves the following steps:

> **(M1)** Generate m simulated random point patterns $\mathbf{x}^{(1)}, \ldots, \mathbf{x}^{(m)}$ from the null hypothesis, where typically m is 19, 39, or 99. These are computer-generated random point patterns, similar to the observed point pattern $\mathbf{x}$, but generated under the assumption that the null hypothesis is true. The random patterns should be independent of each other (in the sense of probability) and independent of the observed data $\mathbf{x}$.
>
> **(M2)** Reduce each point pattern to a single numerical value using a test statistic T. The observed point pattern $\mathbf{x}$ is reduced to the number $t_{obs} = T(\mathbf{x})$. The simulated patterns $\mathbf{x}^{(1)}, \ldots, \mathbf{x}^{(m)}$ are reduced to the numbers $t_1, \ldots, t_m$ where $t_i = T(\mathbf{x}^{(i)})$ for each i.
>
> **(M3)** Assuming that larger values of T are more favourable to the alternative hypothesis, the test rule is to reject H_0 at significance level $1/(m+1)$ if the observed value t_{obs} is larger than all of the simulated values $t_1, \ldots, t_m$.

The basic rationale for Monte Carlo tests is **symmetry**. Assume the null hypothesis is true (and that any parameters of the null model are known). Then the original data and the m simulated patterns are statistically equivalent, so the test statistic value t_{obs} calculated for the original data, and the test statistic values $t_1, t_2, \ldots, t_m$ calculated for the simulated patterns, are statistically equivalent. By symmetry, there is a 1 in $(m+1)$ chance that the test statistic value t_{obs} is the largest of these $m+1$ numbers — that is, that t_{obs} is larger than (each of) the other m values $t_1, t_2, \ldots, t_m$. If this happens, the result is statistically significant at level $\alpha = 1/(m+1)$.

For example, if $m = 19$, the test procedure is to reject H_0 if the value t_{obs} for the observed data is larger than the maximum of the 19 simulated values $t_{max} = \max\{t_1, \ldots, t_{19}\}$. The significance level of the test is $\alpha = 1/(19+1) = 0.05$ (often reported in terms of the p-value as '$p < 0.05$' although in this case the p-value is exactly equal to 0.05).

There is considerable freedom in choosing the test statistic T. Ideally it should be a continuously varying quantity (so that tied values are impossible) and one which tends to give different values under the null and alternative hypotheses.

There are several modifications of this basic form of the test. Instead of a 'one-sided' test which rejects the null hypothesis when t_{obs} is large, we could have a 'two-sided' test which rejects the null hypothesis when t_{obs} is either the largest or smallest of the $m+1$ numbers. This has a significance level of $\alpha = 2/(m+1)$. If we want the standard significance level of 0.05, then we would typically choose $m = 39$ simulations giving a significance level of $2/40 = 0.05$.

Instead of taking the maximum and minimum of the simulated values, a Monte Carlo test can be performed using the k-th largest and k-th smallest values out of m simulations, where k is a chosen rank. For a one-sided test we reject the null hypothesis if the observed value t_{obs} is larger than the k-th largest simulated value. This test has significance level $\alpha = k/(m+1)$. For a two-sided test, we reject the null hypothesis if the observed value t_{obs} is either larger than the k-th largest simulated value, or smaller than the k-th smallest simulated value. This test has significance level $\alpha = 2k/(m+1)$.

Note that many choices of m and k will give a test of significance level $\alpha = 0.05$. Examples include a one-sided test using the 5th largest simulated value out of 99 simulations, and a two-sided test using the 5th largest and 5th smallest simulated values out of 199 simulations. Researchers are free to make their own choices of m and k with an eye to computational cost, standards of evidence, performance, and other factors.

If we prefer to compute a p-value instead of specifying a fixed significance level α, the test procedure is as follows. Count the number of simulated values t_i which are greater than the observed value t_{obs}. If there are j such values then the p-value is $(j+1)/(m+1)$ for a one-sided test, or $2\min(j+1, m+1-j)/(m+1)$ for a two-sided test.

For integer k, the Monte Carlo test with (nominal) significance level $\alpha = k/(m+1)$ rejects H_0 if $t_{obs} > t_{(k)}$, where $t_{(k)}$ is the kth largest value amongst $t_1, \ldots, t_m$. Equivalently, the Monte Carlo p-value is, if large positive values of t are favourable to H_1,

$$p_+ = \frac{1 + \sum_{j=1}^{m} \mathbf{1}\{t_j \geq t_{obs}\}}{m+1}. \tag{10.14}$$

The numerator is the rank of t_{obs} in the set $\{t_1, \ldots, t_m\} \cup \{t_{obs}\}$. If small values (or large negative values) of t are favourable to H_1, the p-value is

$$p_- = \frac{1 + \sum_{j=1}^{m} \mathbf{1}\{t_j \leq t_{obs}\}}{m+1} \tag{10.15}$$

and for two-sided tests $p = 2\min\{p_+, p_-\}$.

10.6.2.2 Code examples

Monte Carlo tests are relatively straightforward to encode in the R language. For example, to test the null hypothesis of CSR for the `cells` data using the Clark-Evans statistic, we would first evaluate

```
> (Tobs <- clarkevans(cells, correction="none"))
[1] 1.67
```

The following code generates 19 simulated point patterns, each containing the same number of points as the `cells` data, and generated according to CSR, and computes the Clark-Evans index for each simulated pattern:

```
> Tsim <- numeric(19)
> for(i in 1:19) {
```

```
    X <- runifpoint(npoints(cells), Window(cells))
    Tsim[i] <- clarkevans(X, correction="none")
}
```

An important point is illustrated by the foregoing code. It might seem 'more natural' to use `X <- rpoispp(intensity(cells),Window(cells))`. However, this would be a conservative test; see Section 10.6.3.

If the alternative is a regular (inhibited) point process, large values of the Clark-Evans index are favourable to the alternative, so the *p*-value is, from (10.14),

```
> (preg <- (1 + sum(Tsim > Tobs))/(1 + length(Tsim)))
[1] 0.05
```

For the alternative of a clustered process the *p*-value is

```
> (pclus <- (1 + sum(Tsim < Tobs))/(1 + length(Tsim)))
[1] 1
```

and for the two-sided alternative

```
> (peither <- 2 * min(pclus, preg))
[1] 0.1
```

For convenience, a few functions in `spatstat` can perform a Monte Carlo test as an option, instead of using asymptotic approximations. These include `quadrat.test` and `clarkevans.test`, which perform tests of CSR.

```
> quadrat.test(redwood, nx=5, alternative="clustered",
               method="MonteCarlo", nsim=999)
> clarkevans.test(redwood, alternative="clustered", nsim=999)
```

Here it should be remembered that the Clark-Evans index is sensitive to inhomogeneity in the point pattern, so a significant result from this test does not always indicate clustering or inhibition.

10.6.2.3 General form

The astute reader will have noticed that the code example in Section 10.6.2.2 above does not strictly follow the procedure laid out in step (**M1**). That would require the simulated point patterns $\mathbf{x}^{(1)}, \ldots, \mathbf{x}^{(m)}$ to be independent of the observed pattern $\mathbf{x}$. However, in Section 10.6.2.2, the simulated patterns were chosen to have the same number of points as the observed data, so the simulated patterns are not independent of the data.

The symmetry principle nevertheless applies to this example. Random numbers $Y_1, \ldots, Y_n$ are called *exchangeable* if any permutation of the list has the same statistical characteristics.[2] We apply the same definition to a list of random point patterns. Then the Monte Carlo test scheme above can be modified by replacing (**M1**) by

> (**M1'**) Given the data pattern $\mathbf{x}$, generate m simulated random point patterns $\mathbf{x}^{(1)}, \ldots, \mathbf{x}^{(m)}$ such that, if the null hypothesis is true, then $\mathbf{x}, \mathbf{x}^{(1)}, \ldots, \mathbf{x}^{(m)}$ are exchangeable.

The test is still valid with this modification because, if the data and simulations are exchangeable, then the summary statistic values $t, t_1, \ldots, t_m$ are exchangeable, and by symmetry, the probability that t is greater than all the simulated values $t_1, \ldots, t_m$ is $1/(m+1)$.

---

[2]The joint distribution of $(Y_1, \ldots, Y_n)$ must be the same as that of $(Y_{i_1}, \ldots, Y_{i_n})$ where $(i_1, \ldots, i_n)$ is any permutation of $(1, 2, \ldots, n)$.

In the example in Section 10.6.2.2, given the observed data, the simulated patterns were generated by placing the same number of points independently and uniformly in the same window. If the null hypothesis of CSR is true, then the data also satisfy this description (by property (PP4) of the Poisson process stated on page 300), so the data and the simulated patterns are exchangeable. Thus, the example in Section 10.6.2.2 is a valid Monte Carlo test.

One can extend the scope of Monte Carlo tests even further, replacing **(M2)** by

> **(M2')** The observed point pattern **x** and simulated patterns $\mathbf{x}^{(1)}, \ldots, \mathbf{x}^{(m)}$ are reduced to the numbers t and $t_1, \ldots, t_m$ which are exchangeable if the null hypothesis is true.

This allows the test statistic T to depend on both the observed and simulated point patterns,

$$t_{obs} = T(\mathbf{x}; \mathbf{x}^{(1)}, \ldots, \mathbf{x}^{(m)}). \tag{10.16}$$

For example, in Section 10.6.2.2, the test statistic t_{obs} could be the Clark-Evans statistic for the `cells` data *minus the average for the simulated patterns*. To preserve symmetry, the statistic t_i for the ith simulated pattern must be calculated using the same rule with $\mathbf{x}^{(i)}$ and $\mathbf{x}$ exchanged.

10.6.3   Conservatism when parameters are estimated

An important caveat about Monte Carlo tests, which is often overlooked, is that they are strictly invalid, and usually *conservative*, if parameters have been estimated from the data [225, p. 89], [40]. The problem arises when the null hypothesis is a model that depends on a parameter or parameters θ which must be estimated from the data (known as a 'composite' hypothesis). The usual procedure is to fit the null model to the observed point pattern, obtaining a parameter estimate $\widehat{\theta}$, and then to generate the simulated point patterns from the model using this fitted parameter value $\widehat{\theta}$. But this violates the essential requirement of symmetry. Under the procedure just described, the simulated point patterns have been generated from the null model with parameter value $\widehat{\theta}$, while (if the null hypothesis is true) the observed point pattern came from the null model with unknown parameter value θ. See Figure 10.4. Since the estimate $\widehat{\theta}$ is usually not exactly equal to the true parameter value θ, the simulated and observed point patterns do not come from the same random process, so they are not exchangeable, and the Monte Carlo test is invalid, strictly speaking (i.e. the Monte Carlo test does not have the claimed significance level).

Figure 10.4 suggests that, if we generate random values from a model fitted to the observed data, the random values are likely to scatter *around* the observed value, so that the observed value is unlikely to lie in the extremes of the simulated values. This causes the test to be *conservative*. A test is called conservative if the true significance level — the true probability of Type I error — is smaller than the reported significance level — the significance level which we quote in reporting the outcome of the test. Equivalently, the reported p-value is higher (less significant) than the true p-value. A conservative test may conclude that the data are not statistically significant when in fact they should be declared statistically significant.

Experiments [40] show that the conservatism effect can be severe in spatial examples, with a test of nominal significance 0.05 having true significance level 0.02 or lower. Conservatism is a small-sample problem in the sense that it disappears in large datasets: if the spatial point pattern dataset is large enough, the parameter estimate $\widehat{\theta}$ will be close to the true parameter value θ, and the conservatism effect will be small. However, conservatism is *not* reduced by taking a larger number of simulations m.

This problem arises even in the simplest case of testing CSR. The null hypothesis of CSR has one free parameter, namely the intensity λ (the average number of points per unit area). Typically we estimate λ from our observed data by $\widehat{\lambda} = n_{obs}/A$ where n_{obs} is the number of points in the observed point pattern and A is the area of the survey region. If we were to generate the simulated

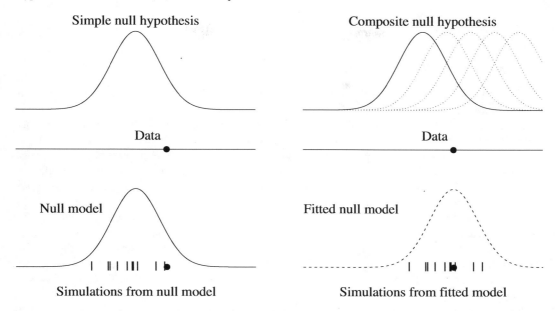

Figure 10.4. *Why Monte Carlo tests are typically conservative when the null hypothesis is composite.* Left column: *simple null hypothesis;* Right column: *composite null hypothesis. The Monte Carlo test compares the observed value (dot) with simulated values (vertical bars). For a simple null hypothesis, the simulated values are generated from the null distribution, and the test is exact. For a composite null hypothesis, simulated values are generated from the* fitted *distribution, which is typically centred around the observed value, and the test is typically conservative.*

point patterns according to a Poisson process with intensity $\widehat{\lambda}$, the Monte Carlo test would be strictly invalid and probably conservative.

The effect can be measured by an experiment like the following. First we define a function that conducts a Monte Carlo test of CSR for a given point pattern using the Clark-Evans statistic, with simulations generated according to CSR.

```
> dotest <- function(X, nsim=19) {
    Tobs <- clarkevans(X, correction="none")
    Tsim <- numeric(nsim)
    for(i in 1:nsim) {
      Xsim <- rpoispp(intensity(X), win=Window(X))
      Tsim[i] <- clarkevans(Xsim, correction="none")
    }
    return(Tobs > max(Tsim))
  }
```

The result of `dotest` is a logical value equal to TRUE if the test rejects the null hypothesis of CSR. Taking `nsim=19` this test has nominal significance level (=probability of rejecting the null hypothesis when it is true) equal to 0.05. To estimate the true significance level, we generate simulated data according to CSR. (This can take a long time with N=10000.)

```
> N <- 10000
> lambda <- 50
> rejected <- logical(N)
> for(i in 1:N) {
```

```
        X <- rpoispp(lambda)
        rejected[i] <- dotest(X)
    }
  > alpha <- mean(rejected)
```

This gives an estimated true significance level of `alpha` = 0.046 with standard error `sqrt(alpha * (1-alpha)/N)` = 0.002 so the Monte Carlo test is only slightly conservative.

In the case of CSR, a solution to the problem of conservatism is to hold the number of points fixed. We generate the simulated patterns with the same number of points as the observed pattern. This exploits the conditional property (PP4) defined on page 300. If the number of points $n(\mathbf{x})$ is known, the locations of the points are independent and uniformly distributed, whatever the value of the intensity λ. It follows that (**M1'**) holds, and so the Monte Carlo test of CSR *conditional on the number of points* is exact (non-conservative).

In the more realistic scenario where the null hypothesis is a model involving spatial trend and interpoint interaction, requiring several parameters to be fitted, the conservatism effect may be greater and yet harder to handle. In general we would need to generate simulations conditionally on the value of the sufficient statistic for the model. This may be quite difficult. The statistic $n(\mathbf{x})$ is typically one component of the sufficient statistic, so conditioning on $n(\mathbf{x})$ may partially alleviate the conservatism effect.

A conservative test may be tolerable in some applications, since it effectively applies a standard of statistical significance that is more stringent than intended. The current usage of Monte Carlo tests in spatial statistics is defensible for some purposes, such as exploratory data analysis. However, for goodness-of-fit testing, a very conservative test is problematic, since a 'non-significant' outcome (where the null hypothesis is not rejected) is often misinterpreted as confirmation of the fitted model. For definitive formal inference and for goodness-of-fit tests, some remedy is needed.

The correct handling of p-values for composite null hypotheses is regarded as an unresolved research problem in statistical inference. See, for example, [133, 85, 582]. Various strategies have been suggested, including adjusting the test statistic so that its mean value is less sensitive to the parameter [582]; using a summary function which is unrelated to the model fitting procedure [225, p. 89]; and adjusting the p-value itself by performing additional simulations [133, 200]. Some of these remedies are discussed below.

10.7 Monte Carlo tests based on summary functions

Monte Carlo tests for spatial point patterns, based on simulation envelopes of summary functions, were pioneered by Ripley [572] and their statistical rationale was set out in detail in [99, 574, 575]; see also [442].

Such tests may be performed using any summary function for the point pattern; typical choices would be the K-function, L-function, or pair correlation function g (Chapter 7), the empty-space function F, or the nearest-neighbour distance distribution function G (Chapter 8).

The Monte Carlo test involves generating simulated realisations from the null hypothesis. If we wish to test whether the pattern is completely random, the null hypothesis is complete spatial randomness, and the simulated realisations are completely random point patterns. However, we can nominate any null hypothesis at all, so long as it is possible to generate simulated realisations from it.

We use the letter S to stand for any chosen summary function. To perform the test, we first calculate the S-function estimate for the observed data, $S_{obs}(r)$. Next we generate m synthetic point patterns generated by simulation from the null hypothesis, and obtain their S-function estimates, $S^{(1)}(r), \ldots, S^{(m)}(r)$.

The Monte Carlo test principle of Section 10.6 cannot be applied to the function $S(r)$ directly, because functions cannot be ranked from smallest to largest. We need to *reduce the function $S(r)$ to a single number T* that will serve as the test statistic. Four possible strategies are described in Sections 10.7.1, 10.7.3, 10.7.4 and 10.7.5 below.

10.7.1 Pointwise envelopes test

One strategy for reducing the summary function $S(r)$ to a single numerical value T is to specify a fixed distance $r = r_0$, and define the test statistic as $T = S(r_0)$, the value of the summary function at the specified distance. Under the null hypothesis the values of $\widehat{S}(r_0)$ for the data and simulated patterns are statistically equivalent random variables and the Monte Carlo test principle applies.

The two-sided Monte Carlo test with significance level $\alpha = 2/(m+1)$ rejects the null hypothesis if the observed value $t_{obs} = S_{obs}(r_0)$ lies outside the range of all the simulated values $t_1 = S^{(1)}(r_0), \ldots, t_m = S^{(m)}(r_0)$, that is, if either $t_{obs} > \max_j t_j$ or $t_{obs} < \min_j t_j$. This rule is equivalent to drawing the graph of the *upper and lower envelopes*

$$U(r) = \max_j S^{(j)} r) = \max\{S^{(1)}(r), \ldots, S^{(m)}(r)\}$$

$$L(r) = \min_j S^{(j)}(r) = \min\{S^{(1)}(r), \ldots, S^{(m)}(r)\}, \tag{10.17}$$

slicing this graph by a vertical line that meets the horizontal axis at the position $r = r_0$, and rejecting the null hypothesis if the observed function value $S_{obs}(r)$ lies outside the interval bounded by these envelopes *at the prespecified distance $r = r_0$*.

Figure 10.5 shows an example using the K-function. The empirical K-function for the `cells` dataset is shown as a thick black line. The grey-shaded region is bounded by the upper and lower envelopes of the empirical K-functions of $m = 39$ completely random patterns. The significance level is $\alpha = 2/(m+1) = 2/40 = 0.05$.

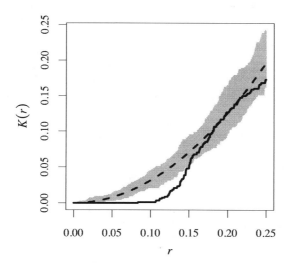

Figure 10.5. *Envelopes of the K-function. Thick black line is empirical K-function for the* `cells` *dataset. Grey shading is bounded by the maximum and minimum values of the K-functions of 39 simulated realisations of CSR with the same number of points as the data.*

The outcome of the pointwise Monte Carlo test depends only on the relative ordering of the values $S_{obs}(r_0)$ and $S^{(1)}(r_0), \ldots, S^{(m)}(r_0)$. Therefore the same test outcome will be obtained if we apply any monotone increasing transformation to the summary function. For example, the K-function could be replaced by the L-function $L(r) = \sqrt{K(r)/\pi}$, the centred functions $K(r) - \pi r^2$ or $L(r) - r$, and so on, without affecting the outcome of the pointwise test.

10.7.2 Misinterpretation of pointwise envelopes

An important caveat about pointwise envelopes is that their formal interpretation as a significance test requires the distance value r_0 to have been fixed in advance of the experiment and analysis.

In Figure 10.5, the empirical K-function for the data crosses the envelopes of the K-functions of 39 completely random patterns. Many writers attach a statistical significance of $2/(39+1) = 0.05$ to this outcome. However, this would be valid only if we had decided in advance to inspect the K-function at one particular interpoint distance. The usual practice is to plot the simulation envelopes without any preconceived idea, and look for excursions outside the area bounded by the envelopes at *any* position. This practice leads to invalid statistical inferences. Assuming H_0 is true, the probability that the K-function will wander outside the region bounded by the envelopes *somewhere* is very much higher than 0.05, so that the results in Figure 10.5 cannot then be declared statistically significant at the 0.05 level. This is a problem of *multiple testing* or *multiple comparison* [336, 347, 612] sometimes known as 'data snooping' or the 'look elsewhere effect'.

This caveat was made in Ripley's pioneering paper [572, p. 181] and has been echoed by many others. Loosmore and Ford [431, p. 1926] also championed it, but their contribution has been widely misinterpreted as meaning that simulation envelopes do not have *any* valid interpretation as a significance test. Some writers in spatial analysis have claimed that "envelope tests should not be thought of as formal tests of significance" [411, p. 619] and have drawn a distinction between invalid 'envelope tests' and valid 'deviation tests' [292]. This cannot be correct since 'envelope tests' and 'deviation tests' have the *same* statistical rationale, namely that of a Monte Carlo test. Simulation envelopes are widely used elsewhere in statistical science without demur [122, 144].

One benefit of graphically displaying a summary function like the K-function is that it contains information from different spatial scales. Plotting the pointwise envelopes over a range of r values enables us to assess, for different distance values r, what the result of the test *would have been* if we had chosen that distance value to perform the test. Figure 10.5 indicates that we would have obtained a statistically significant result if we had chosen the distance value r_0 to be anywhere between 0.05 and 0.15 distance units.

In summary, pointwise envelopes do give a valid significance test when such a test is correctly interpreted and applied. Admittedly they are very prone to misinterpretation. Other types of simulation envelopes which do not suffer from the same risk of misinterpretation are discussed below.

10.7.3 Simultaneous envelopes and the MAD test

Another strategy for reducing the summary function $S(r)$ to a single numerical value T is to compute the maximum deviation between the S-function of the observed data and the theoretical ('expected') S-function of the null model.

In some cases the theoretical value of the S-function under the null model is known. For example under CSR, the K-function takes the form $K_{theo}(r) = \pi r^2$ and so $L_{theo}(r) = r$. In such cases we may take the test statistic T to be the maximum deviation, in absolute value, between $S(r)$ and its theoretical value $S_{theo}(r)$ under the null model, where the maximum is taken over the range of distances from 0 to R units, say, where R is a chosen upper limit on the interaction distance (to be discussed below). That is, we choose

$$T = \max_{0 \le r \le R} |S(r) - S_{theo}(r)|. \tag{10.18}$$

Then T is the maximum vertical separation between the graphs of $S(r)$ and $S_{theo}(r)$ over the chosen range of distances. This procedure reduces the data to a single number T called the 'maximum absolute deviation' (MAD), analogous to the Kolmogorov-Smirnov test statistic (10.10).

To apply a Monte Carlo test we would compute the observed value t_{obs} of the MAD for the point pattern data, and the MAD values $t_1, \ldots, t_m$ for the simulated point patterns, then reject the null hypothesis if t_{obs} is greater than the maximum $t_{max} = \max\{t_1, \ldots, t_m\}$ of all the simulated values. This is the *MAD test* with significance level $\alpha = 1/(m+1)$.

The spatstat function mad.test performs this test. It calls envelope to perform the simulations and to compute the summary functions; the syntax of mad.test is essentially the same as that of envelope.

```
> mad.test(cells, Lest, nsim=99, ginterval=c(0, 0.2),
          fix.n=TRUE, verbose=FALSE)
        Maximum absolute deviation test of CSR
        Monte Carlo test based on 99 simulations with fixed number of points
        Summary function: L(r)
        Reference function: sample mean
        Alternative: two.sided
        Interval of distance values: [0, 0.25]

 data:   cells
 mad = 0.09, rank = 1, p-value = 0.01
```

The result of mad.test is a hypothesis test object (class "htest") with additional attributes storing information about the simulations.

The MAD test rule is equivalent to plotting envelopes bounding a region of *constant* width $2t_{max}$ centred on the theoretical curve $S_{theo}(r)$, that is, the region bounded by the limits

$$L(r) = S_{theo}(r) - t_{max}$$
$$U(r) = S_{theo}(r) + t_{max} \tag{10.19}$$

and rejecting the null hypothesis if the observed S-function $S_{obs}(r)$ *ever* wanders outside this region. The upper and lower bounds of this region are called *global* or *simultaneous* envelopes. Global envelopes were also proposed by Ripley [572, 574, 575], but have not been widely adopted in spatial statistics.

See Figure 10.6 for an example with $m = 19$ so that $\alpha = 0.05$. This was generated by calling envelope with the argument global=TRUE. As distinct from the 'pointwise envelopes' in Figure 10.5, the 'global envelopes' in each panel of Figure 10.6 bound a zone of *constant width*. The width is determined by finding the most extreme deviation (from the theoretical curve) achieved by any of the 39 simulated curves, at any distance r along the horizontal axis. The L-function of the data point pattern wanders outside the region bounded by the global envelopes; this is statistically significant at the level $1/(1+19) = 0.05$, often reported in terms of the p-value as '$p < 0.05$' although the p-value is actually equal to 0.05. Global envelope tests are statistically valid; the problem of multiple testing has been avoided.

Note that, unlike the pointwise test, the outcome of the MAD test *is* affected by transformations of the summary function. In Figure 10.6, different test outcomes are obtained using the K- and L-functions on the same data and the same simulated patterns. A more powerful test is obtained if we (approximately) stabilise the variance, by using the L-function in place of K, as shown in the right panel of Figure 10.6. This clearly rejects the null hypothesis of CSR at the 0.05 significance level, whereas the left panel indicates non-rejection.

In the general case the MAD test rejects the null hypothesis if t_{obs} is greater than the k-th largest of the simulated values, and this has significance level $\alpha = k/(m+1)$.

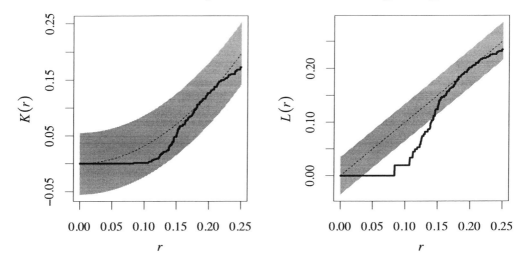

Figure 10.6. *'Global envelopes' based on the K-function (*Left*) or L-function (*Right*). Grey shading shows regions of constant vertical width determined by the maximum deviation (from the expected value for CSR) of the summary function estimates for 19 simulations of CSR.*

If the theoretical value $S_{theo}(r)$ is not known for every r, then it could be estimated from a separate set of simulations of the null model [221], which guarantees the basic requirement of symmetry. Alternatively, using only a single set of simulations, we can replace $S_{theo}(r)$ by

$$\overline{\overline{S}}(r) = \frac{1}{m+1}(S^{(1)}(r) + \ldots + S^{(m)}(r) + S_{obs}(r)), \qquad (10.20)$$

the average of all the simulated and observed S-functions. This preserves symmetry (exchangeability) and ensures that the test has significance level $k/(m+1)$.

10.7.4   DCLF test

A third strategy for reducing the summary function $S(r)$ to a single numerical value T is to compute the integrated squared deviation from the theoretical value,

$$T = \int_0^R (S(r) - S_{theo}(r))^2 \, dr \qquad (10.21)$$

as proposed by Diggle [223, p. 122], [225, eq. (2.7), p. 12]. Here R is a chosen upper limit on the range of distances of interest. The statistic T was subsequently advocated by Cressie [190, eq. (8.5.42), p. 667] and Loosmore and Ford [431]. A Monte Carlo test based on T has come to be known as the *Diggle-Cressie-Loosmore-Ford (DCLF)* test.

The motivation is that in many statistical applications it is better to use an integrated squared error than a maximum absolute error. For example in goodness-of-fit testing, the Kolmogorov-Smirnov test, based on the maximum absolute difference (10.10), typically has less power than the Cramér-Von Mises test, based on a weighted integral of squared difference (10.11). This heuristic argument is borne out by experiments [40].

If the theoretical value S_{theo} has to be estimated then (10.21) is replaced by

$$T = \int_0^R (S(r) - \overline{\overline{S}}(r))^2 \, dr, \qquad (10.22)$$

where again $\overline{\overline{S}}$ is the average of the simulated and observed S-functions as in (10.20).

The test is performed in `spatstat` using `dclf.test`, analogous to `mad.test`. The function `dclf.test` calls `envelope` to perform the simulations and to compute the summary functions; the syntax of `dclf.test` is essentially the same as that of `envelope`.

```
> dclf.test(cells, Lest, nsim=99, fix.n=TRUE, verbose=FALSE)
        Diggle-Cressie-Loosmore-Ford test of CSR
        Monte Carlo test based on 99 simulations with fixed number of points
        Summary function: L(r)
        Reference function: sample mean
        Alternative: two.sided
        Interval of distance values: [0, 0.25]

data:   cells
u = 5e-04, rank = 1, p-value = 0.01
```

The result of `dclf.test` is a hypothesis test object (class `"htest"`) with additional attributes storing information about the simulations. In the printed output, the line 'Reference function: sample mean' indicates that (10.22) was computed rather than (10.21).

The first argument of `dclf.test` can be, amongst other things, an `"envelope"` object, say E, which should have been computed using `savefuns=TRUE` in the call to `envelope`. One can call `dclf.test(E)`, and if the null hypothesis in question is rejected, plot E to gain insight into the nature of the departure from null hypothesis.

Like all Monte Carlo tests, the DCLF test is slightly conservative when the model parameters have to be estimated from the data.

10.7.5   One-sided tests and envelopes

Simulation envelopes may also be used for *one-sided* tests. For example, suppose we wish to test the null hypothesis of complete randomness against the alternative hypothesis of clustering. If the data are believed to be clustered, we expect the K-function of the data to lie *above* (and not below) the K-functions of simulations from the null hypothesis. A test which rejects the null hypothesis only when this occurs would be a one-sided test.

A simple approach is to use the upper and lower pointwise envelopes as the critical boundaries for one-sided tests against upper and lower alternatives. For example, suppose we construct envelopes based on the maximum and minimum of m simulated functions, equation (10.17). For any fixed distance r_0, the one-sided test which rejects H_0 if the observed function value $S_{obs}(r_0)$ *exceeds* the upper limit $U(r_0)$, has significance level $1/(m+1)$. Similarly the test which rejects H_0 if $S_{obs}(r_0) < L(r_0)$ has significance $1/(m+1)$.

For one-sided versions of the MAD and DCLF tests, we need a different approach. A Monte Carlo test can be based on the *positive part* of the deviation

$$(S_{obs}(r) - S_{theo}(r))_+$$

where $x_+ = \max(0, x)$, or on the negative part

$$(S_{obs}(r) - S_{theo}(r))_-$$

where $x_- = \min(0, x)$.

In the one-sided version of the DCLF test, the test statistic is

$$T = \int_a^b [(S(r) - S_{theo}(r))_+]^2 \, dr$$

for the upper one-sided test and similarly for the lower one-sided test. It is performed in `spatstat` by calling `dclf.test` with the argument `alternative` specifying the direction of the alternative hypothesis.

```
> dclf.test(cells, Lest, nsim=19, fix.n=TRUE, alternative="less")
        Diggle-Cressie-Loosmore-Ford test of CSR
        Monte Carlo test based on 19 simulations with fixed number of points
        Summary function: L(r)
        Reference function: sample mean
        Alternative: less
        Interval of distance values: [0, 0.25]

data:  cells
u = 4e-04, rank = 1, p-value = 0.05
```

In the one-sided version of the MAD test, the test statistic (for the positive part) is

$$T = \max_{r \in [a,b]} (S(r) - S_{theo}(r))_+.$$

We reject the null hypothesis if $t_{obs} > \max\{t_1, \ldots, t_m\}$, or equivalently if

$$S_{obs}(r) > S_{theo}(r) + t_{max} \quad \text{for some } r \in [0, R].$$

That is, the upper one-sided MAD test corresponds to plotting a critical boundary $U(r) = S_{theo}(r) + t_{max}$ at a constant vertical height above the theoretical curve $S_{theo}(r)$, and rejecting H_0 if the observed curve $S_{obs}(r)$ ever wanders above this boundary. Similarly the lower one-sided MAD test rejects the null hypothesis if $S_{obs}(r)$ ever wanders below the lower boundary $L(r) = S_{theo}(r) + t^*_{max}$ where t^*_{max} is the maximum absolute value of the negative deviations.

The one-sided MAD test is performed by calling `mad.test` with the appropriate value of `alternative`. Corresponding envelopes are computed by calling `envelope` with `global=TRUE` and with the appropriate value of `alternative`.

Note that the two-sided tests described above are *not* equivalent to performing both of the corresponding one-sided tests.

10.8   Envelopes in `spatstat`

The `spatstat` function `envelope` performs the calculations required for envelopes. It computes the summary function for a point pattern dataset, generates simulated point patterns, computes the summary functions for the simulated patterns, and computes the envelopes of these summary functions. The result is an object of class `"envelope"` and `"fv"` which can be printed and plotted and manipulated using the tools for `"fv"` objects described in Section 7.5. (See Table 7.2 on page 223.) The print method gives a lot of detail: For example, Figure 10.7 was generated by the commands

```
> plot(envelope(redwood, Lest, nsim=39))
> plot(envelope(redwood, Lest, nsim=39, global=TRUE))
```

The function `envelope` is generic, with methods for class `"ppp"`, `"ppm"`, and several other classes. If X is a point pattern, then by default `envelope(X, ...)` constructs pointwise envelopes for the null hypothesis of CSR. If M is a fitted point process model, then by default `envelope(M, ...)` constructs pointwise envelopes for the null hypothesis represented by the model M. The default

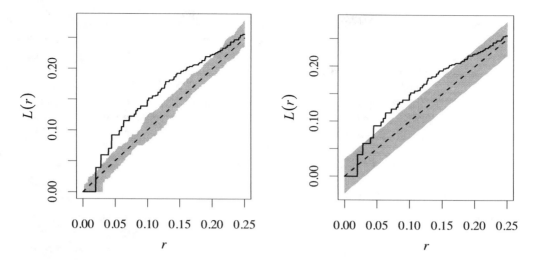

Figure 10.7. *Examples of the* envelope *command. Pointwise (Left) and global (Right) envelopes of Besag's L-function based on 39 simulations of CSR, and observed L-function for the* redwood *data.*

behaviour is to construct pointwise envelopes from the *K*-functions of 99 simulations. Different summary functions, different types of envelopes, different null hypotheses, different alternative hypotheses, and different simulation procedures are selected using additional arguments listed in Table 10.1.

ARGUMENT	DEFAULT	MEANING
Y		point pattern data, or fitted model
fun	Kest	summary function
nsim	99	number of simulations *m* for test
nrank	1	rank for statistical significance
global	FALSE	global envelopes
VARIANCE	FALSE	envelopes based on sample variance
simulate		procedure for generating simulated patterns
alternative	two.sided	alternative hypothesis
fix.n	FALSE	fix number of points
fix.marks	FALSE	fix number of points of each type
correction		edge correction for summary function
transform		transformation of summary function
ginterval		range of distance values for global envelopes
nsim2	nsim	number of simulations to estimate mean
savepatterns	FALSE	store point patterns
savefuns	FALSE	store summary functions

Table 10.1. *Common arguments for* envelope.

10.8.1 Specification of the null hypothesis

The most important consideration, and also the most prone to mistakes, is the specification of the null hypothesis for the envelope test.

By default, envelope.ppp generates envelopes for testing Complete Spatial Randomness (CSR).

A test of CSR is useful in initial investigation of the data [225, p. 12], [30] where it serves as a 'dividing hypothesis' that guides the choice of strategy for further analysis [179, pp. 51–52].

For definitive formal inference, CSR is not usually the appropriate null hypothesis. The null hypothesis should not be chosen naively, but should correspond to a scientifically meaningful scenario in which 'nothing is happening' [644], [431, p. 1929].

The envelope function can be used to test essentially any null hypothesis. The only requirement is that we must be able to simulate realizations of the point process that is specified by the null hypothesis. To direct envelope to generate envelopes based on a particular null hypothesis, the user must specify that hypothesis either as (a) a fitted point process model ("ppm", "kppm", or "lppm" object) representing the null hypothesis; or (b) a recipe for generating simulated realizations of the null hypothesis; or (c) a list of point patterns, taken to be realizations of the null hypothesis.

10.8.1.1   Envelopes for testing CSR

If Y is an *unmarked* point pattern, by default the simulated point patterns are realisations of complete spatial randomness (the uniform Poisson point process) with the same intensity as the pattern Y and in the same window as Y. Each simulated pattern has a random number of points, according to a Poisson distribution. The resulting envelopes support a test of CSR. However, this test is slightly conservative, as explained in Section 10.6.3.

If fix.n=TRUE, simulations are generated from CSR conditionally on having the same number of points as the observed data. That is, each simulated pattern consists of n independent random points distributed uniformly in the same window as $\mathbf{x}$, where $n = n(\mathbf{x})$ is the original number of data points. As explained on page 390, this avoids the conservatism of the previous test.

If Y is a multitype point pattern, then the simulated patterns are also given independent random marks. The simulated patterns are realisations of Complete Spatial Randomness and Independence (the uniform Poisson point process with independent random marks). By default the probability distribution of the random marks is determined by the relative frequencies of marks in Y. If fix.marks=TRUE, the number of points of each type is held fixed. This reduces or removes conservatism of the test. In either case, the resulting envelopes support a test of CSRI.

10.8.1.2   Envelopes for any fitted model

If the first argument of envelope is a fitted point process model (object of class "ppm", "kppm" or "lppm"), then the simulated patterns will be generated according to this fitted model. The original data point pattern, to which the model was fitted, is stored in the fitted model object; the original data are extracted and the summary function for the data is also computed.

If fix.n=TRUE the number of points in each simulated realisation is held fixed; for multitype point process models, if fix.marks=TRUE, the number of points of each type is held fixed. Other arguments are available to control the simulation procedure: for a full list see help(envelope).

The following code fits an inhomogeneous Poisson process to the full Japanese Pines data of Numata and Ogata, then generates simulation envelopes of the L function by simulating from the fitted inhomogeneous Poisson model.

```
> numata <- residualspaper$Fig1
> fit <- ppm(numata ~ polynom(x,y,3))
> E <- envelope(fit, Lest, nsim=19, global=TRUE, correction="border")
> plot(E)
```

For another example we turn to the breakdown spots data bdspots described in Section 16.6.2. We fit the Gibbs hard core point process model (Section 13.3.2) to this pattern:

```
> B    <- bdspots[[1]]
> fit <- ppm(B ~ 1, Hardcore())
```

Then simulation envelopes based on this fitted model can be computed simply by

```
> EH <- envelope(fit, Lest, nsim=39, global=TRUE)
```

10.8.1.3 Envelopes based on any simulation procedure

Envelopes can also be computed using any user-specified procedure to generate the simulated realisations. This allows us to perform randomisation tests, for example.

The simulation procedure should be encoded as an R expression which, when evaluated, yields a point pattern. An `expression` in the R language represents a command that has not yet been executed. It is created by typing `expression(.....)` where the dots represent any command in R. The expression can then be executed or 'evaluated' using `eval`. Typically the expression should involve a random generator, so that the result of evaluating the expression is different each time it is evaluated. For example if we type

```
> e <- expression(rpoispp(100))
```

then each time the expression e is evaluated, it will yield a different random outcome of the Poisson process with intensity 100 in the unit square. Try this by typing `eval(e)` several times.

The expression should be passed to the `envelope` function as the argument `simulate`. Then simulated point patterns will be generated by evaluating the expression repeatedly.

For the breakdown spots data `bdspots` mentioned above, analysis in [472, 598] suggests that a Simple Sequential Inhibition model (Section 5.5.3) is appropriate. We could proceed as follows (this is somewhat slow):

```
> n <- npoints(B)
> W <- Window(B)
> r <- min(nndist(B))*n/(n+1)
> e <- expression(rSSI(r, n, W))
> E <- envelope(B, Lest, nsim=39, global=TRUE, simulate=e)
```

Note that since global envelopes are requested, e is evaluated 39 times to estimate the theoretical *L*-function for the SSI model, and then another 39 times to form the simulation envelopes.

Specifying the simulation procedure as an R expression makes it possible, for instance, to perform a *randomisation test* in which each simulated point pattern is a random alteration of the original data. For example, the function `rlabel` randomly permutes the mark values of a marked point pattern. The following code generates and plots simultaneous envelopes for the cross-type *L*-function based on random labellings of the `amacrine` data.

```
> e <- expression(rlabel(amacrine))
> E <- envelope(amacrine, Lcross, nsim=19, global=TRUE, simulate=e)
> plot(E)
```

The corresponding null hypothesis is that the pattern has the *'random labelling property'*: given the locations of the points, the marks are independent random variables with a common distribution. See Chapter 14 starting at page 567 for a full discussion.

10.8.1.4 Envelopes based on a set of point patterns

Envelopes can also be computed from a user-supplied list of point patterns, instead of the simulated point patterns generated by a chosen simulation procedure. The argument `simulate` can be a list of point patterns:

```
> Xlist <- runifpoint(42, nsim=99)
> envelope(cells, Kest, nsim=99, simulate=Xlist)
```

The argument `simulate` can also be an `"envelope"` object. This improves efficiency and consistency if, for example, we are going to calculate the envelopes of several different summary statistics.

```
> EK <- envelope(cells, Kest, nsim=99, savepatterns=TRUE)
> Ep <- envelope(cells, pcf,  nsim=99, simulate=EK)
```

In the first call to `envelope`, the argument `savepatterns=TRUE` indicates that we want to save the simulated point patterns. These are stored in the object EK. Then in the second call to `envelope`, the simulated patterns are extracted from EK and used to compute the envelopes of the pair correlation function.

The method `envelope.envelope` allows envelope calculations to be based on an existing `"envelope"` object. The previous example could have been done by typing

```
> EK <- envelope(cells, Kest, nsim=99, savepatterns=TRUE)
> Ep <- envelope(EK, pcf)
```

The same principle applies to the functions `mad.test` and `dclf.test`, which have essentially the same syntax as `envelope`. For example, to perform both the MAD and DCLF tests of CSR using the same simulations for both tests:

```
> A <- mad.test(cells, Lest, nsim=99, savefuns=TRUE)
> B <- dclf.test(A)
```

10.8.2   The summary function and its arguments

The second argument of `envelope` specifies the summary function `fun` to be used in constructing the envelopes. The default is `fun=Kest` giving envelopes of the K-function.

Typically `fun` is one of the standard `spatstat` summary functions such as `Kest`, `Gest`, `Fest`, `Jest`, `pcf`, or one of their variants. It may also be a character string containing the name of one of these functions. Optionally, `fun` could be a function created by the user.

Additional arguments to `fun` may be given in the call to `envelope`. For example, to compute the translation edge correction estimate of the K-function, we would call `Kest` with the argument `correction="trans"`. To compute envelopes based on the translation-corrected K-function, we simply type `envelope(X, Kest, correction="trans")`. Note that `correction` is not a formal argument of `envelope` or `envelope.ppp`. Here we are using the 'pass-through' argument '`. . .`' explained on page 82. What happens to these arguments depends on the particular function. The help for `envelope.ppp` states that '`. . .`' arguments are passed to the summary function `fun`, in this case `Kest`. Thus `Kest` will be called with `correction="trans"` as we intended.

The list of available edge corrections is different for each summary function, and may also depend on the kind of window in which the point pattern is recorded. However, all the summary functions in `spatstat` recognise the options `correction="best"` and `correction="good"` which indicate, respectively, the most accurate edge correction, and the most accurate *fast* edge correction. In a call to `envelope`, if `fun` is one of the summary functions provided in `spatstat`, then the default is `correction="best"`. This means that *by default, the envelopes will be computed using the 'best' available edge correction*. To increase speed, the user could specify `correction="good"` or choose a specific edge correction.

Envelopes based on inhomogeneous summary functions like `Kinhom` need careful attention. Recall that `Kinhom(X, lambda)` computes an estimate of the inhomogeneous K-function of the process which generated X, assuming that the intensity function is given by `lambda`. Typically `lambda` should be estimated from the data X somehow. If `lambda` is missing, it is estimated by kernel smoothing, so that `Kinhom(X)` produces a valid estimate. The following is therefore a valid way to build envelopes:

```
> D <- density(X)
> envelope(X, Kinhom, simulate=expression(rpoispp(D)))
```

The null hypothesis is an inhomogeneous Poisson process; the intensity of the process is estimated by kernel smoothing the data to yield the image D; realisations of this fitted model are generated by evaluating the expression. For each simulated pattern, since `lambda` is not given, the intensity will be estimated by kernel-smoothing the simulated pattern, and the inhomogeneous *K*-function for the simulated pattern will be estimated. The procedure used to compute an inhomogeneous *K*-function is exactly the same for the simulated patterns as it is for the data.

However, the following is **not valid**:

```
> D <- density(X)
> envelope(X, Kinhom, lambda=D, simulate=expression(rpoispp(D)))
```

The null hypothesis and the simulation procedure are the same as above, but in this case the function `Kinhom` will be called with `lambda = D` each time. That is, the intensity function estimate D obtained from the data will be used to compute the inhomogeneous *K*-functions of the simulated patterns as well as the original data. The simulated patterns are not treated in the same way as the data. The corresponding envelopes do not support a valid Monte Carlo test of the null hypothesis. The `spatstat` package will usually detect this kind of problem and issue a warning.

For a parametric model, the following is **valid**, but depends on a special feature of `Kinhom`:

```
fit <- ppm(.....)
envelope(fit, Kinhom, lambda=fit)
```

Here the null hypothesis is the Poisson point process model specified by the trend formula in `fit`. The simulated point patterns will be generated from this Poisson process, using the parameters estimated in `fit`. The special feature is that when `lambda` is a fitted point process model, `Kinhom(XX, lambda)` will re-fit the model to the new point pattern `XX`, before computing the predicted intensity function, and using this intensity to compute the inhomogeneous *K*-function estimate. The procedure is exactly the same for the simulated patterns as it is for the data, and the test is valid.

10.8.3 Types of envelopes in `spatstat`

10.8.3.1 Pointwise envelopes

By default, `envelope` computes pointwise envelopes of the summary function. The argument `nsim` controls the number *m* of simulated point patterns. The argument `nrank` specifies the rank *k* of the critical value amongst the simulated values: the envelopes are based on the *k*th largest and *k*th smallest simulated values. The default is $k = 1$, meaning that the minimum and maximum simulated values will be used. For a two-sided test, the significance level is $2k/(m+1)$.

10.8.3.2 Simultaneous envelopes

To obtain simultaneous (global) envelopes, set `global=TRUE`:

```
> envelope(redwood, Lest, global=TRUE)
```

The significance level is $k/(m+1) = 1/(1+99) = 0.01$.

As noted in Section 10.7.3, the outcome of the test is affected by transforming the summary function. In order to achieve the maximum power of the test, it is usually recommended to apply a *variance-stabilising* transformation to the summary function, that is, a transformation $H(r) = f(S(r))$ such that the variance of $H(r)$ is approximately constant as a function of *r*. The global measure of deviation (10.18) treats all fluctuations of the summary function *S* equally; this could be inappropriate if fluctuations of $S(r)$ at some distance r are more informative than at other distances. After a variance-stabilising transformation, fluctuations in the summary function at different distances r have equal scale, so can be treated equally.

An approximate variance-stabilising transformation for the K-function is the square root, as explained on page 207. Indeed this was one of the motivations for Besag's [103] original proposal of the L-function, $L(r) = \sqrt{K(r)/\pi}$. Thus we would typically use Lest rather than Kest for global envelopes.

For the summary functions F and G, an approximate variance-stabilising transformation is Fisher's arcsine transformation $f(x) = \arcsin\sqrt{x}$ as explained on page 269. Rather than writing a new function for the variance-stabilised versions of Fest and Gest, we can specify the transformation using the argument transform:

```
> fisher <- function(x) { asin(sqrt(x)) }
> envelope(redwood, Fest, global=TRUE,
          transform=expression(fisher(.)))
```

To allow very flexible specification of the transformation, envelope uses with.fv to evaluate the transform expression. In the syntax of with.fv, the symbol '.' represents the function value, in this case the value of the estimate $\widehat{F}(r)$.

10.8.3.3 Envelopes based on sample mean & variance

Envelopes can be constructed using the sample mean and sample variance of the simulations. By default the envelopes consist of the sample mean plus or minus 2 times the sample standard deviation. This is useful for understanding the variability of the summary function. Be aware that these envelopes do not have the same significance interpretation as the global envelopes produced from ranking absolute deviations.

```
> envelope(cells, Kest, nsim=100, VARIANCE=TRUE)
```

10.8.3.4 One-sided envelopes

In envelope methods, the argument alternative is a character string determining whether the envelopes correspond to a two-sided test (alternative="two.sided", the default) or a one-sided test with a lower critical boundary (alternative="less") or a one-sided test with an upper critical boundary (alternative="greater").

10.8.4 Re-using envelope data

Computing new envelopes

The method envelope.envelope allows new envelope commands to be applied to a previously computed "envelope" object, provided it contains the necessary data.

In the original call to envelope, if the argument savepatterns=TRUE was given, the resulting "envelope" object contains all the simulated point patterns. Alternatively if the argument savefuns=TRUE was given, the resulting object contains the individual summary functions for each of the simulated patterns. This information is not saved, by default, for efficiency's sake.

Envelopes created with savepatterns=TRUE allow any kind of new envelopes to be computed using the same simulated point patterns:

```
> E1 <- envelope(redwood, Kest, savepatterns=TRUE)
> E2 <- envelope(E1, Gest, global=TRUE,
                transform=expression(fisher(.)))
```

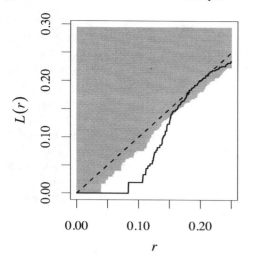 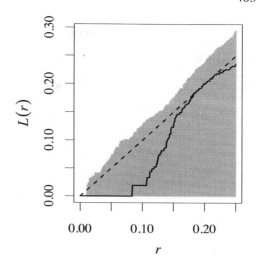

Figure 10.8. *One-sided envelopes. Grey shading shows the pointwise acceptance region for a one-sided test of CSR against the alternative of regularity (*Left*) or clustering (*Right*) based on the L-functions of 19 simulations of CSR. Thick line shows empirical L-function for the* `cells` *data.*

Envelopes created with `savefuns=TRUE` allow the user to switch between pointwise and global envelopes of the same summary function, to apply different transformations of the summary function, and to change some parameters:

```
> A1 <- envelope(redwood, Kest, nsim=39, savefuns=TRUE)
> A2 <- envelope(A1, global=TRUE, nsim=19,
                 transform=expression(sqrt(./pi)))
```

Pooling several envelopes

It is also possible to combine the simulation data from several envelope objects and to compute envelopes based on the combined data. This is done using `pool.envelope`, a method for the `spatstat` generic `pool`. The envelopes must be compatible, in that they are envelopes for the same function, and were computed using the same options. The individual summary functions must have been saved.

```
> E1 <- envelope(cells, Kest, nsim=10, savefuns=TRUE)
> E2 <- envelope(cells, Kest, nsim=20, savefuns=TRUE)
> E <- pool(E1, E2)
```

The method for pooling summary functions is detailed in Section 16.8.1.

10.9 Other presentations of envelope tests

10.9.1 Significance traces

Many statistical tests require us to choose a value for smoothing bandwidth or a similar quantity. The *significance trace* [122] is a plot of the *p*-value obtained from the test against this quantity. It is used to check whether the outcome of the test is sensitive to the value of this quantity.

The MAD test and DCLF test require us to choose the range of distance values over which the deviations in the summary function will be maximised or summed. It can be quite useful to plot a significance trace for these tests against the length of the range of distances. The functions `mad.sigtrace`, `dclf.sigtrace`, and `mctest.sigtrace` generate significance traces for the corresponding tests. Figure 10.9 shows the result of

```
> dclf.sigtrace(cells, Lest, nsim=19)
```

The black line is the Monte Carlo p-value based on 19 simulations for the DCLF test over the distance range $[0, R]$ plotted against R. The plot also shows the desired significance level $\alpha = 0.05$, and a pointwise 95% confidence band for the *'true'* p-value corresponding to an optimal Neyman-Pearson test. The confidence band is based on the Agresti-Coull [3] confidence interval for a binomial proportion. The plot indicates that this test result is quite robust to the choice of different ranges of distance values, except at very small distances.

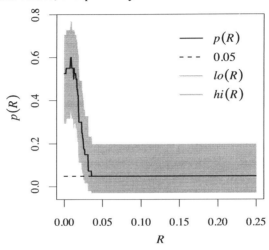

Figure 10.9. *Significance trace for the DCLF test of CSR, based on the L-function, for the* `cells` *data.*

10.9.2 Progress plots

At face value, the DCLF test does not appear to have a graphical representation in terms of simulation envelopes. However, such an interpretation *does* exist, and was pointed out by A. Hardegen [40]. The test statistic T in (10.21) or (10.22) depends on the choice of the upper limit on distances, R. For any specified value of R, suppose $t_{obs}(R)$ is the observed value of the test statistic and $t_i(R)$ is the ith simulated value. Then the DCLF test at significance level $\alpha = 1/(m+1)$ based on the interval $[0, R]$ is to reject the null hypothesis if $t_{obs}(R) > t_{max}(R)$ where $t_{max}(R) = \max_i t_i(R)$ is the largest simulated value. Thus by plotting $t_{obs}(R)$ and $t_{max}(R)$ against R, we can represent the outcome of the DCLF test for each R. We call this the *progress plot* for the DCLF test.

Figure 10.10 shows the progress plot for the DCLF test applied to the Swedish Pines data using the L-function. The left panel shows the outcome based on 19 simulations. The DCLF test statistic $t_{obs}(R)$ and the Monte Carlo critical value $\max_i t_i(R)$ are plotted against R. The DCLF test rejects the null hypothesis at the 0.05 level when R is greater than about 0.7 metres.

The same principle applies when we take the kth largest of the simulated values. The right panel of Figure 10.10 is based on $m = 1999$ simulations and rank $k = 100$, giving the same significance level $\alpha = 0.05$. The DCLF test statistic $t_{obs}(R)$ is plotted against R, and the grey shading shows the

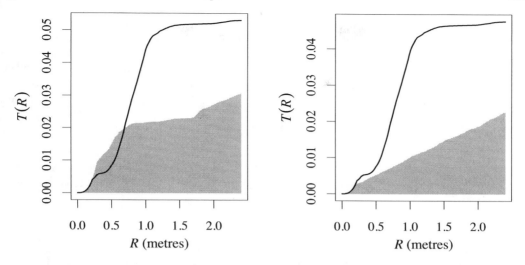

Figure 10.10. *Envelope representation (progress plot) for the DCLF test of CSR with* α = *0.05, applied to the Swedish Pines data. DCLF test statistic T (solid lines) and Monte Carlo acceptance/non-rejection region (shaded) are plotted as a function of the length R of the distance interval* [0, R]. Left: *based on 19 simulations of CSR; critical value is the maximum of the simulated values.* Right: *based on 1999 simulations of CSR; critical value is the 100th largest of the simulated values.*

acceptance region (non-rejection region) delimited by the Monte Carlo critical value, the kth largest of the m values $t_1(R), \ldots, t_m(R)$. Based on this larger suite of simulations, the DCLF test now rejects the null hypothesis at the 0.05 level at values of R greater than about 0.2 metres.

The progress plot is generated by the function `dclf.progress` which has essentially the same syntax as `envelope`. The left panel of Figure 10.10 was generated by

```
> X <- rescale(swedishpines)
> plot(dclf.progress(X, Lest, nsim=19))
```

The analogous plot for the MAD test is generated by `mad.progress`.

10.10 Dao-Genton test and envelopes

Dao and Genton [200] proposed a modification of the Monte Carlo test that reduces or eliminates the problem of conservatism, described in Section 10.6.3. We have extended this approach to the construction of global envelopes [41]. This section includes the result of joint work with A. Hardegen, T. Lawrence, R.K. Milne, G.M. Nair, and S. Rakshit.

10.10.1 Dao-Genton test and modifications

Suppose the null hypothesis is a point process model with parameter θ whose value is unknown and must be estimated. The key problem is that the Monte Carlo p-value $\hat{p}$ is not uniformly distributed under H_0.

In the test proposed by Dao and Genton [200], and slightly modified in [41], the usual Monte Carlo test procedure applied to the data **x** is also applied to each of the simulated point patterns $\mathbf{x}^{(j)}$, yielding p-values $\hat{p}_1, \ldots, \hat{p}_m$. The observed distribution of $\hat{p}_1, \ldots, \hat{p}_m$ is then taken as an approximation to the true null distribution of $\hat{p}$ with the unknown true value of θ.

For each $j = 1, \ldots, m$, we apply the model-fitting procedure to the simulated pattern $\mathbf{x}^{(j)}$, yielding a parameter estimate $\hat{\theta}^{(j)} = \hat{\theta}(\mathbf{x}^{(j)})$. From the fitted model with this parameter value, we generate m' independent simulated realisations $\mathbf{x}^{(j,1)}, \ldots, \mathbf{x}^{(j,m')}$ where typically $m' = m - 1$, then compute $t_{jk} = T(\mathbf{x}^{(j,k)})$ for $k = 1, \ldots, m'$, and compute the Monte Carlo p-value from this set of simulations,

$$\hat{p}_j = \frac{1 + \sum_{k=1}^{m'} \mathbf{1}\left\{t_{jk} \geq t_j\right\}}{m' + 1}, \tag{10.23}$$

where $t_j = T(\mathbf{x}^{(j)})$ are the summary statistics at the first level, and for simplicity we assume that large values of t are favourable to the alternative hypothesis. The p-value associated with the Dao-Genton test is [41]

$$p^* = \frac{1 + \sum_{j=1}^{m} \mathbf{1}\left\{\hat{p}_j \leq \hat{p}\right\}}{m + 1}. \tag{10.24}$$

Unlike the ordinary Monte Carlo p-value, which is biased no matter how many simulations we perform, the adjusted p-value p^* is asymptotically exact, in the sense that as the number of simulations $m \to \infty$, the null distribution of p^* approaches the uniform distribution [200, 41]. The approximation can be further improved by smoothing the t values [200, 41].

It is useful to express the Dao-Genton test with fixed significance level α as a rejection rule in terms of the ranks

$$R = 1 + \sum_{j=1}^{m} \mathbf{1}\left\{t_j \geq t_{obs}\right\} \tag{10.25}$$

$$U_j = 1 + \sum_{k=1}^{m'} \mathbf{1}\left\{t_{jk} \geq t_j\right\}. \tag{10.26}$$

Then H_0 is rejected at significance level α if

$$\sum_{j=1}^{m} \mathbf{1}\left\{R \leq U_j\right\} \leq \alpha(m + 1). \tag{10.27}$$

If $\alpha(m + 1) = k$ is a positive integer, let $\ell = U_{(\alpha m)}$ be the kth largest value among $U_1, \ldots, U_m$. Assuming there are no ties, $p^* \leq \alpha$ iff $R \leq \ell$. Equivalently, the Dao-Genton test rejects H_0 at level α if $t_{obs} \geq t_{(\ell)}$, the ℓth largest of the simulated values in the original suite of simulations.

This procedure is implemented in the `spatstat` function `dg.test`.

```
> dg.test(cells, Lest, nsim=19)

        Dao-Genton adjusted goodness-of-fit test
        based on Diggle-Cressie-Loosmore-Ford test of CSR
        Monte Carlo test based on 19 simulations
        Summary function: L(r)
        Reference function: sample mean
        Alternative: two.sided
        Interval of distance values: [0, 0.25]

data: cells
p0 = 0.05, p-value = 0.05
```

For the test based on maximum absolute deviation, instead of the integrated squared deviation, we would set the exponent to infinity:

```
> dg.test(cells, Lest, nsim=19, exponent=Inf)
```

A one-sided test is specified using the argument `alternative` as in Section 10.7.5.

10.10.2 Dao-Genton correction for simultaneous envelopes

Because simulation envelopes are associated with Monte Carlo tests, the Dao-Genton test has an analogue for simulation envelopes, which was developed in [41]. This technique is only practical for the Dao-Genton version of the MAD test. Assume that $\mathbf{x}$ is a point pattern dataset and $\hat{S}(r) = \hat{S}(\mathbf{x}, r)$, $r > 0$ is a summary function. The statistic t is taken to be

$$t = T(\mathbf{x}) = \max_{r \in [0,R]} |\hat{S}(\mathbf{x}, r) - S_{theo}(r)|$$

where $S_{theo}(r)$ is the reference value of $S(r)$ under the null hypothesis. Again we estimate the parameter θ by $\hat{\theta} = \hat{\theta}(\mathbf{x})$, generate m simulated point patterns $\mathbf{x}^{(1)}, \ldots, \mathbf{x}^{(m)}$ from $P_{\hat{\theta}}$, and evaluate $t_j = T(\mathbf{x}^{(j)})$ for $j = 1, \ldots, m$. The Monte Carlo test of size $k/(m+1)$ rejects H_0 if $t \geq t_{(k)}$, or, equivalently, if the graph of $\hat{S}(\mathbf{x}, r)$ lies outside the region bounded by $L(r) = S_{theo}(r) - t_{(k)}$ and $U(r) = S_{theo}(r) + t_{(k)}$ for any $r \in [0, R]$.

For the Dao-Genton test, for each $\mathbf{x}^{(j)}$ we compute the parameter estimate $\hat{\theta}^{(j)} = \hat{\theta}(\mathbf{x}^{(j)})$ and generate m' simulated patterns $\mathbf{x}^{(j,1)}, \ldots, \mathbf{x}^{(j,m')}$ from $P_{\hat{\theta}^{(j)}}$, evaluating $t_{jh} = T(\mathbf{x}^{(j,h)})$ for $h = 1, \ldots, m'$. We evaluate the ranks S_j as in (10.26). The Dao-Genton test of size $k/(m+1)$ rejects H_0 if $t \geq t_{(\ell)}$ where ℓ is again the kth largest among the ranks S_j. Equivalently, we reject H_0 if the graph of $\hat{S}(\mathbf{x}, r)$ lies outside the region bounded by $L(r) = S_{theo}(r) - t_{(\ell)}$ and $U(r) = S_{theo}(r) + t_{(\ell)}$ for any $r \in [0, R]$. Figure 10.11 shows the envelopes associated with the two-sided and one-sided Dao-Genton tests of CSR for the Swedish Pines data, generated by

```
> swp <- rescale(swedishpines)
> E <- dg.envelope(swp, Lest)
> Eo <- dg.envelope(swp, Lest, alternative="less")
```

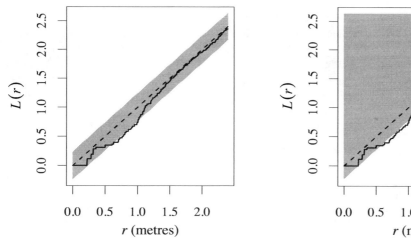

Figure 10.11. *Envelope versions of the Dao-Genton test of complete spatial randomness for the Swedish Pines data, based on the L-function.* Left: *two-sided;* Right: *one-sided, regular alternative.*

These envelopes largely avoid the problem of conservatism that arises when model parameters must be estimated from the data. They should typically be used in small-sample situations, when the conservatism effect is largest and the cost of additional simulations is low.

10.11   Power of tests based on summary functions

The performance of a test is measured by its *power*, the probability of making the correct decision if the null hypothesis is false. Here we discuss various strategies that can improve the power of a Monte Carlo test based on summary functions.

The power of a test is the probability $1 - \beta$ of rejecting the null hypothesis when the null hypothesis is false and the specified alternative hypothesis is true. The test power depends on the alternative, and can be regarded as a measure of the sensitivity of the test to the specified alternative. A test may have strong power against one alternative and weak power against another alternative.

All the choices involved in designing a Monte Carlo test affect the power of the test. Power typically increases if we increase the number of simulations m, for a fixed significance level α. Power is affected by the choice of summary function S and the test statistic T.

The choice of summary function affects the sensitivity of the test to different types of spatial pattern. For example an envelope test based on the nearest-neighbour distance distribution G is usually very sensitive to the presence of inhibition between points, but less sensitive to clustering. The cell process (page 211) has the same K-function as CSR, so a test of CSR based on the K-function has no utility against this alternative.

The power of a test based on a summary function is usually improved substantially by stabilising the variance. For this reason, Besag [103] (in the discussion of [572]) proposed transforming $K(r)$ to $L(r) = \sqrt{K(r)/\pi}$ before applying the MAD test. For the nearest-neighbour distance distribution function G and the empty-space function F the appropriate variance-stabilising transformation is due to Fisher [259] (see [8]) as discussed in Chapter 8.

A detailed description of experiments to measure the power of a test has been given by [391, sec. 3.5]. The power of the DCLF and MAD tests using different summary functions was measured in a series of simulation experiments in [40, 41]. The power of each test is maximised when the interval length R is slightly larger than the range of spatial interaction. The DCLF test is quite sensitive to the choice of R while the MAD test is insensitive to R. The DCLF test is typically (but not always) more powerful than the MAD test, as might be expected by analogy with the the Cramér-Von Mises and Kolmogorov-Smirnov tests of goodness-of-fit. Consequently the recommendation of [40] is to use the DCLF test, provided the range of spatial interaction is known approximately (and the interval length is chosen accordingly). If there is no information about the range of spatial interaction, then it may be prudent to use the MAD test, choosing the interval length to be as large as practicable.

10.12   FAQs

- *I applied two tests to the same data, and obtained different answers: one test says the result is significant and the other says it is not. What is wrong? What do I do?*

  There's probably nothing wrong. Two different tests of the same null hypothesis will sometimes give different answers for the same data. (If they didn't sometimes disagree, they would be the same test.)

- *Why are several different tests available for the same purpose?*

  Usually there is no single test that is always better than the others. Each test is typically designed to have high power (sensitivity) against a particular alternative hypothesis, at the expense of lower power against other alternatives. Sometimes a test that is good in theory (e.g., the likelihood ratio test) is computationally expensive to perform, and another test (e.g., the score test) is much easier to perform. Different tests also depend on different underlying assumptions.

- *I have plotted pointwise envelopes of the L-function, and also performed a DCLF test using the L-function. The empirical L-function stays entirely inside the pointwise envelopes (although it lies very close to the upper envelope) while the DCLF test gives me a very significant result of $p = 0.007$. Aren't these results contradictory?*

No. The DCLF test statistic accumulates evidence against the null hypothesis from each distance r in the specified range. Essentially the DCLF test statistic is something like the area between the empirical and theoretical L-functions. In the situation described above, although there is insufficient evidence against the null hypothesis at any fixed distance r, the evidence accumulated over a range of distances is strong enough to reject the null hypothesis.

- *I'm confused about how to report the results of a hypothesis test. In this chapter you speak of statistical significance at the level 0.05, or a p-value less than 0.05, whereas some journals seem to report results as being '95% significant' or having '95% confidence'.*

Interpreting a test result in terms of 'confidence' is erroneous. See page 370.

- *I have heard that 'envelope tests' are invalid and should be replaced by 'deviation tests'.*

'Envelope tests' and 'deviation tests' have the *same* statistical rationale, that of a Monte Carlo test [249, 79, 338], so it cannot be true that envelope tests are invalid while deviation tests are valid. See Section 10.7 or [40].

- *What null hypotheses can be tested in* spatstat *apart from the basic ones like CSR, random labelling, and independence?*

Almost any null hypothesis that you can think of. It can be any fitted point process model (homogeneous or inhomogeneous, Poisson, Cox, Gibbs, or cluster type) fitted by the model-fitting functions ppm and kppm. Alternatively the null model can be specified by giving a rule in the R language for generating simulated realisations from the null model. Alternatively you can supply several point patterns which you have generated according to the null model. See Section 10.8 for details.

- *A referee said that it is naive to perform a test of CSR, when we already know that CSR is false. The referee says that ecologists "long ago abandoned naive choices of the null model", quoting Strong [644].*

This depends on the scientific objectives and the stage of analysis. In initial investigation, CSR often serves as a 'dividing hypothesis' although it is *known to be false* [179, pp. 51–52]: the test outcome guides the direction of the subsequent data analysis [225, p. 12], [30] and serves as a cross check that the data are roughly consistent with our cherished theoretical assumptions.

In the final, definitive stages of analysis, the null hypothesis should correspond to a scientifically meaningful scenario in which 'nothing is happening' [644]. Rejection of the null hypothesis allows us to conclude that 'something is happening', for example, that a particular variable does have an effect. In this context, CSR is often not the most appropriate scientific hypothesis. For example, in ecological applications, the appropriate null hypothesis might involve spatial inhomogeneity due to known environmental factors, and spatial clustering and regularity due to known processes of dispersal and competition.

- *Which null model do you recommend: CSR or restricted randomisation?*

The appropriate choice of null hypothesis depends on the scientific context. It should be a statement that means 'nothing is happening' in that context.

- *Editorial policy of the journal requires that the conclusions should be summarised in a hypothesis test, but I don't see why this is necessary for my case.*

- *Editorial policy of the journal is that hypothesis tests are invalid or unreliable, and that everyone should now be quoting confidence intervals for the* effect size, *rather than p-values.*

- *Editorial policy of the journal is that frequentist inferential methods like hypothesis tests are unreliable, and that everyone should now be using Bayesian inference.*

Most statisticians disagree with the imposition of any hard-and-fast rules for statistical methodology. The appropriate choice of statistical technique depends greatly on the application context.

There are many polemics against hypothesis tests [163, 73, 560]. Some of these have valid points to make. However, the blanket dismissal of hypothesis testing is unjustified. Used correctly and appropriately, hypothesis testing is a useful tool.

The fiercest critics of statistical methods are statisticians themselves. For example, statisticians have argued for more than 70 years that confidence intervals are preferable to hypothesis testing in some applications [179, 707]. Statisticians understand the weaknesses of these methods and the limitations on their use. All techniques have weaknesses; this does not mean that they must be completely abandoned.

- *The printout from* `envelope(X, nsim=99)` *says the significance level is* $\alpha = 0.02$, *but I want significance level* $\alpha = 0.01$. *According to the help file for* `envelope`, *the significance level is* $\alpha = 2k/(n+1)$ *where n is the number of simulations specified by the argument* `nsim`, *and k is the rank specified by the argument* `nrank`. *So by this formula I should set* $k = 1/2$ *to get* $\alpha = 0.01$. *But when I do* `envelope(X, nsim=99, nrank=1/2)` *it gives an error. Why?*

Fractional ranks are meaningless. Rank 1 means the largest value, rank 2 means the second largest, etc. Only some values of α are achievable (namely multiples of $2/(m+1) = 0.02$ in this case). If you want $\alpha = 0.01$ you could set `nsim=199`, and leave `nrank` at its default value of 1.

- *Why is hypothesis testing such a confusing topic?*

Oh, for the good old days when the guilt or innocence of an accused could be determined simply by drowning them! Statistical inference is similar to the proceedings in a modern court of law, not a show trial. In order to give both sides a fair hearing, the court must apply the rules of evidence, and that can be a complicated process.

It is naive to expect statistics to be easy. Statistical inference is designed to be applicable to all fields of research. It must be flexible enough to handle the realities in many different applications.

Hypothesis testing is not terribly complicated. Unfortunately, many classes in data analysis are taught by people who are not formally trained in statistics, and this tends to show when they are trying to explain hypothesis tests. For a friendly introduction we highly recommend [489].

- *How do I ensure that two envelope objects* A *and* B *are plotted with the same axis scales?*

Use `plot(anylist(A,B), equal.scales=TRUE)`. Alternatively, determine the axis limits using `lapply(list(A,B), plot, limitsonly=TRUE)`, and calculate the scale appropriately.

11

Model Validation

After fitting a point process model to a spatial point pattern dataset, we need to check that the model is faithful to the data. Techniques for checking or *validating* a point process model are introduced in this chapter. Here we shall deal only with Poisson point process models; validation for other point process models is covered in Chapter 13.

11.1 Overview of validation techniques

11.1.1 Principles

Model validation is important because a fitted model is not like a fitted shoe. A shoe must conform to the shape of the wearer's foot reasonably well, before we will say it has been fitted; but a model that we have 'fitted' to data does not necessarily conform to the pattern of the real data *at all*.

For example, linear regression software, which 'fits' a straight line through a cloud of data points, will perform this task even if the data follow a non-linear relationship, or follow no pattern at all. The software effectively assumes that the linear regression model is true. Software outputs such as confidence intervals and *p*-values also assume the model is true, and may give us a false sense of security[1] about the validity of the model.

Similarly when we fit a Poisson point process model to a spatial point pattern dataset as described in Chapter 9, the model makes several assumptions which could be challenged. It assumes that the points are independent of each other, and often assumes that the intensity is a loglinear function of the specified covariates. The fitting procedure does not examine whether these assumptions are appropriate. This is a problem because an incorrect model could undermine the validity of the entire analysis [190, Section 8.5], [30].

A test of *goodness-of-fit* can be used to decide whether the data appear to follow the fitted model. An example is the χ^2 test based on quadrat counts (Sections 6.4.2, 6.7.1, and 10.5.3). Unfortunately, goodness-of-fit tests are rather uninformative: if the test finds significant evidence of 'lack of fit', we do not know which of the model assumptions is incorrect (see Sections 6.4.3 and 10.5.4).

A more searching way to validate the model is by using a suite of *diagnostics*, each designed to focus on one specific component ('assumption') of the model. Diagnostics are well developed for linear regression and for generalized linear models [26, 164, 170, 325]. Diagnostics are mostly built from the *residuals*

$$(\text{residual}) = (\text{observed}) - (\text{fitted})$$

which contain essentially all information about the discrepancies between the model and the data. Residuals for point processes can be defined in the same way [55, 45, 504, 505, 506, 508, 512, 712, 711]. The residuals can be difficult to interpret on their own, but we can extract diagnostic information by combining and plotting the residuals in particular ways. Important regression diagnostics are

---

[1] "If you give people a linear model function, you give them something dangerous." John Fox, UseR! 2004, Vienna.

leverage, *influence*, *partial residual*, and *added variable* plots, all of which can be extended from linear models to spatial point process models [35, 36, 49]. See also [673, 690, 633, 594, 413, 668, 602], [355, sec. 4.6].

A great advantage of model diagnostics is that they help us to *improve* the model. Each specialised diagnostic assesses whether a particular assumption of the model appears to be correct, and, if not, suggests how it may be corrected. This supports a cyclic process of statistical modelling in which we formulate tentative models, fit them to data, criticise the fit, and update the model [181, 180].

11.1.2　Tools

Table 11.1 lists some diagnostic tools, discussed in this chapter, for validating specific components of a fitted point process model.

TARGET	TOOL	IN spatstat	SECTION
Fitted intensity	Relative intensity	`density.ppp` with weights	11.2.1
	Residuals	`residuals.ppm`	11.3.1
	Smoothed residuals	`Smooth.msr`	11.3.2
	Residual plots	`diagnose.ppm`	11.3.4
Form of covariate effect	Relative distribution	`rhohat.ppm`	11.2.2
	Partial residual plot	`parres`	11.4
Presence of covariate effect	Lurking variable plot	`lurking`	11.3.3
	Added variable plot	`addvar`	11.5
Independence	QQ plot of residuals	`qqplot.ppm`	11.6.2
	Residual K-function	`Kres`	11.6.4
	Residual G-function	`Gres`	11.6.4
Sensitivity to data	Leverage	`leverage.ppm`	11.7
	Influence	`influence.ppm`	11.7
	Parameter influence	`dfbetas.ppm`	11.7

Table 11.1. *Diagnostic tools for validating a fitted Poisson point process model.*

To understand how these tools should be applied, a useful guide is the analogy between fitting a Poisson point process model and fitting a regression model, explained in Chapter 9. A regression model has a 'systematic' part (the regression line or regression curve) and a 'random' part (the assumed variability of the observations around the regression line).

Systematic part

The 'systematic' part of a Poisson point process model is its intensity function $\lambda(u)$. If there is a systematic relationship between the points of a Poisson process and a particular covariate $\mathbf{Z}(u)$, this relationship is expressed through the dependence of $\lambda(u)$ on $\mathbf{Z}(u)$.

A fitted loglinear intensity model $\log \lambda(u) = \boldsymbol{\theta}^\top \mathbf{Z}(u)$ is analogous to a fitted loglinear regression. We can validate the model on the scale of the mean (intensity) or on the scale of the linear predictor (log intensity).

On the scale of the intensity, a useful diagnostic is a smooth nonparametric estimate of the intensity $\lambda(u)$, such as the kernel smoothing estimate (Section 6.5.1) computed by `density.ppp`, or a smooth estimate of the function ρ in the relationship $\lambda(u) = \rho(\mathbf{Z}(u))$, such as the relative distribution estimate (Section 6.6.3) computed by `rhohat.ppp`.

On the scale of the log intensity, the appropriate diagnostic is the *partial residual plot* described in Section 11.4 below, and computed by the spatstat function `parres.ppm`. Deviations from a straight line in this plot suggest that the effect of the covariate is not linear on the log scale, and may suggest what shape it does take.

The *residuals* from a fitted point process model can also be defined (Section 11.3 below), and can be computed in spatstat by residuals.ppm. As in elementary statistics, so also for point processes, the residuals capture essentially all information about the discrepancy between the model and the data, but are used or presented in many different ways, to answer different questions. In regression models it is common to smooth the residuals; the analogue for point processes is a smoothed residual plot (Section 11.3.2). It can be computed from the residuals using Smooth.msr or plotted directly from the fitted model by diagnose.ppm.

Random part

The 'random' component of a Poisson point process model is the assumption that the points are independent of each other. To check this assumption we rely on having a good estimate of the intensity function. A *Q–Q plot of the smoothed residuals* is an effective tool (Section 11.6.2). It is produced by qqplot.ppm.

Summary statistics such as the *K*-function can of course be used to check the independence assumption, but they are only defined for models which are stationary or at least correlation-stationary. *Residual summary functions* (Section 11.6.4) avoid this limitation. The residual *K*-function with respect to a fitted model is a modification of the *K*-function which has mean equal to zero if the model is true. If the fitted model is CSR, the residual *K*-function is $K(r) - \pi r^2$, and the residual *G*-function is $G(r) - F(r)$, ignoring some details.

Sensitivity to data

Section 11.7 introduces the *leverage* and *influence* of a fitted point process model. The leverage is a function $h(u)$ expressing the potential effect that observing a point at location u would have on the fitted model. It indicates the relative 'importance' of data from different parts of the spatial domain to the final fitted model. The influence is a numerical value s_i for each data point x_i, expressing the actual importance of that data point in determining the fitted model.

11.2   Relative intensity

We start by describing methods based on the ratio of observed and fitted intensities. These will form a basis for more sophisticated methods.

For demonstration purposes we will use a fitted model which is probably not a good fit:

```
> fit2e <- ppm(bei ~ polynom(elev,2), data=bei.extra)
```

The model states that the *Beilschmiedia* forest density is a log-quadratic function of terrain elevation.

11.2.1   Inverse-lambda weighting

If a point process **X** with true intensity function $\lambda(u)$ is modelled by a point process with intensity $\lambda_0(u) > 0$, the *relative intensity* $r(u) = \lambda(u)/\lambda_0(u)$ measures agreement between the truth and the model. Values $r(u)$ close to 1 indicate agreement while values greater than 1 indicate that the model underestimates the true intensity. The relative intensity can be estimated by kernel smoothing. One estimator is

$$\hat{r}(u) = \frac{1}{e(u)} \sum_i \frac{1}{\lambda_0(x_i)} \kappa(u - x_i) \tag{11.1}$$

where κ is a smoothing kernel on the two-dimensional plane and $e(u)$ is the edge correction factor (6.10) defined on page 168. This is a weighted kernel intensity estimate, with weights equal to the reciprocal of the reference intensity λ_0. By Campbell's formula (6.11) the expected value of $\hat{r}(u)$ is

$$\mathbb{E}\hat{r}(u) = \frac{1}{e(u)} \int_W \frac{1}{\lambda_0(v)} \kappa(u-v)\lambda(v)\,dv = \frac{1}{e(u)} \int_W r(v)\kappa(u-v)\,dv,$$

a smoothed version of the relative intensity $r(u)$.

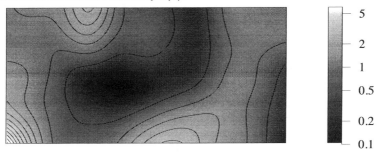

Figure 11.1. *Estimate of relative intensity* $r(u)$ *of* Beilschmiedia *data relative to a fitted model in which the intensity is a log-quadratic function of terrain elevation. Logarithmic greytone scale.*

The estimate $\hat{r}(u)$ can be computed in spatstat by density.ppp using the argument weights. The weights $w_i = 1/\lambda_0(x_i)$ are computed from the fitted model using fitted.ppm. For the model fitted to the *Beilschmiedia* data:

```
> lam0 <- fitted(fit2e, dataonly=TRUE)
> rel2e <- density(bei, weights=1/lam0)
```

The estimate is plotted in Figure 11.1. Values of $\hat{r}(u)$ have quite a wide range:

```
> range(rel2e)
[1] 0.1048 6.1246
```

which suggests the model fit2e is a poor fit.

11.2.2 Relative intensity as function of covariate

There are several goodness-of-fit tests (such as the Kolmogorov-Smirnov test and Berman's tests, Sections 10.5.3 and 10.3.5) which compare the observed distribution of the values of a spatial covariate Z to the expected distribution under a model. Another way to perform this comparison is to assume that the true intensity of the point process is

$$\lambda(u) = \rho(Z(u))\lambda_0(u) \tag{11.2}$$

where $\lambda_0(u)$ is the intensity under the fitted model, and ρ is an unknown function; then we may estimate the function ρ using 'relative distribution' methods (Section 6.6.3).

Equation (11.2) is a special case of equation (6.24) on page 181, where the 'baseline' function $B(u)$ is the intensity of the fitted model $\lambda_0(u)$. The function ρ can be estimated using the methods described there.

A direct way to perform this calculation is to use predict.ppm to compute the fitted intensity function $\lambda_0(u)$, then apply rhohat to the point pattern using the argument baseline:

```
> lambda0 <- predict(fit2e)
> grad <- bei.extra$grad
> rh1 <- rhohat(bei, grad, baseline=lambda0)
```

This uses the method `rhohat.ppp`. There is an alternative method `rhohat.ppm` for fitted models, which performs the same calculation:

```
> rh2 <- rhohat(fit2e, grad)
```

The results `rh1` and `rh2` are identical up to numerical error, and are shown in Figure 11.2. The value $\rho = 1$ corresponds to agreement with the fitted model depending only on terrain elevation, so Figure 11.2 suggests that this model overestimates the intensity on flat parcels of land (low values of `grad`) and underestimates it on steep slopes (high values of `grad`).

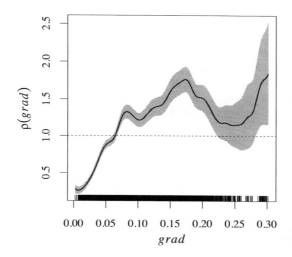

Figure 11.2. *Estimated relative density for* Beilschmiedia *trees against terrain slope, relative to the fitted model in which forest density is a log-quadratic function of terrain elevation. Dashed line at* $\rho \equiv 1$ *corresponds to agreement with fitted model.*

11.3 Residuals for Poisson processes

Residuals from the fitted model are a fundamental building block for diagnostics in applied statistics. They are ubiquitous because they are closely connected to the likelihood.

Residuals can be examined for diagnostic purposes in their own right, although they can be difficult to interpret. More often, the residuals are used to construct specialised diagnostic plots [26, 325].

11.3.1 Residual measure

Residuals from a fitted spatial point process model have been studied and used for decades, mainly for Poisson processes or space-time processes where the theory is simpler. Residuals for *non-Poisson* point processes have only recently been developed in a coherent way [504, 633], [594, pp.

49–50], [55]. Here we present the simplest case, of residuals for a Poisson point process. Residuals for Gibbs models are presented in Chapter 13.

A residual is obtained by subtracting the fitted mean from the observed values. For a spatial point process, the 'mean' is the intensity function, while the 'observed values' are (naively speaking) the data points of the point pattern. How can these be subtracted from each other?

For a fitted Poisson process model, with fitted intensity $\widehat{\lambda}(u)$, the predicted number of points falling in any region B is $\int_B \widehat{\lambda}(u)\,du$. The **point process residual** [55] for the region B is defined to be the *observed minus expected* number of points falling in B:

$$\mathscr{R}(B) = n(\mathbf{x} \cap B) - \int_B \widehat{\lambda}(u)\,du \tag{11.3}$$

where $\mathbf{x}$ is the observed point pattern, $n(\mathbf{x} \cap B)$ the number of points of $\mathbf{x}$ in the region B, and $\widehat{\lambda}(u)$ is the intensity of the fitted model.

This residual is closely related to the residuals for quadrat counts that were mentioned above. Taking the set B to be one of our quadrats, the 'observed' quadrat count is $n(\mathbf{x} \cap B)$. The 'expected' quadrat count is $\widehat{\lambda}\,|B|$ if the model is CSR, or more generally $\int_B \widehat{\lambda}(u)\,du$ if the model is an inhomogeneous Poisson process. Hence the 'raw residual' for quadrat B is (11.3).

However, unlike the residuals associated with quadrats, the residuals (11.3) are defined for **any** region B. The equation (11.3) is the definition of a *measure* $\mathscr{R}$. As explained on page 162, a measure is a function M which assigns a value $M(B)$ to *any* region B, intuitively representing the amount of 'stuff' contained in B.

There has been some confusion over this in the literature, so we want to emphasise that the definition (11.3) does not force anyone to choose a particular set B or choose a particular size of set B. It is incorrect to describe $\mathscr{R}(B)$ as a 'pixel residual' requiring a choice of pixel size, or a 'quadrat residual' requiring a specification of quadrats. On the contrary, (11.3) is meant to be applied to *all* possible regions B, large and small. This is the mathematician's roundabout way of defining a spatial distribution of mass: when we know the amount of 'stuff' contained in each region B, we effectively know the exact spatial distribution of the 'stuff'.

A good physical analogy is to think of the residual measure as a spatial distribution of 'electric charge'. Each of the data points carries a positive electric charge of 1 unit. The background space carries a diffuse negative charge, with a spatially varying density of $\widehat{\lambda}(u)$ units of negative charge per unit area. The total electric charge in a region B is the sum of the positive and negative charges as shown in (11.3). If the fitted model is true, then the positive and negative charges should approximately cancel and the total charge should be approximately zero.

The mathematical theory shows the right way to think about point patterns. It is not completely true that the 'observed values' of a point process are the data points of the point pattern. Information is also contained in the fact that points were *not* observed at other locations in the observation window W. The true 'observed values' in a point pattern $\mathbf{x}$ are the numbers $N(B) = n(\mathbf{x} \cap B)$ of points falling in *any* region B. These values together form a measure N, the *counting measure* of the point pattern. Given complete knowledge of $n(\mathbf{X} \cap B)$ for all B, we can recover the locations of all data points.

The residual measure of a fitted model is computed by `residuals.ppm`:

```
> res2e <- residuals(fit2e)
> res2e
Scalar-valued measure
Total mass:
discrete = 3604    continuous = -3604    total = -2.1159e-10
```

The result is an object of class `"msr"` representing a measure.

Note that in this example *the total residual $\mathscr{R}(W)$ is zero*, up to numerical error. This will be familiar to users of simple linear regression, where the residuals always sum to zero. The same holds for any loglinear Poisson model that contains an intercept term, because the fitted model is a solution of the score equation, and the score equation for the intercept is equivalent to saying that the total raw residual must be zero. Thus, the *total* residual $\mathscr{R}(W)$ is not useful as a measure of 'total discrepancy' between data and model. Rather, we should identify subregions B of W where the residual $\mathscr{R}(B)$ is large, and try to interpret these discrepancies.

To visualise the electric charge in a direct way, we plot the positive charges as points, and the negative charge density as a greyscale map, as shown in Figure 11.3, produced by the plot method `plot.msr` for measures. The black dots represent positive unit charges. The background greyscale image represents a density of negative charge: lighter shades of grey represent stronger negative charge.

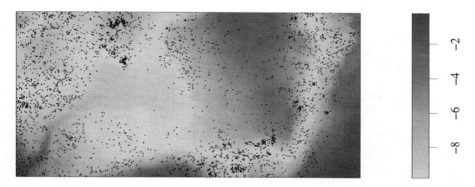

Figure 11.3. *Point process residual measure for the model in which* Beilschmiedia *intensity is a log-quadratic function of terrain elevation. Generated by* `plot(res2e)`. *Residual values multiplied by 1000.*

The class `"msr"` has methods for `plot`, `print`, `Smooth`, `Window` (giving the window where the measure is defined), `"["` (restricting the measure to a sub-region), and geometric operations such as `shift`. Components of the measure can be extracted using `with.msr`.

The residual $\mathscr{R}(B)$ for any spatial region B can be computed using `integral.msr`, a method for the generic function `integral`. For the residual measure `res2e` computed above, if B is a window object, `integral(res2e, B)` computes the residual $\mathscr{R}(B)$, yielding a numerical value:

```
> integral(res2e, square(200))
[1] 56.75
```

If A is a tessellation, then `integral(res2e, A)` computes the values of the residual $\mathscr{R}(B)$ for each tile B of the tessellation, yielding a numeric vector:

```
> qua <- quadrats(bei, 4, 2)
> resQ <- integral(res2e, qua)
```

The values can be plotted in the appropriate spatial position using `plot.tess`:

```
> plot(qua, do.labels=TRUE, labels=signif(resQ, 2))
```

The results are shown in Figure 11.4.

11.3.2 Smoothed residual field

The usual way to interpret Figure 11.3 is that, if the model fits well, the darker and lighter colours should balance out. The darker colours are either positive charges or weak negative charge densi-

84	110	−170	84
69	−380	160	2.8

Figure 11.4. *Raw residual values* $\mathscr{R}(B)$ *for a* 4×2 *array of quadrats.*

Figure 11.5. *Smoothed residual field for model in which* Beilschmiedia *intensity is a log-quadratic function of terrain elevation. Generated by* plot(Smooth(res2e)). *Residual values multiplied by 1000.*

ties, while the lighter colours are strong negative charge densities. Areas with lighter background (stronger negative charge) should have more data points (positive charges) to balance out. If we were to smear the ink in Figure 11.3, the result should be a neutral grey if the model fits well.

This suggests that we *smooth* the residuals: a smoothed version of Figure 11.3 is shown in Figure 11.5. If we apply a two-dimensional smoothing kernel κ to the positive and negative charges, we obtain the 'smoothed residual field'

$$s(u) = \frac{1}{e(u)} \left[\sum_{i=1}^{n(\mathbf{x})} \kappa(u - x_i) - \int_W \kappa(u - v) \lambda_{\hat{\theta}}(v) \, dv \right] \qquad (11.4)$$

where $e(u)$ is the edge correction defined in equation (6.10) on page 168. The smoothed residual field is a spatially weighted average of the positive residual masses and the negative residual density. We can recognise (11.4) as the difference of two terms

$$s(u) = \widetilde{\lambda}(u) - \lambda^{\dagger}(u) \qquad (11.5)$$

where $\widetilde{\lambda}(u)$ is a *nonparametric* estimate of intensity, namely the uniformly-corrected kernel estimate defined in equation (6.8) on page 168, and

$$\lambda^{\dagger}(u) = \frac{1}{e(u)} \int_W \kappa(u - v) \lambda_{\hat{\theta}}(v) \, dv$$

is a correspondingly smoothed version of the *parametric* estimate of the intensity according to the fitted model, The difference (11.5) should have mean value close to zero if the model is true. Positive values of $s(u)$ suggest that there is an overabundance of data points relative to the prediction of the model, that is, they suggest that the model underestimates the true intensity. Negative values of $s(u)$ suggest that the model overestimates the true intensity.

Figure 11.5 shows the smoothed residual field for the *Beilschmiedia* example, calculated by `Smooth(res2e)`. There are several regions of over- and underestimation of the intensity. The scale of values depends on the unit of length, because $s(u)$ has dimensions $length^{-2}$.

The distribution of $s(u)$ is not known exactly, except when the model is CSR [45, Appendix]. However, its mean and variance can be calculated [45, 162].

11.3.3 Lurking variable plot

If there is a spatial covariate $Z(u)$ that could play an important role in the analysis, it may be useful to display a *lurking variable plot* [55] of the residuals against Z. For each possible covariate value z plotted on the horizontal axis, the lurking variable plot shows on the vertical axis the total residual for the region of space where Z is less than or equal to z. That is, the lurking variable plot is a line plot of $C(z) = \mathscr{R}(B(z))$ against z, where

$$B(z) = \{u \in W : Z(u) \le z\}$$

is the sub-level set, the region of space where the covariate value is less than or equal to z.

The left panel of Figure 11.6 shows the lurking variable plot for the terrain slope variable, generated by the command

```
> lurking(fit2e, grad, type="raw")
```

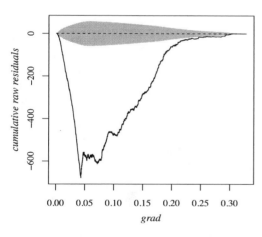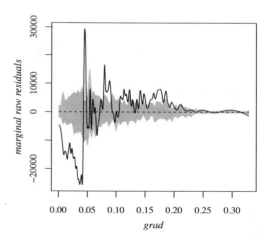

Figure 11.6. *Lurking variable plot for terrain slope, for model depending only on terrain elevation.* Left: *cumulative residual, with two-standard-deviation limits calculated from fitted model.* Right: *non-cumulative residual, with envelopes of 39 simulations from fitted model.*

For each value of terrain slope s on the horizontal axis, the solid line on the graph in the left panel of Figure 11.6 shows the residual for the parts of the survey with slopes no steeper than s, that is, the cumulative number of *Beilschmiedia* trees growing on slopes less than or equal to s, minus the expected number according to the model.

Note that the lurking variable plot typically starts and ends at the horizontal axis, since $\mathscr{R}(W) = 0$ as explained above.

The plot also shows an approximate 5% significance band for the cumulative residual function $C(z)$, obtained from the asymptotic variance under the model [45].

It can be helpful to display the derivative $C'(z)$, which often indicates which values of z are associated with a lack of fit. This is shown in the right panel of Figure 11.6 and is generated by

```
> lurking(fit2e, grad, type="raw", cumulative=FALSE, envelope=TRUE)
```

The derivative is estimated using a smoothing spline and you may need to tweak the smoothing parameters (argument `splineargs`) to get a useful plot. The package currently calculates significance bands for $C'(z)$ by Monte Carlo simulation when the argument `envelope=TRUE` is given.

Both panels of Figure 11.6 suggest that the model substantially overrestimates the intensity on level terrain (slope less than 0.05) and underestimates intensity on steeper slopes. To assess whether this discrepancy is important, we inspect the distribution of slope values, using `hist(grad)` or `plot(ecdf(grad))` or `ecdf(grad)(0.05)`, to find that slopes less than 0.05 constitute about 40% of the survey region. This indicates either that terrain slope is an important variable in determining intensity, or, at the very least, that another unknown variable spatially correlated with terrain slope is important in determining intensity; and that this effect has not been captured by the current model depending only on terrain elevation. One candidate for such a variable is water availability.

11.3.4   Four-panel residual plot

Especially when there are no spatial covariates available, it is often convenient to use the command `diagnose.ppm` to generate four different plots of the residuals, as shown in Figure 11.7, generated by `diagnose.ppm(fit2e)`. This has proved to be a useful quick indication of departure from the model.

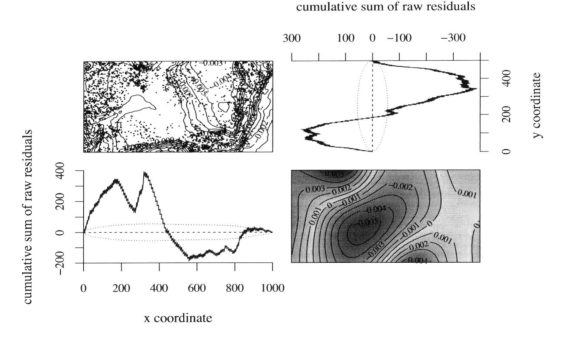

Figure 11.7. *Four-panel diagnostic plot generated by* `diagnose.ppm`.

The top left panel is a direct representation of the residual measure as an 'electric charge' as in Figure 11.3, with circles representing the data points (positive residuals) and a colour scheme or contour map representing the fitted intensity (negative residuals). This carries all information about discrepancies between the data and the model, but is often difficult to interpret — especially in large datasets or where the fitted intensity surface is complicated.

The bottom right panel is an image of the smoothed residual field as in Figure 11.5. This simplifies the information in the top left panel, drawing attention to parts of the window where there is the greatest discrepancy between data and model. Positive values of the smoothed residual field would suggest that the fitted intensity is an underestimate of the true intensity; negative values suggest overestimation of the true intensity.

The two other panels are lurking variable plots against one of the cartesian coordinates. The bottom left panel is a lurking variable plot for the x-coordinate. Imagine a vertical line which sweeps from left to right across the window. The progressive total residual to the left of the line is plotted against the position of the line. Shading shows the pointwise two-standard-deviation limits assuming the model is true. Although these are only pointwise significance bands, the deviations shown in Figure 11.7 are clearly significant.

11.3.5 Scaled or weighted residuals

Interpreting the magnitudes of the residuals can be complicated. In regression analysis, it is common to rescale the residuals so that they satisfy some desired statistical property.

For a Poisson point process we can define the *weighted* residual measure with any desired weight function $w(u)$,

$$\mathscr{R}_w(B) = \sum_{x_i \in \mathbf{X} \cap B} w(x_i) - \int_B w(u)\hat{\lambda}(u)\,\mathrm{d}u \qquad (11.6)$$

for each spatial region B. Choosing the constant weight $w \equiv 1$ gives the raw residual measure $\mathscr{R}$ defined above. Other common choices are the *Pearson residual* measure with $w(u) = \hat{\lambda}(u)^{-1/2}$,

$$\mathscr{R}^{(\mathrm{P})}(B) = \sum_{x_i \in \mathbf{X} \cap B} \hat{\lambda}(x_i)^{-1/2} - \int_B \hat{\lambda}(u)^{1/2}\,\mathrm{d}u \qquad (11.7)$$

and the *inverse-lambda residual* measure with $w(u) = 1/\hat{\lambda}(u)$,

$$\mathscr{R}^{(\mathrm{I})}(B) = \sum_{x_i \in \mathbf{X} \cap B} \frac{1}{\hat{\lambda}(x_i)} - |B|. \qquad (11.8)$$

Values of the Pearson residuals and inverse-lambda residuals have dimensions $length^1$ and $length^2$, respectively.

If it is possible for the fitted intensity $\hat{\lambda}(x_i)$ at a data point to be zero, then the corresponding weight $w(x_i)$ would be infinite in these two cases. To avoid this, we redefine the weight $w(u)$ to be equal to zero whenever $\hat{\lambda}(u) = 0$. This means that the sums in (11.7) and (11.8) must be restricted to data points x_i where $\hat{\lambda}(x_i) > 0$, and the region B in (11.7)–(11.8) must be replaced by the subregion of B where the fitted intensity satisfies $\hat{\lambda}(u) > 0$.

The idea of weighting comes from Stoyan and Grabarnik [633] who proposed a diagnostic equivalent to the inverse-lambda residuals. The inverse-lambda weights have the big computational advantage that they only require calculation at the data points. They also have the practical advantage that we can draw a picture of them, simply by drawing a circle of radius $1/\hat{\lambda}(x_i)$ around each data point x_i. However, the inverse-lambda residuals can have quite large variance, and their variance depends on the model [55].

In `spatstat` the weighted residual measures are computed using `residuals.ppm` with the argument `type` set to `"pearson"`, `"inverse"`, or `"score"`. Figure 11.8 shows a simple example, generated by the following code:

```
> Pat <- rpoispp(function(x,y){50 * exp(2*x)})
> fitpat <- ppm(Pat ~ x)
> resI <- residuals(fitpat, type="inverse")
> resP <- residuals(fitpat, type="Pearson")
```

The simulated inhomogeneous Poisson pattern `Pat` is shown in the left panel of Figure 11.8 The middle panel, generated by `plot(resI)`, shows the inverse-lambda residual measure: circle diameters are proportional to the residual masses $w(x_i) = 1/\hat{\lambda}(x_i)$ at the data points. Large circles at the left of the panel represent large weights corresponding to small values of the fitted intensity at these data points. The background colour is uniform, because the negative density of this measure is constant and equal to 1.

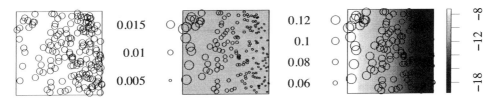

Figure 11.8. *Weighted residuals.* Left: *inhomogeneous Poisson pattern.* Middle: *inverse-lambda residual measure.* Right: *Pearson residual measure. Residuals computed for a fitted inhomogeneous Poisson model of the correct form. Circle diameters represent residual masses at data points. Background colour represents residual density.*

The Pearson residual measure was developed in [55, 45]. Its main advantage is that it is standardised to have unit variance per unit area: $\mathscr{R}^{(P)}(B)$ has mean 0 and variance $|B|$ if the model is true.

The right panel of Figure 11.8 shows the Pearson residual measure for the synthetic example. It was generated by `plot(resP)`. The circle diameters are proportional to the residual masses $w(x_i) = 1/\sqrt{\hat{\lambda}(x_i)}$ which are the reciprocals of the square roots of the fitted intensities. The background greyscale is spatially varying in accordance with the square root of the fitted intensity.

> Warning: The Pearson residual measure is not the same as the Pearson residuals from quadrat counting.

Note that there is a big difference between the Pearson residual measure defined in (11.7) above, and the Pearson residuals used in quadrat counting, defined in equation (10.7) on page 379. The Pearson residuals of quadrat counts are meaningful only for the quadrats used in the counting; these residuals are scaled to have mean 0 and variance 1, approximately, for each quadrat, regardless of the size of the quadrats. The Pearson residual measure $\mathscr{R}^{(P)}(B)$ is defined for *any* region B, and is scaled differently, to have mean 0 and variance $|B|$ if the model is true. The Pearson residual measure for a region with *unit* area has variance 1.

Like all measures, the Pearson residual measure is additive. If A and B are two disjoint sets, then $\mathscr{R}^{(P)}(A \cup B) = \mathscr{R}^{(P)}(A) + \mathscr{R}^{(P)}(B)$. This is not true for the Pearson residuals of quadrat counts: if we merge two quadrats, the merged quadrat count is the sum of the counts in the two quadrats, but the Pearson residual of the merged quadrat count is *not* the sum of the Pearson residuals of the two quadrat counts.

The value of the inverse-lambda residual measure or the Pearson residual measure for any subregion B can be obtained using `integral.msr`:

```
> B <- owin(c(0.1, 0.3), c(0.4, 0.6))
> integral(resI, B)
[1] -0.01345
> integral(resP, B)
[1] -0.1261
```

The total value of the weighted residual for the entire window is not necessarily equal to zero:

```
> integral(resI)
[1] -0.003548
> integral(resP)
[1] -0.015
```

In the fine-pixel limit of logistic regression (Section 9.9), the raw residuals of logistic regression converge to the raw residual measure. The Pearson residuals of logistic regression, rescaled by dividing by the square root of pixel area, converge to the Pearson residual measure. These facts can be used to compute the residual measures approximately from pixel data.

Applying kernel smoothing to these residual measures, we get the *smoothed Pearson residual field*

$$s^{(P)}(u) = \sum_{x_i \in \mathbf{X} \cap B} \hat{\lambda}(x_i)^{-1/2} \kappa(u - x_i) - \int_B \hat{\lambda}(v)^{1/2} \kappa(u - v) \, \mathrm{d}v \qquad (11.9)$$

which has dimensions *length*$^{-1}$, and the *smoothed inverse-lambda residual field*

$$s^{(I)}(u) = \sum_{x_i \in \mathbf{X} \cap B} \frac{1}{\hat{\lambda}(x_i)} \kappa(u - x_i) - \int_B \kappa(u - v) \, \mathrm{d}v \qquad (11.10)$$

which is dimensionless. The smoothed inverse-lambda residual field is almost identical to $\hat{r}(u) - 1$ where $\hat{r}(u)$ is the smoothing estimate (11.1) of the relative intensity $r(u)$ defined in Section 11.2.1. The smoothed inverse-lambda residual field was first used for space-time analysis of earthquake aftershocks [504].

Assuming the model is true, the smoothed Pearson residual field has mean $\mathbb{E}s(u) = 0$ and variance

$$\mathrm{var}\, s(u) \approx \int_W \frac{1}{\left(\lambda(u)^{1/2}\right)^2} \kappa(u - v)^2 \lambda(v) \, \mathrm{d}v = \int_W \kappa(u - v)^2 \, \mathrm{d}v. \qquad (11.11)$$

If $\kappa(x)$ is the isotropic Gaussian kernel with standard deviation σ, then using some algebra, $\kappa(x)^2$ is $1/(4\pi\sigma^2)$ times the isotropic Gaussian kernel with standard deviation $\sigma/2$, so that (11.11) is approximately $1/(4\pi\sigma^2)$. That is, the *standard deviation of $s(u)$ is approximately* $1/(2\sigma\sqrt{\pi})$ when the model is true, where σ is the bandwidth of the smoothing kernel.

Returning to the *Beilschmiedia* example, Figure 11.9 shows the Pearson residual measure

```
> pres2e <- residuals(fit2e, type="pearson")
```

The top panel was generated by `plot(pres2e)`. The sizes of the '+' signs are proportional to the positive charges $\hat{\lambda}(x_i)^{-1/2}$ associated with the data points x_i, and again the greyscale image represents the negative charge density $\hat{\lambda}(u)^{1/2}$.

When reading the smoothed Pearson residual field, we must remember that the Pearson residuals are scaled so that the total Pearson residual in a region *of unit area* has unit variance. In Figure 11.9 the numerical scale might suggest that the deviations are small, with smoothed Pearson residuals of the order of ± 0.15. However the nominal standard deviation of the smoothed Pearson residuals (assuming the model is true) is even smaller: applying the calculations below equation (11.11),

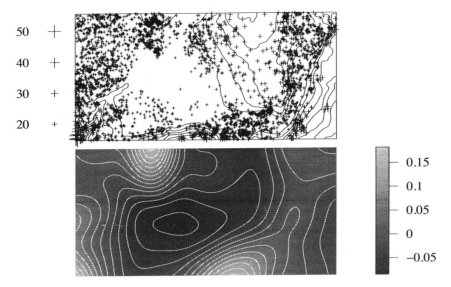

Figure 11.9. *Pearson residuals for model in which* Beilschmiedia *intensity is a log-quadratic function of terrain elevation.* Top: *Pearson residual measure* residuals(fit2e, type="pearson"); Bottom: *smoothed Pearson residual field* Smooth(residuals(fit2e, type="pearson")). *Two-sigma critical values are* ±0.009.

```
> psmo2e <- Smooth(pres2e)
> 1/(2 * sqrt(pi) * attr(psmo2e, "sigma"))
[1] 0.004514
```

Taking the rule of thumb that a residual is 'large' if its absolute value exceeds 2 standard deviations, the threshold for 'large' values of the smoothed Pearson residual field is 2 * 0.004514 = 0.009027. These are substantial deviations, and indicate that the model is a poor fit.

The small numbers arise because the *Beilschmiedia* coordinates are given in *metres*; the survey area is 500,000 square metres. Let us rescale to kilometres:

```
> beikm <- rescale(bei, s=1000, unitname="km")
> beikm.extra <- lapply(bei.extra, rescale,
                         s=1000, unitname="km")
> fit2eKM <- ppm(beikm ~ polynom(elev, 2), data=beikm.extra)
> pres2eKM <- residuals(fit2eKM, type="Pearson")
```

The result (not shown) would give smoothed Pearson residuals of the order of ±150. The unit of length has been divided by 1000 so the smoothed Pearson residuals are multiplied by 1000. The nominal standard deviation would be rescaled to 4.514.

To assess the magnitude of values of the smoothed residual field, it is preferable to generate the four-panel plot for the Pearson residuals, by calling diagnose.ppm with type="Pearson". The printed output then includes the standard deviation of the smoothed Pearson residual field assuming the model is true, using the calculations below equation (11.11), enabling us to assess 'significance' of the smoothed residuals.

11.3.6   Score residual measure

For a loglinear Poisson point process model (9.6) we can also define the vector-valued *score residual* measure with $w(u) = \mathbf{Z}(u)$,

$$\mathscr{R}^{(s)}(B) = \sum_{x_i \in B} \mathbf{Z}(x_i) - \int_B \mathbf{Z}(u)\hat{\lambda}(u)\,du. \tag{11.12}$$

Equivalently, this is a sequence of signed measures, with weight functions equal to the canonical covariates $Z_j(u)$ in the model.

Figure 11.10 shows the score residuals for the synthetic example in Figure 11.8, computed by `residuals(fitpat, type="score")`.

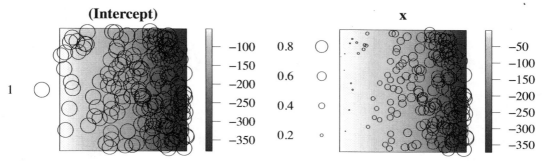

Figure 11.10. *Score residual measure for the simple example in Figure 11.8.*

The score residuals are intimately related to the maximum likelihood estimation procedure, because (assuming regularity conditions) the MLE is the parameter value for which the total score residual $\mathscr{R}^{(s)}(W)$ is zero. They arise in many other contexts, because of this connection.

11.4   Partial residual plots

11.4.1   Example

The model `fit2e`, in which forest density is a function of terrain elevation, has proved to be inadequate in the analysis above, so we shall abandon it and consider models depending on terrain *slope*. We start with a loglinear relationship:

```
> fit1g <- ppm(bei ~ grad, data=bei.extra)
```

This is a much better fit to the data, as we could verify by applying the techniques described above.

Once we have fitted a reasonably good tentative model in which we seem to have chosen the relevant covariates, the next step is to validate the form of the systematic relationship.

A *partial residual plot* shows the fitted effect of a particular covariate, on the scale of the log intensity, together with a smoothed estimate of the true effect. The theory is explained in Section 11.4.3 below.

The left panel of Figure 11.11 shows the partial residual plot for the loglinear model `fit1g` as a function of the slope covariate, produced by

```
> par1g <- parres(fit1g, "grad")
> plot(par1g, xlim=c(0,0.25))
```

The horizontal axis is the covariate value, and the vertical axis is effectively the logarithm of the intensity. The dot-dash line shows the fitted relationship, which is a straight line on this scale. The solid line is the smoothed partial residual, which is roughly speaking an estimate of the true relationship. The grey shading indicates a pointwise 95% confidence interval for the estimated true relationship.

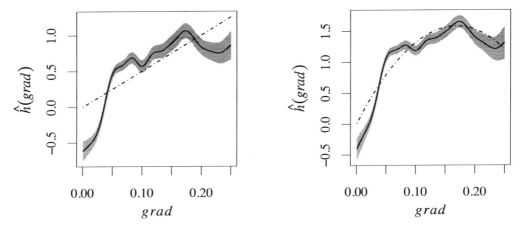

Figure 11.11. *Partial residual plots for the terrain slope covariate, for models where the intensity is a loglinear function* (Left) *or log-quadratic function* (Right) *of terrain slope. Shading shows pointwise 95% confidence intervals based on asymptotic standard errors.*

The partial residual plot for `fit1g` is definitely not a straight line, and looks vaguely quadratic, so we might try fitting a quadratic relationship:

```
> fit2g <- update(fit1g, ~polynom(grad,2))
```

The corresponding partial residual plot is shown in the right panel of Figure 11.11. This suggests a much better fit.

The quadratic term is also significant according to the likelihood ratio test (Section 10.3.2):

```
> anova(fit2g, test="LRT")
Analysis of Deviance Table
Terms added sequentially (first to last)
          Df Deviance Npar Pr(>Chi)
NULL                    1
grad       1      382    2  <2e-16 ***
I(grad^2)  1      198    3  <2e-16 ***
---
Signif. codes:  0 '***' 0.001 '**' 0.01 '*' 0.05 '.' 0.1 ' ' 1
```

The partial residual plot is a useful way to detect relatively small deviations from a model which is already reasonably good. For an application, see [145].

11.4.2   **Implementation in** spatstat

The spatstat command parres computes a partial residual plot. Its syntax is

$$\text{parres}(\text{model, covariate, } \ldots)$$

where model is a fitted Poisson point process model and covariate is a character string giving the name of a covariate used in the model. The result of `parres` is an object that belongs to the classes `"parres"` and `"fv"`. This object is plotted to generate the partial residual plot.

In normal use, `covariate` is the name of one of the *canonical* covariates in the model. These are the functions Z_j that appear in the loglinear intensity (9.6). Type `names(coef(model))` to see the names of the canonical covariates in `model`. If the selected canonical covariate is Z_j, then the diagnostic plot concerns the model term $\beta_j Z_j(u)$. The plot shows a smooth estimate of a function $h(z)$ that should replace this linear term, that is, $\beta_j Z_j(u)$ should be replaced by $h(Z_j(u))$. The linear function is also plotted as a dotted line.

Alternatively `covariate` may be one of the *original* covariates that were supplied as data when fitting the model, and from which the canonical covariates are derived. There may be several canonical covariates which depend on the specified `covariate`. Then the covariate effect is computed using all these canonical covariates. For example in a log-quadratic model which includes the terms `x` and `I(x^2)`, the effect involving both these terms will be computed.

By default, the bandwidth for the smoothing kernel κ is selected automatically, by the one-dimensional bandwidth selection rule `bw.nrd0` from the `stats` package. This can be overridden by specifying the argument `bw`. The argument `adjust` can also be used to adjust the automatically selected bandwidth by a specified adjustment factor.

To scrutinize outliers, the analyst may *restrict* the domain of computation to a subregion $V \subset W$, by restricting the sum in (11.17) to the points of $\mathbf{y} \cap V$ and restricting the integrals in (11.17)–(11.18) to the domain V. The argument `subregion` allows the user to specify the subregion V.

In Figure 11.11, the plots have been restricted to slope values less than 0.25 because the plots beyond this limit show much greater deviations and inflated standard errors. This is attributable to the paucity of data with slopes greater than 0.25, as shown by `hist(grad)` or `plot(ecdf(grad))` or `quantile(grad, 0.99)` or `plot(eval.im(grad >= 0.25))`. This is a common artefact.

11.4.3 Theory*

Partial residuals in regression

In regression models, a simple diagnostic for nonlinearity of covariate effects is the *partial residual plot* (component-plus-residual plot) [255, 403, 400], [164, pp. 135–137], [325, sec 10.4, p. 230 ff.]. Suppose we fit a linear relationship $\mathbb{E}y = \beta^\top \mathbf{x} + \gamma^\top \mathbf{z}$ between responses y_i and vector-valued covariates $\mathbf{x}_i$ and $\mathbf{z}_i$ for $i = 1, \ldots, n$, but we suspect the true relationship is $\mathbb{E}y = h(\mathbf{x}) + \gamma^\top \mathbf{z}$ where $h(\mathbf{x})$ is a nonlinear function of $\mathbf{x}$. The partial residual [255] for $\mathbf{x}$ is the fitted effect for $\mathbf{x}$ plus the residual, $s_i = \widehat{\beta}^\top \mathbf{x}_i + r_i$ where $r_i = y_i - \widehat{y}_i = y_i - (\widehat{\beta}^\top \mathbf{x}_i + \widehat{\gamma}^\top \mathbf{z}_i)$ is the residual. If $h(\mathbf{x})$ is close[2] to a linear function, the delta method gives $\mathbb{E}s_i \approx h(\mathbf{x}_i)$, so a scatterplot of s_i against $\mathbf{x}_i$ suggests the true shape of the function h.

Similarly, in a generalised linear model (such as logistic regression), suppose we fit a relationship $\mathbb{E}(y) = g(\beta^\top \mathbf{x} + \gamma^\top \mathbf{z})$ where g is the inverse link function, but we suspect the true relationship is $\mathbb{E}(y) = g(h(\mathbf{x}) + \gamma^\top \mathbf{z})$. The partial residual is defined on the scale of the linear predictor $g^{-1}(\mathbb{E}(y))$ as

$$s_i = \widehat{\beta}^\top \mathbf{x}_i + r_i^W \qquad (11.13)$$

where r_i^W is the *working residual*

$$r_i^W = \frac{y_i - \widehat{y}_i}{g'(\widehat{y}_i)}. \qquad (11.14)$$

The derivative g' is related to the *variance* of the response.

---

* Starred sections contain advanced material, and can be skipped by most readers.

[2]To be precise, for some value of β, the discrepancy $\varepsilon(\mathbf{x}) = h(\mathbf{x}) - \beta\mathbf{x}$ should be small enough to justify a Taylor expansion of $\mathbb{E}s_i$ in terms of $\varepsilon(\mathbf{x}_i)$.

If the function h is close to a linear function, then $\mathbb{E}s_i \approx h(\mathbf{x}_i)$. A scatterplot smoother of s_i against $\mathbf{x}_i$ should suggest the correct shape of $h(\mathbf{x})$. See [400, 325].

Partial residual for Poisson point process

Partial residual plots for spatial point processes were developed in [35]. Suppose we have fitted a loglinear Poisson point process model with intensity

$$\lambda(u) = \exp(\beta_0 + \beta_1 Z(u) + \gamma \mathbf{V}(u)) \tag{11.15}$$

where $Z(u)$ is a real-valued spatial covariate function and $\mathbf{V}(u)$ is a vector-valued covariate; however, we suspect that the true relationship may be

$$\lambda(u) = \exp(h(Z(u)) + \gamma \mathbf{V}(u)) \tag{11.16}$$

where h is an unknown function. (The simpler model without the term $\gamma \mathbf{V}(u)$ is embraced by taking $\mathbf{V}(u) \equiv 0$.)

If we approximate the model by a logistic regression, then in the fine-pixel limit, the smoothed partial residual curve (using Nadaraya-Watson [494, 694] smoothing, Section 6.9) converges to the *smoothed partial residual* for the point process,

$$\hat{h}(z) = \hat{\beta}_1 z + \frac{\sum_i \frac{1}{\hat{\lambda}(x_i)} \kappa(Z(x_i) - z)}{\int_W \kappa(Z(u) - z)\, \mathrm{d}u} - 1 \tag{11.17}$$

for all real z where the denominator is positive. Here $(\hat{\beta}_0, \hat{\beta}_1, \hat{\gamma})$ are the MLE's of the parameters of (11.15), $\hat{\lambda}(u) = \exp(\hat{\beta}_0 + \hat{\beta}_1 Z(u) + \hat{\gamma} \mathbf{V}(u))$ is the fitted intensity, and κ is a smoothing kernel on the real line. The numerator is a sum over all points of the point pattern dataset $\mathbf{x}$.

A plot of $\hat{h}(z)$ against z is a diagnostic for nonlinearity of the covariate effect of Z. Kernel smoothing is not obligatory; other alternatives could be applied. Standard error calculations for $\hat{h}(z)$ are described in [35].

To understand why (11.17) is a suitable diagnostic, notice that the terms in the sum have weights $1/\hat{\lambda}(x_i)$. In brief, this is because the working residuals (11.14) are inversely scaled by the variance of the response, and a Poisson random variable has variance equal to its mean. The ratio term in (11.17) can be recognised as an inverse-lambda weighted estimate (Section 11.2) of the relative intensity, that is, an estimate of the ratio of the true intensity to the fitted model intensity. If the model fits well, this ratio should be approximately 1, so that (11.17) should be approximately equal to the fitted effect $\hat{\beta}_1 z$.

The partial residual is also intimately connected to the inverse-lambda point process residuals of Section 11.3. We may rewrite (11.17) as

$$\hat{h}(z) = \hat{\beta}_1 z + \frac{\int_W \kappa(Z(u) - z)\, \mathrm{d}\mathscr{R}^{(\mathrm{I})}(u)}{\int_W \kappa(Z(u) - z)\, \mathrm{d}u} \tag{11.18}$$

where $\mathscr{R}^{(\mathrm{I})}$ is the inverse-lambda residual measure (11.8). Note that the residuals are smoothed before the covariate effect is added.

11.5   Added variable plots

Added variable plots [176, sec. 4.5], [688] are commonly used in linear models and generalized linear models, to decide whether a model with response y and predictors $\mathbf{x}$ would be improved by including another predictor z. Essentially the same technique can be used for point processes [35].

11.5.1 Example

We now have a reasonably good model for the *Beilschmiedia* data as a function of terrain slope. One unanswered question is whether the terrain elevation covariate should be included in the model. One tool for this purpose is the *added variable plot*, explained in Section 11.5.3. Roughly speaking, this is a plot of the smoothed Pearson residuals for the current model against the residuals from a linear regression of the new covariate on the existing covariates.

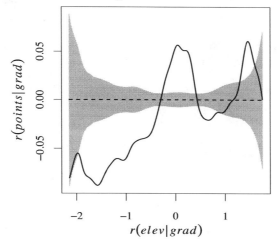

Figure 11.12. *Added-variable plot for terrain elevation covariate, for the model with log-quadratic effect of terrain slope.*

The added variable plot for adding the variable `elev` to the current fitted model `fit2g` is calculated by `addvar(fit2g, elev)` and is plotted in Figure 11.12. The broadly increasing trend in the plot is a strong suggestion that the terrain elevation should be added, but is not a reliable estimate of the form of the effect for this covariate. Once it has been decided to include the terrain elevation into the model, other methods (such as the partial residual plot) should be used to validate the form of the relationship. See [145] for an example.

11.5.2 Implementation in `spatstat`

The `spatstat` function `addvar` computes an added variable plot for a fitted point process model. Its syntax is `addvar(model, covariate, ...)`. The argument `model` is again a fitted point process model.

The argument `covariate` identifies the covariate that is to be considered for addition to the model. It should be either a pixel image (object of class `"im"`) or a `function(x,y)` giving the values of the covariate at any spatial location. Alternatively `covariate` may be a character string, giving the name of a covariate that was supplied (in the `data` argument to `ppm`) when the model was fitted, but was not used in the model.

The result of `addvar(model, covariate)` is an object belonging to the classes `"addvar"` and `"fv"`. Plot this object to generate the added variable plot. The plot method shows the pointwise significance bands for a test of the *null* model, i.e. the null hypothesis that the new covariate has no effect.

11.5.3 Theory∗

11.5.3.1 Added variable plot for regression

In a linear model with response y, predictors $\mathbf{x}$, and i.i.d. Normal errors, the added variable plot [176, sec. 4.5] for a new covariate z is a scatterplot of the raw residuals $r = y - \hat{y}$ from the original model against the residuals $t = z - \hat{z}$ from a regression of z on $\mathbf{x}$. If this plot suggests a linear relationship between r and t with nonzero slope, then the new covariate z is needed. The slope of the added variable plot is equal to the coefficient of z in a regression of y on $(\mathbf{x}, z)$ and is related to the score test [25, 168].

Added variable plots for generalised linear models [688, 689] are similar, based on the approximating linear model [170], [268, p. 412 ff.], [164, pp. 133–134]. The added variable plot for a new covariate z is a plot of the smoothed *Pearson* residuals from the original model against the scaled residuals from a *weighted* linear regression of z on x. If this plot has nonzero slope, then the new covariate z is needed [170, 325]. The plot is not expected to give accurate estimates of the nonlinear effect of z; it should only be used to decide whether or not to add the covariate z to the model [26].

11.5.3.2 Added variable plot for point processes

Consider a Poisson point process $\mathbf{X}$ with intensity of the form

$$\lambda(u) = \exp(A(u) + \boldsymbol{\theta}^{\mathsf{T}} \mathbf{Z}(u)) \tag{11.19}$$

where $A(u)$ is a known baseline, $\mathbf{Z}$ is a *vector-valued* covariate function, and $\boldsymbol{\theta}$ is a vector parameter. Let $Y(u)$ be an additional, *real-valued* covariate function.

First consider weighted linear regression of $Y(u)$ on $\mathbf{Z}(u)$ with weights proportional to $\lambda(u)$. The Pearson residuals for this linear regression are

$$T_{\widehat{\boldsymbol{\theta}}}(u) = \widehat{\lambda}(u)^{1/2} \left(Y(u) - \mathbf{Z}(u)^{\mathsf{T}} \boldsymbol{I}(\widehat{\boldsymbol{\theta}})^{-1} \int_W \mathbf{Z}(v) Y(v) \widehat{\lambda}(v)^{1/2} \, dv \right) \tag{11.20}$$

where $\widehat{\lambda}(u) = \lambda_{\widehat{\boldsymbol{\theta}}}(u)$ is the fitted intensity and $\boldsymbol{I}(\boldsymbol{\theta})$ is the Fisher information matrix (9.29).

The Pearson residual measure of the point process, equation (11.7) on page 421, contains discrete atoms and a continuous component; it should be smoothed for visualisation.

The *added variable plot*, for adding the real-valued covariate $Y(u)$ to the model (11.19), is the plot of the smoothed Pearson point process residual

$$\tilde{r}(t) = \sum_i \lambda_{\widehat{\boldsymbol{\theta}}}(x_i)^{-1/2} \kappa(T_{\widehat{\boldsymbol{\theta}}}(x_i) - t) - \int_W \lambda_{\widehat{\boldsymbol{\theta}}}(u)^{1/2} \kappa(T_{\widehat{\boldsymbol{\theta}}}(u) - t) \, du \tag{11.21}$$

against t, where $\widehat{\lambda}(u)$ denotes the fitted intensity for the model (11.19), $T_{\widehat{\boldsymbol{\theta}}}(u)$ is the linear regression residual (11.20), and κ is a smoothing kernel on $\mathbb{R}$. Pointwise standard error calculations are described in [35].

Again, the added variable plot is not expected to be accurate at estimating the effect of the new variable; it should only be used to decide whether or not to add the covariate $Y(u)$ to the model.

∗ Starred sections contain advanced material, and can be skipped by most readers.

11.6 Validating the independence assumption

11.6.1 Caveat on residual plots

The residual plots described in Sections 11.2–11.5 above are useful for detecting misspecification of the *intensity* in a fitted Poisson process model. They are not very useful for checking the *independence* assumption of the Poisson process (assumption (PP3) on page 300).

An extreme example is provided by the `cells` dataset which gives the locations of the centres of biological cells observed in a histological section [572]. The four-panel diagnostic plot for a uniform Poisson process fitted to the cells data, in Figure 11.13, shows no evidence against the model.

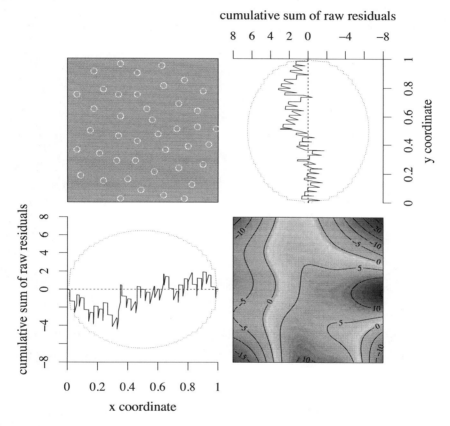

Figure 11.13. *Four-panel diagnostic plot for uniform Poisson model fitted to the* `cells` *data.*

However, the empirical *K*-function (plotted in Figure 10.5 on page 391) shows that the `cells` dataset is clearly not a Poisson pattern, and has strong inhibition.

11.6.2 Q–Q plot of residual field

Interaction between points in a point process affects the probability distribution of the number of points in a subregion. To detect changes in the distribution of numbers of points, we may consider an analogy with regression. To validate distributional assumptions in regression, a reliable tool is the quantile-quantile plot or *Q–Q plot* in which observed quantiles of the residuals are plotted against corresponding theoretical quantiles of their assumed distribution (or sometimes, against quantiles of the assumed distribution of *errors*).

By analogy, to validate the interaction terms in a point process model, we could use a Q–Q plot of the values of the *smoothed residual field* $s(u)$, equation (11.4). After computing $s(u)$ on a fine grid of pixels u, we effectively take the pixel values $s(u)$ as observations, and sort them into ascending order. The distribution of values of $s(u)$ under the model is complicated [45] so we approximate it by Monte Carlo methods. We simulate realisations from the fitted model, re-fit the model to them, compute the smoothed residual fields of these new fits, and evaluate the quantiles. Following [286] we estimate the *expected* quantiles, by averaging the jth order statistic of s over the different simulations, for each j. Details are given in [55]. The observed quantiles of s are plotted against the estimated expected quantiles of s under the fitted model.

Figure 11.14 shows the Q–Q plot of the smoothed raw residuals for the uniform Poisson model fitted to the `cells` data. Dashed lines show pointwise 5% critical envelopes from simulations of the fitted model. This indicates that the uniform Poisson model is grossly inappropriate for the `cells` data.

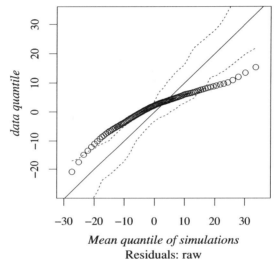

Figure 11.14. *Q–Q plot of smoothed residuals for uniform Poisson model fitted to cells data.*

For some particular models of interpoint interaction, the Q–Q plot is closely related to the summary functions F, G, and K. See [55].

11.6.3   Summary functions

Dependence between points can be measured using summary functions such as the K-function (Section 7.3), pair correlation function (Section 7.6), empty-space function F, and nearest-neighbour distance distribution function G (Section 8.3). However, these summary functions assume the point process is **stationary**, and they can give very misleading results if applied to non-stationary processes. In the context of model validation, these summary functions should only be used if the fitted model is stationary.

Summary functions can be modified to 'adjust' for inhomogeneity. One strategy is to introduce weights into the summary function estimator, to compensate for the spatially varying intensity. Examples are the inhomogeneous K-function and inhomogeneous pair correlation function (Section 7.10.2) and inhomogeneous F- and G-functions (Section 8.7). In model validation, these can be used to discriminate between a fitted inhomogeneous Poisson model and an inhomogeneous process exhibiting interpoint dependence.

An often-overlooked problem with the inhomogeneous K-function is that it also imposes an assumption on the point process, which may or may not be satisfied in a real application. It assumes that the process is 'correlation-stationary'[3] meaning that the pair correlation function depends only on the distance between points. See Section 7.10.2. Other kinds of spatial inhomogeneity, such as local scaling (Section 7.10.3), require different kinds of 'adjustments' to the K-function. See also [668, 712].

11.6.4 Residual summary functions

Overview

For model validation, there is an alternative strategy for compensating for spatial inhomogeneity. We can compute a *residual summary function* which would have expected value zero if the fitted model was true. This approach was developed in [49] using a connection to the score residuals.

In a nutshell, for certain summary functions $T(r)$ it is possible to define a *model compensator* $\mathbf{C}T(r)$ depending on the fitted model and the observed data, such that $T(r) - \mathbf{C}T(r)$ has mean zero if the model is true. The difference $\mathbf{R}T(r) = T(r) - \mathbf{C}T(r)$ is called the *residual summary function* of $T(r)$.

In simple cases, the residual summary function takes a very sensible form. If the fitted model is CSR, the compensator of the K-function is $\mathbf{C}K(r) = \pi r^2$, so the residual is $\mathbf{R}K(r) = K(r) - \pi r^2$, which is a natural way to measure discrepancy between the data and CSR. If the fitted model is CSR, the compensator of the nearest-neighbour function $G(r)$ is the empty-space function $F(r)$, apart from some details, so the residual is $\mathbf{R}G(r) = G(r) - F(r)$ apart from some details. The difference $D(r) = G(r) - F(r)$ has long been used [218, eq. (5.7)] as a measure of discrepancy from CSR: see Section 8.6, page 275.

Implementation and examples in `spatstat`

The residual K-function is computed in `spatstat` by the command `Kres`. If `fit` is a point process model fitted by `ppm`, then `Kres(fit)` is the residual K-function of the fitted model. Further arguments to `Kres` specify the edge correction to be used, and other options for the calculation. For a point pattern X, typing `Kres(X)` gives the residual K-function of CSR fitted to X. By specifying additional arguments, another point process model can be fitted to X instead.

Similarly `Kcom` computes the compensator of the fitted model or the compensator of CSR. For the G-function, `Gres` computes the residual and `Gcom` the compensator.

Figure 11.15 shows the residual K- and G-functions for CSR fitted to the `cells` data, computed by

```
> cellKr <- Kres(cells, correction="best")
> cellGr <- Gres(cells, correction="best")
```

Dotted lines in Figure 11.15 show the two-standard-deviation limits based on the 'Poincare variance' [49], a rough approximation to the variance of the residual, which is easy to compute in this context. The residual K- and G-functions wander substantially outside these limits, strongly suggesting that CSR is not a satisfactory model for the `cells` data, and suggesting that the `cells` data show greater regularity than a Poisson process.

Figure 11.16 shows the residual K-function for a Poisson model with intensity a log-cubic function of the coordinates, fitted to the full Japanese Pines data, computed by

---

[3]or 'second order intensity reweighted stationary' [46] if we do not wish to assume the existence of the pair correlation.

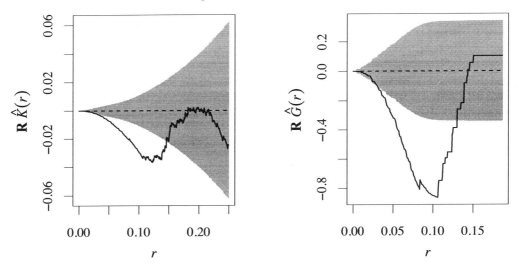

Figure 11.15. *Residual K-function* (Left) *and G-function* (Right) *for the uniform Poisson point process fitted to the* cells *data. Solid lines: residual function. Grey shading: pointwise two-standard-deviation limits. Dashed lines: expected residual (zero) under fitted model.*

```
> jfit <- ppm(residualspaper$Fig1 ~ polynom(x,y,3))
> jKr <- Kres(jfit, correction="best")
```

The conclusion here is marginal: there may be slight evidence against the Poisson assumption, after accounting for spatial inhomogeneity.

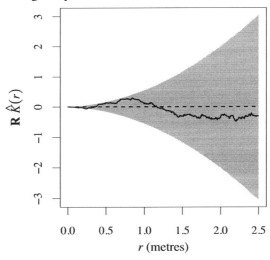

Figure 11.16. *Residual K-function for the Poisson point process with intensity a log-cubic function of the coordinates, fitted to the full Japanese Pines data.*

Our practical recommendation is to use the residual K-function and G-function when the dataset is large or the fitted model is complex. In these situations it is much faster to calculate the residual summary function than the simulation envelopes of the original summary function. Residual summary functions are particularly useful for Gibbs models (Chapter 13).

Theory

We start with a summary statistic or summary function T that is a sum of contributions from each point in $\mathbf{x}$:

$$T(\mathbf{x}) = \sum_i t(x_i, \mathbf{x} \setminus x_i) \tag{11.22}$$

for any function t, where $\mathbf{x} \setminus x_i$ is the set $\mathbf{x}$ excluding the point x_i. Examples include the empirical K-function

$$\widehat{K}_{\mathbf{x}}(r) = \frac{1}{\widehat{\lambda^2}(\mathbf{x})|W|} \sum_i \sum_{j \neq i} e_K(x_i, x_j) \mathbf{1}\left\{ \|x_i - x_j\| \leq r \right\} \tag{11.23}$$

and the empirical nearest-neighbour distance distribution function G,

$$\widehat{G}_{\mathbf{x}}(r) = \frac{1}{n(\mathbf{x})} \sum_i e_G(x_i, \mathbf{x} \setminus x_i, r) \mathbf{1}\left\{ d(x_i, \mathbf{x} \setminus x_i) \leq r \right\} \tag{11.24}$$

where $e_K(u,v), e_G(u,v,r)$ are edge correction weights, and $\widehat{\lambda^2}(\mathbf{x}) = n(\mathbf{x})(n(\mathbf{x}) - 1)/|W|^2$ is an unbiased estimator of λ^2 for CSR. These are of the general form (11.22) where the function t is, respectively,

$$t_K(u, \mathbf{y}) = \frac{1}{\widehat{\lambda^2}(\mathbf{y} \cup \{u\})|W|} \sum_j e_K(u, y_j) \mathbf{1}\left\{ \|u - y_j\| \leq r \right\} \tag{11.25}$$

$$t_G(u, \mathbf{y}) = \frac{1}{n(\mathbf{y} \cup \{u\})} e_G(u, \mathbf{y}, r) \mathbf{1}\left\{ d(x_i, \mathbf{y}) \leq r \right\} \tag{11.26}$$

for any given value of r.

For a Poisson point process $\mathbf{X}$ with intensity function $\lambda(u)$,

$$\mathbb{E} \sum_i t(x_i, \mathbf{X} \setminus x_i) = \mathbb{E} \int_W t(u, \mathbf{X}) \lambda(u) \, du \tag{11.27}$$

under minimal conditions on the function t, by the Campbell-Mecke formula (8.50). Therefore the difference

$$\mathbf{R}T(\mathbf{x}) = \sum_i t(x_i, \mathbf{x} \setminus x_i) - \int_W t(u, \mathbf{x}) \lambda(u) \, du \tag{11.28}$$

satisfies $\mathbb{E}\,\mathbf{R}T(\mathbf{X}) = 0$ for the Poisson process with intensity function $\lambda(u)$. Notice that the integral on the right-hand side of (11.28) is data-dependent: it is not equal to the expected value of $T(\mathbf{X})$, but has the same expected value as $T(\mathbf{X})$. Define the *compensator* [49]

$$\mathbf{C}T(\mathbf{x}) = \int_W t(u, \mathbf{x}) \lambda(u) \, du \tag{11.29}$$

so that the *residual* is $\mathbf{R}T(\mathbf{x}) = T(\mathbf{x}) - \mathbf{C}T(\mathbf{x})$.

In applications we usually have to estimate the intensity $\lambda(u)$. Then, for example, the *residual K-function* is

$$\mathbf{R}K_{\mathbf{x}}(r) = \widehat{K}_{\mathbf{x}}(r) - \int_W t_K(u, \mathbf{x}) \widehat{\lambda}(u) \, du. \tag{11.30}$$

The residual K-function depends on the model, and both terms on the right side of (11.30) are data-dependent. If the model is a good fit, then $\mathbf{R}K_{\mathbf{x}}(r)$ should have approximately zero mean, for all values of r.

Suppose the fitted model is CSR. Ignoring edge effects, for the K-function we find [49]

$$\mathbf{R}K_{\mathbf{x}}(r) \approx \frac{n(\mathbf{x})^2}{2|W|} \left[\widehat{K}_{\mathbf{x}}(r) - \pi r^2 \right]. \tag{11.31}$$

The term in brackets is a commonly used measure of departure from CSR, and is a sensible diagnostic because $K(r) = \pi r^2$ under CSR. Similarly for the residual G-function, when the fitted model is CSR,

$$\mathbf{R}\,G_{\mathbf{x}}(r) \approx n(\mathbf{x})(\widehat{G}_{\mathbf{x}}(r) - \widehat{F}_{\mathbf{x}}(r)) \tag{11.32}$$

where $\widehat{F}_{\mathbf{x}}(r)$ is the empirical estimate of the empty-space function F. The right-hand side of (11.32) is a reasonable diagnostic for departure from CSR, since $F \equiv G$ under CSR. This argument lends support to Diggle's [218, eq. (5.7)] proposal to judge departure from CSR using the quantity $\max_r |\hat{G}(r) - \hat{F}(r)|$.

Residual summary functions can be justified (and indeed derived) as score residuals for a non-Poisson alternative to the Poisson null model. See Chapter 13.

11.7   Leverage and influence

11.7.1   Overview

Standard procedures for fitting parametric models can be highly sensitive to anomalies in the data. An important step in model validation is to check whether the fitted model was likely to have been influenced by such anomalies.

In regression, two standard tools are *leverage* and *influence*. They measure the sensitivity of the model-fitting procedure to changes in the input data. The *leverage* of a data point is the change in the predicted response that would be caused by a change in the observed response, expressed in standard deviation units. The *influence* of a data point is a measure of the change in the fitted model if we delete the data point in question.

Leverage is a measure of the *potential* ability of each data point to 'pull' the regression line toward itself. The leverage of a data point depends only on its covariate value (and the fitted parameters), not directly on the response value. The data points with highest leverage are those with the most extreme values of the covariate.

Influence is a measure of the *actual* impact of the data point on the final result. The data points with greatest influence are those which are both anomalous in their response value (have a large residual) and extreme in their covariate values (have a relatively high leverage).

For Poisson point processes, leverage and influence were developed in [36]. For a loglinear intensity model $\lambda(u) = \exp(\theta^{\top}\mathbf{Z}(u))$ fitted to data, the *leverage* is the function

$$\widehat{h}(u) = \widehat{\lambda}(u)\mathbf{Z}(u)\mathbf{I}_{\widehat{\theta}}^{-1}\mathbf{Z}(u)^{\top} \tag{11.33}$$

where $\mathbf{I}_{\theta}$ is the Fisher information matrix. A relatively large value of $\widehat{h}(u)$ indicates a part of the space where the data may have a strong effect on the result. The *influence* of data point x_i is

$$s_i = \frac{1}{p}\mathbf{Z}(x_i)\mathbf{I}_{\widehat{\theta}}^{-1}\mathbf{Z}(x_i)^{\top}. \tag{11.34}$$

A relatively large value of s_i indicates a data point whose removal would cause a large change in the fitted model.

The *parameter influence measure* is a weighted residual measure (11.6) with weight function $w(u) = \mathbf{I}_{\widehat{\theta}}^{-1}\mathbf{Z}(u)$. Equivalently, it is the score residual measure (11.12) multiplied by $\mathbf{I}_{\widehat{\theta}}^{-1}$. Each data point x_i has mass $\mathbf{I}_{\widehat{\theta}}^{-1}\mathbf{Z}(x_i)$, indicating the effect on each of the fitted parameters of including the data point x_i (that is, equal to minus the effect of deleting x_i). Each background location u has a density of $-\mathbf{I}_{\widehat{\theta}}^{-1}\mathbf{Z}(u)\widehat{\lambda}(u)$ essentially indicating the effect on each of the fitted parameters of adding a new data point at location u, multiplied by the probability that such a point will occur.

11.7.2 Implementation in `spatstat`

For a fitted point process model `fit`, the commands

```
lev <- leverage(fit)
inf <- influence(fit)
dfb <- dfbetas(fit)
```

calculate the leverage, influence, and parameter influence, respectively. The functions `leverage`, `influence`, and `dfbetas` are generic and have methods for linear models and generalised linear models. The commands above invoke `leverage.ppm`, `influence.ppm`, and `dfbetas.ppm`, respectively, which are defined in `spatstat` for point process models.

The result of `leverage.ppm` is an object belonging to the special class `"leverage.ppm"` representing the leverage function $\widehat{h}(u)$. It can be plotted (by the plot method `plot.leverage.ppm`) or converted to a pixel image by `as.im`.

The result of `influence.ppm` is an object of class `"influence.ppm"` representing the influence masses s_i. It can be plotted (by `plot.influence.ppm`), or converted to a marked point pattern by `as.ppp` (see `as.ppp.influence.ppm`).

The result of `dfbetas.ppm` is a signed measure (object of class `"msr"`). There are methods for `print`, `plot`, `"["`, and `integrate` for this class.

If the point process intensity has irregular parameters ('covariate function arguments' that were fitted using `ippm`) then the leverage/influence calculation requires the first and second derivatives of the log intensity with respect to the irregular parameters. The argument `iScore` should be a list, with one entry for each irregular parameter, of R functions that compute the partial derivatives of the log intensity with respect to each irregular parameter. The argument `iHessian` should be a list, with p^2 entries where p is the number of irregular parameters, of R functions that compute the second-order partial derivatives of the log intensity with respect to each pair of irregular parameters.

Two worked examples are given below; another is available in [684, 145].

11.7.3 Example: Murchison gold data

The Murchison gold data were introduced in Section 9.3.2.2. They consist of the spatial locations of gold deposits (`gold`), a line segment pattern of geological faults (`faults`), and the polygonal boundary of greenstone outcrop (`greenstone`) in a 330 by 400 km survey region.

```
> mur <- lapply(murchison, rescale, s=1000, unitname="km")
> attach(mur)
> green <- greenstone
> dfault <- distfun(faults)
```

Here we consider the full loglinear model with interaction between the variables `green` and `dfault`. That is, we allow the log intensity to depend on distance to the nearest fault according to two completely different linear relationships holding inside and outside the greenstone.

```
> murfit1x <- ppm(gold ~ green * dfault, eps=1)
> murlev1x <- leverage(murfit1x)
> murinf1x <- influence(murfit1x)
> murdfb1x <- dfbetas(murfit1x)
```

Figure 11.17 shows a perspective plot of the leverage function for this model, generated by `persp(as.im(murlev1x))`. Sharp peaks indicate small regions of space (in fact, small regions of greenstone outcrop) with very high leverage. The tallest peak in this plot, at upper left, is associated with a small region of greenstone outcrop at a relatively large distance from the nearest fault.

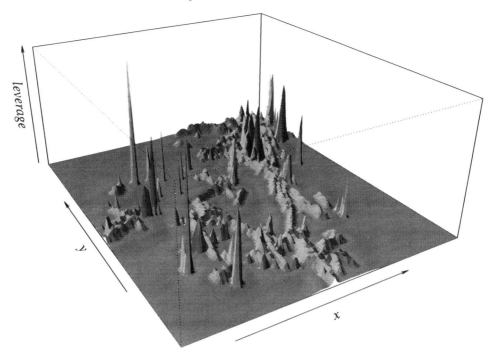

Figure 11.17. *Perspective plot of leverage function for full loglinear model of Murchison data.*

The high leverage indicates that the presence or absence of gold deposits inside this region has a relatively large effect on the fitted intensity.

Figure 11.18 shows the influence measure including a few strikingly large values, indicating that these points (gold deposits) had a substantial effect on the fitted model. To investigate more closely, we selected small neighbourhoods of influential points (by displaying Figure 11.18 in a graphics window and using `clickbox`) and displayed the detailed survey data in those neighbourhoods.

Figure 11.19 shows the neighbourhood of the most highly influential point, in the top right corner of Figure 11.18. The survey data are shown in the left panel, and the corresponding influence values in the right panel. Similarly Figure 11.20 shows the neighbourhood of another highly influential point in the middle left of the survey. It is clear why these gold deposits are so influential: they lie a long way from the nearest recorded fault, and are therefore anomalous with respect to the rest of the survey data. The influence measure has automatically identified these anomalies.

For automatic identification of the most influential points, the influence values s_i can be extracted as `marks(as.ppp(murinf1x))`.

The *direction* in which each data point affects the fitted model can be read from the parameter influence measure `murdfb1x <- dfbetas(murfit1x)`. Figure 11.21 shows a detail of this measure in the same area as Figure 11.20. The influential point causes the coefficient `greenTRUE` to decrease substantially, and the coefficient `greenTRUE:dfault` to increase substantially.

What should be done with highly influential points? This depends on the context. The influential points shown in Figures 11.19 and 11.20 are anomalies in that they lie far away from the nearest fault (if we take the survey data at face value). For the strikingly large value of influence at the top right of Figure 11.18, shown in detail in the left panel of Figure 11.19, there is an additional question of *edge effects*. The gold deposit lies on the boundary of the survey region. The true distance from this gold deposit to the nearest geological fault may be less than the observed distance, since only faults intersecting the survey region are recorded. This can be remedied by applying the border

Figure 11.18. *Influence for Murchison gold deposits (circles) and two regions of high influence (grey shaded rectangles).*

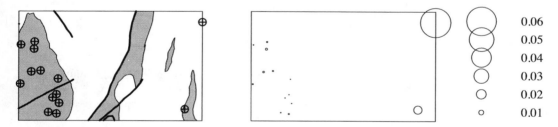

Figure 11.19. *Details of Murchison survey in the vicinity of the most influential point, in the top right corner of the survey. Detail is about 50 by 30 kilometres.* Left: *data, including gold deposits (⊕), faults (—), and greenstone outcrop (grey shading).* Right: *influence values for each gold deposit, represented by circle diameter.*

method, that is, restricting attention to the part of the survey region where the nearest fault is closer than the boundary of the survey:

```
> DL <- as.im(dfault)
> subW <- (DL <= bdist.pixels(Window(DL)))
```

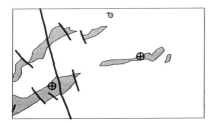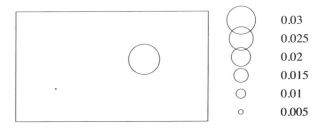

Figure 11.20. *Detail of Murchison survey in the vicinity of a highly influential point at middle left of the survey. Detail is about 60 by 40 kilometres.* Left: *data, including gold deposits ($\oplus$), faults (—), and greenstone outcrop (grey shading).* Right: *influence values.*

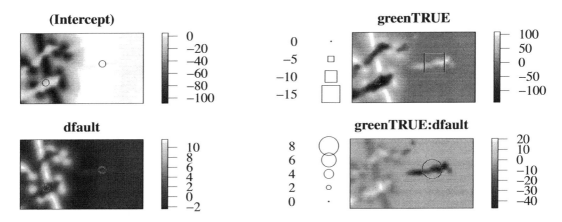

Figure 11.21. *Detail of the parameter influence measure* dfbetas *for the Murchison gold deposits survey in the same area as Figure 11.20. The lack of a symbol legend for the panels on the left indicates that the values at the data points were extremely small. Ribbon scale multiplied by 10^6; symbol scale multiplied by 1000.*

Repeating the analysis with gold replaced by gold[subW] would exclude the data point in question.

11.7.4 Example: Chorley-Ribble cancer data

The Chorley-Ribble data were introduced on page 9 and a Poisson point process model was fitted in Section 9.12.3.3. This analysis, and analysis in [224, 236, 55], concluded there is evidence of an increased disease risk associated with proximity to the incinerator. An important caveat [224, 236] is that there is a cluster of four cases of laryngeal cancer very close to the incinerator. There is concern that these cases may have a strong influence on the fitted model and hence on the evidence for an incinerator effect.

The model chorleyDfit fitted on page 364 is a 'raised incidence' model (9.81)–(9.82) that includes an effect $b_\theta(u)$ due to proximity to the incinerator. The scalar parameters α, β control the magnitude and dropoff rate, respectively, of the incinerator effect. Their MLE's are $\hat{\alpha} = 23.67$ and $\hat{\beta} = 0.91 \text{ km}^{-2}$.

For leverage and influence calculations, since the model (9.81)–(9.82) is *not* loglinear in α and

β, we need the first and second partial derivatives of $\log b_\theta(u)$ with respect to α and β. There are two ways to do this. The first is to write functions by hand which evaluate the partial derivatives, and pass them to `leverage.ppm`, `influence.ppm`, and `dfbetas.ppm` as the arguments `iScore` and `iHessian`. The second way is to use the symbolic calculus features of R.

The first strategy involves differentiating $\log b_\theta(u)$ and coding the result by hand. Using the function `d2incin` defined on page 363, the components of the score are

```
> Zalpha <- function(x,y, alpha, beta) {
    expbit <- exp( - beta * d2incin(x,y))
    expbit/(1 + alpha * expbit)
  }
> Zbeta <- function(x,y, alpha, beta) {
    d2 <- d2incin(x,y)
    topbit <- alpha * exp( - beta * d2)
    - d2 * topbit/(1 + topbit)
  }
> Zscore <- list(alpha=Zalpha, beta=Zbeta)
```

Differentiating again we get the Hessian:

```
> Zaa <- function(x,y, alpha, beta) {
    expbit <- exp( - beta * d2incin(x,y))
    -(expbit/(1 + alpha * expbit))^2
  }
> Zab <- function(x,y, alpha, beta) {
    d2 <- d2incin(x,y)
    expbit <- exp( - beta * d2)
    - d2 * expbit/(1 + alpha * expbit)^2
  }
> Zbb <- function(x,y, alpha, beta) {
    d2 <- d2incin(x,y)
    topbit <- alpha * exp( - beta * d2)
    (d2^2) * topbit/(1 + topbit)^2
  }
> Zhess <- list(aa=Zaa, ab=Zab, ba=Zab, bb=Zbb)
```

Then we compute the new leverage and influence measures:

```
> chorleyDXlev <- leverage(chorleyDfit,
                      iScore=Zscore, iHessian=Zhess)
> chorleyDXinf <- influence(chorleyDfit,
                      iScore=Zscore, iHessian=Zhess)
> chorleyDXdfb <- dfbetas(chorleyDfit,
                      iScore=Zscore, iHessian=Zhess)
> chorleyDXdfbsmo <- Smooth(chorleyDXdfb, sigma=1.5)
```

The second approach uses R's facilities for symbolic calculus. First we write an `expression` for $\log b_\theta(u)$ using only standard mathematical functions:

```
> logbexpr <- expression(log(1 +
      alpha * exp( - beta * ((x - 354.5)^2 + (y-413.6)^2))))
```

Then we use the symbolic differentiation command `deriv`:

```
> logB <- deriv(logbexpr, c("alpha", "beta"),
              c("x", "y", "alpha", "beta"), hessian=TRUE)
```

The result logB is a function with arguments x,y,alpha,beta. The function returns a vector of values of $\log b_\theta(u)$ which also has an attribute "gradient" containing the first derivatives and an attribute "hessian" containing the second derivatives. This function logB is passed to ppm or ippm as a covariate. The presence of the attributes "gradient" and "hessian" is detected, and these are used to compute the leverage and incidence diagnostics. The computation of derivatives also makes ippm faster and more reliable. However, this approach is only valid when the user function (in this case logB) is an additive term in the log intensity: it cannot be used in more general cases.

```
> chorleyDfitI <- ippm(Q ~offset(log(smo) + logB),
                      start=list(alpha=10, beta=1))
> chorleyDfitI
Nonstationary Poisson process
Log intensity:  ~offset(log(smo) + logB)
Fitted trend coefficient:  (Intercept) = -2.896
> chorleyDXlev <- leverage(chorleyDfitI)
> chorleyDXinf <- influence(chorleyDfitI)
> chorleyDXdfb <- dfbetas(chorleyDfitI)
```

Figure 11.22 shows the leverage function for Diggle's model, chorleyDXlev. In order to account for the effect of estimating the irregular parameters α and β, the expression for leverage (11.40) is modified as described on page 445, with $\mathbf{Z}(u)$ replaced by (11.42) and I_θ replaced by (11.43). Figure 11.22 shows that the population close to the incinerator has very high leverage relative to other regions. This is intuitively obvious because the incinerator effect, if present, will only be visible close to the incinerator.

Figure 11.23 shows the influence masses for Diggle's model, again including the effect of estimating the irregular parameters. That is, (11.44) was modified by replacing $\mathbf{Z}(u)$ by (11.42) and I_θ by (11.43). The observed group of four cases very close to the incinerator has very large influence. However, the largest influence is enjoyed by a case slightly to the northwest of the incinerator, at a location where the reference population density $\rho(u)$ is small. This sensitivity is an undesirable consequence of the modelling approach in which we have estimated spatially varying risk by first estimating the population density.

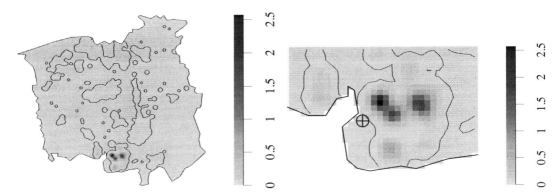

Figure 11.22. *Leverage function for raised incidence model of Chorley-Ribble.* Left: *full data.* Right: *detail close to incinerator ($\oplus$). Contour lines are drawn only at the mean value of leverage.*

Figure 11.24 shows the components of the effect measure $\mathbb{D}\widetilde{\theta}$ for the irregular parameters α and β in Diggle's model. Here (11.45) was modified by replacing $\mathbf{Z}(u)$ by (11.42) and I_θ by

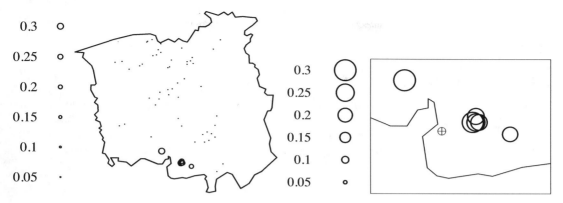

Figure 11.23. *Influence for raised incidence model of Chorley-Ribble.* Left: *full data.* Right: *detail close to incinerator (⊕).*

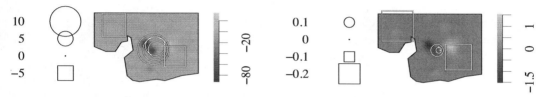

Figure 11.24. *Components of effect measure for parameters α (Left) and β (Right) in Diggle's model of Chorley-Ribble data; detail close to the incinerator. Density of continuous component is depicted as a greyscale image. Masses at data points are shown as circles and squares, for positive and negative masses, respectively.*

(11.43). Remember that the effect measure is the *reverse* or negative of the change in the fitted parameter value that would occur if the relevant data were deleted. Thus a positive value of effect measure means that the corresponding data are 'pulling the parameter estimate upward'. Note the presence of two large squares, representing cases of laryngeal cancer with large negative effect on the parameter estimates, situated to the north-west and south-east of the incinerator. We conclude that the estimated strength α of the incinerator effect is strongly influenced by these two cases (tending to reduce the estimated effect) and by the cluster of four cases close to the incinerator (tending to increase the estimated effect). The estimated dropoff rate β of the incinerator effect is overwhelmingly influenced by the two cases with large negative effect. Since β is the crucial parameter for the effect of proximity to the incinerator, the dramatic sensitivity of $\widehat{\beta}$ to the position of a few controls is important.

Diggle and Rowlingson [236] advocated a different analysis of the same data, analysing conditionally on the spatial locations. Each location has a binary response indicating whether it is a case or control. The conditional model is a binary regression (logistic regression if (α, β) is fixed). The conditional analysis can be validated using the standard diagnostics for leverage and influence in logistic regression [553]. These diagnostics give leverage and influence values for each *data point* (case of cancer of the larynx or lung), in a fashion quite similar to our analysis, but without the effect of kernel smoothing of the reference population density.

Since stochastic clustering of disease cases is a possible alternative explanation, it would be prudent to consider point process models which incorporate positive association between points. Leverage and influence diagnostics then provide information about the evidence for such clustering, as well as the evidence for covariate effects after allowing for clustering. See Section 13.10.7.

11.8   Theory for leverage and influence∗

11.8.1   Leverage and influence in regression models

Leverage and influence are standard tools in applied statistics for validating fitted regression models [26, 169, 325]. Suppose we have data $(\mathbf{x}_1, y_1), \ldots, (\mathbf{x}_n, y_n)$ where $\mathbf{x}_i$ is the covariate value, and y_i is the observed response, for the ith observation. We fit the regression relationship $\mathbb{E}y = f(\beta^{\mathsf{T}}\mathbf{x})$ where β is the parameter vector to be estimated. The *influence* (or likelihood influence) of the ith observation is

$$s_i = \frac{2}{p}\log\frac{L(\hat{\beta})}{L(\hat{\beta}_{(-i)})} \tag{11.35}$$

where L is the likelihood function (based on all the data), $\hat{\beta}$ is the fitted parameter value, and $\hat{\beta}_{(-i)}$ is the fitted parameter value obtained by fitting the model to all of the data except $(\mathbf{x}_i, y_i)$. That is, s_i is a measure of how much the model changes if we delete the ith observation.

Applied statisticians also use the *parameter influence* (charmingly nicknamed '*dfbeta*') which is the effect on the parameter estimate $\hat{\beta}$ of deleting the ith observation,

$$\Delta_i\hat{\beta} = \hat{\beta} - \hat{\beta}_{(-i)}. \tag{11.36}$$

In practice, these influence measures are not computed exactly, but are approximated using first-order Taylor expansions. The key to these approximations is the 'leverage equation',

$$V^{-1/2}(\hat{\boldsymbol{\mu}} - \boldsymbol{\mu}) \approx H^* V^{-1/2}(\mathbf{y} - \boldsymbol{\mu}), \tag{11.37}$$

where $\mathbf{y} = (y_1, \ldots, y_n)$ is the vector of responses, V is the variance matrix of the responses, $\boldsymbol{\mu} = (\mu_1, \ldots, \mu_n)$ is the vector of true mean responses $\mu_i = f(\beta^{\mathsf{T}}\mathbf{x}_i)$, and $\hat{\boldsymbol{\mu}} = (\mu_1(\hat{\beta}), \ldots, \mu_n(\hat{\beta}))$ is the vector of fitted mean responses $\hat{\mu}_i = f(\mathbf{x}_i^{\mathsf{T}}\hat{\beta})$.

The matrix $H^* = H^*(\hat{\beta})$ appearing in (11.37) is called the (standardised) *leverage matrix*. The scaling by $V^{-1/2}$ in (11.37) means that, in the words of McCullagh and Nelder [458, p. 397], the leverage matrix "measures the influence, in Studentized units, of changes in $\mathbf{y}$ on $\hat{\boldsymbol{\mu}}$."

The *leverage* of observation i is defined to be the diagonal entry h_{ii}^* of H^*. That is, the leverage of observation i measures the effect that a change in the observed response y_i will have on the fitted response $\hat{\mu}_i$ for the same observation, where the measurement unit is the standard deviation of the response. The leverage value h_{ii}^* depends only on the covariate values and the fitted model parameters, and not directly on the observed response. A large leverage value can be interpreted as meaning that the observation has a strong *ability* to influence the fitted model. Since $\sum_i h_{ii}^* = \text{tr}\, H^* = p$, the number of model coefficients, observations i with $h_{ii}^* > p/n$ are interpreted as having relatively large leverage. See [208, 270, 703].

11.8.2   Leverage and influence for Poisson point process models

Leverage and influence diagnostics for Poisson point processes were developed in [36]. Roughly speaking, if we view a loglinear Poisson model as the limit of logistic regression on a fine pixel grid, then the new diagnostics are the corresponding limits of the standard leverage and influence diagnostics for logistic regression.

∗ Starred sections contain advanced material, and can be skipped by most readers.

Leverage for Poisson point process models

Given an observed point pattern $\mathbf{x} = \{x_1, \ldots, x_n\}$ in a bounded domain $W \subset \mathbb{R}^d$, suppose we model $\mathbf{x}$ as a realisation of a Poisson point process $\mathbf{x}$ with loglinear intensity function $\lambda_\theta(u) = \exp(B(u) + \theta^{\mathsf{T}} \mathbf{Z}(u))$, where $B(u)$ is a known offset function, $\mathbf{Z}(u)$ is a known, p-dimensional-vector-valued covariate function, and θ is again the parameter vector. If $\hat{\theta}$ is the maximum likelihood estimate of θ, a first-order Taylor approximation gives $\hat{\theta} - \theta \approx \boldsymbol{I_\theta}^{-1} \mathbf{U}(\theta)$ where $U(\theta)$ is the score (9.47) defined on page 344 and $\boldsymbol{I_\theta}$ is the Fisher information (9.29) defined on page 330. This gives

$$\hat{\lambda}(u) - \lambda(u) = \lambda_{\hat{\theta}}(u) - \lambda_\theta(u) = \left[e^{(\hat{\theta} - \theta)^{\mathsf{T}} \mathbf{Z}(u)} - 1 \right] \lambda_\theta(u)$$

$$\approx (\hat{\theta} - \theta)^{\mathsf{T}} \mathbf{Z}(u) \lambda_\theta(u)$$

$$\approx \lambda_\theta(u) \mathbf{Z}(u)^{\mathsf{T}} \boldsymbol{I_\theta}^{-1} \left[\sum_i \mathbf{Z}(x_i) - \int_W \mathbf{Z}(v) \lambda_\theta(v) \, \mathrm{d}v \right] \tag{11.38}$$

where $\lambda(u) = \lambda_\theta(u)$ denotes the true intensity and $\hat{\lambda}(u) = \lambda_{\hat{\theta}}(u)$ is the fitted intensity based on the point pattern $\mathbf{x}$. Dividing both sides by $\lambda(u)^{1/2}$ to 'standardise', we get the analogue of the leverage equation (11.37)

$$\lambda(u)^{-1/2} \left[\hat{\lambda}(u) - \lambda(u) \right] \approx \sum_i h(u, x_i) \lambda(x_i)^{-1/2} - \int_W h(u, v) \lambda(v)^{1/2} \, \mathrm{d}v \tag{11.39}$$

where $h(u, v) = \lambda(u)^{1/2} \lambda(v)^{1/2} \mathbf{Z}(u) \boldsymbol{I_\theta}^{-1} \mathbf{Z}(v)^{\mathsf{T}}$ is the 'leverage kernel', the analogue of the leverage matrix H^*. In practice θ and $\lambda(u)$ will be replaced by their estimates $\hat{\theta}$ and $\hat{\lambda}(u) = \lambda_{\hat{\theta}}(u)$. The *leverage* is the diagonal of this estimated kernel, the function

$$\hat{h}(u) = \hat{\lambda}(u) \mathbf{Z}(u) \boldsymbol{I_{\hat{\theta}}}^{-1} \mathbf{Z}(u)^{\mathsf{T}}. \tag{11.40}$$

The leverage of a spatial point process model is a function of spatial location, and is typically displayed as a colour pixel image or perspective plot. The leverage value $\hat{h}(u)$ at a spatial location u represents the (standardised) change in the intensity of the fitted point process model that would have occurred if a data point were to have occurred at the location u. A relatively large value of $\hat{h}(\cdot)$ indicates a part of the space where the data have a *potentially* strong effect on the fitted model (specifically, a strong effect on the intensity of the fitted model) due to the values of the covariates.

Notice that the leverage function of a point process has dimensions $length^{-2}$ while the leverage matrix of linear regression is dimensionless. The trace of the leverage kernel is again

$$\int_W \hat{h}(u) \, \mathrm{d}u = \int_W \mathbf{Z}(u) \boldsymbol{I_\theta}^{-1} \mathbf{Z}(u)^{\mathsf{T}} \hat{\lambda}(u) \, \mathrm{d}u = p. \tag{11.41}$$

Locations u where $\hat{h}(u) > p/|W|$ can be considered to have relatively large leverage.

The preceding calculations are for the canonical case of a Poisson point process whose intensity is a loglinear function of the parameter θ. For a general model with intensity $\lambda_\theta(u)$, the same results hold with $\mathbf{Z}(u)$ replaced by

$$\mathbf{Z}_\theta(u) = \frac{\partial}{\partial \theta} \log \lambda_\theta(u) \tag{11.42}$$

and the Fisher information $\boldsymbol{I_\theta}$ replaced by the negative Hessian of the loglikelihood

$$J(\theta) = \int_W \mathbf{Z}_\theta(u) \mathbf{Z}_\theta(u)^{\mathsf{T}} \lambda_\theta(u) \, \mathrm{d}u + \int_W \lambda_\theta(u) \frac{\partial}{\partial \theta} \mathbf{Z}_\theta(u) \, \mathrm{d}u - \sum_i \frac{\partial}{\partial \theta} \mathbf{Z}_\theta(x_i). \tag{11.43}$$

The calculations can be generalised to models fitted by maximising a criterion other than the likelihood, including robust maximum likelihood [708, 21, 23]. See [36] for the generalisation.

Influence for Poisson point process models

The influence of a point process model is effectively a value attached to each data point (i.e. each point of the point pattern to which the model was fitted). It is a discrete measure on the data points x_i with masses

$$m_i = \frac{1}{p} \mathbf{Z}(x_i) \mathbf{I}_{\hat{\theta}}^{-1} \mathbf{Z}(x_i)^{\mathsf{T}}. \tag{11.44}$$

The influence value m_i at data point x_i represents the change in the maximised log-likelihood that occurs when the point x_i is deleted. A relatively large value of m_i indicates a data point with a large influence on the fitted model.

Since the influence measure turns out to be an atomic measure on the data points only, it would be sufficient to compute the values $\log(L(\hat{\theta})/L(\hat{\theta}_{-i}))$ exactly for each data point x_i. Taylor series approximation is not necessary on computational grounds unless the dataset is very large. Similar comments apply in classical generalized linear models [400, 703]. However, the result (11.44) provides greater insight into the causes of influence.

Parameter influence for Poisson point process models

The *parameter influence* or *parameter effect* measure ('dfbeta') is

$$\mathbb{D}\widetilde{\theta}(B) = \mathbf{I}_{\hat{\theta}}^{-1} \left[\sum_{x_i \in B} \mathbf{Z}(x_i) - \int_B \mathbf{Z}(u) \lambda_{\hat{\theta}}(u) \, du \right]. \tag{11.45}$$

Since $\mathbf{Z}(u)$ is a vector-valued covariate function, the effect measure $\mathbb{D}\widetilde{\theta}$ is a vector-valued measure. The jth component of the influence measure, corresponding to the jth parameter θ_j, is a signed measure in two-dimensional space. It consists of a discrete mass on each data point (i.e. each point in the point pattern to which the model was originally fitted) and a continuous density at all locations. The mass at a data point represents the *negative* change in the fitted value of the parameter θ that would occur if this data point were to be deleted. The density at other non-data locations represents the *negative* effect (on the fitted value of θ) of deleting these locations (and their associated covariate values) from the input to the fitting procedure.

The measure has an atomic component, with masses $\mathbf{I}_{\hat{\theta}}^{-1} \mathbf{Z}(x_i)$ at the data points x_i, and a continuous component, with density $-\mathbf{I}_{\hat{\theta}}^{-1}(u) \lambda_{\hat{\theta}}(u)$ at location $u \in W$.

The effect measure $\mathbb{D}\widetilde{\theta}$ is also a weighted point process residual, equation (11.6), with weight $w(u) = \mathbf{I}_{\hat{\theta}}^{-1}(u) \mathbf{Z}(u)$.

11.9   FAQs

- *I generated a realisation of CSR, and fitted two models to the pattern: a homogeneous Poisson process, and a Poisson process with intensity a loglinear function of the x-coordinate. The results of* diagnose.ppm *for these two models suggest that the loglinear model is a better fit than the homogeneous model, even though the homogeneous model is correct.*

  A model with more parameters will typically produce a better fit to data, even when it is not correct, because of 'overfitting'.

- *In a model where the intensity depends on a covariate Z through $\lambda(u) = \rho(Z(u))$, if the partial residual* parres *is essentially estimating $h(z) = \log \rho(z)$, then why do* parres *and* rhohat *give different results?*

The interpretation of the partial residual as an estimate of $h(z)$ is approximate: it relies on a Taylor approximation of the residuals in terms of the difference between the true and estimated functions $h(z)$. If the model is misspecified or the fit is poor, the approximation is poor.

- *How can I compute the "pixel residuals (Baddeley et al., 2005)"?*

This is a misunderstanding of the residuals defined in [55] and in Section 11.3.1. The residuals are *not* a collection of numbers defined for the pixels in a grid. They are not even computed on such a grid.

For any spatial region B, the raw residual $\mathscr{R}(B)$ is defined as the observed-minus-expected number of points falling in B. This is defined for any region B. Technically $\mathscr{R}$ is known as a **measure**, a function giving a value for each spatial region.

The distinction becomes even more important when considering the weighted residuals such as the Pearson residual measure. The value of the Pearson residual measure $\mathscr{R}^{(P)}(B)$, defined in (11.7), is completely different from the classical Pearson residual of the number of points in B. This has been a source of confusion in recent literature.

The function `residuals.ppm` returns an object of class `"msr"` representing a measure. For a fitted point process model `fit`, to compute the raw residual $\mathscr{R}(B)$ for any window B, use `integral(residuals(fit), B)`. To compute the Pearson residual measure value $\mathscr{R}^{(P)}(B)$ use `integral(residuals(fit, type="Pearson"), B)`.

- *How can I test whether the leverage or influence values in my analysis are statistically significant?*

Leverage values do not have a significance interpretation. The leverage function depends mainly on the covariates and expresses the *potential* sensitivity of the fitting procedure to random points that might fall at any given location.

Individual values of influence do not have a straightforward interpretation in terms of statistical significance. It is better to use residual plots of various kinds to test significance.

12

Cluster and Cox Models

Until now we have focussed on *Poisson* process models, in which the points are assumed to be independent of each other. In the next two chapters, we introduce models that allow dependence between points. Roughly speaking the models in Chapter 12 have positive dependence between points (clustering) and those in Chapter 13 typically have negative dependence between points (regularity).

This chapter introduces Cox point processes and Neyman-Scott cluster processes, which are convenient and flexible models for clustered point patterns. We explain the basic concepts and properties of these models, and show how to fit such models to data.

12.1 Introduction

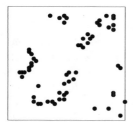 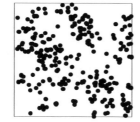 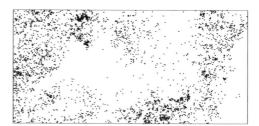

Figure 12.1. *Point patterns exhibiting positive association between points.* Left: *California redwood seedlings and saplings, Ripley's subset, survey plot about 63 feet across.* Middle: *Newly emergent bramble canes in a 9-metre square field.* Right: Beilschmiedia *trees in a* 1000×500 *metre survey plot.*

A crucial assumption of Poisson point process models (Chapter 9) is that points of the process are independent of each other. This assumption is often inappropriate for real data, such as the examples in Figure 12.1. The left panel shows the redwood data giving the locations of 62 seedlings and saplings of California giant redwood. Clusters of points, clearly visible in the plot, are thought to have arisen by propagation from a few parent trees which are no longer alive [643].

The middle panel of Figure 12.1 shows 359 newly emergent bramble cane (blackberry) plants from a larger dataset recorded by Hutchings [352]. Analysis in [221, 235, 664] found that the pattern exhibits clustering, which Hutchings attributes to "vigorous vegetative reproduction".

A different kind of positive association is represented in the *Beilschmiedia* data, shown in the right panel of Figure 12.1. Spatial variation in tree density is evident. This could be attributable to habitat preferences (such as a preference for steep slopes or acidic soil), spatial variation in the availability of water and sunlight, and other factors which may or may not be observable. There is positive association between trees, because the presence of a *Beilschmiedia* tree implies that the habitat is favourable there, which implies that other *Beilschmiedia* trees are likely to be found nearby.

The California redwood data in Figure 12.1 could be modelled as a *cluster process*. In this model, each of the original parent trees (which are unseen) gave rise to a group or cluster of young trees. The observed pattern of young trees is the superposition of clusters originating from different parents. Cluster processes were introduced by Neyman and Scott [499] and others.

The *Beilschmiedia* data could be modelled as a *Cox process*, which is essentially a Poisson process with a random intensity function. The model postulates that there is an underlying, spatially varying, intensity function $\Lambda(u)$ which is *random* because it depends on unobservable external factors as well as observable covariates. If the intensity surface $\Lambda(u)$ were known, then the points would constitute a Poisson process with intensity function $\Lambda(u)$. Cox processes, also known as 'doubly stochastic Poisson processes' or 'modulated Poisson processes', were introduced by Cox [175]; see also [174, 82, 294, 295].

The bramble canes data could be modelled *either* as a cluster process (in which we explicitly postulate the existence of an unseen parent for each newly emergent plant) or as a Cox process (in which unobserved factors such as the unseen pattern of adult plants have determined the spatially varying intensity of newly emergent plants).

Cox and cluster processes are modifications of the Poisson process to incorporate additional random influences — what statisticians would call a *random effect*. They reflect slightly different ways of thinking about the underlying random influences.

A Cox process is effectively a Poisson process in which the intensity itself is random. In applications, the random variation is often attributable to an unobserved spatial covariate, such as soil quality. Spatial variation in the random intensity function can cause points to be more abundant in some areas than in others. This phenomenon can make the point pattern look clustered, especially if there tend to be high peaks ('hot spots') in the random intensity.

In a cluster process, the additional random influence arises from the unseen *parent* points which give rise to *offspring* points which are actually observed. Each observed point's location depends on its parent's location, and this influence is shared by other offspring of the same parent. The pattern looks clustered because points really are organised in clusters.

Deciding whether to use a random effects model (Cox or cluster process) or a fixed effects model (Poisson process) is a strategic choice. In a random effects model we focus on estimating the statistical properties of the unobserved random influences, while in a fixed effects model we try to estimate the actual values of these unobserved influences in the experiment. When we fit a Cox process model to a point pattern, the fitted model parameters tell us the statistical properties (mean, variance, covariance) of the random driving intensity, but do not give estimates of the actual values of the driving intensity. When we fit a Poisson model to the same data, the actual intensity function is estimated.

Both Cox processes and cluster processes (under the assumptions in this chapter) are 'clustered' as measured by the K-function. They combine the variability of the Poisson point process with an additional source of variability, so they are overdispersed (the variance of the number of points falling in a region is greater than the mean). Consequently the K-function of the model always exceeds the theoretical K-function for a Poisson process.

A very attractive feature of Cox and cluster processes is that it is relatively easy to calculate their first and second moments (intensity, pair correlation, K-function). Since we know how to estimate these quantities from data (Chapters 6 and 7), we can match theoretical values of these quantities for Cox and cluster models with estimates obtained from observed data. This can be used to fit models to data.

Cox and cluster process models are fitted to data in `spatstat` using the function `kppm`, which is closely analogous to ppm. For example, to fit a cluster model with Gaussian clusters to the redwoods data we use:

```
> kppm(redwood ~ 1, "Thomas")
```

To fit a Cox model to the *Beilschmiedia* data, where the log intensity is a Gaussian random field with a mean value that depends on terrain slope, and a spatial covariance that is exponentially decreasing as a function of distance, the appropriate command is:

```
> kppm(bei ~ grad, "LGCP", model="exp", data=bei.extra)
```

12.2 Cox processes

12.2.1 Basic concepts

As foreshadowed above, a *Cox point process* is effectively defined as a Poisson process with a random intensity function. To generate a realisation of the Cox process, we first generate a realisation of the underlying random function $\Lambda(u)$, called the 'driving intensity'. Given this function, we generate a realisation of a Poisson process with intensity $\Lambda(u)$. The driving intensity is hidden from us; only the points are observed. See Figure 12.2.

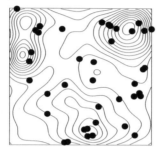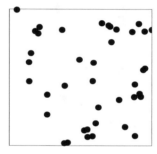

Figure 12.2. *Generation of a Cox process.* Left: *a realisation of a random function $\Lambda(u)$ is generated.* Middle: *a realisation of a Poisson point process with intensity function $\Lambda(u)$ is generated.* Right: *only the points are observed.*

The random function $\Lambda(u)$ could be generated by any random mechanism that we choose, so long as it produces a valid intensity function: the values of $\Lambda(u)$ must be non-negative, and every realisation of $\Lambda(u)$ must be *integrable* over bounded sets (the right-hand side of (9.1) must be finite).

12.2.2 The mixed Poisson process

The simplest example of a Cox process is the *mixed Poisson* process in which we generate a random *number* Λ and, given the value of Λ, generate a uniform Poisson process with intensity Λ. Figure 12.3 shows eight different realisations *of the same process* in the unit square, with the driving intensity Λ following a negative exponential distribution with mean 100. Each panel was effectively generated by typing `lambda <- rexp(1, 1/100)` and `X <- rpoispp(lambda)`.

What is the intensity of this point process? Given the value of Λ, the expected number of points falling in a region B is $\mathbb{E}[n(\mathbf{X} \cap B) \mid \Lambda] = \Lambda|B|$. Averaging over different possible outcomes for the driving intensity Λ, we get

$$\mathbb{E}[n(\mathbf{X} \cap B)] = \mathbb{E}\left(\mathbb{E}\left[n(\mathbf{X} \cap B) \mid \Lambda\right]\right) = \mathbb{E}\left(\Lambda |B|\right) = (\mathbb{E}\Lambda) |B|. \tag{12.1}$$

Therefore the intensity of the mixed Poisson process is $\lambda = \mathbb{E}\Lambda$, the mean value of the random driving intensity. In the example in Figure 12.3 we have $\lambda = \mathbb{E}\Lambda = 100$.

We can also find the *variance* of the number of points falling in a region B, using the analysis of

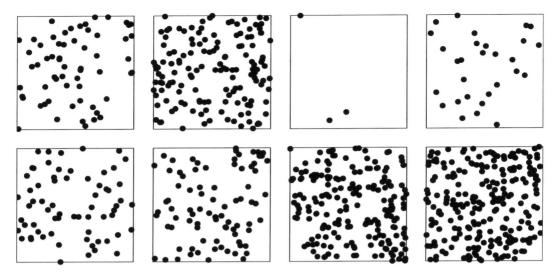

Figure 12.3. *Independent realisations of the mixed Poisson process in the unit square, where the driving intensity* Λ *follows a negative exponential distribution with mean 100.*

variance principle

$$\operatorname{var} n(\mathbf{X} \cap B) = \mathbb{E}\left(\operatorname{var}\left[n(\mathbf{X} \cap B) \mid \Lambda\right]\right) + \operatorname{var}\left(\mathbb{E}\left[n(\mathbf{X} \cap B) \mid \Lambda\right]\right) \tag{12.2}$$

where on the right-hand side, the first term is the variability due to the Poisson process, and the second term is the variability due to Λ. For a given value of Λ, the number of points has a Poisson distribution, with variance equal to the mean, $\operatorname{var}\left[n(\mathbf{X} \cap B) \mid \Lambda\right] = \mathbb{E}\left[n(\mathbf{X} \cap B) \mid \Lambda\right] = \Lambda|B|$, so we get

$$\operatorname{var} n(\mathbf{X} \cap B) = (\mathbb{E}\Lambda)|B| + (\operatorname{var}\Lambda)|B|^2. \tag{12.3}$$

The variance of $n(\mathbf{X} \cap B)$ exceeds its mean, so the point counts are *overdispersed*. In the example in Figure 12.3 the overdispersion is extreme: we have $\mathbb{E}\Lambda = 100$ and $\operatorname{var}\Lambda = 100^2$; if B is the unit square, the variance of the number of points is $100 + 100^2 = 10100$ whereas the Poisson process with the same mean would have a variance of only 100.

A difficulty with applying the mixed Poisson model in practice is that, very often, we observe only one realization of the process. In such cases, in the absence of further information, we have in effect observed a realisation of CSR. There is no way for us to know that what we are seeing is underlain by a Cox process and that if we were to observe another realization it would have a different (random) intensity. Only by observing multiple realizations can we discern that the intensity is random and try to estimate its distribution. In technical terms, the mixed Poisson process is not 'ergodic'.

The mixed Poisson model does become useful if we have replicated observations (repetitions of the same experiment, each resulting in a different realisation of the point process) as discussed in Chapter 16. In that case, we can discern the difference between the mixed Poisson model and CSR, and we can estimate the distribution of the driving intensity Λ.

12.2.3 Random fields

In most applications we need a model in which the random driving intensity $\Lambda(u)$ is spatially varying. A few simple examples are discussed here.

One simple-minded way to generate a random, undulating surface is to create a pixel image with random values in each pixel, and then smooth the result. The left panel of Figure 12.4 shows a pixel image containing independent, uniformly distributed random pixel values, generated by

```
> Y <- rnoise(runif, square(1), max=100)
```

The middle panel shows the result of kernel smoothing the left panel using

```
> Z <- Smooth(Y, sigma=0.07, normalise=TRUE, bleed=FALSE)
```

This randomly generated image Z could now serve as the intensity for a Poisson point process,

```
> X <- rpoispp(Z)
```

which is shown in the right panel.

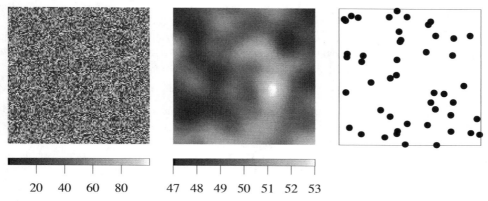

Figure 12.4. *A simple construction of a Cox process.* Left: *Pixel values are independent random numbers.* Middle: *Kernel-smoothed version of left panel.* Right: *Poisson process with intensity given by middle panel.*

One might envisage this phenomenon as arising from the relevant quantity (e.g., soil nutrients or moisture) mixing or seeping between locations.

Another simple model for a random intensity function is the 'mosaic field' shown in Figure 12.5. We start by dividing the window randomly into tiles $T_1, T_2, \ldots$ according to some random mechanism. For example we could use the Dirichlet tessellation generated by random points. The left panel of Figure 12.5 shows a random tessellation generated by random straight lines. Next, a random numerical value Λ_i is associated with each tile T_i. The values associated with different tiles are independent, and are drawn from the same probability distribution. We define the driving intensity $\Lambda(u)$ to be equal to Λ_i inside tile T_i. That is, if the location u lies inside tile T_i, then its value is $\Lambda(u) = \Lambda_i$. A Poisson process with intensity function $\Lambda(u)$ is called a mosaic Cox process.

The point patterns in the right-hand panels of Figures 12.4 and 12.5 show the kind of behaviour we are trying to achieve, in which there are spatial variations in the density of points. Apparent 'hot spots' or clusters of points, and 'cold spots' where points are scarce, are both attributable to spatial variation in the driving intensity. Different realisations of these models would show hot and cold spots, but in different places in each realisation.

12.2.4 Log-Gaussian Cox processes

We saw in Chapter 9 that a very natural and appealing way to build a Poisson point process model is to specify the *logarithm* of the intensity. Similarly for Cox processes, the most natural approach is to build models for $G(u) = \log \Lambda(u)$, the logarithm of the driving intensity. Modelling on the logarithmic scale also ensures that the intensity values are positive.

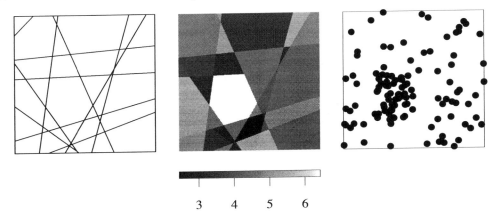

3 4 5 6

Figure 12.5. *Cox process generated by a mosaic random field.* Left: *random tessellation.* Middle: *each tile of the tessellation is assigned a different random value, giving the mosaic random field.* Right: *Poisson process with intensity equal to the mosaic random field.*

In many applications it makes sense to assume that $G(u)$ has a normal (Gaussian) distribution for each location u. Additionally for each pair of locations u, v the pair of values $(G(u), G(v))$ should have a bivariate Gaussian distribution. In general, for any chosen set of locations $u_1, \ldots, u_n$ the corresponding values $G(u_1), \ldots, G(u_n)$ should have a multivariate Gaussian joint distribution. A random function $G(u)$ with these properties is called a *Gaussian random field.*

The random field illustrated in Figure 12.4 is essentially a Gaussian random field. Each pixel value in the smoothed field Z is an average of contributions from many pixels in the random noise image Y, so the distribution of pixel values in Z is approximately Gaussian. For any pair of pixels, the corresponding pixel values of Z are approximately bivariate Gaussian distributed, and so on.

However, the mosaic random field illustrated in Figure 12.5 is **not** a Gaussian random field. Even if the random values assigned to each tile are Gaussian random variables, the joint distribution of values at two different pixels is not bivariate Gaussian. This is explained in Section 12.2.5.

Since Gaussian random variables can take negative values, a true Gaussian random field cannot constitute a model for intensity. However, the log-Gaussian random field obtained by exponentiating a Gaussian random field is always non-negative and hence can be used as the intensity of a point process. A *log-Gaussian Cox process (LGCP)* is a Cox process whose driving intensity is of the form

$$\Lambda(u) = \exp G(u)$$

where $G(u)$ is a Gaussian random field [484, 483].

A very great advantage of a Gaussian random field is that it is completely specified by its first and second moments. Since the Gaussian distribution is completely specified by giving its mean and variance, a Gaussian random field $G(u)$ is uniquely identified by its **mean function**

$$\mu(u) = \mathbb{E}G(u) \tag{12.4}$$

and its **covariance function**

$$C(u, v) = \text{cov}(G(u), G(v)) = \mathbb{E}[G(u)G(v)] - \mu(u)\mu(v). \tag{12.5}$$

Two Gaussian random fields with the same mean function and the same covariance function must be identical with respect to their probability distribution.

Gaussian random fields can be adjusted and rescaled in the same way as Gaussian random variables. If $G_0(u)$ is a Gaussian random field with mean 0 and covariance function $C_0(u, v)$, and

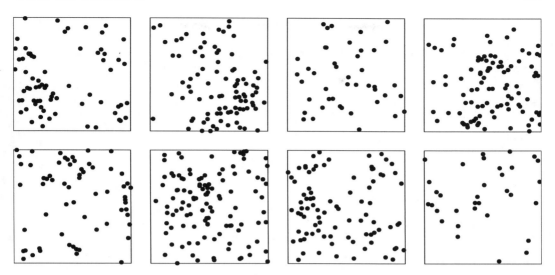

Figure 12.6. *Independent realisations of the log-Gaussian Cox process with log mean* $\mu = 4$ *and exponential covariance with scale* $\alpha = 0.2$ *and log variance* $\sigma^2 = 0.5$.

if $a(u)$ and $b(u)$ are non-random functions, then $G(u) = a(u)G_0(u) + b(u)$ is a Gaussian random field with mean function $\mu(u) = b(u)$ and covariance function $C(u, v) = a(u)a(v)C_0(u, v)$. This makes it possible to build a wide range of inhomogeneous Gaussian random fields, and therefore inhomogeneous log-Gaussian Cox processes, starting from a small collection of stationary models.

However, one cannot simply write down a mean function and a covariance function and assume that there exists a Gaussian random field with these characteristics. The mean and covariance must satisfy technical conditions (most importantly that the covariance is 'positive definite') in order that a Gaussian random field exists [484, section 5.6.1, p. 73]. In practice, we usually select our covariance function from a 'menu' of standard models, which are known to be mathematically valid, and for which the relevant algorithms have been written.

We will exclusively study models where C is isotropic, i.e. only depends on the distance $r = \|u - v\|$ between locations u and v. Furthermore, the models that we study will always be of the form

$$C(u, v) = C_0(r) = \sigma^2 R(r/\alpha) \tag{12.6}$$

where $\alpha > 0$ is a scale parameter, $\sigma^2 > 0$ is a variance parameter, and R is a known 'template' covariance function with $R(0) = 1$ (making it a correlation function). The contributed R package RandomFields supports a wide range of Gaussian random field models, and we rely on this excellent resource in spatstat. The covariance function C_0 can be any of those implemented in RandomFields (see the help for RFmodel). The default template is $R(r) = \exp(-r)$ which gives rise to the *exponential covariance function*:

$$C_0(r) = \sigma^2 \exp(-r/\alpha). \tag{12.7}$$

This function has only two parameters, α and σ^2. In the literature on random fields the more flexible Whittle-Matérn covariance model (also known as the Matérn model) is often recommended [627]. This model has an additional positive shape parameter ν. The template covariance for this model is

$$R(r) = 2^{1-\nu}\Gamma(\nu)^{-1}(\sqrt{2\nu}r)^\nu K_\nu(\sqrt{2\nu}r) \tag{12.8}$$

where K_ν is the modified Bessel function of second kind of order ν. Whenever this model is used in `spatstat` the shape parameter ν has to be specified by the user (i.e. it is not estimated from the data).

Figure 12.6 shows eight independent realisations of a log-Gaussian Cox process with exponential covariance function.

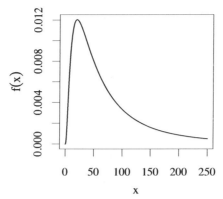

Figure 12.7. *Probability density of the lognormal distribution with mean (on log scale) 4 and standard deviation (on log scale) 1. The median is $e^4 = 54.6$, the mean is $e^{4+1/2} = 90.0$, and the standard deviation $e^{4+1/2}\sqrt{e^1 - 1} = 118.0$.*

Since the 'bell curve' is symmetric about its mean, positive and negative fluctuations in a Gaussian random field $G(u)$ are equally likely. In a log-Gaussian Cox process, 'hot spots' and 'cold spots' of equal size *on a log scale* occur with the same frequency. However, the distribution of the driving intensity value $\Lambda(u) = \exp G(u)$ is asymmetric: it is a log-Gaussian or *lognormal* distribution, which has a heavy right tail. See Figure 12.7. If $Z \sim N(\mu, \sigma^2)$ and $\Lambda = e^Z$ then Λ has median e^μ but has mean $\exp(\mu + \sigma^2/2)$ and standard deviation $\exp(\mu + \sigma^2/2)\sqrt{\exp(\sigma^2) - 1}$. Hence, in a LGCP, the overdensity of points in a hot spot is vastly greater than the underdensity of points in a cold spot.

12.2.5   More on mosaic random fields

Suppose that, in the mosaic model (Figure 12.5), each tile is given an intensity value $\Lambda_i = \exp(V_i)$ where V_i are Gaussian random variables with mean μ and standard deviation σ, generated independently for each tile. Then $G(u) = \log \Lambda(u)$ has a Gaussian distribution, for each location u.

However, the joint distribution of $(G(u), G(v))$ for two locations u, v is not bivariate Gaussian. If the two locations u, v happen to fall in different tiles, then $G(u), G(v)$ are independent Gaussian variables; but if they fall in the same tile, then $G(u) = G(v)$. The joint distribution of $(G(u), G(v))$ is a mixture of these two components, which is not Gaussian. Hence, *such a mosaic model is not a Gaussian random field*. This mosaic model is actually quite difficult to analyse. It is not uniquely characterised by its first and second moments, and it has complicated long-range dependence, induced by the infinite random lines.

To remedy this problem, suppose we generate several independent realisations of the random mosaic field, and average them. That is, suppose $G_1(u), G_2(u), \ldots, G_n(u)$ are independent realisations of the random mosaic field based on a Poisson line process. (See Section 12.2.6 for an explanation of why 'Poisson line process' is important.) We take the average $\overline{G}(u) = \frac{1}{n}\sum_{i=1}^{n} G_i(u)$ and rescale this to $Y(u) = \sqrt{n}(\overline{G}(u) - \mu) + \mu$ to have the same mean and variance as the original. The middle panel of Figure 12.8 shows the rescaled average $Y(u)$ based on $n = 10$ independent realisations. If we increase the number of repetitions n, then the distribution of $Y(u)$ approaches a Gaussian distribution (according to the Central Limit Theorem). The joint distribution of $(Y(u), Y(v))$

for two locations u, v also approaches a bivariate Gaussian distribution; and similarly the joint distribution of values at any number of locations approaches a multivariate Gaussian. Consequently, the result is a Gaussian random field. A realisation is shown in the right panel of Figure 12.8. This random field has the same mean and variance as the original mosaic random field. The covariance is $C(u,v) = \sigma^2 \exp(-2\tau\|u-v\|)$ where $\sigma^2 = \operatorname{var} G(u)$. We have just provided a constructive procedure for creating a Gaussian random field with exponential covariance.

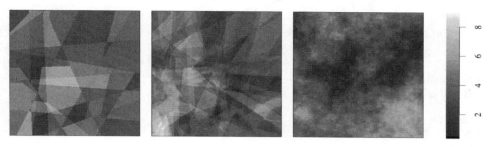

Figure 12.8. *Rescaled average of several copies of the mosaic random field.* Left: *two copies.* Middle: *ten copies.* Right: *an infinite number of copies.*

12.2.6 Moments of Cox processes

The moments of any Cox process can be calculated in the same way as we demonstrated in Section 12.2.2 for the mixed Poisson model. The Cox process with driving intensity $\Lambda(u)$ has intensity function

$$\lambda(u) = \mathbb{E}\Lambda(u) \tag{12.9}$$

and pair correlation function

$$g_2(u,v) = \frac{\mathbb{E}[\Lambda(u)\Lambda(v)]}{\lambda(u)\lambda(v)} = 1 + \frac{\operatorname{cov}(\Lambda(u),\Lambda(v))}{\lambda(u)\lambda(v)}. \tag{12.10}$$

This is the 'general' pair correlation function $g_2(u,v)$ defined in equation (7.46) in Section 7.10, since a Cox process is not necessarily correlation-stationary as defined in Section 7.10.2.

For a log-Gaussian Cox process the pair correlation function becomes

$$g_2(u,v) = \exp(C(u,v)) \text{ where } C(u,v) = \operatorname{cov}(G(u),G(v))$$

where in turn $G(u) = \log \Lambda(u)$. This can easily be calculated from the moment generating function for the Gaussian distribution.

For the mosaic model, illustrated in Figure 12.5, the intensity is $\lambda = \mathbb{E}\Lambda_1$ and the pair correlation is $g(u,v) = 1 + p(u,v)(\operatorname{var}\Lambda_1)/\lambda^2$ where Λ_1 is the random value associated with a typical tile, and $p(u,v)$ is the probability that the points u and v lie inside the same random tile. This expression can be made more explicit when the random lines are generated by a 'Poisson line process' (see for example [636, section 8.4.2]), as they were in the given example. For this tessellation, the probability $p(u,v)$ that two given points u and v lie in the same tile is equal to the probability that u and v are not separated by any random line, which equals $\exp(-2\tau\|u-v\|)$. Thus for the mosaic Cox process, based on a mosaic random field generated by the Poisson line process, the pair correlation function is

$$g(u,v) = 1 + \exp(-2\tau\|u-v\|)\frac{\operatorname{var}\Lambda_1}{\lambda^2}$$

(where $\lambda = \mathbb{E}\Lambda_1$) which decays exponentially as a function of distance.

12.2.7   Simulating Cox processes in `spatstat`

In one sense it is easy to simulate any desired Cox process. The user simply has to generate a realisation of the random driving intensity function $\Lambda(u)$, and then use `rpoispp` to generate the Poisson random points. The random intensity $\Lambda(u)$ can be generated either as a single number (thus leading to a mixed Poisson process), a pixel image, or a function in the R language. All these formats are acceptable as the argument `lambda` to `rpoispp`.

The catch is that, except for a few special cases, it falls to the user to write code to generate the random driving intensity $\Lambda(u)$, which can be challenging. The task is easy in the case of a mixed Poisson process. Simply use the appropriate random number generator provided by R to generate a Poisson mean, and then use `rpoispp` to simulate a realization of a Poisson process having that intensity. For example, to create a realization, in a window `A`, of a mixed Poisson process with a χ^2_5 mixing distribution the user would type:

```
> Lambda <- rchisq(1, df=5)
> X       <- rpoispp(Lambda, W=A)
```

Other cases require more ingenuity. For generating Cox processes driven by a mosaic random field (Figure 12.5) `spatstat` provides the specialized function `rMosaicField` which can be used to assign random values to each tile of the tessellation. The Poisson line tessellation can be generated using `rpoislinetess`. Once the random field has been generated it is trivial to use `rpoispp` to generate the points:

```
> P       <- rpoislinetess(4)
> Lambda <- rMosaicField(P, rchisq, rgenargs=list(df=5))
> X       <- rpoispp(Lambda)
```

The most important case, that of simulating a log-Gaussian Cox process, is implemented in the specialised function `rLGCP`, generously contributed by R. Waagepetersen and A. Jalilian. As mentioned above, the simulation of the underlying Gaussian random field is performed by the `RandomFields` package. A typical call to `rLGCP` is

```
> X <- rLGCP(model="exp", mu=4, var=0.2, scale=0.1, win = square(1))
```

where `model` specifies the covariance model, `mu` is the mean or mean function $\mu(u)$, `var` is the variance parameter σ^2, `scale` is the scale parameter α, and `win` is the window of simulation. Keep in mind that all the parameters are for the *logarithm* of the random intensity since it is $\log \Lambda$ which is a Gaussian random field. The covariance model can be any of those implemented in the `RandomFields` package (see `RFmodel` in `RandomFields` and `help(rLGCP)`). The default is `model="exp"`, the exponential covariance (12.7), so it need not actually have been specified in the foregoing example. Other popular choices include `model="matern"` for the Whittle-Matérn covariance (12.8), and `model="gauss"` which gives the Gaussian covariance template $R(r) = \exp(-r^2)$.

The mean function `mu` can be a numeric constant, an image of class `"im"`, or a `function`. The window needs to be an `"owin"` object. If it is missing or `NULL` it defaults to `Window(mu)` if `mu` is a pixel image, and it defaults to the unit square otherwise. The return value is of course a point pattern (object of class `"ppp"`). By default the generated random driving intensity function $\Lambda(u)$ is attached as an attribute extractable by `attr(X,"Lambda")`. Figure 12.9 shows both the generated random driving intensity function (left panel) and the generated point pattern (right panel).

Some covariance models have additional 'shape' parameters, beyond the mandatory variance and scale parameters, which must be specified. For example, for the Matérn covariance model, we need to specify the shape parameter ν, as in:

```
> X <- rLGCP(model="matern", mu=4, nu = 1, var=0.2, scale=0.1)
```

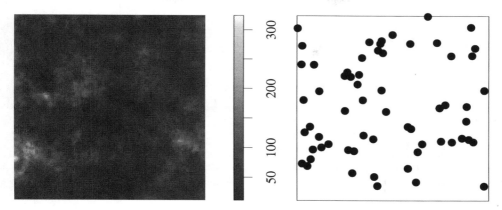

Figure 12.9. *Log-Gaussian Cox Process.* Left: *random driving intensity function* $\Lambda(u)$ *such that* $\log \Lambda(u)$ *is a Gaussian random field with mean* $\mu = 4$ *and exponential covariance* $C_0(r) = 0.2\exp(-10r)$. Right: *realised point process* $\mathbf{X}$.

The `var` and `scale` default to 1, but this default would rarely be sensible.

If the `RandomFields` package is not available, the user *could* with a bit of effort simulate a log-Gaussian Cox process from scratch using the constructive procedure described in Section 12.2.5. However, this possibility only allows for generating log-Gaussian Cox processes with an *exponential* covariance structure.

12.3 Cluster processes

12.3.1 Basic concepts

A *cluster process* is formed by a two-step procedure, illustrated in Figure 12.10. In the first step, a point process $\mathbf{Y}$ of 'parent' points is generated. In the second step, each 'parent' point y_i gives rise to a random pattern of 'offspring' points x_{ij}. Only the offspring points are observed; we may as well assume that the parent point was *replaced* by its offspring. The set consisting of all the offspring points, regardless of their parentage, forms the cluster point process $\mathbf{X}$.

For example, a cluster process could serve as a simplified model of the spatial clustering of plants due to seed dispersal or propagation. Imagine that in a previous generation of plants (which is no longer observable) each plant dispersed its seeds into the space around itself, giving rise to the current generation of plants, which we can observe. This might be a good model for the `redwood` data in the left panel of Figure 12.1.

Cluster processes were developed in a classic paper by Neyman and Scott [499] as models for the spatial pattern of galaxies in the distant universe. Their rationale was that galaxies are observed to lie in groups or clusters, so that a good statistical model of the universe should take a *galaxy cluster* as its basic unit.

In principle we could allow reproduction to continue through several generations: the offspring would have second-generation offspring, and so on. Multi-generational clustering was envisaged by Neyman and Scott [499] and studied by Kingman [382]. For practical reasons, we shall not consider

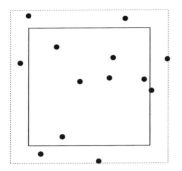 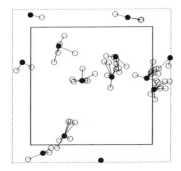 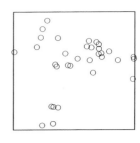

Figure 12.10. *Formation of a cluster process.* Left: *parent points are generated.* Middle: *each parent (filled dot) gives rise to a cluster of offspring (open circles).* Right: *the offspring constitute the cluster point process.*

multi-generational clustering, and we consider only the single-generation cluster models sketched in Figure 12.10.

12.3.2   Model assumptions

To make progress, we need to make simplifying assumptions about the cluster formation process. The most popular items on the 'menu' of assumptions are:

(CLP1) **Poisson parents:** the parent points constitute a Poisson process.

(CLP2) **independent clusters:** different clusters are independent of each other.

(CLP3) **identically distributed clusters:** different clusters, shifted to the same parent location, have the same distribution.

(CLP4) **offspring independent within a cluster:** the locations of the offspring of a given parent point are independent and identically distributed (given the parent location).

(CLP5) **Poisson number of offspring:** for each parent point, the number of offspring is a Poisson random variable.

(CLP6) **isotropic clusters:** the probability density of the offspring of a given parent point depends only on distance from offspring to parent.

Table 12.1 lists various common choices for the set of assumptions, and the standard names for these choices (although the literature is not completely consistent). A point process which satisfies (CLP1)–(CLP4) is a *Neyman-Scott* process. The techniques described in this chapter are applicable to Neyman-Scott processes.

PARENTS	CLUSTERS	OFFSPRING LOCATIONS	NUMBER OF OFFSPRING	NAME OF PROCESS
any	independent	any	any	independent cluster process
Poisson	independent	any	any	Poisson cluster process
Poisson	i.i.d.*	independent	any	Neyman-Scott cluster process
Poisson	i.i.d.*	independent	Poisson	Neyman-Scott Cox process†

Table 12.1. *Nomenclature for cluster processes.* (* *i.i.d.* = *independent and identically distributed, properties* (CLP2) *and* (CLP3); † = *not widely established terminology*).

12.3.3 Specific cluster models

Examples of cluster models are presented in this section. For these examples, we assume *all* of the properties (CLP1)–(CLP6) hold. Consequently, each example can be generated by the following simple procedure:

1. The parent points are generated by a homogeneous Poisson process with intensity κ.

2. Each parent point gives rise to a random number of offspring, according to a Poisson distribution with mean μ offspring per parent.

3. For each parent point y_i, the offspring points x_{ij} are independent and identically distributed, with a spatial probability density or 'kernel' $h(x \mid y) = h(\|x - y\|)$ depending only on the distance from offspring to parent.

The resulting cluster process has intensity $\lambda = \kappa\mu$. Different choices for the offspring probability density $h(x \mid y)$ are discussed below, and depicted in Figure 12.16.

Matérn cluster process

In a *Matérn cluster process* [452, pp. 46–47], [454], the offspring of each parent point are uniformly distributed in a disc of radius R centred around the parent. See Figure 12.11. The spatial scale of the clusters is controlled by the radius R. (Note that R is denoted by `scale` in the argument list of `rMatClust`; see Section 12.3.6.) The parameter μ determines how many points (on average) are in a cluster.

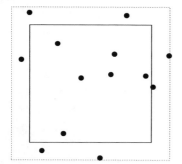 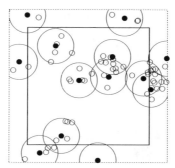 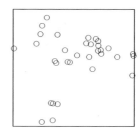

Figure 12.11. *Formation of a Matérn cluster process.* Left: *parent points are generated according to a homogeneous Poisson process with intensity* κ. Middle: *for each parent, a random number of offspring are generated according to a Poisson distribution with mean* μ, *and are placed independently and uniformly in a circle of radius R around the parent.* Right: *the offspring constitute the Matérn cluster process.*

Notice that some of the offspring points inside a window W could have come from parents lying outside W. To generate a realisation of the Matérn cluster process in W, we would need to generate parent points in the larger domain $W_{\oplus R}$.

Figure 12.12 shows eight realisations of the same Matérn cluster process.

The relation between the parent points and offspring points can be understood in another way. For each parent point, there is a Poisson random number of offspring, at independent random locations, so that the offspring of a particular parent constitute a finite Poisson point process. That is, *under assumptions* (CLP4) *and* (CLP5)*, each cluster is itself a Poisson process.* For the Matérn cluster process just described, each cluster contains on average μ points spread uniformly over a disc of radius R, so the offspring cluster is a Poisson process with intensity $\mu/(\pi R^2)$ inside the circle of radius R centred on the parent point, and intensity 0 outside the circle. When we superimpose the offspring of different parents, we are superimposing independent Poisson processes, so the

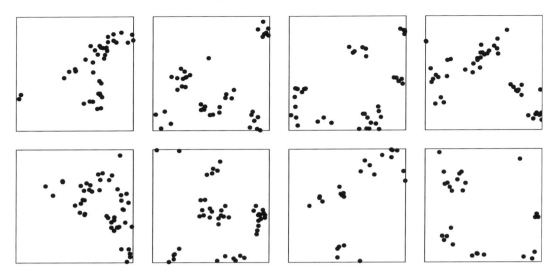

Figure 12.12. *Independent realisations of the stationary Matérn cluster process with parent intensity* $\kappa = 8$, *mean cluster size* $\mu = 5$, *and cluster radius* $R = 0.1$.

result is again a Poisson process. That is, *under assumptions* (CLP1)–(CLP5), *the cluster process is a Cox process*. Given the locations of all the parent points y_i, the offspring process is a Poisson process whose intensity is the sum of the intensities of the individual clusters. For the Matérn cluster process the driving intensity is

$$\Lambda(u) = \sum_i h(u - y_i) \tag{12.11}$$

where $h(z) = \mu/(\pi R^2)$ if $\|z\| \le R$ and $h(z) = 0$ otherwise. This is depicted in Figure 12.13.

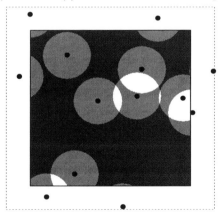

Figure 12.13. *Driving intensity* $\Lambda(u)$ *of the Matérn cluster process. Black areas: zero intensity; grey areas: intensity* $\mu/(\pi R^2)$; *white areas: intensity* $2\mu/(\pi R^2)$.

The representation of the Matérn cluster process as a Cox process is also helpful for calculating the moments. The intensity is $\lambda(u) = \mathbb{E}\Lambda(u) = \mathbb{E}\sum_i h(u - y_i)$; using Campbell's formula (6.11) we get $\mathbb{E}\sum_i h(u - y_i) = \int \kappa h(u - v)\,\mathrm{d}v = \kappa\mu$ so that the Matérn cluster process has homogeneous intensity $\lambda = \kappa\mu$. Using a similar technique (spelt out in Section 12.6) we can calculate the pair correlation function.

Thomas cluster process

In the ('modified') *Thomas cluster process* [653, 217], the probability density of offspring locations is an isotropic Gaussian density. Effectively, each offspring is randomly displaced from its parent, with the displacement vectors having an isotropic Gaussian distribution $N(0, \sigma^2 I)$ with standard deviation σ along each coordinate axis. See Figure 12.14. The spatial scale of the clusters is controlled by σ.

In the original model of Thomas [653], clusters were constrained to be non-empty, that is, the number of offspring in each cluster was a Poisson random variable N conditioned on $N > 0$. Diggle [217] pointed out that the algebra is simpler if this constraint is removed, and called this the 'modified' Thomas process. The original Thomas process with parent intensity κ is equivalent to the modified Thomas process with a higher parent intensity κ/p where $p = 1 - e^{-\mu}$. Because the two formulations are equivalent, the terminology is often blurred and 'Thomas' usually means 'modified Thomas'.

Again the Thomas process is a Cox process, with random driving intensity $\Lambda(u)$ equal to the superposition of Gaussian densities centred at each of the parent points. The middle panel of Figure 12.14 depicts this function $\Lambda(u)$ (the 'shot noise field') for the pattern of parent points shown in the left panel.

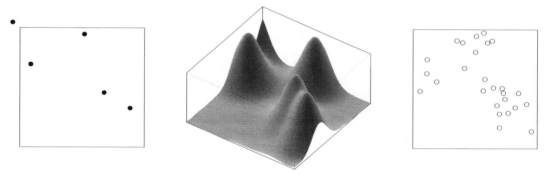

Figure 12.14. *Formation of Thomas process.* Left: *parent points.* Middle: *total offspring density (sum of Gaussian densities centred at each parent point).* Right: *offspring points constitute the Thomas process.*

Cauchy cluster process

In a *Cauchy cluster process* [360] the probability density of the offspring is a bivariate Cauchy distribution,

$$h(u) = \frac{1}{2\pi\omega^2} \left[1 + \frac{\|u\|^2}{\omega^2} \right]^{-3/2} \tag{12.12}$$

where ω is the scale parameter, playing essentially the same rôle as σ for the Thomas process. The spatial extent of a cluster is proportional to ω. The Cauchy density is extremely heavy-tailed so that offspring can be very distant from their parents. A realisation is shown in the left panel of Figure 12.15.

Variance-gamma cluster process

In a *variance-gamma cluster process* [360] the probability density of the offspring is

$$h(u) = \frac{1}{2^{\nu+1}\pi\eta^2\Gamma(\nu+1)} \frac{\|u\|^\nu}{\eta^\nu} K_\nu\left(\frac{\|u\|}{\eta}\right) \tag{12.13}$$

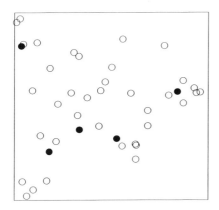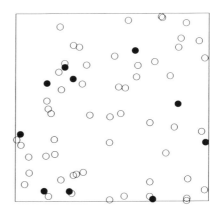

Figure 12.15. *Realisation of a Cauchy process with scale* $\omega = 0.1$ *(Left) and a variance-gamma process with* $\nu = 2$ *and scale* $\eta = 0.1$ *(Right) in the unit square. Open circles: offspring points. Filled dots: parent points.*

where η is the scale parameter and ν is an extra parameter controlling the 'shape' of the density, which must satisfy $\nu > -1/2$. Here Γ is the gamma function and K_ν is the modified Bessel function of the second kind of order ν. The tails of the variance-gamma density are also heavy. A realisation is shown in the right panel of Figure 12.15.

Comparison of cluster shapes

Figure 12.16 compares the shapes of the kernels (bivariate probability density functions) considered in this section. The left-hand panel shows vertical cross-sections of these two-dimensional probability density surfaces, taken along the x-axis. For example, the Matérn cluster process has an offspring density which is uniform over the disc of radius R, so the vertical section of this density is a rectangular shape. Since the kernels are isotropic (circularly symmetric), all cross-sections through the centre of the kernel are the same.

The right panel of 12.16 shows the corresponding probability density *of the distance from parent to offspring* for each model, derived from the two-dimensional densities h_0 by $h_1(t) = 2\pi t h_0(t\mathbf{e})$ where $\mathbf{e}$ is a unit vector. This may give better intuition about the tails of the densities.

Shot noise Cox models

The models considered here belong to the class of 'shot noise Cox processes'. They are indeed Cox processes with driving intensity equal to the superposition of the kernels centered on the parent points. The intensity surface in question is a *random* field since it is determined by the random locations of the parent process. In this setting it is referred to as a *shot noise field*.

This concept arose in physics in the study of electromagnetic phenomena, particularly photons [605]. The points of the parent process may be thought of as the impact points of 'shots' with the effect of the impact dying away as the distance to the point of impact increases. For a general discussion of shot noise processes see [484, section 5.4].

Useful work on this class of models includes [128, 129, 476, 482, 479, 555, 678, 679, 680, 511, 649].

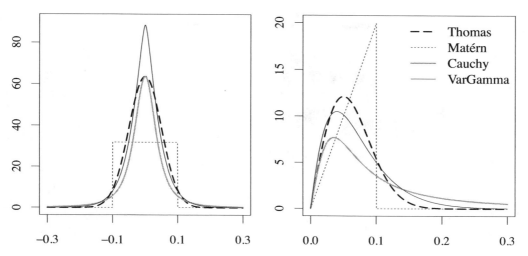

Figure 12.16. *Comparison of cluster models.* Left: *Cross-section of the kernel (bivariate probability density function).* Right: *Probability density of distance from cluster centre to offspring. Thick dashed lines: Thomas model, scale* = 0.05. *Thin dotted lines: Matérn cluster model, R* = 0.1. *Thick grey lines: Cauchy model, scale* = 0.05. *Thin solid lines: variance-gamma model, scale* = 0.03, *shape* $\nu = 1$.

12.3.4   Inhomogeneous cluster models

The inhomogeneous Poisson process was introduced in Section 5.4 and covered in Section 9.2. We can also introduce spatial inhomogeneity into the Neyman-Scott models described in the previous section. This can be done in several distinct ways: by making the parent process inhomogeneous, making the offspring processes inhomogeneous, or both. Various kinds of inhomogeneous models are discussed in [558, 491].

12.3.4.1   Inhomogeneous parents

The first option is to make the *parents* inhomogeneous. We simply generate the parent points from an inhomogeneous Poisson process with some intensity function $\kappa(u)$.

It is straightforward to generate realisations of this model. The `spatstat` functions for simulating cluster processes will accept a parent intensity `kappa` which is a `function(x,y)` or a pixel image with the result that the parents are Poisson with inhomogeneous intensity `kappa`.

However, statistical inference for such models would be relatively complicated. The intensity function is

$$\lambda(u) = \int \kappa(v)h(u \mid v)\,\mathrm{d}v$$

which is not proportional to the parent intensity $\kappa(u)$. Parents occurring in a region where the parent intensity is high may contribute offspring to a nearby region where the parent intensity is low. In simple cases, $h(u \mid v) = h(u - v)$, and the intensity function $\lambda(u)$ of the cluster process is the convolution of the parent intensity $\kappa(u)$ with the offspring kernel h. The second-moment structure (the pair correlation function and K-function) is also complicated because these models are not correlation-stationary. Algorithms for fitting these models are proposed in [493]; they are not currently implemented in `spatstat`.

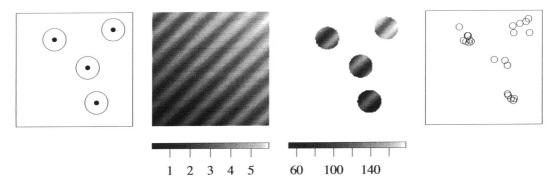

Figure 12.17. *Generation of an inhomogeneous Matérn cluster process.* Left to Right: *parent points and associated cluster radii; reference intensity $\mu(u)$; driving intensity; and offspring points.*

12.3.4.2 Inhomogeneous offspring

To make the process inhomogeneous by generating clusters in an inhomogeneous manner, we use a clever construction due to Waagepetersen [678]. Parent points are first generated from a *homogeneous* Poisson process. For a parent point at location y_i, the offspring are generated from a Poisson process with intensity $\beta_i(u) = \mu(u)h(u - y_i)$, where $h(v)$ is a cluster kernel (offspring density), and $\mu(u)$ is the *reference* or *modulating* intensity.

This is an inhomogeneous generalisation of the stationary cluster processes described above, because if $\mu(u)$ is constant and equal to μ, the model reduces to a stationary cluster process where μ is the mean number of offspring per parent.

The inhomogeneous Matérn cluster process is illustrated in Figure 12.17. In the left panel are the parent points, generated according to a homogeneous Poisson process. The kernels $h(u - y_i)$ are sketched for each parent point y_i. The second panel shows the modulating intensity function $\mu(u)$ which we have chosen arbitrarily. The third panel shows the total offspring intensity $\beta(u) = \sum_i \beta_i(u)$. Finally the right panel shows the offspring points, generated according to a Poisson process with intensity $\beta(u)$. In this simple model, a 'cookie-cutter' effect is applied to the modulating intensity $\mu(u)$ to determine the total offspring intensity $\beta(u)$.

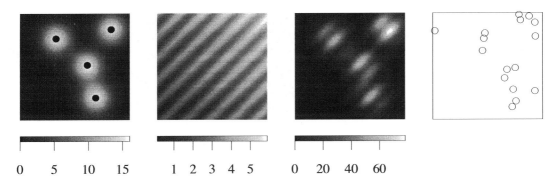

Figure 12.18. *Generation of an inhomogeneous Thomas cluster process.* Left to Right: *parent points and associated cluster densities; reference intensity $\mu(u)$; driving intensity; and offspring points.*

For the Thomas process the 'cookie-cutter' is replaced by a smooth version of the same operation, shown in Figure 12.18.

For any inhomogeneous cluster process of this kind, the intensity is

$$\lambda(u) = \mathbb{E}\beta(u) = \mu(u)\mathbb{E}\sum_i h(u - y_i) = \mu(u)\int h(u - v)\kappa\,\mathrm{d}v = \kappa\mu(u)$$

using Campbell's formula. That is, this model has intensity proportional to the reference intensity $\mu(u)$, as desired.

Note that different clusters have different mean numbers of points. The number of offspring from a given parent y_i is a Poisson random variable with mean

$$B(y_i) = \int \beta_i(u)\,\mathrm{d}u = \int h(u - y_i)\mu(u)\,\mathrm{d}u.$$

The inhomogeneous cluster process can also be described as an inhomogeneous thinning of a homogeneous cluster process [678]. Suppose that M is the maximum value of the reference intensity function $\mu(u)$, or any value greater than this, so that $M \geq \mu(u)$ for all points u in the window in which we are generating patterns. Let us start by generating a *stationary* cluster process with the same parent intensity κ and the same offspring density $h(u \mid y) = h(u - y)$, but where the mean number of offspring per parent is equal to M. The cluster of offspring from parent y_i is a Poisson process with intensity $\zeta_i(u) = Mh(u - y_i)$. Now suppose we *thin* the resulting pattern (see Sections 5.3.3 and 9.2.2) so that the probability of retaining a point at location u is $p(u) = \mu(u)/M$. By this procedure, the offspring of parent y_i are thinned to become a Poisson process with intensity $\zeta_i(u)p(u) = Mh(u - y_i)\mu(u)/M = h(u - y_i)\mu(u)$. That is, this procedure is equivalent to generating the clusters using intensity $\beta_i(u)$ given above. This trick is used by the simulation functions in spatstat.

This thinning interpretation is the key to understanding the properties of the inhomogeneous cluster process. Since an independent random thinning of a stationary process is correlation-stationary, this means that our inhomogeneous cluster process is correlation-stationary, and has the same pair correlation function as the original stationary cluster process. The inhomogeneous pair correlation function and inhomogeneous K-function of the model are well defined, and take the same form as in the stationary model.

12.3.5 Other cluster processes

This chapter concentrates on Neyman-Scott processes because they are tractable and they represent what a modeller would normally understand by a 'cluster model'.

However, there do exist more general kinds of cluster process. The nomenclature is summarised in Table 12.1. In this more general context, many of the statements we make in this chapter do not hold. We give two examples here.

In the *Gauss-Poisson process*, the parent points are drawn from a Poisson process. For each parent, with probability $1 - p_2$ a single offspring point is produced at exactly the parent location; with probability p_2 two offspring points separated by distance r are produced. In the latter case the line joining the two points has uniform random orientation and passes through the parent location. This model satisfies properties (CLP1)–(CLP3) and (CLP6). According to Table 12.1 this belongs to the class of *Poisson cluster processes*. We hasten to add that this does not mean it is a Poisson process! Rather, it means essentially that the parents are Poisson. Realisations of the Gauss-Poisson process can be generated in spatstat by rGaussPoisson for readers who care to experiment.

The Baddeley-Silverman cell process (page 211) is also technically a kind of cluster process. The points within a tile can be treated as the offspring of a parent lying at the centre of the tile. The patterns in different tiles are independent, so the process satisfies properties (CLP2) and (CLP3), so it is an 'independent cluster process' according to Table 12.1. But as we saw in Chapter 7, this process is not 'clustered' in the conventional sense: it has the same first and second moments as a Poisson process.

12.3.6   Simulating cluster processes in `spatstat`

Table 12.2 lists `spatstat` functions for generating simulated realisations of various Neyman-Scott cluster processes. For example, a realisation of the Matérn cluster process with $\kappa = 5, R = 0.1, \mu = 10$ in the square of side length 2 is generated by

```
> rMatClust(kappa = 5, scale = 0.1, mu = 10, win = square(2))
```

A completely general Neyman-Scott cluster process can be simulated using `rNeymanScott`. The other functions in the table use `rNeymanScott` to perform the actual simulation.

`rNeymanScott`	General Neyman-Scott process
`rMatClust`	Matérn cluster process
`rThomas`	Thomas process
`rVarGamma`	Variance gamma cluster process
`rCauchy`	Cauchy cluster process

Table 12.2. *Functions for simulating Neyman-Scott cluster processes.*

To simulate these processes, some parent points must be generated *outside* the specified window of simulation W, giving rise to offspring that fall *inside* W. The simulation algorithm currently used in `spatstat` simulates parents up to a maximal range R_{max} (called `expand` in `rNeymanScott`) away from the target window. The simulation functions first generate the parents in the dilated window $W_{\oplus R_{max}}$ obtained by extending W by a border region of width R_{max}. They then generate clusters of offspring in this extended window, and finally retain only those offspring that fall inside the original window W. See Figure 12.10.

Normally the user does not need to specify or know the value of R_{max}. All the functions in Table 12.2 handle this automatically except `rNeymanScott`. The user needs to be concerned with R_{max} only when calling `rNeymanScott` directly, or when simulating a model in which the parent points are inhomogeneous. In the latter case, if the parent intensity `kappa` is specified by a pixel image, the domain of this image must be larger than the containing rectangle of W; it must contain the dilated window $W_{\oplus R_{max}}$.

The appropriate value of R_{max} is easy to determine for `rMatClust`. It is clear that each cluster has a well-defined finite range $R_{max} = R$, the radius of the cluster disc. For the other functions listed in Table 12.2 the appropriate value is less obvious since the true range is infinite. In `spatstat` a suitable practical finite range is calculated automatically and used for simulation. For example the Thomas process is given a value of R_{max} equal to 4σ. Parents lying further away from the simulation window have fewer than 0.004 percent of their offspring falling within the simulation window.

The function `clusterradius` calculates R_{max} for standard models. For example

```
> clusterradius("Thomas",scale=1)
[1] 4
> clusterradius("VarGamma",scale=1,nu=2)
[1] 11.53
```

All of the Neyman-Scott simulation functions (listed in Table 12.2) allow the offspring mean parameter `mu` to be given as a `function(x,y)` or a pixel image. This specification determines an inhomogeneous reference density for the clusters. For example:

```
> mu <- as.im(function(x,y){ exp(2 * x + 1) }, owin())
> X <- rMatClust(10, 0.05, mu)
```

The result is shown in Figure 12.19.

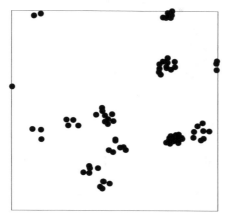

Figure 12.19. *Example of a simulated inhomogeneous Matérn cluster process with reference inten-sity $\mu(x,y) = exp(2x+1)$ in the unit square.*

The function `rNeymanScott` can simulate a general Neyman-Scott process. The parents are Poisson with intensity `kappa`, where `kappa` is given as a numeric value, a function, or a pixel im-age. The cluster mechanism is specified by the argument `rcluster` which can be either a function that generates random point patterns, or a `list(mu, f)` where `mu` is the mean number of off-spring per parent (or the reference intensity), and `f` is a function that generates the random vector displacements of the offspring from their parent. More details are given in the help file. For effi-ciency, the code does not check the validity of the argument `rcluster` if it is a function, so it is currently possible to 'trick' `rNeymanScott` into generating any Poisson cluster process (see Sec-tion 12.3.5). For honesty and clarity of code, Poisson cluster processes should be simulated using `rPoissonCluster`.

12.4 Fitting Cox and cluster models to data

12.4.1 Modelling data as a Cox process or cluster process

Modelling a point pattern as a realisation of a Cox process or Neyman-Scott cluster process involves formulating a model for the intensity, and choosing the type of clusters or driving random field.

Cluster models

A Neyman-Scott cluster process model is sometimes chosen because it explicitly postulates a mech-anism for 'scattering' of points. It was designed for this purpose in astronomy [499] and is often used for modelling seed dispersal in plants. The fitted cluster parameters describe the range of scattering and the productivity of individual parents.

A Neyman-Scott model is also useful for statistical inference when we want to test for the *presence* of clusters. Unlike many other mechanisms for cluster detection, fitting a Neyman-Scott model does not require us to estimate the locations of the individual clusters, which is usually difficult.

One can also adopt a Neyman-Scott model because it exhibits *clustering* or positive dependence between points. The fitted cluster parameters measure the range and strength of clustering, and can be used (for example) to compare the strength of clustering in different point patterns.

While there are many methods for detecting clusters and for measuring the strength of spatial clustering, fitting an explicit point process model has the great advantage that we can easily adjust for spatial inhomogeneity and covariate effects, and still assess statistical significance of clustering in the presence of these effects (and *vice versa.*)

Simulation of a Neyman-Scott model can be used to generate patterns with 'the same kind of clustering' as the data (for example in Monte Carlo methods) or to perform spatial sampling.

Examples of applications of Neyman-Scott cluster process models include [129, 130, 132, 481, 500, 610, 614].

Cox models

A Cox process model is often chosen when we believe that the point pattern is influenced by unobserved covariates. The decision whether to use a Poisson model or Cox model comes down to deciding whether to treat these unobserved influences as fixed or random, respectively. Fixed effects must be estimated; for random effects we need only estimate the variability. In a Cox model, unobserved factors cause spatially varying random fluctuations in the density of points.

By fitting a Cox model we are able to identify and measure different *sources of variability* [360]. This enables us to answer questions like "how much of the variation in the spatial distribution of trees is explained by variation in terrain slope?"

Fitting a Cox model allows us to account for different sources of variability when estimating the effects of given covariates, when calculating the amount of uncertainty, and when assessing the statistical significance of covariate effects.

Examples of application of Cox point process models include [129, 130, 132, 479, 481, 614].

12.4.2   Methods for fitting Cox and cluster models

In Chapter 9 we used the standard technique of *maximum likelihood* to fit Poisson point process models to point pattern data. This is implemented in the spatstat function ppm. Unfortunately this technique is not readily applicable to Cox models and cluster process models, for which the likelihood is intractable [484, sec. 10.3].

Instead, to fit a Cox or cluster process, it is better to exploit the fact that the first and second moments (intensity, pair correlation function, K-function) of the model are known in simple form. A natural way to fit these models to point pattern data is to simply *find the values of the model parameters which give the best match between the corresponding moments of data and model.* For a stationary model, this means that we estimate the intensity in the usual way, and find the values of the cluster parameters or random field parameters to give the best match between the theoretical K-function of the model and the empirical K-function from the data.

This basic principle (essentially the 'method of moments') can be implemented in different ways, according to the definition of the 'best match'. Three popular techniques are sketched below with more details given in Section 12.6.

Method of minimum contrast [544, 231]: This is a kind of nonlinear curve-fitting. We define a measure of discrepancy ('contrast') between the K-function of the model, $K_\theta(r)$, and the estimated K-function from the data, $\widehat{K}(r)$. The discrepancy is something like the area between the two curves or the integrated squared difference between curves. The model parameters θ are chosen to minimise the discrepancy. Alternatively, the pair correlation function could be used instead of the K-function.

Palm likelihood [511, 649, 557]: In effect this fits a Poisson process model to the Fry plot. If the Fry plot is considered as a point process, its intensity is closely related to the pair correlation function of the original point process (see Section 7.9.3). For a Cox or cluster model, the pair correlation function is known, so we effectively know the predicted intensity function of the Fry plot. The Fry plot is approximately a Poisson process [649], so we could fit a Poisson process model *to the Fry plot* and obtain estimates of the parameters of the Cox or cluster process.

Composite likelihood [299]: The K-function of the model predicts the relative frequencies of different distances between pairs of points. After computing the observed distances d_{ij} between all pairs of points, we pretend that these distances are independent and identically distributed random variables, and find the model parameter values which best fit the observed frequencies.

Bayesian inference is very natural for Cox and cluster process models, because the observable point process depends on a hidden random process (the point process of parents or the driving random intensity) in a simple way, allowing us to reverse the relationship and draw Bayesian inferences about the hidden process from the observed process. For Cox processes, Bayesian inference would furnish results such as the (spatially varying) posterior mean of the driving random intensity $\Lambda(u)$, a 95% credible interval for $\Lambda(u)$, and a posterior probability density for the correlation parameter. For log-Gaussian Cox processes, efficient techniques for Bayesian inference include Markov Chain Monte Carlo (MCMC) and Integrated Nested Laplace Approximation (INLA). The contributed R package lgcp supports Bayesian inference for spatial and spatio-temporal LGCP models using MCMC techniques described in [652]. The restricted-licence R package INLA, downloadable from r-inla.org, supports Bayesian inference on a very wide range of spatial statistical problems, as described in [592]; its application to LGCP models is explained in [356, 358]. Overviews and comparisons of the two approaches are given in [652, 567]. Roughly speaking INLA is faster to compute but possibly less accurate unless the right choices are made.

12.4.3 Fitting stationary Cox and cluster models in spatstat

A Cox process or cluster process model can be fitted to point pattern data by the spatstat function kppm. Its usage is very similar to ppm. Examples of its use are given in Section 12.1. The first argument to kppm is a formula in the R language, with exactly the same interpretation as the formula argument for ppm, as discussed in Chapter 9. The left side of the formula gives the point pattern dataset, and the right side specifies the predictors in a model for the log intensity.

The second argument of kppm is the name of the cluster process or Cox process model to be fitted. The currently available options are "Thomas", "MatClust", "Cauchy", "VarGamma", and "LGCP".

```
> fitM <- kppm(redwood ~ 1, "MatClust")
> fitM
Stationary cluster point process model
Uniform intensity:        62
Cluster model: Matern cluster process
Fitted cluster parameters:
   kappa    scale
24.55790  0.08654
Mean cluster size:  2.525 points
```

The resulting object fitM is a fitted model of class "kppm". See Tables 12.3 and 12.4 below for the most commonly used functions for handling "kppm" objects.

The output shown above is generated by print.kppm and shows the estimated intensity of parents $\hat{\kappa} = 24.56$, cluster radius $\hat{R} = 0.087$, and number of offspring per cluster $\hat{\mu} = 2.525$. Note that $\hat{R}$ is referred to as 'scale' in the output. The estimated intensity of the process is $\hat{\kappa}\hat{\mu} = 62$. This is exactly equal to the average intensity of the `redwood` pattern, due to the nature of the estimation algorithm.

It is worth mentioning that although we get exactly the same point estimate of the intensity from the cluster model as we get from fitting a Poisson model, the *inferential* considerations are different. The estimated variance of $\log \lambda$, given by `vcov(fitM)=0.053`, is larger for the cluster model than it is for the Poisson model, given by `vcov(ppm(redwood))=0.0161`. Thus the confidence interval for $\log \lambda$ from the cluster model, given by `confint(fitM)=[3.68, 4.58]`, is wider than that for the corresponding Poisson model, `[3.88, 4.38]`. Of course the fact that the Poisson model yields a narrower confidence interval does not make it a *better* model. Confidence intervals rely on the assumption that the model is true; the `redwood` dataset clearly does not arise from a Poisson process. The wider confidence interval from the cluster model properly reflects the fact that there is greater uncertainty in the estimate of intensity when the data are generated by a cluster process. As a general rule, positive correlation between observations increases uncertainty — intuitively because there are fewer independent observations.

We can similarly fit a Thomas model to the redwood data by means of

```
> fitT <- kppm(redwood ~ 1, "Thomas")
```

The result of `plot(fitT)` is shown in Figure 12.20. It shows the theoretical K-function of the fitted Thomas process (*fit*), a nonparametric estimate of the K-function using Ripley's isotropic correction (*iso*), and the Poisson K-function (*pois*).

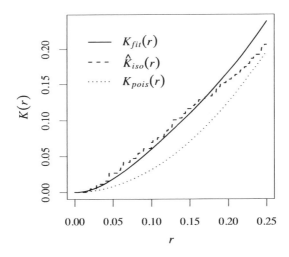

Figure 12.20. *Plot of K-functions corresponding to a fitted Thomas model for the* `redwood` *data.*

To fit an LGCP model (with exponential covariance function) one would use:

```
> fitL <- kppm(redwood ~ 1, "LGCP")
```

To specify a covariance function other than the default exponential covariance, the user should give the name of the covariance function as the argument `model`, and give values for any additional 'shape' parameters of that function. For example the Matérn covariance is specified by `model="matern"` and requires the extra shape parameter `nu`. To identify the covariance function, `spatstat` prefixes the string `model` by `"RM"`, and a function of this name is sought in the `RandomFields` package. For example `model="matern"` specifies the function `RMmatern`. For a list of currently available models see `help(RMmodel, package=RandomFields)`. Extra parameters for a particular covariance function (other than the standard `var` and `scale`) are listed in the help for the model function. For example `help(RMmatern, package=RandomFields)` shows that the parameter nu is required.

In the examples above, the model was fitted using the method of minimum contrast, by matching the K-function of the model to the K-function of the data. To change this default behaviour we can give the argument `statistic="pcf"` which specifies that the model should be fitted by matching the pair correlation function; or give the argument `method="palm"` or `method="clik2"` to specify the use of Palm likelihood or second-order composite likelihood respectively. For example:

```
> fitTp <- kppm(redwood ~1, "Thomas", statistic="pcf")
> fitTp
Stationary cluster point process model
Uniform intensity:         62
Cluster model: Thomas process
Fitted cluster parameters:
    kappa      scale
25.29263    0.03968
Mean cluster size:   2.451 points
```

12.4.4 Fitting inhomogeneous Cox and cluster models in `spatstat`

Specifying a model

Inhomogeneous Cox and cluster process models can also be fitted to point pattern data using kppm. For example we can fit an inhomogeneous Thomas cluster model to the *Beilschmiedia* data as follows:

```
> fitBeiThom <- kppm(bei ~ elev + grad, "Thomas", data=bei.extra)
> fitBeiThom
Inhomogeneous cluster point process model
Log intensity:   ~elev + grad
Fitted trend coefficients:
(Intercept)         elev         grad
   -8.55862      0.02141      5.84104
Cluster model: Thomas process
Fitted cluster parameters:
    kappa      scale
5.024e-05  2.738e+01
Mean cluster size:   [pixel image]
```

Note that the formula given to kppm always specifies the form of the logarithm of the **intensity**, just as it does in ppm. The intensity model is of the loglinear form

$$\log \lambda (u) = B(u) + \theta^\top \mathbf{Z}(u) \tag{12.14}$$

where $B(u)$ is an optional offset or baseline function, θ is a parameter vector, and $\mathbf{Z}(u)$ is a known vector-valued covariate function. Both $B(u)$ and $\mathbf{Z}(u)$ are determined by the terms on the right-hand side of the model formula.

For a cluster process, the parent points are assumed to be a homogeneous Poisson process with intensity κ. The reference intensity (offspring mean) for the cluster process is therefore $\mu(u) = \lambda(u)/\kappa$ so that

$$\log \mu (u) = - \log \kappa + B(u) + \theta^\top \mathbf{Z}(u). \tag{12.15}$$

For a log-Gaussian Cox process, we need to remember that $\lambda(u) = \mathbb{E}\Lambda(u) = \mathbb{E}\exp(G(u))$ is equal to $e^{\mu(u)+\sigma^2/2}$, where $\mu(u)$ is now the mean function of the Gaussian random field and $\sigma^2 = C(0)$ is its variance. Therefore the mean function of the Gaussian random field is not $\log \lambda(u)$ but

$$\mu(u) = -\frac{\sigma^2}{2} + B(u) + \boldsymbol{\theta}^\top \mathbf{Z}(u). \tag{12.16}$$

Notice the absence of log on the left side.

Note that inhomogeneous models of this kind are 'correlation-stationary' as defined in Section 7.10.2. The formal test for correlation-stationarity described in Section 16.8.5 could be used to decide whether such models are appropriate for the data.

Fitting procedure

In normal usage, kppm fits the model by a two-step procedure: first the intensity is fitted, and then the cluster parameters or correlation parameters. In the first step, a Poisson model with the same model formula is fitted to the point pattern data using ppm. This determines the estimates of the coefficients of all terms in the model formula. The fitted intensity is computed. In the second step, the fitted intensity is treated as the true intensity, and the cluster parameters or correlation parameters are estimated by one of the methods described above. For example in the minimum contrast method, the empirical (estimated) inhomogeneous K-function is matched to the theoretical K-function of the corresponding *homogeneous* model. See the last paragraph of Section 12.3.4.2 for the explanation of why this makes sense. See also [678].

In the model fitBeiThom fitted above, kppm first estimates the intensity by fitting the *Poisson* model:

```
> ppm(bei ~ elev + grad, data=bei.extra)
```

Then predict.ppm is used to compute the predicted intensity at the data points, and the results are passed to Kinhom to calculate the inhomogeneous K-function. The parameters of the Thomas process are estimated by matching this estimated inhomogeneous K-function to the theoretical K-function of the Thomas process, by minimum contrast. Just as with ppm the intensity is specified on the log-scale, so the foregoing call to kppm stipulates that the reference intensity of offspring is of the form

$$\mu(x,y) = \exp(\beta_0 + \beta_1 E(u) + \beta_2 S(u))$$

where $\beta_0, \beta_1, \beta_2$ are parameters to be estimated, $E(u)$ is the terrain elevation in metres at location u, and $S(u)$ is the terrain slope, a dimensionless gradient. This is exactly the same expression as for the intensity of the Poisson model (9.16) in Chapter 9.

Because of the two-step fitting procedure described above, the estimates of the coefficients of the model formula returned by kppm are exactly the same as those that are returned by ppm. (This is true unless the improve.type — see below — argument of kppm is changed from its default value of none.) However, since clustering is allowed for (instead of the points being assumed to be independent), the uncertainty of the coefficient estimates is larger than it would be for a Poisson model. The confidence intervals for the coefficients are correspondingly wider (and more trustworthy, inasmuch as the cluster model is almost certainly more appropriate for these data than is the Poisson model). For the model fitBeiThom fitted above, the 95% confidence intervals for the coefficients are

```
> confint(fitBeiThom)
              2.5 % 97.5 %
(Intercept) -15.37  -1.75
elev         -0.02   0.07
grad          0.24  11.44
```

whereas when a Poisson model is fitted they are:

```
> confint(ppm(bei ~ elev + grad, data=bei.extra))
            2.5 % 97.5 %
(Intercept) -9.23  -7.89
elev         0.02   0.03
grad         5.34   6.34
```

The choice of fitting method for the second step is controlled by the argument `method`: options are `"mincon"`, `"palm"`, and `"clik2"` for minimum contrast, Palm likelihood, and composite likelihood, respectively. The default is the minimum contrast method.

More flexible models of the intensity, and greater control over the intensity fitting procedure, are possible using the arguments `covfunargs`, `use.gam`, `nd`, `eps` passed to ppm.

Several auxiliary functions perform individual steps in this fitting procedure. The function `clusterfit` performs the second stage of the fitting procedure, estimating the cluster parameters or correlation parameters. It only requires a K-function, which can be useful when we have estimated the K-function by some unusual means, for example, when we have pooled the K-function estimates from several point patterns.

The very low-level function `mincontrast` performs minimum-contrast fitting, and could be used to fit models which are not currently supported in `spatstat`.

Improving the intensity estimate

The procedure described above is statistically inefficient because the intensity is fitted using a Poisson likelihood (or 'first order composite likelihood' [603]), which ignores the dependence between points. Better fitting techniques have been developed [492, 304, 303].

In advanced usage, kppm can deploy these methods to improve the estimate of the intensity, by using the second moment information to calculate optimal weights for the data points and dummy points of the quadrature scheme (see Section 2.4.4 and Section 9.8.1) and re-fitting the intensity model using these weights. This can substantially reduce the mean squared error of estimation of the intensity. The arguments `improve.type` and `improve.args` to kppm control this procedure. Alternatively the auxiliary function `improve.kppm` can be used to apply the improvement procedure to a fitted model.

There are three improvement methods: `clik1`, first order composite likelihood; `wclik1`, a weighted version of the previous method [304]; and `quasi`, a quasi-likelihood method [303]. The method `clik1` is included for testing purposes and is not recommended for normal use. Methods `wclik1` and `quasi` incorporate the dependence structure between the points in the estimation procedure. This is done by making use of the fitted pair correlation function in the estimating equation.

This estimating equation is constructed by counting the number of points in each pixel of a digital mask representation of the observation window. The resolution of this representation is controlled by the argument `dimyx` which is passed to the function `improve.kppm` and thence to `as.mask`. Finer resolutions yield more accurate estimates, but the required computational time grows with the *cube* of the number of pixels. Consequently fine grids may result in very long computation times and very large memory usage.

12.4.5   Operations on fitted models

The result of kppm is a fitted model of class `"kppm"`. Table 12.3 lists some of the standard operations on fitted models in R which can be applied to Cox and cluster point process models. That is, these generic operations have methods for the class `"kppm"`. For information on these methods, consult the help for `print.kppm`, `summary.kppm`, `plot.kppm`, etc.

Important omissions from this table, compared to the corresponding Table 9.3 for "ppm", are the generics `anova`, `influence`, `leverage`, and `dfbetas`. At the time of writing, the theory underpinning these operations is not yet developed.

`print`	print basic information
`summary`	print detailed summary information
`plot`	plot the empirical and fitted summary functions
`predict`	compute the fitted intensity
`simulate`	generate simulated realisations of model
`update`	re-fit the model
`coef`	extract the fitted coefficient vector $\widehat{\theta}$
`vcov`	variance-covariance matrix of $\widehat{\theta}$
`formula`	extract the model formula
`terms`	extract the terms in the model
`fitted`	compute the fitted intensity at data points
`residuals`	compute residuals
`model.matrix`	compute the design matrix
`AIC, extractAIC`	compute Akaike Information Criterion

Table 12.3. *Standard* R *generic operations which have methods for Cox and cluster point process models (class* "kppm"*).*

Table 12.4 lists other functions defined in `spatstat` for "kppm" objects.

The functions `model.matrix` and `model.images` compute the *canonical* covariates, that is, the covariates which appear in loglinear form in the intensity model (9.6). To relate these to the *original* covariates supplied in the call to kppm, one can use the functions `model.depends`, `model.covariates`, `model.is.additive`, and `has.offset` described on page 339.

`as.fv`	extract summary function of data and model
`as.owin, Window, domain`	extract window of data or covariates
`as.ppm`	extract Poisson model for intensity
`parameters`	extract all model parameters
`clusterfield`	compute driving intensity given parents
`clusterkernel`	offspring density
`clusterradius`	cluster radius (for simulation)
`confint`	confidence intervals for parameters
`envelope`	envelopes based on simulating fitted model
`is.kppm`	check whether object is a "kppm"
`is.stationary`	check whether model is stationary
`Kmodel`	*K*-function of fitted model
`pcfmodel`	pair correlation function of fitted model
`model.images`	images of canonical covariates
`objsurf`	objective function surface
`unitname`	name of unit of distance

Table 12.4. *Additional functions for* "kppm" *objects. All except* `confint` *are generics defined in* `spatstat`.

Note that `vcov.kppm` calculates the variance-covariance matrix for the estimates of the *trend parameters only*, that is, for the parameters θ in the loglinear intensity model (12.14). It is harder to quantify uncertainty in the parameters of the correlation function of the Cox process or the offspring distribution of the cluster process. Theory is given in [680] but the asymptotic variance-covariance matrices are not easy to compute. They could be estimated by a parametric bootstrap based on simulations of the fitted model. Another trick is to calculate the objective function (the function that was minimised to obtain the parameter estimates) for parameter values close to the optimal values. The spatstat function objsurf calculates these values and produces a plottable result:

```
> fit <- kppm(redwood ~ 1, "Thomas")
> contour(objsurf(fit))
```

The result, shown in Figure 12.21, suggests that there is strong correlation between the estimates of parent intensity κ and cluster scale σ.

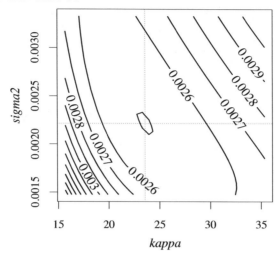

Figure 12.21. *Objective function surface for the cluster parameters of the Thomas model fitted to the* redwood *data.*

The spatstat generic function parameters can be used to extract all the model parameters, including the trend coefficients and the cluster parameters. The result of parameters is a named list which can usually be compressed by unlist:

```
> unlist(parameters(fit))
   trend    kappa    scale       mu
62.00000 23.55114  0.04705  2.63257
```

12.4.6 Simulating fitted models

A model can be simulated directly from the fitted model object returned by kppm.

```
> fit <- kppm(redwood ~ 1, "Thomas")
> plot(simulate(fit, nsim=4))
```

The command simulate is generic; here it dispatches to the method simulate.kppm. This function handles all the gory details of extracting the relevant parameters from the fitted model object and then calling the appropriate simulation function (in this case rThomas) with the extracted parameters supplied appropriately as arguments.

Monte Carlo tests of a fitted Cox or cluster model can be performed using envelope.kppm, dclf.test, or mad.test. For example, Figure 12.22 displays the envelope of L-functions from 39 simulations from the Thomas model fitted to the redwood data. It was computed by

```
> envLT <- envelope(fit, Lest, nsim=39)
```

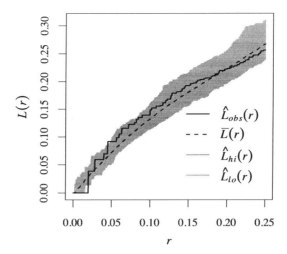

Figure 12.22. *L-function envelopes for fit of Thomas model to the* redwood *data.*

It suggests no obvious inadequacies in the fit.

Note that when a simulation envelope or Monte Carlo test is based on an inhomogeneous summary function such as Kinhom and the simulations are drawn from a fitted model, the default behaviour of envelope.kppm, dclf.test, or mad.test is to re-fit the model to each simulated pattern before the fitted intensity is computed and passed to Kinhom. This avoids a substantial bias in the test. If you like bias, this behaviour can be disabled by setting update=FALSE.

12.4.7 Clustering strength, and nonsensical parameter values

The clustering in a cluster process or Cox process is controlled by the cluster parameters or correlation parameters ψ. For example, for the stationary Matérn cluster process, $\psi = (\mu, R)$ controls the strength and range of clustering.

The fitted parameters $\widehat{\psi}$ may sometimes be physically nonsensical, very large or very small numbers. This occurs frequently when the clustering in the data is weak.

```
> X <- rpoispp(500)
> kppm(X ~ 1, "Thomas")
Stationary cluster point process model
Uniform intensity:        509
Cluster model: Thomas process
Fitted cluster parameters:
 kappa  scale
7506.6  197.1
Mean cluster size:  0.06781 points
> kppm(X ~ 1, "LGCP")
Stationary Cox point process model
Uniform intensity:        509
Cox model: log-Gaussian Cox process
       Covariance model: exponential
Fitted covariance parameters:
      var       scale
```

```
7.752e-01 1.429e-08
Fitted mean of log of random intensity: 5.845
```

A cluster process with an average of 0.0678 points per cluster spread over a scale of 197 units, and a Cox process with a tiny correlation range of 1.429e-08 units, are not physically meaningful.

This anomaly is caused by a quirk of the parametrisation: the Poisson process does not correspond to a particular finite value of ψ. A Matérn cluster process in the unit square with huge cluster radius is effectively a Poisson process. A log-Gaussian Cox process with tiny correlation range is also effectively a Poisson process. In theory, the Poisson process is obtained in the asymptotic limit as the cluster radius increases to infinity, or as the correlation scale decreases to zero, or as the parent intensity κ increases with the process intensity λ held fixed.

A strategy proposed in [33] is to measure the 'strength' of clustering by $p = s(0)$ where

$$s(r) = \frac{g(r) - 1}{g(r)} = 1 - \frac{1}{g(r)} \tag{12.17}$$

where $g(r)$ is the pair correlation function. For a cluster process, $s(r)$ is the probability that two given points, separated by a distance r, are members of the same cluster ('siblings'). The value $p = s(0)$ is the probability that two very close points are siblings. For example, the Thomas process has sibling probability $p = 1/(1 + 4\pi\kappa\sigma^2)$. In the example above, the sibling probability is 2.73e-10, so small that the fitted model is effectively Poisson.

A more far-reaching problem is that — unlike loglinear Poisson point process models, which can be fitted to *any* point pattern dataset — a Cox model or Neyman-Scott model *cannot always be fitted* in a sensible way to any point pattern. All Cox and Neyman-Scott models are overdispersed, so a point pattern dataset which exhibits underdispersion (regularity or inhibition) lies outside the range of data that can sensibly be fitted. Particular Cox models and Neyman-Scott models assume a particular form for the second moments: for example the K-function of a Matérn cluster process has a particular shape. A point pattern whose K-function looks nothing like the assumed shape cannot be fitted adequately. In these situations, the fitting algorithm in kppm may fail to converge, or may converge happily to a nonsensical result. The three choices of fitting method may give different, bizarre results.

All three methods used by kppm are 'minimization' estimators of irregular parameters (i.e. parameters that do not appear in the model in 'exponential family form'). Such estimators usually perform well if the model is true, but can perform poorly in other settings. The objective function that is being minimized may have several local minimum points, and the iterative minimization algorithm may give different results depending on the initial parameter estimate.

The moral of the story is that one should not attempt to fit a Cox or cluster model to a dataset unless you have *some* reason to believe that such a model is appropriate. As always it is highly advisable to follow the cyclic modelling paradigm of 'explore the data; model the data; criticize the fit of the model; repeat'. In criticizing the model, keep a lookout for extreme or anomalous estimates of parameters.

12.5 · Locally fitted models

Local likelihood or 'geographically weighted regression' methods for Poisson process models were described in Section 9.13. The same idea can be extended to Cox and cluster process models [33], at least for the Palm likelihood method.

A simple version of this idea is that, at each spatial location u, we consider only the data inside the disc $b(u, R)$ where R is a fixed radius. We fit a Cox or cluster model to these data and record the fitted model parameters $\widehat{\theta}(u)$. More generally the disc is replaced by a smoothing kernel so that the model fitted at location u involves contributions from all of the original observations, with spatially-dependent weights that favour observations close to u.

This is implemented in the function `lockppm` which will shortly be included in `spatstat`. For the full redwood data (Figure 8.3 on page 260):

```
> redTom <- lockppm(redwoodfull ~ 1, "Thomas", sigma=bw.lockppm)
```

The smoothing bandwidth σ was chosen by cross-validation. Figure 12.23 shows the cluster strength index p for each of the locally fitted models, computed by `psib(redTom)`. It indicates strongly that the left side of the dataset is clustered while the right side is not. This procedure has automatically identified the boundary between the two spatial patterns sketched in Figure 8.3; knowledge of that boundary was not used in the fitting procedure.

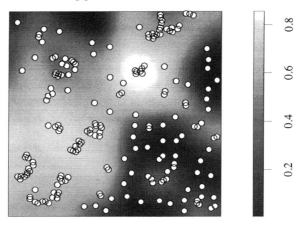

Figure 12.23. *Cluster strength index p for the local composite likelihood fits of the Thomas model to the full redwood data.*

12.6   Theory∗

12.6.1   Likelihood

First we explain why the method of maximum likelihood, used successfully for Poisson models in Chapter 9, is not easily applicable to Cox process models.

Let $\Lambda(u)$, $u \in W$ be a random field such that $\int_W \Lambda(u)\,du$ is finite with probability 1. Conditional on the realisation of Λ, let $\mathbf{Y}$ be a Poisson point process with intensity function Λ. Then $\mathbf{Y}$ is a Cox process and the probability density is

$$f(\mathbf{y}) = \mathbb{E}\left[\exp\left(-\int_W \Lambda(u)\,du\right) \prod_i \Lambda(y_i)\right]. \tag{12.18}$$

This is typically not expressible in simple form so maximum likelihood methods are difficult to apply.

∗ Starred sections contain advanced material, and can be skipped by most readers.

12.6.2   Pair correlation functions and *K*-functions

General formula for pair correlation of cluster models

The pair correlation function for a stationary Neyman-Scott Cox process can easily be calculated. Given the parent points y_i, the process is Poisson with intensity $\beta(u) = \mu \sum_i h(u - y_i)$. The second moment intensity of this Poisson process is $\Lambda_2(u, v) = \beta(u)\beta(v)$. Taking the expectation over the positions of the parent points, the second moment intensity of the cluster process is $\lambda_2(u, v) = \mathbb{E}\Lambda_2(u, v) = \mathbb{E}[\beta(u)\beta(v)]$. To evaluate this expectation we expand out the product:

$$\beta(u)\beta(v) = \left(\mu \sum_i h(u - y_i) \right) \left(\mu \sum_j h(v - y_i) \right)$$
$$= \mu^2 \sum_i \sum_{j \neq i} h(u - y_i)h(v - y_j) + \mu^2 \sum_i h(u - y_i)h(v - y_i). \tag{12.19}$$

The sum on the far right of (12.19) is a sum over all parent points y_i. Applying Campbell's formula (6.11), its expectation over the positions of the parent points is

$$\mathbb{E} \sum_i h(u - y_i)h(v - y_i) = \kappa \int h(u - x)h(v - x)\,\mathrm{d}x.$$

The first sum on the right-hand side of (12.19) is a sum over all distinct *pairs* of parent points. Applying the second order Campbell formula (7.44), its expectation is

$$\mathbb{E} \sum_i \sum_{j \neq i} h(u - y_i)h(v - y_j) = \int \int h(u - x)h(v - y)\kappa^2 \,\mathrm{d}x\,\mathrm{d}y = \kappa^2.$$

Consequently the second moment intensity of the cluster process is

$$\lambda_2(u, v) = \mu^2 \kappa^2 + \mu^2 \kappa G(u, v)$$

where

$$G(u, v) = \int h(u - x)h(v - x)\,\mathrm{d}x. \tag{12.20}$$

Dividing by $\lambda^2 = \mu^2 \kappa^2$, the pair correlation is

$$g(u, v) = 1 + \frac{1}{\kappa} G(u, v). \tag{12.21}$$

Intuitively we can interpret $h(u - x)h(v - x)$ as the joint probability [density] that a parent at location x would give rise to offspring at both locations u and v. Summing this probability over all possible locations of the parent x gives the probability that two offspring of the same parent will be found at the locations u and v.

Pair correlation of standard cluster processes

For the stationary Matérn process with parameters $\theta = (\mu, \kappa, R)$ the pair correlation function is, from (12.21),

$$g_\psi(r) = 1 + \frac{1}{\pi^2 R^2 \kappa} a \left(\frac{r}{2R} \right) \tag{12.22}$$

where $\psi = (\kappa, R)$ and

$$a(z) = 2 \left[\cos^{-1} z - z\sqrt{1 - z^2} \right]$$

for $z \leq 1$, and $a(z) = 0$ for $z > 1$. Here $a(z)$ is the area of overlap between two unit discs whose centres are separated by a distance of $2z$. The K-function is

$$K_\psi(r) = \pi r^2 + \frac{1}{\kappa} f\left(\frac{r}{2R}\right) \quad (12.23)$$

where

$$f(z) = 2 + \frac{1}{\pi}[(8z^2 - 4)\arccos(z) - 2\arcsin(z) + 4z\sqrt{(1-z^2)^3} - 6z\sqrt{1-z^2}]$$

for $z \leq 1$, and $f(z) = 1$ for $z > 1$.

For the stationary Thomas process with parameters $\theta = (\mu, \kappa, \sigma)$ the pair correlation function is

$$g_\psi(r) = 1 + \frac{1}{4\pi\kappa\sigma^2} \exp\left(-\frac{r^2}{4\sigma^2}\right) \quad (12.24)$$

where $\psi = (\kappa, \sigma)$. The K-function is

$$K_\psi(r) = \pi r^2 + \frac{1}{\kappa}\left\{1 - \exp\left(-\frac{r^2}{4\sigma^2}\right)\right\}. \quad (12.25)$$

Analytic expressions for the pair correlation function of the Cauchy and the variance-gamma cluster models are given in [360].

Pair correlation of log-Gaussian Cox processes

For the log-Gaussian Cox process the pair correlation function and the K-function depend, of course, on the particular covariance model that is used. However, the dependence is remarkably simple: as we saw in Section 12.2.6, the pair correlation function is just the exponential of the covariance function of the underlying Gaussian random field. We are focussing exclusively on the setting in which the covariance is isotropic whence we may write $g(r) = \exp(C_0(r))$. Using (7.25) on page 227,

$$K(r) = \int_0^r 2\pi s g(s)\,\mathrm{d}s = \int_0^r 2\pi s \exp(C_0(s))\,\mathrm{d}s.$$

The integration must be carried out numerically; this is done in `spatstat` using either the function `integrate` or a homegrown implementation of Simpson's rule.

Recall that $C_0(r) = \sigma^2 R(r/\alpha)$ where R is a known 'template' covariance function and where σ^2 and α are the variance and scale parameters. The parameter vector to be estimated here is $\psi = (\sigma^2, \alpha)$. In particular for the default exponential model the pair correlation function is

$$g_\psi(r) = \exp(\sigma^2 \exp(-r/\alpha)) \quad (12.26)$$

and the K-function is

$$K_\psi(r) = \int_0^r 2\pi s \exp(\sigma^2 \exp(-s/\alpha))\,\mathrm{d}s. \quad (12.27)$$

It may be of interest to estimate the mean of the underlying Gaussian random field for an LGCP model. This mean is not directly estimated via the method of minimum contrast; rather we apply the fact that

$$\lambda(u) = \exp\left(\mu(u) + \frac{C_0(0)}{2}\right)$$

where $\mu(u) = \mathbb{E}(G(u))$, $G(u)$ being the underlying Gaussian random field, and where $\lambda(u) = \mathbb{E}(\Lambda(u))$ is the intensity of the process. Here we are focussing on the homogeneous setting so $\mu(u) = \mu$ and $\lambda(u) = \lambda$ are constants. We can then estimate λ using the unbiased estimator $\overline{\lambda}$, and then estimate μ via $\hat{\mu} = \log\overline{\lambda} - \hat{C}_0(0)/2$.

> WARNING: There is an unfortunate notational conflict over the symbol μ, which is used to denote two different quantities: the mean number of points per cluster in a cluster process, and the mean of the log intensity in a log-Gaussian Cox process.

Two-step estimation

In applications it is usual to work with models in which the intensity and interaction parameters are 'separable' in the sense that

$$\lambda_2(u, v; \theta) = \lambda(u; \varphi)\lambda(v; \varphi)g_2(u, v; \psi) \tag{12.28}$$

where $\theta = (\varphi, \psi)$ and where $g_2(\ldots)$ is the (general) pair correlation function. That is, the parameter vector θ may be separated into two components φ and ψ where the intensity $\lambda()$ depends only on φ and the pair correlation function depends only on ψ.

In a separable model we can use a *two-step estimation* procedure [680], first estimating the intensity $\lambda(u; \varphi)$ via maximising a Poisson likelihood and then estimating the interaction parameters ψ by other means.

Examples include log-Gaussian Cox processes and certain inhomogeneous Neyman-Scott processes [678, 680]. Consider a stationary Thomas cluster process with parent intensity κ, a mean number μ of offspring, and scale (standard deviation) parameter σ. This process has intensity $\lambda = \kappa\mu$ and pair correlation function

$$g_2(u, v) = 1 + \frac{1}{4\pi\kappa\sigma^2} \exp\left(-\frac{\|u - v\|^2}{4\sigma^2}\right).$$

Taking $\varphi = \lambda = \kappa\mu$ and $\psi = (\kappa, \sigma)$ we get $g_2(u, v; \psi) = 1 + (4\pi\kappa\sigma^2)^{-1}\exp(-\|u - v\|^2/(4\sigma^2))$ so that the model is separable.

12.6.3 Minimum contrast estimation

Formulae for the first and second moments (intensity, pair correlation function, and K-function) of a Cox or cluster process were derived above. These formulae can be used to fit Cox and cluster process models to data, effectively by matching the moments of the model to the corresponding moments of the data. This is an application of the classical principle called the *method of moments* which has some theoretical justification. However, there are some peculiarities in this context.

In the case of constant intensity, we fit our model to a point pattern dataset **x** by matching the model intensity λ_φ to the estimated intensity $\overline{\lambda}$, and the model K-function $K_\psi(r)$ to the empirical K-function, $\widehat{K}(r)$. For the intensity we can of course achieve an exact match $\lambda_\varphi = \overline{\lambda}$, but this is impossible for the K-functions; no choice of parameters would make the theoretical K-function exactly equal to the estimate. In particular, $K_\psi(r)$ is a smooth function while $\widehat{K}(r)$ has discrete jumps.

In practice the cluster parameters ψ are chosen by a kind of *curve-fitting*. The 'data' are the values of the empirical K-function. The task is to find the curve of best fit, that is, a curve of the form $K_\psi(r)$ which has the best agreement with the empirical data. A popular approach would be to fit the curve by nonlinear least squares, where we find the value of ψ that minimises the sum of squared errors

$$D(\psi) = \sum_{i=1}^{N} (K_\psi(r_i) - \widehat{K}(r_i))^2 \tag{12.29}$$

where $r_1, \ldots, r_N$ is a sequence of distance values. That is, the value of ψ which achieves the minimum of $D(\psi)$ would become our estimate $\widehat{\psi}$. This ensures that the fitted curve $K_{\widehat{\psi}}(r)$ 'goes through' the empirical curve $\widehat{K}(r)$, rather than exactly matching it. In the present context, D is called the *contrast* between the two curves; estimating ψ by minimising D is the *method of minimum contrast*.

Taking a finely spaced sequence of distances would effectively replace the sum (12.29) by the integral

$$D(\psi) = \int_a^b (K_\psi(r) - \widehat{K}(r))^2 \, \mathrm{d}r \tag{12.30}$$

where a and b are the limits of the range of distances considered. This measures the separation between the two curves $K_\psi(r)$ and $\widehat{K}(r)$ over the distance interval $[a,b]$. The squared error in (12.29)–(12.30) could be replaced by the absolute error

$$D(\psi) = \int_a^b \left| K_\psi(r) - \widehat{K}(r) \right| \, \mathrm{d}r$$

which is the total *area* between the two curves. We could also pre-transform the K-function, for example by taking the square root to convert it to the L-function, and in general we could raise the error to any chosen power, so the contrast could be

$$D(\psi) = \int_a^b \left| \widehat{K}(r)^q - K_\psi(r)^q \right|^p \, \mathrm{d}r \tag{12.31}$$

where $0 \le a < b$, and where $p, q > 0$ are indices. This method was developed by Pfanzagl [544] in a general statistical context, and independently developed for spatial point processes by Diggle *et al.* [231]. See [484, sec. 10.1]. In the `spatstat` implementation the user may choose the interval $[a,b]$ and the indices p and q but there are sensible defaults.

Minimum contrast can be extended to other point process models for which an analytic formula for $K_\psi(r)$ is not available [231] by using Monte Carlo simulation to estimate $K_\psi(r)$ for any given parameter values ψ. The contrast $D(\psi)$ could then be estimated. This is of course computationally expensive, since the simulation procedure has to be repeated for each candidate value of the parameters.

There are advantages and disadvantages to using the pair correlation function rather than the K-function. An advantage is that the theoretical pair correlation function is somewhat easier to calculate than is the theoretical K-function. This is especially true for the LGCP model as we have explained above. A disadvantage is that the pair correlation function is harder to estimate from data than is the K-function and the quality of estimates of the pair correlation function may be lower.

12.6.4 Palm likelihood

Stationary case

Another approach can be motivated by the Fry plot (Section 7.2.2). Considered as a point process in its own right, the Fry plot has an intensity, which is determined by the first and second moments of the original point process and by the geometry of the observation window. If $\mathbf{X}$ is the original point process, assumed to be stationary, with second moment intensity $\lambda_2(u,v) = \lambda_2(u-v)$ then the expected number of points *of the Fry plot* falling in a region B is

$$\int_W \int_W \mathbf{1}\{u - v \in B\} \lambda_2(u-v) \, \mathrm{d}u \, \mathrm{d}v = \int_B \lambda_2(x) |W \cap (W-x)| \, \mathrm{d}x$$

where W is the observation window for $\mathbf{X}$. So the Fry plot has intensity function

$$\beta(u) = \lambda_2(u) |W \cap (W-u)|. \tag{12.32}$$

For sufficiently large samples, the Fry plot is approximately a Poisson process except in regions very close to the origin or very close to the border of the Fry plot [649]. Consequently, we could fit a stationary Cox or cluster process model to point pattern data by fitting the Poisson process model with intensity (12.32) *to the Fry points*.

An almost equivalent approach is to consider, for each data point x_i, the pattern of Fry points $v_{ij} = x_j - x_i$ representing the vectors joining x_i to other data points x_j. The intensity function of this pattern is $\gamma(u) = \beta(u)/\lambda = \lambda_P(u)|W \cap (W-u)|$ where $\lambda_P(u) = \lambda_2(u,0)/\lambda$ is the *Palm intensity* (intensity of the Palm distribution) of the stationary point process $\mathbf{X}$ given there is a point of $\mathbf{X}$ at the origin.

Ogata and Katsura [511] defined the *Palm likelihood* $L_P(\theta)$, for a stationary point $\mathbf{X}$ governed by a parameter vector θ, by

$$\log L_P(\theta) = \sum_i \sum_{j \neq i} Q(x_i - x_j) \log \lambda_P(x_j - x_i; \theta) - n(\mathbf{X} \cap W) \int_{\mathbb{R}^2} Q(u) \lambda_P(u; \theta) \, du \qquad (12.33)$$

where W is the observation window, the double sum is over all pairs of distinct points $x_i, x_j \in \mathbf{X} \cap W$, and typically $Q(u) = \mathbf{1}\{\|u\| \leq R\}$ for a fixed distance $R > 0$. These ideas have been further studied in [649] and [557].

12.6.4.1 Non-stationary processes

We now extend the idea of the Palm likelihood to non-stationary processes [33, 555]. Suppose that $\mathbf{X}$ is not (necessarily) stationary and assume that $\mathbf{X}$ has intensity function $\lambda(u; \theta)$ and second moment intensity $\lambda_2(u, v; \theta)$. The Palm distribution of $\mathbf{X}$ given a point at $v \in \mathbb{R}^2$ then has intensity

$$\lambda_P(u \mid v; \theta) = \frac{\lambda_2(u, v; \theta)}{\lambda(v; \theta)}.$$

We propose the following generalisation of (12.33):

$$\log L_P(\theta) = \sum_i \sum_{j \neq i} Q(x_i, x_j) \log \lambda_P(x_j \mid x_i; \theta) - \sum_i \int_W Q(x_i, u) \lambda_P(u \mid x_i; \theta) \, du \qquad (12.34)$$

where typically $Q(u, v) = \mathbf{1}\{\|u - v\| \leq R\}$. If $\mathbf{X}$ is stationary, then $\lambda_P(v \mid u; \theta) = \lambda_0(v - u; \theta)$ so that (12.34) reduces to (12.33).

12.6.5 Second-order composite likelihood

For fitting clustered point process models, Guan [299] proposed a composite loglikelihood based on knowledge of the second moment intensity $\lambda_2(u, v; \theta)$ (see equation (7.45) in Section 10.3):

$$\log \mathrm{CL}(\theta) = \sum_i \sum_{j \neq i} w(x_i, x_j) \left[\log \lambda_2(x_i, x_j; \theta) - \log \int_W \int_W w(u, v) \lambda_2(u, v; \theta) \, du \, dv \right] \qquad (12.35)$$

where w is a weight function, designed to simplify computation and optimize statistical properties, typically taken to be $w(u, v) = \mathbf{1}\{\|u - v\| \leq R\}$ where $R > 0$ is an upper limit on the correlation distance of the model. The corresponding composite score function is

$$\frac{\partial}{\partial \theta} \log \mathrm{CL}(\theta) = \sum_i \sum_{j \neq i} w(x_i, x_j) \frac{\kappa_2(x_i, x_j; \theta)}{\lambda_2(x_i, x_j; \theta)} - \sum_i \sum_j w(x_i, x_j) \frac{\langle \kappa_2, w \rangle}{\langle \lambda_2, w \rangle} \qquad (12.36)$$

where $\kappa_2(u, v; \theta) = \partial/(\partial \theta) \lambda_2(u, v; \theta)$ and where

$$\langle f, w \rangle = \int_W \int_W w(u, v) f(u, v; \theta) \, du \, dv$$

for any integrand $f(u, v; \theta)$. Using the second-order Campbell formula (7.44) on page 242, it is easy to show that (12.36) is an unbiased estimating function for any point process with the correctly specified second moment intensity λ_2.

The second-order composite likelihood method can be applied in `spatstat` by calling `kppm` with the `method` argument specified to be `clik2`. For example:

```
> fit <- kppm(redwood ~ 1, "Thomas", method="clik2")
```

12.7 FAQs

- *Why are Cox processes and cluster processes fitted using the same command* `kppm`*?*

  Cox and cluster models are amenable to the same kind of analysis, and in fact the two classes of models overlap considerably. They are both fitted by matching the first and second moments of the model to the estimated first and second moments of the data pattern.

- *What is the essential difference between Cox processes and cluster processes?*

  Cox and cluster processes are different ways to model a point pattern which is influenced by unobserved random processes. The main distinction in their behaviour is that a cluster process always has 'hot spots against a cold background', that is, regions of high intensity around a parent point, against a background where points are absent; while a Cox process often has a mixture of 'hot spots' and 'cool spots' arising from fluctuations in the driving intensity.

- *I get different results when I fit the same model to my data by different methods (minimum contrast, Palm likelihood, second-order composite likelihood).*

  This usually means that the data are not clustered, or the evidence for clustering is weak (e.g., there are not enough data points). See Section 12.4.7.

- *When I try to simulate the fitted cluster model, I get errors or warnings about trying to simulate a large number of points.*

  This means that the estimate of the parent intensity κ, or the mean number of offspring per parent μ, is ridiculously large. In turn this usually means that the original data were not clustered.

- *Is it possible to generate point patterns according to a cluster process or Cox process but with a fixed number of points?*

  This is possible, but not straightforward. It involves specialised algorithms for 'conditional simulation' [484, sec. 10.2] which are not currently implemented in `spatstat`.

  If the intention is to hold the number of points fixed, so as to avoid problems with the conservatism of a Monte Carlo test (Section 10.6.3), note that this strategy is only valid for Poisson models, not for Cox or cluster models.

- *Are there tools for model validation for Cox and cluster processes, like those in Chapter 11 for Poisson processes?*

  These have not yet been developed at the time of writing.

13

Gibbs Models

This chapter introduces Gibbs (including Markov) point process models, which are flexible and natural models for point patterns with dependence between points. We explain the basic concepts, and show how to build a Gibbs model appropriate to the data, to fit the model to the data, to interpret the results, and to identify weaknesses in the analysis.

13.1   Introduction

Dependence between the points in a spatial point process is often loosely called *interaction*. But the word 'interaction' suggests a very specific kind of behaviour, in which individual points 'act' on each other, causing them to stay away from each other ('inhibition' or 'repulsion') or to come closer together ('attraction').

Interaction in this stricter sense is a very important concept in many sciences. In spatial ecology, it is important to learn how individual organisms interact with each other, for example by competing for resources or territory. The spatial pattern of organisms will be influenced by such interactions.

However, interactions are not directly observed. Summary statistics such as the K-function *do not measure interaction* in this stricter sense. The K-function measures correlation, not causal interaction (see Section 7.3.5.2).

To draw inferences about the interactions that gave rise to a spatial point pattern, we need to fit a model to the pattern. A *Gibbs point process model* explicitly postulates that interactions occur between the points of the process. It is the only kind of spatial point process model that can truly be said to involve explicit 'interaction' between the points.

Gibbs processes are very useful and natural models for spatial point patterns in many different contexts including forestry [639], anatomy [227], and materials science [519]. They can produce a wide range of spatial patterns, ranging from strongly regular to strongly clustered patterns, and can easily combine repulsion and attraction at different scales.

A Gibbs process is the model of choice for a point pattern with inhibition between points. There is a widespread belief that Gibbs processes are only useful as models of inhibition, and are not able to produce strong clustering, e.g., [355, pp. 138, 157, 171]. This cannot be true in general, because all finite point process models (under reasonable conditions) can be represented mathematically as Gibbs models. However, this is not useful in practice; for example the Gibbs representation of a Neyman-Scott cluster process is prohibitively complicated [71]. In practice, many of the popular Gibbs models only produce inhibition or weak clustering, and it may be true that Gibbs models which are easy to write down are usually not strongly clustered.

Gibbs processes are very convenient for statistical model-building and inference. Models for specific applications can be built using simple tools, based on the concept of conditional intensity. Gibbs models can be fitted to data by pseudolikelihood (composite likelihood) methods, broadly similar to the likelihood methods used in Chapter 9. A simulated realisation of any Gibbs process can be generated by Markov chain Monte Carlo methods.

Although many researchers say they prefer not to assume a model when analysing data, the use of a summary function is almost equivalent to fitting a Gibbs model. The estimates of the K-function, empty-space function F, and nearest-neighbour distance function G are essentially the 'sufficient statistics' for Gibbs point process models called the Strauss process, area-interaction process, and Geyer saturation process, respectively [22, 70, 49]. Statistical theory says that estimation and inference for these Gibbs models should be based on the summary statistics. Using the K-function is therefore almost equivalent to fitting a Strauss model.

As we have seen in previous chapters, summary functions do not support a very sophisticated statistical analysis, because they cannot easily be adjusted to account for features such as spatial inhomogeneity. Using an explicit model gives us access to a much larger toolbox of statistical methods, makes it much easier to account for spatial inhomogeneity and covariate effects, and allows us to validate the analysis.

The statistical methods for Poisson processes presented in Chapters 9 and 11 are extended to Gibbs point processes in this chapter. We show how to formulate a Gibbs model appropriate to the data, fit the model to the data, interpret the results, and identify weaknesses in the analysis. Some of these techniques have only recently been developed.

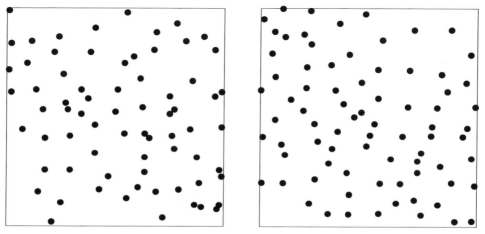

Figure 13.1. Left: *Swedish Pines data.* Right: *simulated realisation of a fitted Gibbs point process model involving spatial inhomogeneity and interpoint interaction. (Log-quadratic trend, hybrid of Diggle-Gates-Stibbard and area-interaction models.)*

Figure 13.1 shows the Swedish Pines data and a simulated realisation from a Gibbs model fitted to these data. The model includes both spatial inhomogeneity and interpoint interaction. Using the tools in `spatstat` it is possible to build and fit such models in a flexible way, to conduct formal hypothesis tests for the inhomogeneity after accounting for interaction, and *vice versa*, and to validate each component of the model.

13.2 Conditional intensity

A Poisson point process is completely specified by its intensity function $\lambda(u)$. The intensity is the ideal tool for modelling: its meaning is clear, we can easily formulate different models, and the models can be fitted efficiently to point pattern data, as we saw in Chapter 9.

For a Gibbs point process, the intensity function is not so convenient or useful as a modelling tool. In technical terms, the intensity does not completely specify a Gibbs process, and has a complicated relationship to other model parameters. For modelling purposes, the intensity is a complicated quantity that is the outcome of several competing effects (for example, increasing fertility and increasing competition have opposite effects on forest density).

However, there is a related concept of *conditional intensity* which is very useful. For a point pattern **x** and a location u, the conditional intensity $\lambda(u \mid \mathbf{x})$ is roughly speaking 'the intensity at u given the rest of the point pattern **x**'.

A heuristic definition is sketched in Figure 13.2. Suppose we divide our spatial window W into a fine grid of pixels, each of area a. Focus on one pixel, centred on a location u. Suppose we have full knowledge of the point pattern *outside* this pixel, and we want to know the conditional probability that the pixel in question contains a random point, *given* the rest of the point pattern **x**.

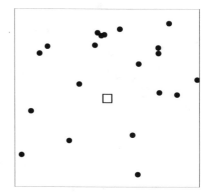

Figure 13.2. *Heuristic definition of conditional intensity $\lambda(u \mid \mathbf{x})$ as the intensity of points inside a pixel ('□') at location u given the point pattern **x** outside the pixel.*

If the point process is CSR, the points are independent; the contents of different pixels are independent, so this conditional probability does not depend on the rest of the pattern **x**. The conditional probability is equal to the probability of at least one point falling in the pixel, which is approximately λa where λ is the intensity. (Here we are assuming the pixels are extremely small). For an inhomogeneous Poisson process with intensity function $\lambda(u)$, the same reasoning applies, and the conditional probability would be $\lambda(u) a$.

Now if the point process exhibits dependence between points, then we expect that the probability of observing a point in the pixel *will* depend on the rest of the pattern. Suppose that the conditional probability of a point in the pixel, given the rest of the pattern **x**, is equal to $\lambda(u \mid \mathbf{x}) a$, where $\lambda(u \mid \mathbf{x})$ is a function that deserves to be called the *conditional* intensity. It behaves like the intensity except that it depends on the rest of the pattern. It is the intensity of the process inside the pixel, given the configuration of points outside the pixel.

For a simple example, imagine a forest where each tree has a circular crown of diameter R which captures all the rain falling on it. No tree trunk can grow within a distance R of an existing tree trunk. Then it makes sense to assume that the conditional intensity for this forest is $\lambda(u \mid \mathbf{x}) = 0$ for any location u lying within a distance R of an existing point of **x**, and $\lambda(u \mid \mathbf{x}) = \beta$ otherwise, where $\beta > 0$ is a constant. This is like CSR except that trees are forbidden to grow too close together. This is the Gibbs *hard core* model. The left panel of Figure 13.3 shows an image of the hard core model's conditional intensity $\lambda(u \mid \mathbf{x})$ as a function of the spatial location u, given the spatial point pattern **x** indicated. Conventionally the hard core distance is denoted by h rather than R.

More generally we could imagine that each tree crown captures a fraction — say a half — of all rain falling on it. Translating this into a probability we could then assume that each existing tree crown reduces, by half, the chance of a new tree growing underneath it. A location u not covered by any tree crowns will have conditional intensity $\lambda(u \mid \mathbf{x}) = \beta$, say; a location covered by one crown will then have $\lambda(u \mid \mathbf{x}) = \beta/2$; a location covered by two crowns will have $\lambda(u \mid \mathbf{x}) = \beta/4$; and in general a location covered by k tree crowns will have $\lambda(u \mid \mathbf{x}) = \beta/2^k$. In this model, trees can grow close together but this is less likely than for CSR. This is the *Strauss model* and its conditional intensity is shown in the right panel of Figure 13.3.

The conditional intensity is the main tool for modelling using Gibbs processes, and the discussion above is representative of the way we think when we build Gibbs models for real data.

However, a picture of the conditional intensity like Figure 13.3 does not tell us what a typical realisation of the Gibbs point process will look like. To answer that question, we need some additional tools, which are surveyed in the next section.

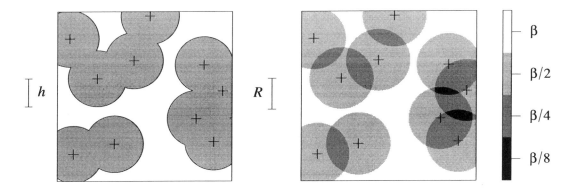

Figure 13.3. *Conditional intensity* $\lambda(u \mid \mathbf{x})$ *as a function of spatial location u given a point pattern* $\mathbf{x}$ *(marked by '+'). Left: hard core with distance h: new points are forbidden in the shaded region.* Right: *Strauss model with interaction distance R: new points are less likely in the darker regions.*

13.3 Key concepts

This section introduces many of the key concepts for Gibbs point process models, using the simple examples of the hard core process and the Strauss process. For technical details and references, see Section 13.12.1.

13.3.1 Probability density

A statistical distribution is often studied using its *probability density* $f(x)$, such as the famous 'bell curve' of the normal distribution. A spatial point process can also be studied using its probability density $f(\mathbf{x})$. Here $\mathbf{x}$ is one of the possible *point patterns* that could be produced, and we can intuitively think of $f(\mathbf{x})$ as 'the relative chance of getting the point pattern $\mathbf{x}$'.

Technical details are postponed to Section 13.12.1, but we need to mention two key facts. First, probability densities can only be defined for *finite* point processes, that is, processes which always contain a finite number of points. We shall assume all the point processes discussed in this chapter are observed in a bounded window W so that they are finite. (The observed pattern may arise either from Window Sampling, page 143, or from the Small World model, page 145; for the moment we assume the latter.)

Second, the probability density for a point process is expressed *relative to CSR* with intensity 1. The value of the probability density $f(\mathbf{x})$ can be interpreted as the probability of getting the point pattern $\mathbf{x}$ from the point process in question, *divided by* the same probability for CSR. Imagine that the window W is divided into tiny pixels. For a given point pattern $\mathbf{x}$, assign each pixel the value 1 or 0 according to whether it does or does not contain any points of $\mathbf{x}$. Now calculate the probability that this particular sequence of 0's and 1's would occur, for the point process $\mathbf{X}$. Also calculate the probability of the sequence if the process was CSR with intensity 1. The ratio of these probabilities is the density $f(\mathbf{x})$.

The probability densities of some point processes can be calculated by hand. Complete spatial randomness, the homogeneous Poisson process with intensity β in W, has probability density

$$f(\mathbf{x}) = c\,\beta^{n(\mathbf{x})} \tag{13.1}$$

where $n(\mathbf{x})$ is the number of points in $\mathbf{x}$, and c is a constant, namely $\exp((1-\beta)|W|)$. The inhomogeneous Poisson process with intensity function $\beta(u)$ in W has probability density

$$f(\mathbf{x}) = c_1\,\beta(x_1)\cdots\beta(x_n) = c_1 \prod_{i=1}^{n}\beta(x_i) \tag{13.2}$$

where $\mathbf{x} = \{x_1,\dots,x_n\}$, and c_1 is another constant, namely $\exp(\int_W (1-\beta(u))\,du)$. That is, the probability density of the Poisson process for the pattern $\mathbf{x}$ is calculated by multiplying together a contribution from each of the points $x_1,\dots,x_n$ in the pattern.

The mathematical form of the probability density determines the behaviour of the point process. For CSR, the probability density (13.1) does not depend on the locations of the points x_i, so that all possible patterns of n points are equally likely, different points are independent, and each point is uniformly distributed. For the inhomogeneous Poisson process, the probability density (13.2) *does* depend on the locations of points x_i, so there is a preference for particular locations. However, the contribution to $f(\mathbf{x})$ from a point x_i does not depend on the other points x_j, so the points are independent of each other.

Other point processes can be *defined* simply by writing down their probability densities $f(\mathbf{x})$. By choosing the mathematical form of the probability density, we can control the behaviour of the point process. This is the approach taken in Gibbs models.

13.3.2 Hard core process

The simplest kind of interaction between points is one in which they are forbidden to come too close together. Imagine placing circular coins, of equal diameter h, on a table so that the coins do not overlap: the locations of the centres of the coins form a point process in which each pair of points is at least h units apart. This is called a *hard core* constraint. See Figure 13.4.

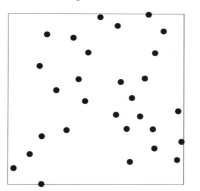 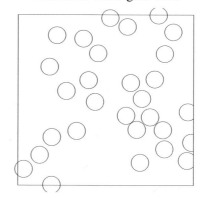

Figure 13.4. Left: *Realisation of Gibbs hard core process with* $\beta = 100$ *and hard core distance* $h = 0.1$ *in the unit square. Generated by* `rHardcore(100, 0.1)`. *Right: circles of diameter h around each point do not overlap.*

Point process models with a hard core — that is, models which always respect the hard core constraint — are often needed for analysing spatial point pattern data when the points are the locations of the centres of objects with appreciable size, such as mature trees, biological cells, or cities.

For modelling purposes it would be useful to have a point process model that is *'completely random, subject to the hard core constraint'*. We can easily define this using a probability density.

Take the probability density of CSR, equation (13.1), and modify it by requiring that $f(\mathbf{x}) = 0$ if the pattern violates the hard core condition. The new model has the probability density

$$f(\mathbf{x}) = \begin{cases} c_2 \, \beta^{n(\mathbf{x})} & \text{if hard core is satisfied} \\ 0 & \text{if hard core is violated} \end{cases} \qquad (13.3)$$

where c_2 is another constant (which is difficult to calculate). This is the probability density of the *Gibbs hard core process*, usually just called 'the' hard core process. One way to interpret this model is that

'the hard core process is the Poisson process *conditional on* the event that it satisfies the hard core constraint.'

The intuitive meaning of this statement is sketched in Figure 13.5. Imagine generating many realisations of CSR, and *deleting* any realisation that does not respect the hard core constraint. Then the remaining point patterns are realisations of the hard core process.

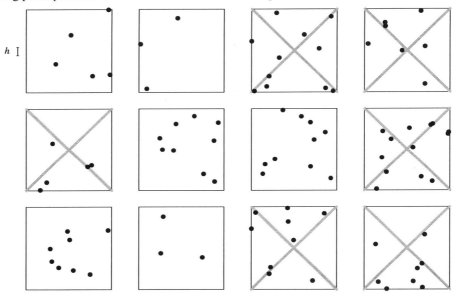

Figure 13.5. *Intuitive definition of the Gibbs hard core process in a bounded window W. Generating many realisations of CSR inside W, we delete those realisations which violate the hard core condition (shown as crossed-out). The remaining point patterns are realisations of the Gibbs hard core process. Parameters for the example are $\beta = 8$ and $h = 0.1$, and the observation window is the unit square.*

We emphasise that the acceptance-rejection procedure sketched in Figure 13.5 is not intended as a description of the real world. Fitting a hard core point process model to a forest survey does not mean we believe in a forest goddess who tried many random patterns of trees until she found a pattern that she liked. A hard core process can also be produced by several other, completely different, random mechanisms. For biological applications, a much more natural and realistic scenario which leads to the hard core model is a spatial birth-and-death process, discussed in Section 13.3.5.

Figure 13.4 shows a simulated realisation of the Gibbs hard core process in the unit square with $\beta = 100$ and $h = 0.1$. Note that the parameter β is the intensity of the original Poisson process; the intensity λ of the Gibbs hard core process is smaller, in this case $\lambda \approx 34.2$. This is an important general fact about Gibbs processes which we discuss in Section 13.3.6.

The acceptance-rejection procedure described in Figure 13.5 is usually not a practical way to generate realisations of a Gibbs hard core process in a computer. The simulated realisation in Figure 13.4 was not generated this way: that would have been completely impractical. For the parameter settings $\beta = 100$ and $h = 0.1$ used in Figure 13.4, the probability that a completely random pattern in the unit square will satisfy the hard core constraint is[1] about $p = 10^{-68}$, so that we could expect to generate about $1/p = 10^{68}$ completely random patterns before finding one that satisfied the hard core constraint. At a rate of one point pattern per millisecond, we would be waiting until the end of the universe to generate one picture like Figure 13.4. More practical Markov chain Monte Carlo techniques for simulating Gibbs models are described in Section 13.9.

13.3.3 Conditional intensity

For a point pattern $\mathbf{x}$ with probability density $f(\mathbf{x})$, we can now define the conditional intensity at a location u given $\mathbf{x}$, where u is not in $\mathbf{x}$, by

$$\lambda(u \mid \mathbf{x}) = \frac{f(\mathbf{x} \cup \{u\})}{f(\mathbf{x})} \tag{13.4}$$

where $\mathbf{x} \cup \{u\} = \{x_1, \ldots, x_n, u\}$ is the result of adding the new point u to the point pattern $\mathbf{x}$. It is also possible to define the conditional intensity at a data point x_i, by

$$\lambda(x_i \mid \mathbf{x}) = \frac{f(\mathbf{x})}{f(\mathbf{x} \setminus x_i)} \tag{13.5}$$

where $\mathbf{x} \setminus x_i = \{x_1, \ldots, x_{i-1}, x_{i+1}, \ldots, x_n\}$ is the result of deleting the point x_i from the point pattern $\mathbf{x}$. Some writers use the more compact definition $\lambda(v \mid \mathbf{x}) = f(\mathbf{x} \cup \{v\})/f(\mathbf{x} \setminus v)$ which applies to all locations v whether they are data points or not.

The conditional intensity always satisfies $\lambda(v \mid \mathbf{x}) = \lambda(v \mid \mathbf{x} \setminus v)$, that is, the conditional intensity at v is conditional on information about the point pattern at all locations *other than* v. The presence or absence of data points at all locations other than v informs the conditional intensity at v.

In this book, for better clarity, we write expressions for the conditional intensity $\lambda(u \mid \mathbf{x})$ always assuming that u is not in $\mathbf{x}$. The conditional intensity at a data point x_i can be recovered using $\lambda(x_i \mid \mathbf{x}) = \lambda(x_i \mid \mathbf{x} \setminus x_i)$.

For many Gibbs models it is straightforward to calculate the conditional intensity using (13.4) directly. For CSR, the homogeneous Poisson process with intensity β and probability density (13.1), the conditional intensity (13.4) is $\lambda(u \mid \mathbf{x}) = \beta$ for all u and all $\mathbf{x}$. This makes sense: since the points of a Poisson process are independent of each other, the conditional intensity is the same as the 'unconditional' intensity.

Similarly for the inhomogeneous Poisson process with intensity function $\beta(u)$ and probability density (13.2), the conditional intensity is $\lambda(u \mid \mathbf{x}) = \beta(u)$, the same as the unconditional intensity function.

For the Gibbs hard core process with probability density (13.3), we get

$$\lambda(u \mid \mathbf{x}) = \begin{cases} \beta & \text{if } u \text{ is permissible} \\ 0 & \text{if } u \text{ is not permissible} \end{cases} \tag{13.6}$$

where u is 'permissible' if $\mathbf{x} \cup \{u\}$ would satisfy the hard core constraint, that is, if the distance from u to each existing point x_i is greater than the hard core diameter h. That is, the conditional intensity is the same as for CSR, provided the hard core is respected, and otherwise the conditional intensity is zero. The left panel of Figure 13.3 is a picture of the conditional intensity $\lambda(u \mid \mathbf{x})$ as a

---

[1]The hard core constraint is equivalent to stipulating that the minimum nearest-neighbour distance in the pattern is greater than h. Using the approximation described in the footnote on page 269, the desired probability is about $\exp(-\pi h^2 \beta^2 |W|/2) = 6 \times 10^{-69}$.

function of u for a fixed point pattern $\mathbf{x}$, with the shaded regions having the value $\lambda(u \mid \mathbf{x}) = 0$, and the unshaded regions having $\lambda(u \mid \mathbf{x}) = \beta$.

Notice that the difficult normalising constants c, c_1, c_2 in the probability densities have been cancelled out in the conditional intensity because this is a ratio of densities (13.4). This is vitally important for applications since it enables us to calculate the conditional intensity explicitly.

13.3.4 Physics context

Gibbs models were first developed in physics, for studying the behaviour of a gas or fluid consisting of molecules. The molecules are constantly in motion; when two molecules come close together, they exert forces on each other, which deflect them away from each other. When the gas is in equilibrium at a macroscopic scale, it is still random at a microscopic scale; an instantaneous snapshot of the spatial positions of all molecules follows the probability law of a Gibbs point process.

The Gibbs law says essentially that the probability of a particular point pattern $\mathbf{x}$ is proportional to $e^{-U(\mathbf{x})/T}$, where $U(\mathbf{x})$ is the total potential energy of the configuration $\mathbf{x}$, and T is the temperature. Potential energy is the total amount of work (= force exerted $\times$ distance moved) required to overcome the molecular forces and move the molecules so that they form the pattern $\mathbf{x}$. The harder it is to push molecules into a particular configuration, the less likely it is that they will be found in that configuration in an instantaneous snapshot.

The Gibbs hard core process is the equilibrium law of a molecular gas in which molecules have a physical core which cannot be penetrated (overlapping molecules have infinite potential energy). Molecules rebound off each other elastically, according to 'billiard ball dynamics'. When equilibrium is reached, a snapshot of the molecules follows the distribution of the hard core process.

We mentioned the widespread belief that Gibbs models cannot produce strongly clustered patterns. Intuitively this is plausible — at least for the simple models discussed here — because if all the forces or interactions between particles were attractive, the particles would collapse together, or the population size would explode, so there would be no equilibrium state.

13.3.5 Birth-and-death process

Of course, trees in a forest do not behave like molecules in a gas. Individual trees do not travel through the forest, bouncing off other trees. A better model for forest succession would involve the death of existing trees at random times, and the germination of new seedlings at random places and random times. The probability that a seedling is viable could be assumed to depend on the spatial pattern of existing trees around the seedling, through some 'interaction' between the seedling and the adult trees. Similarly the viability of an existing tree could be assumed to depend on the pattern of surrounding trees. It turns out that, if this birth-and-death model of forest succession is run for many generations until it reaches equilibrium, an instantaneous snapshot of the spatial pattern of trees also follows the probability law of a Gibbs point process. In this model, although there are no physical 'forces' between trees, there is a quantity $U(\mathbf{x})$ analogous to total potential energy, and the probability of a particular point pattern $\mathbf{x}$ is again proportional to $e^{-U(\mathbf{x})}$.

Indeed this is a modified version of the random birth-and-death scenario described in Section 9.2.2, page 303. Suppose that, in each small interval of time Δt, each existing tree has probability $m \Delta t$ of dying, independently of other trees, where m is the 'mortality rate' per tree per unit time. In the same time interval, in any small region of area Δa, a new tree germinates with probability $g \Delta a \Delta t$, *provided the new tree lies more than h units away* from the nearest living tree. Here g is the 'germination rate' per unit area per unit time. Effectively, around each existing tree there is a circular zone of *radius h* where new trees are forbidden to germinate. See the left panel of Figure 13.3. Outside this forbidden region, new seedlings germinate at a uniform rate g.

No matter what the initial state of the forest, over long time scales this *spatial birth-and-death process* [554] reaches an equilibrium in which any snapshot of the forest is a realisation of the Gibbs hard core process with parameter $\beta = g/m$ and hard core diameter h. Figure 13.6 shows a sequence of states of the spatial birth-death process for the hard core model, with $\beta = 200$ and $h = 0.07$ in the unit square. The last state in the sequence might be taken as a realisation of the hard core model.

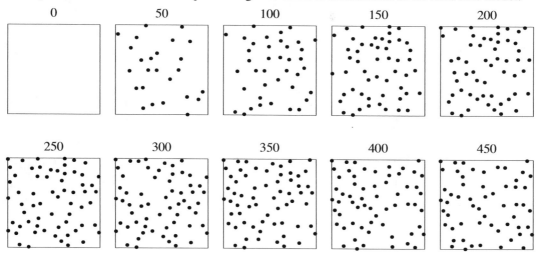

Figure 13.6. *The state of the spatial birth-death process after every 50 transitions, starting from an empty configuration.*

A spatial birth-death process is a realistic scenario for many natural and physical processes. It is also very convenient from a modelling viewpoint. We often find it easier to think about the changes in a spatial pattern over time than to think about an instantaneous snapshot of the pattern. In modelling a forest, it is easier to think about the morbidity of an individual tree and the competitive pressure experienced by a new seedling, rather than trying to predict the spatial pattern that will result from these effects over many generations.

Spatial birth-and-death processes also provide a practical method for computer simulation of the hard core process. Figure 13.4 was generated by simulating such a spatial birth-death process.

In practice, one would usually run the spatial birth-death process for many more generations than are depicted in Figure 13.6, to ensure that the sequence has converged. Note that 'convergence' does not mean that the snapshots will ultimately all look the same; the pattern is eternally changing. Instead, the spatial birth-death process exhibits 'convergence in distribution', which means that the snapshots ultimately follow the exact probability law of the Gibbs hard core process.

If we need more than one realisation of the Gibbs hard core process, we can either (a) re-run the entire spatial birth-death process each time, or (b) first run the spatial birth-death process for a large number of generations until convergence is assured, and then take a snapshot every N generations, where N is large enough to ensure there is negligible dependence between the snapshots.

The key concept for spatial birth-death processes is the *equilibrium* between birth and death. As sketched in Figure 13.7, birth (the instantaneous addition of a new point) and death (the instantaneous deletion of an existing point) can be considered as the reverse of each other. It is the *ratio* of the birth rate to the death rate which determines the final probability distribution of the point process.

Suppose we want to generate a realisation of a hard core process. We run a spatial birth-death process with death rate m (per point per unit time) and birth rate b (per unit area per unit time) subject to the hard core with diameter h. When the spatial birth-death process has converged, births and deaths are in equilibrium. If the point pattern $\mathbf{x}$ is the current state, consider the birth which adds a new point u, so that $\mathbf{x}$ becomes $\mathbf{x} \cup \{u\}$. The reverse transition is a death from $\mathbf{x} \cup \{u\}$ to $\mathbf{x}$. In equilibrium, the frequency with which we find ourselves in state $\mathbf{x}$ and jump to $\mathbf{x} \cup \{u\}$ should

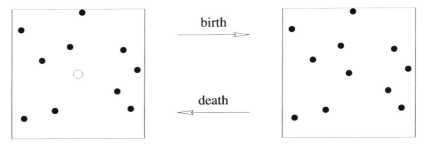

Figure 13.7. *Birth and death transitions in a spatial birth-and-death process are the reverse of each other. In equilibrium the frequency of transitions in each direction will be equal.*

equal the frequency with which we find ourselves in state $\mathbf{x} \cup \{u\}$ and jump back to $\mathbf{x}$. That is,

$$f(\mathbf{x})b = f(\mathbf{x} \cup \{u\})m \tag{13.7}$$

so that

$$\frac{b}{m} = \lambda(u \mid \mathbf{x}). \tag{13.8}$$

If we rescale the unit of time so that the mortality rate m is equal to 1, then the birth rate of the spatial birth-and-death process is equal to the conditional intensity for the hard core process.

13.3.6 Inhomogeneous models, intensity, and 'trend'

Spatially inhomogeneous models, including models involving covariate effects, are needed for many applications. Suppose that in Figure 13.5, instead of generating a large collection of realisations of CSR, we generate many realisations of an inhomogeneous Poisson process, with intensity function $\beta(u)$. As before, we delete any pattern which violates the hard core condition. The remaining patterns are realisations of an *inhomogeneous hard core process*, with 'first-order trend' $\beta(u)$. It has probability density

$$f(\mathbf{x}) = \begin{cases} c_3 \, \beta(x_1) \ldots \beta(x_n) & \text{if hard core is satisfied} \\ 0 & \text{if hard core is violated} \end{cases} \tag{13.9}$$

where c_3 is another constant (difficult to calculate), and conditional intensity

$$\lambda(u \mid \mathbf{x}) = \begin{cases} \beta(u) & \text{if } u \text{ is permissible} \\ 0 & \text{if } u \text{ is not permissible} \end{cases} \tag{13.10}$$

where again u is 'permissible' if the distance from u to the nearest point of $\mathbf{x}$ is greater than the hard core diameter h.

Figure 13.8 shows a simulated realisation of the inhomogeneous hard core process with $\beta(x,y) = 500\exp(-2(x+y))$ and hard core diameter $h = 0.05$ in the unit square.

The right panel is a kernel estimate of the intensity function $\lambda(u)$ of this point process. Clearly, the qualitative behaviour of $\lambda(u)$ is strongly influenced by the qualitative behaviour of $\beta(u)$, which drops exponentially from bottom left to top right of the picture. However, $\lambda(u)$ is not proportional to $\beta(u)$. The original Poisson intensity $\beta(u)$ varies by a factor of $\exp(4) \approx 50$, while the estimated intensity $\widehat{\lambda}(u)$ varies by a factor of about 30. A heuristic explanation is that the effect of inhibition between points is stronger at higher intensities (because there are more points to inhibit each other) so that the ratio $\lambda(u)/\beta(u)$ will be smaller for larger values of $\beta(u)$.

It is a mistake to treat $\beta(u)$ as the 'intensity' of the model. We should rather think of $\beta(u)$ as

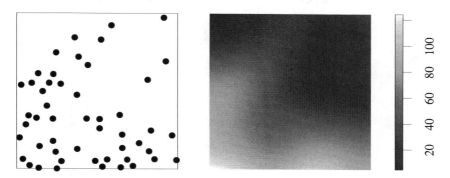

Figure 13.8. *Inhomogeneous hard core process.* Left: *Simulated realisation of inhomogeneous hard core process with $\beta(x,y) = 500\exp(-2(x+y))$ and hard core diameter $h = 0.05$ in the unit square.* Right: *kernel smoothed intensity estimate using Scott's bandwidth selector.*

the spatially varying 'fertility' which is counterbalanced by the 'competitive' effect of the hard core to give the final intensity. In `spatstat` documentation we refer to $\beta(u)$ as the *(first-order) trend*. Other terms used in the literature include 'chemical activity' and 'fugacity'. When analysing spatial point patterns of plants, for example, it is typically more relevant to study the effect of a spatial covariate on the underlying fertility (expressed through the first-order trend) than on the intensity, which is the product of fertility and competition.

Another useful way to think of the first-order trend $\beta(u)$ is that it is the conditional intensity for an *empty* point pattern, $\beta(u) = \lambda(u \mid \emptyset)$.

A completely different way to achieve spatial inhomogeneity is to take a homogeneous point process and apply a non-uniform geometrical transformation [362, 502, 556, 363], [484, sec. 6.5.3]. We do not explore this option here.

13.3.7 Strauss process

The hard core process is an appropriate model when it is physically impossible for two points to lie closer than a distance h apart. If close pairs of points are not impossible, but are unlikely to occur, an appropriate model is the *Strauss process* shown in Figure 13.9, in which close pairs of points are 'penalised' rather than 'forbidden'.

We can explain the Strauss process intuitively along the lines of Figure 13.5. Suppose we generate a realisation of a homogeneous Poisson process in the desired window W. Inspecting this realisation, we determine whether any pair of points is closer than a threshold distance R. If there are no close pairs, the realisation is accepted as before. Otherwise, for each close pair of points, we flip a coin to decide whether to accept or reject this pair. If *any* pair of close points is rejected, the *entire* realisation is rejected. That is, the realisation is accepted only if all close pairs of points are accepted (i.e. all coin flips are successful). The coin may be biased, with probability γ of acceptance, where $0 \leq \gamma \leq 1$. If $s(\mathbf{x}, R)$ is the number of close pairs of points in the configuration $\mathbf{x}$, then the probability that the realisation $\mathbf{x}$ will be accepted is $\gamma^{s(\mathbf{x},R)}$.

Figure 13.10 shows this scheme applied to the Poisson point patterns from Figure 13.5 using a fair coin, $\gamma = \frac{1}{2}$. Numerals show the number of close pairs $s(\mathbf{x}, R)$ for each pattern. A pattern with $s(\mathbf{x}, R) = 0$ respects the hard core constraint and is accepted always. A pattern with $s(\mathbf{x}, R) = 1$ is accepted with probability $\gamma^1 = \frac{1}{2}$; so in this scheme, about half of the patterns with $s(\mathbf{x}, R) = 1$ will be crossed out. A pattern with $s(\mathbf{x}, R) = 2$ is accepted with probability $\gamma^2 = \frac{1}{4}$; so about three

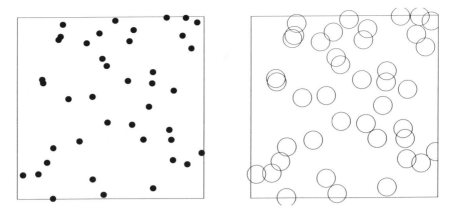

Figure 13.9. Left: *Realisation of Strauss process with* $\beta = 100$, *interaction parameter* $\gamma = 1/2$, *and interaction distance* $R = 0.1$ *in the unit square. Generated by* `rStrauss(100, 0.5, 0.1)`. *Right: circles of diameter R around each point sometimes overlap.*

quarters of the patterns with $s(\mathbf{x}, R) = 2$ are crossed out, and so on. The remaining point patterns can be regarded as realisations of the Strauss point process with $\beta = 8$ and interaction parameter $\gamma = \frac{1}{2}$. These patterns do not always respect the hard core constraint, but they tend to have fewer close pairs of points than would be expected for a completely random pattern.

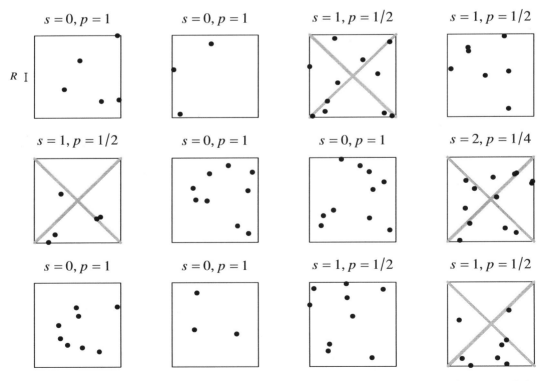

Figure 13.10. *Generating realisations of the Strauss process. Starting with realisations of the Poisson process, we count the number* $s(\mathbf{x}, R)$ *of close pairs of points in each realisation* $\mathbf{x}$. *Each realisation is accepted with probability* $\gamma^{s(\mathbf{x}, R)}$. *In this example* $\gamma = 1/2$.

The Strauss process is intermediate between the hard core process and complete spatial randomness. Its behaviour depends on the value of the interaction parameter γ, ranging between 0 and 1. If γ is chosen to be equal to 1, then all realisations are accepted, and the Strauss point process is a homogeneous Poisson process. If γ is chosen to be equal to 0, then a realisation is only accepted if it has no close pairs of points, and the Strauss point process is the hard core process.

Note again that β is the intensity of the original Poisson process; the intensity λ of the Strauss process is smaller. In Figure 13.10 we have $\beta = 8$ and $\lambda \approx 7.2$; in Figure 13.9 we have $\beta = 100$ and $\lambda \approx 47.5$, using an approximation explained in Section 13.12.6.

Figure 13.11 shows simulated realisations of the Strauss process with $\gamma = 0, \frac{1}{3}, \frac{2}{3}$, and 1, with the original Poisson intensity β adjusted so that each of the Strauss patterns has approximately the same intensity $\lambda \approx 50$.

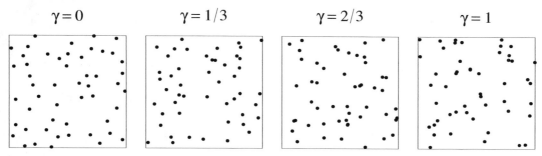

Figure 13.11. *Simulated realisations of Strauss processes in the unit square with different values of γ ranging from 0 to 1 and fixed interaction distance $R = 0.08$. Parameter β adjusted so that all patterns have intensity approximately equal to 50.*

The Strauss process has probability density

$$f(\mathbf{x}) = c_4 \, \beta^{n(\mathbf{x})} \, \gamma^{s(\mathbf{x},R)} \tag{13.11}$$

where c_4 is a constant, $\beta > 0$ is the 'abundance' ('fertility', 'activity') parameter, $0 \leq \gamma \leq 1$ is the interaction parameter, and $s(\mathbf{x},R)$ is the number of unordered pairs of points in $\mathbf{x}$ which are closer than the interaction distance R. According to (13.11), a pattern with many pairs of close points is much less likely than a pattern with the same number of points but only few close pairs.

The conditional intensity of the Strauss process at a location u not in $\mathbf{x}$ is

$$\lambda(u \mid \mathbf{x}) = \beta \, \gamma^{t(u,R,\mathbf{x})} \tag{13.12}$$

where $t(u,R,\mathbf{x}) = s(\mathbf{x} \cup \{u\}, R) - s(\mathbf{x}, R)$ is the number of points in $\mathbf{x}$ which lie within distance R of the location u. For a data point x_i in $\mathbf{x}$ this gives

$$\lambda(x_i \mid \mathbf{x}) = \beta \, \gamma^{t(x_i, R, \mathbf{x} \setminus x_i)} \tag{13.13}$$

where $t(x_i, R, \mathbf{x} \setminus x_i) = s(\mathbf{x}, R) - s(\mathbf{x} \setminus x_i, R)$ is the number of neighbours of x_i in $\mathbf{x}$, not counting x_i itself. According to (13.12), a point is much less likely to be found in an environment where there are many other points close by. The right panel of Figure 13.3 shows an example of the conditional intensity of the Strauss process.

Although the Strauss process was originally intended as a model of clustering [643], it can only be used to model inhibition, because the parameter γ cannot be greater than 1. If we take $\gamma > 1$, the function (13.11) is not integrable [371], so it does not define a valid probability density.

Realisations of the Strauss process can also be generated by a spatial birth-death process similar to that described in Section 13.3.5. In each small interval of time Δt, each existing tree has probability $m \Delta t$ of dying, independently of other trees, where m is the mortality rate per tree per unit time.

In the same time interval, in any small region of area Δa, a new tree germinates with probability $b\gamma^k\Delta a\Delta t$, where k is the number of existing trees which lie closer than a distance R to this location. Here b is the 'base germination rate' per unit area per unit time.

No matter what the initial state of the forest, over long time scales this spatial birth-and-death process reaches an equilibrium in which any snapshot of the forest is a realisation of the Strauss process with $\beta = b/m$, interaction distance R, and interaction parameter γ.

An inhomogeneous version of the Strauss process is straightforward following the idea of Section 13.3.6. In Figure 13.10 we replace CSR by the inhomogeneous Poisson process with intensity function $\beta(u)$. The probability density is

$$f(\mathbf{x}) = c_5\,\beta(x_1)\cdots\beta(x_n)\,\gamma^{s(\mathbf{x},R)} \tag{13.14}$$

and the conditional intensity is

$$\lambda(u\mid\mathbf{x}) = \beta(u)\,\gamma^{t(u,R,\mathbf{x})}. \tag{13.15}$$

13.3.8 Interactions and the potential

In summary, a Gibbs model is defined by the interactions that occur between points. We can specify a Gibbs model by writing down a formula for the probability density $f(\mathbf{x})$ as a product of terms associated with each interaction. The *logarithm* of the probability density

$$V(\mathbf{x}) = \log f(\mathbf{x}) \tag{13.16}$$

is called the (negative) *potential* of the model. (For consistency with other techniques we have removed the minus sign so that $U(\mathbf{x}) = -V(\mathbf{x})$ is what physicists would call the potential energy.) Because a Gibbs density is a product of interaction terms, the potential function $V(\mathbf{x})$ is a sum of terms associated with interactions. For example, CSR with intensity β has potential

$$V(\mathbf{x}) = \log c + n(\mathbf{x})\log\beta. \tag{13.17}$$

The inhomogeneous Poisson process with intensity function $\beta(u)$ has potential

$$V(\mathbf{x}) = \log c_1 + \log\beta(x_1) + \cdots + \log\beta(x_n) = \log c_1 + \sum_{i=1}^{n}\log\beta(x_i). \tag{13.18}$$

The homogeneous Strauss process has potential

$$V(\mathbf{x}) = \log c_4 + n(\mathbf{x})\log\beta + s(\mathbf{x},R)\log\gamma. \tag{13.19}$$

The terms $n(\mathbf{x})\log\beta$ and $\sum_i\log\beta(x_i)$ are called 'first-order' potential terms because they are sums of contributions from each individual point. The term $s(\mathbf{x},R)\log\gamma$ is called a 'second-order' potential or interaction term, because it is a sum of contributions from each pair of points x_i,x_j.

One can also have higher-order interactions: a third-order interaction term would be a sum of contributions from each *triple* of points x_i,x_j,x_k. A full discussion of interaction models is postponed to Sections 13.6–13.8.

Note that the labels 'first-order' and 'second-order' refer to the degree of interaction; these are not the same as the classification of summary statistics into 'first-order' summaries (estimates of the first moment, the intensity) and 'second-order' (estimates of the correlation, covariance or K-function). As we saw above, the first-order trend $\beta(u)$ is not the same as the 'first-order' intensity $\lambda(u)$, unless interaction is very weak. Likewise the second-order interaction $c(d)$ is not the same as the pair correlation function $g(d)$, although they are connected if the interaction is weak.

13.4   Statistical insights

This section provides some additional, statistical insights about Gibbs models. It can be skipped at a first reading.

13.4.1   Connection to summary statistics

An important statistical insight about the Strauss model [22, 52, 70] is that it is intimately connected to the K-function. For the Strauss process with interaction range R, the statistic $s(\mathbf{x}, R)$ is the number of pairs of distinct points lying closer than R apart. This is closely related to the K-function for the distance R. If $\widehat{K}_{\mathbf{x}}(r)$ is the empirical K-function for the point pattern $\mathbf{x}$ as defined in equation (7.3), ignoring the edge correction terms gives

$$s(\mathbf{x}, R) \approx \frac{n(n-1)}{2|W|} \widehat{K}_{\mathbf{x}}(R) \tag{13.20}$$

where the factor 2 arises because $s(\mathbf{x}, R)$ counts unordered pairs. In technical terms the 'sufficient statistic' for the Strauss model is the pair of numbers $(n(\mathbf{x}), s(\mathbf{x}, R))$ and ignoring edge effects the sufficient statistic could be taken as $(n(\mathbf{x}), \widehat{K}_{\mathbf{x}}(R))$.

This gives the insight that the Strauss model could be defined as a point process in which the probability of getting a configuration $\mathbf{x}$ depends only on the number of points and on the value of $\widehat{K}(R)$, with smaller values of $\widehat{K}(R)$ being more likely.

This also has many implications for statistical analysis of point patterns. Fitting the Strauss model to a point pattern dataset by maximum likelihood would involve calculating the empirical K function at the distance R. The theoretically optimal (Neyman-Pearson) hypothesis test of the null hypothesis of CSR, against the alternative of a Strauss process with interaction range R, must depend only on the statistics $n(\mathbf{x})$ and $\widehat{K}_{\mathbf{x}}(R)$.

The good news, then, is that the K-function has some theoretical support as a tool for formal inference about spatial point patterns, as well as being intuitively reasonable. Indeed, to a first approximation due to Penttinen [539], the maximum likelihood estimate of the Strauss interaction parameter γ is approximately

$$\widehat{\gamma} \approx \frac{\widehat{K}(R)}{\pi R^2} \tag{13.21}$$

provided γ is not too far from 1. This is also the mean field approximation [47].

However, there are also several caveats. If we do not assume the point process is a Strauss process, then the use of the K-function does not have the same theoretical support.

Another caveat is that, if we have fitted a Strauss model to point pattern data, we cannot assess the model's goodness-of-fit using envelopes of the K-function: this would be circular logic since $\widehat{K}(r)$ is intimately connected to the Strauss model. Another summary function unrelated to the K-function should be used [225, pp. 10, 89]. Alternatively the data may be split into separate halves, and a model fitted to one half may be validated using the other half [225, p. 93].

13.4.2   Conditioning

Unlike Cox and cluster processes, Gibbs point processes are very well behaved under conditioning.

13.4.2.1   Conditioning on the presence of a point

Many calculations in spatial statistics (for example, the definition of the K-function) involve the conditional probability or conditional expectation of some quantity, given that the point process includes a specified point u (i.e. given that u is a point of the point process).

For a finite Gibbs process $\mathbf{X}$ with probability density $f(\mathbf{x})$, if we condition on the presence of a point at location u, then $\mathbf{X}$ is no longer a Gibbs process, because it contains a fixed point. However, the *rest* of the process, say $\mathbf{Y} = \mathbf{X} \setminus u$, is a finite Gibbs process with probability density $h(\mathbf{y}) = cf(\mathbf{y} \cup \{u\}) = cf(\mathbf{y})\lambda(u \mid \mathbf{y})$ where c is a normalising constant. In technical terms, the reduced Palm distribution of a Gibbs model is a Gibbs model.

For example, in the Gibbs hard core process with parameters β and h in a window W, if we condition on the presence of a point at location u, then $\mathbf{Y}$ has probability density proportional to $f(\mathbf{y} \cup \{u\})$ where f was defined in (13.3). No point is allowed to fall inside the disc $b(u,h)$ of radius h centred at u. The conditional process $\mathbf{Y}$ is just a hard core process with the same parameters, inside the smaller window $W \setminus b(u,h)$. See the left panel of Figure 13.12.

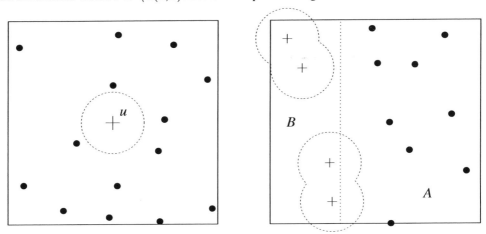

Figure 13.12. *Finite Gibbs hard core process conditioned on the presence of a point* (Left) *or conditioned on the realisation in a subregion* (Right). *Fixed (conditioning) points:* +. *Free points:* •.

13.4.2.2 Conditioning on the realisation in a subdomain

Suppose we divide the window W into two disjoint parts A and B, and condition on the realisation inside the region B. If $\mathbf{X}$ is the point process then we consider the conditional distribution of $\mathbf{Y} = \mathbf{X} \cap A$ given that $\mathbf{X} \cap B = \mathbf{z}$ where $\mathbf{z}$ is a point pattern in B. Again it turns out that $\mathbf{Y}$ is conditionally a finite Gibbs process with probability density $h(\mathbf{y}) = cf(\mathbf{y} \cup \{\mathbf{z}\})$ where c is another normalising constant depending on $\mathbf{z}$.

For example, in the Gibbs hard core process above, if we condition on $\mathbf{X} \cap B = \mathbf{z}$, then $\mathbf{Y}$ is a hard core process in A which is constrained not to put any points in any of the discs $b(z_i, h)$ centred at the points $z_i \in \mathbf{z}$. Equivalently $\mathbf{Y}$ is a hard core process in the smaller window $W \setminus \bigcup_i b(z_i, h)$. See the right panel of Figure 13.12.

13.4.2.3 Markov property

In many Gibbs point processes, the conditional intensity $\lambda(u \mid \mathbf{x})$ depends only on the points of $\mathbf{x}$ that lie within a distance R^* of the location u. If so, R^* is called the *'interaction range'*, and the process is called a *Markov* point process [579].

For example, this is true for the Gibbs hard core process: the conditional intensity at u only depends on knowing whether there are any points of $\mathbf{x}$ within a distance h of u. It is true for the Strauss process: the conditional intensity at u depends only on the number of points of $\mathbf{x}$ within a distance h of u.

Markov point processes have a conditional independence property which is analogous to the conditional independence of Markov chains. Suppose we divide the window W into disjoint parts A and B as before. Now consider the border region $C = A \cap B_{\oplus R^*}$, the part of A that lies within a distance R^* of the region B. See Figure 13.13. If we condition on the realisation of $\mathbf{X}$ in this border region, $\mathbf{X} \cap C = \mathbf{z}$, then the patterns inside A and B are *conditionally independent* [579, 67].

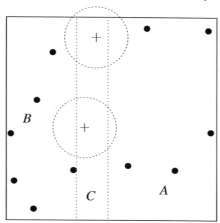

Figure 13.13. *Markov property for Markov point processes. Given the points in the region C (points +), the points in A are independent of the points in B.*

13.4.3 Infinite Gibbs processes

We have focused on *finite* Gibbs processes until now. There also exist *infinite* Gibbs processes, such as the stationary hard core process, and the stationary Strauss process. These infinite Gibbs models cannot be defined so easily, and indeed they involve some theoretical problems which are beyond the scope of this book [279].

Infinite *Markov* point processes are relatively easy to describe. Suppose that, in any bounded window W, we have defined a finite Markov point process with probability density $f_W(\mathbf{x})$ and interaction range R^*. Taking one such window W, we divide *the two-dimensional plane* into the parts inside and outside the window, $A = W$ and $B = W^c$. Then we stipulate that the conditioning property described in Section 13.4.2.2 must hold. That is, conditional on the realisation of $\mathbf{X}$ *outside* the window, $\mathbf{X} \cap B$, the realisation inside the window, $\mathbf{X} \cap A$, must be a finite Gibbs process. Let $C = A \cap B_{\oplus R^*}$ be the border region. The requirement is that, conditional on $\mathbf{X} \cap C = \mathbf{z}$, the process $\mathbf{Y} = \mathbf{X} \cap A = \mathbf{X} \cap W$ must be a finite Gibbs process with probability density $h(\mathbf{y}) = c f_{A \cup C}(\mathbf{y} \cup \{\mathbf{z}\})$ where c is a normalising constant depending on $\mathbf{z}$ and W. If this stipulation holds for all choices of W, then $\mathbf{X}$ is an infinite Markov point process. For more explanation, see [484, sec. 6.4].

If an infinite Gibbs point process is observed under Window Sampling, then edge corrections are required, to deal with the fact that points outside the observation window could have influenced the observed points. See Section 13.5.3 and [228, sec. 8.1.4], [50].

13.5 Fitting Gibbs models to data

13.5.1 Formulating and fitting a model

In `spatstat`, Gibbs point process models are fitted to a point pattern data set using the same function ppm as for Poisson models, introduced in Section 9.3. The syntax is simply extended with an extra argument `interaction`:

$$\texttt{ppm(X ~ trend, interaction)}$$

or in general

$$\texttt{ppm(X ~ trend, interaction, ..., data)}$$

The first argument is still a formula in the R language with the left-hand side specifying the name of the point pattern dataset X. The right-hand side of this formula now specifies the form of the *logarithm* of the *first-order term* of the conditional intensity for the model, $\log \beta(u)$.

The second argument `interaction` is an object of the special class `"interact"` which describes the interpoint interaction component of the model. See Table 13.2 on page 516 and Table 13.3 on page 519 for the interactions implemented in `spatstat`. To fit a homogeneous Strauss process to the `swedishpines` dataset:

```
> ppm(swedishpines ~ 1, Strauss(r=5))
Stationary Strauss process
First order term:  beta = 0.01237
Interaction distance:      5
Fitted interaction parameter gamma:          0.2726
```

Here `Strauss` is a special function that creates the interaction structure of the Strauss process. Notice that we had to specify the value of the interaction range R, which is denoted by r in `spatstat`.

The result of `ppm` is a fitted point process model (object of class `"ppm"`). The output above[2] is from the print method, `print.ppm`, and tells us that the estimated values of the parameters β and γ of the Strauss process are 0.012 and 0.273, respectively.

In this model the first-order term β is constant (not dependent on spatial location), which may often be unrealistic; it implies that we should expect approximately the same number of points in regions of equal size. In Chapter 6 it was concluded that the Swedish Pines data have inhomogeneous intensity as well as inhibition: see Figures 6.8, 6.10, and 6.15.

To allow for spatial variation we simply replace the constant β by a function $\beta(u)$. Inhomogeneity, and dependence on spatial covariates, may then enter the model through the first-order term. Inhomogeneity can easily be incorporated into our model by specifying a formula for the trend, as explained in Section 9.3.1. For example, we might fit an inhomogeneous Strauss process with first-order ('trend') term depending on the spatial coordinates:

```
> ppm(swedishpines ~ polynom(x,y,2), Strauss(5))
Nonstationary Strauss process
Log trend:   ~x + y + I(x^2) + I(x * y) + I(y^2)
Fitted trend coefficients:
(Intercept)             x             y       I(x^2)       I(x * y)       I(y^2)
  -5.9882056     0.0314702     0.0399793   -0.0001157     -0.0003459   -0.0002371
Interaction distance:         5
Fitted interaction parameter gamma:          0.2558
```

This is often a pragmatic choice, in the absence of covariate data or background information to guide the choice of model.

To interpret these models it helps to remember that the arguments to ppm specify the form of the *conditional intensity*, on a *logarithmic* scale. The first argument (the formula) specifies the first-order terms, while the second argument (the interaction) specifies the second-order and higher-order terms. The command `ppm(X~1, Strauss(5))` fits the model with log conditional intensity

$$\log \lambda(u \mid \mathbf{x}) = \theta_0 + \theta_1\, t(u, 5, \mathbf{x}) \tag{13.22}$$

---

[2]Printed output has been shortened to save space, by setting `options(digits=4)` and `spatstat.options(terse=3)`.

where θ_0, θ_1 are parameters to be estimated, and $t(u, 5, \mathbf{x})$ is the number of points of $\mathbf{x}$ lying closer than 5 units (0.5 metres) away from the location u. This is the homogeneous Strauss process with parameters β, γ where $\theta_0 = \log \beta$ and $\theta_1 = \log \gamma$.

The parameters which are actually estimated by ppm are the coefficients θ_0, θ_1 in the log conditional intensity, known as the 'canonical parameters'. They can be extracted using the coef method:

```
> fit0 <- ppm(swedishpines ~ 1, Strauss(5))
> coef(fit0)
(Intercept) Interaction
     -4.393      -1.300
```

It is the print method for "ppm" which converts the canonical parameters into the more conventional parameters like γ. The spatstat generic function parameters can be used to extract all the model parameters, including the trend coefficients and the interaction parameters (fitted and irregular):

```
> unlist(parameters(fit0))
trend.beta            r     gamma
   0.01237      5.00000   0.27260
```

Thus the command ppm(X ~ polynom(x,y,2), Strauss(5)), which we used above, fitted the model with log conditional intensity

$$\log \lambda(u \mid \mathbf{x}) = \theta_0 + \theta_1 u_1 + \theta_2 u_2 + \theta_3 u_1^2 + \theta_4 u_2^2 + \theta_5 u_1 u_2 + \theta_6\, t(u, 5, \mathbf{x}) \qquad (13.23)$$

where (u_1, u_2) are the spatial coordinates of the location u, and $\theta_0, \ldots, \theta_6$ are parameters to be estimated. This is the inhomogeneous Strauss process with log-quadratic first-order trend

$$\beta(u) = \exp(\theta_0 + \theta_1 u_1 + \theta_2 u_2 + \theta_3 u_1^2 + \theta_4 u_2^2 + \theta_5 u_1 u_2)$$

and interaction parameter $\gamma = \exp \theta_6$, that is, $\theta_6 = \log \gamma$. To extract the fitted parameters θ_i use coef as before.

Dependence on a covariate can also be included in the model. For example the vesicles pattern (Figure 3.6 on page 68) appears to have higher intensity near the active zone. The pattern also appears to be inhibited, although this may simply be a result of the fact that the vesicles have appreciable size and do not overlap. We could fit a Strauss model in which the first-order term depends on distance to the active zone (output not shown):

```
> Dactive <- distfun(vesicles.extra$activezone)
> ppm(vesicles ~ Dactive, Strauss(40))
```

In general the model fitted by ppm has loglinear conditional intensity

$$\log \lambda(u \mid \mathbf{x}) = V(u \cup \{\mathbf{x}\}) - V(\mathbf{x}) = \eta^{\mathsf{T}} Z(u) + \varphi^{\mathsf{T}} T(u, \mathbf{x}) \qquad (13.24)$$

where $(\eta, \varphi) = \theta$ is the vector of 'canonical parameters', and the vector-valued functions $Z(u)$ and $T(u, \mathbf{x})$ together comprise the 'canonical covariates'. The function $Z(u)$ is specified by the trend formula and $T(u, \mathbf{x})$ is specified by the interaction argument. The trend parameters η determine the first-order trend $\beta(u) = \exp(\eta^{\mathsf{T}} Z(u))$, while the interaction parameters φ determine the interaction term $\exp(\varphi^{\mathsf{T}} T(u, \mathbf{x}))$. Further examples of loglinear conditional intensities are given in Section 13.5.1.

We stress again that the first-order trend $\beta(u) = \exp(\eta^{\mathsf{T}} Z(u))$ does not directly determine the intensity of the model. When $\beta(u)$ increases, the intensity typically increases; but the expected number of points in a Gibbs model with $\beta = 2$ is not twice that in the same model with $\beta = 1$. We return to this discussion in Section 13.12.6.

Smooth functions in the *trend* terms are available by setting use.gam=TRUE and including appropriate terms in the trend formula. See Section 13.13.

13.5.2   Choice of fitting method

Additional arguments "..." in the call to ppm control the fitting algorithm, including the choice of several fitting methods, edge corrections, and the density of 'dummy' (non-data) sample points.

The default values of these algorithm parameters are set so that *computation is reasonably fast*, rather than aiming for high accuracy. The default settings are useful at the beginning of an analysis when many different candidate models are tried out. Later, when formal conclusions are desired, it is strongly recommended to change these parameters to increase accuracy, for example, setting the number of dummy points to nd=256 or higher. See help(ppm.ppp) for details.

Fitting a Gibbs model is much harder than fitting a Poisson model. The likelihood $L(\theta)$ is the probability density $f(\mathbf{x})$ of the observed data $\mathbf{x}$ considered as a function of the model parameters θ. The formulae given above for the probability densities of Gibbs models are quite simple in form, except that they involve a normalising constant $c, c_1, c_2, \ldots$ which is an intractable function of θ. For example, for the hard core process (13.3), the constant c_2 is the reciprocal of the probability that a Poisson process of intensity β in the window W satisfies the hard core condition at distance h: this is not known in simple form.

Methods for fitting Gibbs processes are described in detail in Section 13.13. The default method is maximum *pseudolikelihood* (Section 13.13.4), a surrogate for maximum likelihood invented by Besag [98], and now often called 'composite likelihood'. The spatstat package uses the Berman-Turner-Baddeley [90, 50] approximation to pseudolikelihood (Section 13.13.5).

Alternative fitting methods are available in spatstat. Conditional logistic regression for Poisson models (Section 9.10) can be extended to Gibbs models as logistic composite likelihood [158, 38] and is available in ppm by setting method="logi". See Section 13.13.7. A Monte Carlo approximation to maximum likelihood due to Huang and Ogata [348] is selected by method="ho". See Section 13.13.6.

Generalized additive models [327] allow terms in the model to include very flexible functions of the original covariates. The models are then fitted by maximising a penalised likelihood formed by adding a roughness penalty to the likelihood. Such terms can be included in the *trend* of a point process model by setting use.gam=TRUE. See Section 13.13.

Variational Bayes estimation for Poisson models (Section 9.11) also extends to Gibbs models (Section 13.13.9) and is available in ppm by setting method="VBlogi".

Local likelihood or geographically weighted regression methods (Section 9.13) also extend to Gibbs models, using separate code described in Section 13.11.

13.5.3   Edge correction

Edge corrections may be needed because the conditional intensity $\lambda(u \mid \mathbf{x})$ at a point u close to the edge of the window W may depend on unobserved points of $\mathbf{x}$ lying outside W. The need for edge corrections depends on the sampling context (Section 5.6.2). In 'Window Sampling' (page 143) an edge correction is required, while in the 'Small World Model' (page 145) it is not required.

The default behaviour in spatstat is to use the *border correction* (correction="border"), in which the pseudolikelihood or composite likelihood is a sum of contributions from $W_{\ominus r}$, the window W eroded by a distance r, treating the data in the border region $W \setminus W_{\ominus r}$ as if it were fixed. Here r is taken as the range of interaction of the Gibbs model, unless specified by the argument rbord. This approach is valid in any of the sampling contexts discussed in Section 5.6.2.

Alternatives are correction="translation" and correction="isotropic", which apply edge-correction weights to each contribution to the pseudolikelihood or composite likelihood [228, sec. 8.1.4], [50]; correction="periodic" which treats a rectangular window as if opposite sides were glued together; and correction="none" which avoids all edge correction. In the 'Small World Model' the most appropriate choice is correction="none", and weighted edge corrections would be inappropriate.

13.5.4 Validity of the model

The printout of a fitted model may indicate that it is *'invalid'*. By default, the canonical parameters θ are fitted without any constraint. However, most interpoint interactions have constraints on the parameters. For example in the Strauss process, the interaction strength parameter γ must be less than or equal to 1 in order to obtain a well-defined point process. Equivalently the canonical parameter $\theta_2 = \log \gamma$ should be less than or equal to 0. This constraint is **not** imposed (by default) by ppm when it fits a Strauss model. Consider the redwood data, which are known to be very clustered:

```
> ppm(redwood ~ 1, Strauss(0.1))
Stationary Strauss process
First order term:  beta = 20.39
Interaction distance:       0.1
Fitted interaction parameter gamma:       1.4363
*** Model is not valid ***
*** Interaction parameters are outside valid range ***
```

The fitted parameter γ is greater than 1, so the model is invalid: a Strauss process with these parameters simply does not exist. The printed output reports this. To force ppm to fit a valid model, use the argument project:

```
> ppm(redwood ~ 1, Strauss(0.1), project=TRUE)
Stationary Poisson process
Intensity: 71.53
```

The result is a Poisson process because the best-fitting (or least-poorly-fitting) Strauss model is one with $\gamma = 1$, corresponding to a Poisson process. To test whether a fitted model is valid in this sense, use valid.ppm. To convert an invalid fitted model to a valid one, use project.ppm.

By default, invalid models are not projected, because the un-projected model can often convey useful information, and because this policy ensures that models fitted by the same command always have the same format (which is important in simulation experiments).

The model is also invalid if any of the fitted interaction coefficients are NA, meaning that they cannot be estimated. This can occur, for example, if we try to fit a Strauss model with a very small interaction range r:

```
> ppm(cells ~ 1, Strauss(0.00001))
Stationary Strauss process
First order term:  beta = 42.08
Interaction distance:       1e-05
Fitted interaction parameter gamma:       NA
*** Model is not valid ***
*** Some coefficients are NA or Inf ***
```

This happens because models are fitted using a finite set of sample points (the 'quadrature scheme', Section 13.13.5). No inferences can be drawn about interactions at a distance smaller than the minimum interpoint distance in the quadrature scheme. Again the argument project=TRUE could be used to force the model to be valid.

13.5.5 Methods for fitted Gibbs models

The result of ppm is a fitted point process model object of class "ppm". As we saw in Chapter 9, there are many methods available for this class, including the usual tasks like printing, summarising, plotting, and predicting the model, but also for simulating the fitted model, calculating confidence intervals, performing stepwise model selection, testing hypotheses, and so on. Tables 9.3 and 9.4 in

Section 9.4.1 list the standard R generic operations which have methods for the class "ppm". These methods now apply to Gibbs models as well, after a decade of research [18, 55, 45, 37, 36, 35, 48, 47, 18, 162].

`as.interact`	interaction component of model
`fitin`	*fitted* interaction component of model
`coef`	fitted coefficients (trend and interaction)
`parameters`	all model parameters
`predict`	fitted trend, conditional intensity, intensity
`intensity`	fitted intensity
`effectfun`	fitted trend as function of covariate
`model.matrix`	design matrix
`model.covariates`	identify covariates used in a model
`model.images`	pixel images of the canonical covariates
`Kmodel`	approximate K-function of fitted model
`pcfmodel`	approximate pair correlation function of fitted model
`nobs`	number of data points
`reach`	interaction range of model
`rmh`	generate simulated realisations of model
`simulate`	generate simulated realisations of model
`rmhmodel`	Convert to a model that can be simulated
`vcov`	variance-covariance matrix of $\widehat{\theta}$

Table 13.1. *Generic operations with methods for class* "ppm" *for extracting components or predictions of a fitted model. (The generics* `predict`, `model.matrix`, `simulate`, *and* `vcov` *are defined in the base package* `stats`; *all others are defined in* `spatstat`.*)*

Additionally, `spatstat` defines many generic operations with methods for "ppm" objects. Generics for extracting predictions from a model are listed in Table 13.1 here. Generics for validating a fitted model are listed in Table 13.5 on page 536. Type `methods(class="ppm")` to see a list of all methods applicable to class "ppm".

13.5.6   Updating the model

Gibbs models can be changed using `update.ppm`:
$$\text{newmodel <- update(oldmodel, ...)}$$
where the optional additional arguments "`...`" determine how the model will be changed. These arguments can provide a new formula for the trend, a new point pattern dataset, a new interaction, or new values of the algorithm parameters:

```
> fit0 <- ppm(swedishpines ~ 1, Strauss(r=5))
> fit2  <- update(fit0, . ~ polynom(x,y,2))
> fit9 <- update(fit0, Strauss(9))
> fit0fine <- update(fit0, nd=256)
```

Using `update.ppm` is often safer and more convenient than using `ppm` when fitting many candidate models to the same data.

If external data have changed, then simply type `update(oldmodel)` to re-fit the model to the new data.

13.5.7   Variance, standard errors, and confidence intervals

Standard errors and confidence intervals for the canonical parameter estimates $\widehat{\theta}$ can be obtained using `vcov` or `confint`:

```
> sqrt(diag(vcov(fit0)))
(Intercept) Interaction
     0.2132      0.5075
> confint(fit0)
              2.5 % 97.5 %
(Intercept) -4.811 -3.975
Interaction -2.294 -0.305
```

or together with a significance test by

```
> coef(summary(fit0))
             Estimate   S.E. CI95.lo CI95.hi Ztest     Zval
(Intercept)    -4.393 0.2132  -4.811  -3.975   *** -20.601
Interaction    -1.300 0.5075  -2.294  -0.305     *  -2.561
```

The latter table will be printed automatically, whenever a Gibbs model is printed, if the user sets `spatstat.options(print.ppm.SE="always")`. This does not happen by default, to save computation time.

All these calculations rely on estimating the variance-covariance matrix of $\widehat{\theta}$ using `vcov.ppm`, the method for the generic `vcov`.

```
> vcov(fit0)
            (Intercept) Interaction
(Intercept)     0.04547    -0.06819
Interaction    -0.06819     0.25759
```

The algorithm for calculating variance estimates is quite recent [45, 162]. The default algorithm has been chosen for computational speed, and requires only summation over data points and pairs of data points. If numerical problems are encountered, the slower but more reliable estimate, involving numerical integration over the domain, should be selected, using the call `vcov(fit0, fine=TRUE)`.

13.5.8 Prediction

In Section 9.4.3 we explained how to calculate the fitted intensity of a Poisson point process model using `predict.ppm`. For a Gibbs point process model there are three fitted quantities of interest which can be computed by `predict.ppm`: the first-order term β or $\beta(u)$, the conditional intensity $\lambda(u \mid \mathbf{x})$, and the intensity λ or $\lambda(u)$. These are equivalent for a Poisson process.

The argument `type` determines which quantity will be calculated. If `type="cif"` the fitted conditional intensity $\widehat{\lambda}(u \mid \mathbf{x})$ is calculated by substituting the estimates of the canonical parameters θ into the formula for the conditional intensity (13.24). If `type="trend"` the fitted first-order term is calculated as $\widehat{\beta}(u) = \exp(\widehat{\eta}^\mathsf{T} Z(u))$ where $\widehat{\eta}$ is the vector of trend (non-interaction) parameters. If `type="intensity"` then an **approximation** to the intensity of the fitted model is computed using the recently developed Poisson saddlepoint approximation (Section 13.12.6), currently implemented only for pairwise interaction models.

By default, predictions are calculated at a regular grid of locations u, covering the observation window of the original point pattern data, yielding a pixel image.

Figure 13.14 shows the trend, conditional intensity, and (approximate) intensity for the Strauss model with interaction range 5 units and log-quadratic trend, fitted to the Swedish Pines data. They were computed by `predict(fit2, type="trend")`, `predict(fit2, type="cif")`, and `predict(fit2, type="intensity")`, respectively. There are many useful ways to display these images: see Section 4.1.4 of Chapter 4. Diagonal transects of the three images (bottom left corner to top right corner) obtained using `transect.im` are shown in Figure 13.15.

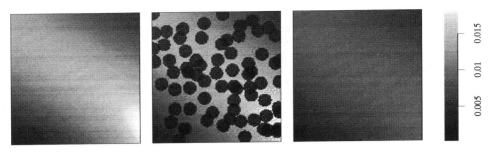

Figure 13.14. *Fitted first-order trend (Left), conditional intensity (Middle), and intensity (Right) for the Strauss model with log-quadratic trend fitted to the Swedish Pines data.*

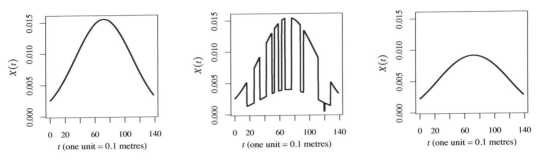

Figure 13.15. *Diagonal transects of the three images in the previous figure, using equal scales.*

The conditional intensity $\lambda(u \mid \mathbf{x})$ depends on the point pattern $\mathbf{x}$. The default is to compute $\widehat{\lambda}(u \mid \mathbf{x})$ using the original point pattern dataset $\mathbf{x}$. To replace $\mathbf{x}$ by another point pattern when calculating the conditional intensity, use the argument X:

```
> scrambled <- runifpoint(ex=swedishpines)
> ll <- predict(fit2, type="cif", X=scrambled)
```

A plot of the fitted conditional intensity for the *original* data is useful for qualitative understanding of the fitted model, and also for diagnostic purposes (Section 13.10).

Predictions can also be calculated in another window by giving the argument window, or only at specific locations by giving the argument locations. For full details see help(predict.ppm).

If the model is homogeneous, the fitted trend and fitted intensity are constant, so a pixel image of these functions will be uninteresting. The numerical value of the fitted trend β is printed by print.ppm and can be extracted as summary(fit)$trend$value. The numerical value of the fitted (approximate) intensity λ is computed by intensity.ppm via the generic function intensity:

```
> intensity(fit0)
[1] 0.007883
```

Thus the fitted homogeneous Strauss model is expected to produce approximately 0.00788 points per square decimetre. This corresponds to intensity(fit0)*area(swedishpines)=75.7 points in the window, which is slightly more than the observed number 71.

Standard errors for the predictions are available by setting se=TRUE.

13.5.9 Plotting the fitted model and fitted interaction

For a fitted model object in R, the convention is that the `plot` method displays a 'sensible' selection of graphics that portray the fitted model and support a visual assessment of validity of the model. For a fitted Gibbs point process model, the method `plot.ppm` plots the fitted first-order trend and the fitted conditional intensity (described above), the fitted interaction, and a selection of diagnostics for model validation (described in Section 13.10).

The *fitted interaction* can be extracted from the model by the function `fitin`.

```
> fitin(fit2)
Strauss process
Interaction distance:          5
Fitted interaction parameter gamma:        0.2558
```

The result is an object of class `"fii"` ('fitted interpoint interaction') describing the interaction part of the model, including the fitted interaction parameters. It can be printed and plotted: the left panel of Figure 13.19 shows an example.

Note that a fitted interaction is different from an object of class `"interact"`. They both describe the interaction, but an `"fii"` object includes the fitted parameters such as γ, while an `"interact"` object does not. We can think of `Strauss(5)` as 'the Strauss interaction ready to be fitted to data.' After fitting the model, the original un-fitted interaction can be extracted by `as.interact`:

```
> as.interact(fit2)
Interaction:Strauss process
Interaction distance:          5
```

13.5.10 Simulation

Another way to study the fitted model is to generate simulated realisations from it:

```
> s <- simulate(fit2, nsim=100)
```

The result is a list of 100 point patterns. The result also belongs to the class `"solist"` so that it can be plotted immediately. Figure 13.16 shows the first four simulated realisations, plotted by `plot(s[1:4])`.

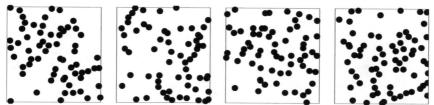

Figure 13.16. *Four simulated realisations from the inhomogeneous Strauss model fitted to the Swedish Pines data.*

Executing the `simulate` command for a fitted Gibbs model is noticeably slower than for a fitted Poisson model (cf. Chapter 9). To generate each simulated realisation of a Gibbs point process, we essentially run a spatial birth-and-death process for a large number of iterations to approach equilibrium. The actual simulation procedure is somewhat different, and is discussed in more detail in Section 13.9. If you use simulation as part of the analysis of a point pattern dataset it is highly recommended to read that material.

Visualising simulated realisations is a useful way to understand the fitted model, and to decide whether the fitted model is producing patterns that are qualitatively similar to the observed data.

Simulation also makes possible a wide variety of techniques for validating the model. At a simple level we can assess whether the simulated patterns produce roughly the same number of points as the data:

```
> npoints(swedishpines)
[1] 71
> np <- sapply(s, npoints)
> mean(np)
[1] 71.58
> sd(np)
[1] 6.703
```

However, 'validation' of this kind can give a false sense of security. For a Poisson point process model, under typical conditions, the expected total number of points generated by the fitted model must be equal to the observed number of points. (This is implied by the score equations, if the model includes an intercept term and has been fitted by maximum likelihood.) For a Gibbs process, this is only approximately true, for several reasons, including the use of surrogate likelihoods and edge corrections; nevertheless the criticism still holds.

A wide range of techniques for simulation-based inference is supported in spatstat, using simulate.ppm to generate the simulated realisations. For example, the function envelope has a method for class "ppm" making it possible to generate simulation envelopes of any summary function based on any fitted Gibbs model:

```
> E0 <- envelope(fit0, Lest, nsim=39)
```

The function dclf.test likewise has a method for "ppm" allowing us to perform a goodness-of-fit test of any fitted Gibbs model.

13.5.11   Model selection and hypothesis testing

Section 9.4.6 described the R commands step, drop1, and add1 for model selection. The Akaike Information Criterion AIC is calculated for each candidate model; drop1 and add1 show a comparison table of AIC values, while step recursively chooses the model with the smallest AIC.

These commands can be applied to "ppm" objects representing Gibbs models. Since the AIC (defined in equation (9.33) on page 335) involves the intractable likelihood, our implementation uses the *composite AIC*

$$\text{AIC}^* = -2\log\text{CL}_{max} + 2m \tag{13.25}$$

where $\text{CL}_{max} = \text{CL}(\widehat{\theta})$ is the maximised value of the composite likelihood used to fit the model (either the pseudolikelihood or the logistic composite likelihood) and m is the Takeuchi penalty [667, eq. (5)]. For a Poisson process, m is equal to the number of fitted parameters. Strictly speaking, the statistical theory supporting the Takeuchi criterion has only been established for finite collections of random variables [667] but we believe it can be extended to spatial point processes.

When applied to a Gibbs model, step performs stepwise model selection on the terms in the trend formula only, keeping the same interaction in all candidate models.

```
> step(fit2, trace=0)
Stationary Strauss process
First order term:  beta = 0.01237
Interaction distance:        5
Fitted interaction parameter gamma:        0.2726
```

Section 10.3.2 described the Likelihood Ratio Test and the related Analysis of Deviance, for Poisson point process models. The function `anova.ppm`, a method for the generic `anova`, performs these procedures. For Gibbs models, since the likelihood is intractable, we replace the likelihood ratio test statistic Γ defined in equation (10.1) by an analogue based on composite likelihood:

$$\Gamma^* = 2A \log \frac{CL_1}{CL_0} = 2A(\log CL_1 - \log CL_0) \tag{13.26}$$

where CL_0, CL_1 are the maximised values of composite likelihood for the null and alternative hypotheses, respectively, and A is a data-dependent adjustment factor [56] introduced to ensure that the null distribution of Γ^* is approximately χ^2. The factor A is determined by 'moment matching' [532, 56].

To decide whether there is evidence for a Strauss interaction in the Swedish Pines pattern after allowing for the spatial inhomogeneity, we can apply `anova.ppm`:

```
> fit2P <- update(fit2, Poisson())
> anova(fit2P, fit2, test="LR")
Analysis of Deviance Table
Model 1: ~x + y + I(x^2) + I(x * y) + I(y^2)          Poisson
Model 2: ~x + y + I(x^2) + I(x * y) + I(y^2)          Strauss
  Npar Df AdjDeviance Pr(>Chi)
1    6
2    7  1          43  5.3e-11 ***
---
Signif. codes:  0 '***' 0.001 '**' 0.01 '*' 0.05 '.' 0.1 ' ' 1
```

The Strauss interaction is highly significant. On the other hand,

```
> anova(fit0, fit2, test="LR")
Analysis of Deviance Table
Model 1: ~1              Strauss
Model 2: ~x + y + I(x^2) + I(x * y) + I(y^2)          Strauss
  Npar Df AdjDeviance Pr(>Chi)
1    2
2    7  5        2.23     0.82
```

the inhomogeneous trend is not significant.

13.6 Pairwise interaction models

13.6.1 Pairwise interactions

In the Strauss model, two points interact if they lie closer than the threshold distance R apart. Each pair of close points contributes a factor γ to the overall probability of the pattern.

Instead of this threshold behaviour, we could assume that the contribution from each pair of points is a more general function of the distance between them. For a pair of points lying d units apart, the contribution is $c(d)$, where c is the 'pairwise interaction function'. Assume for a moment that $0 \le c(d) \le 1$ for all distances d. Following a scheme similar to those in Figures 13.5 and 13.10, we first generate many realisations from a Poisson process with intensity β. For each realisation $\mathbf{x}$, we consider each pair of points x_i, x_j, calculate the distance $d_{ij} = \|x_i - x_j\|$ between them, and calculate the contribution $c(d_{ij})$. This is the acceptance probability for the 'coin flip' associated with the pair x_i, x_j. Multiplying the contributions $c(d_{ij})$ together for all pairs of points x_i, x_j gives the overall probability of accepting the realisation $\mathbf{x}$. Using computer-generated random numbers we accept or delete each realisation $\mathbf{x}$ according to the calculated probability of acceptance. The undeleted patterns are realisations of a 'pairwise interaction' Gibbs process with interaction function c.

A value of $c(d) = 1$ would mean that a pair of points d units apart effectively does not interact, because the contribution from this pair of points does not change the overall probability (i.e. multiplication by 1 does not change the result). A value $c(d) = 0$ would mean that an interpoint distance of d is forbidden, and this would cause the *entire* point pattern $\mathbf{x}$ to be rejected (i.e. multiplication by 0 yields a value of 0).

More generally still we could assume the contribution from each pair of points x_i, x_j is a function $c(x_i, x_j)$ depending on the positions of both points. The probability density of this *pairwise interaction* model is

$$f(\mathbf{x}) = c_6 \, \beta(x_1) \cdots \beta(x_n) c(x_1, x_2) c(x_1, x_3) \cdots c(x_{n-1}, x_n)$$

$$= c_6 \left[\prod_{i=1}^{n} \beta(x_i) \right] \left[\prod_{i=1}^{n} \prod_{\substack{j=1 \\ j>i}}^{n} c(x_i, x_j) \right] \tag{13.27}$$

and the conditional intensity is

$$\lambda(u \mid \mathbf{x}) = \beta(u) c(u, x_1) \cdots c(u, x_n) = \beta(u) \prod_{i=1}^{n} c(u, x_i). \tag{13.28}$$

If c depends only on relative position, $c(u, v) = c(u - v)$, then we shall call the interaction *homogeneous*. If c depends only on distance, $c(u, v) = c(\|u - v\|)$, we shall call it *homogeneous and isotropic*. The hard core and Strauss models are special cases of the pairwise interaction model. For the Strauss process, the interaction function is

$$c(d) = \begin{cases} \gamma & \text{if } d \leq R \\ 1 & \text{if } d > R \end{cases} \tag{13.29}$$

and the hard core process is the special case of (13.29) where $\gamma = 0$.

Figure 13.17 shows several pairwise interaction functions proposed in the literature which can be used in `spatstat`. The hybrid *Strauss-hard core* model combines a hard core effect with a Strauss-like effect: it has the pairwise interaction function

$$c(d) = \begin{cases} 0 & \text{if } d \leq h \\ \gamma & \text{if } h < d \leq R \\ 1 & \text{if } d > R \end{cases} \tag{13.30}$$

where $0 < h < R$. Figure 13.18 shows a simulated realisation of the Strauss-hard core process and its conditional intensity.

A pairwise interaction function is called *purely inhibitory* if it stays below the value 1, that is, if $c(d) \leq 1$ for all distances d. In this case, the pairwise interaction process involves inhibition at all scales. Adding more points to the pattern $\mathbf{x}$ will either reduce the conditional intensity $\lambda(u \mid \mathbf{x})$ or leave it unchanged.

In the case of the Strauss process, we needed to impose the constraint $\gamma \leq 1$ to ensure that the model is well defined (a Strauss process with $\gamma > 1$ simply does not exist). There are no technical problems about the existence of a purely inhibitory point process. For any function c satisfying $c(d) \leq 1$, and any first-order term $\beta(u)$ which is integrable, the pairwise interaction process with density (13.27) is well defined.

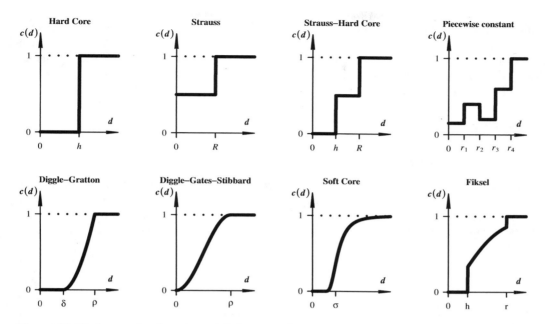

Figure 13.17. *Interaction functions $c(d)$ for several pairwise interaction models used in* spatstat.

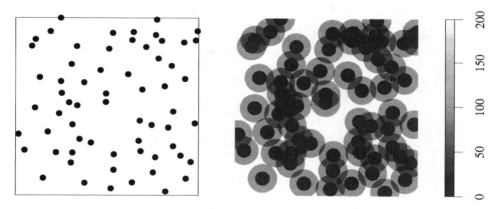

Figure 13.18. *Strauss-hard core process.* Left: *simulated realisation with $\beta = 200$, hard core diameter $h = 0.04$, and Strauss interaction radius $R = 0.08$ in the unit square.* Right: *conditional intensity.*

There do exist valid pairwise interaction models where $c(d) > 1$ for *some* values of d. This is a more delicate matter. One way to ensure that the process exists is to include a hard core, that is, to make $c(d) = 0$ for values of d below a threshold. Any pairwise interaction function $c(d)$ that includes a hard core gives rise to a well-defined point process. For example, the hybrid Strauss-hard core interaction (13.30) can have any parameter value $\gamma > 0$ assuming the hard core radius h is greater than 0.

Other pairwise interactions that are considered in spatstat include the *soft core* interaction with scale $\sigma > 0$ and index $0 < \kappa < 1$, with pair potential $c(d) = (\sigma^2/d^2)^{1/\kappa}$; the *Diggle-Gates-*

Stibbard interaction with interaction range ρ, defined by $c(d) = \sin((\pi/2)d/\rho)^2$ if $d \leq \rho$, and $c(d) = 1$ if $d > \rho$; the *Diggle-Gratton* interaction with hard core distance δ, interaction distance ρ, and index κ, defined by $c(d) = 0$ if $d \leq \delta$, $c(d) = ((d - \delta)/(\rho - \delta))^{\kappa}$ if $\delta < d \leq \rho$, and $c(d) = 1$ if $d > \rho$; the *Fiksel* double exponential interaction with hard core distance h, interaction distance r, interaction strength a, and interaction decay rate κ, defined by $c(d) = 0$ if $d \leq h$, $c(d) = \exp(a\exp(-\kappa d))$ if $h < d \leq r$, and $c(d) = 1$ if $d > r$.

There is also the general *piecewise constant* interaction in which $c(d)$ is a step function of d. See [221, section 4.9], [225, p. 109], [333, 94].

13.6.2   Pairwise interactions in `spatstat`

When fitting a Gibbs point process model using ppm, the user specifies the interaction structure of the model using the argument `interaction`. For example, `interaction=Strauss(5)` specifies the interaction structure of the Strauss process with interaction range 5 units. The function `Strauss` returns an object of class `"interact"` that describes the properties of the Strauss model needed for calculations. Table 13.2 lists functions provided in `spatstat` for creating an interaction object that represents a pairwise interaction. Detailed information about each interaction model is available in the help files for these functions.

FUNCTION	MODEL
DiggleGratton	Diggle-Gratton potential
DiggleGatesStibbard	Diggle-Gates-Stibbard potential
Fiksel	Fiksel double exponential potential
Hardcore	hard core process
Hybrid	hybrid of several interactions
LennardJones	Lennard-Jones potential
PairPiece	pairwise interaction, piecewise constant
Pairwise	pairwise interaction, user-supplied potential
Softcore	soft core potential
Strauss	Strauss process
StraussHard	hybrid Strauss/hard core point process
MultiHard	multitype hard core process
MultiStrauss	multitype Strauss process
MultiStraussHard	multitype Strauss-hard core process
HierStrauss	hierarchical Strauss process
HierHard	hierarchical hard core process
HierStraussHard	hierarchical Strauss-hard core process

Table 13.2. *Functions in* `spatstat` *for creating an interpoint interaction object representing a pairwise interaction. (Interactions with names beginning with* `Multi` *or* `Hier` *are presented in Chapter 14.)*

Table 13.2 includes some simple interaction models for immediate practical use, and also some very flexible general models, `Pairwise` and `PairPiece`, which effectively allow any pairwise interaction at all. Additionally the `Hybrid` command allows us to combine several interactions to make a new one: see Section 13.8.

Most of the functions in Table 13.2 have arguments, which must be given by the user, specifying scale parameters such as the Strauss interaction distance R. These are the *irregular parameters* or *nuisance parameters* of the model. The irregular parameters are not estimated directly by ppm. Indeed the statistical theory for estimating such parameters is often unclear.

An exception is the hard core diameter h in the hard core model and other models with a hard core. These parameters can be estimated immediately from data. The maximum likelihood estimator of the hard core distance [581] is the minimum nearest-neighbour distance $d_{min} = \min_i \min_{j \neq i} d_{ij}$. In `spatstat` we use the modified estimator $h^* = (n(\mathbf{x})/(n(\mathbf{x}) + 1))d_{min}$ which has smaller bias and avoids computational problems. If the user calls the function `Hardcore` without specifying the hard core diameter h, then h will be estimated by calculating h^* for the point pattern data, and the model-fitting will proceed with that value. Thus, the user has the option to fit the hard core model with an automatically estimated hard core radius h^*, or with an estimate of h chosen by the user. This 'self-starting' feature is also available in the functions `DiggleGratton`, `DiggleGatesStibbard`, `Fiksel`, `StraussHard`, `MultiHard`, and `MultiStraussHard`.

13.6.3   Determining the range of pairwise interaction

It is often claimed that the 'scale of interaction' of a point process can be identified by inspecting the graph of the K-function or L-function, and finding the distance value (on the horizontal axis) where the observed function achieves its maximum deviation from the theoretical function expected for CSR. This is false, in general, as we saw in Section 7.3.5.4.

To draw inferences about the range of interaction, we need to assume a *model* of the interaction. The interaction range can then be defined unambiguously in terms of the model parameters, and it may become clear how to estimate the range.

The use of the K-function to identify the interaction range *is* justified by statistical theory if we are willing to assume that the process is a homogeneous Strauss process. Firstly the likelihood of the Strauss process, with interaction range R unknown, depends only on the number of points and the empirical K-function (ignoring edge effects, as explained in Section 13.4.1). This implies that theoretically optimal statistical inference can be performed using only these summary statistics. Secondly, for a Strauss model with interaction range R, the largest value of the deviation $|K(r) - \pi r^2|$ of the *true* K-function does occur when $r = R$. It is then reasonable to estimate R by the distance r^* of maximal deviation of the empirical K-function [634].

The pair correlation function could be used in the same way, but does not perform as well (neither in practice nor in theory). Intuitively this is because the Strauss model postulates a threshold or changepoint in the behaviour of pairwise distances, and $\widehat{K}(r)$ accumulates all available evidence for a threshold at distance r. This is analogous to other statistical procedures for changepoint models.

For other pairwise interaction models, it remains true that the likelihood depends only on the number of points and on the empirical K-function. The second-order potential term is related to the empirical K-function by

$$\prod_{i=1}^{n} \prod_{\substack{j=1 \\ j>i}}^{n} c(d_{ij}) \approx \exp\left(\frac{n(n-1)}{|W|} \int_0^\infty \log c(r) \, d\widehat{K}(r) \right)$$

so there is theoretical support for performing inference about the model parameters using only the statistics $n(\mathbf{x})$ and $\widehat{K}(\cdot)$. However, in this general case it is not quite as easy to use the graph of the K-function to judge the best values of the irregular parameters. A practical alternative is to fit a highly flexible pairwise interaction model such as a step function, plot the fitted interaction function, and use this to guide the selection of irregular parameters.

The left panel of Figure 13.19 shows a step-function pairwise interaction [221, section 4.9], [225, p. 109], [333] fitted to the Swedish Pines data by

```
> mod <- ppm(swedishpines ~ polynom(x,y,2), PairPiece(1:15))
> f <- fitin(mod)
```

It suggests the data could better be modelled using a pairwise interaction function which increases gradually from 0 to 1, such as the Diggle-Gratton model.

Other methods for selecting irregular parameters were described in Section 9.12. For threshold parameters such as the Strauss interaction range R, Newton's method (`ippm`) is not useful, but the method of maximum profile likelihood (`profilepl`) is applicable. It can be adapted to Gibbs models by replacing the likelihood by the pseudolikelihood, logistic composite likelihood, or Akaike Information Criterion. If $\theta = (\phi, \eta)$ where ϕ denotes the nuisance parameters and η the regular parameters, define the profile log-pseudolikelihood by

$$\text{PPL}(\phi) = \text{PPL}(\phi, \mathbf{x}) = \max_{\eta} \log \text{PL}\left((\phi, \eta); \mathbf{x}\right).$$

The right-hand side can be computed, for each fixed value of ϕ, by ppm. Then we just have to maximise $\text{PPL}(\phi)$ over ϕ. This is done by the function `profilepl`: similarly for the logistic composite likelihood and the AIC. The following code compares different trial values of the Strauss interaction range R using AIC:

```
> rr <- data.frame(r=seq(3,14,by=0.05))
> p <- profilepl(rr, Strauss, swedishpines~polynom(x,y,2), aic=TRUE)
> p
Profile log pseudolikelihood for model:
ppm(swedishpines ~ polynom(x, y, 2), aic = TRUE, interaction = Strauss)
Optimum value of irregular parameter:  r = 6.95
```

The result p is an object of class `profilepl` containing the profile AIC function, the optimised value of the irregular parameter R, and the final fitted model. There are methods for print, summary, and plot for these objects. The result of plot(p) is shown in the right panel of Figure 13.19 and suggests that $r = 7$ is roughly the best choice.

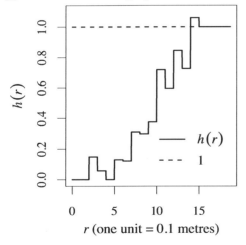
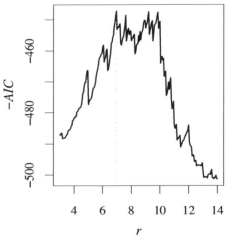

Figure 13.19. *Investigating interaction range in the Swedish Pines data.* Left: *fitted interaction function (assumed to be a step function) for an inhomogeneous pairwise interaction model.* Right: *profile AIC for the inhomogeneous Strauss model as a function of Strauss interaction range parameter.*

To extract the final fitted model, use as.ppm:

```
> as.ppm(p)
Nonstationary Strauss process
Log trend:  ~x + y + I(x^2) + I(x * y) + I(y^2)
Fitted trend coefficients:
```

```
(Intercept)              x              y      I(x^2)      I(x * y)      I(y^2)
  -5.4090458   -0.0309246    0.1297984   0.0005028    -0.0002593   -0.0012422
Interaction distance:             6.95
Fitted interaction parameter gamma:            0.1359
```

An algorithm for estimating the interaction $c(d)$ nonparametrically from $\widehat{K}(r)$ using the Percus-Yevick integral equation (another approximate relationship between K and c) was proposed by Diggle *et al.* [229] but this does not seem to work reliably in practice, for reasons which are not well understood.

Bayesian estimation of the interaction $c(d)$ has also been proposed [333, 114].

13.7 Higher-order interactions

There are some useful Gibbs point process models which exhibit interactions of higher order, that is, in which the probability density has contributions from groups of three or more points.

Table 13.3 shows higher-order interactions implemented in spatstat.

FUNCTION	MODEL
AreaInter	area-interaction process
BadGey	multiscale Geyer saturation process
Concom	connected component interaction
Geyer	Geyer saturation process
Hybrid	hybrid of several interactions
Ord	Ord model, user-supplied potential
OrdThresh	Ord model, threshold potential
Saturated	saturated model, user-supplied potential
SatPiece	multiscale saturation process
Triplets	Geyer triplet interaction process

Table 13.3. *Functions in* spatstat *for creating an interpoint interaction object of order 3 or higher.*

The area-interaction model and the Geyer saturation model are quite handy, as they can be used to model both clustering and regularity. Hybrid interactions are extremely versatile.

13.7.1 Area-interaction

The Widom-Rowlinson penetrable sphere model [699] or *area-interaction* process [70] was developed in physics, as a model for interactions between molecules. It has probability density

$$f(\mathbf{x}) = c\kappa^{n(\mathbf{x})}\gamma^{-A(\mathbf{x})} \tag{13.31}$$

where c is a constant, $\kappa > 0$ is an intensity parameter, and $\gamma > 0$ is an interaction parameter. Here $A(\mathbf{x})$ denotes the area of the region obtained by drawing a disc of radius r centred at each point x_i, and taking the union of these discs,

$$A(\mathbf{x}) = \left| W \cap \bigcup_{i=1}^{n(\mathbf{x})} b(x_i, r) \right|. \tag{13.32}$$

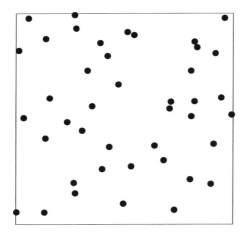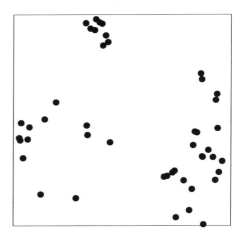

Figure 13.20. *Simulated realisations of the area-interaction process exhibiting inhibition* (Left) *and clustering* (Right). *Scale-free parameters: left panel* $\beta = 250$, $\eta = 0.02$; *right panel* $\beta = 5$, $\eta = 50$; *disc radius* $r = 0.07$ *in the unit square.*

The density (13.31) is integrable for all values of γ. The value $\gamma = 1$ again corresponds to a Poisson process. If $\gamma < 1$, the density favours configurations where $A(\mathbf{x})$ is large, so that the process is regular or inhibitory; while if $\gamma > 1$, the density favours configurations where $A(\mathbf{x})$ is small, so that the process is clustered or attractive. This process has interactions of all orders. It can be used as a model for moderate regularity or clustering. Figure 13.20 shows simulated realisations.

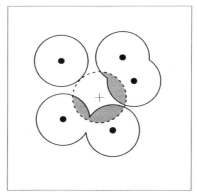

Figure 13.21. *Area-interaction model. Sketch of the 'contested zone' (dark grey) associated with the location u (+) given the point pattern $\mathbf{x}$ ($\bullet$).*

Each isolated point of the pattern $\mathbf{x}$ contributes a factor $\kappa \gamma^{-\pi r^2}$ to the probability density. It is better to rewrite (13.31) in the *canonical scale-free form*

$$f(\mathbf{x}) = c\beta^{n(\mathbf{x})}\eta^{-C(\mathbf{x})} \tag{13.33}$$

where $\beta = \kappa \gamma^{-\pi r^2} = \kappa/\eta$ is the true first-order term, $\eta = \gamma^{\pi r^2}$ is the interaction parameter, and

$$C(\mathbf{x}) = \frac{A(\mathbf{x})}{\pi r^2} - n(\mathbf{x})$$

is the interaction potential term. This is a true interaction, in the sense that $C(\mathbf{x}) = 0$ if the point pattern $\mathbf{x}$ does not contain any points that lie close together (closer than $2r$ units apart).

In the canonical scale-free form, the parameter η can take any nonnegative value. The value $\eta = 1$ again corresponds to a Poisson process, with intensity β. If $\eta < 1$ then the process is regular, while if $\eta > 1$ the process is clustered. The value $\eta = 0$ corresponds to a hard core process with hard core distance $h = 2r$, so that circles of radius r centred at the data points do not overlap.

The conditional intensity is

$$\lambda(u \mid \mathbf{x}) = \beta \eta^{-C(u,\mathbf{x})} \tag{13.34}$$

where $C(u,\mathbf{x}) = C(\mathbf{x} \cup \{u\}) - C(\mathbf{x})$ is the fraction of *contested area*, the fraction of area of the disc of radius r centred on u that is also covered by one or more of the discs of radius r centred at the other points $x_i \in \mathbf{x}$. This is illustrated in Figure 13.21. If the points represent trees, we may imagine that each tree takes nutrients and water from the soil inside a circle of radius r. Then we may interpret $C(u,\mathbf{x})$ as the fractional area of the 'contested zone' where a new tree at location u would have to compete with other trees for resources. For $\eta < 1$ we can interpret (13.34) as saying that a random point is less likely to occur when the contested area is large.

The area-interaction model is fitted in `spatstat` using the function `AreaInter`:

```
> fitSA <- ppm(swedishpines ~ polynom(x,y,2), AreaInter(4))
> cifSA <- predict(fitSA, type = "cif", ngrid=512)
> fitSA
Nonstationary Area-interaction process
Log trend:  ~x + y + I(x^2) + I(x * y) + I(y^2)
Fitted trend coefficients:
(Intercept)           x          y       I(x^2)      I(x * y)      I(y^2)
 -4.599e+00   9.228e-03   2.878e-02   2.562e-04    -6.357e-04   -2.394e-05
Disc radius:            4
Fitted interaction parameter eta:         0.00353333
```

The fitted value $\eta = 0.0035$ is very small, meaning that the model behaves like a hard core at short distances. The left panel of Figure 13.22 shows a plot of the fitted interpoint interaction, generated by `plot(fitin(fitSA))`. It shows the interaction term $\eta^{-C(u,\{v\})}$ between a *pair* of points u,v as a function of the distance between them. Although this is not a pairwise interaction model, the plot gives a sense of the strength and scale of the interaction. The right panel of Figure 13.22 shows the fitted conditional intensity `cifSA`.

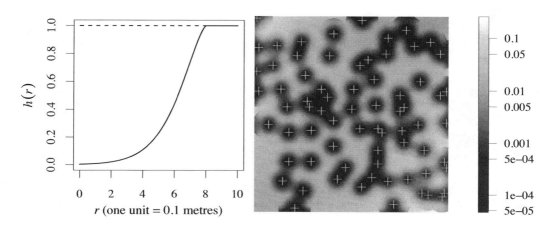

Figure 13.22. *Area-interaction model fitted to the* `swedishpines` *dataset. Left: Fitted interaction. Right: Conditional intensity. Note greyscale is logarithmic.*

The sufficient statistic for the area-interaction model is effectively the empty-space function

$\widehat{F}(r)$ as explained in [70]. The theoretically optimal (Neyman-Pearson) test of CSR against an area-interaction model with known r is based on $n(\mathbf{x})$ and $\widehat{F}(r)$.

13.7.2   Geyer saturation model

In Section 13.3.7 it was explained that a Strauss process with $\gamma > 1$ does not exist, because the probability density is not integrable. In terms of the conditional intensity, $\lambda(u \mid \mathbf{x}) = \beta \gamma^{t(u,r,\mathbf{x})}$ where $t(u,r,\mathbf{x})$ is the number of points in the configuration $\mathbf{x}$ which lie within a distance r of the location u. If we were to take γ greater than 1, then this conditional intensity could take arbitrarily large values, which is unreasonable from a practical standpoint.

To avoid this problem, Geyer [282] proposed the *saturation process*, a modification of the Strauss process in which the overall contribution from each point is trimmed to never exceed a maximum value. The probability density is

$$f(\mathbf{x}) = c \, \beta^{n(\mathbf{x})} \prod_{i=1}^{n(\mathbf{x})} \gamma^{\min(s,\, t(x_i,r,\mathbf{x}))} \tag{13.35}$$

where c is a constant, β, γ, r, s are parameters, and $t(x_i, r, \mathbf{x}) = t(x_i, r, \mathbf{x} \setminus x_i)$ is the number of *other* data points x_j lying within a distance r of the point x_i.

The parameter $s \geq 0$ is a saturation threshold which ensures that each term in the product is never larger than γ^s, so that the product is never larger than $\gamma^{sn(\mathbf{x})}$. If $s > 0$, then the interaction parameter γ may take any positive value (unlike the case of the Strauss process), with values $\gamma < 1$ describing an 'ordered' or 'inhibitive' pattern, and values $\gamma > 1$ describing a 'clustered' or 'attractive' pattern.

We can also allow the values $s = 0$ (when the model reduces to a Poisson process) and $s = \infty$ (when it reduces to the Strauss process with interaction parameter γ^2). The value of s may be fractional rather than integer.

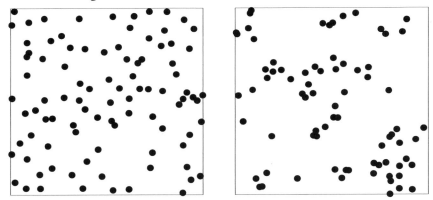

Figure 13.23. *Simulated realisations of Geyer saturation process exhibiting inhibition* (Left) *and clustering* (Right). *Parameters: left panel* $\beta = 300$, $\gamma = 0.5$; *right panel* $\beta = 20$, $\gamma = 2$; *distance threshold* $r = 0.07$ *in the unit square; saturation parameter* $s = 2$.

The task of calculating the conditional intensity of the saturation process seems to be extremely prone to human error. Using (13.4) the conditional intensity is

$$\lambda(u \mid \mathbf{x}) = \beta \prod_{i=1}^{n(\mathbf{x})} \gamma^{\min(s,\, t(u,r,\mathbf{x}))} \prod_{i=1}^{n(\mathbf{x})} \gamma^{\min(s,\, t(x_i,r,\mathbf{x} \cup \{u\})) - \min(s,\, t(x_i,r,\mathbf{x}))}. \tag{13.36}$$

The first product in (13.36) is simply the extra factor introduced into (13.35) associated with the additional point u. The second product is the impact of adding the point u on the other terms in (13.35) associated with existing data points x_i.

Note that the interaction range is $2r$, rather than r, because each term $t(x_i, r, \mathbf{x})$ in (13.36) depends on other points of $\mathbf{x}$ lying up to r units away from x_i and hence up to $2r$ units away from u.

The saturation model is fitted in `spatstat` using the function `Geyer`:

```
> ppm(redwood ~ 1, Geyer(r=0.05, sat=2))
Stationary Geyer saturation process
First order term:   beta = 20.43
Interaction distance:         0.05
Saturation parameter:         2
Fitted interaction parameter gamma:        2.205
```

To choose suitable values for r and s we could minimise AIC using `profilepl`:

```
> df <- expand.grid(r=seq(0.02, 0.1, by=0.001), sat=c(1,2))
> pG <- profilepl(df, Geyer, redwood ~ 1, correction="translate",
                   aic=TRUE)
> as.ppm(pG)
Stationary Geyer saturation process
First order term:   beta = 19.54
Interaction distance:         0.045
Saturation parameter:         2
Fitted interaction parameter gamma:        2.3949
```

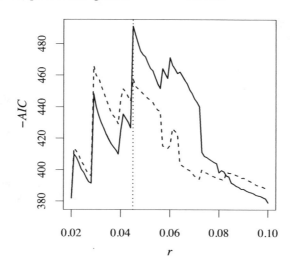

Figure 13.24. *AIC profile for Geyer saturation model fitted to* `redwood` *dataset. Solid lines: saturation $s = 2$. Dashed lines: $s = 1$.*

Figure 13.24 shows the AIC as a function of r for the Geyer saturation model fitted to the `redwood` dataset. Note that we have only considered the two integer values $s = 1$ and $s = 2$.

The case $s = 1$ of the Geyer saturation model is interesting because

$$\min(1, t(x_i, r, \mathbf{x})) = \mathbf{1}\{t(x_i, r, \mathbf{x}) > 0\} = \mathbf{1}\{d(x_i, \mathbf{x}) \le r\}$$

where $d(x_i, \mathbf{x})$ is the nearest-neighbour distance for point x_i. The probability density (13.35) becomes

$$f(\mathbf{x}) = c\,\beta^{n(\mathbf{x})}\gamma^{M(\mathbf{x})} \tag{13.37}$$

where

$$M(\mathbf{x}) = \sum_{i=1}^{n(\mathbf{x})} \mathbf{1}\{d(x_i, \mathbf{x}) \leq r\} \approx n(\mathbf{x})\widehat{G}(r) \qquad (13.38)$$

ignoring edge corrections. That is, for $s = 1$ and if r is known, the sufficient statistic for the Geyer saturation model is effectively $(n(\mathbf{x}), \widehat{G}(r))$. The theoretically optimal (Neyman-Pearson) test of CSR against a Geyer saturation model with known r and known $s = 1$ is based on $n(\mathbf{x})$ and $\widehat{G}(r)$. To estimate the distance parameter r, it would make intuitive sense to inspect the graph of $\widehat{G}$ (rather than $\widehat{K}$) and determine the distance where the deviation from CSR is greatest; however, we do not know whether this is theoretically justified or experimentally supported.

13.7.3   Triplet interaction

The stationary triplet interaction process [282] with parameters β, γ, r has probability density

$$f(\mathbf{x}) = c_7 \beta^{n(\mathbf{x})} \gamma^{\nu(\mathbf{x})} \qquad (13.39)$$

where $\nu(\mathbf{x}) = \nu(\mathbf{x}, r)$ is the number of unordered triplets (x_i, x_j, x_k) of distinct points in $\mathbf{x}$ in which each pair of points is closer than r units: $\|x_i - x_j\| \leq r$, $\|x_i - x_k\| \leq r$, and $\|x_j - x_k\| \leq r$. That is, $\nu(\mathbf{x})$ is the number of triangles formed by the data points such that the longest side of the triangle is less than or equal to r.

The parameter γ (not the same as the Strauss interaction parameter) may take any value between 0 and 1. If $\gamma = 1$ the model reduces to a Poisson process. If $\gamma = 0$ then we get a rather interesting process in which *triangles* of maximum side length less than r are forbidden, but *pairs* of points may come closer than a distance r. For intermediate values $0 < \gamma < 1$ the process is inhibitory.

Figure 13.25 shows simulated values of the triplets process for $\gamma = 0$ and $\gamma = 0.5$, and the conditional intensity for $\gamma = 0$. Although this model is mainly of theoretical interest [282], it has been fitted to real data, as a component in a hybrid model [54, Sec. 11.8].

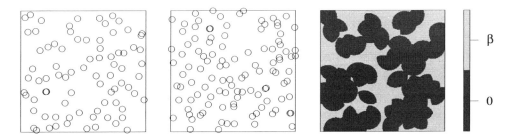

Figure 13.25. *Triplets process.* Left: *simulated realisation with* $\gamma = 0$. Middle: *simulated realisation with* $\gamma = 0.5$. Right: *conditional intensity for left panel. Unit square window, interaction distance* $r = 0.1$.

The sufficient statistic for the triplets model is $(n(\mathbf{x}), \nu(\mathbf{x}))$. Ignoring edge corrections, we notice that $\nu(\mathbf{x})$ is equivalent to $\widehat{T}(r)$, the triangle counting summary statistic introduced in Section 7.12. In order to choose the interaction distance r we should inspect the graph of $\widehat{T}(r)$ using a similar rationale to that used for the Strauss model.

13.7.4   Connected component interaction

The left panel of Figure 13.26 shows the full California redwoods dataset `redwoodfull` with a line segment joining each pair of points that lie closer than 0.1 units apart. The lines were generated by

```
> clo <- closepairs(redwoodfull, 0.1)
> crx <- with(clo, psp(xi, yi, xj, yj, window=Window(redwoodfull)))
```

A connected component of this graph is a set of vertices that can all be reached from one another. The method `connected.ppp` identifies the connected components and gives each point a mark specifying the component to which it belongs:

```
> redcon <- connected(redwoodfull, 0.1)
```

This is shown in the right panel of Figure 13.26.

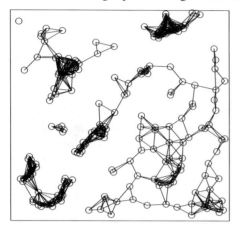 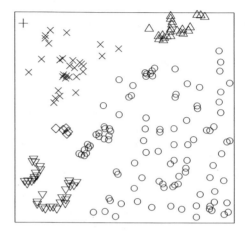

Figure 13.26. Left: *lines join each pair of points in the full redwoods dataset that lie closer than 0.1 units apart.* Right: *points labelled according to the connected component to which they belong.*

The *connected component process* [67, 475, 148] with connection radius r, first-order parameter κ, and interaction parameter γ has probability density

$$f(\mathbf{x}) = c\,\beta^{n(\mathbf{x})}\eta^{n(\mathbf{x})-C(\mathbf{x})} \tag{13.40}$$

where c is a constant and $C(\mathbf{x}) = c(\mathbf{x}, r)$ is the number of connected components in the graph described above.

When a new point u is added to an existing point pattern $\mathbf{x}$, the potential $n(\mathbf{x}) - C(\mathbf{x})$ increases by zero or a positive integer. The increase is zero if u is not close to (within distance r of) any point of $\mathbf{x}$. The increase is a positive integer k if there are k different connected components of $\mathbf{x}$ which contain members that lie close to u. The conditional intensity is $\lambda(u \mid \mathbf{x}) = \beta\eta^{k}$.

Simulation of the connected component process is not yet implemented in `spatstat` at the time of writing, but the model can be fitted to data using the function `Concom`.

The Gordon Square data (Figure 5.21 on page 145) were collected as part of a study on human social interaction [54]. This pattern is a likely candidate for a connected-component interaction model because it is plausible that members of a social group 'interact' with all group members when they choose where to sit on the grass. To select a value for the threshold distance r, we exploit the relationship between this model and a hierarchical clustering procedure called *single link clustering*. The command

```
> plot(hclust(dist(coords(gordon)), "single"))
```

produces the 'dendrogram' shown in Figure 13.27. Vertical lines at the bottom of the diagram represent individual data points. As distance r increases — along the y-axis — connected components gradually fuse together, as shown by the horizontal crossbars. In this example the dendrogram

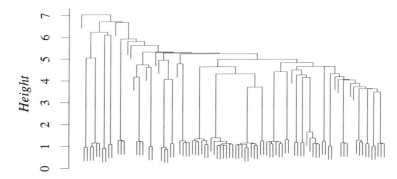

Figure 13.27. *Dendrogram for single-linkage hierarchical clustering based on pairwise distances in Gordon Square data.*

strongly suggests that at a distance of about 1.5 to 2 metres, most of the points have been grouped into sensible components. A plot of `connected(gordon, 2)` would show that these components correspond to the social groups that we see in the Gordon Square data: a few lone individuals, several pairs of people, and some large circles. Taking $r = 2$ as the interaction distance we fit the connected component model:

```
> ppm(gordon ~ 1, Concom(2), rbord=0)
Stationary Connected component process
First order term:  beta = 0.009959
Distance threshold:        2
Fitted interaction parameter eta:        12.4485
```

The estimate of η corresponds to very strong clustering. To assess whether this is significant, we could perform a simple Monte Carlo test using the fitted value of η as the test statistic.

```
> etahat <- function(X) {
    mod <- ppm(X ~ 1, Concom(2), rbord=0)
    return(parameters(mod)$eta)
  }
> (etaobs <- etahat(gordon))
[1] 12.45
> Xsims <- rpoispp(ex=gordon, nsim=99)
> etasim <- sapply(Xsims, etahat)
> (pval <- mean(etaobs <= c(etaobs, etasim)))
[1] 0.01
```

The result appears to be significant.

The connected component interaction is the simplest example of a *nearest-neighbour Markov point process model* [67, 71, 96, 95, 211, 212, 213].

13.7.5   Example of an invalid conditional intensity function

It may be tempting to invent a new point process model by writing down a formula for the conditional intensity function. For example, it is a surprisingly common mistake in spatial statistics to propose a new model with the conditional intensity function

$$\lambda(u \mid \mathbf{x}) = \begin{cases} \beta_1 & \text{if } \|u - x_i\| < r \text{ for some } x_i \in \mathbf{x} \\ \beta_2 & \text{otherwise} \end{cases} \tag{13.41}$$

where β_1 and β_2 are non-negative constants.

The problem here is that *the conditional intensity cannot be chosen at will*; the values $\lambda(u \mid \mathbf{x})$ for different locations u and different patterns $\mathbf{x}$ must be logically consistent.

The conditional intensity $\lambda(u \mid \mathbf{x})$ of a point process is related to its probability density $f(\mathbf{x})$ through equation (13.4). Rearranging this equation gives $f(\mathbf{x} \cup \{u\}) = f(\mathbf{x})\lambda(u \mid \mathbf{x})$. If we know the probability density for $\mathbf{x}$ and the conditional intensity at a new point u given $\mathbf{x}$, we can calculate the probability density for the augmented pattern $\mathbf{x} \cup \{u\}$. Applying this idea repeatedly, we can determine the probability density for any point pattern $\mathbf{x}$ by starting with an empty point pattern $\emptyset$ and building up the pattern one point at a time:

$$f(\mathbf{x}) = f(\emptyset) \times \lambda(x_1 \mid \emptyset) \times \lambda(x_2 \mid \{x_1\}) \times \ldots \times \lambda(x_n \mid \{x_1, \ldots, x_{n-1}\}) \qquad (13.42)$$

where $\mathbf{x} = \{x_1, x_2, \ldots, x_n\}$. Now the left-hand side of (13.42) does not depend on the ordering of the points $x_1, \ldots, x_n$ in the point pattern. Consequently the right-hand side of (13.42) must not depend on the ordering of the points. This is the logical consistency requirement for a conditional intensity function.

For the proposed conditional intensity function (13.41), consider three points, a, b, and c, such that $||a - b|| < r$, $||b - c|| < r$, but $||a - c|| > r$. Adding the points in the order a, b, c gives

$$\lambda(a \mid \emptyset) \times \lambda(b \mid \{a\}) \times \lambda(c \mid \{a, b\}) = \beta_2 \times \beta_1 \times \beta_1,$$

but adding the points in a different order gives

$$\lambda(a \mid \emptyset) \times \lambda(c \mid \{a\}) \times \lambda(b \mid \{a, c\}) = \beta_2 \times \beta_2 \times \beta_1.$$

These are not equal. Consequently, (13.41) is not a valid conditional intensity function. Thus, a quite reasonable-looking candidate for the conditional intensity may well turn out to be invalid.

13.8 Hybrids of Gibbs models

The interpoint interaction models described above have relatively simple mathematical structure, which simplifies theoretical study and software coding, but makes them unrealistic in many applications. In particular, the most commonly used Gibbs models exhibit spatial interaction at only a single spatial scale, whereas most natural processes exhibit dependence at multiple scales. This has motivated statisticians to construct 'multi-scale' generalisations of the classical Gibbs models [14, 333, 547], [221, section 4.9], [225, p. 109].

Despite Geyer's assertion that it is "easy to invent new processes and do statistical inference for them" [282, p. 110], this is not easy for the average user. If we write down a formula for the probability density $f(\mathbf{x})$ we must first check that the density is integrable (a famous example of non-integrability is the Strauss process for $\gamma > 1$). The density must satisfy other technical requirements such as 'local stability' (Section 13.12.5) in order that we may simulate realisations of the model using standard techniques [283, 484]. Then the algorithms for parameter estimation and simulation must be implemented in software — usually also requiring further research to ensure the validity and efficiency of these algorithms. Meaningful interpretation of the model is a further challenge.

There is one very easy way to build new Gibbs interaction models, and that is to form *hybrids*. A hybrid [54] is a point process model created by combining two or more existing point process models. Following is a brief introduction; see [54] for details and a case study.

The hybrid of two point processes, with probability densities $f_1(\mathbf{x})$ and $f_2(\mathbf{x})$, respectively, is the point process with probability density

$$f(\mathbf{x}) = c f_1(\mathbf{x}) f_2(\mathbf{x}) \qquad (13.43)$$

where c is a normalising constant (assuming the right-hand side is integrable). Equivalently, the hybrid of two point processes with conditional intensities $\lambda_1(u \mid \mathbf{x})$ and $\lambda_2(u \mid \mathbf{x})$ is the point process with conditional intensity

$$\lambda(u \mid \mathbf{x}) = \lambda_1(u \mid \mathbf{x}) \, \lambda_2(u \mid \mathbf{x}). \tag{13.44}$$

The hybrid of any number of point processes is defined in a similar way.

Thanks to the simple relationship (13.44), the technical properties of a hybrid can be deduced from the known properties of the original models [54, Sect. 4]. In particular, it is fairly easy to determine whether the hybrid is a valid point process. We can easily fit hybrid interaction models to data, and generate simulated realisations of hybrid models, using existing code for calculating the conditional intensities of the original models.

The appropriate description of a hybrid interaction in `spatstat` is yielded by the function `Hybrid()`.

```
> Hybrid(Strauss(0.1), Geyer(0.2, 3))
```

The arguments of `Hybrid` will be interpreted as interpoint interactions, and the result will be the hybrid of these interactions. Each argument must either be an interpoint interaction (object of class `"interact"`), or a point process model (object of class `"ppm"`) from which the interpoint interaction will be extracted. Arguments may also be given in the form `name=value` to attach simple mnemonic names to the component interactions.

```
> M <- Hybrid(H=Hardcore(), G=Geyer(0.2, 3))
> fit <- ppm(redwood ~ 1, M, correction="translate")
> fit
Stationary Hybrid interaction
First order term:  beta = 77.11
Hybrid of 2 components: 'H' and 'G'
H:
Interaction:Hard core process
Hard core distance:         0.01968
G:
Interaction:Geyer saturation process
Interaction distance:       0.2
Saturation parameter:       3
Fitted G interaction parameter gamma:        0.9434
```

Here we used the translation edge correction because the interaction range of the model is too large for the border correction.

In the example just given, the hybrid model has a straightforward interpretation. The hybrid of a hard core model with a Geyer saturation model is simply a Geyer model *conditioned* on the event that the hard core is respected. The same principle applies to the hybrid of a hard core with any Gibbs model. In this way we can introduce hard core effects into any of the Gibbs models described in this chapter.

Hybrids have a straightforward interpretation from the viewpoint of statistical physics: a hybrid is created simply by adding the potential energy functionals of several models.

The statistical interpretation of a hybrid requires a little care. Each factor in (13.43) serves as a *relative* term which modifies the overall probability of the point pattern $\mathbf{x}$. The interpretation of the parameters of the component models becomes slightly different when they are part of a hybrid. See [54] for further details.

In some cases we can regard one of the factors in (13.43) or (13.44) as an offset or baseline that should be included in any sensible model, while the other factor describes the interesting part of the model [54].

The software makes it easy to incrementally add new terms into a hybrid model:

```
> Mplus <- Hybrid(fit, g=Geyer(0.1,1))
> update(fit, Mplus)
```

The fitted interaction `fitin(m)` of a hybrid model `m` can be plotted. By default the combined interaction is plotted; if `separate=TRUE` the separate components are shown.

```
> fit2 <- ppm(swedishpines ~ 1,
               Hybrid(DG=DiggleGratton(2,10), S=Strauss(5)))
> plot(fitin(fit2))
> plot(fitin(fit2), separate=TRUE, mar.panel=rep(4,4))
```

Hybrids are particularly useful for modelling interactions at multiple scales. A hybrid of several point process densities, each having a different characteristic 'scale' of interaction, will typically exhibit 'multi-scale' interaction. The multi-scale generalisation of the Strauss process — namely, a pairwise interaction process with a step function potential [221, section 4.9], [225, p. 109], [333], — is a hybrid of several Strauss processes. Ambler and Silverman's [14, 15, 547] multi-scale generalisation of the area-interaction model [699, 70] is a hybrid of area-interaction potentials.

13.9 Simulation

Gibbs point process models can be simulated by Markov chain Monte Carlo algorithms: indeed, these algorithms were originally invented to simulate Gibbs processes [465, 329, 573].

The basic idea is that a Gibbs process is the equilibrium state of a spatial birth-and-death process (introduced in Section 13.3.5). By running the birth-and-death process for a very long time, we can expect to come close to equilibrium, so that the state of the birth-and-death process after many generations can be taken as a simulated realisation of the desired Gibbs point process [554, 474]. The spatial birth-and-death process evolves in continuous time; it would be enough to simulate the transitions (births or deaths) at the successive instants when transitions occur. The `spatstat` package includes a flexible implementation of this approach, in the function `rmh`, which implements the *Metropolis-Hastings algorithm* for Gibbs point processes [283], [484, chap. 7]. This is currently also used in `simulate.ppm`, the method for the generic `simulate`.

A weakness of the Metropolis-Hastings algorithm is that it only achieves approximate convergence — although the approximation is usually very good. *Exact simulation* is an alternative approach in which convergence is guaranteed to have occurred when the algorithm terminates. The `spatstat` package also provides some exact simulation algorithms for specific models, listed in Table 13.4. See Section 13.9.3. A weakness of these algorithms is that they may fail to terminate in a reasonable time when interaction is very strong.

13.9.1 Simulation using `rmh`

The `spatstat` function `rmh` simulates Gibbs processes using a Metropolis-Hastings algorithm.

```
rmh(model, start, control, ...)
```

The argument `model` determines the point process model to be simulated; `start` determines the initial state of the Markov chain; and `control` specifies control parameters for running the Markov chain and for saving results.

The function `rmh` is generic, with methods `rmh.default` and `rmh.ppm`. The default method accepts a simple specification of the Gibbs model that can easily be put together by hand. The method `rmh.ppm` makes it possible to simulate a fitted Gibbs point process model directly.

If the model is a Poisson process, then it is still permissible to simulate it using `rmh`, but the

Metropolis-Hastings algorithm is not used; the Poisson model is generated directly using `rpoispp` or `rmpoispp`.

13.9.1.1 Simulation of a fitted model: `rmh.ppm`

The method `rmh.ppm` generates simulated realisations from a fitted Gibbs model.

```
> fit <- ppm(swedishpines ~ 1, Strauss(8))
> Y   <- rmh(fit)
```

The result `Y` is a simulated realisation from the fitted Strauss model `fit` in the same window. The current implementation enables simulation from a fitted Gibbs model involving almost any[3] of the interactions described in this chapter, including hybrid models.

As explained in Section 13.5.4, it is possible that the fitted coefficients of a point process model may be 'invalid', i.e. that there may not exist a mathematically well-defined point process with the given parameter values. For example, we saw that a Strauss process with interaction parameter $\gamma > 1$ does not exist, but the model-fitting procedure used in ppm will sometimes produce values of γ greater than 1. The default behaviour of `rmh.ppm` is to find the nearest legal model using `project.ppm` and simulate this model instead.

Additional arguments to `rmh.ppm` are passed to `rmh.default` and these are discussed below.

The result of `rmh` is a point pattern. It carries additional information about the simulation parameters, which can be revealed using `print` or `summary`.

13.9.1.2 Simulation of a chosen model: `rmh.default`

The default method `rmh.default` accepts a simple specification of the Gibbs point process model. To generate a simulated realisation of the Strauss process with parameters $\beta = 2, \gamma = 0.2, r = 0.7$ in a square of side 10,

```
> mo <- list(cif="strauss",
            par=list(beta=2,gamma=0.2,r=0.7), w=square(10))
> X <- rmh(model=mo,
            start=list(n.start=42), control=list(nrep=1e6))
```

The other arguments specify a random initial state of 42 points, and that the algorithm shall be run for a million iterations.

The argument `model` specifies the model to be simulated. It should be a list including the entries `cif` (a character string identifying the interpoint interaction), `par` (a list of model parameters), `w` (the spatial window in which the simulation should be generated), and optionally `trend` (a specification of the inhomogeneous first-order trend) and `types` (the vector of possible types, for a multitype point process). This loosely structured information will be checked by the function `rmhmodel` and converted into an object of class "rmhmodel" ready for simulation. Details of the requirements for these arguments are given in the help for `rmhmodel.default`. We recommend calling `rmhmodel` explicitly to check the model parameters:

```
> mo <- rmhmodel(cif="strauss",
            par=list(beta=2,gamma=0.2,r=0.7), w=square(10))
```

The `lookup` conditional intensity function permits approximate simulation of any pairwise interaction process in which the interaction between a pair of points depends only on the distance between them.

---

[3]At the time of writing, the exceptions are `OrdThresh`, `Concom`, and `SatPiece`.

A hybrid model is specified by a vector of character strings giving the names of the component interactions, and a list containing the parameter lists for these interactions. For example a hybrid of the hard core model with hard core distance 0.03, and the Strauss model with interaction range 0.07 and interaction parameter $\gamma = 0.5$, could be specified by

```
> rmhmodel(cif = c('hardcore', 'strauss'),
          par = list(list(beta = 10, hc = 0.03),
                     list(beta = 1, gamma = 0.5, r = 0.07)),
          w   = square(1))
```

13.9.1.3 Control of the simulation

The optional argument `start` of `rmh` specifies the initial state. There is a sensible default. Although the Metropolis-Hastings algorithm converges to the same equilibrium from any initial state, convergence will be faster from a well-chosen initial state. If `start` is given, it should be a list including *either* the entry `n.start` specifying the initial number of points, *or* the entry `x.start` specifying the initial point pattern. This information will be checked by the function `rmhstart`: see `help(rmhstart.default)`.

The optional argument `control` controls the simulation procedure, including proposal distributions, conditional simulation, iterative behaviour, tracking, and termination. It should be a list with optional components p (probability of proposing a 'shift' transition, Section 13.9.2, as opposed to a birth or death), q (conditional probability of proposing a death given that the proposal is either a birth or death), `nrep` (number of iterations of algorithm), `expand` (data specifying that the model will be simulated on a larger window and clipped to the original window), `periodic` (data specifying that spatial coordinates should be treated as periodic so that opposite edges of the window are joined), `ptypes` (probability distribution to be used in assigning a random type to a new point), `fixall` (logical value indicating whether to condition on the number of points of each type), and `nverb` (number of iterations between progress reports). If the entry `x.cond` is present, then conditional simulation will be performed, and `x.cond` specifies the conditioning points and the type of conditioning (see below). If the entries `nsave`, `nburn` are present, then intermediate states of the simulation algorithm will be saved every `nsave` iterations after an initial burn-in period of `nburn` iterations. For full information see the help for `rmhcontrol`.

Entries in the arguments `model`, `start`, `control` can also be passed individually to `rmh` if their names are unambiguous. For example:

```
> X4 <- rmh(model=mo, nrep=1e4)
```

Default values of the parameters p, q, and `nrep` can also be set in `spatstat.options` using the arguments `rmh.p`, `rmh.q`, `rmh.nrep`, respectively.

13.9.1.4 Conditional simulation

It is sometimes desired to simulate a Gibbs model conditionally on some constraint or information. There are several kinds of conditional simulation.

Simulation *conditional upon the number of points* means that we hold the number of points fixed. To do this in `rmh`, set `control$p` (the probability of a shift transition) equal to 1. The number of points is then determined by the starting state, which may be specified either by setting `start$n.start` to be a single number, or by setting the initial pattern `start$x.start`.

In the case of multitype processes, it is possible to simulate the model *conditionally upon the number of points of each type*, i.e. holding the number of points of each type to be fixed. To do this, set `control$p` equal to 1 and `control$fixall` to be TRUE. The number of points is then determined by the starting state, which may be specified either by setting `start$n.start` to be an integer vector, or by setting the initial pattern `start$x.start`.

We can also simulate *conditional on the configuration observed in a sub-window*, that is, requiring that, inside a specified sub-window V, the simulated pattern should agree with a specified point pattern **y**. To do this, set `control$x.cond` to equal the specified point pattern y, making sure that it is an object of class `"ppp"` and that the window `Window(control$x.cond)` is the conditioning window V.

We can simulate *conditional on the presence of specified points*, that is, requiring that the simulated pattern should include a specified set of points. This is simulation from the Palm distribution of the point process given a pattern **y**. To do this, set `control$x.cond` to be a `data.frame` containing the coordinates (and marks, if appropriate) of the specified points.

13.9.1.5   Window expansion

If the parameter `expand` is given, the point process model will be simulated on a window larger than the original data window. This is appropriate in the context of Window Sampling (page 143) where the point process is thought to exist on a larger domain.

The argument `expand` specifies the rule for expansion of the window, or gives the expanded window itself. The expansion rules are parsed by `rmhexpand`. For example, we can specify that the window should be expanded to double its area (`expand=2` or `expand=list(area=2)`), double the spatial scale (`expand=list(length=2)`), or extended by a distance of 10 units on all sides (`expand=list(distance=10)`).

Of course, the model must be capable of extrapolation to this larger window. This is usually not possible for models which depend on external covariates, because the domain of a covariate image is usually the same as the domain of the fitted model. Nor is it possible when we simulate conditionally on the number of points, or conditionally on the number of points of each type.

If the model has a trend, then in order for expansion to be feasible, the trend must be given either as a function, or an image whose bounding box is large enough to contain the expanded window.

If `start$x.start` is specified then `expand` is set equal to 1 and simulation takes place in `Window(start$x.start)`. Any specified value for `expand` is simply ignored.

For point process models which have a trend depending only on x and y, the simulation window is taken to be the same as the original window containing the data (by default). That is, 'expansion' does not take place, by default. This avoids the common problem that the extrapolation of such models often leads to huge values of the extrapolated intensity, so that the algorithm fails to converge in the time available.

13.9.1.6   Monitoring the Metropolis-Hastings algorithm

Monitoring the behaviour of the Metropolis-Hastings algorithm is sometimes important, for example, for checking convergence and for improving performance. After convergence, the rate of acceptance of proposals should stabilise. For greatest efficiency the acceptance rate should lie somewhere between 0.1 and 0.5.

If `control$track=TRUE`, the history of transitions of the algorithm will be saved, and returned as an attribute of the resulting point pattern.

```
> fit <- ppm(swedishpines ~ 1, Strauss(8))
> Y <- rmh(fit, track=TRUE)
> h <- attr(Y, "history")
> with(h, mean(accepted))
[1] 0.3935
> with(h, prop.table(table(proposaltype, accepted), 1))
            accepted
proposaltype    FALSE      TRUE
       Birth  0.10377   0.89623
```

```
Death 0.90176 0.09824
Shift 0.58255 0.41745
```

The acceptance rate is about 39 percent which is good overall. However, the very asymmetric acceptance rates for births and for deaths may suggest that the proposal probabilities for these transitions should be adjusted.

If `control$nsave` is given, then the state of the Metropolis-Hastings algorithm will be saved every `nsave` iterations. A list of these intermediate states will be returned as an attribute `"saved"` of the resulting point pattern. This was used to generate Figure 13.6.

Figure 13.28. *Screenshot of graphical debugger for Metropolis-Hastings algorithm.*

If `snoop = TRUE`, an interactive debugger is activated. The debugger displays the current state of the Metropolis-Hastings algorithm together with the proposed transition to the next state. Figure 13.28 shows a screenshot of the debugger display. The user can interact with the graphical display, print detailed information, dump the current state to a file, accept or reject the proposed transition, jump to the next iteration or skip some iterations, and so on.

13.9.2 Theory for the Metropolis-Hastings algorithm

Markov chain methods for simulating spatial point process models are surveyed in [484, chap. 7], [477], [275, chap. 4]. For Gibbs models, simulation algorithms typically require only computation of the Papangelou conditional intensity.

The standard birth-death Metropolis-Hastings algorithm for point processes [283], [484, chap. 7] is a discrete-time Markov chain whose states are point patterns $\mathbf{x}$. At each time step, a 'proposal' is made to change the current state $\mathbf{x}$ to a new state $\mathbf{y}$. The proposal is either 'accepted' (the state is changed to $\mathbf{y}$) or 'rejected' (the state remains as $\mathbf{x}$) according to the outcome of a random coin-flip with probability $A(\mathbf{x}, \mathbf{y})$ of accepting the proposal.

Each proposal is either a 'death' or a 'birth' with probabilities q and $1 - q$, respectively. In a 'birth' proposal $\mathbf{x} \mapsto \mathbf{x} \cup \{u\}$ the existing configuration $\mathbf{x}$ is augmented by adding a random point u with proposal density $b(u \mid \mathbf{x})$ in the simulation window W. A 'death' proposal $\mathbf{x} \mapsto \mathbf{x} \setminus x_i$ is the deletion of one of the existing points of the configuration $\mathbf{x}$, chosen with equal probability $1/n(\mathbf{x})$. For a target point process density $f(\mathbf{x})$ the Metropolis-Hastings acceptance probabilities are $A(\mathbf{x},\mathbf{y}) = \min(1,R(\mathbf{x},\mathbf{y}))$ where $R(\mathbf{x},\mathbf{y})$ is the *Hastings ratio*,

$$R(\mathbf{x},\mathbf{x}\cup\{u\}) = \frac{f(\mathbf{x}\cup\{u\})q/n(\mathbf{x}\cup\{u\})}{f(\mathbf{x})(1-q)b(u\mid\mathbf{x})} = \frac{\lambda(u\mid\mathbf{x})}{b(u\mid\mathbf{x})}\frac{q}{(1-q)(n(\mathbf{x})+1)},$$

$$R(\mathbf{x},\mathbf{x}\setminus x_i) = \frac{f(\mathbf{x}\setminus x_i)(1-q)b(x_i\mid\mathbf{x}\setminus x_i)}{f(\mathbf{x})q/n(\mathbf{x})} = \frac{b(x_i\mid\mathbf{x}\setminus x_i)}{\lambda(x_i\mid\mathbf{x})}\frac{n(\mathbf{x})(1-q)}{q}.$$

Computation of the acceptance probabilities requires only the ratio $\lambda(u\mid\mathbf{x})/b(u\mid\mathbf{x})$ of the conditional intensity to the birth proposal density. However, convergence rates depend on the choice of proposal density $b(u\mid\mathbf{x})$.

The simulation algorithm implemented in `spatstat` has birth, death, and *shift* proposals. In a shift proposal, one of the points x_i in the current configuration is selected at random and moved to another location u, so that the pattern $\mathbf{x}$ is changed to $\mathbf{x}\setminus x_i\cup\{u\}$. Currently the new location is generated uniformly at random in the simulation window (or with probability density proportional to the first-order trend). The Hastings ratio is

$$R(\mathbf{x},\mathbf{x}\setminus x_i\cup\{u\}) = \frac{f(\mathbf{x}\setminus x_i\cup\{u\})p}{f(\mathbf{x})p} = \frac{\lambda(u\mid\mathbf{x}\setminus x_i)}{\lambda(x_i\mid\mathbf{x})}.$$

There is never a guarantee that the Metropolis-Hastings algorithm has converged to its limiting distribution. The Metropolis-Hastings algorithm may have poor convergence in some situations. It has the great advantage of being computationally reliable and applicable to a wide range of models.

13.9.3 Exact simulation

Exact simulation is another class of simulation procedures, in which convergence is guaranteed to have occurred when the algorithm terminates.

A simple metaphor (due to P. Diaconis) may help. Imagine that we shuffle a pack of cards, one card at a time, by picking the top card and replacing it in a randomly chosen position in the pack. The Ace of Spades was at the bottom of the pack. When the Ace of Spades finally reaches the top of the pack and is randomly shuffled, we know that every card has been shuffled, and the pack is *guaranteed* to be in a completely random order. We do not have to wait an infinite time to ensure the pack is shuffled: the procedure terminates after a finite time. The drawback is that it may take a very long time.

rDiggleGratton	Diggle-Gratton model
rDGS	Diggle-Gates-Stibbard model
rFiksel	Fiksel model
rHardcore	hard core process
rStrauss	Strauss process
rStraussHard	Strauss-hard core process

Table 13.4. *Functions for exact simulation of Gibbs processes in* `spatstat`.

Exact simulation techniques such as coupling-from-the-past (CFTP) have been developed [307, 93, 375] for point process densities f which are purely inhibitory, i.e. for which $\mathbf{x}\subset\mathbf{y}$ implies $f(\mathbf{x})\geq f(\mathbf{y})$. In such techniques the spatial birth-death process associated with a Poisson point process is coupled to a sampler for the target process. The `spatstat` package contains CFTP samplers for the point processes listed in Table 13.4 based on original code for the Strauss model by K.K. Berthelsen. By default the point pattern is generated on a larger window than desired, and trimmed to the original window, to reduce edge effects.

Experience with these algorithms is that they are very fast in most cases, but that the computation time increases dramatically as the strength of inhibition increases. For example, `rHardcore` becomes dramatically slower as the hard core distance R increases. In these extreme cases the algorithm may run out of memory before it has terminated. Therefore in extreme cases the user is better advised to fall back on rmh, and to monitor convergence as described above.

For further information on exact simulation of point processes, see [661, p. 89 ff.], [484].

13.9.4 Simulated tempering

For Gibbs processes with very strong inhibition, *simulated tempering* is the method of choice for generating simulated realisations. If $f(\mathbf{x})$ is the target probability density and $T > 0$, let $f_T(\mathbf{x}) = c_T f(\mathbf{x})^{1/T}$ be the modified probability density, where c_T is a constant needed to normalise the density. In the statistical physics analogy, T corresponds to the temperature of the system. If $T = 1$ we have the target density. If we increase T, the range of values of $f_T(\mathbf{x})$ becomes narrower, and the point process with density $f_T(\mathbf{x})$ has weaker interaction. Intuitively, at higher temperatures, interactions between molecules become less important.

To generate a realisation from the desired point process with density $f(\mathbf{x})$, we may start by generating a realisation $\mathbf{X}_1$ from the process with density $f_T(\mathbf{x})$ with a large temperature $T = T_1$, using (say) the Metropolis-Hastings algorithm. Then reducing the temperature to $T_2 < T_1$, and taking $\mathbf{X}_1$ as the initial state, we generate a realisation $\mathbf{X}_2$ from the process with density $f_T(\mathbf{x})$ with $T = T_2$. Continuing to reduce the temperature we eventually arrive at $T = 1$ and obtain a realisation from the desired process. This is analogous to the metallurgical process of tempering, in which a bar of metal is heated to a high temperature and worked while the temperature is progressively reduced.

Simulated tempering for hard core processes has been studied in [432, 451]. The spatstat function temper provides a simulated tempering algorithm based on the Metropolis-Hastings sampler.

13.10 Goodness-of-fit and validation for fitted Gibbs models

For a fitted Poisson point process model, tests of goodness-of-fit were described in Chapter 10, and diagnostics for validating the model were described in Chapter 11. These techniques have been extended to Gibbs point process models [55, 28, 45, 36, 49, 37, 35, 2, 157, 668, 712, 711, 710].

13.10.1 Overview

Table 13.5 lists generic functions with methods for class "ppm" that are particularly useful for validating the fitted model.

For example, envelope has a method for class "ppm", so that it is easy to generate simulation envelopes of a summary function based on simulation from a fitted Gibbs model. Other methods make it possible to perform specialised hypothesis tests (dclf.test, berman.test), compute diagnostics for model validation, extract the fitted interaction, and compute the *predicted K*-function or pair correlation function of the model.

13.10.2 Goodness-of-fit testing for Gibbs processes

Standard goodness-of-fit tests based on the cumulative distribution function, such as the Kolmogorov-Smirnov test, are not directly applicable to a fitted Gibbs point process model. The points in a Gibbs process are dependent, violating the assumption of independence on which the test is based. The

`berman.test`	Berman's tests
`cdf.test`	spatial CDF tests
`quadrat.test`	χ^2 test based on quadrat counts
`envelope`	simulation envelopes of summary function
`residuals`	compute residuals
`leverage`	leverage diagnostic
`influence`	likelihood influence diagnostic
`dfbetas`	parameter influence diagnostics
`relrisk`	relative risk predicted by model
`rhohat`	relative intensity as a function of covariate
`Kmodel`	predicted *K*-function of model
`pcfmodel`	predicted pair correlation function of model

Table 13.5. *Generic operations with methods for class* `"ppm"` *for validating a fitted model.*

test statistic itself is difficult to calculate, because it involves the intensity of the process, which is not known exactly for Gibbs models.

A practical solution is to replace the intensity $\lambda(u)$ by the conditional intensity $\lambda(u \mid \mathbf{x})$ in the definition of the test statistic, and to perform a Monte Carlo test of goodness-of-fit. This is implemented in `cdf.test` as the default behaviour if the fitted model is not a Poisson process. It is justified at least when the model has weak interaction, when it will be approximately equivalent to the classical goodness-of-fit test.

In the Queensland copper data (see Sections 1.1.4 and 6.6.4) we could assume there is clustering of deposits as well as dependence on distance to the nearest lineament:

```
> dlin <- distfun(copper$SouthLines)
> copfit <- ppm(copper$SouthPoints ~ dlin, Geyer(1, 1))
```

The analogue of the Kolmogorov-Smirnov test for the distribution of distance to the nearest lineament under the model — based on the conditional intensity, and using Monte Carlo simulations — is performed by:

```
> coptest <- cdf.test(copfit, dlin, nsim=39)
> coptest

        Monte Carlo spatial Kolmogorov-Smirnov test of Gibbs process in two
        dimensions

data:  covariate 'dlin' evaluated at points of 'copper$SouthPoints'
     and transformed to uniform distribution under 'copfit'
D = 0.069, p-value = 0.2
alternative hypothesis: two-sided
```

The result suggests no evidence against the fitted model, based on comparing the observed and expected distributions of distance-to-nearest-lineament. This is primarily an assessment of the trend component of the model, after allowing for the interaction.

Another standard approach to goodness-of-fit testing is to use simulation envelopes of summary functions such as the *K*-function, or the related tests such as the Diggle-Cressie-Loosmore-Ford test. The `spatstat` command `envelope` will accept as its first argument a fitted Gibbs model, and will simulate from this model to determine the critical envelopes.

For example, let us fit an inhomogeneous Strauss model to the Swedish Pines data:

```
> swedfit <- ppm(swedishpines ~ polynom(x,y,2), Strauss(9))
```

Now we compute the envelopes of the *L*-function based on simulations from the fitted model:

```
> swedenv <- envelope(swedfit, Lest, nsim=39, global=TRUE,
                      savepatterns=TRUE)
```

where we set `savepatterns=TRUE` to save the simulated point patterns for re-use in other calculations. The centred *L*-function is plotted in Figure 13.29 using the plot formula `. ~ . - r`. This does not suggest significant departure from the fitted model.

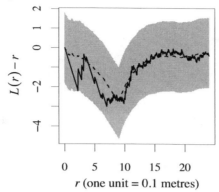

Figure 13.29. *Centered L-function for the copper data (solid black line) and global envelopes of the same function from 39 simulations of the fitted inhomogeneous Geyer model.*

Readers may object that this is not appropriate because the fitted model is inhomogeneous, and the definition of the *L*-function assumes a stationary process. The formal hypothesis test of goodness-of-fit is still valid here: the rationale for Monte Carlo tests (Section 10.6) does not require stationarity. However, if the test had rejected the fitted model, it would have been difficult to interpret the deviations of the *L*-function outside the envelopes, since these deviations could be caused either by a lack of fit to the trend in the model or a lack of fit to the interaction. The main weakness of the goodness-of-fit test based on the envelopes in Figure 13.29 is the usual problem of conservatism, explained in Section 10.6.3, which may lull us into a false sense of security. For Gibbs processes the need for validation diagnostics is even greater than for Poisson processes.

13.10.3 Residuals for Gibbs processes

13.10.3.1 Definition

Point process residuals for a general Gibbs model were first defined in [55]. See also [45, 161]. The raw residual for any region $B \subset \mathbb{R}^2$ is defined as

$$\mathscr{R}(B) = n(\mathbf{x} \cap B) - \int_B \widehat{\lambda}(u \mid \mathbf{x}) \, du \tag{13.45}$$

where again $n(\mathbf{x} \cap B)$ is the observed number of points in the region B, and $\widehat{\lambda}(u \mid \mathbf{x})$ is the **conditional intensity** of the fitted model, *evaluated for the data point pattern* $\mathbf{x}$. If the fitted model is correct, the residual $\mathscr{R}(B)$ has mean zero, by the Campbell-Mecke formula (8.50).

Equation (13.45), applied to all regions B, defines a 'measure' $\mathscr{R}$ on two-dimensional space, as explained in Section 11.3.1.

This definition is similar to the definition of residuals for Poisson processes (Section 11.3) except that the intensity $\widehat{\lambda}(u)$ of the fitted Poisson process has been replaced by the *conditional* intensity $\widehat{\lambda}(u \mid \mathbf{x})$ of the fitted Gibbs process evaluated for the data point pattern $\mathbf{x}$. By using the conditional intensity, the calculation of the residual includes an adjustment for the fitted interaction between points as well as the fitted spatial trend.

For example, a hard core process (Section 13.2) has conditional intensity $\lambda(u \mid \mathbf{x}) = \beta$ if u is 'permissible' and $\lambda(u \mid \mathbf{x}) = 0$ if u is not 'permissible', where u is 'permissible' if it lies further than h units away from any point in $\mathbf{x}$. Here h is the hard core distance and β is the activity parameter. The residual for a region B is

$$\mathscr{R}(B) = n(\mathbf{x} \cap B) - \widehat{\beta} |B \setminus \mathbf{x}_{\oplus h}|, \tag{13.46}$$

in which $B \setminus \mathbf{x}_{\oplus h}$ is the 'permissible' part of B. See the left panel of Figure 13.3. The mean value of $\mathscr{R}(B)$ is equal to zero if the model is true, by the Campbell-Mecke formula (8.50), or by appealing to the equilibrium principle (Figure 13.7).

The definition of the residuals is derived [55] from statistical theory applied to the pseudolikelihood; essentially, the residuals are increments of the pseudolikelihood score. Consequently the residuals arise frequently in different contexts in statistical analysis of point patterns: they encapsulate *all* discrepancies between the data and the fitted model. Although there are many other quantities which have the property that their mean is equal to zero when the model is true, these do not have the same fundamental connection to model-fitting as the residuals do.

A common misunderstanding of equation (13.45) is that it presupposes a particular choice of the region or regions B. It is incorrect to describe $\mathscr{R}(B)$ as a 'pixel residual' requiring a choice of pixel size, or a 'quadrat residual' assuming a particular division of space into quadrats. On the contrary, (13.45) is meant to be applied simultaneously to *all possible sets B* in the window W. As explained in Section 11.3.1, equation (13.45) defines a signed **measure** on the window W, which we can think of as a spatial distribution of 'electric charge'. Each of the data points carries a positive electric charge of 1 unit. The background space carries a diffuse negative charge, with a spatially varying density of $\widehat{\lambda}(u \mid \mathbf{x})$ units of negative charge per unit area. If the fitted model is true, then the positive and negative charges *in any region* should approximately cancel, and the total charge *in any region* should be approximately zero. For further explanation see Section 11.3.1.

The raw residual measure for a fitted Gibbs model is computed by `residuals.ppm`:

```
> swedR <- residuals(swedfit)
```

The result is an object of class `"msr"` representing a signed measure. Methods for this class are listed in Section 11.3.1. The result of `plot(swedR)` is shown in the left panel of Figure 13.30.

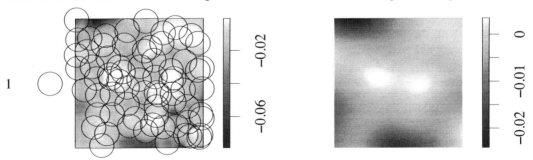

Figure 13.30. *Raw residuals for inhomogeneous Strauss model fitted to Swedish Pines data.* Left: *raw residual measure* (13.45). *Each circle represents a positive mass of 1 unit. Greyscale represent negative residual density.* Right: *smoothed raw residual field* (13.47).

13.10.3.2 Smoothed residuals

A smoothed version of the residuals was defined for Poisson models in Section 11.3.2. Similarly for a Gibbs model, the smoothed raw residual field is

$$s(u) = \frac{1}{e(u)} \left[\sum_{i=1}^{n(\mathbf{x})} \kappa(u - x_i) - \int_W \kappa(u - v)\widehat{\lambda}(v \mid \mathbf{x})\,dv \right] \tag{13.47}$$

where κ is a two-dimensional smoothing kernel, and $e(u)$ is the edge correction defined in equation (6.10) on page 168. The smoothed residual field is a spatially weighted average of the positive residual masses and the negative residual density. The right-hand side of (13.47) is the difference $s(u) = \widetilde{\lambda}(u) - \lambda^{\dagger}(u \mid \mathbf{x})$ between a nonparametric kernel estimate of intensity $\widetilde{\lambda}(u)$ and a smoothed version of the conditional intensity, similar to (11.5).

The function `Smooth.msr` applies kernel smoothing to a measure. The right panel of Figure 13.30 shows the smoothed raw residual field $s(u)$ for the model fitted to Swedish Pines, computed by `Smooth(swedR)`.

13.10.3.3 Inverse-lambda residuals and exponential energy marks

As in Section 11.3.5 we can define scaled (or weighted) versions of the residual measure $\mathcal{R}$. In the *inverse-lambda* residual measure, each data point is weighted by the reciprocal of its fitted conditional intensity. In the simple case where the conditional intensity is always positive,

$$\mathcal{R}^{(\mathrm{I})}(B) = \sum_i \frac{1}{\widehat{\lambda}(x_i \mid \mathbf{x})} - |B|. \tag{13.48}$$

In the general case where the conditional intensity could be zero (typically because the interaction includes a hard core or extremely strong inhibition), the sum and integral are restricted to locations where the conditional intensity is positive:

$$\mathcal{R}^{(\mathrm{I})}(B) = \sum_i \frac{\mathbf{1}\left\{ \widehat{\lambda}(x_i \mid \mathbf{x}) > 0 \right\}}{\widehat{\lambda}(x_i \mid \mathbf{x})} - \int_B \mathbf{1}\left\{ \widehat{\lambda}(u \mid \mathbf{x}) > 0 \right\}\,du. \tag{13.49}$$

The inverse-lambda residual measure for a fitted Gibbs model is computed by `residuals.ppm` with `type="inverse"`:

```
> swedI <- residuals(swedfit, type="inverse")
```

The result is an object of class `"msr"`. If the conditional intensity is always positive, so that the simpler expression (13.48) holds, we might as well calculate only the masses $1/\widehat{\lambda}(x_i \mid \mathbf{x})$ at the data points, known as the *exponential energy marks* [633]. These can be computed using `eem`:

```
> swedE <- eem(swedfit)
```

and the result is a numeric vector with one entry for each point in the point pattern. The interpretation is that the expected total exponential energy weight of the data points in *any* given region B should be equal to the area of the region [633].

The left panel of Figure 13.31, generated by `plot(swedI)`, shows the inverse-lambda residual measure. Circle diameters are proportional to the exponential energy marks. This is equivalent to plotting the data points marked by the exponential energy marks, `plot(swedishpines %mark% swedE)`. There is one data point with a relatively large value of the exponential energy mark, i.e. a relatively small value of conditional intensity.

Variances associated with the exponential energy marks were studied in [633] and can be analysed using the general methods of [45].

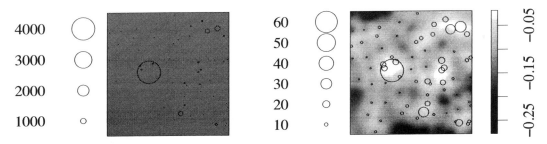

Figure 13.31. *Weighted residual measures.* Left: *inverse-lambda residual measure.* Right: *Pearson residual measure. Inhomogeneous Strauss model for Swedish Pines data.*

Kernel smoothing of the inverse-lambda residual measure gives the smoothed inverse-lambda residual field

$$s^{(\mathrm{I})}(u) = \frac{1}{e(u)} \sum_{i=1}^{n(\mathbf{x})} \frac{\kappa(u - x_i)}{\widehat{\lambda}(x_i \mid \mathbf{x})} - 1 \tag{13.50}$$

(assuming the conditional intensity is always positive) with the same notation as in (13.47). This is equivalent to kernel-smoothing the point pattern with weights $1/\widehat{\lambda}(x_i \mid \mathbf{x})$ and subtracting the expected value 1.

13.10.3.4 Pearson residuals

In the *Pearson* residual measure of a fitted Gibbs model, each data point is weighted by the reciprocal of the square root of its fitted conditional intensity:

$$\mathscr{R}^{(\mathrm{P})}(B) = \sum_i \frac{1}{\widehat{\lambda}(x_i \mid \mathbf{x})^{1/2}} - \int_B \widehat{\lambda}(u \mid \mathbf{x})^{1/2}\, du \tag{13.51}$$

or in the general case where the conditional intensity can be zero,

$$\mathscr{R}^{(\mathrm{P})}(B) = \sum_i \frac{\mathbf{1}\left\{\widehat{\lambda}(x_i \mid \mathbf{x}) > 0\right\}}{\widehat{\lambda}(x_i \mid \mathbf{x})^{1/2}} - \int_B \widehat{\lambda}(u \mid \mathbf{x})^{1/2}\, du. \tag{13.52}$$

The right-hand panel of Figure 13.31 shows a plot of `residuals(swedfit, type="Pearson")`.

For a Poisson model, the rationale for dividing by the square root of fitted intensity is that the resulting residuals have approximately constant variance per unit area — that is, for a Poisson model, var $\mathscr{R}^{(\mathrm{P})}(B)$ is approximately equal to $|B|$. For a Gibbs model this is a much less accurate approximation: it remains approximately true for models with weak to moderate interaction. Exact formulae for the variance of $\mathscr{R}^{(\mathrm{P})}(B)$ can be obtained from results of [45]. Future versions of `spatstat` will incorporate such variance calculations.

Kernel smoothing of the Pearson residual measure gives the smoothed Pearson residual field

$$s^{(\mathrm{P})}(u) = \frac{1}{e(u)} \left[\sum_{i=1}^{n(\mathbf{x})} \frac{\kappa(u - x_i)}{(\widehat{\lambda}(x_i \mid \mathbf{x}))^{1/2}} - \int_W \kappa(u - v)\widehat{\lambda}(v \mid \mathbf{x})\, dv \right] \tag{13.53}$$

with the same notation as in (13.47).

At the time of writing, the distribution of the smoothed Pearson residuals for a Gibbs process is not known exactly; an estimator of the variance is under development, based on [45, 162]. Provisional results suggest that, in most cases, the interpretation of values of the smoothed Pearson residual field is *roughly* the same as described under Figure 11.9 on page 424.

13.10.3.5 Residual plots

Residuals for Gibbs processes can be plotted using the same techniques as in Section 11.3. Figure 13.32 shows the four-panel plot of the Pearson residual measure, for the inhomogeneous Strauss model fitted to the Swedish Pines data in Section 13.10.2. It was generated by

```
> diagnose.ppm(swedfit, type="Pearson", envelope=TRUE, nsim=39)
```

or more efficiently by

```
> diagnose.ppm(swedfit, type="Pearson", envelope=swedenv, nsim=39)
```

since `swedenv` already contains simulated point patterns.

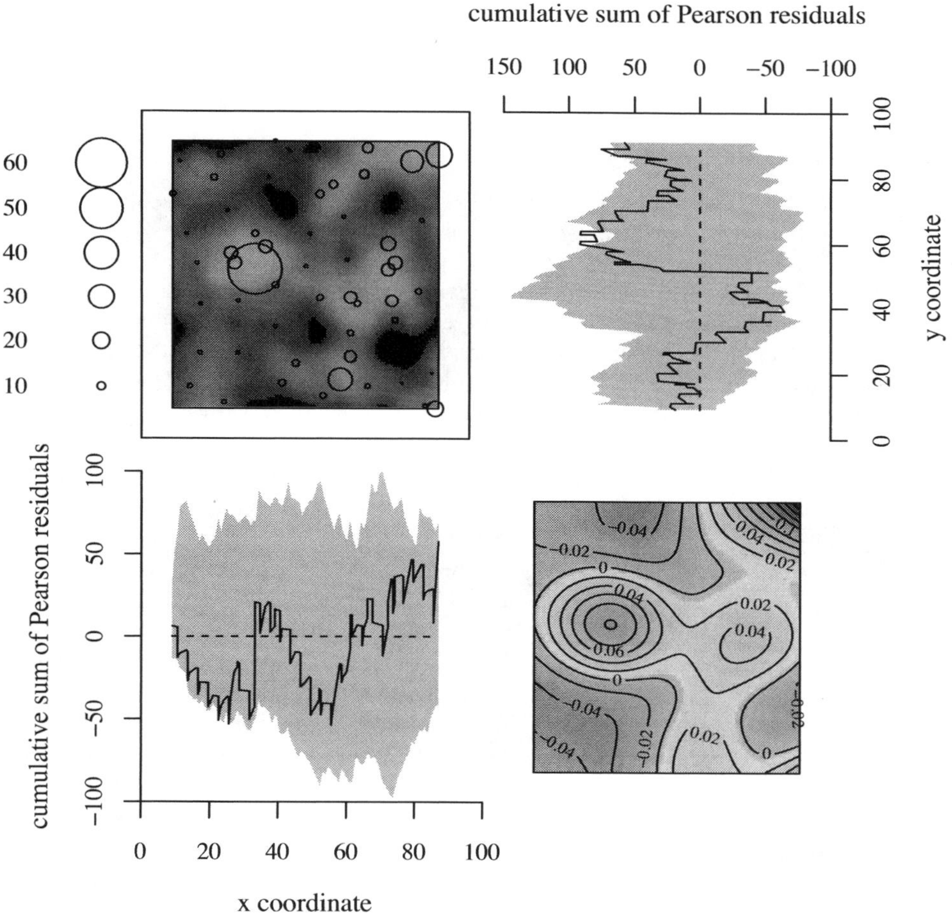

Figure 13.32. *Four-panel diagnostic plot for the inhomogeneous Strauss model fitted to the Swedish Pines data. Top left: mark plot of Pearson residual measure. Bottom right: smoothed Pearson residual field. Top right and bottom left: lurking variable plots for the x- and y-coordinates: cumulative Pearson residual (solid black line) with pointwise 5% significance bands (grey shading) based on 39 simulations of fitted model.*

These plots have several new features in the case of Gibbs processes. In the top left and bottom right panels, a margin has been removed from the border of the window. The width of the margin is the range of interaction of the model, 9 units (0.9 metres), and corresponds to the use of the border edge correction when the model was fitted. Proper comparison of goodness-of-fit should use this eroded region. The raw residuals sum to zero over this eroded region, rather than over the full window.

The lurking variable plots include pointwise 5% significance bands shown in grey shading. These are currently based on Monte Carlo simulation rather than theoretical standard errors, in the case of Gibbs processes (this may change in future versions of `spatstat`). They are selected by setting `envelope=TRUE`. Choosing `nsim=39` gives the significance level $1/(39+1) = 0.05$.

Applying the rule of thumb explained on page 424, the threshold for 'large' values is about

```
> SWP <- Smooth(residuals(swedfit, type="Pearson"))
> 2/(2 * sqrt(pi) * attr(SWP, "sigma"))
[1] 0.04702
```

(this value is also printed in the output from `diagnose.ppm`) suggesting that some parts of the smoothed residual field are moderately large.

13.10.4   Partial residual and added variable plots

Once we have fitted a reasonably good tentative model, the next step is to validate the form of the systematic relationship. A *partial residual plot* (introduced in Section 11.4) shows the fitted effect of a particular covariate, on the scale of the log intensity, together with a smoothed estimate of the true effect. Partial residual plots are also available for fitted Gibbs models [35], using essentially the same definition (11.17)–(11.18) based on the inverse-lambda residuals. Roughly speaking, the smooth curve in the partial residual plot is a smoothed version of the fitted effect plus the inverse-lambda residuals. If the fitted loglinear model is approximately true, the smooth curve is an approximately unbiased estimate of the smoothed true effect of the covariate.

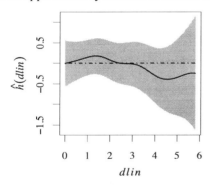

Figure 13.33 shows the partial residual plot for the distance covariate `dlin` in the fitted inhomogeneous Geyer model for the Queensland copper data, computed by `parres(copfit, dlin)` where `copfit` and `dlin` were computed in Section 13.10.2. The pointwise 5% significance bands (boundaries of the shaded region) are currently based on a Poisson approximation which may be inaccurate for very strong interactions. Nevertheless the plot suggests the true relationship is that there is essentially no dependence on distance from the nearest fault.

Figure 13.33. *Partial residual plot for inhomogeneous Geyer model fitted to the Queensland copper data.*

Added variable plots (Section 11.5) can also be used for fitted Gibbs models [35]. Roughly speaking, this is a plot of the smoothed Pearson residuals for the current model against the residuals from a linear regression of the new covariate on the existing covariates. To decide whether a new covariate Z should be added into an existing fitted model `fit`, plot the result of `addvar(fit, Z)`.

13.10.5   Validating the interaction model

13.10.5.1   Q–Q plots

As we noted in Section 11.6.1, the four-panel residual plot and the lurking variable plot are useful for detecting misspecification of the *trend* in a fitted model. They are not very useful for checking misspecification of the *interaction* in a fitted model. A more appropriate tool for this purpose is the Q–Q plot described in Section 11.6.1. In a residual Q–Q plot the observed quantiles of the smoothed residual field s are plotted against the expected quantiles of s under the fitted model, estimated by simulation. Figure 13.34 shows the residual Q–Q plot for the inhomogeneous Strauss model fitted to the Swedish Pines data, with pointwise 5% critical envelopes, generated by `qqplot.ppm(swedfit, nsim=39)`. The plot suggests that this model is a marginally inadequate fit to the interaction in the Swedish Pines.

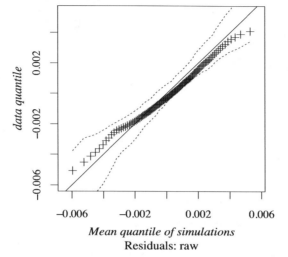

Figure 13.34. *Residual Q–Q plot for inhomogeneous Strauss model fitted to the Swedish Pines data.*

These validation techniques generalise and unify many existing exploratory methods. For particular models of interpoint interaction, the Q–Q plot is closely related to the summary functions F, G, and K. See [55].

13.10.5.2 Residual K- and G-functions

Residual summary functions were introduced in Section 11.6.4 for Poisson processes. Now that the reader is familiar with Gibbs processes, we can explain that the residual K-function for a fitted Poisson model, evaluated at a specific distance r, is the *score residual* used for testing the Poisson model against the alternative of a Strauss process with the same trend and with interaction distance r. Similarly the residual G-function for a fitted Poisson model, evaluated at a given r, is the score residual used for testing the Poisson model against the alternative of a Geyer saturation model with saturation parameter 1 and interaction radius r. For details, see [49].

The definitions of residual summary functions and compensators in Section 11.6.4 can be extended to Gibbs processes by replacing the fitted intensity $\widehat{\lambda}(u)$ by the fitted conditional intensity $\widehat{\lambda}(u \mid \mathbf{x})$. See [49]. They are computed using the same `spatstat` functions `Kcom`, `Kres`, `Gcom`, `Gres`, `psst`, `psstA`, `psstG`, and `compareFit`, which apply to any fitted Gibbs model.

Figure 13.35 shows the residual K-function for the inhomogeneous Strauss model fitted to the Swedish Pines data, computed by `Kres(swedfit, correction="iso")`. The acceptance region shaded in Figure 13.35 is based on a very rough variance approximation (with the advantage that it can be computed quickly) so this should not be used as a basis for formal inference. Note also that the residual K-function value at distance $r = 9$ is *constrained to equal zero* (approximately) in this example, because the K-function is the sufficient statistic for the Strauss model, ignoring edge corrections (so that $\widehat{K}(9)$ is approximately the total score residual for the interaction parameter).

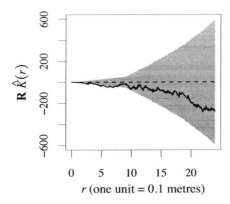

Figure 13.35. *Residual K-function for the inhomogeneous Strauss model fitted to the Swedish Pines data. Shading shows two-standard-deviation limits based on a rough approximation.*

13.10.6   Leverage and influence diagnostics

Leverage and influence diagnostics are used to check whether the fitted model was likely to have been influenced by anomalies in the data. These diagnostics were explained in Section 11.7 for the case of Poisson point process models; they have been extended to apply to Gibbs models [36].

In a Gibbs model of the loglinear form $\lambda(u \mid \mathbf{x}) = \exp(\boldsymbol{\theta}^\top \mathbf{Z}(u \mid \mathbf{x}))$ where $\mathbf{Z}(u \mid \mathbf{x})$ is a known function satisfying $\mathbf{Z}(u \mid \mathbf{x}) = \mathbf{Z}(u \mid \mathbf{x} \setminus u)$, the leverage is the scalar-valued function

$$\widehat{h}(u \mid \mathbf{x}) = \widehat{\lambda}(u \mid \mathbf{x})\mathbf{Z}(u \mid \mathbf{x})\boldsymbol{I_{\hat{\theta}}}^{-1}\mathbf{Z}^*(u \mid \mathbf{x}, \hat{\boldsymbol{\theta}})^\top \tag{13.54}$$

where $\boldsymbol{I_\theta}$ is the Fisher information matrix and

$$\mathbf{Z}^*(u \mid \mathbf{x}, \hat{\boldsymbol{\theta}}) = \mathbf{Z}(u \mid \mathbf{x}) - \int_W \left[\widehat{\lambda}(v \mid \mathbf{x} \cup \{u\})\mathbf{Z}(v \mid \mathbf{x} \cup \{u\}) - \widehat{\lambda}(v \mid \mathbf{x})\mathbf{Z}(v \mid \mathbf{x})\right] \mathrm{d}v \tag{13.55}$$

is the effect on the pseudolikelihood score of adding the point u to the dataset $\mathbf{x}$. A relatively large value of $\widehat{h}(u \mid \mathbf{x})$ indicates a part of the space where the data may have a strong effect on the result.

The *influence* of data point x_i is the scalar value

$$s(x_i \mid \mathbf{x}) = \frac{1}{p}\mathbf{Z}(x_i \mid \mathbf{x})\boldsymbol{I_{\hat{\theta}}}^{-1}\mathbf{Z}^*(x_i \mid \mathbf{x}, \hat{\boldsymbol{\theta}})^\top \tag{13.56}$$

where

$$\mathbf{Z}^*(x_i \mid \mathbf{x}, \hat{\boldsymbol{\theta}}) = \mathbf{Z}(x_i \mid \mathbf{x}) - \int_W \left[\widehat{\lambda}(v \mid \mathbf{x})\mathbf{Z}(x_i \mid \mathbf{x}) - \widehat{\lambda}(v \mid \mathbf{x} \setminus x_i)\mathbf{Z}(v \mid \mathbf{x} \setminus x_i)\right] \mathrm{d}v. \tag{13.57}$$

A relatively large value of $s(x_i \mid \mathbf{x})$ indicates a data point whose removal would cause a large change in the fitted model.

Finally the *parameter influence measure* is the weighted residual measure with weight function $w(u) = \boldsymbol{I_{\hat{\beta}}}^{-1}\mathbf{Z}(u \mid \mathbf{x})$.

13.10.7   Chorley-Ribble example

We continue the analysis of the Chorley-Ribble data from Section 11.7.4. Since clustering of disease cases is a possible alternative explanation for the finding of elevated risk near the incinerator, it would be prudent to consider point process models which incorporate positive association between points. Diagnostics then provide information about the evidence for such clustering, as well as the evidence for covariate effects after allowing for clustering.

Here we add a Geyer interaction term to Diggle's model and assess the evidence for an incinerator effect in the presence of clustering (and vice versa).

```
> rGey <- 0.25
> chorleyDGfit <- ippm(larynx ~ offset(log(smo) + log(raisin)) ,
                Geyer(rGey, 1), eps = 0.5, iScore=Zscore,
                start=list(alpha=20, beta=1),
                nlm.args=list(steptol=0.001))
> unlist(parameters(chorleyDGfit))
trend.(Intercept)            alpha             beta                 r
          -2.8712          20.7406           0.8946            0.2500
              sat            gamma
           1.0000           0.9420
> chorleyDGopt <- parameters(chorleyDGfit)[c("alpha", "beta")]
```

Next we calculate the leverage and influence:

```
> chorleyDGlev <- leverage(chorleyDGfit,
                iScore=Zscore, iHessian=Zhess, iArgs=chorleyDGopt)
> chorleyDGinf <- influence(chorleyDGfit,
                iScore=Zscore, iHessian=Zhess, iArgs=chorleyDGopt)
> chorleyDGdfb <- dfbetas(chorleyDGfit,
                iScore=Zscore, iHessian=Zhess, iArgs=chorleyDGopt)
```

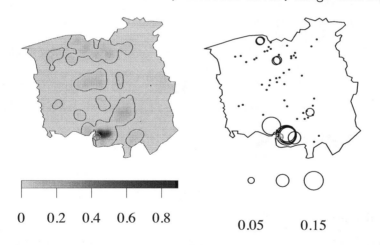

Figure 13.36. *Leverage function (*left*) and influence measure (*right*) for Geyer model with inciner-ator effect, fitted to the Chorley-Ribble data. Location of incinerator is marked by $\oplus$.*

The leverage function, plotted in the left panel of Figure 13.36, shows the overwhelming importance of data close to the incinerator. The influence measure for the Geyer model with incinerator effect, plotted in the right panel of Figure 13.36, is quite similar to the influence for the corresponding Poisson model in Figure 11.23. The most influential points are actually two points located a little further away from the incinerator than the suspect cluster of larynx cases.

The parameter influence measure (Figure 13.37) shows that these two most influential points are lung cancer cases, represented as squares in the bottom two plots in Figure 13.37, indicating that they tend to reduce the estimates of the parameters α and β.

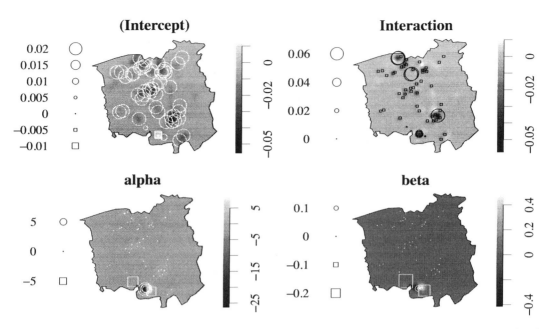

Figure 13.37. *Parameter influence measure* dfbetas *for Geyer model with incinerator effect, fitted to the Chorley-Ribble data.*

The top right panel of Figure 13.37 also suggests that the strongest support for the Geyer interaction comes from several pairs of cases of cancer of the larynx which are far away from the incinerator, while there is relatively little support near the incinerator. Thus, while spatial clustering may or may not be present, spatial clustering artefacts are *not* a good explanation for the apparent increased risk of laryngeal cancer near the incinerator.

Figure 13.38 shows a smoothed version of the parameter influence measure, computed by

```
> chorleyDGdfbsmo <- Smooth(chorleyDGdfb, sigma=0.8)
```

This reinforces the same conclusions.

13.11   Locally fitted models

Local likelihood or 'geographically weighted regression' methods were described in Section 9.13 for Poisson process models and in Section 12.5 for Cox and cluster models. They also apply to Gibbs models [33]. The spatstat implementation in the function locppm is based on local pseudolikelihood, defined by applying a smoothing kernel weight to each contribution to the log-pseudolikelihood.

The full redwood data (Figure 8.3 on page 260) contain patches that are clearly clustered, while other patches appear to be regular. To capture both kinds of behaviour we choose a Geyer saturation process model, and fit it locally.

```
> RedGey <- locppm(redwoodfull ~ 1, Geyer(0.05, 2), sigma=bw.locppm,
                   correction="isotropic")
```

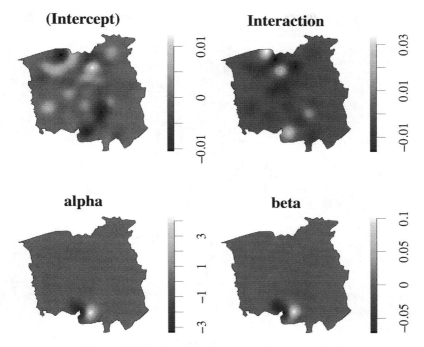

Figure 13.38. *Smoothed version of previous figure.*

Figure 13.39 shows the spatially varying values of the fitted canonical interaction parameter $\log \gamma$ of the Geyer model. Like the analysis in Section 12.5, this has automatically detected the transition between clustered and regular textures. For further details see [33].

13.12 Theory: Gibbs processes*

This section summarises essential theory for *finite* Gibbs point process models including their probability densities and conditional intensities, and model-fitting techniques. For more detail, see [661], [484, chap. 6], [355, sec. 3.6], [576, 577, 477].

For infinite Gibbs point processes, see [279, 661], [638, sec 5.5.3], [484, sec. 6.4], [199, sec. 10.4]. For spatial birth-and-death processes, see [554, 434, 474, 480] and [484, chap. 11 and appendix G].

13.12.1 Probability densities for finite point processes

One way to construct a statistical model (in any field of statistics) is to write down its probability density. This is useful because the functional form of the density reflects its probabilistic properties; terms or factors in the density often have an interpretation as 'components' of the model; and it is easy to introduce terms that represent the dependence of the model on covariates, etc. This approach is useful provided the density *can* be written down, and provided the density is tractable.

Roughly speaking, a point process with an *infinite* number of points cannot have a probability density. For example, we cannot define a probability density for the homogeneous Poisson process on the entire two-dimensional plane. This is a consequence of the strong law of large numbers.

---

* Starred sections contain advanced material, and can be skipped by most readers.

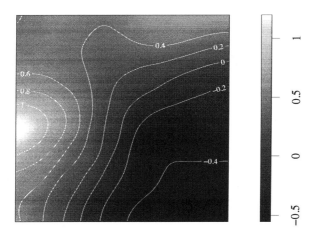

Figure 13.39. *Canonical interaction parameter* $\log\gamma$ *in the locally fitted Geyer saturation model for the full redwood data.*

It is possible to define probability densities for *finite* spatial point processes. For convenience we suppose the points must lie inside a bounded window W. The probability density will be a function $f(\mathbf{x})$ defined for each finite configuration $\mathbf{x} = \{x_1, \ldots, x_n\}$ of points $x_i \in W$ for any $n \geq 0$. Notice that the number of points n is not fixed, and may be zero. Apart from this peculiarity, probability densities for point processes behave much like probability densities in more familiar contexts.

A point process $\mathbf{X}$ inside W is defined to have probability density f if and only if, for any nonnegative integrable function h,

$$\mathbb{E}[h(\mathbf{X})] = e^{-|W|}\left(h(\emptyset)f(\emptyset) + \sum_{n=1}^{\infty}\frac{1}{n!}\int_W \cdots \int_W h(\{x_1, \ldots, x_n\})f(\{x_1, \ldots, x_n\})\,\mathrm{d}x_1 \cdots \mathrm{d}x_n\right) \quad (13.58)$$

where $|W|$ denotes the area of W. In particular, the probability that $\mathbf{X}$ contains exactly n points is

$$p_n = \mathbb{P}\{n(\mathbf{X}) = n\} = \frac{e^{-|W|}}{n!}\int_W \cdots \int_W f(\{x_1, \ldots, x_n\})\,\mathrm{d}x_1 \cdots \mathrm{d}x_n$$

for $n \geq 1$ and $p_0 = \mathbb{P}\{n(\mathbf{X}) = 0\} = e^{-|W|}f(\emptyset)$. Given that there are exactly n points, the conditional joint density of the locations $x_1, \ldots, x_n$ is $f(\{x_1, \ldots, x_n\})/p_n$.

The definition (13.58) has a simple interpretation. Suppose $\mathbf{Y}$ is a Poisson process with intensity $\lambda = 1$ in W. The expectation of any function $h(\mathbf{Y})$ can be written out by first principles:

$$\mathbb{E}[h(\mathbf{Y})] = \mathbb{E}\left[\mathbb{E}[h(\mathbf{Y}) \mid n(\mathbf{Y})]\right] \quad (13.59)$$

$$= \sum_{n=0}^{\infty}\mathbb{P}\{n(\mathbf{Y}) = n\}\,\mathbb{E}[h(\mathbf{Y}) \mid n(\mathbf{Y}) = n]. \quad (13.60)$$

The number of points $n(\mathbf{Y})$ is Poisson-distributed with mean $|W|$, and given $n(\mathbf{Y}) = n$, the points are independent and uniformly distributed in W. Hence the sum above can be written

$$\mathbb{E}[h(\mathbf{Y})] = e^{-|W|}h(\emptyset) + e^{-|W|}\sum_{n=1}^{\infty}\frac{|W|^n}{n!}\int_W \cdots \int_W h(\{x_1, \ldots, x_n\})\frac{1}{|W|^n}\,\mathrm{d}x_1 \cdots \mathrm{d}x_n. \quad (13.61)$$

Comparing (13.61) with (13.58) we find that, for a point process $\mathbf{X}$ with probability density $f(\mathbf{x})$,

$$\mathbb{E}[h(\mathbf{X})] = \mathbb{E}[h(\mathbf{Y})f(\mathbf{Y})] \quad (13.62)$$

where $\mathbf{Y}$ is the Poisson process with unit intensity. In other words, the probability density f defined by (13.58) is a density with respect to the unit rate Poisson process.

13.12.2 Gibbs representation

A finite point process density $f(\mathbf{x})$ is said to have *hereditary positivity* (or just to be *hereditary*) if

$$f(\mathbf{x}) > 0 \text{ and } \mathbf{y} \subset \mathbf{x} \text{ imply } f(\mathbf{y}) > 0. \tag{13.63}$$

That is, if $\mathbf{x}$ is a possible outcome, then any subset $\mathbf{y}$ of $\mathbf{x}$ cannot be impossible. To put it another way, if $f(\mathbf{y}) = 0$ and $\mathbf{y} \subset \mathbf{x}$ then $f(\mathbf{x}) = 0$: a point pattern which contains an impossible configuration must be impossible.

For example, the probability density of the hard core process (13.3) is hereditary. The probability density of the binomial point process with n points is not hereditary, because a point pattern with fewer than n points is impossible.

Theorem 13.1 (Hammersley-Clifford-Ripley-Kelly). *[579] If the probability density $f(\mathbf{x})$ of a finite point process is hereditary, then it has a* Gibbs representation

$$f(\mathbf{x}) = \exp\{V_0 + \sum_{x \in \mathbf{x}} V_1(x) + \sum_{\{x,y\} \subset \mathbf{x}} V_2(x,y) + \ldots\} \tag{13.64}$$

where V_0 is a constant, and $V_k : W^k \to \mathsf{R} \cup \{-\infty\}$ is called the potential of order k. *The representation is unique.*

Definition 13.1. *Any finite point process with probability density of the form* (13.64) *is called a* finite Gibbs process. *It has* interaction order m *if, in the Gibbs representation* (13.64), *the potentials V_k for $k > m$ are identically zero. It has* interaction range R^* *if all the potentials satisfy $V_k(\mathbf{y}) = 0$ whenever $\mathbf{y}$ contains two points $y_1, y_2 \in \mathbf{y}$ such that $\|y_1 - y_2\| > R^*$. A point process with a finite interaction range is called a* Markov point process *[661, 484].*

For example, the Poisson processes have interaction order 1 and interaction range 0. The pairwise interaction processes (13.27) have interaction order 2, ignoring trivial cases which reduce to a Poisson process. The Strauss process (13.11) with parameters β, γ, r has interaction order 2 and interaction range $R^* = r$. The triplet interaction process (13.39) has interaction order 3 and interaction range $R^* = r$. The Widom-Rowlinson model (13.31) and the Geyer saturation process (13.35) have infinite interaction order, and interaction range $R^* = 2r$. All the above examples are Markov point processes.

The higher-order potentials V_k for $k \geq 2$ introduce dependence between points of the process (interpoint interaction). The point process $\mathbf{X}$ is Poisson if and only if the higher-order potentials all vanish, $V_k \equiv 0$ for all $k \geq 2$. In this case the process is Poisson with intensity $\beta(u) = \exp(V_1(\{u\}))$.

It is often convenient to pool all the terms of order $k \geq 2$, yielding the representation

$$V(\mathbf{x}) = V_0 + V_1(\mathbf{x}) + V_{[\geq 2]}(\mathbf{x}) \tag{13.65}$$

where we regard V_1 as controlling 'spatial trend' while $V_{[\geq 2]}(\mathbf{x})$ controls 'spatial interaction'. This interpretation is correct provided $V_{[\geq 2]}(\mathbf{x})$ is invariant under translations of the point pattern $\mathbf{x}$, as we would normally assume in practice.

For the conditional intensity we have correspondingly

$$\lambda(u \mid \mathbf{x}) = \exp(\Delta_u V(\mathbf{x})) \tag{13.66}$$

where

$$\Delta_u V(\mathbf{x}) = V(\mathbf{x} \cup \{u\}) - V(\mathbf{x}). \tag{13.67}$$

The decomposition of V above yields

$$\Delta_u V(\mathbf{x}) = V_1(\{u\}) + \Delta_u V_{[\geq 2]}(\mathbf{x}) \tag{13.68}$$

where again V_1 may be interpreted as 'spatial trend', while $\Delta_u V_{[\geq 2]}$ determines 'interpoint interaction'.

13.12.3 Constructing finite Gibbs models

Any finite Gibbs model has a probability density of the form given by Theorem 13.1. To build Gibbs models we can simply write down the interaction functions $V_1, V_2, \dots$. For example, to define a pairwise interaction model, we would define the log trend $V_1(u)$ and the pair potential $V_2(u, v)$ for all points u and v.

In any probability distribution, the total probability must equal 1. The total probability for a Gibbs process is equal to the right-hand side of (13.58) when we set $h(\mathbf{x}) \equiv 1$. A total probability equal to 1 can be achieved by adjusting the value of the constant V_0 in (13.64), *provided* the infinite sum on the right-hand side of (13.58) is a finite number. The density is called *integrable* if the right-hand side of (13.58) is finite when $h(\mathbf{x}) \equiv 1$.

Care must be taken to ensure that the probability density $f(\mathbf{x})$ is integrable [282, 371]. This places a constraint on the choice of the interaction functions V_k. An example is the Strauss process [643, 371] discussed in Section 13.3.7. Suppose $V_1(u) = a$ for all locations u, and $V_2(u, v) = b$ if $\|u - v\| \le r$ while $V_2(u, v) = 0$ if $\|u - v\| > r$. If $b > 0$ then a point pattern $\mathbf{x}$ with many pairs of close points will have a very large value of $f(\mathbf{x})$, and the total on the right side of (13.58) is *infinite* — the density $f(\mathbf{x})$ is not integrable; a point process with these characteristics does not exist. If $b < 0$ then the density is integrable, and we get the Strauss process.

For loglinear models of the form $f(\mathbf{x}) \propto \exp(\theta^\mathsf{T} A(\mathbf{x}))$, with potential $V(\mathbf{x}) = \theta^\mathsf{T} A(\mathbf{x})$, integrability is equivalent to requiring that θ is in the domain of the Laplace transform of the distribution of the statistic A under the unit rate Poisson process [282]. Note that integrability depends only on the interaction terms: provided $\int_W \exp V_1(\{u\}) \, du < \infty$, the unnormalised density

$$h_{\varphi, \psi}(\mathbf{y}) = \exp\{\varphi^\mathsf{T} V_1(\mathbf{y}) + \psi^\mathsf{T} V_{[\ge 2]}(\mathbf{y})\}$$

is integrable if and only if $h_{0, \psi}$ is integrable. The practical implication is that we may formulate Gibbs point process models with an arbitrary first-order spatial trend term $V_1(\{u\}) = Z(u)$, but we will usually want to choose the interaction term $V_{[\ge 2]}(\mathbf{y})$ from a list of well-studied higher-order potentials whose integrability properties are known.

13.12.4 Conditional intensity

The (Papangelou) conditional intensity of a point process is formally defined in [197, pp. 580–590]. Here we present a very simplified explanation of the conditional intensity, omitting many technical conditions. See [27, 368], [370, sec. 2.6] for more details.

First a heuristic sketch. Suppose $\mathbf{X}$ is a point process in a bounded window W. Divide the window into tiny pixels, and take one of the pixels U. Suppose that we know the contents of all pixels *except* U, so that we know the point pattern $\mathbf{X} \setminus U$. Given this information, let

$$M(U \mid \mathbf{X} \setminus U) = \frac{1}{|U|} \mathbb{E}[n(\mathbf{X} \cap U) \mid \mathbf{X} \setminus U]$$

be the expected number of points falling in pixel U, divided by the area of U, that is, the average intensity of $\mathbf{X}$ inside U given complete information about $\mathbf{X}$ outside U.

Now let the pixel grid become increasingly fine. If the point process is well behaved, then the values of $M(U \mid \mathbf{X} \setminus U)$ will converge to a function $\lambda(u \mid \mathbf{X})$ defined at all spatial locations u, and this is the conditional intensity function. Roughly speaking, $\lambda(u \mid \mathbf{X}) \, du$ is the conditional probability that there will be a point of $\mathbf{X}$ in an infinitesimal neighbourhood of u, given the location of all points of $\mathbf{X}$ outside this neighbourhood.

For example, if $\mathbf{X}$ is CSR, then points inside and outside U are independent, so the conditional expectation is equal to the usual expectation, and $M(U \mid \mathbf{X} \setminus U) = \lambda$ is the usual intensity. In the fine-pixel limit we obtain $\lambda(u \mid \mathbf{x}) = \lambda$.

Definition 13.2. *For a point process* **X** *in two dimensions, suppose there is a function* $\lambda(u \mid \mathbf{x})$ *such that*

$$\mathbb{E}\left[\sum_{x_i \in \mathbf{X} \cap B} h(x_i, \mathbf{X} \setminus x_i)\right] = \int_B \mathbb{E}[\lambda(u \mid \mathbf{X})h(u, \mathbf{X})]\,du \qquad (13.69)$$

for any function $h(u, \mathbf{x})$ *and any spatial region B for which both sides of the equation are well defined. Then* $\lambda(u \mid \mathbf{x})$ *is called the* Papangelou *conditional intensity of* **X**.

For a Poisson process with intensity function $\lambda(u)$, the independence properties can be used to show that $\lambda(u \mid \mathbf{x}) = \lambda(u)$.

The structure of equation (13.69) is interesting. On the left-hand side, we visit all random points x_i of the point process within the region B, apply the function h to the point process *after deleting* x_i to get the contribution $h(x_i, \mathbf{X} \setminus x_i)$, and sum these contributions. On the right-hand side, we visit every spatial location u in B, apply the function to the point process without modification to get the contribution $h(u, \mathbf{X})$, and integrate these contributions over all possible locations u with weight $\lambda(u \mid \mathbf{X})$.

Suppose, for example, that $h(u, \mathbf{x}) = (u, \mathbf{x}) = \min_j \|u - x_j\|$ is the shortest distance from the location u to a point of $\mathbf{x}$. Then on the left side of (13.69), the contribution $h(x_i, \mathbf{X} \setminus x_i)$ is the distance from x_i to the nearest *other* point of $\mathbf{X}$, that is, the *nearest-neighbour* distance for x_i. On the right side of (13.69), the contribution $h(u, \mathbf{X})$ is the *empty-space* distance from u to $\mathbf{X}$. Equation (13.69) relates these two quantities through the conditional intensity.

Taking $h \equiv 1$ in (13.69) gives

$$\mathbb{E}n(\mathbf{X} \cap B) = \int_B \mathbb{E}[\lambda(u \mid \mathbf{X})]\,du.$$

Since $\mathbf{X}$ has intensity function $\lambda(u)$, we have $\mathbb{E}n(\mathbf{X} \cap B) = \int_B \lambda(u)\,du$. This implies that

$$\mathbb{E}\lambda(u \mid \mathbf{X}) = \lambda(u) \qquad (13.70)$$

for every location u (ignoring a set of measure zero).

In Section 8.12.1 we defined the Palm distribution of a point process by a formula (8.48) similar to the left-hand side of (13.69). Applying (8.48) or (8.50) to the left-hand side of (13.69) we find that

$$\int_B \mathbb{E}^u[h(u, \mathbf{X} \setminus u)]\lambda(u)\,du = \int_B \mathbb{E}[\lambda(u \mid \mathbf{X})h(u, \mathbf{X})]\,du$$

where $\mathbb{E}^u$ denotes expectation with respect to the Palm distribution at u. Equivalently using the reduced Palm distribution (Section 8.12.4)

$$\int_B \mathbb{E}^{!u}[h(u, \mathbf{X})]\lambda(u)\,du = \int_B \mathbb{E}[\lambda(u \mid \mathbf{X})h(u, \mathbf{X})]\,du.$$

Consequently we have the *Georgii-Nguyen-Zessin (GNZ)* formula [278, 279, 501]

$$\mathbb{E}^{!u}[h(u, \mathbf{X})] = \mathbb{E}\left[\frac{\lambda(u \mid \mathbf{X})}{\lambda(u)}h(u, \mathbf{X})\right] \qquad (13.71)$$

for every location u (ignoring a set of measure zero). This states that the reduced Palm distribution (left side) and the ordinary distribution (right side) of a point process are related through the conditional intensity. We may think of (13.71) as a form of Bayes' Theorem, because the Palm distribution at u is essentially the conditional probability given that there is a random point at location u, while the conditional intensity is the reverse, the conditional probability that there is a random point at location u given the rest of the process.

As an example, the J-function of a stationary point process (Section 8.6) can be written explicitly in terms of the conditional intensity: since

$$\mathbb{P}^0\{d(0,\mathbf{X}\setminus 0) > r\} = \mathbb{P}^{!0}\{d(0,\mathbf{X}) > r\} = \mathbb{E}\left[\frac{\lambda(0\mid\mathbf{X})}{\lambda}\mathbf{1}\{d(0,\mathbf{X}) > r\}\right]$$

where λ is the intensity, we have

$$J(r) = \frac{\mathbb{P}^0\{d(0,\mathbf{X}\setminus 0) > r\}}{\mathbb{P}\{d(0,\mathbf{X}) > r\}} = \frac{\mathbb{E}[\frac{\lambda(0\mid\mathbf{X})}{\lambda}\mathbf{1}\{d(0,\mathbf{X}) > r\}]}{\mathbb{P}\{d(0,\mathbf{X}) > r\}} = \mathbb{E}\left[\frac{\lambda(0\mid\mathbf{X})}{\lambda}\Big| d(0,\mathbf{X}) > r\right]. \quad (13.72)$$

This representation can often be evaluated, while F and G often cannot be evaluated explicitly.

For a Cox process with random driving intensity $\Lambda(u)$, assuming $\mathbb{E}\int_W \Lambda(u)\,du < \infty$, the conditional intensity has the implicit representation

$$\lambda(u,\mathbf{y}) = \mathbb{E}[\Lambda(u)\mid\mathbf{Y} = \mathbf{y}]. \quad (13.73)$$

13.12.5 Stability

For simulation and statistical inference about Gibbs processes we often need the model to satisfy an additional 'stability' property.

A finite point process with probability density $f(\mathbf{x})$ is called *Ruelle stable* if there exists a constant $M \geq 1$ such that $f(\mathbf{x}) \leq M^{n(\mathbf{x})}$ for all patterns $\mathbf{x}$. It is called *locally stable* if there exists a constant $M \geq 1$ such that $\lambda(u \mid \mathbf{x}) \leq M$ for all patterns $\mathbf{x}$ and all $u \in W$. Local stability implies Ruelle stability. See [484, 282] for more detail.

If a process is locally stable, then the Metropolis-Hastings algorithm will converge to the desired steady state distribution [484, Chapter 7].

For the Strauss process, Kelly and Ripley [371] showed that the Strauss density (13.11) is integrable iff $\gamma \leq 1$. In this case, $f(\mathbf{x})$ is Ruelle stable, since $f(\mathbf{x}) \leq c_4\beta^{n(\mathbf{x})}$. The conditional intensity (13.12) satisfies $\lambda(u \mid \mathbf{x}) \leq \beta$ so that $f(\mathbf{x})$ is also locally stable.

Returning for a moment to the general pairwise interaction density (13.27) with conditional intensity (13.28), a sufficient condition for local stability is that the pairwise interaction function c is bounded above by 1. However, this condition is quite restrictive, since it implies that the process must exhibit regularity (inhibition) at all spatial scales. The Lennard-Jones [418] pairwise interaction process is Ruelle stable but not locally stable [484, p. 88].

Gibbs processes are commonly used to model inhibitory (regular) patterns, as it is more difficult to construct locally stable Gibbs models which exhibit attraction (clustering). Notable exceptions that are locally stable even in the attractive case include the Widom-Rowlinson [699, 591] penetrable spheres model or 'area-interaction' process [70], the Geyer [282] saturation process, continuum random cluster process [385], quermass-interaction processes [376], and shot-noise weighted processes [665].

The Geyer saturation model density is integrable, Ruelle stable, and locally stable for all values of $\gamma > 0$, by a geometrical argument [282].

13.12.6 Intensity of a Gibbs model

The intensity of a Gibbs model is not a straightforward function of the model parameters. This creates difficulties for beginners and experts alike. The intensity is related to the conditional intensity through the Georgii-Nguyen-Zessin formula (13.71), which gives

$$\lambda(u) = \mathbb{E}\lambda(u \mid \mathbf{x}). \quad (13.74)$$

A simple example is the stationary Strauss process, with conditional intensity $\lambda(u \mid \mathbf{x}) = \beta \gamma^{t(u,r,\mathbf{x})}$ where $t(u,r,\mathbf{x})$ is the number of points of $\mathbf{x}$ lying within a distance r of the location u. As we have noted many times, β is **not** the intensity of the Strauss process. According to (13.74), the intensity λ is

$$\lambda = \beta \mathbb{E} \gamma^{t(u,r,\mathbf{x})} \tag{13.75}$$

where on the right-hand side, $\mathbf{X}$ is the stationary Strauss process, and the expectation is taken with respect to the probability distribution of the Strauss process. The right-hand side cannot easily be evaluated. Since $0 \leq \gamma \leq 1$, we can certainly say that $\gamma^{t(u,r,\mathbf{X})} \leq 1$ so that $\lambda \leq \beta$.

A useful *approximation* to λ can be obtained by pretending that the expectation on the right-hand side of (13.75) is taken with respect to a Poisson process $\mathbf{Y}$, of the same (unknown) intensity λ as the Strauss process. For the approximating Poisson process, $t(u,r,\mathbf{Y})$ is a Poisson random variable, with mean $\lambda \pi r^2$, and using the generating function of the Poisson distribution, we get $\mathbb{E}_\lambda \gamma^{t(u,r,\mathbf{Y})} = \exp(\lambda \pi r^2 (\gamma - 1))$. This gives us a 'self-consistency' or 'closure' relation

$$\lambda = \beta \exp(\lambda \pi r^2 (\gamma - 1)) \tag{13.76}$$

which can be solved to obtain an approximate value λ^* which we call the *Poisson-saddlepoint approximation* [48, 47, 18], [30, p. 360]. It is closely related to the Poisson-Boltzmann-Emden [384, 84] approximations for the density of a molecular gas and the Percus-Yevick [541] approximation for the pair correlation. See also [446, 115, 308]. The approximation is surprisingly good in many examples, though not in all [606, 645].

The `spatstat` function `intensity.ppm` computes the Poisson-saddlepoint approximation to the intensity, if the fitted model is a Gibbs process. The approximation is also available for inhomogeneous models [47, 18].

13.13 Theory: Fitting Gibbs models∗

For overviews of techniques for fitting Gibbs models, see [576, chap. 4], [577], [484, chap. 9], [355, chap. 7], [275, sec. 5.5], [228, 567].

13.13.1 Parameters of a Gibbs model

The algorithms implemented in ppm assume an 'exponential family' model, that is, a point process model with probability density

$$f(\mathbf{x}) = c_\theta \exp(B(\mathbf{x}) + \theta^\mathsf{T} S(\mathbf{x})) \tag{13.77}$$

where θ is a vector of 'canonical parameters', $S(\mathbf{x})$ is a known, vector-valued function called the 'canonical sufficient statistic', and c_θ is a constant depending on θ. The optional baseline function $B(\mathbf{x})$ must be of the form $B(\mathbf{x}) = \sum_i B(x_i)$. This model is equivalent to assuming that the conditional intensity is a loglinear function of the parameters:

$$\lambda(u \mid \mathbf{x}) = \exp(B(u) + \theta^\mathsf{T} Z(u, \mathbf{x})) \tag{13.78}$$

where $Z(u, \mathbf{x}) = S(\mathbf{x} \cup \{u\}) - S(\mathbf{x})$.

∗ Starred sections contain advanced material, and can be skipped by most readers.

13.13.2   Maximum likelihood for Gibbs models

The likelihood function is a cornerstone of statistical science: it is used to build models, develop statistical tools, and analyse statistical performance. We saw its importance in Chapter 9, which presented some powerful tools for model-building and statistical inference for spatial point processes, all based on the likelihood function for the Poisson point process.

The likelihood function of a *Gibbs* point process can also be written down in simple form, as it is simply the probability density $f(\mathbf{x})$ regarded as a function of the model parameters.

The main difficulty with performing maximum likelihood inference for Gibbs models is that the normalising constant V_0 in equation (13.64) is a complicated function of the model parameters.

For example, the likelihood of the Gibbs hard core process is simply the likelihood of a Poisson process conditional on the event that the hard core constraint is respected:

$$f_{hard}(\mathbf{x}) = \begin{cases} \alpha\,\beta^{n(\mathbf{x})} e^{(1-\beta)|W|} & \text{if hard core is respected} \\ 0 & \text{otherwise} \end{cases}$$

where α is the normalising constant, $\alpha = 1/p$ where p is the probability that the Poisson process respects the hard core. The normalising constant α depends on the parameters β and h in a complicated way. Clearly $p = p(\beta, W, h)$ is decreasing as a function of h and as a function of β. For fixed β, the likelihood $L(\beta, h) = \beta^{n(\mathbf{x})} e^{(1-\beta)|W|}/p$ if $d_{min} = \min_{i \neq j} d_{ij} > h$, and zero otherwise. For fixed β this is maximised at $h = d_{min}$. That is, the maximum likelihood estimate of the hard core distance r is $\widehat{h} = d_{min}$. However, the maximum likelihood estimate of β is much harder to calculate.

Techniques developed for fitting Gibbs models by approximate maximum likelihood include numerical approximation of the normalising constant [513, 514, 515, 516, 517, 510, 512], [576, chap. 4], Monte Carlo MLE [539, 282, 283], [484, chap. 9], stochastic approximation [490], and one-step Monte Carlo MLE [348]. The asymptotic theory of maximum likelihood for Gibbs models is studied in [447].

13.13.3   Alternatives to maximum likelihood

Maximum likelihood estimation is intractable for most point process models. At the very least it requires Monte Carlo simulation to evaluate the likelihood (or the score and the Fisher information): see [282, 484, 483]. Ripley [576] argued convincingly that maximum likelihood is not necessarily desirable in the analysis of spatial data.

Alternatives to maximum likelihood estimation include, of course, Bayesian inference using Markov chain simulation methods to generate realisations from the posterior [100, 92], [484, sec. 9.3]. There are also Bayesian techniques for estimating the pairwise interaction potential [333, 114] and variational Bayes estimation of the model parameters (Section 13.13.9).

13.13.4   Maximum pseudolikelihood

A workable alternative fitting technique, at least for investigative purposes, is to maximise the log-*pseudolikelihood* [97, 98, 594, 595]

$$\log \mathrm{PL}(\boldsymbol{\theta}; \mathbf{x}) = \sum_i \log \lambda_{\boldsymbol{\theta}}(x_i \mid \mathbf{x}) - \int_W \lambda_{\boldsymbol{\theta}}(u \mid \mathbf{x})\, du. \tag{13.79}$$

This is formally similar to the loglikelihood (9.40) of the Poisson process. In general it is not a loglikelihood, but the analogue of the score equation, $\frac{\partial}{\partial\boldsymbol{\theta}} \log \mathrm{PL}(\boldsymbol{\theta}) = 0$, is an unbiased estimating equation. The maximum pseudolikelihood estimator is asymptotically unbiased, consistent, and asymptotically normal under appropriate conditions [365, 364, 448, 450, 105]. See [484, sec. 9.2].

The main advantage of maximum pseudolikelihood over maximum likelihood is that, at least for popular Gibbs models, the conditional intensity $\lambda(u \mid \mathbf{x})$ is easily computable, so that the pseudo-likelihood is easy to compute and to maximise. The main disadvantage is the statistical inefficiency of maximum pseudolikelihood in small samples. Both maximum likelihood and maximum pseudolikelihood estimates are biased; there are reports that maximum pseudolikelihood has greater bias, but this may be due to the software implementation [53]. For technical investigation, the spatstat function exactMPLEstrauss can be used to compute the maximum pseudolikelihood fit of the stationary Strauss model, without numerical artefacts.

13.13.5   Fitting Gibbs models in spatstat

The model-fitting function ppm contains an implementation of the algorithm of Baddeley and Turner [50] for maximum pseudolikelihood (which extends the Berman-Turner device for Poisson processes to a general Gibbs process). The conditional intensity of the model, $\lambda_\theta(u \mid \mathbf{x})$, must be loglinear in the parameters θ:

$$\log \lambda_\theta(u \mid \mathbf{x}) = B(u) + \theta^\mathsf{T} Z(u, \mathbf{x}), \qquad (13.80)$$

generalising (9.6), where $B(u)$ is an optional, real-valued function representing a baseline or offset, $Z(u, \mathbf{x})$ is a real-valued or vector-valued function of location u and configuration $\mathbf{x}$. Parameters θ appearing in the loglinear form (13.80) are called 'regular' parameters, and all other parameters are 'irregular' parameters. For example, the Strauss process conditional intensity (13.12) can be recast as

$$\log \lambda(u \mid \mathbf{x}) = \log \beta + (\log \gamma) t(u, r, \mathbf{x})$$

so that $\theta = (\log \beta, \log \gamma)$ are regular parameters, but the interaction distance r is an irregular parameter or 'nuisance parameter'.

In spatstat we split the conditional intensity into first-order and higher-order terms as explained in Section 13.12.3:

$$\log \lambda_\theta(u \mid \mathbf{x}) = B(u) + \eta^\mathsf{T} Z(u) + \varphi^\mathsf{T} T(u, \mathbf{x}). \qquad (13.81)$$

The 'first-order term' $Z(u)$ describes spatial inhomogeneity and/or covariate effects. The 'higher-order term' $T(u, \mathbf{x})$ describes interpoint interaction.

The model with conditional intensity (13.81) is fitted by calling ppm in the form

```
ppm(X ~ terms, interaction)
```

The first argument X ~ terms is a model formula, with the left side specifying the point pattern dataset, and the right side specifying the first-order term $Z(u)$ and optional offset $B(u)$ in (13.81) in the manner described in Section 9.7.4. Thus the first-order term $Z(u)$ in (13.81) may take very general forms.

The second argument interaction is an object of the special class "interact" which describes the interpoint interaction term $T(u, \mathbf{x})$ in (13.81). It may be compared to the 'family' argument which determines the distribution of the responses in a linear model or generalised linear model. Only a limited number of canned interactions are available in spatstat, because they must be constructed carefully to ensure that the point process exists. However, they include very flexible classes of interactions.

The first and second arguments, taken together, specify the form of the *logarithm* of the conditional intensity of the model. For example, ppm(swedishpines ~ 1, Strauss(9)) specifies that

$$\log \lambda(u \mid \mathbf{x}) = \theta_1 + \theta_2 t(u, r, \mathbf{x})$$

where θ_1, θ_2 are parameters to be estimated, and $t(u, r, \mathbf{x})$ is the number of points of $\mathbf{x}$ lying within a distance r of the location u where in this case we have chosen $r = 9$ units, i.e. 0.9 meters. The model is the homogeneous Strauss process with parameters β, γ, r where $\theta_1 = \log \beta$ and $\theta_2 = \log \gamma$.

13.13.6　Huang-Ogata approximate maximum likelihood

Huang and Ogata [348] proposed a 'one-step' approximation to maximum likelihood. Starting from the maximum pseudolikelihood fit, we simulate M independent realisations of the model with the fitted parameters, evaluate the canonical sufficient statistics, and use them to form estimates of the score and Fisher information. Then we take one step of Fisher scoring, updating the parameter estimate. The rationale is that the log-likelihood is approximately quadratic in a neighbourhood of the maximum pseudolikelihood estimator, so that one update step is almost enough.

Consider a Gibbs model with probability density of loglinear form $f(\mathbf{x}) \propto \exp(\boldsymbol{\theta}^{\mathsf{T}} S(\mathbf{x}))$. The likelihood score is $U(\boldsymbol{\theta}, \mathbf{x}) = S(\mathbf{x}) - \mathbb{E}_{\boldsymbol{\theta}} S(\mathbf{X})$ and the Fisher information is $I_{\boldsymbol{\theta}} = \mathrm{var}_{\boldsymbol{\theta}} S(\mathbf{X})$. Given data $\mathbf{x}$, we compute the maximum pseudolikelihood estimate $\widehat{\boldsymbol{\theta}}_{\mathrm{MPL}}$, then generate M independent realisations $\mathbf{x}^{(1)}, \ldots, \mathbf{x}^{(M)}$ of the model with parameters $\widehat{\boldsymbol{\theta}}_{\mathrm{MPL}}$. The likelihood score $U(\widehat{\boldsymbol{\theta}}_{\mathrm{MPL}}, \mathbf{x})$ is then approximated by $\overline{U} = S(\mathbf{x}) - \overline{S}$ and the Fisher information $I_{\widehat{\boldsymbol{\theta}}_{\mathrm{MPL}}}$ by $\overline{I} = \sum_{i=1}^{M} S(\mathbf{x}^{(i)}) S(\mathbf{x}^{(i)})^{\mathsf{T}} - \overline{S}\overline{S}^{\mathsf{T}}$ where $\overline{S} = \frac{1}{M} \sum_{i=1}^{M} S(\mathbf{x}^{(i)})$. The parameter estimate is updated by Newton-Raphson (equivalent here to Fisher scoring) to give $\widehat{\boldsymbol{\theta}} = \widehat{\boldsymbol{\theta}}_{\mathrm{MPL}} + \overline{I}^{-1} \overline{U}$.

To use the Huang-Ogata method instead of maximum pseudolikelihood, call ppm with the argument method="ho".

```
> fit <- ppm(simdat ~ 1, Strauss(r=0.275), method="ho")
> fit
Stationary Strauss process
First order term:  beta = 2.377
Interaction distance:        0.275
Fitted interaction parameter gamma:        0.5597
> vcov(fit)
          [,1]      [,2]
[1,]   0.008337 -0.008654
[2,]  -0.008654  0.032317
```

For models fitted by the Huang-Ogata method, the variance-covariance matrix returned by vcov is computed from the simulations.

13.13.7　Logistic composite likelihood for Gibbs models

Gibbs models can also be fitted by a form of conditional logistic regression. This was introduced for Poisson models in Section 9.10. The extension to Gibbs models was developed in [38]; see also [158, 101].

Suppose $\mathbf{x}$ is a realisation of a Gibbs process $\mathbf{X}$ with conditional intensity of the loglinear form $\lambda_{\boldsymbol{\theta}}(u \mid \mathbf{x}) = \exp(\boldsymbol{\theta}^{\mathsf{T}} Z(u, \mathbf{x}))$. In addition to the observed point pattern $\mathbf{x}$, let us also generate a random pattern of 'dummy' points $\mathbf{d}$ by simulation from a Poisson point process $\mathbf{D}$ with intensity function $\delta(u)$. Treating $(\mathbf{x}, \mathbf{d})$ as a multitype point pattern with two types (data and dummy), let $\mathbf{y} = \mathbf{x} \cup \mathbf{d}$ be the pattern of all points regardless of type, and $I_i = \mathbf{1}\{y_i \in \mathbf{x}\}$ the indicator that equals 1 if y_i was a data point, and 0 if it was a dummy point. Now condition on the pattern of locations $\mathbf{y}$ and consider only the data/dummy indicators. The conditional probability that $I_i = 1$ (i.e. that y_i is a data point) given all the other indicators $I_j, j \neq i$, equals

$$p_i = \mathbb{P}\{I_i = 1 \mid I_j, j \neq i\} = \frac{\lambda_{\boldsymbol{\theta}}(y_j \mid \mathbf{x})}{\delta(y_i) + \lambda_{\boldsymbol{\theta}}(y_j \mid \mathbf{x})}. \tag{13.82}$$

Consequently the log odds are

$$\log \frac{p_i}{1 - p_i} = \boldsymbol{\theta}^{\mathsf{T}} Z(y_j, \mathbf{x}) - \log \delta(y_i). \tag{13.83}$$

That is, the indicators I_i satisfy a conditional autologistic regression. The model can be fitted by maximising the logistic composite log-likelihood

$$\text{LRL}(\boldsymbol{\theta}) = \sum_{x_i \in \mathbf{x}} \log \frac{\lambda_{\boldsymbol{\theta}}(x_i \mid \mathbf{x})}{\delta(x_i) + \lambda_{\boldsymbol{\theta}}(x_i \mid \mathbf{x})} + \sum_{d_j \in \mathbf{d}} \log \frac{\delta(d_j)}{\delta(d_j) + \lambda_{\boldsymbol{\theta}}(d_j \mid \mathbf{x})} \qquad (13.84)$$

which is Besag's discrete log-pseudolikelihood for the indicators (I_i) given the locations $\mathbf{y}$. This can be maximised using standard software for logistic regression. The loglikelihood is a concave function of $\boldsymbol{\theta}$, and conditions for existence and uniqueness of the maximum are well known [616]. The pseudoscore $\partial/(\partial\boldsymbol{\theta})\text{LRL}(\boldsymbol{\theta})$ is an unbiased estimating function, so that the estimates are consistent and asymptotically normal under reasonable conditions [38]. Instead of a Poisson pattern of dummy points, one may use a binomial or stratified binomial pattern. This does not change the estimating function (13.84), but the expression for the variance depends on the choice of dummy process as detailed in [38] with stratified binomial giving the smallest variance.

To use logistic composite likelihood in `spatstat` the additional argument `method = "logi"` should be given to `ppm` in the same manner as described in Section 9.10.5.

13.13.8 Other estimating equations for Gibbs models

Numerous other techniques for parameter estimation in Gibbs models are available. Many are based on an unbiased estimating equation, that is, the parameter estimate $\hat{\boldsymbol{\theta}}$ is the solution of $H(\mathbf{x}, \boldsymbol{\theta}) = 0$ where $\mathbf{x}$ is the observed data and H is an 'estimating function' with the 'unbiasedness' property that

$$\mathbb{E}_{\boldsymbol{\theta}} H(\mathbf{X}, \boldsymbol{\theta}) = 0. \qquad (13.85)$$

Note this does not imply that $\hat{\boldsymbol{\theta}}$ is an unbiased estimator of the true $\boldsymbol{\theta}$. Under reasonable conditions $\hat{\boldsymbol{\theta}}$ will be asymptotically consistent.

A *Takacs-Fiksel* estimating function [646, 647, 648, 256, 258, 228, 160] takes the form

$$H(\mathbf{x}, \boldsymbol{\theta}) = \sum_i T(x_i, \mathbf{x}) - \int_W \lambda_{\boldsymbol{\theta}}(u \mid \mathbf{x}) T(u, \mathbf{x}) \, du \qquad (13.86)$$

where $T(u, \mathbf{x})$ is any chosen function. This satisfies (13.85) because of the Campbell-Mecke formula (8.50), under integrability conditions. Maximum pseudolikelihood for a loglinear Gibbs model is a special case of Takacs-Fiksel estimation, where T is chosen to be the canonical statistic for the model. Estimating equations can also be derived from variational principles [39].

There is an intimate connection between simulation and estimation of a Gibbs process. For any simulation algorithm which is a Markov chain, the generator of the chain has mean zero under the equilibrium distribution. This fact can be used to derive unbiased estimating functions, called *time-invariance* estimating functions [60], which subsume all the estimation methods discussed in this chapter. For example, the Takacs-Fiksel estimating equation (13.86) can be derived from a spatial birth-and-death process with birth rate 1 and death rate $\lambda_{\boldsymbol{\theta}}(u \mid \mathbf{x})$ by applying the generator of the birth-death process to the function T.

13.13.9 Approximate Bayesian inference for Gibbs models

Rajala's [561] approximate Bayesian inference presented in Section 9.11 carries directly over from Poisson to Gibbs models. The setup and assumptions are the same as in Section 13.13.7 above and then (13.84) is taken to be the true likelihood of the model and a multivariate Gaussian prior for the parameter vector $\boldsymbol{\theta}$ is assumed. Using the same approximation as in Section 9.11 a variational Bayes approximation to the posterior is obtained. The implementation in `spatstat` is the same as in the Poisson case with the addition of extra interaction parameters. When specifying priors it is important to be aware of the ordering of parameters in ppm objects: Trend parameters come before interaction parameters. Thus,

```
> p.mean <- rep(0,7)
> p.var <- diag(c(rep(10000,6), 0.001))
> swedfitVB <- ppm(swedishpines ~ polynom(x,y,2), Strauss(9),
                 method = "VBlogi",
                 prior.mean = p.mean, prior.var = p.var)
```

would fit the same model as in Section 13.10.2 using the variational Bayes approximation with a 'flat prior' for the six trend parameters and a strong prior belief that the interaction parameter $\theta_7 = \log \gamma$ is close to zero (corresponding to $\gamma = 1$, i.e. the Poisson process).

Fully Bayesian inference for Gibbs models [484, sec. 9.3] is not yet available in spatstat.

13.14 Determinantal point processes

Determinantal point processes (DPPs) constitute an interesting class of point processes which was recently studied from a statistical viewpoint in [410]. There is a rich literature on the probability theory of DPPs; see in particular [345].

From a statistical viewpoint the most important property of a DPP is that after accounting for spatial inhomogeneity it is inhibitory at all scales. Thus DPPs can only be used to model data with repulsive interactions between points, and any apparent clustering can only be attributed to an inhomogeneous intensity.

While DPPs are in fact Gibbs point processes, they are usually not studied from this point of view, since they are very different from all the other Gibbs point processes we have considered thus far. Some useful properties of DPPs which are unusual for Gibbs processes are: they have an explicitly known intensity function as well as pair correlation function and higher-order moments, and they can be simulated by a sequential algorithm which avoids using Markov chain Monte Carlo methods such as the spatial birth-death algorithm.

A DPP is defined by a so-called kernel (which is effectively a covariance function) $C(u,v)$ defined for any pair of locations u and v. Here we will assume it to be of the special form

$$C(u,v) = \sqrt{\lambda(u)\lambda(v)}R(\|u-v\|/\alpha) \tag{13.87}$$

where $\lambda(u)$ is a non-negative function (the intensity of the process), $\alpha > 0$ is a scale parameter, and R is a known 'template' kernel (correlation function) with $R(0) = 1$. Here we will only use the template $R(r) = \exp(-r^2)$ which gives rise to the *Gaussian kernel*:

$$C(u,v) = \sqrt{\lambda(u)\lambda(v)}\exp(-\frac{\|u-v\|^2}{\alpha^2}). \tag{13.88}$$

A DPP defined by a kernel of the form (13.87) is always correlation-stationary as defined in (7.47) in Chapter 7. The function $\lambda(u)$ is the intensity of the process and the pair correlation function is

$$g(r) = 1 - R(r/\alpha)^2.$$

From this relation it is clear that the pair correlation function is at most 1 and the behaviour of $g(r)$ is completely determined by the behaviour of $R(r)$, and α appears as a scale parameter. An important fact is that the parameter α and the function $\lambda(u)$ cannot be chosen independently of each other. Intuitively this is because α determines how far points have to be separated to appear with a reasonable probability. Thus a large value of α forces the points to be separated by a large distance, but that naturally imposes a bound on the intensity of points. For example the Gaussian kernel (13.88) yields a valid DPP when

$$\lambda(u) \leq 1/(\pi\alpha^2)$$

for all locations u.

13.14.1 Implementation in `spatstat`

Support for DPP models has recently been added to `spatstat`. To define a determinantal point process model with Gaussian kernel we will use the function `dppGauss`. For example

```
model <- dppGauss(lambda=100, alpha=0.05, d=2)
```

defines a Gaussian DPP with constant intensity 100 and scale parameter 0.05 in two-dimensional space (model specification and simulation allows for any number of dimensions). The resulting `model` is of class `"dppmodel"` and it is possible to extract its kernel (`dppkernel`), intensity (`intensity`), pair correlation function (`pcfmodel`), and K-function (`Kmodel`). Furthermore, there is a method for `simulate` to generate realisations from the model:

```
simulate(model)
```

To fit a DPP model to a given dataset, `dppm` can be used, with a formula as first argument, and an empty `"dppmodel"` object as second argument:

```
X <- residualspaper[["Fig1"]]
jfit <- dppm(X ~ polynom(x,y,3), dppGauss())
```

The command `dppGauss()` generates a `dppmodel` where none of the parameters are fixed, and thus all need to be estimated. It is also possible, for example, to fix the scale parameter and only estimate the inhomogeneous intensity by setting `dppGauss(alpha=0.25)`.

For inhomogeneous models the default behaviour of `dppm` is to use two-step minimum contrast estimation like `kppm`. In principle it should be straightforward to implement composite likelihood and Palm likelihood for DPPs as well, but this has not been attempted yet. However, for homogeneous models (i.e. models with right-hand formula ~1) it is possible to (approximately) calculate the likelihood of the model. By default for homogeneous models the intensity is estimated by the non-parametric estimate $\overline{\lambda} = n/|W|$, where n is the number of points and $|W|$ is the area of the study region, and then the remaining parameters are estimated via maximum likelihood.

13.15 FAQs

• *I don't understand the different Gibbs models. Which one should I use for my data?*

A Gibbs model postulates an 'interaction' between the points. It is characterised by the conditional intensity $\lambda(u \mid \mathbf{x})$, which expresses the influence that the point pattern $\mathbf{x}$ would have on a new point added at the spatial location u. The choice of Gibbs model really depends on scientific insight into the kinds of interactions that may be occurring.

• *What is the 'trend' in a Gibbs model? Is it the same as the 'intensity'?*

Trend and intensity are different. The intensity or intensity function of a point process was explained in Chapter 6: it gives the expected number of points per unit area. In a Gibbs model, the 'trend' is the first-order term in the conditional intensity. For example in an inhomogeneous Strauss model with conditional intensity $\lambda(u \mid \mathbf{x}) = \beta(u)\gamma^{t(u,r,\mathbf{x})}$, the 'trend' function is $\beta(u)$. In this model the intensity function $\lambda(u)$ satisfies $\lambda(u) \leq \beta(u)$. See Sections 13.3.6 and 13.5.8.

- *When I fit a Gibbs model to my data, the output says that* `"the model is invalid"`. *What does this mean?*

It means the fitted model does not determine a point process that could be simulated. Some of the fitted parameter values may be `NA` (typically because there was insufficient data to estimate them). Alternatively the estimated interaction parameters may lie outside the range where the model is properly defined — for example if the estimated interaction parameter γ in a Strauss process is greater than 1. See Section 13.5.4 for advice.

- *I have heard that maximum pseudolikelihood is biased and should not be used.*

Maximum pseudolikelihood estimation is theoretically consistent and asymptotically normal, so this can only be a problem in small samples. There seems to be relatively little *published* evidence of bias in maximum pseudolikelihood estimation in small samples: most of the evidence is anecdotal, and may depend on the software implementation used [53]. Nevertheless it is true that the estimates of Gibbs model parameters obtained from ppm, with the default settings, are biased toward the Poisson process (i.e., tend to underestimate the strength of interaction). The default settings are chosen for speed rather than accuracy. For formal analysis it may be wise to change the default 32×32 grid of dummy points to a finer grid using the arguments nd or eps to ppm, or by setting the value of `spatstat.options("ndummy.min")`.

- *When I try to fit an area-interaction model using* `profilepl`, *the algorithm runs well until it prints the message* `"Fitting optimal model..."`, *then stops or hangs.*

Fitting the area-interaction model involves much more computation than other models. If this happens then you probably have a lot of data. It may be worth considering an alternative model that can be calculated faster, such as Geyer's saturation model.

The explanation for the slowdown is that `profilepl` uses a shortcut to fit the candidate models, which reduces the amount of computation required. After selecting the optimal parameter values, the algorithm has to fit the optimal model in full, without this shortcut. In the case of the area-interaction model, this can be a much greater task.

This is a topic of current research, so please check for updates at `www.spatstat.org`.

- *My study region is a country which has both land borders and coastline. The point pattern extends into neighbouring countries (where they are not observed in my study) but not into the sea. Is this an instance of the Small World Model, or Window Sampling? Which edge correction should I use?*

This is not an instance of either the Small World or Window Sampling scenarios. We recommend using the border correction for fitting Gibbs models.

14

Patterns of Several Types of Points

This chapter deals with the analysis of point patterns consisting of several different types of points, called *multitype* point patterns. Such patterns are usually presented as marked point patterns with categorical marks.

14.1 Introduction

A *multitype* spatial point pattern is one that consists of several different types of points. Examples include a map of trees in a wood, with labels showing the species classification of each tree; a map of recent crimes, labelled by the type of crime; and a map of the sky positions of galaxies, classified by colour.

Figure 14.1 shows the spatstat dataset mucosa, a multitype point pattern which gives the locations of two types of cells in a cross-section of the gastric mucosa of a rat.

Multitype point patterns were introduced briefly in Chapter 1. It was seen that, by classifying points into different types, we introduce a new set of scientific questions, requiring new kinds of statistical analysis. These matters have been postponed until now, apart from a few basics. Tools for handling multitype point pattern data were explained in Section 3.3.2, and plotting multitype point patterns in Section 4.1.2. This chapter covers everything else: the exploratory and formal analysis of multitype point patterns, using all the tools developed in the book so far.

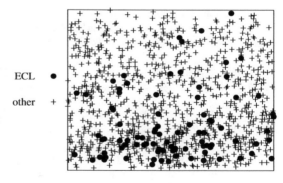

Figure 14.1. *Gastric mucosa data: entero-chromaffin-like cells (•) and other cells (+) in a microscope field of view.*

14.2 Methodological issues

Suppose we have data giving the spatial locations $x_1, \ldots, x_n$ of points, and corresponding labels $m_1, \ldots, m_n$ indicating the type of each point. Without background information, it is unclear how such data should be analysed. It may or may not be appropriate to treat them as a realisation of a point process. The right choice of analysis also depends on the relationship between the different types of points, the number of types, and especially on the scientific context. These choices are discussed below.

14.2.1   Should the data be treated as a multitype point process?

As discussed in Section 5.6.1, treating a spatial point pattern as a realisation of a spatial point process effectively assumes that the locations of points are not fixed, and that the point pattern is the *response* or observation of interest. The analyst should consider whether this is appropriate in the scientific context.

Scenario 14.1. *A weather map for Europe displays a symbol for each major city indicating the expected type of weather (e.g., sunny, cloudy, storms).*

This should *not* be treated as a multitype point process because the cities are fixed locations. Methods from random fields or geostatistics would be appropriate.

Scenario 14.2. *An optical astronomy survey records the sky position and qualitative shape (elliptical, spiral, etc.) of each galaxy in a nearby region of space.*

This can be treated as a multitype point process. Questions of interest could include the spatial distribution of galaxies, the clustering of galaxies, the relative abundance of galaxies of different shapes, whether different shapes of galaxy have the same spatial distribution, and whether galaxies of the same type are clustered together.

Scenario 14.3. *Trees in an orchard are examined and their disease status (infected/not infected) is recorded. We are interested in the spatial characteristics of the disease, such as contagion between neighbouring trees.*

These data probably should *not* be treated as a point process. The response is 'disease status'. We can think of disease status as a label applied to the trees after their locations have been determined. Since we are interested in the spatial correlation of disease status, the tree locations are effectively fixed covariate values. It would probably be best to treat these data as a discrete random field (of disease status values) observed at a finite known set of sites (the trees). Appropriate methods are described in [111] and supported by the contributed R package spdep [112, 110].

Scenario 14.4. *A police map shows the home addresses of the last 100 people fined for traffic offences (violations), labelled by the type of offence.*

This can be treated as a multitype point process, provided the spatial resolution is not too fine (otherwise it becomes similar to the previous scenario). The fact that most people live at a fixed residential address does not mean that the spatial pattern of offenders is fixed. The fixed number of points in the sample is not an obstacle to point process methods, but should be taken into account in the formal analysis (for example, by simulating patterns with exactly 100 points when constructing simulation envelopes).

There are some 'grey areas' which permit several alternative choices of analysis, one of which might be to treat the data as a realisation of a point process. One such grey area occurs when the locations are random, but may be ancillary for the parameters of interest.

Scenario 14.5. *Case-control study of cancer. The domicile locations of all new cases of a rare cancer are mapped. To allow for spatial variation in the density of the susceptible population, domicile locations are recorded for a random sample of (matched) controls.*

This dataset can be analysed either as a multitype point pattern (where the type is the case/control label), or (by conditioning on locations) as a discrete random field of case/control values attached to the known domicile locations. See [224] and [236], respectively, for a full analysis of each kind applied to the Chorley-Ribble data (page 9).

14.2.2 Responses and covariates

As in any statistical analysis, it is vital to decide which quantities to treat as *response* variables, and which as *explanatory* variables (e.g., [181, 180]).

Point process methods are appropriate only if the spatial locations x_i are 'response' variables. *Multitype* point process methods are appropriate only if both the spatial locations and the associated types are part of the 'response'.

Scenario 14.6. *In an intensive geological survey of a mineral province, the locations x_i of all natural deposits of a rare mineral are recorded. Deposits are effectively points at the scale of the survey. For each deposit location x_i, we record whether the surrounding rock is volcanic ($v_i = 1$) or non-volcanic ($v_i = 0$). We wish to determine whether deposits are more likely to occur in volcanic rock.*

The deposit locations x_i are the 'response' of interest, and should be interpreted as a point process. However the rock type values v_i are clearly intended to serve as a *covariate* (explanatory variable), since we wish to determine whether the abundance of deposits depends on rock type v.

A major difficulty with Scenario 14.6 is that these data are inadequate for the purpose. It is not sufficient to record covariate values only at the points of the point pattern. The covariate must also be observed at least at *some* other locations in the study region. The relative frequencies of the rock types $v = 0, 1$ observed only at the deposit locations are not sufficient to estimate the relative frequencies of deposits in the two rock types. In schematic terms, $P(v \mid \text{deposit})$ does not determine $P(\text{deposit} \mid v)$. Bayes' formula indicates that we would need additional information about the relative frequencies $P(v)$ of the two rock types in the study area.

Marks and covariates play different statistical roles. Marks are attributes of the individual points in the pattern, and are part of the 'response' in the experiment, while covariates are 'explanatory' variables. A covariate must be observable at spatial locations other than the points of the pattern, and really should be observable at *any* location, while the mark variable is usually meaningful only at a point of the pattern.

It may be difficult in practice to decide whether a variable should be treated as a response or as a covariate [181]. For example, the Longleaf Pines data (Figure 1.8 on page 7) give the location and diameter at breast height (*dbh*) of each tree in a forest survey. We could make this a multitype pattern by classifying *dbh* into size ranges: the usual practice is to label trees as adults if *dbh* ≥ 30 *cm*, and as juveniles if *dbh* < 30 *cm*. This label is clearly an attribute of the tree, rather than a quantity that could be measured at any spatial location. However, it is clear from Figure 1.8 that the age distribution is not spatially homogeneous. The survey contains some areas where most trees are relatively small. It is known that some parts of the survey region were cleared, decades ago, and such areas would now contain relatively young trees. A more sophisticated analysis of the Longleaf Pines data might use a spatially smoothed trend surface of the *dbh* values as a covariate — effectively a surrogate for the history of the forest. This would be unnecessary if we had a detailed map showing the disturbance history.

14.2.3 The nature of 'types'

The classification of points into types could be based on any criterion. It may be an intrinsic attribute, such as species of tree, sex of plant, type of crime, cause of fire, type of cell, or type of vehicle. A continuously varying attribute may be divided into discrete classes such as size ranges (small/ medium/ large), age ranges, compass directions, colour of galaxy (red/blue), or shape of galaxy (elliptical, spiral, etc.). Alternatively, type may indicate status, such as disease status (susceptible/ infected/ recovered; case/control), disease stage, legal status (major/minor accident), quality of information (verified/not verified), or sampling status (selected/not selected).

Just as in the analysis of any categorical data, the relationship *between* the categories is important. The different types of points may simply be different possibilities, to which we want to give equal treatment, e.g., when studying the spatial distribution of species in a woodland. Alternatively, one type may be the 'control' or 'reference' type, to which other types are compared, e.g., in a case/control study. The types may have a natural ordering, for example when sizes are divided into small, medium, and large ranges. The relationship may be asymmetric, e.g., in the relationship between predator and prey species. Different types of statistical analysis are appropriate in each case.

Categories can also be defined by external or artificial criteria. If the points represent locations where a sample was taken, and if we observe a categorical variable at each location (rock type, vegetation type) then this variable can be taken as the 'type' of the point. In random thinning, the points of the original point pattern before thinning can be labelled according to whether they were retained or deleted. We can label each point in a point pattern according to whether it is or is not a reflexive nearest neighbour (the nearest neighbour of its nearest neighbour). In these artificial cases, the type of each point depends both on the spatial location and on other external information; the interpretation of such data requires great care.

14.2.4   Number of types

In practice, the analysis of multitype patterns depends heavily on the number of possible types, and on their relative abundance.

Most of the examples presented in this book are multitype point patterns with only a few types of points, for which plotting and analysis are easily manageable. We can, for example, compute a multitype version of the K function, $K_{ij}(r)$, based on distances from points of type i to points of type j, and do this for each possible pair of types i and j. We can fit a multitype counterpart of the Strauss model in which a pair of points, of types i and j, respectively, interact if they are closer than a distance r_{ij}.

For multitype point patterns involving a larger number of distinct types, separate visualisation and analysis of each type — and comparison of each pair of types — rapidly become unwieldy. The summary functions K_{ij} cannot be displayed for all pairs of types i and j, but we may pool the 'diagonal' functions K_{ii} for all i, giving us a count of the number of pairs of points *of the same type* within a given distance. Summaries of dependence 'within' and 'between' types are described in Section 14.6.7. Similarly, the multitype Strauss model, with a separate interaction term between each pair of types, has too many parameters and is too difficult to interpret in such cases, but we can easily fit a pairwise interaction that depends on whether the points are of the same type or of different type. These are effectively multivariate data reduction techniques; other multivariate techniques such as principal component analysis may be useful [354], [355, sec. 4.9].

Some types of points may occur with such low frequency that a separate analysis of each type is unreliable, and it is appropriate to pool some of the types. For example, pooling of this nature was done with the Lansing Woods survey (see Section 14.3.3) before the data were released. The `misc` category combines several other species which occur in the woods but are comparatively rare.

Some multitype datasets involve an *a priori* unlimited number of types, for example, a forestry survey of a rainforest with high biodiversity. Such datasets often empirically satisfy 'Zipf's Law': the frequency of the kth most frequent type is approximately proportional to $1/k$. Practical strategies include pooling the lowest-ranking types, or mapping the types to continuous numerical values. It may then be possible to apply methods applicable to continuous marks (see Chapter 15).

14.2.5   Multitype and multivariate viewpoints

There are two, closely related, ways to view a spatial point pattern that consists of several different types of points. One is to record the spatial locations $x_1, \dots, x_n$ of all points, along with labels $m_1, \dots, m_n$ indicating the type of each point. This is a *multitype* point pattern.

Figure 14.1 on page 561 shows the standard presentation of the dataset mucosa (see Section 6.1) as a multitype point pattern, with the two types of cells represented by two different symbols, generated by plot(mucosa). This is an abstraction of the pattern that would have been visible under the microscope.

The other approach is to separate the points according to their type, yielding M distinct point patterns, where M is the number of possible types. We then have a *multivariate* observation $\mathbf{y} = (\mathbf{x}^{(1)}, \ldots, \mathbf{x}^{(M)})$ where $\mathbf{x}^{(m)}$ is the pattern of locations of points of type m for each value of $m = 1, 2, \ldots, M$. Figure 14.2 shows the multivariate presentation of the gastric mucosa data, in which different types of cells are shown in different panels, generated by plot(split(mucosa)). Figure 14.2 is what might have been observed if the different types of cells had to be visualised using different staining or illumination.

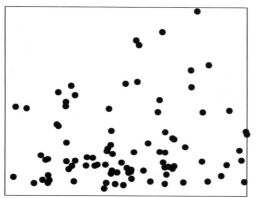 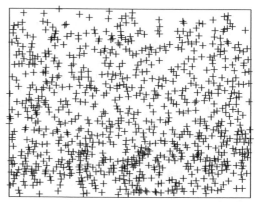

Figure 14.2. *Multivariate presentation of gastric mucosa data.* Left: *enterochromaffin-like cells;* Right: *other cells.*

Data in spatstat are organised using the 'multitype' approach, in which all the points are collected together in one point pattern. Some other packages, importantly including splancs, use the 'multivariate' approach. It is simple to reorganize data to conform to either paradigm, as explained in Section 14.3.2.

The multitype and multivariate viewpoints enable slightly different statistical analyses of the data, as explained below.

14.2.6 Strategy for analysis

Multitype point pattern datasets raise new and interesting questions about the appropriate choice of statistical analysis. They offer more choices — and more traps.

In a statistical analysis of two response variables X and Y, we have the choice of analysing the joint distribution $P(X, Y)$, or the conditional distribution $P(Y \mid X)$ of Y given X, or the marginal distribution $P(X)$ together with $P(Y \mid X)$, or other options. These strategies are mathematically related through $P(X, Y) = P(X) P(Y \mid X)$, but they are not statistically equivalent.

Similarly, in the analysis of multitype point patterns, there are several possible strategies, which permit different exploratory tools, typically lead to different stochastic models, have different inferential interpretations, and suggest different choices for the 'natural' null hypothesis. Some questions are much easier to formulate using one strategy than using another. The appropriate choice depends on the scientific context and objectives.

Adopting the **'multitype'** viewpoint to start with, suppose $X = (x_1, \ldots, x_n)$ denotes the locations of the points, and $M = (m_1, \ldots, m_n)$ the marks (types) of these points. Three possible schemes for analysis are:

joint distribution $P(X,M)$**:** regard the locations and types as having been generated at the same time.

A typical example is the joint analysis of a mixed woodland. Each individual tree belongs intrinsically to a species; the viability of each tree depends on its neighbours' locations and their species; hence the spatial pattern of locations and species should be analysed jointly. The most general model is a multitype point process. We might assume that the process is stationary, allowing us to estimate a multitype version of the K-function, for example. The simplest null model is a homogeneous *Poisson* multitype point process (dubbed *'complete spatial randomness and independence'*, *CSRI* below).

conditional distribution $P(M \mid X)$**:** regard the locations X as fixed or given, and study only the types M attached to these locations.

A typical example is the conditional analysis of a spatial case-control study such as the Chorley-Ribble data (page 9) where the spatial locations are home addresses and the 'type' label identifies whether the person is a disease case or a control. In a conditional analysis [236] the home address locations are treated as fixed covariates, and the disease status is treated as a random variable. The most general model is a discrete random field for the disease status labels. We could use kernel smoothing to estimate the spatially varying disease risk, and neighbourhood-based correlation indices such as Moran's I to investigate disease clustering. See [111] and the spdep package. The simplest null model is that disease status of each person is random, independently of other people, with a constant risk of disease (called *'random labelling'* below).

marginal and conditional distributions $P(X)P(M \mid X)$**:** regard the locations X as having been generated first, and subsequently annotated with types M.

This approach is often used in an initial analysis of multitype point pattern data, where we first focus on the spatial locations only, applying methods for spatial point processes described in the previous chapters, then study the marks attached to the points. The most general model is a point process for X together with a discrete random field for M given X. The simplest null model is a homogeneous Poisson process (CSR) for the locations, and random labelling for the marks given the locations, which together are equivalent to CSRI.

Now adopting the **'multivariate'** viewpoint, suppose for simplicity that there are only two types of points, a and b. The multivariate point pattern is $Y = (A,B)$ where A and B are the patterns of points of type a and b, respectively. Three possible schemes for analysis are:

joint distribution $P(A,B)$**:** regard the data as a pair of spatial point patterns.

A typical application is the amacrine cells data (Figure 1.5) in which the two types of cells lie in different layers of the retina. Questions about the dependence between the two layers are best investigated by treating them as two point processes. The most general model is equivalent to a multitype point process. The simplest null model states that A and B are independent point processes (*'independence of components'*).

conditional distribution $P(B \mid A)$**:** regard the locations of the points of type a as fixed or given, and study the points of type b.

A typical application is the ants' nests data (see Section 1.1.2, page 6), giving the locations of nests of two species of ants, one of which (*Messor*, A) is potential food for the other (*Cataglyphis*, B). To investigate how B colonies place their nests, we could treat the pattern of A nests as an explanatory covariate. The most general model is a point process for B in which one of the covariates is the pattern of A nests. The simplest null model is that B does not depend on A (equivalent to assuming independence of components).

Spatial case-control data such as the Chorley-Ribble data can also be analysed by modelling only the cases, treating the controls as a covariate. The locations of the controls are viewed as a sample from the spatially varying density of the susceptible population. Kernel smoothing the controls gives an estimate of the population density (up to a constant factor related to the sampling fraction) which can then serve as a covariate in a point process model for the cases. See [224]. The usual null model in that context is that the cases are Poisson with intensity proportional to the population density (the model of *'constant disease risk'*).

marginal and conditional distributions $P(A)P(B \mid A)$: regard the data as having been built by first determining the pattern A and then the pattern B given A.

In analysing the ants' nests data we could model the spatial pattern of A (*Messor*) nests, then the pattern of B (*Cataglyphis*) nests conditional on A. The most general model is a *hierarchical point process model* (Section 14.9) for A and B with B depending on A. The simplest null model is a homogeneous Poisson process (CSR) for A and a separate homogeneous Poisson process model for B independent of A, which together are equivalent to CSRI.

In a nutshell, if we wish to test whether *'the marks depend on the locations'*, there are at least three different choices for the null hypothesis:

complete spatial randomness and independence (CSRI): the locations of points are completely random; the locations of the points of a given type are completely random; points of different types are independent of each other; each point is randomly allocated to one of the possible types independently of other points and independently of its location.

random labelling: each point is randomly allocated to one of the possible types independently of other points and independently of its location.

independence of components: points of different types are independent of each other.

To see, for instance, that random labelling does not imply independence of components, take a hard core process, and randomly label the points. The resulting multitype point process still obeys the hard core constraint; it does not have the independence of components property because if points with different marks were independent, they would not obey the hard core constraint.

The appropriate choice of null hypothesis depends crucially on the scientific context. We discuss this further below.

14.3 Handling multitype point pattern data

In this section we explain how to create multitype point patterns in `spatstat` and how to perform useful operations on them.

14.3.1 Creating multitype point pattern datasets

A multitype point pattern is represented in `spatstat` by a `"ppp"` object with categorical mark values, that is, a point pattern X for which `marks(X)` is a `factor`. Such data can be created from scratch, or from other types of data, using the tools listed in Table 14.1. See also Sections 3.3.2.1, 4.2.4, and sections below.

The syntax for using ppp to create a multitype point pattern from raw data is

```
ppp(x, y, ..., marks=m)
```

ppp	create point pattern dataset
as.ppp	convert other data to point pattern
superimpose	combine several point patterns
scanpp	read point pattern data from text file
clickppp	create a pattern using point-and-click interface
marks	extract marks
marks<-	attach marks (assignment operator)
%mark%	attach marks (binary operator)
unmark	remove marks
cut.ppp	classify points into types

Table 14.1. *Tools useful for making a multitype point pattern.*

where x and y are vectors of equal length containing the spatial coordinates, m is a vector of categorical values (a factor) of the same length as x specifying the type of each point, and "..." are arguments that determine the window for the point pattern.

As it is quite common to make errors when creating a factor, vigilance is strongly recommended. Ensure that the marks argument m is stored as a **factor**, by typing is.factor(m) or by printing the values. When the point pattern X has been created, check that it is indeed multitype, using is.multitype(X) or printing X. Check that the factor levels are as you intended, using levels(marks(X)). Check for NA values in the factor, which arise when a data value did not match any of the specified levels. Check the ordering of the factor levels: by default they are sorted alphabetically; see Section 2.1.9 for ways to reorder the levels. Finally, when performing equality comparisons involving a factor, check that the result is as intended; this is particularly prone to human error when the factor levels are strings that represent integers.

The function ppp issues a warning if two data points are identical, meaning that they occupy the same spatial location *and* have the same mark value. It does not warn if two data points with different marks occupy the same spatial location. To detect duplication of the spatial locations, use duplicated(unmark(X)) or duplicated(X, rule="unmark"). Such duplication might occur, for instance, if the spatial coordinates were coarsely discretised. See Section 3.4.4.

The function as.ppp is used to convert data from another format to a point pattern object of class "ppp". This is normally used to convert spatial data from another package, such as a "SpatialPointsDataFrame" object obtained from a shapefile (see Section 3.10.2). It also applies to simple data structures. If the input is a matrix or data frame, the first two columns will be interpreted as spatial coordinates, and subsequent columns as marks:

```
> mydata <- data.frame(x=runif(10), y=runif(10),
            m=factor(sample(letters[1:3], 10, replace=TRUE)))
> X <- as.ppp(mydata, square(1))
```

For a multitype pattern, the input mydata needs to be a data frame rather than a matrix, in order that the third column can be a factor.

The usual way to convert experimental data into a spatstat point pattern object is to read the data from files into R as vectors of numerical or categorical data, then create the point pattern using ppp. A shortcut is available for data stored in a text file in table format (Section 3.2.1). The function scanpp will read the data and create a point pattern object.

```
> copyExampleFiles("amacrine")
> W <- owin(c(0,1060/662), c(0,1))
> X <- scanpp("amacrine.txt", factor.marks=TRUE, window=W)
```

The argument factor.marks determines whether the mark values are to be interpreted as a factor.

If there is more than one column of marks, `factor.marks` can be a logical vector prescribing a different fate for each column.

Recall from Chapter 4 that marks can also be attached to an existing point pattern X by using the assignment function `"marks<-"` as in `marks(X) <- m`, or by using the binary operator `%mark%` as in `Y <- X %mark% m`, where m is a vector or factor with one entry for each point, or a data frame with one row for each point. These are convenient for assigning new marks to a dataset or randomising the existing marks:

```
> marks(X) <- sample(marks(X), npoints(X))
```

Note that if X is already a marked point pattern, then `marks(X) <- m` will *replace* the marks by m. Marks can be removed (resulting in an unmarked point pattern) by setting `marks(X) <- NULL` or assigning `Y <- X %mark% NULL` or `Y <- unmark(X)`. The function `superimpose` can be used to combine several individual point patterns into a single multitype point pattern, as explained in the next section.

A multitype point pattern can also be created interactively using `clickppp`, using the argument `types` to specify the possible types. See the help for `clickppp` or Section 3.9.

The function `cut.ppp` classifies the points of a point pattern into categories. By default the classification is based on the existing mark values, so that if X is a point pattern with numeric marks, then `cut(X, breaks=3)` will divide the range of mark values into three bands, and classify each point accordingly. Alternatively the classifier can be other data such as a tessellation or image, so that points can be classified according to where they fall, or according to the value of a covariate. See Section 4.2.4 or the help for `cut.ppp`.

Multitype point patterns can also be generated randomly using functions listed in Table 14.2.

`rmpoint`	generate *n* random multitype points
`rmpoispp`	simulate Poisson multitype point process
`rNeymanScott`	simulate Neyman-Scott process
`rmh.default`	simulate Gibbs point process
`rmh.ppm`	simulate fitted Gibbs point process model
`simulate.ppm`	simulate fitted Gibbs point process model

Table 14.2. *Functions in* `spatstat` *for generating random multitype point patterns.*

14.3.2 Split, superimpose, and unsplit

The generic function `split` divides a dataset into subsets according to a grouping factor. The method `split.ppp` (see Section 4.2.9.1) takes a point pattern and separates it into a list of sub-patterns. If the point pattern is multitype and the grouping factor is omitted, then by default, the result is a list of unmarked point patterns containing the points of each given type:

```
> Y <- split(mucosa)
> Y
Point pattern split by factor

ECL:
Planar point pattern: 89 points
window: rectangle = [0, 1] x [0, 0.81] units

other:
Planar point pattern: 876 points
window: rectangle = [0, 1] x [0, 0.81] units
```

This effectively changes the representation of the data from the 'multitype' to the 'multivariate' viewpoint.

To pass back from the 'multivariate' to the 'multitype' viewpoint, one can use the function `superimpose` which combines several point patterns into a single pattern. If the input patterns are unmarked, and are specified using 'name=value' syntax, they will be interpreted as patterns of different types of points, and the resulting pattern will be multitype, with the types given by the names.

```
> superimpose(A=runifpoint(5), B=runifpoint(7))
Marked planar point pattern: 12 points
Multitype, with levels = A, B
window: rectangle = [0, 1] x [0, 1] units
```

Note that the names are needed in order to obtain a multitype point pattern:

```
> superimpose(runifpoint(5), runifpoint(7))
Planar point pattern: 12 points
window: rectangle = [0, 1] x [0, 1] units
```

The arguments to `superimpose` should be unmarked point patterns; otherwise the existing marks will also be retained, and the result will have several columns of marks, as explained in the help for `superimpose`.

If Y is a named list of unmarked point patterns, such as the result of `split.ppp`, then

```
X <- do.call(superimpose, Y)
```

will produce a multitype point pattern, with the types given by the names of the list elements of Y.

Note that `superimpose` is not exactly the reverse of `split.ppp` because the ordering of points may be changed. When we split a multitype point pattern, all the points of a given type are extracted and stored in consecutive order, even if they were not contiguous in the original dataset. When we superimpose point patterns A, B, ..., the resulting point pattern contains all the points of A followed by the points of B, and so on. Thus `do.call(superimpose, split(X))` is equivalent to X but possibly with a different ordering.

The dataset `betacells` is a marked point pattern giving the locations, types ('on' or 'off'), and projected areas of beta ganglion cells in the retina, originally from Figure 6(a) of [692], scanned by S.J. Eglen. The point pattern has a data frame of marks, with two columns: a factor named `type` and a numeric variable `area`. First we will re-mark this pattern using only the first (factor) column of the data frame:

```
> X <- betacells
> marks(X) <- marks(X)[,1]
```

Next we compare X with the result of splitting and superimposing:

```
> XX <- do.call(superimpose, split(X))
> identical(X, XX)
[1] FALSE
> max(nncross(X, XX)$dist)
[1] 0
```

If it is important to preserve the original ordering after such operations, then the generic function `split<-` can be used. For example, suppose we wish to apply random displacements to only one type of point in a multitype point pattern. We can split the pattern into components, apply the random displacement to the desired component, and un-split:

```
> X <- amacrine
> Y <- split(amacrine)
> Y$on <- rjitter(Y$on, 0.1)
> split(X) <- Y
```

14.3.3 Inspecting a multitype point pattern

Basic tools for inspecting a multitype point pattern include the methods for `print`, `summary`, `plot`, and `intensity`, together with the interactive plotting command `iplot`.

If X is a multitype point pattern, `plot(X)` plots the data using the multitype presentation, with all points displayed in a single window, using a different symbol for each type of point. See Figure 14.1. The mapping of types to symbols is shown in an explanatory legend, and is also returned invisibly from the plot command as an object of class `"symbolmap"`, allowing the same mapping to be re-used or modified. See Section 4.1.2 for further details.

Alternatively `plot(split(X))` would plot the data using the multivariate presentation, with a separate panel for each type of point. See Figure 14.2. Since `split(X)` is a list belonging to the class `"solist"`, the plotting is performed by `plot.solist`.

The multitype plot is often useful for detecting small-scale dependence between different types. The multivariate plot is often more useful for detecting spatial inhomogeneity of the points of one type, and large scale dependence between types.

The interactive plotting function `iplot` is very useful in initial examination of the data. It allows the user to switch between the multitype and multivariate presentations of a multitype point pattern, tweak the symbol map, zoom in to inspect fine detail, and so on. See Section 4.1.4.

Figure 14.3 shows the `lansing` dataset, the result of a survey of a 20-acre tract in Lansing Woods, Michigan [280]. The data give the locations of 2251 trees and their botanical classification into hickories, maples, red oaks, white oaks, black oaks, and miscellaneous trees. One of the main questions is whether the species are *segregated*, in the sense that there are regions of the woods where one species predominates [280, 182, 219]. Figure 14.3 is the default 'multitype' display, generated by `plot(lansing)`. This monochrome display is not useful for assessing evidence for segregation of species, although the colour version is better. The interactive plot function `iplot` makes it easy to tweak the symbols, their sizes and colours, and to see fine detail.

Figure 14.4 shows the same data separated by species, using `plot(split(lansing))`. Spatial variation in the intensity of trees of a given species, and evidence for segregation of species, is much easier to visualise in this 'multivariate' presentation.

Factor data are often part of a more complicated data structure. For example the New Brunswick forestfires dataset `nbfires` has nine columns of data, one of which is a factor named `fire.type`. To display only this information as if it were a multitype point pattern, use `which.marks`:

```
> plot(nbfires, which.marks="fire.type")
```

The subset method for point patterns is useful for restricting attention to a subset defined by the other columns of marks:

```
> plot(subset(nbfires, year == "1996"), which.marks="fire.type")
```

or equivalently

```
> plot(subset(nbfires, year == "1996", select=fire.type))
```

The result is shown in Figure 14.5.

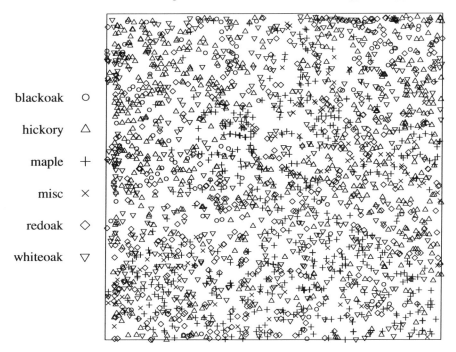

Figure 14.3. *Lansing Woods data: locations and botanical classification of 2251 trees in a survey region 924 feet (282 metres) across.*

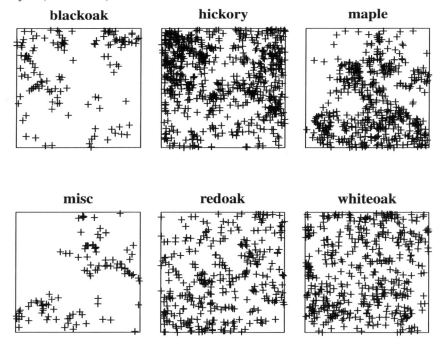

Figure 14.4. *Lansing Woods data separated by species.*

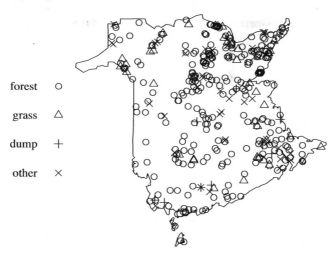

forest   ○

grass   △

dump   +

other   ×

Figure 14.5. *The 1996 New Brunswick fires, marked by fire type. Note the apparent symbol* ∗ *is an overprinting of the symbols for* dump *and* other.

The print and summary methods provide basic summaries:

```
> lansing
Marked planar point pattern: 2251 points
Multitype, with levels = blackoak, hickory, maple, misc, redoak, whiteoak
window: rectangle = [0, 1] x [0, 1] units (one unit = 924 feet)
> summary(lansing)
Marked planar point pattern:  2251 points
Average intensity 2251 points per square unit (one unit = 924 feet)
*Pattern contains duplicated points*
Multitype:
          frequency proportion intensity
blackoak      135     0.05997     135
hickory       703     0.31230     703
maple         514     0.22830     514
misc          105     0.04665     105
redoak        346     0.15370     346
whiteoak      448     0.19900     448
Window: rectangle = [0, 1] x [0, 1] units
Window area = 1 square unit
Unit of length: 924 feet
```

The output shows the number of trees of each species, their relative proportions, and the average intensity of each species of tree. The summary indicates there are duplicated points; as discussed above, this means there exist two trees of the same species at the same spatial location.

The function intensity.ppp, applied to a multitype point pattern, computes the average intensities for each type of point:

```
> intensity(rescale(lansing))
  blackoak   hickory     maple     misc    redoak  whiteoak
 0.0001581 0.0008234 0.0006020 0.0001230 0.0004053 0.0005247
```

Here we used rescale to convert the coordinates to comprehensible units (feet, rather than multiples of 924 feet) so that the intensity is expressed as the number of trees per square foot. Alternatively in trees per square mile (where 1 mile is 5280 feet):

```
> intensity(rescale(lansing, 5280/924))
 blackoak  hickory    maple     misc  redoak whiteoak
    4408    22955    16784     3429   11298    14629
```

We could alternatively have written `rescale(rescale(lansing), 5280)`.

It is often useful to inspect the coordinates and marks together, by converting them to a data frame:

```
> head(as.data.frame(amacrine), 3)
       x      y marks
1 0.0224 0.0243    on
2 0.0243 0.1028    on
3 0.1626 0.1477    on
```

The `head` function was used here to show only the first few lines of data. In an interactive session one can use `page`, as in `page(as.data.frame(amacrine),method="print")`.

14.4 Exploratory analysis of intensity

Normally the first major step in exploratory analysis of a multitype point pattern dataset is to study its intensity. This conveys information about the abundance and density of the points (with the interpretations of fertility, productivity, accident rate, etc as described in Chapter 6), but also about the relative abundance of points of different types (with interpretations of relative frequency, relative risk, or probability distribution). It may provide evidence of an association between spatial location and type of point (suggesting there is spatial segregation of types, spatially varying disease risk, etc.). Only minimal statistical assumptions are needed to estimate some properties of the intensity using nonparametric techniques such as kernel smoothing.

14.4.1 Homogeneous intensity

There are no additional theoretical difficulties in defining the intensity of a multitype point process: we simply consider the intensity of each type of point separately.

Adopting the multivariate viewpoint, we split a multitype point process $\mathbf{Y}$ into sub-processes $\mathbf{X}^{(1)}, \ldots, \mathbf{X}^{(M)}$ where $\mathbf{X}^{(m)}$ is the point process of points of type m for each $m = 1, 2, \ldots, M$. We also write $\mathbf{X}^{(\bullet)}$ for the 'unmarked' point process consisting of the locations only.

The multitype process $\mathbf{Y}$ will be called *first-order stationary* if each sub-process $\mathbf{X}^{(m)}$ is first-order stationary, that is, if the expected number of points of $\mathbf{X}^{(m)}$ falling in any region B is proportional to the area of B:

$$\mathbb{E}[n(\mathbf{X}^{(m)} \cap B)] = \lambda_m |B| \tag{14.1}$$

analogous to equation (6.1) on page 159, where λ_m is a nonnegative constant, the intensity of the process of points of type m.

If the multitype point process $\mathbf{Y}$ is first-order stationary, then the unmarked process $\mathbf{X}^{(\bullet)}$ is also first-order stationary, with intensity

$$\lambda_\bullet = \sum_{m=1}^{M} \lambda_m \tag{14.2}$$

because $\mathbf{X}^{(\bullet)}$ is the superposition of all the processes $\mathbf{X}^{(m)}$ for $m = 1, 2, \ldots, M$. Hence it is really appropriate to borrow the traditional symbol $\bullet$ from classical statistics, which often denotes the 'margin' or 'sum' over an index.

Assuming our multitype point pattern dataset **y** in window W comes from a first-order stationary process, an unbiased estimator of λ_m is

$$\overline{\lambda}_m = \frac{n(\mathbf{x}^{(m)})}{|W|}, \tag{14.3}$$

the observed number of points of type m divided by the area of the observation window. This calculation is performed by `intensity.ppp` or `summary.ppp` as demonstrated above.

The corresponding unbiased estimator of $\lambda_\bullet$ is the sum of these intensity estimates, $\overline{\lambda}_\bullet = \sum_m \overline{\lambda}_m$, which is the usual intensity estimate for the unmarked point pattern. It can be computed by summing the individual intensity estimates, or by unmarking the data:

```
> sum(intensity(lansing))
[1] 2251
> intensity(unmark(lansing))
[1] 2251
```

Standard errors or confidence intervals for these estimates would require further assumptions. For a stationary multitype Poisson process the estimator $\overline{\lambda}_m$ has standard error $(\lambda_m/|W|)^{1/2}$. The plug-in standard error is

```
> sqrt(intensity(lansing)/area(Window(lansing)))
 blackoak  hickory    maple     misc   redoak whiteoak
    11.62    26.51    22.67    10.25    18.60    21.17
```

An alternative is to use the sample variance of quadrat counts, assuming stationarity and independence between sufficiently large quadrats. We demonstrate with the Lansing Woods data although the stationarity assumption is clearly inappropriate:

```
> Nx <- Ny <- 4
> QC <- lapply(split(lansing), quadratcount, nx=Nx, ny=Ny)
> QC <- lapply(QC, as.vector)
> v <- sapply(QC, var)
> (se <- sqrt(Nx * Ny * v)/area(Window(lansing)))
 blackoak  hickory    maple     misc   redoak whiteoak
    27.32    92.00    91.24    27.32    34.53    32.46
```

14.4.2 Intensity function and kernel estimation

The multitype point process **Y** has *intensity function* $\lambda(u,m)$ for locations u and marks m, if

$$\mathbb{E}[n(\mathbf{X}^{(m)} \cap B)] = \int_B \lambda(u,m)\,\mathrm{d}u. \tag{14.4}$$

That is, the process of points of type m has intensity function $\lambda_m(u) = \lambda(u,m)$.

If **Y** has intensity function $\lambda(u,m)$, then the unmarked process $\mathbf{X}^{(\bullet)}$ has intensity function

$$\lambda_\bullet(u) = \sum_{m=1}^{M} \lambda_m(u) = \sum_{m=1}^{M} \lambda(u,m), \tag{14.5}$$

the sum of the individual intensity functions. Campbell's formula (page 169) applies to multitype point processes, so that if $f(u,m)$ is a function of location u and mark m, then

$$\mathbb{E}\left[\sum_i f(x_i, m_i)\right] = \int \sum_{m=1}^{M} f(u,m)\,\lambda(u,m)\,\mathrm{d}u. \tag{14.6}$$

Kernel estimation of the intensity function of a point process was described in Section 6.5.1. For multitype point processes, the intensity function $\lambda(u,m)$ can also be estimated by kernel smoothing, justified by Campbell's formula (14.6). In effect kernel smoothing is applied separately to each of the sub-patterns $\mathbf{x}^{(m)}$ of points of type m, to produce the estimated intensity functions $\widehat{\lambda}_m(u)$ for each type.

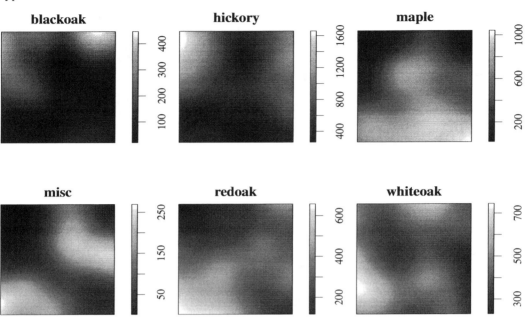

Figure 14.6. *Kernel estimates of intensity for each species in Lansing Woods.*

In `spatstat` this calculation is performed by `density.splitppp`, the `density` method for objects of class `"splitppp"`. For example, we saw in Section 14.3.3 that the Lansing Woods data appear to be inhomogeneous, and that the different species probably have different intensity functions. Kernel estimates of the intensity of each species are calculated and plotted by

```
> plot(density(split(lansing)))
```

See Figure 14.6. If the smoothing bandwidth `sigma` is not specified, or if it is given a numerical value, then all intensity estimates will be based on the same smoothing bandwidth. However, if `sigma` is a function for bandwidth selection, such as `sigma=bw.diggle`, then this function will be applied separately to each sub-pattern, yielding different smoothing bandwidths for each sub-pattern. The use of different smoothing bandwidths may be sensible when the frequencies of different types are very unequal, but this also makes it harder to compare the intensities of different types.

Standard errors are available under the additional assumption that the process is Poisson, as discussed in Section 6.5.1.5:

```
> plot(density(split(lansing), se=TRUE)$SE)
```

(The plot is not shown.)

To investigate evidence for segregation between types, we need to establish what would be expected if the types are *not* segregated. We will say that the intensity is *separable* if $\lambda(u,m) = a_m\beta(u)$, where $\beta(u)$ is a function of spatial location and $a_1,\ldots,a_M$ are constants. Then the intensity functions of each type of point are proportional to $\beta(u)$, and hence proportional to one another; the points of different types share the same 'form' of spatial inhomogeneity.

14.4.3 Probabilities of each type

In a first-order stationary process, we can speak of the probability distribution of types. The probability that a typical point belongs to type m is

$$p_m = \frac{\lambda_m}{\lambda_\bullet}. \tag{14.7}$$

The natural estimator of p_m is $\overline{p}_m = n_m/n_\bullet = \widehat{\lambda}_m/\widehat{\lambda}_\bullet$, where n_m is the observed number of points of type m, and $n_\bullet$ is the total number of points. As a ratio of unbiased estimators, $\overline{p}_m$ is not exactly unbiased, in general. For a stationary multitype Poisson process (Section 14.5.1), $\overline{p}_m$ is exactly unbiased, given the pattern is not empty. These estimates can be obtained using `intensity.ppp`:

```
> lambda <- intensity(lansing)
> probs <- lambda/sum(lambda)
> probs
 blackoak  hickory    maple     misc   redoak whiteoak
  0.05997  0.31231  0.22834  0.04665  0.15371  0.19902
```

For an inhomogeous multitype point process, the probability distribution of types is spatially varying. Assuming there is an intensity function $\lambda(u,m) = \lambda_m(u)$, the conditional probability, given that there is a point at location u, that the point is of type m, is

$$p(m \mid u) = \frac{\lambda_m(u)}{\lambda_\bullet(u)} = \frac{\lambda_m(u)}{\sum_{m'=1}^{M} \lambda_{m'}(u)} \tag{14.8}$$

provided that $\lambda_\bullet(u) \neq 0$. The probability is undefined if $\lambda_\bullet(u) = 0$.

Notice that if the intensity is separable, $\lambda_m(u) = a_m \beta(u)$, then $p(m \mid u) = a_m/a$ is constant. That is, spatial variation in the probability distribution of types $p(m \mid u)$ is equivalent to segregation of types.

Nonparametric estimates of the probabilities $p(m \mid u)$ could be obtained by substituting kernel smoothing estimates of $\lambda_m(u)$ and $\lambda_\bullet(u)$ into (14.8). If we use the same smoothing bandwidth to estimate $\lambda_m(u)$ and $\lambda_\bullet(u)$, then with the 'uniform' edge correction, equation (6.8) on page 168, the estimate is

$$\widehat{p}(m \mid u) = \frac{\sum_i \mathbf{1}\{m_i = m\} \kappa(u - x_i)}{\sum_i \kappa(u - x_i)} \tag{14.9}$$

where the sum is over all multitype points in the dataset, and the term $\mathbf{1}\{m_i = m\}$ effectively restricts the sum in the numerator to be taken over the points of type m only. Similarly if we use Diggle's edge correction (6.9), the estimate of the distribution of types is

$$\widehat{p}(m \mid u) = \frac{\sum_i \mathbf{1}\{m_i = m\} \kappa(u - x_i)/e(x_i)}{\sum_i \kappa(u - x_i)/e(x_i)} \tag{14.10}$$

where $e(x_i)$ is defined in (6.10). Notice that the uniform edge correction term $1/e(u)$ in (6.8) cancels out in the numerator and denominator of (14.9), while the data weights $1/e(x_i)$ in the Diggle correction (6.9) remain in (14.10).

It is instructive to see that (14.9), for example, is the Nadaraya-Watson smoother (6.26) of the 0-1 values $z_i = \mathbf{1}\{m_i = m\}$ attached to the data points x_i. That is, in estimating the spatially varying distribution of types, we are effectively applying spatial smoothing to the indicator variables for each type.

In `spatstat` these calculations are performed by `relrisk.ppp`, a method for the generic `relrisk`:

```
> ProbU <- relrisk(lansing)
> ProbD  <- relrisk(lansing, diggle=TRUE)
```

Note that `relrisk.ppp`, unlike `density.ppp`, performs automatic bandwidth selection by default: this is discussed below. The result is a list of pixel images, of class `"imlist"` and `"solist"`. Figure 14.7 shows the Diggle-corrected estimates, plotted by `plot(ProbU)`.

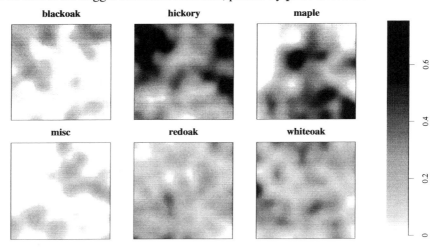

Figure 14.7. *Estimates of spatially varying proportions of each species in Lansing Woods, obtained using* `relrisk` *with Diggle's edge correction.*

It would be possible to calculate the estimates (14.9) and (14.10) by 'hand':

```
> lami <- density(split(lansing))
> lamdot <- Reduce("+", lami)
> probi <- solapply(lami, "/", lamdot)
```

Note the use of `Reduce` to compute the sum of a list of images. The function `relrisk.ppp` does essentially the same calculation but also deals with extreme cases and with numerical errors.

If we require estimates $\widehat{p}(m \mid x_i)$ of the probabilities at the data locations themselves, then, as explained in Section 6.5.1.3, to avoid substantial bias we should compute the 'leave-one-out' intensity estimates (6.12) or (6.13), by summing over all points x_j other than the data point x_i in question. In `spatstat` this is the default when estimation is performed at the data points:

```
> ProbX <- relrisk(lansing, at="points")
> ProbX[1:5]
[1] 0.008269 0.017932 0.017053 0.033044 0.041233
```

Standard errors for the probabilities $\widehat{p}(m \mid u)$ or $\widehat{p}(m \mid x_i)$ are available assuming a Poisson process and using the delta-method approximation to the variance of a ratio of random variables:

$$\operatorname{var} \widehat{p}(m \mid u) \approx \frac{1}{\lambda_\bullet(u)^2} \left[\operatorname{var} \widehat{\lambda}_m(u) + p(m \mid u)^2 \operatorname{var} \widehat{\lambda}_\bullet(u) - 2p(m \mid u) \operatorname{cov}(\widehat{\lambda}_m(u), \widehat{\lambda}_\bullet(u)) \right]. \quad (14.11)$$

When `relrisk` is called with `se=TRUE`, the result is a list of length 2, containing both the intensity estimates and the standard errors.

14.4.4 Bandwidth selection

By default, `relrisk.ppp` selects the smoothing bandwidth automatically by calling `bw.relrisk`. This implements the bandwidth selection techniques of [372]. Consider the indicator variables $z_{im} = \mathbf{1}\{m_i = m\}$ which equal 1 when the ith data point belongs to type m, and equal 0 otherwise. Then the bandwidth can be chosen to minimise either the negative loglikelihood

$$L = -\sum_i \log \widehat{p}(m_i \mid x_i) \tag{14.12}$$

or the sum of squared errors

$$L = \sum_i \sum_m (z_{im} - \widehat{p}(m \mid x_i))^2 \tag{14.13}$$

or the sum of standardised squared errors

$$L = \sum_i \sum_m \frac{(z_{im} - \widehat{p}(m \mid x_i))^2}{\widehat{p}(m \mid x_i)(1 - \widehat{p}(m \mid x_i))} \tag{14.14}$$

where $\widehat{p}(m \mid x_i)$ is the leave-one-out estimator. Each of these criteria measures how well the smoother is able to predict the type of x_i, using information from all points other than x_i. The default choice is the likelihood criterion (14.12). After typing

```
> b <- bw.relrisk(lansing)
```

simply printing b yields the numerical value of the optimal bandwidth, 0.0488, while typing `plot(b)` generates the display in the left panel of Figure 14.8 showing the cross-validation criterion plotted against smoothing bandwidth. The right panel is obtained by adding the argument `xlim` to the plot command to narrow the range of bandwidth values. The plotted function itself can be extracted as `as.fv(b)`. Having decided that a reasonable choice of bandwidth has been made, we could then compute the spatially varying probabilities by

```
> Prob <- relrisk(lansing, sigma=b)
```

or use the argument `adjust` to change this value. Other bandwidth selection rules could be applied instead of `bw.relrisk`.

We analysed the Lansing Woods data in order to investigate segregation between species. Figure 14.7 does strongly suggest that the species are segregated: for example, hickories and maples are strongly segregated from each other. The intensity of the hickories tends to be high where that of the maples is low, and vice versa.

The following code determines, for each spatial location u, which species of tree has the highest probability at this location — effectively dividing the forest into regions where different species predominate.

```
> dominant <- im.apply(ProbD, which.max)
> species <- levels(marks(lansing))
> dominant <- eval.im(factor(dominant, levels=1:6, labels=species))
```

The result is shown in Figure 14.9. This is a pixel image with factor values, plotted using the command `textureplot` to show each level of the factor as a different texture.

The term 'segregation' may suggest that the points of different types were separated by a causal mechanism. However, Dixon [239] warns that observed segregation of plant species may have various explanations, such as different habitat preferences, patchy history of disturbance, and natural reproductive processes. Lack of observed segregation may be attributable to intervention such as forest logging. Correlation is not causation.

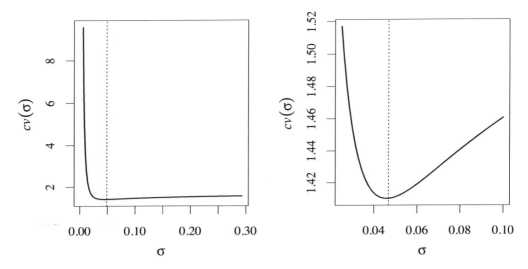

Figure 14.8. *Plots of the likelihood cross-validation criterion cv(σ) against bandwidth σ for the Lansing Woods data, computed by* bw.relrisk. *Left:* full range of bandwidths. *Right:* narrower range of bandwidths around the optimal value.

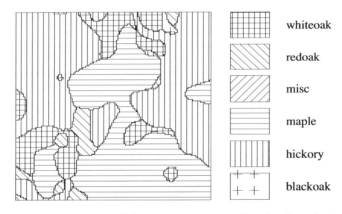

Figure 14.9. *Most likely species at each location in Lansing Woods, plotted using* textureplot.

Kelsall and Diggle [373] proposed a Monte Carlo test of spatial variation in disease risk, and Diggle *et al.* [237] generalised this to a Monte Carlo test of segregation in multitype point patterns. The null hypothesis is the more specific assumption of *random labelling*: given the locations x_i, the type labels m_i are random variables, independent of each other, with common probability distribution $\mathbb{P}\{m_i = m\} = p_m$. This null hypothesis *implies* that $p(m \mid u) = p_m$ is constant as a function of u, for each m, so that the types are not segregated. The test statistic is a measure of constancy of the distribution of types:

$$T = \sum_i \sum_m \left(\widehat{p}(m \mid x_i) - \overline{p}_m\right)^2$$

where $\widehat{p}(m \mid x_i)$ is the leave-one-out kernel estimate of $p(m \mid x_i)$, and $\overline{p}_m$ is the observed relative frequency of type m. Randomisation for the Monte Carlo test is performed by randomly permuting the type labels m_i while holding the locations u_i fixed. Under the null hypothesis, the original pattern and its randomised versions are exchangeable, justifying the Monte Carlo test. The test is performed in spatstat by segregation.test:

```
> segregation.test(lansing, nsim=19)
        Monte Carlo test of spatial segregation of types

data:  lansing
T = 0.014, p-value = 0.05
```

By default, automatic bandwidth selection will be performed separately for each randomised pattern. This computation can be time-consuming, but is strictly necessary for the validity of the test. Automatic bandwidth selection for the randomised patterns may also generate warnings, because the optimal bandwidth for such patterns can be infinite (or at least larger than the default range of bandwidths). To avoid both problems, one can simply specify a fixed value for the smoothing bandwidth sigma.

For the gastric mucosa data (Figure 14.1), the spatial locations of cells are clearly inhomogeneous, but we are interested in the relative proportions of the two cell types. Figure 14.10 shows the nonparametric estimates of these spatially varying proportions.

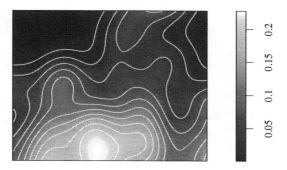

Figure 14.10. *Nonparametric estimate of spatially varying proportion of enterochromaffin-like cells in the gastric mucosa data.*

See the contributed package spatialsegregation, written by T. Rajala, for a comprehensive set of tools for measuring and testing spatial segregation, using spatstat data types.

14.4.5   Relative risk

It may be the *relative* probability of the different types that is of interest. A very important example is a spatial *case-control* study, giving the spatial locations of a set of disease cases, and of a separate set of controls (notionally a random sample from the population at risk of the disease). Then the ratio of the intensities of the two point processes is an indication of the spatially varying risk of disease.

In the context of a case-control study, let D be the point process of disease cases and C the point process of controls. We define the *relative risk* of disease as

$$\rho(u) = \frac{\lambda_D(u)}{\lambda_C(u)} \tag{14.15}$$

where $\lambda_D(u)$ and $\lambda_C(u)$ are the intensity functions of the disease cases and the controls, respectively. The rationale for this definition is as follows. Assume that C is an independent random sample (with unknown sampling fraction f) from the population at risk. Assuming the population at risk is large, and has spatially varying density $h(u)$ people per square kilometre, the locations of the controls C are approximately a Poisson point process with intensity $\lambda_C(u) = f h(u)$.

The null hypothesis of *constant disease risk* states that each person in the population has the same probability p of contracting the disease. Under this hypothesis, D has intensity function $\lambda_D(u) = p h(u)$. The alternative hypothesis of *spatially varying risk* states that a person living at spatial location u has probability $p(u)$ of contracting the disease, where $p(u)$ is some function to be estimated. In this case, D has intensity function $\lambda_D(u) = p(u) h(u)$. Comparing expressions for λ_D and λ_C we find that

$$\rho(u) = \frac{1}{f} p(u) \tag{14.16}$$

that is, the ratio of the intensity of cases to the intensity of controls is proportional to the spatially varying probability of contracting the disease.

Note that we did not say the pattern of *disease* cases D is a Poisson process; this would only be true if different people's health outcomes are stochastically independent (in particular, not dependent on the disease status of family members, etc.) and is probably not true for infectious diseases. The beauty of analysing the intensity is that it is a first-order (mean value) property of the point process, so that the equations for λ_D and λ_C above remain true even when disease cases are spatially correlated.

The customary estimator of the relative risk $\rho(u)$ is the 'plug-in' estimator obtained by substituting the kernel estimators of the numerator and denominator. A related, but slightly different, approach is to combine the cases and controls into a single point process X, and to study

$$q(u) = \frac{\lambda_D(u)}{\lambda_D(u) + \lambda_C(u)}.$$

Then $q(u)$ is the probability of a case (rather than a control) given that there is a point of the combined process X at the location u. The two are related by $q(u) = \rho(u)/(1 + \rho(u))$ and $\rho(u) = q(u)/(1 - q(u))$.

For a multitype point pattern with M types, we designate one of the possible types m_0 as the 'control' or 'reference' type. Then the *relative risk* of another type m with respect to the control type m_0 is

$$\rho(m \mid u) = \frac{\lambda(u, m)}{\lambda(u, m_0)} = \frac{p(m \mid u)}{p(m_0 \mid u)}, \tag{14.17}$$

equal to the ratio of the spatially varying probabilities of the types m and m_0. Spatial segregation between types occurs if and only if $\rho(m \mid u)$ is not constant as a function of location u.

Diggle *et al.* [237] studied outbreaks of bovine tuberculosis classified by the genotype of the tuberculosis bacterium. The relative risk of different genotypes of the disease gives clues about the mechanism of disease transmission (e.g., strong segregation of types would suggest that contagion is localised) and clues to appropriate management of new cases (e.g., if there is strong segregation and a new case does not belong to the locally predominant type, the infection is more likely to be the result of importation of infected animals).

The `spatstat` function `relrisk` computes estimates of *relative* risk when `relative=TRUE`. For the Chorley-Ribble data (page 9)

```
> plot(relrisk(chorley, relative=TRUE, control="lung"))
```

The result, shown in the left panel of Figure 14.11, does not suggest there is any large-scale spatial variation in the risk of cancer of the larynx. This is confirmed by the right panel of Figure 14.11, which shows the standard error, computed by

```
> plot(relrisk(chorley, relative=TRUE, control="lung", se=TRUE)$SE)
```

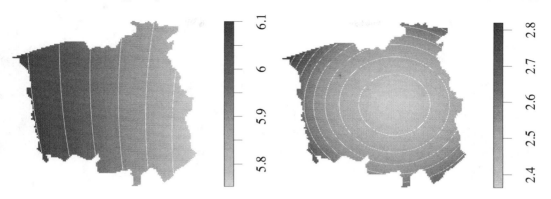

Figure 14.11. *Relative risk of laryngeal cancer in the Chorley-Ribble data.* Left: *estimate of relative risk.* Right: *standard error. Scale graduations multiplied by 100.*

The fluctuations in estimated risk are smaller than the standard error.

Standard error in relative risk is again computed using the delta-method approximation to the variance of a ratio:

$$\mathrm{var}\,\widehat{\rho}(m \mid u) \approx \frac{1}{\lambda_{m_0}(u)^2}\left[\mathrm{var}\,\widehat{\lambda}_m(u) + \rho(m \mid u)^2\,\mathrm{var}\,\widehat{\lambda}_{m_0}(u)\right] \tag{14.18}$$

More advanced, adaptive estimation of spatially varying relative risk is described in [331] and supported by the package `sparr` [202].

14.5 Multitype Poisson models

This section covers multitype Poisson process models: their basic properties, simulation, and fitting models to data.

Multitype Poisson point process models are the most basic and yet most pivotal stochastic models for multitype point patterns. After exploratory analysis of the intensity, the next step in data analysis would usually be to fit a tentative Poisson model, giving us access to a large toolbox of techniques for estimation and inference. The adequacy of the Poisson model should be evaluated. Discrepancies between the data and the predictions of a Poisson model are a useful guide to building better models.

14.5.1 Homogeneous multitype Poisson process

In a multitype point process, suppose that the points of a given type m constitute a homogeneous Poisson process (CSR) with intensity λ_m, and that the points of different types are independent. Then we have a *homogeneous multitype Poisson process*. Alternatively, suppose that the locations of all points, regardless of their type, constitute a homogeneous Poisson process with intensity $\lambda_\bullet$, and that each point is randomly allocated to type m with probability p_m, independently of other points. Then again we have a homogeneous multitype Poisson process with intensity λ_m for points of type m.

Since the established term CSR ('complete spatial randomness') is used to refer to the homo-

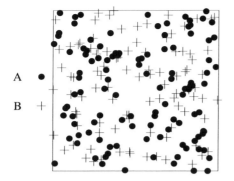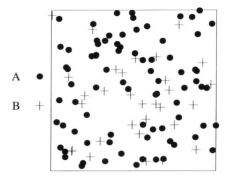

Figure 14.12. *Simulated realisations of CSRI with two types of points.* Left: *Intensities* $\lambda_A =$ $100, \lambda_B = 100$; Right: *Intensities* $\lambda_A = 100, \lambda_B = 20$.

geneous Poisson point process, it would seem appropriate that the homogeneous *multitype* Poisson point process be called *Complete Spatial Randomness with Independence (CSRI)*.[1]

More formally, the *homogeneous multitype Poisson process*, also known as *Complete Spatial Randomness and Independence (CSRI)*, is characterised by the following properties:

(MPP1) **Poisson points:** the unmarked locations x_i form a homogeneous Poisson process (CSR);

(MPP2) **completely random labels:** the type labels m_i are independent and identically distributed;

(MPP3) **Poisson points of each type:** the sub-process $\mathbf{X}^{(m)}$ of points of type m is a homogeneous Poisson process (CSR);

(MPP4) **independent components:** the sub-processes $\mathbf{X}^{(1)}, \dots, \mathbf{X}^{(M)}$ of points of each type are independent point processes.

Notice in particular that the point process is *spatially* homogeneous but the distribution of types is not uniform: the different types of points will usually have different probabilities of occurrence. If λ_m is the intensity of $\mathbf{X}^{(m)}$, then the intensity of $\mathbf{X}^{(\bullet)}$ is $\lambda_\bullet = \sum_m \lambda_m$, and the probability that a point belongs to type m is $p_m = \lambda_m / \lambda_\bullet$. Figure 14.12 shows two simulated realisations of CSRI: both patterns are spatially homogeneous, but the pattern on the right contains unequal proportions of the two types of points.

The list of properties above is redundant. Properties (MPP1) and (MPP2) would be sufficient; these describe CSRI from the 'multitype' viewpoint. Alternatively (MPP3) and (MPP4) would be sufficient, and they describe CSRI from the 'multivariate' viewpoint.

However, it is very important in applications that the random labelling property (MPP2) and the independence of components property (MPP4) are *not* equivalent, in general. These are two complementary aspects of 'randomness' for multitype point patterns.

Lemma 14.1 (Characterisations of CSRI). *For a multitype point process:*

　　1. If (MPP1) *and* (MPP2) *hold, then the process is CSRI.*

　　2. If (MPP3) *and* (MPP4) *hold, then the process is CSRI.*

　　3. If the process is stationary *and satisfies* (MPP2) *and* (MPP4), *then it is CSRI.*

The last statement in the Lemma says that, in a sense, random labelling and independence of components are sufficient to characterise a Poisson process. See [383, 24].

---

[1] Another name is 'completely independent marked point process' [197, p. 205].

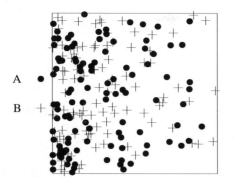 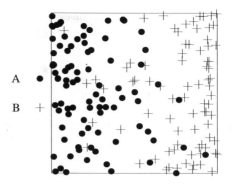

Figure 14.13. *Plots of simulated multitype patterns with spatially varying intensity.* Left: *Same intensity function for both types.* Right: *Different intensity functions.*

14.5.2   Inhomogeneous multitype Poisson process

In an *inhomogeneous* multitype Poisson process, the points are still independent, but may have a spatially varying intensity; the type labels of the points are still independent, but may have a spatially varying probability distribution; and the patterns of points of different types are still independent point processes, but may not be homogeneous.

The *inhomogeneous multitype Poisson process* is characterised by the following properties:

(MPP1′) **Poisson points:** the unmarked locations x_i form a Poisson process;

(MPP2′) **independent labels:** given the locations x_i, the type labels m_i are independent of each other, and m_i depends only on x_i;

(MPP3′) **Poisson points of each type:** the sub-process $\mathbf{X}^{(m)}$ of points of type m is a Poisson process;

(MPP4′)=(MPP4) **independent components:** the sub-processes $\mathbf{X}^{(1)}, \ldots, \mathbf{X}^{(M)}$ of points of each type are independent point processes.

Again the list is redundant: (MPP1′) and (MPP2′) would be sufficient, or (MPP3′) and (MPP4).

If the multitype process has intensity function $\lambda(u,m) = \lambda_m(u)$, then the unmarked process $\mathbf{X}^{(\bullet)}$ has intensity function $\lambda_\bullet(u) = \sum_m \lambda_m(u)$, and the probability that a point at location u belongs to type m is $p(m \mid u) = \lambda_m(u)/\lambda_\bullet(u)$. Figure 14.13 shows simulated realisations of two inhomogeneous multitype Poisson processes.

Notice that the 'independent labelling' property (MPP2′) of the inhomogeneous Poisson process is weaker than the 'completely random labelling' property (MPP2) enjoyed by CSRI. In the inhomogeneous case, the distribution of types may be spatially varying. We can only say that the type of each point is random, independent of the types of other points, and depends only on the location of the point.

If an inhomogeneous Poisson process does satisfy the stronger 'completely random labelling' property (MPP2), then $p(m \mid u) = p_m$ is constant as a function of u for each type m, so that $\lambda_m(u) = p(u \mid m)\lambda_\bullet(u) = p_m\lambda_\bullet(u)$. The intensity functions of the different types of points are proportional to each other, so the different types have *constant relative risk*.

Lemma 14.2 (Characterisations of inhomogeneous multitype Poisson process). *For a multitype point process:*

> *1. If* (MPP1′) *and* (MPP2′) *hold, then the process is a multitype Poisson process.*
>
> *2. If* (MPP3′) *and* (MPP4) *hold, then the process is a multitype Poisson process.*

3. *If the process has constant relative risk, i.e.* $\lambda_m(u) = p_m\lambda_\bullet(u)$, *then independent labelling* (MPP2$'$) *implies completely random labelling* (MPP2).

4. *If* (MPP2) *and* (MPP4) *hold [sic], then the process is a multitype Poisson process.*

The last statement says that if a multitype point process has completely random labelling, and independent components, then it is a multitype Poisson process. It must necessarily also have constant relative risk, $\lambda_m(u) = p_m\lambda_\bullet(u)$. Thus, the properties of completely random labelling and independence of components, if we could verify them both, are enough to establish that the point process is Poisson. However, an inhomogeneous multitype Poisson process may fail to satisfy the 'completely random labelling' property (MPP2).

The book by Kingman [383] gives a thorough and elegant discussion of these ideas at a fairly non-technical level. A more succinct account is [484, sec. 3.3].

14.5.3   Simulation

Simulation of a multitype Poisson point process is straightforward, because the points of different types are independent. For each possible type of point, we simply generate a Poisson point process with the required intensity, using the methods described in Sections 5.3.4 and 5.4.2.

Simulated realisations of a multitype Poisson point process are generated in spatstat by the function rmpoispp. The first argument of this function specifies the intensities λ_m or the intensity function $\lambda(u,m)$. It can be a constant, a vector of constants, an R function of the form function(x,y,m, ...), a pixel image, a list of functions of the form function(x,y, ...), or a list of pixel images. Some examples of the use of rmpoispp are as follows:

```
> X1 <- rmpoispp(100, types=c("A","B"))
> X2 <- rmpoispp(c(100,20), types=c("A","B"))
> X3 <- rmpoispp(function(x,y,m){300*exp(-3*x)}, types=c("A","B"))
> ll <- function(x,y,m){ 300*exp(-4*ifelse(m=="A", x, (1-x))) }
> X4 <- rmpoispp(ll, types=c("A","B"))
```

The patterns X1 and X2 are shown in Figure 14.12, while X3 and X4 are shown in Figure 14.13.

When a multitype Poisson process model has been fitted to a dataset (see Section 14.5.4 below) patterns can be simulated from the fitted model automatically using rmh.ppm or simulate.ppm.

14.5.4   Fitting Poisson models

A Poisson multitype point process model is completely determined by its intensity; fitting the model is equivalent to estimating its intensity. Thus the exploratory techniques of Section 14.4 for estimating the intensity of a multitype point process also provide nonparametric techniques for fitting Poisson models.

As argued in Chapter 9, it is also important to be able to fit *parametric* models for multitype Poisson processes to point pattern datasets. Parametric model-fitting give us access to a wide variety of statistical tools, for example for model selection, prediction, simulation, standard errors, confidence intervals, measuring the effect of a covariate, testing whether a covariate has an effect, testing whether different types of points are segregated, testing goodness-of-fit, and validating the fitted model. Even if we do not believe a particular model is true, fitting this model is an essential step in demonstrating that the model is not adequate.

Poisson models for multitype point processes may be fitted to point pattern data using ppm. By default, models are fitted by the method of maximum likelihood using the Berman-Turner [90] device extended to multitype patterns [50]. Alternatives include the logistic composite likelihood [38] selected by method="logi". The logistic method often has better statistical performance (reduced bias, at the cost of slightly increased variability) than approximate maximum likelihood, and is much faster to compute when the number of possible types is large.

The general syntax for calls to ppm is discussed in Section 9.3.1. For multitype point pattern data, the trend formula in the call to ppm is a formula for the logarithm of the multitype intensity function $\lambda(u, m) = \lambda_m(u)$. In addition to the reserved names x and y representing the spatial coordinates, the reserved name marks may also appear in the trend formula. This refers to the marks (type labels) of the points. Since the marks are categorical, marks is treated as a factor variable for modelling purposes.

If a point pattern dataset X is multitype, then ppm(X ~ trend, ...) will always fit a multitype point process model, either Poisson or Gibbs depending on the arguments. To fit the homogeneous multitype Poisson process (CSRI), Section 14.5.1, we call

```
ppm(X ~ marks)
```

The formula ~ marks indicates that the trend depends only on the marks, and not on spatial location; since marks is a factor, the trend has a separate constant value for each level of marks. This is CSRI. For the Lansing Woods data:

```
> ppm(lansing ~ marks)
Stationary multitype Poisson process
Possible marks: 'blackoak', 'hickory', 'maple', 'misc', 'redoak' and 'whiteoak'
Log intensity:   ~marks
Intensities:
beta_blackoak   beta_hickory    beta_maple      beta_misc     beta_redoak
          135            703           514            105             346
beta_whiteoak
          448
```

The printout shows the estimated intensities $\widehat{\lambda}_m$ for each type m. In this simple model, the parameter estimates $\widehat{\lambda}_m$ coincide with those obtained from summary.ppp above. That is a consequence of the fact that the maximum likelihood estimates (obtained by ppm) are also the method-of-moments estimates (obtained by summary.ppp).

Warning: If X is a multitype point pattern, then ppm(X ~ 1) is a multitype point process model. The fitted intensity value is the intensity of each possible type.

For example for the Lansing Woods data

```
> ppm(lansing ~ 1)
Stationary multitype Poisson process
Possible marks: 'blackoak', 'hickory', 'maple', 'misc', 'redoak' and 'whiteoak'
Uniform intensity for each mark level:      375.2
```

This has fitted a multitype point process, namely the special case of CSRI where the intensities λ_m are equal, $\lambda_m \equiv \beta$ say, for all possible marks. That model is only appropriate if we believe that all mark values are equally likely — which they clearly are not in this case. The fitted intensity value $\widehat{\beta} = 375.17 = 2251/6$ is the intensity for each possible type.

Tip: If you really want to estimate a single overall intensity, treating the data as if they were not marked, then unmark the data! That is, do something like ppm(unmark(X) ~ 1).

```
> ppm(unmark(lansing) ~ 1)
```

```
Stationary Poisson process
Intensity: 2251
```

As explained in Section 9.3 on page 316, the interpretation of the fitted coefficients depends on which *contrasts* are in force. The model fit <- ppm(X ~ marks) includes an intercept term (by default) and uses the treatment contrasts (by default), so the fitted coefficients coef(fit) include an intercept and coefficients for each species *except* the first species blackoak. Under the same conditions, the model fit1 <- ppm(X ~ marks-1) would not include an intercept term, and coef(fit1) includes a fitted coefficient for each species. The two models are equivalent, and give identical predictions.

An example of a fit with spatially varying intensity is:

```
> ppm(lansing ~ marks + x)
Nonstationary multitype Poisson process
Possible marks: 'blackoak', 'hickory', 'maple', 'misc', 'redoak' and 'whiteoak'
Log intensity:   ~marks + x
Fitted trend coefficients:
   (Intercept)   markshickory      marksmaple      marksmisc    marksredoak
       4.94295        1.65008         1.33695       -0.25131        0.94116
 markswhiteoak              x
       1.19952       -0.07582
```

This is the multitype Poisson process whose intensity function $\lambda(u,m)$ at location $u = (x,y)$ and mark m satisfies

$$\log \lambda(u,m) = \alpha_m + \beta x \qquad (14.19)$$

where $\alpha_1, \dots, \alpha_6$ and β are parameters. The intensity is loglinear in x, with a different intercept for each species, but the same 'slope' β (analogous to 'parallel loglinear regression'). In the printout above, the fitted slope parameter β is $\hat{\beta} = -0.07582$. As discussed in Section 9.3 on page 316, the fitted coefficients α_m for the categorical mark are interpreted in the light of the 'contrasts' in force. The default is the treatment contrasts, and the first level of the mark is blackoak, so in this case the fitted coefficient for m=blackoak is 4.943, while the fitted coefficient for m=hickory is $4.943 + 1.65 = 6.593$ and so on.

The relationship (14.19) is an 'additive' model because the effects of the mark m and of the spatial coordinate x are added. In the spatial context this implies that the model has *constant relative risk*, because the probability that a tree at location u belongs to species m is predicted to be

$$p(m \mid u) = \frac{\lambda(u,m)}{\sum_{m'} \lambda(u,m')} = \frac{\exp(\alpha_m + \beta x)}{\sum_{m'} \exp(\alpha_{m'} + \beta x)} = \frac{\exp(\alpha_m)\exp(\beta x)}{\sum_{m'} \exp(\alpha_{m'})\exp(\beta x)} = \frac{\exp(\alpha_m)}{\sum_{m'} \exp(\alpha_{m'})} \quad (14.20)$$

which is constant as a function of location u.

We can allow the intensity to vary in a different way for each type, in the following manner:

```
> ppm(lansing ~ marks * x)
Nonstationary multitype Poisson process
Possible marks: 'blackoak', 'hickory', 'maple', 'misc', 'redoak' and 'whiteoak'
Log intensity:   ~marks * x
Fitted trend coefficients:
    (Intercept)      markshickory       marksmaple         marksmisc     marksredoak
         5.2378            1.4425           0.6796           -0.8483          0.6916
  markswhiteoak                 x    markshickory:x      marksmaple:x      marksmisc:x
         1.0902           -0.7064           0.4511            1.3243          1.2138
  marksredoak:x  markswhiteoak:x
         0.5380            0.2421
```

The symbol * in the formula is an 'interaction' in the usual sense for linear models, as discussed in Section 9.3.9. The fitted model is the multitype Poisson process with

$$\log \lambda(u,m) = \alpha_m + \beta_m x \qquad (14.21)$$

where $u = (x,y)$ and $\alpha_1,\ldots,\alpha_6$ and $\beta_1,\ldots,\beta_6$ are parameters. The intensity is loglinear in x with a different slope β_m and intercept α_m for each mark m. For example, for m="hickory", we have $\alpha_m = 5.238 + 1.442 = 6.68$ and $\beta_m = 0.4511$.

To plot the fitted intensity and conditional intensity of the fitted model, use `predict` to compute the intensity or conditional intensity, and `plot` to display them. For a multitype point process there will be a separate plot panel for each possible type of point.

A more complicated example is illustrated in Figure 14.14. In the model that was fitted to produce this figure, each species of tree was assumed to have an intensity of log-cubic form in the Cartesian coordinates, that is, $\log \lambda(u,m)$ was assumed to be a cubic function of x,y (where $u = (x,y)$) with coefficients depending on m. This model has 60 parameters. Confidence intervals for these parameters can be obtained from the output of `ppm` using the base R function `confint`. This function makes use of the asymptotic normal distribution of the maximum likelihood estimator of the parameter vector. We do not display this vast collection of confidence intervals to save trees. To reduce the number of parameters we could have replaced `polynom` by `harmonic`.

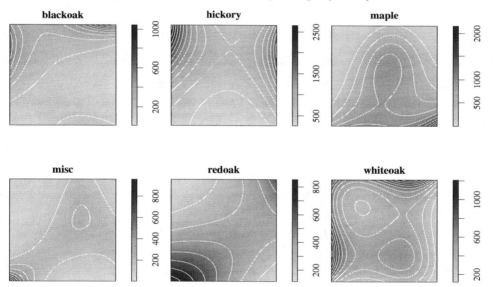

Figure 14.14. *Maximum likelihood estimates of intensity for each species of tree in Lansing Woods assuming a separate log-cubic intensity function for each species. Compare with Figure 14.6.*

A parametric test for segregation in multitype patterns can be performed by fitting models and testing whether certain terms in the model are nonzero. A Poisson model is separable (non-segregated) if the log intensity is a sum $\log \lambda(u,m) = A(u) + B(m)$ of terms $A(u)$ depending only on location, and terms $B(m)$ depending only on the type of point. The presence of any terms that depend on both u and m would imply segregation.

For example we can fit two models to the Lansing Woods data and perform a likelihood ratio test as follows:

```
> f0 <- ppm(lansing ~ polynom(x,y,3) + marks)
> f1 <- ppm(lansing ~ polynom(x,y,3) * marks)
> anova(f0, f1, test="LR")
```

The output of the foregoing calculation is not shown, to save space. Model f0 corresponds to the null hypothesis of no segregation. In this model only the intercept term depends on species, while other coefficients are common to all species. Model f1 is segregated. The likelihood ratio test yielded a test statistic of 613 on 45 df and a resulting *p*-value of '< 2.2e-16' which is extremely significant.

The ants' nests data (Figure 1.7) provide a factor-valued covariate side indicating whether each location belongs to the 'scrub' or 'field' area of the survey. An interaction term marks * side would mean that each species has a different intensity on each side of the survey:

```
> Ants <- ants
> levels(marks(Ants)) <- c("C", "M")
> ppm(Ants ~ marks * (side - 1), data=ants.extra)
Nonstationary multitype Poisson process
Possible marks: 'C' and 'M'
Log intensity:   ~marks * (side - 1)
Fitted trend coefficients:
          marksC            marksM         sidescrub marksM:sidescrub
         -9.3510           -8.8311           -0.7790           0.9682
```

A fitted multitype Poisson process model can be interrogated and manipulated using any of the methods available for the class "ppm", listed in Tables 9.3 and Tables 9.4 on page 328 and Table 9.5 on page 338.

14.5.5   Probabilities and relative risks predicted by a model

Nonparametric estimates of the spatially varying distribution of marks, and the relative risk of each type, were described in Section 14.4.3. We can also obtain parametric estimates of these quantities by fitting a point process model to data.

Consider a Poisson multitype point process model with intensity of the form

$$\lambda(u,m) = \exp(B(u,m) + \theta^{\mathsf{T}} \mathbf{Z}(u,m))$$

at spatial location u for mark m where $m = 1, 2, \ldots, M$, where θ is a p-dimensional vector of parameters, $B(u,m)$ is an optional baseline or offset function, and $\mathbf{Z}(u,m)$ is a p-dimensional vector-valued covariate. The intensity of the unmarked process is

$$\lambda(u) = \sum_m \lambda(u,m) = \sum_m \exp(B(u,m) + \theta^{\mathsf{T}} \mathbf{Z}(u,m)). \tag{14.22}$$

The spatially varying probability of mark m is

$$p(m \mid u) = \frac{\lambda(u,m)}{\lambda(u)} = \frac{\exp(B(u,m) + \theta^{\mathsf{T}} \mathbf{Z}(u,m))}{\sum_{m'} \exp(B(u,m') + \theta^{\mathsf{T}} \mathbf{Z}(u,m'))}. \tag{14.23}$$

The spatially-varing relative risk of mark m, relative to another mark value m_0 designated as the 'control', is

$$r(m \mid u) = \frac{p(m \mid u)}{p(m_0 \mid u)} = \exp(B(u,m) - B(u,m_0) + \theta^{\mathsf{T}}(\mathbf{Z}(u,m) - \mathbf{Z}(u,m_0))). \tag{14.24}$$

When the model is fitted to data and θ is estimated by $\widehat{\theta}$, estimates $\widehat{p}(m \mid u)$ and $\widehat{r}(m \mid u)$ are obtained by substituting $\widehat{\theta}$ for θ above. Standard errors are also easy to obtain using the delta method: we have

$$\operatorname{var} \widehat{p}(m \mid u) \approx P(u,m) \mathbf{I}_\theta^{-1} P(u,m)^{\mathsf{T}} \tag{14.25}$$

$$\operatorname{var} \widehat{r}(m \mid u) \approx R(u,m) \mathbf{I}_\theta^{-1} R(u,m)^{\mathsf{T}} \tag{14.26}$$

where I_θ is the Fisher information matrix and

$$P(u,m) = p(m \mid u) \left[\mathbf{Z}(u,m) - \sum_{m'} \mathbf{Z}(u,m') p(m' \mid u) \right]$$
$$R(u,m) = r(m \mid u) [\mathbf{Z}(u,m) - \mathbf{Z}(u,m_0)].$$

These calculations are performed by `relrisk.ppm`, the method for the generic `relrisk` for fitted Poisson or Gibbs point process models. For example, for the gastric mucosa data (Figure 14.1 on page 561), we might consider the model

```
> fit <- ppm(mucosa ~ marks * polynom(x,y,3))
```

Then the spatially varying probabilities of each type are computed by

```
> probs <- relrisk(fit, casecontrol=FALSE)
```

Figure 14.15 shows the estimated proportion of ECL cells. The spatially varying relative 'risk' of ECL cells relative to other cells is computed by

```
> rusk <- relrisk(fit, relative=TRUE, control="other")
```

To obtain standard errors, set `se=TRUE` in the calls to `relrisk`.

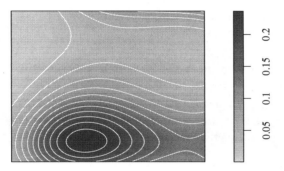

Figure 14.15. *Parametric estimate of spatially varying proportion of enterochromaffin-like cells amongst all cells in the gastric mucosa data.*

14.6 Correlation and spacing

In this section we discuss some tools for exploratory data analysis of the dependence between points in a multitype point pattern. These tools include multitype versions of summary functions such as the K-, L-, F- and G-functions and their inhomogeneous counterparts. Comparisons can be made between each pair of possible types, or between each type and the rest of the pattern, or 'within' and 'between' types.

14.6.1 Distances and nearest neighbours

The `spatstat` package has many functions for 'low-level' calculation with the distances between points in a point pattern. These functions are useful for developing new software, and occasionally for inspecting quirks in a dataset. However, for everyday data analysis, it is much easier to use the 'higher-level' statistical commands presented in subsequent sections.

Table 8.1 on page 256 lists the basic functions in `spatstat` for measuring distances in a point pattern and for identifying the nearest neighbour of each point. The most common commands are `pairdist(X)` for the distances between all pairs of points, `nndist(X)` for the nearest-neighbour distances, `crossdist(X,Y)` for distances from each point of X to each point of Y, and `nncross(X,Y)` for the distance from each point of X to the nearest point of Y.

For a multitype point pattern, one can use the argument 'by' to `nndist` and `nnwhich` to find nearest neighbours of a given type. For example `nndist(X, by=marks(X))` finds the distance from each point of X to the nearest point of type *i*, for each possible type *i*. The result is a matrix with one row for each data point and one column for each possible type. Similarly `nnwhich(X, by=marks(X))` returns a matrix giving for each point in X the index of the nearest point of type *i*, for each possible type *i*.

```
> d <- nndist(amacrine, by=marks(amacrine))
> head(d, 3)
         off       on
[1,]  0.04968  0.07852
[2,]  0.08452  0.07852
[3,]  0.05622  0.06054
```

For example, the distance from the first data point to the nearest point of type "on" is 0.0785. The argument k can also be used to select the *k*th-nearest neighbour.

Matrix operations can be used to compile summary statistics from these basic distances. To find the minimum distance from any point of type *i* to any point of type *j* in the amacrine cells data, for each pair of types *i* and *j*,

```
> d <- nndist(amacrine, by=marks(amacrine))
> a <- aggregate(d, by=list(from=marks(amacrine)), min)
```

To find the mark of the nearest neighbour of a point, invoke `nnwhich` to identify the nearest neighbour by its serial number (index in the point pattern) and use this to index the marks vector. For example, to study the association between the mark of a point and the mark of its nearest neighbour in the amacrine cells data:

```
> m <- marks(amacrine)
> n <- nnwhich(amacrine)
> m1 <- m[n]
> table(m, m1)
      m1
m       off  on
  off   17  125
  on   126   26
```

Several convenient 'wrapper' functions are provided in `spatstat` to perform calculations of this kind. They include `marktable`, `markstat`, and `applynbd`. For each point in the dataset, these functions retrieve either the N nearest neighbours, or the points closer than a distance R, and apply a user-supplied function to these neighbouring points. The results for each point neighbourhood are returned. See the help files for these functions.

14.6.2 Nearest-neighbour correlation

Summary statistics which take the form of a single number, rather than a summary function, are useful for exploratory analysis and in hypothesis tests. For multitype point patterns a very useful summary is the *nearest-neighbour correlation*.

For a stationary multitype point process, the 'unnormalised' nearest-neighbour correlation is the probability that a typical point and its nearest neighbour have the same type:

$$\tilde{n} = \mathbb{P}\{M = M^*\} \tag{14.27}$$

where, for a typical point in the process, M is the mark of the point and M^* is the mark of its nearest neighbour. Values of $\tilde{n}$ are probabilities, between 0 and 1. The 'normalised' nearest-neighbour correlation is

$$\tilde{m} = \frac{\mathbb{P}\{M = M^*\}}{\mathbb{P}\{M = M'\}} \tag{14.28}$$

where M' is an independent random mark value with the same probability distribution as M. That is, the denominator is the value of $\tilde{n}$ that would be obtained if the process had the random labelling property. Values of $\tilde{m}$ are positive numbers, and the value 1 is consistent with random labelling.

To estimate the nearest-neighbour correlation one needs to apply edge correction: for example, the border correction would ignore all points for which the distance to the nearest neighbour is greater than the distance to the window boundary. An edge corrected estimate of the nearest-neighbour correlation is computed by `nncorr`. For the amacrine cells data:

```
> nncorr(amacrine)
unnormalised   normalised
      0.1339       0.2674
```

A value much less than 1 implies that neighbouring cells are usually of different types.

Although the formal definition assumes the process is stationary, the nearest-neighbour correlation is also meaningful for many non-stationary processes, and can be used as a fairly robust measure of dependence between types. Correlation indices based on the k-th nearest neighbours can also be computed using `nncorr`.

The nearest-neighbour correlation is closely related to the 'join count' statistics first proposed by Moran [487] comparing the label of a point with the label of its nearest neighbour. Cuzick and Edwards [195] developed statistical tests for disease clustering in case-control studies, based on the disease status of the k-th nearest neighbours for $k = 1, \dots, K$. See also [102, 143, 201, 210, 276, 334, 415, 444, 607, 659, 704]. The contributed R package `spdep` [112, 110] supports many of these techniques.

Nearest-neighbour correlations can also be defined for point processes with numeric marks, as described in Section 15.2.4.5.

14.6.3 Contingency tables of marks of nearest neighbours

Pielou [548] proposed that spatial segregation of plant species could be analysed using a contingency table of the species of each plant and its nearest neighbour. For each possible combination of species i and j, the contingency table cell on row i, column j counts the number of plants of species i whose nearest neighbour was of species j. This table can be compiled by `marktable`:

```
> marktable(amacrine, N=1, collapse=TRUE)
      neighbour
point off  on
  off  17 125
   on 126  26
```

Statistical analysis of the nearest-neighbour contingency table was developed by Dixon [238, 239]. The contributed R package `dixon` supports these methods. For Lansing Woods, a nearest-neighbour contingency table provides strong evidence of segregation:

```
> library(dixon)
> dixon(as.data.frame(lansing))$tablaC
                      df  Chi-sq P.asymp  P.rand
Overall segregation   30 378.42  0e+00        0
From  blackoak         5  42.58  0e+00        0
From  hickory          5 123.63  0e+00        0
From  maple            5 142.05  0e+00        0
From  misc             5  65.43  0e+00        0
From  redoak           5  42.32  0e+00        0
From  whiteoak         5  23.04  3e-04        0
```

The corresponding result for `amacrine` is also statistically significant:

```
> dixon(as.data.frame(amacrine))$tablaC
                      df  Chi-sq P.asymp  P.rand
Overall segregation    2  85.99       0        0
From  off              1  66.08       0        0
From  on               1  66.56       0        0
```

However, in the present case this does not indicate segregation, rather the opposite: nearest neighbours are likely to be of *different* types.

14.6.4 Summary functions for pairs of types

We start by considering *pairs* of types i and j. Assume the multitype point process **Y** is **stationary**.

14.6.4.1 Cross-type K-function and its relatives

For any pair of types i and j, the *multitype K-function* $K_{ij}(r)$, also called the bivariate or cross-type K-function, proposed by [433, 324], is the expected number of points of type j lying within a distance r of a typical point of type i, standardised by dividing by the intensity of points of type j:

$$K_{ij}(r) = \frac{1}{\lambda_j} \mathbb{E}\left[t(u, r, \mathbf{X}^{(j)}) \mid u \in \mathbf{X}^{(i)} \right] \tag{14.29}$$

for $r \geq 0$, where $\mathbf{X}^{(j)}$ is the sub-process of points of type j, with intensity λ_j, and $t(u, r, \mathbf{x})$ is the count of r-neighbours, defined in equation (7.5) on page 206. Dividing by λ_j allows us to compare point patterns with different intensities.

If the types i and j are equal, then $K_{ij}(r) = K_{ii}(r)$ is the usual K-function of the process $\mathbf{X}^{(i)}$ of points of type i. This has the same interpretation as the usual K-function: for example, the benchmark value $K_{ii}(r) = \pi r^2$ (in two dimensions) is consistent with a Poisson process for the points of type i.

However if $i \neq j$, the interpretation of $K_{ij}(r)$ is quite different: it now measures the dependence or association between types i and j. Suppose the points of types i and j are *independent* (i.e. $\mathbf{X}^{(i)}$ and $\mathbf{X}^{(j)}$ are independent point processes). Then the expected number of points of type j falling within a distance r of a point of type i is simply the expected number of points of type j falling in any disc of radius r, namely $\lambda_j \pi r^2$, so that $K_{ij}(r) = \pi r^2$. That is, the benchmark value $K_{ij}(r) = \pi r^2$ is consistent with *independence* between the points of types i and j. It is even consistent with the weaker property that the points of types i and j are *uncorrelated*. It does not imply that the points are Poisson.

Estimation of $K_{ij}(r)$ from data is very similar to the estimation of the original K-function, but is based on measuring the pairwise distances from all points of type i to all points of type j. In `spatstat` the estimates of $K_{ij}(r)$ are computed by `Kcross`. For example, to compute $K_{ij}(r)$ for the amacrine cells data with i and j being the types `"on"` and `"off"`, respectively,

```
> K <- Kcross(amacrine, "on", "off")
```

Apart from the need to specify i and j, this function is very similar to `Kest`. It offers the same choice of edge corrections, and returns an object of class `"fv"`, very similar to the result of `Kest` except for the labels displayed when it is plotted. It recognises either the arguments `i` and `j` or `from` and `to` as specifying the types of points.

To compute estimates of *all* the functions $K_{ij}(r)$ for all pairs of types i and j, use the `spatstat` function `alltypes`.

```
> Kall <- alltypes(rescale(amacrine), Kcross)
```

The result of `alltypes` is an object of class `"fasp"` ('function array for spatial patterns') which behaves like a two-dimensional array of functions. These objects are discussed further in Section 14.6.6.

Figure 14.16 shows the result of `plot(Kall)`. The rows of the array correspond to the first type i, and the columns to the second type j. The diagonal panels show the ordinary K-functions for the layers of `"on"` and `"off"` cells, respectively. The discrepancies between the estimates $\widehat{K}_{ii}(r)$ and the benchmark πr^2 strongly suggest that these component patterns are not Poisson. However, the off-diagonal panels show substantial agreement between $\widehat{K}_{ij}(r)$ and πr^2, which is consistent with independence between the `"on"` and `"off"` layers.

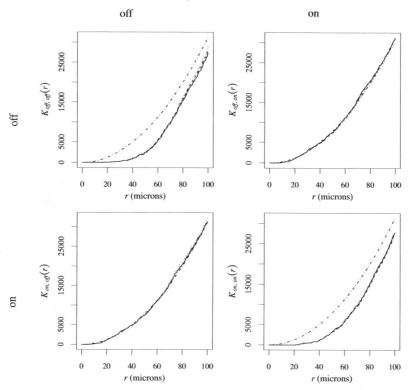

Figure 14.16. *The cross-type K-functions $\widehat{K}_{ij}(r)$ for each pair of types i and j in the amacrine cells data, computed using* `alltypes`.

Simulation envelopes for each function $K_{ij}(r)$ can be generated by setting `envelope=TRUE` in the call to `alltypes`. By default, these envelopes are based on simulated realisations of CSRI. The diagonal panels then correspond to a test of CSR for each type; however, the off-diagonal panels correspond to a test of CSRI, not to the more general hypothesis of independence between components.

We can also define analogues of Besag's *L*-function (Section 7.3.3) and the pair correlation function (Section 7.6). For any types i and j, the cross-type *L*-function is defined as

$$L_{ij}(r) = \sqrt{\frac{K_{ij}(r)}{\pi}}. \tag{14.30}$$

This transformation approximately stabilises the variance of the estimate when the process is Poisson, and also transforms the benchmark value $K_{ij}(r) = \pi r^2$ to $L_{ij}(r) = r$.

The cross-type pair correlation function is defined as

$$g_{ij}(r) = \frac{K'_{ij}(r)}{2\pi r} \tag{14.31}$$

where $K'_{ij}(r)$ is the derivative of K_{ij}. This has an interpretation similar to that described on page 226 for the pair correlation function. If U and V are 'infinitesimal' regions with area dU and dV, respectively, which are separated by a distance r, then

$$\mathbb{P}\{\text{point of type } i \text{ in } U, \text{ point of type } j \text{ in } V\} \doteq \lambda_i \lambda_j g_{ij}(r) \, dU \, dV. \tag{14.32}$$

The benchmark value $g_{ij}(r) = 1$ is consistent with a Poisson process when $i = j$, and consistent with independence between types i and j (or at least, lack of correlation) when $i \neq j$.

Estimates of the bivariate *L*-function and pair correlation function for a given pair of types i and j are computed by `Lcross` and `pcfcross`, respectively, using the same syntax as for `Kcross`. To compute these functions for all pairs of types i and j one can use `alltypes(X, Lcross)` or `alltypes(X, pcfcross)` as explained above.

These summary functions can also be used to conduct a test of random labelling. Under the null hypothesis of completely random labelling (MPP2), we would have $K_{ij}(r) = K_{\bullet\bullet}(r)$, where $K_{\bullet\bullet}(r)$ is the *K*-function of the unmarked point process $\mathbf{X}^{(\bullet)}$.

Note that (14.32) is symmetric in i and j, so that $g_{ij}(r) = g_{ji}(r)$ and so also $K_{ij}(r) = K_{ji}(r)$ and $L_{ij}(r) = L_{ji}(r)$. However, estimates of these functions may not be exactly symmetric in i and j, because of the effect of edge corrections. Some software packages force the estimates to be symmetric by replacing both $\widehat{K}_{ij}(r)$ and $\widehat{K}_{ji}(r)$ by the average of the two. This is not done in `spatstat`, because asymmetry of the estimates could be an important clue in some applications, and because it is easy to compute the symmetrised version when needed, using `eval.fv`.

Since *K*-functions are essentially counts of pairs of points, they satisfy some simple relationships. Suppose there are only two types of points, 1 and 2. Then the *K* function of the unmarked point process $\mathbf{X}^{(\bullet)}$ is related to the cross-type *K* functions by

$$\lambda_{\bullet}^2 K_{\bullet\bullet}(r) = \lambda_1^2 K_{11}(r) + 2\lambda_1 \lambda_2 K_{12}(r) + \lambda_2^2 K_{22}(r). \tag{14.33}$$

Similarly for any number of types. A trick for computing the multitype *K*-function in software packages which only support the 'ordinary' *K*-function, is to compute the *K*-functions $K_{11}(r), K_{22}(r)$, and $K_{\bullet\bullet}(r)$, respectively, by applying the software to the type 1 points, type 2 points, and all points regardless of type, and then to use (14.33) to obtain $K_{12}(r)$.

14.6.4.2 Mark connection function

Interesting relationships between the different types can be revealed by the *mark connection function*

$$p_{ij}(r) = \frac{\lambda_i \lambda_j g_{ij}(r)}{\lambda_{\bullet}^2 g(r)} \tag{14.34}$$

where $g_{ij}(\cdot)$ is the cross pair correlation function between types i and j and $g(\cdot)$ is the pair correlation function for the unmarked process. This is motivated by comparing the heuristic interpretations of the bivariate pair correlation g_{ij}, equation (14.32) above, and the unmarked pair correlation g, equation (7.24) from Chapter 7. Taking the ratio of these two equations gives

$$\frac{\mathbb{P}\{\text{point of type } i \text{ in } U, \text{ point of type } j \text{ in } V\}}{\mathbb{P}\{\text{point in } U, \text{ point in } V\}} \doteq p_i p_j \frac{g_{ij}(r)}{g(r)}$$

where $p_i = \lambda_i/\lambda_\bullet$ is the probability of type i. The left-hand side can be interpreted as the conditional probability, given that there are points of the process in U and V, that the type labels of these points are i and j, respectively. In other words $p_{ij}(r)$ can be interpreted as the conditional probability, given that there is a point of the process at a location u and another point of the process at a location v separated by a distance $\|u - v\| = r$, that the first point is of type i and the second point is of type j:

$$p_{ij}(r) = \mathbb{P}^{u,v}\{m(u) = i, m(v) = j\} \tag{14.35}$$

where $\mathbb{P}^{u,v}$ is the conditional probability given that there are points of the process at the locations u and v (Section 8.12).

The spatstat function markconnect can be used to compute estimates of the mark connection function:

```
> ma <- markconnect(amacrine, "on", "off")
```

We can use alltypes to compute p_{ij} for all pairs of types i and j. Figure 14.17 shows the estimates of $p_{ij}(r)$ for the amacrine cells data, for each pair of types i and j, generated by
```
plot(alltypes(amacrine, markconnect))
```
The horizontal dashed lines show the reference values $p_i p_j$ that would be expected under random labelling. This shows that two points lying close together are more likely to be of different types than we would expect under random labelling. Although this could be termed a positive association between the cells of different types, it does not necessarily indicate dependence between the cell types; it could also be explained as an artefact of the negative association between cells of the same type.

14.6.4.3 Multitype F-, G- and J-functions

Multitype analogues of F, G and J were proposed in [664]. For each type i, we define $F_i(r)$ to be the empty-space function for the points of type i. That is, $F_i(r)$ is the cumulative distribution function of the distance from an arbitrary location u to the nearest point of type i:

$$F_i(r) = \mathbb{P}\left\{d(u, \mathbf{X}^{(i)}) \le r\right\}, \tag{14.36}$$

defined for all distances $r \ge 0$. This assumes the process is stationary. If the points of type i constitute a Poisson process, then $F_i(r) = 1 - \exp(-\lambda_i \pi r^2)$.

The multitype (or bivariate or cross-type) nearest-neighbour function $G_{ij}(r)$ is the cumulative distribution function of the distance from a point of type i to the nearest point of type j:

$$G_{ij}(r) = \mathbb{P}\left\{d(u, \mathbf{X}^{(j)} \setminus u) \le r \,\big|\, \mathbf{X}^{(i)} \text{ has a point at } u\right\}. \tag{14.37}$$

Again, this assumes that the process is stationary. When $i = j$, we get $G_{ii}(r)$, the nearest-neighbour function of the process of points of type i. If the points of type i are Poisson, then $G_{ii}(r) = 1 - \exp(-\lambda_i \pi r^2)$. However, when $i \ne j$, the function $G_{ij}(r)$ measures the association between types i and j. If the points of types i and j are independent of each other, then $G_{ij}(r) = F_j(r)$.

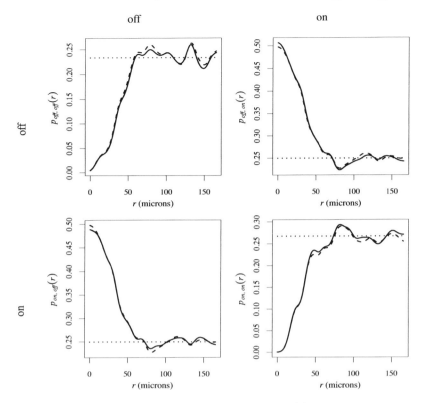

Figure 14.17. *Array of estimated mark connection functions* $p_{ij}(r)$ *of the amacrine cells data, for each pair of types* i, j.

The multitype J-function is then defined as

$$J_{ij}(r) = \frac{1 - G_{ij}(r)}{1 - F_j(r)}. \tag{14.38}$$

When $i = j$, we get $J_{ii}(r)$, the J-function of the process of points of type i. The benchmark value $J_{ii}(r) = 1$ is consistent with CSR. However when $i \neq j$, the benchmark value $J_{ij}(r) = 1$ is consistent with *independence* between the points of types i and j.

These functions are not symmetric in i and j, that is, $G_{ij}(r) \neq G_{ji}(r)$ and $J_{ij}(r) \neq J_{ji}(r)$ in general. Under *CSRI* we would have

$$F_i(r) = 1 - \exp(-\lambda_i \pi r^2), \quad G_{ij}(r) = 1 - \exp(-\lambda_j \pi r^2), \quad J_{ij}(r) = 1. \tag{14.39}$$

Under the hypothesis of *independent components* (the processes of points of each type are independent), we would have

$$G_{ij}(r) = F_j(r), \qquad J_{ij}(r) = 1 \tag{14.40}$$

and the empty-space function F of the unmarked process $\mathbf{X}^{(\bullet)}$ would satisfy

$$1 - F(r) = \prod_i [1 - F_i(r)]. \tag{14.41}$$

Under the hypothesis of random labelling these functions take a complicated form [664].

Estimation of $G_{ij}(r)$ is very similar to the estimation of $G(r)$ except that it is based on measuring the distances from each point of type i to the nearest point of type j. Estimates of the functions $G_{ij}(r)$ and $J_{ij}(r)$ are computed in spatstat by Gcross and Jcross, respectively:

```
> AG <- Gcross(amacrine, "on", "off")
> AJ <- Jcross(amacrine, "on", "off")
```

To compute these functions for all pairs of types i and j one can use alltypes as described above. Figure 14.18 shows the result of plotting alltypes(amacrine, Gcross). The top right panel shows the cumulative distribution function of the distance from an 'off' cell to the nearest 'on' cell. The diagonal panels are strong evidence against a Poisson process for each separate layer of cells.

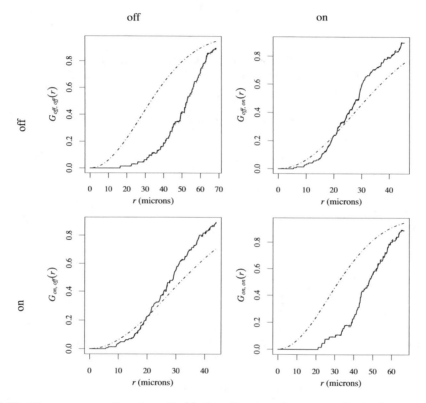

Figure 14.18. *The cross-type functions $G_{ij}(r)$ for all pairs of types i and j in the amacrine cells data.*

There is no separate function in spatstat for computing $F_i(r)$ since this is straightforward to do using split and Fest. However, alltypes(amacrine, Fest) will compute the functions $F_i(r)$ for all types i, and return the result as a "fasp" array containing one row for each possible type, and only one column.

Note that the results of Kcross and Gcross include a 'theoretical' or 'benchmark' value but their interpretation is different. For Kcross the benchmark $K_{ij}(r) = \pi r^2$ is consistent with independence between the points of types i and j. For Gcross, the benchmark $G_{ij}(r) = 1 - \exp(-\lambda_j \pi r^2)$ is consistent with the points of type j being *Poisson* in addition to being independent of the points of type i.

An alternative would be to use the benchmark $G_{ij}(r) = F_j(r)$ which would then be consistent with independence between the points of types i and j. This is most easily done by superimposing the plots:

```
> plot(Gcross(amacrine, "on", "off"), km ~ r)
> plot(Fest(split(amacrine)$off), km ~ r, add=TRUE, lty=2)
```

Alternatively one can use `cbind.fv` to combine the functions.

Tests of independence between types are discussed in Section 14.7.

14.6.5   Summary functions for distances to any point

One may also generalise the classical summary functions in a different way, based on measuring distances from points of type i to points of *any* type [664]. These functions allow us to investigate the dependence of one type of point on the rest of the process.

14.6.5.1   Dot K-function and its relatives

The 'one type to any type' or 'dot' version of the K-function, $K_{i\bullet}(r)$, is the expected number of points of *any* type lying within a distance r of a typical point of type i, standardised by dividing by the intensity of unmarked points:

$$K_{i\bullet}(r) = \frac{1}{\lambda_\bullet}\mathbb{E}\left[t(u,r,\mathbf{X}^{(\bullet)}) \mid u \in \mathbf{X}^{(i)}\right] \tag{14.42}$$

for $r \geq 0$. The corresponding L-function and pair correlation function are

$$L_{i\bullet}(r) = \sqrt{\frac{K_{i\bullet}(r)}{\pi}}, \qquad g_{i\bullet}(r) = \frac{K'_{i\bullet}(r)}{2\pi r}. \tag{14.43}$$

It is easy to see that

$$\lambda_\bullet K_{i\bullet}(r) = \sum_j \lambda_j K_{ij}(r) \tag{14.44}$$

which we could re-express as

$$K_{i\bullet}(r) = \sum_j p_j K_{ij}(r) \tag{14.45}$$

where $p_j = \lambda_j/\lambda_\bullet$ is the proportion of points that belong to type j.

Under the hypothesis of independent components (MPP4) (that is, if the points of different types are independent of each other) we will have

$$K_{i\bullet}(r) = p_i K_{ii}(r) + (1-p_i)\pi r^2 \tag{14.46}$$

$$g_{i\bullet}(r) = 1 + p_i(g_{ii}(r) - 1) \tag{14.47}$$

where $K_{ii}(r)$ and $g_{ii}(r)$ are the K-function and pair correlation function of the point process $\mathbf{X}^{(i)}$ consisting of the points of type i. Under completely random labelling (MPP2)

$$K_{i\bullet}(r) = K_{\bullet\bullet}(r)$$

$$g_{i\bullet}(r) = g_{\bullet\bullet}(r)$$

the corresponding K-function and pair correlation function of the unmarked process $\mathbf{X}^{(\bullet)}$.

14.6.5.2   Dot G- and J-functions

The 'one type to any type' or 'i-to-any' version of the nearest-neighbour function, $G_{i\bullet}(r)$, is the distribution function of the distance from a point of type i to the nearest other point of any type:

$$G_{i\bullet}(r) = \mathbb{P}\left\{d(u,\mathbf{X}^{(\bullet)} \setminus u) \leq r \mid \mathbf{X}^{(i)} \text{ has a point at } u\right\}. \tag{14.48}$$

This is not related to the functions $G_{ij}(r)$ in a simple way. The corresponding version of the J-function is

$$J_{i\bullet}(r) = \frac{1 - G_{i\bullet}}{1 - F(r)}. \tag{14.49}$$

Under the null hypothesis of independent components (MPP4),

$$G_{i\bullet}(r) = 1 - (1 - G_{ii}(r)) \prod_{j \neq i} (1 - F_j(r)) \tag{14.50}$$

$$J_{i\bullet}(r) = J_{ii}(r). \tag{14.51}$$

Under completely random labelling (MPP2), a typical point of type i is just a typical point of $\mathbf{X}^{(\bullet)}$, so that

$$G_{i\bullet}(r) = G_{\bullet\bullet}(r) \tag{14.52}$$

$$J_{i\bullet}(r) = J_{\bullet\bullet}(r), \tag{14.53}$$

the G- and J-functions of the unmarked process $\mathbf{X}^{(\bullet)}$.

The functions $G_{i\bullet}(r)$ and $J_{i\bullet}(r)$ are computed by Gdot and Jdot, respectively. The command alltypes can be used with them, and produces an $M \times 1$ array of functions. Figure 14.19 shows the estimates of $G_{i\bullet}(r)$ and $J_{i\bullet}(r) - J_{ii}(r)$ for type $i =$ 'on' in the amacrine cells data.

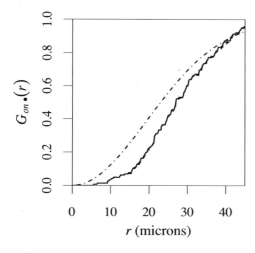
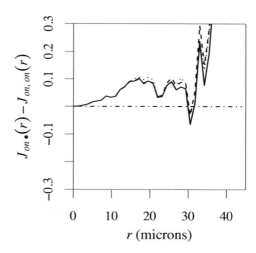

Figure 14.19. *Assessing random labelling in the amacrine data.* Left: *nearest-neighbour distance distribution $G_{i\bullet}(r)$ for type $i =$ 'on'.* Right: *Estimated discrepancy function $J_{i\bullet}(r) - J_{ii}(r)$ for $i =$ 'on'.*

14.6.6 Plotting and manipulating function arrays

Here we give more detail about handlng the "fasp" objects produced by the function alltypes mentioned in Sections 14.6.4 and 14.6.5. Objects of this class can be plotted simply by calling plot which dispatches to plot.fasp. There is also a print method for this class and there are various ways of manipulating objects of this class.

The plot.fasp method is similar to plot.fv and allows the function values to be transformed:

```
> aG <- alltypes(rescale(amacrine), "G")
> Phi <- function(x) asin(sqrt(x))
> plot(aG, Phi(.) ~ Phi(theo))
```

The result is shown in Figure 14.20.

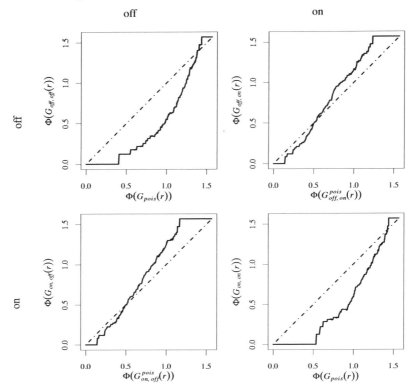

Figure 14.20. *Variance-stabilised estimates of G functions for each pair of types in the amacrine cells data.*

The function array can be indexed and subsetted using array subscripts. This is particularly useful when there are many possible types.

```
> lansL <- alltypes(lansing, Lcross)
> subL <- lansL[1:3, 1:3]
> hickL <- lansL[2, ]
> oaks <- c("redoak", "blackoak", "whiteoak")
> oakL <- lansL[oaks, oaks]
```

Individual functions in the array can also be extracted, as "fv" objects, by setting drop=TRUE to remove the array structure:

```
> Lmh <- lansL[3,2,drop=TRUE]
```

Calculations can be performed on all the functions in the array using eval.fasp. For example Fisher's variance stabilizing transformation can be applied to all of the Gcross function estimates in aG as follows:

```
> aGfish <- eval.fasp(Phi(aG))
```

The function eval.fasp is similar to eval.fv. The expression that is to be evaluated by eval.fasp should involve the *names* of objects of class "fasp".

14.6.7 Comparisons within and between types

When a point pattern has a large number of possible types, it becomes impractical to compute a summary function for each pair of possible types. Furthermore, some of the possible types will occur rarely, so that summary functions for these types will have large variability and bias.

One practical solution is to merge some of the rarer types together — for example, grouping rare species together by their genus or family. An alternative is to collapse the summary functions into two comparisons, within and between types.

14.6.7.1 Mark equality function

The *mark equality function* or *type equality function* $e(r)$ of a stationary multitype point process $\mathbf{Y}$ is a measure of the dependence between the types of two points of the process lying a distance r apart [638]. It is defined as the sum of the mark connection functions $p_{ii}(r)$ between all pairs of points of the same type:

$$e(r) = \sum_i p_{ii}(r) \tag{14.54}$$

for any distance $r \geq 0$, where $p_{ii}(r)$ is defined in (14.34). It can be interpreted as the conditional probability, given that there is a point of the process at a location u and another point of the process at a location v separated by a distance $\|u - v\| = r$, that the two points have the *same* type.

Under random labelling, $p_{ii}(r) = p_i^2$ where $p_i = \lambda_i/\lambda_\bullet$ is the probability of type i, so that $e(r) = \sum_i p_i^2$. The usual practice is to normalise $e(r)$ by dividing by this expected value:

$$k_e(r) = \frac{\sum_i p_{ii}(r)}{\sum_i p_i^2}.$$

Under random labelling, $k_e(r) \equiv 1$. Values $k_e(r) < 1$ indicate that for points separated by a distance r there is a 'lower than expected' probability that the points will belong to the same type. Conversely $k_e(r) > 1$ indicates that for points separated by a distance r there is a 'higher than expected' probability that the points will belong to the same type.

The standardised mark equality function $k_e(r)$ is a special case of the *mark correlation function*

$$k_f(r) = \frac{\mathbb{E}\left[f(m(u), m(v)) \middle| u, v \in \mathbf{X} \right]}{\mathbb{E}[f(M, M')]}$$

defined for a stationary marked point process, where $m(u)$ and $m(v)$ are the marks attached to two points of the process at locations u and v separated by a distance r, while M and M' are independent realisations of the marginal distribution of marks, and $f(m, m')$ can be any function defined for pairs of marks that returns a nonnegative real value. The mark equality function is obtained when $f(m, m') = \mathbf{1}\{m = m'\}$. Note that the mark correlation function, like the pair correlation function, is not a 'correlation' in the usual statistical sense. It can take any nonnegative real value. The value 1 suggests 'lack of correlation'.

In `spatstat` the mark correlation function is calculated using `markcorr`. The user may specify the function f. For multitype point patterns, the default for the function f is $f(m, m') = \mathbf{1}\{m = m'\}$ so that the standardised mark equality function is computed. The left panel of Figure 14.21 shows the standardised mark equality function for the amacrine cells data, computed by `markcorr(amacrine)`. It indicates that nearby points tend to have different types.

14.6.7.2 *I*-function

The *I*-function is defined [664] as

$$I(r) = \sum_{i=1}^{m} p_i J_{ii}(r) - J_{\bullet\bullet}(r) \tag{14.55}$$

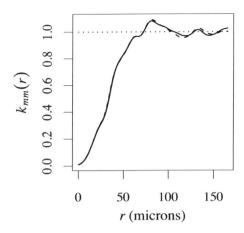 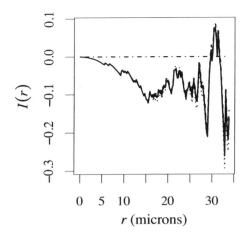

Figure 14.21. *Mark equality function* (Left) *and I-function* (Right) *for the amacrine data.*

where $J_{\bullet\bullet}(r)$ is the J-function for the entire point process ignoring the marks, while J_{ii} is the J-function for the process consisting of points of type i only, and p_i is the proportion of points which are of type i. Again this assumes the process is stationary.

The I-function is designed to measure dependence between points of different types. If the processes $\mathbf{X}^{(i)}$ are independent of each other, then the I-function is identically equal to 0. Deviations $I(r) < 1$ or $I(r) > 1$ typically indicate negative and positive association, respectively, between types. Explicit formulae are available [664].

The `spatstat` function `Iest` calculates estimates of the I-function for a multitype point pattern. See the right panel of Figure 14.21.

14.6.8 Summary functions for non-stationary patterns

The exploratory summary functions described in Sections 14.6.4–14.6.7 rest on the assumption that the point process is stationary. If this is not true, there is a risk of misinterpretation of the summary functions. This is the problem of confounding between inhomogeneity and clustering, explained in Sections 5.7.4 and 7.3.5.

The inhomogeneous K-function [46] (see Sections 7.10 and 7.10.2) can easily be generalised to multitype point processes [484, definition 4.8]. Inhomogeneous analogues of the functions $K_{ij}(r)$ and $K_{i\bullet}(r)$ are obtained by weighting each point by the reciprocal of the appropriate intensity function, and weighting the contribution from each pair of points by the product of these weights. For example, the inhomogeneous cross-type K-function is

$$K_{ij}^{inhom} = \mathbb{E}\left[\sum_{x \in \mathbf{X}^{(j)}} \frac{\mathbf{1}\{\|x - u\| \le r\}}{\lambda_j(x)} \mid u \in \mathbf{X}^{(i)} \right] \tag{14.56}$$

and an estimator of $K_{ij}(r)$ is

$$\widehat{K}_{ij}^{inhom} = \frac{1}{|W|} \sum_{x_\ell \in \mathbf{X}^{(i)}} \sum_{x_k \in \mathbf{X}^{(j)}} \frac{e(x_k, x_\ell)\mathbf{1}\{\|x_k - x_\ell\| \le r\}}{\lambda_i(x_\ell)\lambda_j(x_k)} \tag{14.57}$$

where $e(u, v)$ is an edge correction weight. The `spatstat` functions for calculating $\widehat{K}_{ij}^{inhom}$ and $\widehat{K}_{i\bullet}^{inhom}$ are named `Kcross.inhom` and `Kdot.inhom`.

Inhomogeneous versions of the multitype *L*-functions and pair correlation functions are also defined using the standard relationships $L = \sqrt{K/\pi}$ and $g = K'/(2\pi r)$. Their implementations in spatstat are Kcross.inhom, Ldot.inhom, pcfcross.inhom, and pcfdot.inhom.

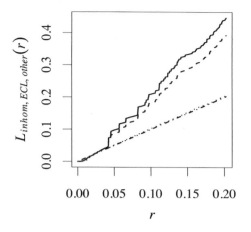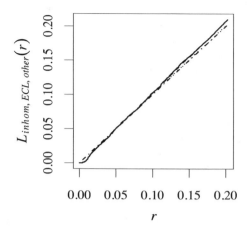

Figure 14.22. *Inhomogeneous cross-type L-function from ECL cells to other cells in the gastric mucosa data. Intensity estimated by kernel smoothing with bandwidth selected by cross-validation for the unmarked points* (Left, $\sigma = 0.0503$) *or by likelihood cross-validation for the multitype points* (Right, $\sigma = 0.0858$).

The cross-type functions Kcross.inhom, Lcross.inhom, pcfcross.inhom require arguments lambdaI and lambdaJ giving estimates of the intensity functions $\lambda_i(u)$ and $\lambda_j(u)$ of the points of types i and j, respectively. The values of these intensities are required only at the data points of types i and j, respectively. Similarly the 'type-i-to-any-type' functions Kdot.inhom, Ldot.inhom, pcfdot.inhom require arguments lambdaI and lambdadot giving estimates of $\lambda_i(u)$ and $\lambda_\bullet(u)$, respectively. The arguments lambdaI, lambdaJ, and lambdadot may be numeric vectors, pixel images, or functions. If these arguments are missing, the intensities will be estimated by kernel smoothing from the same data. Note the warning about possible bias induced by the use of kernel smoothing estimates, given in Section 7.10.2. Before using kernel smoothing it is wise to inspect the resulting intensity surface. An interesting experiment is:

```
> LMD <- Lcross.inhom(mucosa, "ECL", "other", sigma=bw.diggle(mucosa))
> LMR <- Lcross.inhom(mucosa, "ECL", "other", sigma=bw.relrisk(mucosa))
```

Figure 14.22 shows the results for these two automatically selected bandwidths, the first bandwidth $\sigma = 0.0503$ suggesting clustering, while the second bandwidth $\sigma = 0.0858$ does not. Plots (not shown) of the kernel-smoothed density images or kernel-smoothed probabilities relrisk(mucosa, sigma) for these two values of the bandwidth suggest that the smaller value is undersmoothing, so that the left panel of Figure 14.22 is probably an artefact.

14.7 Tests of randomness and independence

For unmarked point patterns, the most important reference model is complete spatial randomness. Sometimes the key scientific question is whether or not a point pattern appears to have been generated by CSR. More often, CSR is not scientifically plausible, but is still an important 'benchmark' or 'dividing hypothesis' [179, pp. 51–52] that separates spatial clustering from spatial regularity, as sketched in Figure 7.1.

For multitype point patterns, the analogous reference model is Complete Spatial Randomness and Independence (CSRI). In some studies, the key scientific question is whether the data could have been generated by CSRI. However, CSRI does not serve quite the same role as a 'dividing hypothesis'. For multitype patterns there are at least two natural choices for a dividing hypothesis:

random labelling (MPP2)**:** the type labels are randomly allocated to the points;

independent components (MPP4)**:** the points of different types are independent.

If **both** of these statements are true, for a stationary point process, then the pattern is CSRI. See [383, 24]. However, there are many contexts where one of these statements is obviously false, and the other statement is a good choice for a dividing hypothesis. For example, in the amacrine cells data, the key question is whether the two layers of cells are independent patterns; the random labelling hypothesis is obviously false for the amacrine cells, because each layer of cells is regular.

This section discusses several tests of random labelling and independence of components, based on envelopes of the summary functions.

It is important to remember when drawing conclusions from these envelopes that the summary functions can have a different interpretation in the multitype setting from that which they have in the unmarked setting. See the remarks on pages 594 and 597 in Section 14.6.4. These interpretations are summarised in Tables 14.3 and 14.4.

14.7.1 Tests for independence of components

A test for independence of components is most appropriate when the different types of points represent different sub-populations which may or may not interact. For example, in the amacrine cells data, the 'on' and 'off' layers of cells are anatomically distinct, and the question is whether the two layers are independent.

For testing independence of components, the 'i-to-j' summary functions are most useful. For a stationary multitype point process, if the sub-processes of types i and j are independent, then $K_{ij}(r) = \pi r^2$, $G_{ij}(r) = F_j(r)$ and $J_{ij}(r) \equiv 1$.

In a randomisation test of the independence-of-components hypothesis, the simulated patterns X are generated from the dataset by splitting the data into the sub-patterns of points of each type, and randomly shifting each of these sub-patterns, independently of each other. Assuming stationarity, any shifted version of the data is statistically equivalent to the original. Under the null hypothesis of independent components, a different shift can be applied to the points of each type.

The random shifting is performed by `rshift`. If we specify the argument `radius=150`, then a shift vector with maximum length 150 will be randomly selected (by choosing a random point in the disc of radius 150) and applied to the points of a given type.

```
> Amac <- rescale(amacrine)
> E <- envelope(Amac, Lcross, nsim=39, i="on", j="off",
                simulate=expression(rshift(Amac, radius=150)))
```

Notice that the arguments i and j here do not match any of the formal arguments of `envelope`, so they are taken to be components of the . . . argument and passed to `Lcross`. This executes `Lcross(X, i="on", j="off")` for each of the simulated point patterns X.

The result E is plotted in Figure 14.23 using the plot formula `. - r ~ r` to bring out the interesting detail. The overall impression is that the independence-of-components hypothesis should not be rejected. Another pair of envelopes (not shown) calculated using `global=TRUE` (so that the conclusions would be unambiguous) also failed to reject the independence-of-components hypothesis.

FUNCTION	THEORETICAL VALUE	CONSISTENT WITH			
		CSRI	CSR(i)	INDEPT.	RANDOM LABELLING
K_{ij}	πr^2	●	○	●	○
L_{ij}	r	●	○	●	○
g_{ij}	1	●	○	●	○
K_{ii}	πr^2	●	●	○	○
L_{ii}	r	●	●	○	○
g_{ii}	1	●	●	○	○
$K_{i\bullet}$	πr^2	●	○	○	○
$L_{i\bullet}$	r	●	○	○	○
$g_{i\bullet}$	1	●	○	○	○
G_{ij}	$1-\exp(-\lambda_j \pi r^2)$	●	○	○	○
J_{ij}	1	●	○	●	○
G_{ii}	$1-\exp(-\lambda_i \pi r^2)$	●	●	○	○
J_{ii}	1	●	●	○	○
F_i	$1-\exp(-\lambda_i \pi r^2)$	●	●	○	○
I	0	●	○	●	○
k_{mm}	1	●	○	○	●

Table 14.3. *Interpretation of the 'theoretical' value of each multitype summary function. Filled dot means true, open circle indicates false. For each summary function h(r) and its theoretical value $h_0(r)$, the table indicates whether $h(r) = h_0(r)$ is (always) consistent with the stated assumptions about the point pattern. Meaning of the columns is:* CSRI = *homogeneous multitype Poisson process;* CSR(i) = *points of type i constitute a homogeneous Poisson process;* INDEPT. = *independent components;* RANDOM LABELLING = *random labelling.*

FUNCTION MATCH	CONSISTENT WITH			
	CSRI	CSR(i)	INDEPT.	RANDOM LABELLING
$K_{i\bullet} = K_{\bullet\bullet}$	●	○	○	●
$L_{i\bullet} = L_{\bullet\bullet}$	●	○	○	●
$g_{i\bullet} = g_{\bullet\bullet}$	●	○	○	●
$K_{ii} = K_{\bullet\bullet}$	●	○	○	●
$J_{i\bullet} = J_{\bullet\bullet}$	●	○	○	●
$J_{i\bullet} = J_{ii}$	●	○	●	○
$G_{ij} = F_j$	●	○	●	○

Table 14.4. *Interpretation of agreement between multitype summary functions. Symbols as in Table 14.3.*

Testing for independence of components by random shifting is somewhat unsatisfactory: stationarity is assumed, and there is a need to handle edge effects associated with the random shifts.

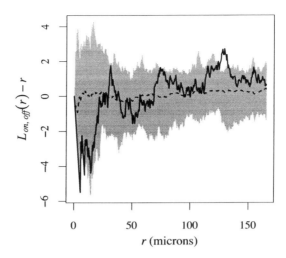

Figure 14.23. *Centred L-function* $L_{ij}(r) - r$ *for* `amacrine` *with* $i = $ *"on" and* $j = $ *"off" (solid line) and the envelopes of 39 simulations generated by random shifts (grey shading). Results suggest "on" and "off" components are independent.*

14.7.2 Tests of random labelling

A test of random labelling is most natural when the type label of each point represents its status, such as disease status or a case-control label. The null hypothesis of completely random labelling is that the status of each point is determined at random, independently of other points, with fixed probabilities. In a case-control study the null hypothesis of completely random labelling implies constant disease risk — the probability of contracting the disease does not depend on spatial location.

A simple test for random labelling is a *permutation test* — a Monte Carlo test based on randomly relabelled versions of the original data. The spatial locations of the observed data points are held fixed, and the marks (type labels) attached to these locations are randomly permuted, with equal probability for each possible permutation. If the null hypothesis of random labelling is true, then these randomly relabelled datasets are statistically equivalent to the original data, and the rationale of the Monte Carlo test applies.

Almost any test statistic could be used, but it is advisable to choose one which has a clear interpretation in the context. For envelope-based tests, the 'dot' functions are the most useful for assessing random labelling. Assuming a *stationary* multitype point process, completely random labelling implies that $K_{i\bullet}(r) = K(r)$, $G_{i\bullet}(r) = G(r)$, and $J_{i\bullet}(r) = J(r)$ where K, G, J are the summary functions for the point process without marks. This suggests using a statistic like $K_{i\bullet}(r) - K(r)$ or $J_{i\bullet}(r) - J(r)$. Although these choices are not strictly necessary for the validity of the test, they are useful for interpreting departures from the null hypothesis.

Here is a worked example using the *J*-function. We use the dataset `paracou` giving the locations of Kimboto trees in a 400×525 metre survey plot at Paracou, French Guiana [264, 547]. Trees are classified as either juvenile or adult.

First we write a function to evaluate $\widehat{J}_{i\bullet}(r) - \widehat{J}(r)$:

```
> Jdif <- function(X, ..., i) {
    Jidot <- Jdot(X, ..., i=i)
    J <- Jest(X, ...)
    dif <- eval.fv(Jidot-J)
```

```
        return(dif)
    }
```

This function will be called by `envelope` to compute simulation envelopes. Here it is important that the function should accept unspecified additional arguments "`...`" and pass them to `Jdot` and `Jest`, so that the `envelope` function can control the sequence of distance values r and the choice of edge correction, etc.

The `spatstat` function `rlabel` permutes the marks in a marked point pattern. To generate the envelopes for a test of random labelling:

```
> EPJ <- envelope(paracou, Jdif, nsim=39, i="juvenile",
              simulate=expression(rlabel(paracou)))
```

Notice that the argument `i` does not match any formal argument of `envelope`, so it is treated as one of the "`...`" arguments and passed to `Jdif`. That is, `envelope` calls `Jdif(X, i="juvenile")` for each of the simulated point patterns `X`.

For each of the 39 simulations, the type labels of the `paracou` dataset are randomly permuted, and the function $\widehat{J}_{i\bullet}(r) - \widehat{J}(r)$ for $i =$ "`juvenile`" is evaluated for this randomised dataset. The pointwise envelopes of $\widehat{J}_{i\bullet}(r) - \widehat{J}(r)$ are computed. The result is shown in Figure 14.24. The null hypothesis would be rejected at the 5% level with almost any choice of distance r. The deviations of $\widehat{J}_{i\bullet}(r) - \widehat{J}(r)$ are negative, which suggests that juvenile trees are more likely to be found close to an adult tree than would be expected if tree age was randomly allocated (i.e. if the age of each tree was independent of the ages of its neighbours).

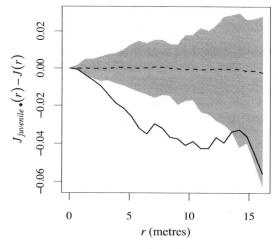

Figure 14.24. *Test of random labelling for the* `paracou` *dataset. Summary function* $\widehat{J}_{i\bullet}(r) - \widehat{J}(r)$ *for the* `paracou` *dataset (solid line) and envelopes based on 39 random labellings of the data (grey shading) with* $i =$ "`juvenile`". *Results emphatically suggest tree ages are not random labels.*

Care must always be taken with the *interpretation* of excursions, of the graph of the estimate of the summary function in question, outside the region bounded by the envelopes. See Chapter 10, in particular Section 10.7.2. For example, the test described above would remain valid even if the `paracou` data were not from a stationary process, because the test is effectively conditional on the observed locations. However, departures from the null value of the statistic would be harder to *interpret* in this case.

14.7.3 Arrays of envelopes

When dealing with multitype patterns one often wishes to investigate relationships amongst all of the types simultaneously. For instance one might wish to look at simulation envelopes for the functions L_{ij} for all pairs of types i and j. This can be done very simply by using `alltypes` with the argument `envelope=TRUE`.

```
> X <- rescale(amacrine)
> aE <- alltypes(X, Lcross, nsim=39, envelope=TRUE, global=TRUE)
```

Figure 14.25 shows the result of `plot(aE, . - r ~ r)`. By default, these envelopes are based on simulations from CSRI; the argument `simulate` specifies how to generate the simulated patterns, as explained in Section 10.8.1.

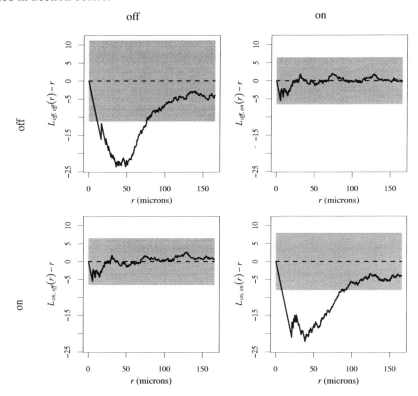

Figure 14.25. *Centred L-function and global envelopes of 39 simulations of CSR, for each combination of types for the amacrine cells data.*

For patterns with a large number of possible types, within- and between-type comparisons (Section 14.6.7) are more practical.

14.7.4 Testing for CSRI

As mentioned at the beginning of Section 14.7, there are some studies where the key scientific question is whether the data could have been generated by a homogeneous Poisson multitype point process (CSRI). This is probably the only context where a test of CSRI is justified (unlike a test of CSR which may often serve as a dividing hypothesis).

The envelopes in Figure 14.25 could serve as the basis for a formal test of CSRI for the `amacrine` data, if that were scientifically appropriate. For formal inference it is advisable to generate simultaneous envelopes by setting `global=TRUE` in the call to `alltypes` to avoid the 'look elsewhere effect'.

Note that CSRI will be accepted only when *all* of the individual plot panels in Figure 14.25 indicate acceptance of CSRI. The null hypothesis will be rejected if *any* of the panels indicate rejection. The significance level of this test is larger (i.e. weaker evidence) than the significance level attached to each individual panel. It is advisable to set `reuse=FALSE` in the call to `alltypes`, so that each panel is based on an independent set of simulated point patterns. Then, since the probability of type I error in each panel is $\alpha = 0.05$ in the example given, the probability of type I error for the test based on all $m = 4$ panels is $\alpha^* = 1 - (1-\alpha)^m = 1 - 0.95^4 = 0.185$. To achieve a test with significance level $\alpha^* = 0.05$ we would need $\alpha = 1 - (1-\alpha^*)^{1/m} = 1 - 0.95^{1/4} = 0.013$. This could be achieved by setting `nsim=156` in the call.

Another approach would be to assume the process is stationary, and to test for CSRI by testing whether the process has *both* the properties of random labelling *and* of independence of components. If randomisation tests are used, then the same kind of calculation can be used (with $m = 2$) to determine the significance level.

14.8 Multitype Gibbs models

Gibbs point process models (introduced in Chapter 13) also apply to multitype point processes [484, sec. 6.6]. A multitype Gibbs model can be fitted to multitype point pattern data using ppm.

A simple example is a multitype version of the hard core process (Section 13.3.2) in which any two points, of types i and j, respectively, are forbidden to lie closer than a distance h_{ij} apart, where the hard core distance h_{ij} can be different for different pairs of types i, j. This model can be fitted to the `amacrine` data (for example) by

```
> ppm(amacrine ~ marks, MultiHard())
```

In this case, the hard core distance parameters h_{ij} will be estimated from the data (essentially by finding the minimum nearest-neighbour distance between specified types of points). However, most other models have nuisance parameters that cannot be estimated so easily, and must be specified by the user, or estimated by other means.

There are no technical difficulties in extending Gibbs models to multitype processes, but there are some important matters to understand when using these models.

14.8.1 Probability densities

A multitype Gibbs point process in a bounded window W can be specified by writing down the form of its *probability density f*.

A typical realisation of a multitype point process in a bounded window W is a set

$$\mathbf{y} = \{(x_1, m_1), \ldots, (x_n, m_n)\}$$

where $x_i \in W$ is the spatial location of the ith point and $m_i \in \{1, 2, \ldots, M\}$ is the type.

The probability density $f(\mathbf{y})$ of a multitype point process $\mathbf{Y}$ is defined using the same approach as in Section 13.3.1. We can interpret $f(\mathbf{y})$ as the probability of getting the pattern $\mathbf{y}$ from the multitype point process $\mathbf{Y}$, *divided by* the corresponding probability for CSRI with intensity $\lambda(u, m) = 1$. The last part is important: the reference model is CSRI with *intensity 1 for each type of point*, so the total intensity of the reference point process is M, the number of types.

This convention ensures that, if the multitype point process $\mathbf{Y}$ has independent components, then the probability density of $\mathbf{Y}$ is the product of the probability densities for the components:

$$f(\mathbf{y}) = f_1(\mathbf{x}^{(1)}) \cdots f_M(\mathbf{x}^{(M)}) \tag{14.58}$$

where $f_m(\mathbf{x})$ is the probability density of $\mathbf{X}^{(m)}$, the sub-process of points of type m, for $m = 1, \ldots, M$.

In particular for CSRI, the stationary multitype Poisson process with intensity $\lambda(u,m) = \lambda_m$, the probability density is (14.58) where $f_m(\mathbf{x})$ is the probability density of CSR with intensity λ_m. That is, the probability density for CSRI is

$$f(\mathbf{y}) = c\, \lambda_1^{n_1} \cdots \lambda_M^{n_M} \tag{14.59}$$

where $n_m = n(\mathbf{x}^{(m)})$ is the number of points of type m, and c is a constant. Each point of type m contributes a factor of λ_m to the probability density.

The inhomogeneous multitype Poisson process with intensity function $\lambda(u,m)$ has probability density

$$f(\mathbf{y}) = c_1 \prod_{i=1}^{n} \lambda(x_i, m_i) \tag{14.60}$$

where c_1 is another constant. Each multitype point (x_i, m_i) contributes a factor $\lambda(x_i, m_i)$ to the probability density.

14.8.2 Conditional intensity

The conditional intensity $\lambda(u \mid \mathbf{x})$ of an (unmarked) point process $\mathbf{X}$ at a location u was defined in Section 13.12.4. Roughly speaking $\lambda(u, \mathbf{x})\,\mathrm{d}u$ is the conditional probability of finding a point near u, given that the rest of the point process $\mathbf{X}$ coincides with $\mathbf{x}$.

For a multitype point process $\mathbf{Y}$ the conditional intensity is a function $\lambda((u,m) \mid \mathbf{y})$ giving a value at a location u for each possible type m. We can interpret $\lambda((u,m) \mid \mathbf{y})\,\mathrm{d}u$ as the conditional probability finding a point *of type m* near u, given the rest of the multitype point process.

The conditional intensity is related to the probability density $f(\mathbf{y})$ by

$$\lambda((u,m) \mid \mathbf{y}) = \frac{f(\mathbf{y} \cup \{(u,m)\})}{f(\mathbf{y})} \tag{14.61}$$

for $(u,m) \notin \mathbf{y}$.

For Poisson processes, the conditional intensity coincides with the intensity function:

$$\lambda((u,m) \mid \mathbf{y}) = \lambda(u,m). \tag{14.62}$$

The important property is that the conditional intensity for a Poisson process does not depend on the configuration $\mathbf{y}$. In particular the homogeneous Poisson multitype point process, CSRI, has conditional intensity

$$\lambda((u,m) \mid \mathbf{y}) = \lambda_m \tag{14.63}$$

where $\lambda_m \geq 0$ is the intensity of the points of type m.

14.8.3 Pairwise interactions

A multitype *pairwise interaction* process has probability density of the form

$$f(\mathbf{y}) = \alpha \left[\prod_{i=1}^{n(\mathbf{y})} \beta_{m_i}(x_i) \right] \left[\prod_{i<j} c_{m_i, m_j}(x_i, x_j) \right] \tag{14.64}$$

where α is a constant, $\beta_m(u) \geq 0$ is the 'first order trend' for points of type m, and $c_{m,m'}(u,v)$ for $m, m' \in \{1, 2, \ldots, M\}$ are the pairwise interaction functions between points of given types m and m'. The interaction functions must be symmetric, $c_{m,m'}(u,v) = c_{m,m'}(v,u)$ and $c_{m,m'} \equiv c_{m',m}$. The conditional intensity of the multitype pairwise interaction process is

$$\lambda((u,m) \mid \mathbf{y}) = \beta_m(u) \prod_{i=1}^{n(\mathbf{y})} c_{m,m_i}(u, x_i). \tag{14.65}$$

14.8.4 Interactions not depending on types

The simplest examples of multitype pairwise interaction processes are those in which the interaction term $c_{m,m'}(u,v)$ does not depend on the types m, m' but only on the locations u, v. For example, we would take any of the pairwise interaction functions $c(u,v)$ described in Section 13.6.2 and use it to construct a multitype point process. This provides a convenient first step in modelling a multitype point pattern.

An equivalent description of such a process is the following [67]. First we generate an *unmarked* Gibbs process with first-order term $\beta_\bullet(u) = \sum_m \beta_m(u)$ and pairwise interaction $c(u,v)$. Then we label each point x_i with a random type m_i, with probability distribution $\mathbb{P}\{m_i = m\} = \beta_m(x_i)/\beta_\bullet(x_i)$, independent of other points.

Such a process always satisfies the independent labelling property (MPP2'). If additionally the first-order terms are proportional, $\beta_m(u) = a_m \beta_\bullet(u)$, then the labels are 'completely random' in the sense of (MPP2). Such models provide an alternative way to test for random labelling.

14.8.5 Interaction only within types

Another simple model arises when pairwise interactions occur only between points of the same type. For example, in a spatial case-control study, we might assume that the controls are a Poisson process (perhaps because of the way they were sampled), but the cases might conceivably be clustered. The appropriate model is a multitype Gibbs process in which interactions occur only between cases.

If pairwise interactions occur only between points of the same type, that is, $c_{m,m'}(u,v) = 1$ whenever $m \neq m'$, then the probability density (13.27) can be factorised in the form (14.58). This process has independent components: the points of different types are independent. The points of type m constitute a Gibbs process with first-order trend $\beta_m(u)$ and pairwise interaction $c_{m,m}(u,v)$. This provides a useful alternative way to test for independence of components, without having to assume stationarity.

14.8.6 Mark-dependent pairwise interactions

In general the pairwise interaction may depend on the marks.

14.8.6.1 Multitype hard core process

A simple example is the (constant intensity) *multitype hard core process* in which $\beta_m(u) \equiv \beta_m$ and

$$c_{m,m'}(u,v) = \begin{cases} 1 & \text{if } \|u-v\| > r_{m,m'} \\ 0 & \text{if } \|u-v\| \leq r_{m,m'} \end{cases} \tag{14.66}$$

where $r_{m,m'} = r_{m',m} > 0$ is the hard core distance that is pertinent to the relationship between points of type m and points of type m'. In this process, two points of type m and m', respectively, can never come closer than the distance $r_{m,m'}$.

The conditional intensity at a location u for type m is equal to β_m if the hard core conditions are satisfied, and equal to 0 otherwise. That is,

$$\lambda((u,m) \mid \mathbf{y}) = \beta_m \mathbf{1} \left\{ t(u, r_{m,m'}, \mathbf{x}^{(m')}) = 0 \text{ for all } m' \right\}. \tag{14.67}$$

Here $t(u, r_{m,m'}, \mathbf{x}^{(m')})$ is the number of points in $\mathbf{y}$ of type m' lying within a distance $r_{m,m'}$ of the location u.

By setting $r_{m,m'} = 0$ for a particular pair of marks m, m' we effectively remove the interaction term between points of these types. If there are only two types, 1 and 2, then setting $r_{1,2} = r_{2,1} = 0$ implies that the sub-processes $\mathbf{X}^{(1)}$ and $\mathbf{X}^{(2)}$, consisting of points of types 1 and 2, respectively, are *independent* hard core processes, as explained above.

If interactions do occur between different types of points, then it becomes much more complicated to describe the behaviour of the points of a given type. In a multitype hard core process, the points of type 1 are not a Gibbs hard core process (although they do respect a hard core constraint) because of the influence of the points of other types.

A well-studied example is the multitype hard core process with two types of points, with $r_{1,1} = r_{2,2} = 0$, known as the bivariate Widom-Rowlinson model. Points of the same type do not interact, while points of different types are forbidden to lie closer than a distance $r_{1,2}$. Then in fact the points of type 1 constitute an *area-interaction* process (Section 13.7.1 and [699, 70]). The points of type 1 are clustered (area-interaction parameter $\eta > 1$) essentially because they gather together in regions that are not forbidden by the presence of points of type 2, and conversely.

One important practical lesson we can draw is that, when building a Gibbs model for multitype data, we cannot infer the form of the interaction between points of type m by studying only the points of type m.

14.8.6.2 Multitype Strauss model

The *multitype Strauss process* has pairwise interaction term

$$c_{m,m'}(u,v) = \begin{cases} 1 & \text{if } \|u - v\| > r_{m,m'} \\ \gamma_{m,m'} & \text{if } \|u - v\| \leq r_{m,m'} \end{cases} \tag{14.68}$$

where $r_{m,m'} > 0$ are interaction radii as above, and $\gamma_{m,m'} \geq 0$ are interaction parameters. Symmetry requires that $r_{m,m'} = r_{m',m}$ and $\gamma_{m,m'} = \gamma_{m',m}$ so there are only $(M^2 + M)/2$ interaction parameters to be fitted rather than M^2.

The conditional intensity at a location u for type m is

$$\lambda((u,m) \mid \mathbf{y}) = \beta_m \prod_{m'=1}^{M} \gamma_{m,m'}^{t(u, r_{m,m'}, \mathbf{x}^{(m')})}. \tag{14.69}$$

In contrast to the unmarked Strauss process, which is well defined only when its interaction parameter γ is between 0 and 1, the multitype Strauss process allows some of the interaction parameters $\gamma_{m,m'}$ to exceed 1 for $m \neq m'$, provided that one of the relevant types has a hard core i.e. that $\gamma_{m,m} = 0$ or $\gamma_{m',m'} = 0$.

Suppose that there are only two types, 1 and 2. If $\gamma_{1,2}$ is equal to 1, then the sub-processes $\mathbf{X}^{(1)}$ and $\mathbf{X}^{(2)}$, consisting of points of types 1 and 2, respectively, are independent Strauss processes. The resulting process is the same as if two independent Strauss processes were generated, marked "1" and "2", respectively, and then superimposed. Thus, a test of independence of components could be performed by testing the null hypothesis $H_0: \gamma_{12} = 1$.

Care should be taken with the interpretation of the interaction parameters in multitype point process models. In a multitype Strauss model, even if all interaction parameters γ_{ij} are less than 1, the marginal behaviour of the component point processes can be spatially *aggregated*. Heuristically this happens because, if points of opposite type have a strong inhibitive effect on each other, while points of the same type have only a weak interaction, then points of the same type will tend to lie close together. See, e.g., [70, 227].

14.8.6.3 Multitype Strauss-hard core process

In a *multitype Strauss-hard core process* each pair of types has both a hard core and a Strauss interaction. The pairwise interaction function is

$$c_{m,m'}(u,v) = \begin{cases} 0 & \text{if } \|u-v\| < h_{m,m'} \\ \gamma_{m,m'} & \text{if } h_{m,m'} \le \|u-v\| \le r_{m,m'} \\ 1 & \text{if } \|u-v\| > r_{m,m'} \end{cases} \qquad (14.70)$$

where $r_{m,m'} > 0$ are interaction distances and $\gamma_{m,m'} \ge 0$ are interaction parameters as above, and $h_{m,m'}$ are hard core distances satisfying $h_{m,m'} = h_{m',m}$ and $0 \le h_{m,m'} < r_{m,m'}$.

14.8.7 Fitting Gibbs models to multitype data

To fit a multitype Gibbs model, ppm is called with the argument `interaction`. For example

```
> ppm(amacrine ~ marks, interaction=MultiHard())
```

fits the multitype hard core process to the amacrine cells data. The argument `interaction` must be an interpoint interaction object (class `"interact"`) created by one of the specialised functions listed in Tables 13.2 and 13.3 on pages 516 and 519, respectively.

14.8.7.1 Interactions not depending on types

In `spatstat` interaction objects designed for unmarked point processes may also be used directly to fit models to a multitype point pattern. The result is a model where the interaction does not depend on the types of the points, only on their locations. Such models were discussed in Section 14.8.4 for the case of pairwise interactions.

These models are a useful starting point for analysis, especially when the number of possible types is larger, or when some of the types are rare. For example the `hyytiala` forest survey data include four species, one of which is represented by just a single individual tree. The following model postulates a hard core between trees, with the same hard core distance for all species (estimated from the data):

```
> ppm(hyytiala ~ marks, Hardcore())
Stationary Hard core process
Possible marks: 'aspen', 'birch', 'pine' and 'rowan'
Log trend:   ~marks
First order terms:
beta_aspen beta_birch  beta_pine beta_rowan
  0.002755   0.046479   0.330434   0.059947
Hard core distance:       0.1542
```

Models in which the interaction does not depend on type have the random labelling property (if the first-order terms are constant) or have constant relative risk (if the first-order terms are proportional), so they may serve as the null hypothesis in a test of random labelling or a test of constant relative risk. For example, the `hamster` data give the locations and types of cells in a histological section of a kidney tumour in a hamster. The locations are slightly regular at a scale of about 5–10 microns (0.02–0.04 units in the dataset). The main question is whether the type labels are randomly allocated, or whether cells of the same type are clustered together. The null hypothesis of random labelling could be represented by the fitted model

```
> ppm(hamster ~ marks, Strauss(0.02))
```

```
Stationary Strauss process
Possible marks: 'dividing' and 'pyknotic'
Log trend: ~marks
First order terms:
beta_dividing beta_pyknotic
       262.55         91.24
Interaction distance:        0.02
Fitted interaction parameter gamma:        0.6314
```

The foregoing code fits the multitype Strauss process with conditional intensity

$$\lambda((u,m) \mid \mathbf{y}) = \beta_m \gamma^{t(u,r,\mathbf{y})}. \tag{14.71}$$

Here β_m are constants which allow for unequal abundance of the different species of tree. The other quantities are the same as in (13.12). The interaction between two trees is assumed to be the same for all species, and is controlled by the interaction parameter γ and interaction radius $r = 0.02$.

The interaction radius r is an *'irregular'* parameter (see Sections 9.12 and 13.13.5). Consequently it is *not* estimated by ppm but must be supplied to the function Strauss when ppm is called. The interaction radius can be estimated by profile methods using profilepl.

14.8.7.2 Interactions depending on types

We now move on to the setting in which the pairwise interaction *does* depend on the marks, i.e. on the types of the points as well as their separation.

In the near future, spatstat will support a long list of models of multitype interactions. At the time of writing, the only examples are the spatstat functions MultiHard, MultiStrauss, and MultiStraussHard which create interaction objects that represent the multitype hard core, multitype Strauss, and multitype Strauss-hard core interactions, respectively. The argument syntax for these functions is

```
MultiHard(hradii, types)
MultiStrauss(radii, types)
MultiStraussHard(iradii, hradii, types)
```

where types is the vector of possible types, and hradii is a matrix of hard core distances between each pair of types, and radii or iradii is a matrix of interaction distances for the Strauss component.

The matrices hradii, iradii, radii must be symmetric, with entries that are either positive numbers or NA. A value of zero or NA indicates that no interaction term should be included for this combination of types. For example, specifying a diagonal matrix of distances (i.e. in which all off-diagonal entries are zero) would fit a model in which different types of points are independent.

The argument types need not be specified in normal use. It will be determined automatically from the point pattern dataset to which the interaction is applied, when the user calls ppm. This is the 'self-starting' feature of interpoint interactions. However, specifying the types will ensure that the rows and columns of the distance matrices are correctly interpreted. If types is not specified, the user needs to be confident that the ordering of rows and columns in the distance matrices corresponds to the ordering of types in levels(marks(X)).

The hard core distances in the matrix hradii and the interaction distances in the matrices iradii, radii are *irregular* parameters, and are not estimated directly by ppm. However, the hard core distance matrix hradii can be estimated easily. The maximum likelihood estimate of each hard core radius is the minimum nearest-neighbour distance between the relevant types of points. In spatstat we reduce the bias by multiplying by $n/(n+1)$ where n is the number of values over which the minimum is taken. If the argument hradii is missing, then it is estimated automatically; for example,

```
> ppm(amacrine ~ marks, MultiHard())
Stationary Multitype Hardcore process
Possible marks: 'off' and 'on'
Log trend:   ~marks
First order terms:
beta_off  beta_on
   106.6    126.3
Hardcore radii:
           off        on
off 0.024690 0.008341
on  0.008341 0.031980
```

Note that the only regular parameters being estimated by the ppm function in this example are the β values, the first-order terms for the two types.

The interaction distance matrix `radii` for the multitype Strauss model, and its counterpart `iradii` for the multitype Strauss-hard core model, cannot be estimated so easily and usually have to be chosen by the user. For an initial analysis, the interaction distances could be chosen by inspecting the cross-type K-functions and finding the distances where the deviation from CSRI is greatest. For a definitive analysis the interaction distances could be optimised using `profilepl`.

The dataset `betacells` described on page 570 has a data frame of marks of which we will again use only the column named `type`, a factor indicating the type of cell:

```
> Beta <- betacells
> marks(Beta) <- marks(Beta)[,"type"]
```

or more slickly

```
> Beta <- subset(betacells, select=type)
```

To fit the stationary multitype Strauss process we must specify the matrix of interaction radii $r_{i,j}$. A plot of `alltypes(Beta, Kcross)` (not shown) suggests that $r_{1,1} = 60$ and $r_{2,2} = 80$ microns would be appropriate, while there does not appear to be any interaction between types. We could try a model with independent components,

```
> R <- diag(c(60,80))
> fit0 <- ppm(Beta ~ marks, MultiStrauss(R))
> fit0
Stationary Multitype Strauss process
Possible marks: 'off' and 'on'
Log trend:   ~marks
First order terms:
 beta_off    beta_on
0.0005566 0.0014877
Interaction radii:
     off on
off   60 NA
on    NA 80
Fitted interaction parameters gamma_ij
       off      on
off 0.033      NA
on     NA 0.0418
```

or perhaps set $r_{1,2} = 60$ microns to allow for a possible interaction between types:

```
> R2 <- matrix(c(60,60,60,80), nrow=2, ncol=2)
> fit1 <- ppm(Beta ~ marks, MultiStrauss(R2))
> fit1
Stationary Multitype Strauss process
Possible marks: 'off' and 'on'
Log trend:   ~marks
First order terms:
 beta_off   beta_on
0.0006044 0.0016147
Interaction radii:
     off on
off  60 60
on   60 80
Fitted interaction parameters gamma_ij
        off       on
off 0.0330 0.9189
on  0.9189 0.0421
```

In the second model the between-types interaction parameter $\gamma_{1,2}$ is close to 1. Using the adjusted composite likelihood ratio test [56] explained in Section 13.5.11, we can assess whether $\widehat{\gamma}_{1,2}$ is significantly different from 1:

```
> anova(fit0, fit1, test="LR")
Analysis of Deviance Table
Model 1: ~marks           MultiStrauss
Model 2: ~marks           MultiStrauss
  Npar Df AdjDeviance Pr(>Chi)
1    4
2    5  1        1.29     0.26
```

This suggests an absence of interaction between types.

Any Gibbs model may have a '*trend*' component and thus have a spatially varying intensity rather than being stationary. (We re-emphasize that the trend *influences* the intensity but is *not* the same as the intensity — except, of course, for Poisson processes.) To fit a non-stationary Gibbs process simply specify the formula argument appropriately.

For example the betacells data appear to be spatially inhomogeneous. We might model the trend for Beta by means of a log-cubic polynomial in the spatial coordinates x and y. We could fit a nonstationary multitype Strauss process having such a trend to Beta as follows:

```
> fitP <- ppm(Beta ~ marks + polynom(x,y,3), MultiStrauss(R))
> fitI <- ppm(Beta ~ marks * polynom(x,y,3), MultiStrauss(R))
```

We could have used harmonic instead of polynom. In the first model fitP the effect of marks is additive, so the first-order trends for each type of cell are proportional. The second model allows completely different first-order trends for the two types of cells.

14.8.7.3 Plotting fitted interactions

Rather than trying to understand the interaction in a fitted point process model by interpreting the fitted interaction parameters in a printout, it is more effective to *plot* the fitted interaction function.

The fitted interaction of a point process model can be extracted by the function fitin, explained in Section 13.5.9. The result is an object of class "fii" ('fitted interpoint interaction'). Methods for print, plot, etc. are provided for handling such objects.

```
> fitBeta <- ppm(Beta ~ marks, MultiStraussHard(iradii=R2))
```

```
> plot(fitin(fitBeta))
```

The result is shown in Figure 14.26. The function $h(r)$ in each panel is the (exponentiated) pairwise interaction between points of the given types. The diagonal panels suggest there is quite strong inhibition between points of the same type, while the off-diagonal panels show very little interaction between points of different type.

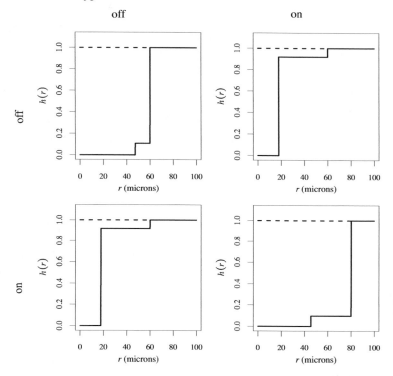

Figure 14.26. *Fitted interpoint interaction functions for a multitype Strauss-hard core model fitted to the* betacells *data marked by '*type*'.*

14.8.7.4 Simulating a Gibbs model

Any of the Gibbs models discussed here can be simulated in spatstat by the function rmh. It is particularly easy to simulate from a *fitted* Gibbs model, which is done using the method rmh.ppm. It may be more convenient to use simulate.ppm, particularly when simulating more than one pattern.

```
> Beta.sim <- rmh(fitBeta)
```

The result of this simulation is shown in the right panel of Figure 14.27.

It is also possible to specify a Gibbs model using rmhmodel.default and then simulate from that model using rmh. For example, the following code specifies a multitype Strauss-hard core model and generates one simulated realisation:

```
> HC <- matrix(c(20,10,10,20),2,2)
> R  <- matrix(c(30,20,20,30),2,2)
> G  <- matrix(0.3,2,2)
> B  <- c(5,10)*1e-5
> M  <- rmhmodel(cif="straushm",
                 par=list(beta=B,gamma=G,iradii=R,hradii=HC),
```

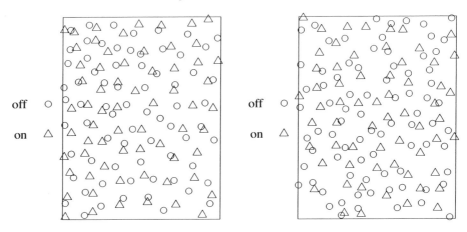

Figure 14.27. *Real (*Left*) and simulated (*Right*) versions of the* betacells *pattern.*

```
                       types=c("A","B"), w=square(1000))
   > X   <- rmh(M)
```

The function `rmhmodel` checks the validity of the model parameters, prepares internal data, and returns an object of class `"rmhmodel"` describing the model to be simulated. The command `rmh(M)` is dispatched to `rmh.rmhmodel` and ultimately to `rmh.default` to perform the simulation.

14.9   Hierarchical interactions

Högmander and Särkkä [337] introduced an alternative class of *hierarchical interaction* models for multitype point patterns.

14.9.1   Motivation

A statistical model is 'hierarchical' if it consists of several stages, each stage depending only on the outcome of the previous stages (or 'higher levels' of the hierarchy). Hierarchical models are particularly common in spatial statistics [191, 192, 77, 414] where the higher levels of the hierarchy usually represent phenomena occurring at larger spatial scales. A Cox point process **X** can be regarded as a hierarchical statistical model with two stages: the first stage or upper level is the generation of the random intensity function $\Lambda(u)$, and the second stage or lower level is the generation of the point process **X** which is Poisson with intensity function $\Lambda(u)$.

Hierarchical *interactions* in a multitype point process model [337, 293, 357] are slightly different in concept: the possible *types of points* are arranged in a hierarchical order, and the model postulates that points of a given type interact only with points of the same type or an earlier (higher) type.

Hierarchical interactions are natural models when the types have a natural ordering (such as age classes or size classes) or a natural asymmetry (such as predator and prey species).

Figure 14.28 shows the locations of 823 bramble cane (blackberry) plants classified according to age as either newly emergent, one year old, or two years old [352, 221, 235, 664]. It would make sense to assume that the spatial pattern of each new generation of plants depends on the pattern of existing plants from previous generations. The 2-year-old plants would be assumed to interact only with each other; the 1-year-old plants would interact with each other and also be influenced by the

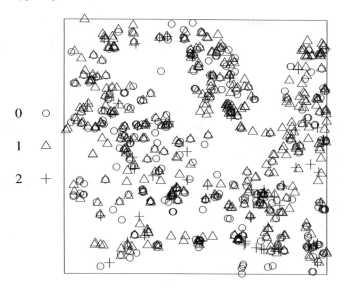

Figure 14.28. *Bramble canes data of Hutchings [352].*

2-year-old plants; the newly-emergent plants would interact with each other and be influenced by the 1-year-old and 2-year-old plants.

In the ants' nests data (Figure 1.7 on page 6) the ecological relationship of the two species is asymmetrical. *Cataglyphis* benefits from living close to populations of other ants, including *Messor*. *Messor* benefits from living close to plant food resources, and does not compete with *Cataglyphis* ants for these resources, nor does it need to avoid *Cataglyphis*.

When there is a natural ordering or asymmetry between types, the pairwise interaction models described in Section 14.8.3 are unsuitable, because they are inherently symmetric. This is the motivation for hierarchical interactions.

A hierarchical model could be imagined as arising one type at a time, with points of 'later' type being placed with reference to how they relate to or are influenced by points of 'earlier' type (higher level in the hierarchy). The influence of earlier types on later types induces non-stationarity or heterogeneity in the later types.

The assumed hierarchy is based on *prior knowledge* about the relationship between the types. It may not be possible to deduce the hierarchy from a single observed point pattern.

14.9.2 Hierarchical models of bivariate point processes

For simplicity we first explain the case where there are two types of points.

If X and Y are random variables, we can *always* write the probability of the outcome (X, Y) as

$$P(X, Y) = P(X)P(Y \mid X) \qquad (14.72)$$

where $P(X)$ is the 'marginal' probability of the outcome X alone, and $P(Y \mid X)$ is the conditional probability of Y given X. In situations where the random variable X somehow 'precedes' the random variable Y, it may be natural to build statistical models by writing down equations for $P(X)$ and $P(Y \mid X)$ because these may have a simple form. Hierarchical modelling is an application of this approach.

Suppose we have a multitype point process $\mathbf{Y}$ with two types of points, A and B, with A taking precedence. Take the 'multivariate' viewpoint by splitting the process into the two types of points, $\mathbf{Y} = (\mathbf{x}^{(A)}, \mathbf{x}^{(B)})$. Assuming the probability densities exist, we can *always* write the probability density of the process as

$$f(\mathbf{y}) = f_{A,B}(\mathbf{x}^{(A)}, \mathbf{x}^{(B)}) = f_A(\mathbf{x}^{(A)}) f_{B|A}(\mathbf{x}^{(B)} \mid \mathbf{x}^{(A)}). \tag{14.73}$$

A *hierarchical model* for $\mathbf{Y}$ (with respect to the hierarchy $A > B$) is a model expressed in the form of the right hand side of (14.73).

In theory, any bivariate point process model could be factorised in the hierarchical form (14.73). In practice, a hierarchical model is one in which the densities $f_A(\mathbf{x}^{(A)})$ and $f_{B|A}(\mathbf{x}^{(B)} \mid \mathbf{x}^{(A)})$ are easy to understand and to compute. A hierarchical modelling approach is advantageous when the conditional distributions are easy to specify, which happens if there is a natural order of precedence or causality.

Assuming that the required probability density functions exist, in a hierarchical model [337, 293] the probability density of the process is defined to be the product of the conditional probability densities for each type, given the types which are above it in the hierarchy. See also [189].

The multitype pairwise interaction models described in Section 14.8.3 can theoretically be factorised in the hierarchical form (14.73), but the marginal probability density $f_A(\mathbf{x}^{(A)})$ is not a pairwise interaction, and may be very complicated, as we saw on page 614.

14.9.3   Hierarchical pairwise interaction for two types

We will now define a hierarchical counterpart of the multitype pairwise interaction (14.64). Take f_A to be the density of a pairwise interaction process,

$$f_A(\mathbf{x}_A) = \alpha_A \prod_{x_i \in \mathbf{x}_A} \beta_A(x_i) \prod_{x_i, x_j \in \mathbf{x}_A} c_{AA}(x_i, x_j) \tag{14.74}$$

where α_A is a normalising constant, $\beta_A(u)$ is the first-order trend for points of type A, and $c_{AA}(u, v)$ is the pairwise interaction between pairs of points which are both of type A. Next, take the conditional density $f_{B|A}$ to be

$$f_{B|A}(\mathbf{x}_B \mid \mathbf{x}_A) = \alpha_{B|A}(\mathbf{x}_A) \prod_{x_i \in \mathbf{x}_B} \beta_B(x_i) \prod_{x_i \in \mathbf{x}_A, x_j \in \mathbf{x}_B} c_{AB}(x_i, x_j) \prod_{x_i, x_j \in \mathbf{x}_B} c_{BB}(x_i, x_j) \tag{14.75}$$

where $\alpha_{B|A}(\mathbf{x}_A)$ is a normalising constant which depends on $\mathbf{x}_A$. Here $\beta_B(u)$ is a first-order term for points of type B, $c_{AB}(u, v)$ reflects the influence of points of type A upon those of type B, while $c_{BB}(u, v)$ is a pairwise interaction between pairs of points of the same type B. Taking $\mathbf{x}_A$ to be fixed in (14.75), we recognise (14.75) as an instance of the general pairwise interaction density (see Section 13.6) with interaction $c_{BB}(u, v)$ and first-order term

$$\beta_{B|A}(u) = \beta_{B|A}(u \mid \mathbf{x}_A) = \beta_B(u) \prod_{x_i \in \mathbf{x}_A} c_{AB}(x_i, u). \tag{14.76}$$

Thus we are assuming that (1) the marginal distribution of $\mathbf{x}_A$ is a pairwise interaction process, and (2) the conditional distribution of $\mathbf{x}_B$ given $\mathbf{x}_A$ is a pairwise interaction process. Note that **neither of these statements is true for the multitype pairwise interaction** (14.64).

Multiplying (14.74) by (14.75), we see that the joint density of the hierarchical pairwise model is

$$f(\mathbf{x}_A, \mathbf{x}_B) = \alpha_A \alpha_{B|A}(\mathbf{x}_A) \left[\prod_{x_i \in \mathbf{x}_A} \beta_A(x_i) \prod_{x_i \in \mathbf{x}_B} \beta_B(x_i) \right] \times$$

$$\left[\prod_{x_i, x_j \in \mathbf{x}_A} c_{AA}(x_i, x_j) \prod_{x_i \in \mathbf{x}_A, x_j \in \mathbf{x}_B} c_{AB}(x_i, x_j) \prod_{x_i, x_j \in \mathbf{x}_B} c_{BB}(x_i, x_j) \right] \tag{14.77}$$

which differs from the multitype pairwise interaction (14.64) by virtue of the presence of a normalising factor $\alpha_{B|A}(\mathbf{x}_A)$.

In applications of this model it is important to understand that the model does *not* mean that points of type A 'do not depend on' points of type B. The two types are not independent in a statistical sense: information about the locations of points of type B does affect our knowledge about points of type A. They are no more independent than 'ordinary' random variables X and Y would be if we wrote their joint probability density function as $f(x,y) = f(x)f(y \mid x)$. Rather, we have constructed the model in such a way that we do not have to explicitly describe the dependence of points of type A on points of type B.

14.9.4   Hierarchical Strauss interaction for two types

Here we spell out the hierarchical version of the (stationary) multitype Strauss process for two types of points. This is the special case of (14.77) where $\beta_A(u) \equiv \beta_A$ and $\beta_B(u) \equiv \beta_B$ are constant first-order terms, and interaction or influence is determined by threshold functions:

$$c_{AA}(u,v) = \begin{cases} \gamma_{AA} & \text{if } \|u-v\| \le r_{AA} \\ 1 & \text{otherwise} \end{cases}$$

$$c_{AB}(u,v) = \begin{cases} \gamma_{AB} & \text{if } \|u-v\| \le r_{AB} \\ 1 & \text{otherwise} \end{cases}$$

$$c_{BB}(u,v) = \begin{cases} \gamma_{BB} & \text{if } \|u-v\| \le r_{BB} \\ 1 & \text{otherwise.} \end{cases}$$

Then (14.74) is

$$f_A(\mathbf{x}_A) = \alpha_A \beta_A^{n(\mathbf{x}_A)} \gamma_{AA}^{s(\mathbf{x}_A, r_{AA})} \tag{14.78}$$

where $s(\mathbf{x},r)$ denotes the number of pairs of points of $\mathbf{x}$ which are closer than r units apart. The marginal distribution of $\mathbf{x}_A$ is a homogeneous Strauss process with parameters $\beta_A, \gamma_{AA}, r_{AA}$. Meanwhile (14.75) is

$$f_B(\mathbf{x}_B \mid \mathbf{x}_A) = \alpha_{B|A}(\mathbf{x}_A) \beta_B^{n(\mathbf{x}_B)} \gamma_{AB}^{s(\mathbf{x}_A, \mathbf{x}_B, r_{AB})} \gamma_{BB}^{s(\mathbf{x}_B, r_{BB})} \tag{14.79}$$

where $s(\mathbf{x},\mathbf{y},r)$ is the number of pairs (x_i, y_j) with $x_i \in \mathbf{x}$ and $y_j \in \mathbf{y}$ which are closer than r units apart. The conditional distribution of $\mathbf{x}_B$ given $\mathbf{x}_A$ is an *inhomogeneous* Strauss process with interaction strength γ_{BB}, interaction range r_{BB}, and first-order trend $\beta_{B|A}(u \mid \mathbf{x}_A) = \beta_B \gamma_{AB}^{s(u, \mathbf{x}_A, r_{AB})}$.

A very important point to note is that, in the hierarchical Strauss model, the parameter γ_{AB} can take values greater than 1, indicating an *attractive* influence of points of type A on points of type B. The 'interaction' parameter between two different types is really an 'influence' parameter — effectively a parameter of the trend for the points of type B given the points of type A. The parameters γ_{AA} and γ_{BB} still have the interpretation of interaction parameters, and must be less than or equal to 1.

14.9.5   Hierarchical interactions for several types

Now suppose there are M different types of points, numbered according to the hierarchy, so that type 1 is the top level of the hierarchy and type M is the bottom level. A hierarchical model expresses the probability density in the form

$$f(\mathbf{y}) = f_1(\mathbf{x}^{(1)}) f_{2|1}(\mathbf{x}^{(2)} \mid \mathbf{x}^{(1)}) f_{3|1,2}(\mathbf{x}^{(3)} \mid \mathbf{x}^{(1)}, \mathbf{x}^{(2)}) \cdots f_{M|1,2,\dots,M-1}(\mathbf{x}^{(M)} \mid \mathbf{x}^{(1)}, \dots, \mathbf{x}^{(M-1)}) \tag{14.80}$$

so that the points of type m are modelled by their dependence on the patterns of all points of types $1, 2, \dots, m-1$.

In a hierarchical *pairwise* interaction model, each of the probability densities on the right hand side of (14.80) is a pairwise interaction density. Each point x of type m contributes a factor $\beta_m(x)$ to the overall density, and each pair of points x, x' of types m and m' contributes a factor $c_{m,m'}(x, x')$ to the overall density. The conditional density $f_{m|1,2,...,m-1}$ for the pattern of points of a particular type m, conditional on the patterns of points of type $1, 2, \ldots, m-1$, collects together all the terms $\beta_m(x)$ for points of that type m, all the terms $c_{m,m'}(x, x')$ for points x, x' of types m, m' **where** m' **equals or precedes** m, and a normalising constant $\alpha_m(\mathbf{x}^{(1)}, \ldots, \mathbf{x}^{(m-1)})$ that depends on the patterns of the preceding types of points.

14.9.6 Fitting hierarchical models in `spatstat`

Hierarchical models can be fitted to multitype point pattern data using ppm, by specifying an appropriate interaction object. In this case ppm maximises the hierarchical pseudolikelihood, the product of the pseudolikelihoods corresponding to each term in the factorisation (14.73) or (14.80).

In the near future, `spatstat` will support many hierarchical interaction models. At the time of writing, the only examples are HierStrauss, HierHard, and HierStraussHard. These create interaction objects that represent the hierarchical Strauss, hierarchical hard core, and hierarchical Strauss-hard core interactions, respectively. The argument syntax is

```
HierStrauss(radii, types, archy)
HierHard(hradii, types, archy)
HierStraussHard(iradii, hradii, types, archy)
```

where types is the vector of possible types, hradii is the matrix of hard core radii, radii or iradii is the matrix of interaction radii, and the new argument archy is a character vector listing the types in their hierarchical order, or an index vector indexing the types in their hierarchical order.

For illustration, we fit the hierarchical Strauss model to the ants' nests data. Note that the levels of the marks in the ants dataset are in alphabetical order: Cataglyphis, Messor. By default, the first level Cataglyphis would be interpreted as the type of higher order in a hierarchical model with Messor having the lower order. This is the opposite of the desired hierarchy. We could make Messor the first level using relevel(ants, "Messor"), or rearrange the levels using factor(...,levels=c("Messor","Cataglyphis")). However, it is easier and safer to use the archy argument of HierStrauss. We will also shorten the names of the levels to shorten the output.

```
> Ants <- ants
> levels(marks(Ants)) <- c("Cat", "Mess")
> rmat <- matrix(c(65,45,45,30), nrow=2, ncol=2)
> fit <- ppm(Ants ~ marks, HierStrauss(rmat, archy=c(2,1)))
> fit
Stationary Hierarchical Strauss process
Possible marks: 'Cat' and 'Mess'
Log trend:  ~marks
First order terms:
 beta_Cat beta_Mess
8.122e-05 2.825e-04
Hierarchy: Mess ~> Cat
Interaction radii:
     Cat Mess
Cat  65
Mess 45   30
Fitted interaction parameters gamma_ij
     Cat    Mess
```

```
Cat  0.2269
Mess 1.3467 0.1382

> coef(summary(fit))
                Estimate   S.E.  CI95.lo CI95.hi Ztest    Zval
(Intercept)     -9.4184   1.3422 -12.049 -6.7876  ***  -7.0169
marksMess        1.2465   1.4107  -1.518  4.0114        0.8836
markCatxCat     -1.4831   1.0726  -3.585  0.6191       -1.3827
markMessxCat     0.2977   0.7749  -1.221  1.8164        0.3842
markMessxMess   -1.9790   0.7583  -3.465 -0.4927   **  -2.6097
```

The radii for the influence matrix in the foregoing code were chosen by means of (a rough) profile pseudolikelihood:

```
> rseq    <- seq(20, 70, by=5)
> s       <- expand.grid(r11=rseq, r12=rseq, r22=rseq)
> pfn <- function(r11, r12, r22) {
      HierStrauss(matrix(c(r11,r12,r12,r22),2,2), archy=2:1)
  }
> prf <- profilepl(s, pfn, ants ~ marks, correction="translate")
```

The fit produced by the foregoing code indicates moderate repulsion between *Cataglyphis* nests ($\widehat{\gamma}_{CC} = 0.2269$) and strong repulsion between *Messor* nests ($\widehat{\gamma}_{MM} = 0.1382$); the *Messor* nests are estimated to attract the *Cataglyphis* nests fairly weakly ($\widehat{\gamma}_{MC} = 1.347$).

However, the printout shows that the standard errors for the coefficients (the logarithms of the interaction parameters γ) in the fitted model are quite large, and the corresponding confidence intervals are wide. If the normal approximation can be trusted, then only the *Messor* by *Messor* interaction is significant. A more reliable test is the adjusted composite likelihood ratio test [56]:

```
> fit0 <- update(fit, HierStrauss(diag(diag(rmat))))
> anova(fit0, fit, test="LR")
Analysis of Deviance Table
Model 1: ~marks          HierStrauss
Model 2: ~marks          HierStrauss
  Npar Df AdjDeviance Pr(>Chi)
1    4
2    5  1        1.43     0.23
```

The idiom `diag(diag(M))` sets all the off-diagonal entries to zero, so `fit0` is the model where the two species of ants are independent Strauss processes. This null hypothesis is not rejected.

There is thus no confirmation of the expected attraction of *Cataglyphis* nests to *Messor* nests. We re-emphasize here a point made earlier (at the end of Section 14.9.4) to the effect that the hierarchical Strauss model *is* capable of modelling attractive influences. That is, obtaining an estimated value of γ_{MC} equal to 1.347 is indeed meaningful, although in this case it was not deemed statistically significant.

14.10 Multitype Cox and cluster processes

Multitype versions of Cox processes and Neyman-Scott cluster processes were envisaged by Cox [175] and Neyman and Scott [499], and by Matérn [452, 454], and have been studied in subsequent literature [235, 132, 131, 649], [197, chap. 8], [484, section 5.8].

At the time of writing, `spatstat` provides code for simulating such processes, but not yet for fitting these models to point pattern data. We expect this to change soon. The `lgcp` package [650, 651] can fit multitype log-Gaussian Cox process models to multitype point pattern data.

14.10.1 Multitype Cox processes

A *multitype Cox process* is effectively a multitype Poisson process whose driving intensity is a random function $\Lambda(u, m)$. If there are M possible marks, a multitype Cox process is equivalent to M Cox processes in $\mathbb{R}^2$ with random driving intensity functions $\Lambda_1(u), \ldots, \Lambda_M(u)$ which may be dependent or independent [484, section 5.8]. A multitype Cox process has independent components if and only if the random driving intensities are independent.

Models proposed in [235] include *linked Cox processes* where the random driving intensities are proportional to each other, say $\Lambda_m(u) = p_m \Lambda_{\bullet}(u)$ where $\Lambda_{\bullet}(u)$ is a random function and p_m is a nonrandom constant for each m; and *balanced Cox processes* where $\sum_m \Lambda(u) = \beta$ is constant. A multitype Cox process has the random labelling property if and only if it is a linked Cox process. See [484, section 5.8] for further theoretical results.

A *multitype log-Gaussian Cox process* is one in which $(\log \Lambda_1(u), \ldots, \log \Lambda_M(u))$ is a vector-valued Gaussian random field. For each location u, the log intensities of the m types have an M-dimensional Gaussian joint distribution. For each pair of locations u and v, the log intensities of the M types at the two locations have a $2M$-dimensional Gaussian joint distribution, and so on for any number of locations. A multitype LGCP has independent components if and only if $\Lambda_1(u), \ldots, \Lambda_M(u)$ are independent random functions. A multitype LGCP does *not* have the random labelling property unless we allow degenerate Gaussian distributions. A multitype LGCP cannot be a linked Cox or balanced Cox process.

Tools in `spatstat` can be used to simulate such models by hand. Figure 14.29 shows examples of linked Cox and balanced Cox processes, generated by

```
> P <- dirichlet(runifpoint(10))
> logLambda <- rMosaicField(P, rnorm,
                            dimyx=512, rgenargs=list(mean=4, sd=1))
> Lambda <- exp(logLambda)
> X <- rpoispp(Lambda)
> Xlinked <- X %mark% factor(sample(c("a","b"), npoints(X),
                            replace=TRUE, prob=c(2,3)/5))
> Y <- rpoispp(100)
> Xbalanced <- Y %mark% factor(ifelse(Lambda[Y] > exp(4), "a", "b"))
```

and between them a multitype log-Gaussian Cox process, generated by

```
> X1 <- rLGCP("exp", 4, var=0.2, scale=.1)
> Z1 <- log(attr(X1, "Lambda"))
> Z2 <- log(attr(rLGCP("exp", 4, var=0.1, scale=.1), "Lambda"))
> Lam1 <- exp(Z1)
> Lam2 <- exp((Z1 + Z2)/2)
> Xlgcp <- superimpose(a=rpoispp(Lam1), b=rpoispp(Lam2))
```

If $\mathbf{Y}$ is a multitype Cox process, then the unmarked process $\mathbf{X}^{(\bullet)}$ is a Cox process. However, if $\mathbf{Y}$ is a multivariate LGCP, then $\mathbf{X}^{(\bullet)}$ is **not** a LGCP, even if the model has independent components, because a sum of log-Gaussian variables is not log-Gaussian in general.

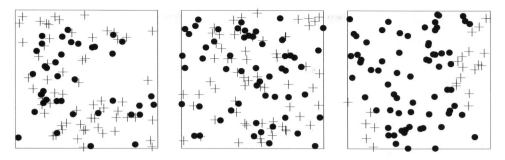

Figure 14.29. *Simulated realisations of (*Left*) linked Cox process, (*Middle*) multitype log-Gaussian Cox process, and (*Right*) balanced Cox process.*

14.10.2 Multitype cluster processes

Similarly a *multitype cluster process* is a cluster process in $\mathbb{R}^2 \times \mathcal{M}$ where $\mathcal{M} = \{1, 2, \ldots, M\}$. That is, a number of parent points are first generated in $\mathbb{R}^2 \times \mathcal{M}$ and then for each parent a cluster of 'offspring' points (in some sense 'centred' on the parent) is generated in $\mathbb{R}^2 \times \mathcal{M}$. The resulting realization of the process is the union or superposition of all of the offspring points. The parents 'disappear', or are 'replaced' by their offspring. It is usually assumed that the parent points form a Poisson process. Each cluster of offspring points is a finite set of multitype points following some stochastic mechanism.

The overlap between Cox processes and Neyman-Scott processes [484, section 5.3] persists into the multitype context. Figure 14.30 shows a realisation of a multitype version of Matérn's cluster process which is both a Neyman-Scott process and a Cox process. Each cluster contains a Poisson number of random points (with mean 5 points) uniformly distributed in a disc of radius $r = 0.1$. The points are independently marked as type 1 or 2 with equal probability. The example was generated as follows:

```
> ftcf <- function(x0, y0, radius, mu) {
    n <- rpois(1, mu)
    Y    <- runifdisc(n, radius=radius, centre=c(x0, y0))
    Z    <- factor(sample(1:2, npoints(Y), replace=TRUE), levels=1:2)
    return(Y %mark% Z)
  }
> Xmc2 <- rNeymanScott(15,0.1,ftcf, radius=0.1, mu=5)
```

This model has the random labelling property, but clearly does not have independent components: the presence of a point of one type implies that there is a cluster at that location, which increases the probability of presence of a point of the other type. This model is also a linked Cox process.

In general a multitype cluster process has its parent points chosen from $\mathbb{R}^2 \times \mathcal{M}$ as well as the offspring points. The mark of the parent point could come into play, for example, by having the type probabilities for the offspring depend on the type of the parent. The generating mechanism of the cluster could also depend on the type of the parent. These intricacies do not arise in the example in Figure 14.30.

Multitype cluster processes are considered in [132].

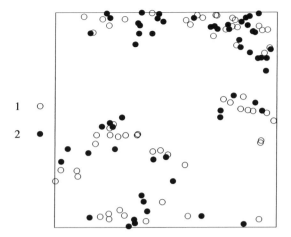

Figure 14.30. *Simulated realisation of multitype Neyman-Scott process which is also a multitype Cox process.*

14.11   Other multitype processes

14.11.1   Dependent thinning

One way to generate non-Poisson multitype point processes is to apply *dependent thinning* to a Poisson multitype point process (Section 5.5.2). Here 'dependent' means that the fates of different points (i.e. whether they are deleted or retained) are stochastically dependent. This can happen, for example, if the probability of retaining a point x_i is not a function $p(x_i)$ of its location only, but depends on other points x_j through a function $p(x_i \mid \mathbf{x})$.

The simplest examples of dependent thinning are those which apply a deterministic *thinning rule* that depends only on the configuration of points. Each point x_i is deleted or retained according to a rule $A(x_i \mid \mathbf{x})$ that gives a yes-or-no answer. Matérn's Model I and Model II are examples (Section 5.5.2). In the multitype case the thinning rule or thinning probability depends on the marks as well as the locations of the points.

The left panel in Figure 14.31 shows a multitype version of Matérn's Model I, obtained by generating a stationary multitype Poisson process, then deleting any point that lay closer than a critical distance r to a point of *different* type.

A slight modification to this model is a *hierarchical* version, in which we first generate points of type 1 according to a Poisson process, and then generate points of type 2 according to a Poisson process conditional on the requirement that no point of type 2 lies within a distance r of a point of type 1. An example is shown in the right panel of Figure 14.31. Code for these examples is available from www.spatstat.org.

Dependent thinning can be used to build models of non-overlapping spatial objects [440]. It can also serve as a model for 'noisy' observation of a true point process [436, 437].

14.11.2   Random field marking

In a *random field marking* model we start with a point process $\mathbf{X}$ in $\mathbb{R}^2$ and a random or non-random function $Z(u)$. We assume only that $\mathbf{X}$ and $Z(u)$ are independent. Then to each point u_i in $\mathbf{X}$ we

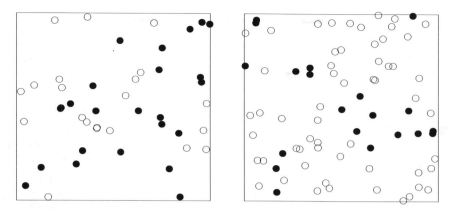

Figure 14.31. *Multitype versions of Matérn Model I.* Left: *simultaneous (annihilation of each type by the other).* Right: *hierarchical (annihilation of type 2 by type 1). Open circles: type 1. Filled circles: type 2.*

attach the mark $m_i = Z(u_i)$ given by the value of the random field at that location. The result is a marked point process **Y**. If the values of $Z(u)$ were categorical values, then **Y** is a multitype point process. The dependence structure of **X** is complicated, even if $Z(u)$ is a deterministic function. This model would be appropriate if the marks are values of a spatial function observed at random locations. Figure 14.32 shows an artificial example of random field marking.

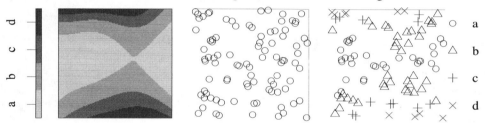

Figure 14.32. *Artificial example of a multitype pattern with marks determined by random field marking.* Left: *random field.* Middle: *point process.* Right: *point process with random field marking.*

More complicated models, where the point process and the random field may be dependent, are studied in [215, 214].

14.12 Theory*

Here we summarise basic statistical theory for multitype point processes. For details, see [31, 484, 355, 199].

---

* Starred sections contain advanced material, and can be skipped by most readers.

14.12.1 Multitype point patterns and point processes

Suppose $\mathscr{M}$ is the set of possible types, which we could take to be $\mathscr{M} = \{1, 2, \ldots, M\}$. A *multitype point pattern* in a window $W \subseteq \mathbb{R}^2$ is a finite unordered set

$$\mathbf{y} = \{(x_1, m_1), \ldots, (x_n, m_n)\}, \quad x_i \in W, \quad m_i \in \mathscr{M}$$

where the x_i are the locations and the m_i are the corresponding marks. In other words, for mathematical purposes, a mark is effectively treated as an extra coordinate. A marked point at location u in W, with mark m from $\mathscr{M}$, is treated as a point (u, m) in the space $W \times \mathscr{M}$. A *multitype point process* $\mathbf{Y}$ in window W with marks in $\mathscr{M}$ is simply defined as a point process in $W \times \mathscr{M}$.

The space $W \times \mathscr{M}$ can be thought of as a stack of M separate copies of W. Another way of thinking about it is that a multitype point pattern with locations in W is equivalent to M point patterns $\mathbf{x}^{(1)}, \ldots, \mathbf{x}^{(M)}$ in W, where $\mathbf{x}^{(m)}$ is the pattern of points with mark m.

Many characteristics of a multitype point process can be expressed by counting the number of points (u, m) whose spatial location u falls in a specified region A *and* whose mark m belongs to a specified set of marks B. If we view the multitype point pattern as a point pattern in $W \times \mathscr{M}$, then this count is the number of points (u, m) falling in the Cartesian product $A \times B$.

14.12.2 Intensity and relative risk

With a few exceptions, it is not necessary to introduce new mathematical definitions for fundamental properties of a multitype point process, such as the intensity measure, moment measures, Palm distribution (Section 8.12), and conditional intensity (Sections 13.3.3 and 13.12.4). Since a multitype point process is a special case of a point process, the existing definitions (see Chapter 6) are sufficient, with a few exceptions.

The *intensity measure* Λ of a multitype process $\mathbf{Y}$ on $\mathbb{R}^2$ with marks in $\mathscr{M}$ is defined by $\Lambda(S) = \mathbb{E}[n(\mathbf{Y} \cap S)]$ for all bounded sets $S \subset \mathbb{R}^2 \times \mathscr{M}$, provided that this expectation is always finite. The measure can be determined from the values $\Lambda(A \times B) = \mathbb{E}[n(\mathbf{Y} \cap (A \times B))]$ for all bounded sets $A \subset \mathbb{R}^2$ and all subsets $B \subseteq \mathscr{M}$.

Specifying the multitype intensity measure Λ is equivalent to specifying intensity measures Λ_m for each of the processes $\mathbf{X}^{(m)}$ of points with mark of type m, for each possible m. That is

$$\Lambda(S) = \sum_{m \in \mathscr{M}} \Lambda_m(S_m)$$

where $S_m = \{u \in \mathbb{R}^2 : (u, m) \in S\}$.

A multitype point process $\mathbf{Y}$ has *homogeneous intensity* if each subprocess $\mathbf{X}^{(m)}$ has homogeneous intensity. Equivalently,

$$\mathbb{E}[\mathbf{X}(A \times B)] = |A| \sum_{m \in B} \lambda_m \tag{14.81}$$

where $\lambda_1, \ldots, \lambda_M$ are the intensities of the processes of points of each type. Note that homogeneity refers only to the spatial distribution of points; it does *not* mean that points of different types are equally abundant. Points of type m represent a fraction $p_m = \lambda_m / \lambda_\bullet$ of the points in the pattern, on average, where $\lambda_\bullet = \sum_m \lambda_m$ is the total intensity.

A multitype point process may have an *intensity function* and for practical purposes this will usually be assumed. For an unmarked point process, the intensity function $\lambda(u)$ is defined to satisfy (6.3). For a multitype point process, the intensity function is a function $\lambda(u, m)$ such that

$$\mathbb{E}[\mathbf{X}(A \times B)] = \int_A \sum_{m \in B} \lambda(u, m) \, du \tag{14.82}$$

for all bounded sets $A \subset \mathbb{R}^2$ and all $B \subseteq \mathscr{M}$.

Note that, in making the definition (14.82), we have effectively assigned equal weight to all possible types m. This is mathematically convenient because (14.82) will reduce to (14.81) when $\lambda(u,m) = \lambda_m$ does not depend on location u. However it requires care when interpreting the results of a fitted model.

14.12.3 Stationarity

A multitype point process is defined to be *stationary* if its distribution is unaffected by shifting *the locations* x_i.

Definition 14.1. *A multitype point process in* $\mathbb{R}^2$ *with marks in* $\mathcal{M}$ *is* stationary *if its distribution is invariant under translations of* $\mathbb{R}^2$, *that is, under transformations* $(u,m) \mapsto (u+v,m)$ *where v is any vector in* $\mathbb{R}^2$.

Under the transformation $(u,m) \mapsto (u+v,m)$ a multitype point pattern simply has its locations shifted by the vector v, with the marks unchanged.

It may help to adopt the multivariate viewpoint (see Section 14.2.5) in which the process $\mathbf{Y}$ consists of M subprocesses $\mathbf{X}^{(1)}, \ldots, \mathbf{X}^{(M)}$ corresponding to the respective types. Stationarity of the multivariate process implies that the subprocesses $\mathbf{X}^{(i)}$ are stationary but additionally implies that they are *jointly stationary*, in the sense that $(\mathbf{X}^{(1)} + v, \mathbf{X}^{(2)} + v, \ldots, \mathbf{X}^{(M)} + v)$ has the same joint probability distribution as $(\mathbf{X}^{(1)}, \ldots, \mathbf{X}^{(M)})$ for any translation vector v.

An important technical result is that, if $\mathbf{Y}$ is a stationary multitype point process, then it has homogeneous intensity in the sense of (14.81).

14.12.4 Likelihoods and pseudolikelihoods

For a multitype Poisson process with intensity function $\lambda(u,m)$ the loglikelihood of a (multitype) pattern $\mathbf{y} = \{(u_1,m_1), \ldots, (u_n,m_n)\}$ is, up to a constant,

$$\log L = \sum_{i=1}^{n} \log \lambda(u_i,m_i) - \sum_{m \in \mathcal{M}} \int_W \lambda(u,m) \, \mathrm{d}u, \tag{14.83}$$

where m_i is the mark attached to data point u_i.

For a multitype Gibbs point process with conditional intensity $\lambda((u,m);\mathbf{y})$, the log pseudolikelihood is

$$\log \mathrm{PL} = \sum_{i=1}^{n(\mathbf{y})} \log \lambda((u_i,m_i);\mathbf{y}) - \sum_{m \in \mathcal{M}} \int_W \lambda((u,m);\mathbf{y}) \, \mathrm{d}u \tag{14.84}$$

where m_i is the mark attached to location u_i,

The two expressions (14.83) and (14.84) have the same functional form. They can be approximated using the Berman-Turner device (Sections 9.8.1 and 13.13.5) applied to the product space $W \times \mathcal{M}$. That is, in addition to the data points (x_i,m_i) we create a set of dummy points (x'_j,m'_j), combine the two sets of points into a single set of marked points, and approximate the integrals in (14.83) and (14.84) by finite sums over these marked points. See [50, 478, 290] for details.

14.13 FAQs

- *I am confused about the difference between marks and covariates.*

  A covariate is an explanatory variable, used to predict the outcome of the experiment — in this context, to predict the abundance of points. We think of the covariate as existing 'before' the points came into existence. A covariate can, in principle, be measured at any spatial location.

Marks are attributes of the points. In most cases they only have a meaning with reference to the points. For example, species and diameter are attributes of each tree in a forest. It makes no sense to measure species or diameter at an arbitrary location where there is no tree. Furthermore, a mark is part of the 'outcome' of the experiment, and is something to be predicted using the covariates.

- *If so, then why is 'random field marking' allowed, where the mark is the value of a spatial covariate at that point?*

We can declare an 'attribute' to be anything. The terrain slope on which a tree is growing can be considered an 'attribute' of the tree. However, this changes the role of the slope variable from an explanatory covariate to a part of the 'response'.

- *How can I manipulate the levels of a factor?*

To rearrange the ordering of the levels, use `relevel` or `reorder`. To change the names of the levels, just reassign them using `levels<-`. To merge several levels into one, use the `spatstat` function `mergeLevels`.

- *How can I reduce the white space around a plot of a* `fasp` *object, or bring the outer labels closer to the plot?*

Tweak the base R graphics parameters documented in `help(par)`. Try `par(mar=c(4,4,0,0))`. See `help(plot.fasp)` for other arguments controlling the plot.

- *Can I fit a multitype version of the Geyer saturation model or the area-interaction model to my data?*

At the time of writing, this is not implemented in `spatstat`. Consult `www.spatstat.org` for updates.

- *I have data from a forest survey with 42 different tree species recorded. I want to fit a multitype Strauss model. How do I determine all the interaction radii for the model?*

This will be difficult, as there are $42 \times 41/2 + 42 = 903$ radii to determine, and presumably some of these parameters will be poorly determined. It may be wiser to fit a simpler model in which (for example) there is one interaction radius between trees of the same type, and another interaction radius between points of different type. The radii can then be estimated by maximum profile likelihood:

```
nt <- length(levels(marks(X)))
Mf <- function(rwithin, rbetween) {
  R <- matrix(rbetween, nt, nt)
  diag(R) <- rwithin
  return(MultiStrauss(R))
}
df <- expand.grid(rwithin=seq(rmin, rmax, length=5),
                  rbetween=seq(rmin, rmax, length=5))
profilepl(df, Mf, X ~ trend)
```

One could also consider functional principal components analysis of $K_{ij}(r)$ [354], [355, sec. 4.9].

- *I am studying the distribution of discrete genetic traits in an ancient cemetery, with the working hypothesis that the graves are clustered by family. How should I test the null hypothesis of random labelling?*

It's unclear whether point process methods are appropriate here, because the graves could be considered as fixed locations. Testing for random labelling would usually imply that we condition on the locations, thereby treating them as if they were fixed. More appropriate techniques are those from spatial neighbourhood analysis. Joint count statistics, such as Moran's I, seem the most appropriate. See [111] and the package `spdep`.

- *I am studying tree mortality in a spatially inhomogeneous forest. After observing the same forest at two different times, I have assembled a dataset showing the locations of all trees that were alive at time 1, marked by a label indicating whether they were still alive at time 2. To test for random mortality I decided to subtract the L-function of all trees (trees alive at time 1) from the L-function of trees still alive at time 2. This is supposed to account for the initial heterogeneity. Is this correct?*

In general the difference $L_1(r) - L_2(r)$, between the L-functions of two point patterns, does not indicate the difference in clustering between the two patterns; an observed difference could simply be explained by a difference in spatial inhomogeneity. To adjust for heterogeneity, instead of subtracting, we should *divide* by the spatially varying intensity, and this is the idea of the inhomogeneous K- and L-functions.

However, in this case, the second point pattern is a subset of the first point pattern, and under the hypothesis of completely random mortality, the expected K-function of the trees at time 2 is equal to the K-function of the trees at time 1, due to the thinning property of the K-function. It does therefore make sense to subtract the two L-functions: the difference should be approximately zero under the random mortality hypothesis.

However, it is unclear how to interpret the difference between the two L-functions if it is found to be significant.

If a Monte Carlo test (envelope test or DCLF test) is performed, it is not necessary to subtract the L-function of the initial pattern at all. The envelope procedure generates a scatter of L-functions obtained from the null hypothesis of random mortality, and these simulated functions automatically include the effect of heterogeneity in the initial pattern. That is, the Monte Carlo test automatically deals with heterogeneity.

Finally the answer depends on the alternative hypothesis. If we suspect that tree deaths are clustered (perhaps due to infectious disease) then the L-function is appropriate. If we are looking for spatially varying risk of tree death (perhaps due to underlying hydrology) it would be better to estimate the spatially varying risk of tree mortality (Section 14.4.3) using `relrisk` or the `sparr` package. If covariates are available then a point process model could be used [684, 145].

Part IV

ADDITIONAL STRUCTURE

15

Higher-Dimensional Spaces and Marks

Part IV covers point patterns with additional 'structure', starting in this chapter with point patterns in higher-dimensional space and space-time, and multivariate marks.

15.1 Introduction

This book has explained the principles and techniques of spatial statistics in the simplest case of a point pattern in two-dimensional space, with no marks or with relatively simple marks. The same principles and techniques can be extended, with some modification, to higher dimensional space and to more complicated kinds of marks.

Point patterns in three-dimensional space, and point patterns in space-time, are becoming a major focus of attention [192, 226] because natural processes occur in three-dimensional space and occur over time. Three-dimensional geometry is particularly important in astronomy, geology, and anatomy. Point patterns of events in space-time are important in epidemiology (e.g., for disease surveillance), seismology, anatomy (e.g., video microscopy), and increasingly in astronomy where new telescopes make it possible to observe transient events across a wide survey volume. There are very important methodological differences between space and space-time, because time has a natural ordering [192, 226].

Marked point patterns were introduced in Sections 1.1.3 and 3.3.2. The special case of categorical marks was covered in Chapter 14. In other cases the mark could be a numerical variable, such as a count, a calendar date or clock time, or a compass direction, There could be several 'mark variables': for each tree in a forest we might record the species, diameter at breast height, number of parasitic insects, a chemical assay of the leaves, and so on. These variables can be regarded as a multivariate value attached to the tree (not to be confused with the multivariate viewpoint on multi-type patterns). Indeed the mark attached to each point may be of any nature: think of an interactive map where clicking on a point will summon a page of information about that point. For example, the mark could be a function (such as the spectral signature of light from a galaxy, attached to the galaxy's sky position), a geometric shape or pixel image (such as an image of each galaxy), another point pattern, and so on. The spatstat data formats allow marks of any kind.

A numerical mark is like an extra spatial coordinate. Indeed the distinction between a numerical mark and a spatial coordinate is sometimes blurred. In many three-dimensional point patterns, one of the spatial coordinates is qualitatively different from the other two coordinates, and might be treated either as a coordinate or as a mark. An earthquake's hypocentre depth is much more difficult to measure than its epicentre longitude and latitude. If the depth estimate is very inaccurate, we might be better advised to treat it as a mark rather than a spatial coordinate. In radioastronomy the redshift of an object is related to, but not equivalent to, its distance from Earth. Redshift can be treated as a spatial coordinate. However, the distance of an object is also related to its age, which might be better treated as a mark.

15.2   Point patterns with numerical or multidimensional marks

First we consider two-dimensional point patterns where the mark attached to each point is a numerical value, or several numerical values.

15.2.1   Data handling and plotting

For a two-dimensional point pattern X of class "ppp", the marks component marks(X) may be null, or a factor, a numeric vector, or a data frame. Different columns of the data frame represent different 'mark variables' recorded. There is currently no mechanism in spatstat for handling the units in which the mark variables are expressed: unitname(X) specifies the scale of the spatial coordinates only.

A marked point pattern can be created from raw data using ppp with the argument marks to specify the mark values (Section 3.3). Marks can be attached to an existing point pattern object X by marks(X) <- value or Y <- X %mark% value (Section 3.3.2). The coordinates and marks can be extracted together as a data frame by as.data.frame(X), for example for printing to a file.

Figure 15.1 shows the dataset spruces giving the locations and sizes of Norwegian spruce (*Picea abies*) trees in a natural forest stand in Saxony, Germany [256]. Each tree is marked with its diameter at breast height (*dbh*) so that marks(spruces) is a numeric vector.

The standard presentation of the data in Figure 15.1 was generated by plot(spruces). The (rescaled) mark value is rendered as the diameter of a circle centred at the data point. This is appropriate for the spruces data because the marks are indeed physical diameters. By default the marks are scaled so that the variation in mark values is visible without too much overlap between circles. In the spruces dataset, according to help(spruces) the spatial coordinates and marks are all expressed in metres, so the diameters could be rendered at the correct physical scale by specifying markscale=1. Figure 15.1 was instead plotted using markscale=4.

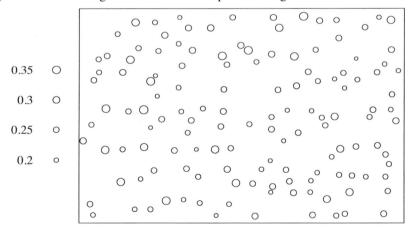

Figure 15.1. *Spruce trees in a* 56 × 38 *metre sampling region, showing tree diameters at breast height. Diameters inflated by a factor of 4 relative to the scale of spatial coordinates by setting* markscale=4.

In other cases, numerical marks should be presented as circles of *area* proportional to the mark value, because the visual weight of a geometric figure is related to its area rather than its diameter [657]. For example, this would apply in a wildlife survey where the location of each bird nest is marked with the number of eggs. It can be achieved by specifying size=function(x){sqrt(x)} or similar, in the `plot` command or as an argument to `symbolmap`. See Section 4.1.2 for further information.

Figure 15.2 shows the Finnish Pines data `finpines` giving the locations of 126 pine saplings in a Finnish forest, marked by their heights and their diameters at breast height. Here `marks(finpines)` is a data frame with two columns named `diameter` and `height`. Figure 15.2 shows the standard presentation produced by `plot(finpines)`, in which the left panel shows the diameters and the right panel the heights. Coordinates and heights are in metres (rounded to the nearest 0.1 metre); diameters are in centimetres (rounded to the nearest centimetre). Specifying `markscale=0.1` reduces the heights by a factor of 10 and inflates the diameters by a factor of 10.

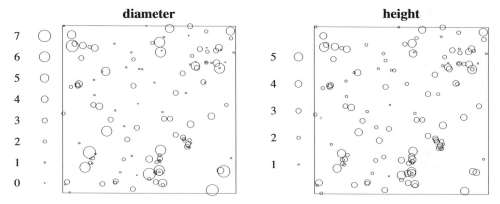

Figure 15.2. *Finnish Pine saplings marked by their diameters at breast height (*Left*) and heights (*Right*). Diameters inflated by a factor of 10, and heights deflated by 1/10, relative to spatial scale. Survey region is* 10×10 *metres.*

See Chapters 3 and 4 for further information about manipulating the marks in a marked point pattern.

15.2.2 Exploratory tools

15.2.2.1 Mark distribution

For a point pattern with marks that are a numeric vector or a data frame containing several columns of numeric values, the mark values can be extracted using the `marks` function and inspected using standard tools for data analysis. Common practice is to inspect each column of marks using the histogram or kernel density estimate, and to compare columns of marks using scatterplots and other standard techniques.

Figure 15.3 shows histograms of the *dbh* values for the Norwegian Spruces and the Longleaf Pines, generated by `hist(marks(spruces))` and `hist(marks(longleaf))`. The histogram for Longleaf Pines strongly suggests there may be two sub-populations of ages.

For multivariate marks, histograms of all mark variables can be generated by an idiom like

```
> Hp <- anylapply(marks(finpines), hist, plot=FALSE)
> plot(Hp)
```

and a scatterplot of each pair of mark variables can be generated using `pairs`:

```
> pairs(marks(finpines))
```

and one could try

```
> pairs(marks(finpines), diag.panel=panel.histogram)
> with(marks(finpines), plot(diameter ~ height))
```

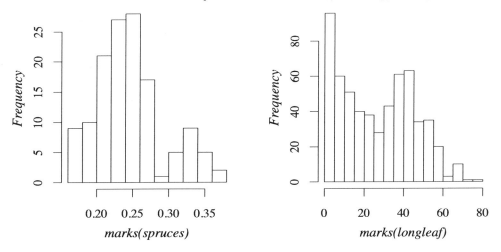

Figure 15.3. *Histograms of dbh values for the Spruces (*Left*) and Longleaf Pines (*Right*).*

Figure 15.4 shows the scatterplot of diameter against height for the Finnish Pines, generated by the last command above. The plot shows immediately that heights and diameters are roughly proportional, but with much more severe discretisation of the diameters.

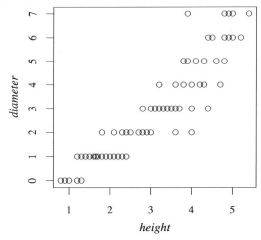

Figure 15.4. *Scatterplot of dbh against height for the Finnish Pines.*

For exploratory purposes it can be helpful to reduce a multivariate mark to a single numerical value. For example

```
> vols <- with(marks(finpines), (pi/12) * height * diameter^2)
> X <- finpines %mark% vols
```

replaces the marks of `finpines` by estimates of the tree volume, treating the tree as a circular cone. Alternatively one could calculate the residuals from a linear or nonlinear regression, or use principal components analysis or other dimension-reduction techniques available in R. See [670] or the functions `princomp` and `cancor` from the base package `stats`.

15.2.2.2 Spatial trend in marks

Spatial trend in the marks — defined as spatial inhomogeneity of the mean mark value or the distribution of marks — can be explored using many tools.

A scatterplot of the mark values against one of the spatial coordinates is often useful. Typing

```
> pairs(as.data.frame(longleaf))
```

generates an array of scatterplots (not shown) for the marks and spatial coordinates of the Longleaf Pines data, strongly suggesting inhomogeneity in the spatial distribution of marks. To supplement the scatterplots with a nonlinear smoothing curve, add the argument `panel=panel.smooth` to the command above. For a point pattern with m columns of marks, the scatterplot will have $m+2$ rows and columns of panels, comparing each mark variable with other mark variables and coordinates.

Converting numerical marks to discrete categories may be useful for exploratory purposes, and is also sensible when the marks only take a few possible values (for example, the count of eggs in each bird nest). As explained in Section 4.2.4, the command `cut` can be used to convert numerical marks to categorical values. The user specifies a series of cut-points on the numerical scale; all mark values between two cut-points are assigned to the same category. For example

```
> Y <- cut(longleaf, breaks=c(0, 30, Inf), labels=c("J","A"))
```

classifies the Longleaf Pines data into juvenile ('J'; $dbh \leq 30$) and adult ('A'; $dbh > 30$) trees. This multitype pattern, shown in Figure 15.5, clearly indicates that juvenile trees are concentrated in the upper right corner of the plot. The `split` command would separate the juvenile and adult trees into different point patterns.

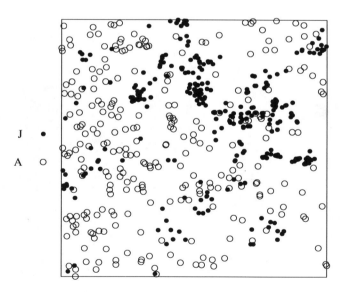

Figure 15.5. *Longleaf Pines classified into 'juvenile' (*J*; dbh < 30, filled circles) and 'adult' (*A*; dbh ≥ 30, open circles).*

In the foregoing call to `cut`, the argument `right=FALSE` was specified so that 'juveniles' are trees whose diameters are *strictly less than* 30 cm (rather than less than or equal to this value).

Cutting can be based on any numerical quantity z given as the second argument of `cut.ppp`. For a point pattern with a data frame of marks, z can be the name of one of the columns of marks. For example the nbfires pattern consists of locations of forest fires, marked by 9 different variables. To classify the fire locations into four groups on the basis of their final fire size:

```
> NB <- cut(nbfires, "fnl.size", breaks=4)
```

Note that setting breaks=4 tells cut to divide (a slight extension of) the range of the fire sizes into four roughly equal intervals.

More complex subdivisions can be accomplished using subset.ppp or arithmetic with the marks:

```
> big.old <- subset(nbfires, fnl.size > 10 & dis.date < "2001-01-01")
```

15.2.3   Intensity for point patterns with real-valued marks

Suppose we have a marked point pattern $\mathbf{y} = \{(x_1, m_1), \ldots, (x_n, m_n)\}$ where the marks m_i are real numbers. In Section 6.9 we mentioned the Nadaraya-Watson smoother of the marks (6.26)–(6.27)

$$\widetilde{m}(u) = \frac{\sum_i m_i \kappa(u - x_i) b(x_i)}{\sum_i \kappa(u - x_i) b(x_i)} \tag{15.1}$$

for each spatial location u, where κ is a smoothing kernel in two-dimensional space, and $b(x_i)$ is an edge correction factor.

For the Longleaf Pines data, using least-squares cross-validation to select the smoothing bandwidth, this smoother would be calculated by

```
> longsmooth <- Smooth(longleaf, bw.smoothppp)
```

The result is shown in the left panel of Figure 15.6: the greyscale value of each pixel u represents the spatially smoothed average $\widetilde{m}(u)$ of tree diameters, in centimetres. The plot strongly suggests a concentration of younger trees in a swath across the upper right quarter of the survey.

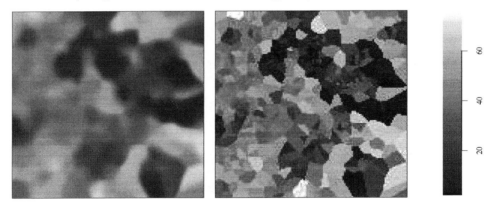

Figure 15.6. *Smoothing the marks of the Longleaf Pines data. Left: Nadaraya-Watson kernel smoothing, Gaussian kernel selected by cross-validation. Right: nearest-neighbour interpolation.*

For very small bandwidths, (15.1) converges to *nearest-neighbour interpolation*, in which the value $m(u)$ is estimated by taking the mark m_i of the data point x_i that is *closest* to the location u. This can be performed in spatstat by nnmark. The result of nnmark(longleaf) for the Longleaf Pines is shown in the right panel of Figure 15.6.

The Nadaraya-Watson smoother is closely related to estimation of intensity. For a marked point process in two dimensions, with real-valued marks, the intensity $\lambda(u, m)$ if it exists is a function of three coordinates (two spatial coordinates and the mark variable). Integrating out the mark variable yields the intensity of the point process of locations,

$$\lambda_\bullet(u) = \int_{-\infty}^{\infty} \lambda(u, m) \, dm.$$

At a given point u, the ratio

$$p(m \mid u) = \frac{\lambda(u, m)}{\lambda_\bullet(u)}$$

is the conditional probability density of the mark m at this location, and the expected mark value at this location is

$$e(u) = \int_{-\infty}^{\infty} m \, p(m \mid u) \, dm = \frac{\int_{-\infty}^{\infty} m \, \lambda(u, m) \, dm}{\int_{-\infty}^{\infty} \lambda(u, m) \, dm}. \tag{15.2}$$

Given a marked point pattern $\{(x_i, m_i)\}$ the intensity function $\lambda(u, m)$ could be estimated by kernel smoothing in *three* dimensions:

$$\hat{\lambda}(u, m) = \sum_i \kappa_3((x_i, m_i) - (u, m))$$

where κ_3 is a probability density on $\mathbb{R}^3$. The intensity of locations $\lambda_\bullet(u)$ would then be estimated by kernel smoothing the locations x_i using the corresponding two-dimensional marginal kernel

$$\kappa((x, y)) = \int_{-\infty}^{\infty} \kappa_3((x, y, z)) \, dz.$$

This ensures that $\hat{\lambda}_\bullet$ is the marginal integral of $\hat{\lambda}$ and that the ratio $\hat{p}(m \mid u) = \hat{\lambda}(u, m) / \hat{\lambda}_\bullet(u)$ is a probability density. The estimator $\hat{e}(u)$ obtained by plugging into (15.2) a kernel estimator of $\lambda(u, m)$ is identical to the Nadaraya-Watson smoother (15.1).

Another useful property of the spatially-varying mark distribution is the *local variance of marks*

$$v(u) = \int_{-\infty}^{\infty} (m - e(u))^2 p(m \mid u) \, dm. \tag{15.3}$$

The kernel smoothing estimate of $v(u)$ is

$$\widetilde{v}(u) = \widetilde{m_2}(u) - (\widetilde{m}(u))^2 \tag{15.4}$$

where $\widetilde{m_2}(u)$ is the Nadaraya-Watson smoother of the squared marks m_i^2, using the same bandwidth as for $\widetilde{m}(u)$. The function $v(u)$ is a measure of variability amongst the marks in the neighbourhood of the location u. If there is a smooth gradient in the mark value, the corresponding standard deviation $\widetilde{\sigma}(u) = \sqrt{\widetilde{v}(u)}$ is an estimate of the gradient. The local mark variance is computed in `spatstat` by `markvar`. To obtain the standard deviation:

```
> msd <- sqrt(markvar(longleaf, bw.smoothppp))
```

The result is plotted in the left panel of Figure 15.7. The highest values of standard deviation appear mainly at the borders of areas where the trees are relatively young.

The residual $r_i = m_i - \widetilde{m}_{-i}(x_i)$ is the difference between the mark value at a data point and the predicted value given by the average of the marks of neighbouring points other than x_i. It may be useful in detecting anomalies. The middle panel of Figure 15.7 shows the Longleaf Pines data marked by their residuals r_i computed by

```
> mfit <- Smooth(longleaf, bw.smoothppp, at="points")
> res <- marks(longleaf) - mfit
```

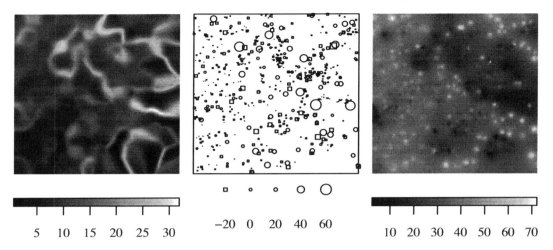

Figure 15.7. *Additional smoothing results for Longleaf Pines data.* Left: *Local standard deviation of tree diameter.* Middle: *Residual diameters. Circles and squares represent positive and negative residuals, respectively.* Right: *Inverse-distance weighted interpolation of diameters.*

Note that `mfit[i]` is the leave-one-out estimate $\widetilde{m}_{-i}(x_i)$ because of the defaults in `Smooth.ppp`.

Inverse-distance weighted smoothing is essentially Nadaraya-Watson smoothing (15.1) with the 'kernel' $\kappa(x) = 1/\|x\|^p$, the inverse pth power of distance. The right panel of Figure 15.7 shows the result of inverse-distance weighted smoothing of the diameters of the Longleaf Pines with the default index value $p = 2$. In this panel the smoothed tree diameter at location u is computed as the weighted average of the diameters of nearby trees, with weights inversely proportional to the squared distance from the tree to the location u. For $p \geq 1$ this kernel has a sharp peak, so that for locations u close to a data point x_i, the mark value m_i at the data point has relatively huge weight, and pictures of the inverse-distance weighted average show a characteristic bright or dark spot around each data point. This kernel is not integrable for any value of p, so it does not enjoy many of the good statistical properties of the Nadaraya-Watson smoother. However, it is relatively fast to compute, and the power p is a kind of smoothing bandwidth, with small values of p giving very smooth pictures, and large values of p giving pictures equivalent to nearest-neighbour interpolation.

If the marks are a data frame, the result of `Smooth.ppp` or `nnmark` or `markvar` or `idw` will be a list of pixel images, one for each mark variable.

15.2.4   Summary functions for numerical marks

Dependence between the marks in a point pattern can be investigated empirically, at least when the process is believed to be stationary. The spruces pattern in Figure 15.1 appears to be approximately stationary. A basic question about this pattern is whether the sizes of neighbouring trees are strongly dependent [256]. One can use various exploratory statistics, which are analogous to the K-function and pair correlation function (Chapter 7) and the nearest-neighbour function (Chapter 8).

15.2.4.1   Mark correlation function

The 'mark correlation function' $k_f(r)$ of a stationary marked point process **Y**, foreshadowed in Section 14.6.7.1, is a measure of the dependence between the marks of two points of the process a distance r apart [638, p. 262]. Choose any 'test function' $f(m, m')$ with two arguments m, m' which are possible marks of the point process, where the value $f(m, m')$ is a nonnegative real number.

Define the mark correlation function with test function f by

$$k_f(r) = \frac{\mathbb{E}\left[f(m(u),m(v))\big|u,v \in \mathbf{X}\right]}{\mathbb{E}[f(M,M')]} \tag{15.5}$$

where M, M' are independent, identically distributed random marks which have the same distribution as the mark of a randomly chosen point. The denominator is chosen so that, under the hypothesis of random labelling, $k_f(r) \equiv 1$.

Note that $k_f(r)$ is not a 'correlation' in the usual statistical sense. Like the pair correlation function, $k_f(r)$ can take any nonnegative real value, and the value 1 suggests 'lack of correlation'. The interpretation of values larger or smaller than 1 depends on the choice of function f.

Common choices of f are, for nonnegative real-valued marks, $f(m,m') = mm'$; for categorical marks (multitype point patterns), $f(m,m') = \mathbf{1}\{m = m'\}$; and for marks representing angles or directions, $f(m,m') = \sin(m - m')$.

In the first case $f(m,m') = mm'$, the mark correlation is

$$k_{mm}(r) = \frac{\mathbb{E}\left[m(u)\,m(v)\big|u,v \in \mathbf{X}\right]}{\mathbb{E}[M]^2}. \tag{15.6}$$

The mark correlation function $k_f(r)$ in (15.5) can be estimated nonparametrically. The denominator $\mathbb{E}[f(M,M')]$ is estimated by the sample average of $f(m_i,m_j)$ taken over all pairs i and j. The numerator

$$c_f(r) = \mathbb{E}\left[f(M(0),M(u))\big|u,v \in \mathbf{X}\right] \tag{15.7}$$

can be estimated by a kernel smoother of the form

$$\hat{c}_f(r) = \frac{\sum_{i<j} f(m_i,m_j)\kappa(||x_i - x_j|| - r)w(x_i,x_j)}{\sum_{i<j} \kappa(||x_i - x_j|| - r)w(x_i,x_j)} \tag{15.8}$$

where κ is a smoothing kernel on the real line and $w(u,v)$ is an edge correction factor. Note that the denominator of $\hat{c}_f(r)$ is the usual kernel estimate of the pair correlation function.

Estimates of the mark correlation function are computed in `spatstat` by `markcorr`. It has the syntax

```
markcorr(X, f, ...)
```

where X is a point pattern and f is an R language function. If f is omitted, then the default choice is $f(m,m') = mm'$ for real-valued marks and $f(m,m') = \mathbf{1}\{m = m'\}$ for categorical marks (multitype point patterns) as discussed in Section 14.6.7.1. Additional arguments control the choice of edge corrections and smoothing algorithms. The default is kernel smoothing (with arguments `kernel` and `bw` to select the smoothing kernel and smoothing bandwidth); alternatives include local polynomial regression smoothing (selected by `method="loess"`, with smoothing bandwidth argument `span`). Note that the estimator assumes the process is stationary.

The result of `markcorr(X,f,...)` is a function value table (object of class `"fv"`) if X has a single column of marks. If the marks are a data frame, the result is a list of `"fv"` objects, one for each column of marks; the result belongs to class `"anylist"`. (See also `markcrosscorr`.)

Figure 15.8 shows the estimated mark correlation function $\widehat{k}_{mm}(r)$ for the spruce trees, which could have been computed by `markcorr(spruces)`, and the pointwise envelopes of the mark correlation function from 39 simulations of random labelling, computed by

```
> envelope(spruces, markcorr, nsim=39,
          simulate=expression(rlabel(spruces)))
```

The estimate $\widehat{k}_{mm}(r)$ suggests there is no dependence between the diameters of neighbouring trees, except for a hint of negative association at short distances (closer than 1 metre apart). The envelopes suggest this is not significant.

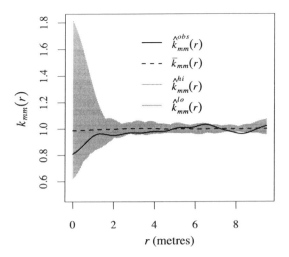

Figure 15.8. *Estimated mark correlation function for the spruce trees (solid line) and the pointwise envelopes (boundary of grey shading) from 39 random relabellings of the spruces data.*

15.2.4.2 Mark-weighted K-function

The mark correlation function is similar to the pair correlation function (Section 7.6) in that they both gather information from pairs of points lying exactly r units apart. This is very convenient for interpretation of results, but it can be statistically inefficient: for hypothesis testing purposes we normally prefer a cumulative function such as Ripley's K-function.

Accordingly Penttinen *et al.* [540] proposed the *mark-weighted K-function*

$$K_f(r) = \frac{1}{\lambda \, \mathbb{E}[f(M,M')]} \mathbb{E}\left[\sum_{x_j \in \mathbf{X}} f(m(u),m(x_j)) \mathbf{1}\left\{ 0 < \|u - x_j\| \le r \right\} \,\middle|\, u \in \mathbf{X} \right] \qquad (15.9)$$

where $m(u)$ and $m(x_j)$ denote the mark values at the points u and x_j, and again M,M' are independent random marks which have the same distribution as the marks in the point process. This is like the definition of Ripley's K-function in equation (7.6) on page 206 except that the contribution from each pair of points is weighted by $f(m,m')$ where m,m' are the marks of the two points. The normalisation is designed so that, under random labelling, $K_f(r) = K(r)$, the usual K-function.

Note that K_f is not exactly the cumulative version of $k_f(r)$. If the mark correlation function $k_f(r)$ exists, then[1]

$$K_f(r) = 2\pi \int_0^r s k_f(s) g(s) \, \mathrm{d}s, \qquad (15.10)$$

where $g(r)$ is the pair correlation function. That is, $K_f(r)$ is essentially the cumulative value of $k_f(s)$ weighted by $s,g(s)$, for all distances less than or equal to r.

The analogue of Besag's L-function is $L_f(r) = \sqrt{K_f(r)/\pi}$ and the empirical estimate $\widehat{L}_f(r) = \sqrt{\widehat{K}_f(r)/\pi}$ should have approximately constant variance for a Poisson marked point process. In spatstat the mark-weighted K-function is estimated by the function Kmark:

```
Kmark(X, f, ...)
```

---

[1]Equation (5.3.36) of [355, p. 351] is incorrect as it omits the factor $g(s)$. Equation (5.3.40) (*op. cit.*) is correct.

The default for the test function f is the same as for `markcorr`, described above. The result of `Kmark` is an `"fv"` object, or a list of `"fv"` objects if X has a data frame of marks. The additional argument `returnL=TRUE` can be used as a shortcut to compute $L_f(r)$ instead of $K_f(r)$.

```
> smki <- Kmark(spruces, returnL=TRUE)
> smke <- envelope(spruces, Kmark, returnL=TRUE,
                   simulate=expression(rlabel(spruces)),
                   nsim=1999, nrank=50, savefuns=TRUE)
```

Figure 15.9 shows the estimate of $L_f(r)$ and the pointwise 5% significance envelopes based on 1999 simulations of random labelling. The apparent deviation should be investigated using a formal test such as `mad.test` or `dclf.test` before concluding statistical significance.

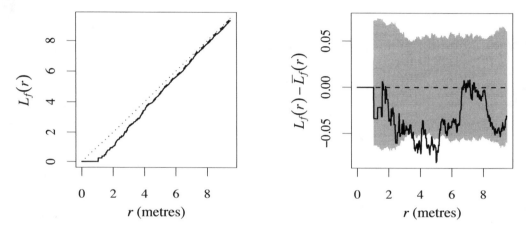

Figure 15.9. Left: *estimated mark-weighted L-function $L_f(r)$ for the spruce trees (solid lines) and the corresponding theoretical value for a Poisson process (dotted line).* Right: *centred estimate of $L_f(r)$ for the spruce trees (solid lines) and the pointwise 5% significance envelopes (boundary of grey shading) based on 1999 simulations of random labelling.*

Note carefully that the 'theoretical' value plotted in the left panel of Figure 15.9 is the value of $L_f(r)$ for a *Poisson* marked point process. This is not useful: Figure 15.1 shows that the pattern is highly regular. The theoretical value of $L_f(r)$ *under random labelling* would be the *L*-function of the unmarked point process. We could compare $L_f(r)$ with $L(r)$ by

```
> Lspruce <- Lest(spruces)
> diffL <- eval.fv(smki - Lspruce)
```

and similarly for the envelopes.

15.2.4.3 Mark variogram

The *mark variogram* [190, 449, 683, 641], for a stationary marked point process with real-valued marks, is

$$\gamma(r) = \frac{1}{2} \, \mathbb{E} \left[(m(u) - m(v))^2 \big| u, v \in \mathbf{X} \right] \tag{15.11}$$

where u and v are arbitrary locations with $\|u - v\| = r$. That is, $2\gamma(r)$ is the expected squared difference between the mark values at two points separated by a distance r.

This definition is *analogous* to the variogram of a random field [190, sec. 2.3.1]. If $Z(u)$ is a stationary random function of spatial location u, the variogram of Z is $\gamma_Z(r) = (1/2)\mathbb{E}[(Z(u) -$

$Z(v))^2]$ where again u and v are arbitrary locations with $\|u - v\| = r$. However, the mark variogram is *not* a variogram in the usual sense of geostatistics, as pointed out by Wälder and Stoyan [683, 641]. It may exhibit properties which are impossible or implausible for a geostatistical variogram. This occurs because the mark variogram is a conditional expectation — the expected squared difference given that there exist two points separated by a distance r — and the conditioning event is different for each value of r.

The mark variogram is of the general form (15.7) with $f(m,m') = \frac{1}{2}(m - m')^2$. It can therefore be estimated using the same nonparametric smoothing methods as before. Roughly speaking, to estimate $\gamma(r)$ for a given distance r, we find pairs of data points x_i, x_j with interpoint distance close to r, extract the corresponding marks m_i, m_j, compute $f(m_i, m_j)$, and average these values.

The `spatstat` function `markvario` computes estimates of the mark variogram, with the same syntax as `markcorr`. It returns an `"fv"` object, or a list of `"fv"` objects if the marks are a data frame.

Figure 15.10 shows the estimated mark variogram of the Norwegian spruces dataset computed by `markvario(spruces)`. There is a hint of increased variance at very short distances. A possible interpretation is that, in order for these rather large trees to be found growing close together, at least one of the trees must be small. The significance of the finding can be assessed using simulation envelopes or deviation tests as described for other summary functions: the significance appears to be marginal.

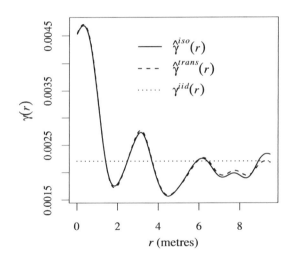

Figure 15.10. *Mark variogram of Norwegian spruces.*

15.2.4.4   Reverse conditional moments

The dependence of mark values on the proximity of other points can be investigated directly using the functions $E(r)$ and $V(r)$ introduced by Schlather *et al.* [601]. These are the conditional mean and conditional variance of the mark attached to a typical random point, given that there exists another random point at a distance r away from it:

$$E(r) = \mathbb{E}\left[m(u)\big|u,v \in \mathbf{X}\right] \tag{15.12}$$

$$V(r) = \mathbb{E}\left[(m(u) - E(r))^2\big|u,v \in \mathbf{X}\right] \tag{15.13}$$

where again u and v are arbitrary locations with $\|u - v\| = r$. Under the random labelling hypothesis, $E(r)$ and $V(r)$ should be constant [601].

The mean mark function $E(r)$ is a special case of the unnormalised mark correlation function (15.7) with $f(m, m') = m$, and can be estimated nonparametrically using smoothing methods. Similarly $V(r)$ can be estimated by smoothing.

The `spatstat` functions `Emark` and `Vmark` compute estimates of $E(r)$ and $V(r)$, respectively. Additional arguments control the smoothing algorithm and edge correction. Figure 15.11 shows estimates of $E(r)$ and $V(r)$ for the spruces data. This plot suggests that a tree with a very close neighbour tends to have a diameter slightly smaller than average, about 10% smaller. The discrepancies are marginally not significant.

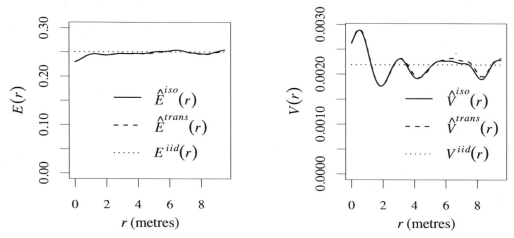

Figure 15.11. *The functions $E(r)$ and $V(r)$ estimated for the spruce trees.*

15.2.4.5 Marks of nearest neighbours

The mark correlation function and its relatives described above are all *second-order* summary statistics, formed by adding contributions from each pair of points. An alternative is to consider only nearest neighbours.

Suppose that the random variable M represents the (random) mark value at a typical point, and M^* is the mark value at its nearest neighbour. The *mean nearest-neighbour mark* is the expected value $\mathbb{E}M^*$, that is, the mean value of the mark at the nearest neighbour of a typical point. The `spatstat` function `nnmean` computes estimates of $\mathbb{E}M^*$ (the 'unnormalised mean') and of $\mathbb{E}M^*/\mathbb{E}M$ (the 'normalised mean'), assuming the process is stationary.

```
> nnmean(spruces)
unnormalised    normalised
   0.2493519     0.9959210
```

The normalised value of 0.99 indicates that the average diameter $\mathbb{E}M^*$ of the nearest neighbour of a typical spruce tree is approximately the same as the average diameter $\mathbb{E}M$ of a typical tree. Under random labelling, these would be equal, and the normalised value would be equal to 1.

The *nearest-neighbour variogram index* is $v_n = (1/2)\mathbb{E}[(M - M^*)^2]$, a measure of discrepancy between the marks of a typical point and its nearest neighbour. Dividing by the variance of a typical mark gives the 'normalised' version $v_n / \mathrm{var} M$. Estimates are computed by the `spatstat` function `nnvario`:

```
> nnvario(spruces)
unnormalised     normalised
 0.002579167    1.168827847
```

These quantities are all special cases of the *nearest-neighbour correlation index* [638, sec. 14.7]

$$\bar{m}_f = \frac{E[f(M, M^*)]}{E[f(M, M')]} \tag{15.14}$$

where f is a test function (as described in the previous sections) and M' is an independent copy of M with the same distribution. Like the other correlation indices we have discussed, the nearest-neighbour correlation $\bar{m}_f$ is not a 'correlation' in the usual statistical sense, but is normalised so that the value 1 suggests 'lack of correlation': under the random labelling hypothesis, $\bar{m}_f = 1$. The interpretation of values larger or smaller than 1 depends on the choice of function f.

If the marks of X are real numbers, we can also compute the **classical** correlation between the mark of a typical point and the mark of its nearest neighbour, that is, the correlation coefficient of the two random variables, M and M^*. The classical correlation has a value between -1 and 1. Values close to -1 or 1 indicate strong dependence between the marks.

The `spatstat` function `nncorr` computes the nearest-neighbour correlation (unnormalised, normalised, and if appropriate also the classical version):

```
> nncorr(spruces)
unnormalised    normalised   correlation
   0.06191204   0.98764221   -0.14855920
> nncorr(finpines)
                diameter      height
unnormalised    5.8240741     7.9618981
normalised      0.9086290     0.9954150
correlation    -0.2173841    -0.1839798
```

The standard error of these estimates is complicated by the dependence between the contributions from different points. If this dependence were ignored, we could use the asymptotic formula for standard error of the classical correlation $\hat{\rho}$, namely se $= ((1 - \hat{\rho}^2)/(n-2))^{1/2}$ where n is the number of independent observations of each variable. Using the heuristic that the nearest-neighbour distances between n points contain roughly $n/2$ independent values (page 269), an approximate standard error is se $= (2(1 - \hat{\rho}^2)/(n-4))^2$.

```
> rr <- nncorr(spruces)[["correlation"]]
> sqrt(2 * (1-rr^2)/(npoints(spruces)-4))
[1] 0.1226584
```

This *ballpark* measure of accuracy is as large in magnitude as the estimated correlation.

15.3 Three-dimensional point patterns

15.3.1 Background

The three-dimensional spatial locations of objects are important in many fields of research, such as astronomy, geology, and anatomy. Spatial patterns of three-dimensional locations are increasingly easy to capture, using sensing technologies such as three-dimensional microscopy and geophysical tomography.

There is also an increasing realisation that, while two-dimensional data may be easier to obtain, the information they carry about three-dimensional space is fundamentally biased. We cannot pretend that space is two-dimensional simply because we observe it that way. This is a crucial problem when cutting two-dimensional *sections* through three-dimensional space [1, 43].

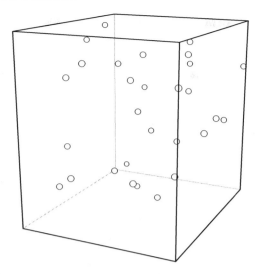

Figure 15.12. *A three-dimensional point pattern (object of class* `"pp3"`*) plotted in isometric projection by* `plot.pp3`.

A dataset giving the locations of points in three-dimensional space may be treated as a realisation of a point process in three-dimensional space (subject to the caveats explained in previous chapters). The statistical methodology for two-dimensional point processes which we described in previous chapters can be extended to three-dimensional point processes without any major obstacles. However, there are some new technical and methodological problems to resolve.

The main methodological question is the status of the third spatial coordinate. It often happens that one of the three spatial coordinates (call it z) is qualitatively different from the other two (call them x and y).

In astronomy, the sky position (x, y) of an object can be established accurately, while its redshift z is related to, but not equivalent to, its distance from Earth (deviations from Hubble's Law are caused by the proper motion of the object). Larger distances are often less accurately measured. Furthermore the probability of observing an object depends on its magnitude (brightness as seen from Earth) which decreases with distance. We have the option of treating redshift either as a spatial coordinate or as a mark.

In three-dimensional modalities of microscopy (such as optical confocal microscopy) the third coordinate ('depth') is different from the first two coordinates, at least with regard to measurement accuracy. The origin of the x, y coordinates is typically chosen to be the bottom left corner of the microscope's field of view, and it would often be appropriate to assume that the point process is stationary with respect to translations of x and y. However the origin of the z coordinates may be fixed, for example, $z = 0$ could represent the upper surface of the material. It may not be appropriate to assume the point process is stationary with respect to translations of z.

15.3.2 Handling 3D point pattern data

Currently `spatstat` provides basic support for three-dimensional point patterns. A 3D point pattern is an object of class `"pp3"`. It is created by the function `pp3` with arguments similar to `ppp`. The spatial 'window' containing the point pattern is currently required to be a 3D rectangular box, stored as an object of class `"box3"`, created by the function `box3`.

The coordinates of the point pattern shown in Figure 15.12 are available in the file `osteo36.txt` installed with `spatstat`. The coordinates, in microns, are stored in tabular format. The sampling

print	print basic information
summary	print detailed summary of data
plot	plot 3D points
domain	extract 3D spatial domain ('window')
npoints	number of points
as.data.frame	extract coordinates and marks
coords	extract coordinates
coords<-	change coordinates
marks	extract marks
marks<-	change marks
unmark	remove marks
unitname	extract name of unit of length
unitname<-	set name of unit of length
split	divide into sub-patterns
pairdist	matrix of distances between all pairs of points
crossdist	distances between all pairs of points in two patterns
nndist	nearest-neighbour distance
nnwhich	identify nearest neighbour
F3est	three-dimensional empty-space function
G3est	three-dimensional nearest-neighbour function
K3est	three-dimensional K-function
pcf3est	three-dimensional pair correlation function
envelope	simulation envelopes of summary functions
nnclean	Byers-Raftery nearest-neighbour cleaning

Table 15.1. *Facilities available for* "pp3" *objects.*

volume was a box of dimensions $81 \times 100 \times 100$ microns. To create a "pp3" object containing these locations,

```
> copyExampleFiles("osteo")
> xyz <- read.table("osteo36.txt", header=TRUE)
> b <- box3(c(0,81),c(0,100),c(-100,0),unitname=c("micron", "microns"))
> X <- with(xyz, pp3(x, y, z, b))
> X
Three-dimensional point pattern
29 points
Box: [0, 81] x [0, 100] x [-100, 0] microns
> summary(X)
Three-dimensional point pattern
29 points
Box: [0, 81] x [0, 100] x [-100, 0] microns
Volume 810000 cubic microns
Average intensity 3.58024691358025e-05 points per cubic micron
```

Facilities available for pp3 objects are listed in Table 15.1. Every "pp3" object also belongs to the class "ppx" of multidimensional point patterns, described in Section 15.4. Many generic functions have methods for class "ppx" which automatically apply to "pp3" objects.

The plot method for 3D point patterns, plot.pp3, produces a perspective view (not an orthogonal projection) of the point pattern and its surrounding box. The plot respects the physical scale of the coordinates: that is, the plot is not stretched to fit the available graphics area. To plot a point pattern X in the style of Figure 15.12,

```
> plot(X, main="", pch=21, bg='white', cex=1.2)
```

Plot character 21 is a filled circle; the argument bg='white' specifies that the interior of the circle is filled in white; this means that a point lying behind another point will be hidden from view. For a transparent view of all points, set pch=1.

In the example dataset used above, a glance at coords(X) shows us that the depth coordinate z is recorded to the nearest micron, while the field-of-view coordinates x, y are recorded to the nearest multiple of $1/11$ micron (due to rescaling of the original record). The z coordinate values are negative, and in general it is important to clarify the meaning of the negative sign, and to clarify whether the value 0 has any special meaning. In this case larger negative values are deeper in the skull (-50 is deeper than -20) and the zero depth *is* special: it was the shallowest possible position of the focal plane that permits observation.

The new generic function domain extracts the spatial domain of a point pattern: the method domain.pp3 currently extracts the three-dimensional rectangular box containing the pattern. Facilities available for box3 objects include print, summary, unitname, unitname<-, diameter, volume, sidelengths, shortside, and eroded.volumes.

Random 3D point patterns can be generated by runifpoint3 and rpoispp3. Currently the only 3D point pattern data installed in spatstat is the dataset osteo, a hyperframe containing 40 point patterns [66, 68].

A version of the model-fitting function ppm for three-dimensional point patterns is under development.

15.3.3 Summary functions

The summary functions introduced in Chapters 7 and 8 can be extended to three dimensions [68]. It is assumed that the point process is stationary with respect to translation in *all* directions.

15.3.3.1 Three-dimensional K-function

The K-function for a stationary three-dimensional point process $\mathbf{X}$ is defined exactly as in equations (7.4)–(7.6) on page 205. That is, $\lambda K(r)$ is the expected number of other points lying within a distance r of a typical point of the process. Here the intensity λ is the expected number of points per unit *volume*. For a completely random (homogeneous Poisson) point process in three dimensions, the K-function is

$$K_{pois}(r) = \frac{4}{3}\pi r^3, \tag{15.15}$$

equal to the volume of the sphere of radius r. Edge-corrected estimators of $K(r)$ were described in [68].

The spatstat function K3est computes estimates of the three-dimensional K-function:

```
K3est(X, ..., rmax, nrval, correction)
```

where rmax is the maximum value of distance r for which the estimates will be computed, nrval is the number of r values (which should be at least 128), and correction is a string indicating which edge corrections should be used. The result is an "fv" object.

The left panel of Figure 15.13 shows the result of plot(K3est(X)) for the example dataset in Figure 15.12.

> WARNING: The cube root of the three-dimensional K-function is not variance-stabilised.

It is not so simple to define an analogue of Besag's L-function in three dimensions. Taking the cube root $K(r)^{1/3}$ would transform (15.15) to a linear function of r, but this would *not* stabilise the variance (contrary to many claims in the literature). For a completely random process, the variance

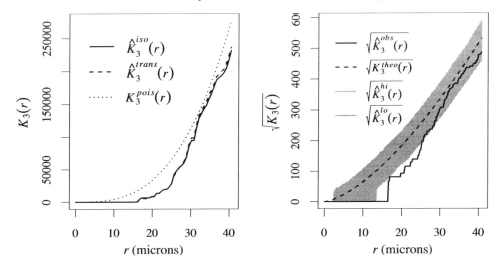

Figure 15.13. Left: *Three-dimensional K-function estimated for the data in Figure 15.12.* Right: **Square root** *of the empirical K-function for the data in Figure 15.12 (solid lines) and pointwise 5% significance envelopes based on simulations of homogeneous Poisson process (boundary of grey shading).*

of $\widehat{K}(r)$ is approximately proportional to $K(r)$ so that, by the delta method, the variance of $\widehat{K}(r)^{1/3}$ is approximately proportional to $K(r)^{-1/3}$, that is, proportional to $1/r$. The variance would be *inflated* at small values of r. When the K-function is used in hypothesis testing and simulation envelopes, we recommend using the *square-root* transformation, which stabilises the variance, and, especially, does not overinflate the variance for small distances r.

Simulation envelopes can be generated by `envelope.pp3`. Unless the argument `simulate` is given, the simulated realisations are generated according to a homogeneous Poisson process with intensity equal to the estimated intensity of the data. The right panel of Figure 15.13 shows the pointwise 5% significance envelopes of the square root of the K-function, generated by

```
> E <- envelope(X, K3est, nsim=1999, nrank=50, nrval=512)
> plot(E, sqrt(.) ~ r)
```

indicating that the variance is well stabilised apart from the familiar 'elephant's foot' feature (see page 269) at small values of r.

15.3.3.2 Three-dimensional pair correlation function

In three dimensions the pair correlation function of a stationary point process is defined as

$$g(r) = \frac{K'(r)}{4\pi r^2} \tag{15.16}$$

analogous to equation (7.22) on page 225. Dividing by $4\pi r^2$, the surface area of the sphere of radius r, ensures that $g(r)$ *is the probability of observing a pair of points of the process separated by a distance r, divided by the corresponding probability for a Poisson process.* Thus, $g(r) = 1$ for a completely random point process.

The pair correlation function is estimated in `spatstat` by `pcf3est`. The current implementation uses fixed-bandwidth kernel smoothing of the interpoint distances d_{ij} with weights $1/(4\pi d_{ij}^2)$. Smoothing is based on the Epanechnikov kernel with a simple default rule-of-thumb for the bandwidth: the half-width of the Epanechnikov kernel is $\delta = 0.26/\overline{\lambda}^{1/3}$ where $\overline{\lambda}$ is the empirical intensity.

Figure 15.14 shows the estimated pair correlation function for the data in Figure 15.12, generated by `plot(pcf3est(X))`. The plot strongly suggests regularity at short distances. Pointwise envelopes (not shown) support this, but also indicate that the variance of the estimated pair correlation (under CSR) is quite large at small distances r.

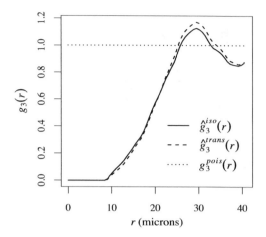

Figure 15.14. *Estimated pair correlation function for the data in Figure 15.12. Solid lines: isotropic correction. Dashed lines: translation correction. Dotted lines: nominal value for completely random pattern.*

15.3.3.3 Three-dimensional nearest-neighbour function G

The nearest-neighbour function G of a stationary point process in three dimensions is the cumulative distribution function of the distance $d(x, \mathbf{X} \setminus x)$ from a typical point of the process to the nearest other point of the process,

$$G(r) = \mathbb{P}\left\{ d(u, \mathbf{X} \setminus u) \leq r \,\middle|\, \mathbf{X} \text{ has a point at } u \right\}, \quad r \geq 0 \tag{15.17}$$

exactly as in equation (8.6). For a homogeneous Poisson process in three dimensions with intensity λ, the nearest-neighbour function is

$$G_{pois}(r) = 1 - \exp(-\frac{4}{3}\lambda\pi r^3). \tag{15.18}$$

Techniques for estimating G in three dimensions are virtually identical to those in two dimensions (Section 8.11) with only minor changes. They require only the nearest-neighbour distances $d(x_i, \mathbf{x} \setminus x_i)$ and the boundary distances $d(x_i, W^c)$ for the data points x_i.

The `spatstat` function `G3est` computes estimates of $G(r)$ for a three-dimensional point pattern. Figure 15.15 shows the result of `plot(G3est(X))` for the data in Figure 15.12. The poor behaviour of the border correction is explained by the fact that in three dimensions this technique discards large proportions of the data:

```
> v <- eroded.volumes(domain(X), c(0, 10, 20, 30))
> f <- v/v[1]
> round(f, 2)
```

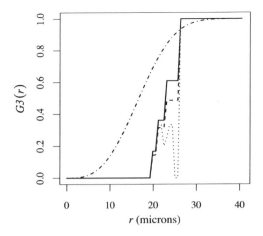

Figure 15.15. *Estimated nearest-neighbour function G for the data in Figure 15.12. Solid lines: normalised Hanisch-type correction. Dashed lines: Kaplan-Meier type correction. Dotted lines: border correction. Dot-dash lines: theoretical value for complete randomness.*

```
[1] 1.00 0.48 0.18 0.04
```

When $r = 20$ microns, only 18% of the original data are used.

15.3.3.4   Three-dimensional empty-space function F

The empty-space function F of a stationary point process in three dimensions is the cumulative distribution function of the distance $d(u, \mathbf{X})$ from a fixed location u to the nearest point of the process,

$$F(r) = \mathbb{P}\{d(u, \mathbf{X}) \le r\}, \quad r \ge 0 \tag{15.19}$$

exactly as in equation (8.5). For a homogeneous Poisson process in three dimensions with intensity λ, the empty-space function is

$$F_{pois}(r) = 1 - \exp(-\frac{4}{3}\lambda \pi r^3). \tag{15.20}$$

Estimation of F in three dimensions is a little more delicate. The usual procedure is to divide the domain into a rectangular grid of *voxels* (three-dimensional pixels) and to evaluate the distances $d(u, \mathbf{x})$ from each voxel u to the dataset $\mathbf{x}$ using a three-dimensional version [117] of the distance transform algorithm [586, 587, 118].

For efficiency, the three-dimensional distance transform algorithm in spatstat compares each pixel only to its immediate neighbours (i.e. the 26 voxels separated from the current voxel by a maximum of one step along each axis) and uses integer arithmetic with step lengths $41, 58, 71$. The values $(41, 58, 71)/41$ are good rational approximations to the true step distances $1, \sqrt{2}, \sqrt{3}$. Estimates of F are then computed using the border correction (reduced sample estimator), the Kaplan-Meier style estimator [64], and the Chiu-Stoyan or Hanisch-style estimator [150] based on the distance transform values, as described in Section 8.11.

WARNING: Empty-space functions are affected by artefacts of the discretisation.

When comparing the estimate $\widehat{F}(r)$ with the benchmark value for a Poisson process, we need to be aware of a digital artefact. The 'sphere' of radius r in the distance transform — that is, the set of voxels that lie at most r units from a given voxel according to the distance transform algorithm — is a polyhedron of fixed shape, even for large values of r. For the `spatstat` algorithm the volume $v(r)$ of the digital sphere of radius r is roughly 0.78 times the volume of the euclidean sphere, $(4\pi/3)r^3$. This is a considerable deficit, so the benchmark value should really be calculated by replacing the ideal volume by the digital volume:

$$F_{pois}(r) = 1 - \exp(-\lambda v(r)). \qquad (15.21)$$

The `spatstat` function `F3est` computes estimates of the three-dimensional empty-space function for a `"pp3"` object. The argument `sphere` determines how the theoretical value for a Poisson process will be calculated. If `sphere="ideal"` then (15.20) will be used; this is not recommended. If `sphere="fudge"` then (15.21) will be used, with $v(r)$ approximated as 0.78 times the volume of the ideal sphere. If `sphere="digital"` then (15.21) is used; the volume $v(r)$ of the sphere in the discretised distance metric is computed exactly using another distance transform. This takes longer to compute, but is exact.

Figure 15.16 shows the estimate of F for the data in Figure 15.12 using the exact digital volume, computed by

```
> F3 <- F3est(X, sphere="digital")
```

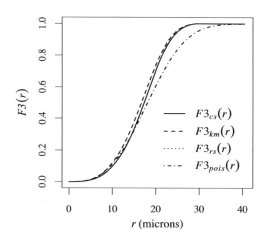

Figure 15.16. *Estimated empty-space function F for the data in Figure 15.12. Solid lines: normalised Hanisch-type correction. Dashed lines: Kaplan-Meier type correction. Dotted lines: border correction. Dot-dash lines: theoretical value for complete randomness using exact digital volume.*

15.4 Point patterns with any kinds of marks and coordinates

An important area of current research is the analysis of *space-time point patterns* where the location and time of each event are recorded. Applications include seismology and disease surveillance. The spatial location could be two-dimensional or three-dimensional.

More generally there are problems involving several spatial coordinates and several time coordinates. For example, pairs of locations giving the origin and destination of a trip taken by a commuter (or a criminal) can be treated as four-dimensional spatial points.

There are also applications where the marks attached to the points need to be very complicated objects. In an astronomical survey ('catalogue') of galaxies, the location (sky position and redshift) of each galaxy could be marked by its spectral signature over a range of optical wavelengths, or its shape [570]. A pattern of randomly placed shapes can be represented as a point pattern of centre locations, marked by shapes [630].

Experimental support for multi-dimensional point patterns, with any kind of marks, has recently been added to `spatstat`. An object of class `"ppx"` represents a marked point pattern in space, time, or space-time. There may be any number of coordinates, designated by the user as 'spatial', 'temporal', or 'local' coordinates, which must be numeric values. Each point is uniquely specified by its spatial and temporal coordinates. 'Local' coordinates provide an alternative coordinate system that may only be applicable in a subset of the space, such as map reference coordinates.

A `"ppx"` object may also contain any number of mark variables. Each individual mark value may be atomic (numeric values, factor values, etc.) or it may be an object of any kind: a function, a window, an image, another point pattern, and so on.

An object of class `"ppx"` is created by `ppx`:

```
ppx(data, domain, coord.type)
```

where `data` is a data frame or hyperframe containing all coordinate and mark values, and `domain` is some kind of data defining the space-time domain of the point pattern, and `coord.type` is a vector of character strings specifying the coordinate type for each column of `data`. Entries of `coord.type` are partially matched to the strings `"spatial"`, `"temporal"`, `"local"`, and `"mark"`.

The following would create a point pattern with three spatial dimensions and one temporal dimension:

```
> df <- data.frame(x=runif(100, max=3),
                   y=runif(100, max=3),
                   z=runif(100, max=2),
                   t=runif(100))
> bb <- boxx(c(0,3), c(0,3), c(0,2), c(0,1))
> X <- ppx(data=df, domain=bb, coord.type=c("s","s", "s", "t"))
> X
Multidimensional point pattern
100 points
3-dimensional space coordinates (x,y,z)
1-dimensional time coordinates (t)
Domain:
        4-dimensional box:
[0, 3] x [0, 3] x [0, 2] x [0, 1] units
```

The space-time domain is currently required to be a rectangular box, stored as an object of class `"boxx"`, created by the function `boxx`.

The point pattern may have marks of any type. They may be included in the `data` argument of `ppx`, or assigned subsequently using `marks<-`.

```
> marks(X) <- with(as.hyperframe(df), disc(centre=c(x,y)))
> X
Multidimensional point pattern
100 points
3-dimensional space coordinates (x,y,z)
```

```
1-dimensional time coordinates (t)
1 column of marks: 'marks'
Domain:
        4-dimensional box:
[0, 3] x [0, 3] x [0, 2] x [0, 1] units
```

Facilities currently available for ppx objects are listed in Table 15.2. Random multidimensional point patterns can be generated by runifpointx and rpoisppx.

print	print basic information
summary	print detailed summary of data
as.hyperframe	extract coordinates and marks
as.data.frame	extract coordinates and *atomic* marks
npoints	number of points
coords	extract coordinates
coords<-	change coordinates
marks	extract marks
marks<-	change marks
unmark	remove marks
unitname	extract name of unit of length
unitname<-	set name of unit of length
duplicated	determine which points are duplicates of others
multiplicity	count points are duplicates of others
split	divide into sub-patterns
pairdist	matrix of distances between all pairs of points
crossdist	distances between all pairs of points in two patterns
nndist	nearest-neighbour distance
nnwhich	identify nearest neighbour

Table 15.2. *Facilities available for "ppx" objects.*

There is also code to compute the theoretical distribution of the nearest neighbour distances, i.e. the distance from a reference point to the k-th nearest neighbour in a uniform Poisson point process in m dimensions. The functions dknn, pknn, qknn, and rknn compute the probability density, cumulative probability, quantiles, and random deviates, respectively.

Functions available for "boxx" objects include print, unitname, unitname<-, diameter, volume, sidelengths, shortside, and eroded.volumes.

Currently the class "ppx" exists mainly to provide basic support for more specific classes of spatial patterns, especially the class "pp3" of three-dimensional point patterns, and the class "lpp" of point patterns on a linear network (Chapter 17). Future versions of spatstat will provide statistical techniques for "ppx" objects.

15.5 FAQs

• *Plots of a three-dimensional point pattern generated by* plot.pp3 *look very different from the standard plots generated by the package* scatterplot3d.

In scatterplot3d the coordinate axes are rescaled separately, to fill the available plotting region. Extra space may also be added at the margins to produce an attractive plot. In plot.pp3 the coordinate axes are maintained at the same scale and an isometric perspective projection is produced. No extra space is added, so a point which lies close to the outer surface of the box will be shown as such. If the viewing angle is rotated, `plot.pp3` produces geometrically consistent views of the pattern, which could be used to make an animation.

- *I have a pattern of trees marked by species and height. How do I analyse the marks?*

Start with exploratory plots such as `plot(X)`, `pairs(marks(X))`, and `plot(Smooth(X))`. Study each column of marks separately: if the marks are called `species` (factor) and `height` (numeric), then `Y <- subset(X, select=species)` can be studied using `relrisk(Y)` and models using `ppm`. Use `Z <- split(X, reduce=TRUE)` to split the pattern by species, yielding a list of sub-patterns with numeric marks, and apply the techniques of this chapter to each sub-pattern. Hypothesis tests of dependence between species and height can be performed using the methods of Section 16.8.3.

- *I have a point pattern with three columns of numerical marks. How do I analyse relationships between these three variables?*

Current methodology is limited. Use exploratory graphics such as `pairs(marks(X))` to identify any strong relationships. Try summary functions such as `Emark`, `Vmark`, `markcorr` and `markcrosscorr` which are applicable to multidimensional marks.

- *I would like to fit a model involving pairwise interaction between trees, where the interaction depends on the trees' diameter. Is this possible?*

Models for spatial point patterns with numerical marks are relatively underdeveloped [355, 484, 601]. This is in stark contrast to the situation for *space-time* point patterns with numerical marks [226, 504, 507, 512, 509].

At the time of writing, `ppm` is not able to fit models to point patterns with numerical marks. Check for updates at `www.spatstat.org`. If this capability has not yet been added, we suggest discretising the diameters and fitting a model to the resulting multitype point pattern.

- *I have a point pattern with numeric marks. How can I test if the marks are independent of the locations?*

It would be prudent to start with exploratory plots such as `pairs(as.data.frame(X))`. We recommend a Monte Carlo test of random labelling: see Sections 10.6, 10.7, and 16.8.3. If the pattern appears to be homogeneous, the test could use any of the summary functions described in Section 15.2.4. See also [601, 302, 310]. For inhomogeneous patterns, the nearest-neighbour correlation (Section 15.2.4.5) can be used.

16

Replicated Point Patterns and Designed Experiments

This chapter deals with the analysis of data consisting of *several* point patterns.

16.1 Introduction

When an observation or experiment can be repeated several times under identical conditions, the results are called *replicated* observations, or *replicates* of the same observation. Replication is vitally important in statistics because it enables us to observe variability, and to separate different sources of variability.

Replicated point patterns are sets of point patterns which can be regarded as realisations of the same point process. For example, the `waterstriders` dataset consists of three point patterns, obtained from three separate photographs of the pattern of insects on the surface of a pond at different times (Figure 1.2 on page 4). The experimental conditions for each photograph were equivalent, so the three patterns can be regarded as replicates of the same observation.

More generally there may be several different groups of point patterns, each group consisting of replicated patterns. For example, the `osteo` dataset consists of 40 point patterns of three-dimensional locations of cells, obtained from 4 different animals, 10 patterns per animal. Patterns obtained from the same animal may be treated as replicates of the same observation. Replication enables us to analyse differences in spatial pattern between the animals and within each animal, and to decide whether there are significant differences between the animals [68, 232].

In the most general situation, we conduct an experiment in which each experimental unit is exposed to different conditions, and we observe a point pattern as the 'outcome' or 'response' from each unit. For example, the `bdspots` dataset contains point patterns of breakdown spots on circular electrodes of three different sizes. The electrode size is an explanatory variable or covariate. In general the covariates could include grouping factors (e.g., control and treatment groups, different materials), numerical covariates such as temperature or voltage, and also spatial covariates such as a pixel image of the spatially varying thickness of the electrode. Some replication would be desirable (say, at least two point patterns observed under each set of experimental conditions). Even without replication, we can fit a model to the data which explains the influence of experimental conditions on the point patterns.

This chapter describes basic methodology for analysing such data, and the capabilities currently available in `spatstat`. We show how to store and handle the data, and apply tools for exploratory data analysis, parametric model-fitting, and model validation.

Techniques for analysing replicated point patterns can also be pressed into service for analysing a *single* point pattern. For example, we could divide the survey region into a few large quadrats, assume that the contents of different quadrats are approximately independent, and test whether the sub-patterns are similar. This is the basis for several statistical techniques for single point patterns, which are presented in this chapter because of their connection to the replicated case.

This chapter includes the results of unpublished research with Leanne Bischof and Ida-Maria Sintorn.

16.2 Methodology

16.2.1 Formulation

We view the experiment as involving a series of *'experimental units'*. Each unit is subjected to a known set of experimental conditions (described by the values of the *covariates*), and each unit yields a *response* which is a spatial point pattern. The value of a particular covariate for each unit can be either a single value (numerical, logical or factor), or a pixel image. Three important cases are:

I. Independent replicates: We observe n different point patterns that can be regarded as independent replicates, i.e. independent realisations of the same point process. The 'responses' are the point patterns; there are no covariates.

II. Replication in groups: there are K different experimental groups (e.g., control, aspirin, ibuprofen). In group k ($k = 1, \ldots, K$) we observe n_k point patterns which can be regarded as independent replicates within this group. We regard this as an experiment with $n = \sum_k n_k$ units. The responses are the point patterns; there is one covariate which is a factor (categorical variable) identifying which group each point pattern belongs to.

III. General case: there are covariates other than factors that influence the response. The point patterns are assumed to be independent, but no two patterns have the same distribution.

Examples of these three cases are given in the datasets `waterstriders`, `osteo`, and `bdspots`, respectively, which are installed in `spatstat`.

The methodological landscape is completely changed when replication is present [539, 346, 68, 232, 234, 87], [355, sec. 4.7]. Independent replication provides a firm basis for non-parametric estimation and inference using summary statistics. For example, the empirical K-function estimates obtained from each replicate point pattern can be treated as i.i.d. estimates of the true K-function of the underlying point process (assuming that the point process is stationary). However, replication also supports *parametric* inference in its fullest form, including parametric standard errors, confidence intervals, and hypothesis tests.

Parametric modelling for replicated spatial patterns (Case I) was first demonstrated by Penttinen [539, Chapter 5]. He fitted the uniform Poisson and Strauss point process models to the water striders data by maximum likelihood, estimating the Strauss normalising constant by a Monte Carlo procedure. Perceived benefits included (a) borrowing strength, when the individual point patterns contain only small numbers of points; (b) standard errors for parameter estimates; and (c) testing the Poisson assumption against overdispersed alternatives.

In Case II (which is also referred to as 'replicated spatial patterns' in the literature), replication is present within groups. This replication again provides a basis for parametric or non-parametric inference. A mainly nonparametric approach was developed in [346, 68] and [232, 234]. The main idea of those papers is to extract a summary statistic from each point pattern, and to perform the analogue of a one-way analysis of variance or analysis of deviance. If the summary statistic for a point pattern is taken to be its total number of points, then analysis of deviance for Poisson counts enables us to fit a Poisson point process model with uniform intensity, where intensity may differ between groups, and to test the goodness-of-fit of the Poisson model against underdispersed and overdispersed alternatives [232]. If the summary statistic is Ripley's K-function [572], then applying ANOVA pointwise (i.e. separately to estimates of $K(r)$ for each value of r) provides a nonparametric pooled estimate of $K(r)$ for each group, a nonparametric pointwise standard error for $K(r)$ within groups, and a simple test for differences in spatial pattern (interpoint interaction) between groups. Diggle *et al.* [234] perform simulation experiments and argue that non-parametric inference based on summary statistics is competitive with parametric inference. Repeated measures analysis or functional data analysis [562] could also be applied (to borrow strength and to handle covariance between $K(r)$ for different r) but has not, to our knowledge. One problem here is that the windows containing the point patterns must be identical if we are to assume that the empirical K functions are i.i.d.

By treating data as replicated observations, we *assume* that they are independent and identically distributed. This assumption may be reasonable in the context of the experiment. Alternatively, it may be only a tentative working assumption for the analysis. Evidence against the assumption would include an apparent trend in the number of points (or other quantity) as a function of the ordering in the sequence of observations.

Parametric modelling for multiple point patterns was developed by Bell and Grunwald [87]. See also [234]. Assuming the point patterns are independent, the likelihood or pseudolikelihood for the entire dataset is the product of the likelihoods or pseudolikelihoods for the individual point patterns. This makes it relatively straightforward to adapt existing techniques to fit parametric models to replicated data, and more generally to designed experiments with point pattern responses. Often we need to include *random effects* in the model.

16.3   Lists of objects

16.3.1   Lists

First we need a convenient way to store the point pattern *responses* from all the units in an experiment. In spatstat the easiest way to store them is to form a *list* of "ppp" objects. The waterstriders dataset is an example:

```
> waterstriders
List of point patterns

Component 1:
Planar point pattern: 38 points
window: rectangle = [0, 48.1] x [0, 48.1] cm

Component 2:
Planar point pattern: 36 points
window: rectangle = [0, 48.8] x [0, 48.8] cm

Component 3:
Planar point pattern: 36 points
window: rectangle = [0, 46.4] x [0, 46.4] cm
```

The object waterstriders is a list, each of whose entries is a point pattern. Note that the observation windows of the three point patterns are **not** identical.

Lists can be interrogated easily using their indexing operator: waterstriders[[2]] is the second point pattern in the water striders data. Functions can be applied to each element of a list using the standard R commands lapply ('list apply'), which returns a list containing the function results, or sapply ('simplified list apply') which simplifies the result to a vector or array where possible. To determine the number of points in each water striders pattern:

```
> sapply(waterstriders, npoints)
[1] 38 36 36
```

To extract the window of each water striders pattern:

```
> wins <- lapply(waterstriders, Window)
```

The result is a list of window objects.

16.3.2   Lists of spatial objects: class "solist"

For convenience, the `waterstriders` dataset also belongs to the class `"solist"` ('spatial object list'). This class provides a simple mechanism to handle the list neatly — for example, there are methods for printing, plotting, and summarising such a list.

The plot of the water striders data in Figure 1.2 was generated by typing `plot(waterstriders)`. This command was dispatched to the method `plot.solist`, which displays each entry of the list in a separate panel. More details are given in Section 16.3.4.

To create a `"solist"` object from raw data, use the function `solist`, analogous to `list`:

```
> solist(rpoispp(100), rpoispp(100))
List of point patterns
Component 1:
Planar point pattern: 106 points
window: rectangle = [0, 1] x [0, 1] units
Component 2:
Planar point pattern: 122 points
window: rectangle = [0, 1] x [0, 1] units
```

To convert an existing list to the class `"solist"`, use `as.solist`. These functions check whether each entry in the list is indeed a spatial object x (meaning that it occupies a specified region of two-dimensional space) by checking whether `Window(x)` is defined. The objects in the list do not have to belong to exactly the same class: the list could contain a mixture of point patterns, windows, pixel images, line segment patterns, tessellations, and other objects.

Another useful function is `solapply`, which is equivalent to `as.solist(lapply(...))` and is often used when the result of calculations should be a list of spatial objects:

```
> Y <- solapply(c(10, 30, 100), rpoispp)
```

16.3.3   Lists of objects: class "anylist"

The class `"anylist"` is another simple mechanism to allow spatstat to handle lists of objects *of any kind*. The objects do not need to belong to the same class. It is assumed only that each object belongs to a class which has methods for `plot` and `print`.

```
> K <- anylist(Kest(cells), Kest(redwood))
```

The entries of K are function objects, of class `"fv"`, for which there is a plot method. The command `plot(K)` would be dispatched to `plot.anylist` which would plot each of the functions in a separate panel. Similarly

```
> K <- as.anylist(lapply(waterstriders, Kest))
```

would calculate the estimate of the K-function for each of the water strider point patterns, and store the three functions in an `"anylist"`, which could be plotted by `plot(K)`. The function `anylapply` is equivalent to `as.anylist(lapply(...))`:

```
> K <- anylapply(waterstriders, Kest)
```

If the list entries may or may not be spatial objects depending on circumstances, one can use `anylist(..., promote=TRUE)` or `solist(..., demote=TRUE)` which will return a `"solist"` if the entries are all spatial objects, and an `"anylist"` otherwise.

16.3.4 Plotting a list of objects

The plot methods plot.solist and plot.anylist attempt to plot each entry of the list in a separate panel.

In plot.anylist the graphics device is divided into a grid of equally sized panels using the R base graphics command layout. Each panel is treated as if it were a separate page of graphics. Each list entry x[[i]] is plotted, effectively by calling plot(x[[i]]). For example, Figure 16.1 shows the result of

```
> plot(K, main="", main.panel=letters[1:3], legend=FALSE)
```

where K is the list computed above. Here main.panel is a recognised argument of plot.anylist giving a header above each panel. The argument legend is not recognised by plot.anylist, so it is passed to the plot call for generating each panel, which dispatches to plot.fv, where it controls the explanatory legend. The most useful argument of plot.anylist to remember is mar.panel, which controls the amount of white space around the margins: it specifies the graphics parameter mar for each panel, giving the number of margin lines in the order bottom, left, top, right, expressed as a multiple of the height of one text character.

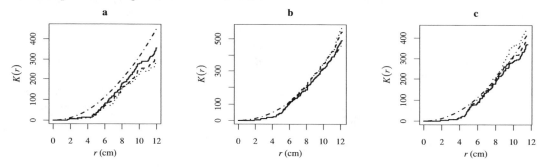

Figure 16.1. *Estimated K-functions for the three water striders patterns.*

In plot.solist the graphics device is treated as a single spatial domain but is notionally divided into a grid of equally sized panels. If equal.scales=TRUE then all objects are plotted at the same physical scale; otherwise each object is rescaled separately (with equal x and y scales) to fill its allotted panel. Objects can be aligned exactly by setting halign=TRUE, valign=TRUE as appropriate. The argument mar.panel again controls the amount of white space around the plot panel, but is now expressed as a multiple of one-sixth (1/6) of the panel width or height.

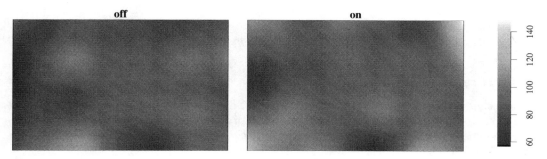

Figure 16.2. *Kernel estimates of intensity for each cell type in the amacrine data, plotted using the same colour map.*

For a list of pixel images, the argument equal.ribbon=TRUE will ensure that all images are plotted using the same colour map, which will be displayed only once, at the margin of the array of panels. Figure 16.2 shows the result of

```
> D <- solapply(split(amacrine), density)
> plot(D, equal.ribbon=TRUE, main="")
```

For a list of `"fv"` objects, the argument `equal.scales=TRUE` will ensure that all plots use the same coordinate axis limits.

16.4   Hyperframes

In statistical analysis we usually think of the results of an experiment as being stored in an array. The array has one column for each variable measured in the experiment, and one row for each experimental unit on which the variables were measured. The 'variables' include both response variables and explanatory variables (covariates). Some of the 'variables' may be factors which describe the organisation of the experiment, for example, specifying how the experimental units are divided into groups which are given different experimental treatments.

In the base R system, data of this kind are stored as a *data frame*. A data frame is like a matrix, in that it is printed as an array, and it can be indexed in the same form `x[i,j]` where `i` is the row index and `j` is the column index. The entries of a matrix must all be values of the same kind (for example they must all be numerical, or all logical). In a data frame, all the entries in a particular *column* must be of the same kind, but different columns may be of different kinds. For example a data frame may contain some columns of numerical data and some columns of categorical (factor) data. All the results from an experiment are stored in a single data frame.

A *hyperframe* is a generalization of a data frame in which the array entries can be *objects* of any class. For example the entries can be point patterns, pixel images, windows, fitted models, ... anything at all. The only constraint is that all the entries in a given column must be objects of the same type. The entries can also be atomic ('ordinary') values such as numbers, categorical values, logical values, and so on.

A hyperframe is designed to store the data from an experiment in which some of the 'variables' are point patterns or other spatial data types. For example, one column of the hyperframe could contain the point patterns observed as the response in the experiment. Other columns could contain categorical or numerical values describing the experimental conditions. Perhaps one column could contain a pixel image representing experimental conditions.

Hyperframes are defined in `spatstat`. They can be created, manipulated, interrogated, and plotted using the functions and operators listed in Table 16.1.

Several examples of hyperframes are installed with `spatstat`, as described in Section 16.6.

16.4.1   Creating hyperframes

Hyperframes can be created using the function `hyperframe`. A call to this function takes the form `hyperframe(...)` where `...` consists of any number of arguments of the form `tag=value`. Each `value` will become a column of the array. The `tag` determines the name of the column: it may be omitted and will default to a name like `V1, V2, ....`

Each `value` can be either: (a) an atomic vector or factor (i.e. numeric vector, integer vector, character vector, logical vector, complex vector or factor); (b) a list of objects which are all of the same class; (c) one atomic value, which will be replicated to make an atomic vector or factor; or (d) one object (of any sort) which will be replicated to make a list of identical objects. The arguments

`hyperframe`	create a hyperframe
`as.hyperframe`	convert other data to a hyperframe
`print.hyperframe`	print a representation of a hyperframe
`dim.hyperframe`	dimensions of hyperframe
`$.hyperframe`	extract column of hyperframe
`[.hyperframe`	extract subset of hyperframe
`[<-.hyperframe`	replace subset of hyperframe
`subset.hyperframe`	extract subset of hyperframe
`as.data.frame.hyperframe`	convert hyperframe to data frame
`cbind.hyperframe`	combine several hyperframes
`rbind.hyperframe`	combine several hyperframes
`with.hyperframe`	compute an expression in each row of hyperframe
`plot.hyperframe`	plot each row of hyperframe

Table 16.1. *Functions available in* `spatstat` *to manipulate hyperframes.*

passed to `hyperframe` (vectors, factors, or lists) must all be of the same length, if their length is greater than 1.

To give an artificial example we construct a hyperframe containing columns that are vectors of numbers or of letters, a list of *functions*, a factor, and lists of point patterns. Note that a column of character strings will be converted to a factor, unless `stringsAsFactors=FALSE` is given in the call to `hyperframe`. This is the same behaviour as for `data.frame`.

```
> G <- hyperframe(X=c(0.23, 1.76, 3.14), T=anylist(sin,cos,tan),
                  Y=letters[1:3], Z=factor(letters[1:3]),
                  W=list(rpoispp(100),rpoispp(100), rpoispp(100)),
                  U=42, V=rpoispp(100), stringsAsFactors=FALSE)
> G
       X          T Y Z     W  U     V
1 0.23 (function) a a (ppp) 42 (ppp)
2 1.76 (function) b b (ppp) 42 (ppp)
3 3.14 (function) c c (ppp) 42 (ppp)
```

This hyperframe has 3 rows. The columns named U and V are constant (all entries in a column are the same). The column named Y is a character vector.

The output above was produced by `print.hyperframe`, which is like `print.data.frame` except that non-atomic objects are replaced by a string indicating their class. In this compact layout it is not practical to print any more information.

Hyperframes are often used to handle replicated point patterns. The `waterstriders` dataset is a list of point patterns, but not a hyperframe. It can easily be converted into one using `hyperframe`:

```
> WS <- hyperframe(Striders=waterstriders)
```

The result is a hyperframe with a single column named `Striders`.

Separate hyperframes may be combined (bound together) using the `"hyperframe"` methods for `cbind` and `rbind`. These functions work in a similar manner to the corresponding `"data.frame"` methods: `rbind` adds new rows, while `cbind` adds new columns. Of course the dimensions along which the hyperframes are being bound must match. When `rbind` is applied, the names of the columns must match. When `cbind` is applied, the row names do not need to match, and indeed will be lost: the row names of the result are always the integers from 1 to n where n is the number of rows of the result. It is also possible to combine hyperframes and data frames together using `cbind` or `rbind` to yield a new hyperframe.

16.4.2   Columns of a hyperframe

Individual columns of a hyperframe can be extracted using $ or getElement:

```
> G$W
List of point patterns
1:
Planar point pattern: 97 points
window: rectangle = [0, 1] x [0, 1] units
2:
Planar point pattern: 102 points
window: rectangle = [0, 1] x [0, 1] units
3:
Planar point pattern: 95 points
window: rectangle = [0, 1] x [0, 1] units
```

The result of $ is a vector or factor if the column contains atomic values; otherwise it is a list of objects (with class "solist" or "anylist" to make it easier to print and plot).

Individual columns can also be assigned (overwritten or created) using $<-:

```
> G$U <- letters[24:26]
> G$animal <- c("horse","dog","deer")
> G
      X            T Y Z     W U      V animal
1 0.23 (function) a a (ppp) x (ppp)  horse
2 1.76 (function) b b (ppp) y (ppp)    dog
3 3.14 (function) c c (ppp) z (ppp)   deer
```

This facility can be used to build up a hyperframe column-by-column:

```
> WS <- hyperframe()
> WS$larvae <- waterstriders
> WS$experiment <- factor(1:3)
> WS
  larvae experiment
1  (ppp)          1
2  (ppp)          2
3  (ppp)          3
```

16.4.3   Subsets of a hyperframe

Subsets of a hyperframe can be extracted with the "hyperframe" method for the subset operator "[". For example G[,2] extracts the second column, G[3,] extracts the third row, G[2:3,] extracts the second and third rows, and G[3,2] extracts the third row, second column.

The result of [.hyperframe is a hyperframe, by default:

```
> G[3,2]
Hyperframe:
           T
3 (function)
```

This behaviour can be overridden using the arguments drop and strip. To extract the element in the third row and second column,

```
> G[3,2,drop=TRUE]
function (x)  .Primitive("tan")
```

The argument `drop` determines whether the hyperframe structure will be discarded where possible. The argument `strip` determines whether the list structure of a *column* will be discarded where possible.

The detailed rules are as follows. If `drop=FALSE` (the default), the result is a hyperframe or data frame. If `drop=TRUE`, and if the selected subset has only one row, or only one column, or both, then the result also depends on the argument `strip` (which defaults to `strip=drop`). If `strip=FALSE`, the result is a list, with one entry for each array cell that was selected. If `strip=TRUE`, then

- if the subset has one row containing several columns, the result is a list;

- if the subset has one column containing several rows, the result is a list or (if possible) an atomic vector;

- if the subset has exactly one row and exactly one column, the result is the object or atomic value contained in this column.

Subsets of a hyperframe can also be assigned new values using `"[<-"`:

```
> G[,2] <- anylist(sqrt,exp,log)
```

Columns of a hyperframe can be indexed by their names as well as by their numbers. For example `G[,"Z"]` is the same as `G[,4]` for the hyperframe G defined on page 667.

The use of `$` to extract columns is equivalent to using `[` with `drop=TRUE`. For example `G$Z` gives the same result as `G[,"Z",drop=TRUE]`.

16.4.4 Subsetting and editing hyperframes

There is also a `subset` method for hyperframes, similar to the `subset` method for data frames, with syntax `subset(x, subset, select)`. The argument `subset` determines which rows of the hyperframe will be retained. It is an expression involving the names of columns of the hyperframe; the expression will be evaluated separately in each row and should return a logical value. For example, the number of points in a point pattern X is returned by `npoints(X)`. The command

```
> B <- subset(G, npoints(W) > 100)
```

selects the subset of the hyperframe in which the point pattern W contains more than 100 points.

The argument `select` determines which columns of the hyperframe will be retained. It is a subset index, or an expression involving the names of columns of the hyperframe, treated as if they were integers:

```
> B <- subset(G, minnndist(W) < 0.02, select=Z:V)
```

The result is the subset of G consisting of the columns Z, W, U, and V, for those rows where the point pattern W has minimum interpoint distance less than 0.02.

One can use `head` and `tail` to extract the first or last few rows of a hyperframe:

```
> head(pyramidal)
  Neurons    group
1   (ppp)  control
2   (ppp)  control
3   (ppp)  control
4   (ppp)  control
5   (ppp)  control
6   (ppp)  control
```

The interactive viewer `View` and interactive text editor `edit` also have methods for hyperframes, so that one can type `View(pyramidal)` or `h <- edit(pyramidal)`. Only the columns of atomic data (numbers, characters, logical values, factors) can be viewed and edited.

The generic function `split`, for dividing a dataset into subsets, has a method for hyperframes. To split the `pyramidal` dataset into the three groups defined by `pyramidal$group` one can do:

```
> pg <- split(pyramidal, pyramidal$group)
```

The result is a list of three hyperframes, with names `control`, `schizoaffective`, and `schizo-phrenic`, the levels of the `group` factor.

16.4.5   Plotting a hyperframe

16.4.5.1   Plotting one column

If `h` is a hyperframe, then the default action of `plot(h)` is to extract the first column of `h` and plot each of the entries in a separate panel on one page, using `plot.solist` or `plot.anylist`. For example, Figure 16.3 shows the result of `plot(simba)`.

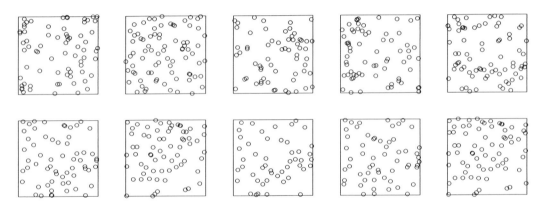

Figure 16.3. *The* `simba` *dataset.*

By default, `plot.hyperframe` tries to plot the objects in the first column of the hyperframe. This is successful if the objects in the first column have a plot method (for example, point patterns, images, or windows). Otherwise the algorithm attempts to plot the second column, and so on. To select a column to be plotted, one can use `$` or `[` or `subset.hyperframe` to extract the relevant column as a list, and `plot.solist` or `plot.anylist` to plot it.

16.4.5.2   Complex plots

More generally, we can display any kind of higher-order plot involving one or more columns of a hyperframe:

```
plot(h, e)
```

where `h` is a hyperframe and `e` is an R language call or expression that must be evaluated in each row to generate each plot panel. For example, Figure 16.4 shows the result of

```
> plot(demohyper,
        quote({ plot(Image, main=""); plot(Points, add=TRUE) }))
```

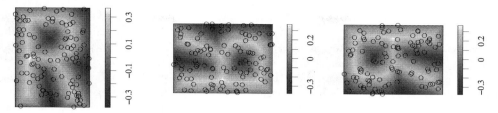

Figure 16.4. *Complicated example of* plot.hyperframe *explained in the text.*

Note the use of quote, which prevents the code inside the braces from being evaluated immediately. Another way to plot the *K*-functions of each of the patterns in the waterstriders dataset is

```
> H <- hyperframe(Bugs=waterstriders)
> plot(H, quote(plot(Kest(Bugs))))
```

This idiom is useful for invoking functions which generate a plot, rather than returning an object which can be plotted. For example, nnclean currently displays a histogram of nearest-neighbour distances but does not return this histogram data in a plottable form. To display the histograms for each of the water strider patterns, we would need to use a command like the one above.

16.5 Computing with hyperframes

16.5.1 with.hyperframe

Often we want to perform some computation on each row of a hyperframe. In a data frame, this can be done using with:

```
> df <- data.frame(A=1:10, B=10:1)
> with(df, A-B)
 [1] -9 -7 -5 -3 -1  1  3  5  7  9
```

Here the expression A-B is effectively evaluated in each row of the data frame, and the result is a vector containing the computed values for each row. The function with is generic, with a default method that works for data frames, lists, and environments.

The hyperframe method for with allows us to perform row-by-row computations on a hyperframe. The general syntax is

```
with(h,e)
```

where h is a hyperframe, and e is an R language construct involving the names of columns in h. For each row of h, the expression e will be evaluated in such a way that each entry in the row is identified by its column name. For example:

```
> H <- hyperframe(Bugs=waterstriders)
> with(H, npoints(Bugs))
 1  2  3
38 36 36
> D <- with(H, distmap(Bugs))
```

The value returned by `with.hyperframe` is a list of objects (of class `"solist"` or `"anylist"`), or a vector or factor if appropriate.

Note that `with.hyperframe` sometimes behaves quite differently from `with.default`. In both cases, the expression is evaluated by matching the names of variables in the expression to the names of columns of the hyperframe. However in `with.hyperframe` the expression is evaluated separately for each row of the hyperframe (substituting the corresponding entries *from the current row*) while in `with.default` the expression is evaluated only once (substituting entire columns of data at once). Thus, for example, `with(flu,table(stain))` is quite different from `with(as.data.frame(flu),table(stain))`.

The reason for this discordance in behaviour is that in a hyperframe, a column is often a *list* of objects, and most operations are not vectorised or adapted to such lists. Consequently `with.hyperframe` has been designed so that the expression does not have to be vectorised. For example, `distmap` expects a single point pattern, and is not vectorised to deal with a list of point patterns. In `with.hyperframe` the expression `distmap(Bugs)` is *evaluated separately in each row of the hyperframe*, giving a sensible result in the example above.

A convenient use of `with.hyperframe` is to create new columns of a hyperframe by computation from the existing columns. For example, we can add a new column to the `simba` dataset that contains pixel images of the distance maps for each of the point pattern responses:

```
> simba$Dist <- with(simba, distmap(Points))
```

16.5.2 Summary statistics

Summary statistics for each row of a hyperframe are easily calculated using `with.hyperframe`. To find the number of points in each pattern in the `Points` column of the `simba` dataset:

```
> with(simba, npoints(Points))
 1  2  3  4  5  6  7  8  9 10
71 80 63 68 75 61 68 49 58 70
```

The summary statistic can be any kind of object. For example, to compute the empirical *K*-functions for each of the patterns in the `waterstriders` dataset:

```
> H <- hyperframe(Gerris=waterstriders)
> K <- with(H, Kest(Gerris))
```

To plot these *K*-functions, one can then just type `plot(K)`. The summary statistic for each row could be a numeric vector:

```
> H <- hyperframe(Gerris=waterstriders)
> m <- with(H, nndist(Gerris))
```

The result `m` is a list, each entry being a vector of nearest-neighbour distances. To find the minimum interpoint distance in each pattern:

```
> with(H, min(nndist(Gerris)))
   1    2    3
1.65 1.89 1.83
```

16.5.3 Simulation

The function `with.hyperframe` can be useful for simulation. For example, to generate Poisson point patterns with different intensities, where the intensities are given by a numeric vector `lambda`:

```
> lambda <- rexp(3, rate=1/50)
> H <- hyperframe(lambda=lambda)
> H$Points <- with(H, rpoispp(lambda))
```

The result could be plotted by:

```
> plot(H, quote(plot(Points, main=lambda)))
```

or more neatly:

```
> H$Title <- with(H, parse(text=paste("lambda==", signif(lambda, 3))))
> plot(H, quote(plot(Points, main=Title)))
```

The result of `parse` is an R language expression. The R base graphics code renders this as a mathematical expression: '==' becomes the mathematical equality sign '=' and `lambda` becomes the Greek letter λ. The result is shown in Figure 16.5.

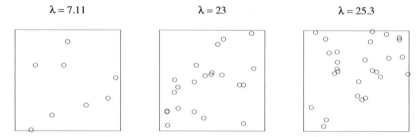

Figure 16.5. *Poisson patterns with random intensity, generated using* `plot.hyperframe`.

It is even simpler to generate independent Poisson point patterns with the *same* intensity 50, say:

```
> H$X <- with(H, rpoispp(50))
```

The expression `rpoispp(50)` is evaluated once in each row, yielding a different point pattern in each row because of the randomness.

16.6 Replicated point pattern datasets in `spatstat`

The following datasets containing replicated point patterns are currently installed in `spatstat`. They are representative of different kinds of replicated data.

16.6.1 Waterstriders

The territorial behaviour of an insect group called water striders (Gerridae) was studied in a series of laboratory experiments by M. Nummelin, University of Helsinki. Larvae of *Limnoporus (Gerris) rufoscutellatus* (larval stage V) were photographed on the surface of a pool. Three photographs obtained at different times were scanned manually, and the locations of the water strider larvae recorded. A border of about 2.5 cm has been deleted at the edge of the photograph, leaving a homogeneous area about 48 cm square. The data were first analysed by A. Penttinen [539]. Professors Nummelin and Penttinen should be acknowledged in any use of these data.

The dataset `waterstriders` is a list of three point patterns giving the locations of the larvae in three windows of slightly different sizes. The patterns can be treated as independent replicates.

It is known that this species exhibits territorialism at older larval stages and at the adult stage. If any deviation from complete spatial randomness exists in these three point patterns, it is expected to be towards inhibition. Penttinen [539, chap. 5] fitted a pairwise interaction model with a Strauss/hard core interaction with hard core radius 1.5 cm and interaction radius 4.5 cm.

16.6.2 Breakdown spots

The application of successive voltage sweeps to the metal gate electrode of a microelectronic capacitor generates multiple breakdown spots on the electrode. The spatial distribution of these breakdown spots in MIM (metal-insulator-metal) and MIS (metal-insulator-semiconductor) structures was observed and analysed by Miranda *et al.* [472, 471] and Saura *et al.* [598, 596, 597]. Please cite [598] in any use of these data.

The dataset `bdspots` gives the breakdown spot patterns for three circular electrodes of different radii, 169, 282, and 423 microns, respectively, in MIM structures analysed in [598]. It is a list (of class `"solist"`) of three spatial point patterns in circular windows. Spatial coordinates are given in microns. Figure 16.6 shows the result of `plot(bdspots, equal.scales=TRUE)`.

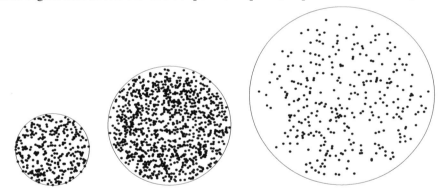

Figure 16.6. *The breakdown spots data. Three point patterns at the same physical scale: largest circle diameter is 846 microns.*

16.6.3 Osteocyte lacunae

Osteocyte lacunae are holes in solid bone which were occupied by osteocytes (bone-building cells) during life. They were amongst the first biological structures ever to be visualised by confocal microscopy, because of their strong optical contrast.

The three-dimensional locations of osteocyte lacunae in bone were visualised in 1984 using a tandem-scanning (confocal) reflected light microscope, operated by Adrian Baddeley [346]. The samples were three intact skulls and one skull cap, all originally identified as belonging to the macaque monkey *Macaca fascicularis*. Later analysis [66, 68] suggested that the skull cap, given here as the first animal, was a different subspecies, and this was confirmed by anatomical inspection. Please cite [346, 66, 68] in any use of these data.

The spatial point patterns in this dataset are *three-dimensional*. Each pattern gives the (x, y, z) coordinates (in microns) of all visible osteocyte lacunae in a three-dimensional rectangular box or 'brick' of dimensions $81 \times 100 \times d$ microns, where d varies. The z-coordinate is depth into the bone (depth of the focal plane of the confocal microscope); the (x, y) plane is parallel to the exterior surface of the bone; the relative orientation of the x and y axes is not important. Three-dimensional point patterns are discussed in Section 15.3.

From each skull or skull cap, observations were collected in 10 separate sampling volumes, giving a total of 40 three-dimensional point patterns. The study thus conformed to a nested sampling design with replication at all levels.

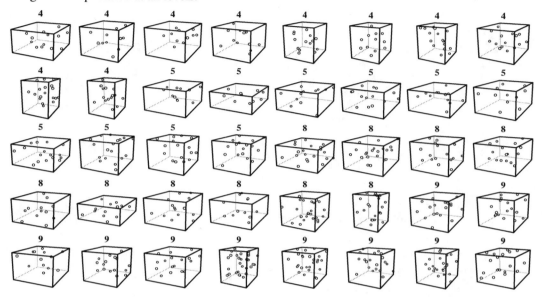

Figure 16.7. *Osteocyte lacunae data. Numerals identify the four animal skulls from which the samples were taken. Each box has base dimensions 81×100 microns.*

The hyperframe `osteo` has 40 rows and 5 columns: `id` is a factor identifying the bone sample by its original record name in the collection; `shortid` is the last digit of the record name; `brick` is the serial number of the sampling volume within the bone sample (ranging from 1 to 10); `pts` is the three-dimensional point pattern (object of class `"pp3"`); and `depth` is the depth of the brick in microns.

```
> head(osteo)
       id shortid brick   pts depth
1 c77za4       4     1 (pp3)    45
2 c77za4       4     2 (pp3)    60
3 c77za4       4     3 (pp3)    55
4 c77za4       4     4 (pp3)    60
5 c77za4       4     5 (pp3)    85
6 c77za4       4     6 (pp3)    90
```

Figure 16.7 shows the 40 point patterns from this dataset.

16.6.4 Pyramidal neurons

The `pyramidal` hyperframe contains data from a study conducted to assess evidence for the existence of an anatomical deficit in the brains of patients with schizophrenia. Histological sections of the cingulate cortex (Brodmann area 24, layer 2) of the brain were obtained from 31 human subjects at death. For each subject, the spatial pattern of neurons in just one histological section is recorded.

The subjects were classified into three groups: controls (12 subjects), schizoaffective (9 subjects), and schizophrenic (10 subjects). The `pyramidal` dataset is a hyperframe with 31 rows and 2 columns, the first `Neurons` consisting of the point patterns and the second `group` being the factor which classifies the subjects into groups. Each point pattern is recorded in a unit square region; the unit of measurement is unknown.

Figure 16.8 shows a plot of the pyramidal neurons data, with panel titles indicating the grouping. The data were introduced and analysed in [232], and re-analysed in [234, 455, 87]. Please cite [232] in any use of these data.

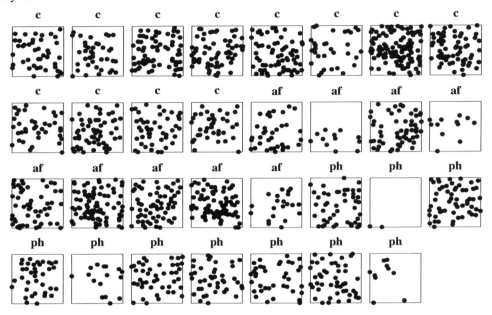

Figure 16.8. *Pyramidal neurons data, with panel titles* c, af, ph *indicating the division into control, schizoaffective, and schizophrenic groups, respectively. Spatial coordinate units unknown.*

One important reservation about this kind of data is that the pattern is a two-dimensional projection of a histological section of brain tissue, that is, a slice of tissue which is perhaps only a few microns thick and a few hundred microns across. This gives rise to sampling effects [43] and changes the context for modelling and analysis. The pair correlation function in three dimensions can be estimated from the observed pair correlation in two dimensions [519, sec. 9.3.2]. The same is not true of the F-, G- and J-functions. A thick section of a three-dimensional Poisson, Cox, or cluster process is a two-dimensional point process of the same type. However, a thick section of a three-dimensional Gibbs process is not a two-dimensional Gibbs process.

16.6.5 Influenza proteins

In research on viral replication, infected cell membranes were stained with antibodies against viral proteins. The antibodies were conjugated (bound) to gold particles so that their location could be visualised by electron microscopy ('immunogold labelling'). Using two different sizes of gold particles, two different viral proteins can be visualised at the same time, giving a spatial pattern of two different types of points. The main question is whether there is evidence of interaction between the two proteins. Four different combinations of experimental conditions were used, and the experiment was replicated about 10 times under each set of conditions.

Canine kidney cells were infected with human influenza, Udorn strain, either the wild type or a mutant which encodes a defective M2 protein. At twelve hours post-infection, membrane sheets were prepared and stained for viral proteins, using two antibodies conjugated to gold particles of two sizes (6 nanometre and 12 nanometre diameter). The 6 *nm* particles were stained for M2 (ion channel protein), while the 12 *nm* particles were stained either for M1 (matrix protein) or for HA (hemagglutinin).

The result is a set of spatial point patterns of gold particle locations with coordinates in *nm*. The patterns are observed in square windows which are 3331 *nm* across. The patterns are marked, and each has two types of points, either M1 and M2, or HA and M2.

The patterns are grouped according to two experimental factors: virustype (either wild type wt or mutant mut1), and stain, which identifies whether the membrane was stained for M2 and M1 (stain="M2-M1") or stained for M2 and HA (stain="M2-HA"). There are four combinations of experimental conditions. The experiment was replicated 11 times under three of the combinations and eight times with the "wt"/"M2-HA" combination. The main question is whether there is evidence of interaction between the two proteins.

Experimental technique and spatial analysis of the membranes stained for M2 and M1 is reported in [146]. Analysis of the membranes stained for M2 and HA is reported in [589]. The M2-HA data show a stronger association between the two proteins, consistent with other research [589].

These data were generously provided by G.P. Leser and R.A. Lamb. Please cite [146] in any use of these data. Figure 16.9 shows two of the 41 point patterns.

The dataset is stored in a hyperframe flu which has 41 rows (one for each membrane sheet) and four columns. The column named pattern contains the spatial point patterns of gold particle locations. There are two types of point in each pattern: either M1 and M2, or HA and M2. The column named stain is a factor identifying whether the membrane was stained for M1 and M2 (stain="M2-M1") or stained for HA and M2 (stain="M2-HA"). The column virustype is a factor identifying the virus: either wild type wt or mutant mut1. The row names of the hyperframe are a succinct summary of the experimental conditions and can be used as labels in plots.

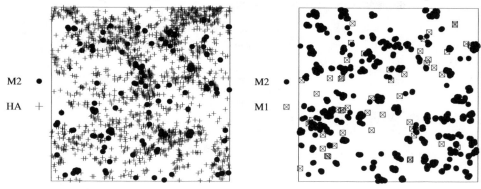

Figure 16.9. *The point patterns in row 12* (left) *and row 24* (right) *of the* flu *dataset. Each panel is 3331 nm across.*

16.6.6 Simulated data

The simba dataset, a hyperframe with 10 rows and 2 columns, contains simulated data from an experiment with a 'control' group and a 'treatment' group, each group containing 5 experimental units. The responses in the control group are independent Poisson point patterns with intensity 80. The responses in the treatment group are independent realisations of a Strauss process with activity parameter $\beta = 100$, interaction parameter $\gamma = 0.5$, and interaction radius $R = 0.07$ in the unit square. The simba dataset is a hyperframe with 10 rows and 2 columns: Points (the point patterns) and group (a factor with levels control and treatment).

The demohyper hyperframe is a very simple artificial example with only three rows and three columns. These columns consist of point patterns (Points), pixel image covariates (Image), and a grouping factor (Group), this last having two levels a and b.

16.7   Exploratory data analysis

Before fitting models to the data, it is prudent to explore the data to detect unusual features and to suggest appropriate models.

16.7.1   Exploring effects of design covariates

A design covariate (as opposed to a spatial covariate) has a single numerical or categorical value for each point pattern. Design covariates specify the experimental conditions for each unit, such as the assignment of units into groups. A simple way to explore the influence of design covariates is to replace the response point patterns by a simple numerical summary, such as the number of points or the Clark-Evans index, then use basic statistical methods.

One question about the pyramidal neurons data is whether the intensity is different between the three groups. To explore this, we reduce the hyperframe to a data frame, adding columns for the number of points and the area of each window:

```
> py <- pyramidal
> py$n <- with(py, npoints(Neurons))
> py$area <- with(py, area(Neurons))
> py <- as.data.frame(py, warn=FALSE)
```

In this example the window for each point pattern is the unit square, but it is still good practice to include a column recording the window area, to prevent erroneous analysis if the code is re-used for other data.

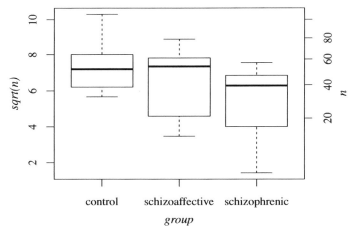

Figure 16.10. *Boxplots (square root scale) of numbers of pyramidal neurons for each group.*

Figure 16.10 shows boxplots of the square root of the number of points in each pattern, broken down by the three groups, generated by plot(sqrt(n) ~ group, data=py). The square root is the variance-stabilising transformation for the Poisson distribution and is commonly used for count data. The tick marks on the right hand side show the original count scale, and were generated by v <- pretty(py$n); axis(4, at=sqrt(v), labels=v). The boxplots give some reason to be cautious about assuming a Poisson distribution. Random effects, due to factors such as randomisation of the orientation of the histological section, are plausible.

If the intensity is estimated from each individual point pattern in the usual way, the group means of these intensity estimates are:

```
> sapply(split(py$n/py$area, py$group), mean)
     control schizoaffective    schizophrenic
        54.6            45.1             33.9
```

If the window areas were unequal, a weighted average might be more appropriate, usually the average with weights proportional to areas:

```
> ntot <- sapply(split(py$n, py$group), sum)
> atot <- sapply(split(py$area, py$group), sum)
> ntot/atot
     control schizoaffective    schizophrenic
        54.6            45.1             33.9
```

giving the same result in this case. The same analysis could be performed using generalised linear modelling:

```
> fitn <- glm(n ~ offset(log(area)) + group,
              family=poisson, data=py)
> newd <- data.frame(area=1, group=levels(py$group))
> predict(fitn, newdata=newd, type="response")
   1    2    3
54.6 45.1 33.9
```

The last line shows the fitted intensities for the three groups, since we asked for predictions of the mean count in a unit area. A simple test for significant difference between the groups is also available:

```
> anova(fitn, test="LRT")
Analysis of Deviance Table
Model: poisson, link: log
Response: n
Terms added sequentially (first to last)
       Df Deviance Resid. Df Resid. Dev Pr(>Chi)
NULL                      30        377
group   2    52.9         28        324  3.3e-12 ***
---
Signif. codes:  0 '***' 0.001 '**' 0.01 '*' 0.05 '.' 0.1 ' ' 1
```

The tentative conclusion is that there is a significantly lower intensity in the schizoaffective and schizophrenic groups. Note that this conclusion refers to the abundance of *cell profiles* of pyramidal neurons observed in a histological section of brain tissue, rather than the density of pyramidal neurons in the three-dimensional tissue. A reduction in intensity on sections can be caused by a loss of cell number, or by a reduction in the average size of cells [43, Chap. 1].

Assuming the point patterns are stationary (which is reasonable because of the sampling protocol for these sections), we could do a similar analysis with the Clark-Evans index.

```
> py$ce <- with(pyramidal, clarkevans(Neurons, correction="Donnelly"))
> sapply(split(py$ce, py$group), mean)
     control schizoaffective    schizophrenic
        1.09            1.03             1.03
```

This calculation suggests that the three groups are not different with regard to their spatial pattern. Under the null hypothesis of CSR, the Clark-Evans index is asymptotically normally distributed, but with a variance inversely proportional to the intensity. The test statistic of the Clark-Evans test is a standardised version of the Clark-Evans index, so it is more appropriate to compare these test statistic values.

```
> py$z <- with(pyramidal, clarkevans.test(Neurons)$statistic)
> anova(lm(z ~ group, data=py))
Analysis of Variance Table
Response: z
          Df Sum Sq Mean Sq F value Pr(>F)
group      2  0.011 0.00543    0.34   0.71
Residuals 28  0.443 0.01582
```

For the synthetic simba dataset the result is quite different:

```
> sdf <- as.data.frame(simba, warn=FALSE)
> sdf$z <- with(simba, clarkevans.test(Points)$statistic)
> anova(lm(z ~ group, data=sdf))
Analysis of Variance Table
Response: z
          Df Sum Sq Mean Sq F value  Pr(>F)
group      1 0.0742  0.0742    27.8 0.00075 ***
Residuals  8 0.0213  0.0027
---
Signif. codes:  0 '***' 0.001 '**' 0.01 '*' 0.05 '.' 0.1 ' ' 1
```

Here there is a significant difference in spatial pattern between the two groups of point patterns.

16.7.2   Exploring spatial trend and spatial covariate effects

Inhomogeneity of the point patterns could be investigated using the kernel smoothed intensity.

```
> plot(pyramidal, quote(plot(density(Neurons), main=group)))
> plot(with(pyramidal, density(Neurons)))
```

Covariate effects due to a real-valued spatial covariate (a real-valued pixel image) can be investigated using rhohat (Section 6.6.3).

```
> plot(with(demohyper, rhohat(Points, Image)))
```

The results of these plots are not shown.

16.8   Analysing summary functions from replicated patterns

16.8.1   Pooling summary functions

Combining or *pooling* several datasets into a single dataset is a common statistical procedure. Pooling can also be applied to summary statistics: if a summary statistic has already been computed for each of several datasets, then the 'pooled' summary statistic is the value that would have been obtained if we had first pooled the datasets and then computed the summary. Often this can be calculated without recourse to the original data. For example, if two groups of people, containing n_1

and n_2 people, respectively, had average heights of h_1 and h_2, respectively, then the pooled average height is $(n_1 h_1 + n_2 h_2)/(n_1 + n_2)$, the weighted average of the group averages with weights proportional to size of group. This is the correct calculation because $n_i h_i$ is the total height in group i, so the numerator $n_1 h_1 + n_2 h_2$ is the total height of all people in the two groups, while the denominator $n_1 + n_2$ is the total number of people.

In general if the estimator of our desired quantity R is a ratio $\widehat{R} = \widehat{Y}/\widehat{X}$, and we calculate estimates $\widehat{R}_1, \ldots, \widehat{R}_m$ from m different datasets, the natural pooled estimate is the *ratio-of-sums* estimator

$$\overline{R} = \frac{\sum_i \widehat{Y}_i}{\sum_i \widehat{X}_i} = \frac{\sum_i \widehat{X}_i \widehat{R}_i}{\sum_i \widehat{X}_i}, \tag{16.1}$$

the weighted average of the individual ratios $\widehat{R}_i$ with weights proportional to $\widehat{X}_i$.

The ratio-of-sums rule (16.1) can be justified as the 'optimal' procedure, under some assumptions. Typically $\widehat{X}_i$ can be interpreted as the 'sample size' of the ith dataset, while $\widehat{Y}_i$ is the sample total of some quantity, so that $\widehat{R}_i = \widehat{Y}_i/\widehat{X}_i$ is the sample average, and the variance of $\widehat{Y}_i$ is proportional to the sample size. Suppose that the pairs $(\widehat{X}_1, \widehat{Y}_1), \ldots, (\widehat{X}_m, \widehat{Y}_m)$ are independent and identically distributed pairs, and that

$$\mathbb{E}[\widehat{Y}_i \mid \widehat{X}_i = x] = Rx \tag{16.2}$$

$$\mathrm{var}[\widehat{Y}_i \mid \widehat{X}_i = x] = cx \tag{16.3}$$

where c is a constant. Then $\overline{R}$ is the minimum variance linear unbiased estimator of R given the denominators $\widehat{X}_1, \ldots, \widehat{X}_m$. See [193, 66, 68, 696].

This pooling principle may also be applied to summary functions such as the K-function. Estimators of the K-function are ratios $\widehat{K}(r) = \widehat{Y}(r)/\widehat{X}(r)$ where the numerator $\widehat{Y}(r)$ and denominator $\widehat{X}(r)$ are sums over all pairs of data points. For the isotropic (7.17) and translation (7.19) corrections, $\widehat{X}(r)$ is simply equal to the number of pairs of distinct points. If we are given the estimates $\widehat{K}_1(r), \ldots, \widehat{K}_m(r)$ of the K-function obtained from m replicated point patterns, the pooled estimate $\overline{K}(r)$ can be obtained by averaging these K-functions with weights proportional to $n_i(n_i - 1)$ where n_i is the number of data points in the ith point pattern.

The ratio-of-sums pooling rule is implemented in spatstat by the generic function pool. It is used to combine the data from several objects of the same type, and to compute statistics based on the combined dataset. It may be used to pool the estimates obtained from replicated datasets. It may also be used in high-performance computing applications, when the objects have been computed on different processors or in different batch runs, and we wish to combine them.

The generic pool has methods for "fv", "fasp", "envelope", and other classes. In order to make use of the ratio estimation procedure described above, the summary functions to be pooled should be computed with the argument ratio=TRUE so that the numerator and denominator are stored. For example, suppose we obtain estimates of the K-function for the three waterstriders patterns,

```
> Keach <- lapply(waterstriders, Kest, ratio=TRUE)
> Keach[[1]]
Function value object (class 'fv')
for the function r -> K(r)
.........................................
        Description
r       distance argument r
theo    theoretical Poisson K(r)
border  border-corrected estimate of K(r)
trans   translation-corrected estimate of K(r)
iso     isotropic-corrected estimate of K(r)
```

```
. . . . . . . . . . . . . . . . . . . . . . . . . . . . . . . . . . . . . . . . . . .
Default plot formula:  .~r
where "." stands for 'iso', 'trans', 'border', 'theo'
Recommended range of argument r: [0, 12.025]
Available range of argument r: [0, 12.025]
Unit of length: 1 cm
[Contains ratio information]
```

Notice the last line of output. To pool these *K*-functions,

```
> K <- pool(Keach[[1]], Keach[[2]], Keach[[3]])
```

or more neatly K <- do.call(pool, Keach) or

```
> K <- pool(as.anylist(Keach))
```

The pooling operation calculates the weighted mean of the individual *K*-function estimates.

16.8.2   Variability in summary functions

Replication also allows us to measure variability, and so to obtain a standard error for the pooled estimate $\overline{R}$. Using the delta-method approximation to the variance of a ratio [159, pp. 154,161], [68]

$$\operatorname{var}\frac{A}{B} \approx \frac{(\mathbb{E}A)^2}{(\mathbb{E}B)^2}\left[\frac{\operatorname{var}A}{(\mathbb{E}A)^2} + \frac{\operatorname{var}B}{(\mathbb{E}B)^2} - 2\frac{\operatorname{cov}(A,B)}{(\mathbb{E}A)(\mathbb{E}B)}\right] \qquad (16.4)$$

and taking $A = \sum_i \widehat{Y}_i$ and $B = \sum_i \widehat{X}_i$, we can estimate the variance $V = \operatorname{var}\overline{R}$ by

$$\widehat{V} = \frac{\overline{R}^2}{m}\left[s_{XX}^* + s_{YY}^* - 2s_{XY}^*\right] \qquad (16.5)$$

where $s_{XX}^*, s_{YY}^*, s_{XY}^*$ are the entries of the empirical covariance matrix of the normalised observations $(X_1^*, Y_1^*), \ldots, (X_m^*, Y_m^*)$ where $X_i^* = m\widehat{X}_i / \sum_{j=1}^m \widehat{X}_j$ and $Y_i^* = m\widehat{Y}_i / \sum_{j=1}^m \widehat{Y}_j$. The square root of (16.5) gives a standard error for $\overline{R}$.

Alternatively, if we are confident that the proportional regression assumptions (16.2)–(16.3) are correct, then the moments on the right hand side of (16.4) could be expressed in terms of R and c, giving an estimator

$$\widetilde{V} = \frac{\widehat{c}\overline{R}^2}{\sum_i \widehat{X}_i} \qquad (16.6)$$

where $\widehat{c}$ is an estimate of the variance parameter c. See [193, 66, 68].

The function pool calculates a standard error based on (16.5) which is then used to make approximate 95% confidence intervals for the true *K*-function. Figure 16.11 shows the result of

```
> plot(K, cbind(pooliso, pooltheo, loiso, hiiso) ~ r,
        shade=c("loiso", "hiiso"))
```

The column named pooliso is the pooled estimate of the *K*-function using the isotropic correction. The columns hiiso and loiso give the upper and lower limits of the pointwise 95% confidence interval for the true value of $K(r)$ based on the estimated standard error.

When the data consist of several different groups of replicated point patterns, we may pool the *K*-functions within each group. For example, the osteo data are replicated patterns from four different animals. Pooled estimates of the *K*-function for each animal are computed by

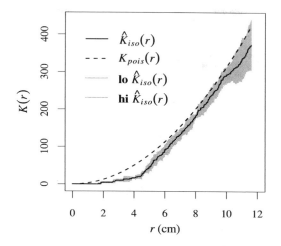

Figure 16.11. *Pooled K-function for the water striders data, with pointwise 95% confidence interval based on sample variance.*

```
> os <- osteo
> os$K <- with(os, K3est(pts, ratio=TRUE))
> Kanimal <- anylapply(split(os$K, os$id), pool)
```

Figure 16.12 shows the results for the intact skulls, `Kanimal[-1]`, plotted in the same style as above.

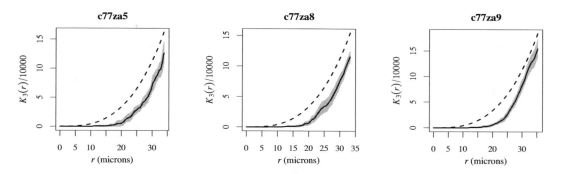

Figure 16.12. *Pooled three-dimensional K-functions for each intact skull in the* `osteo` *data.*

The grey shading in each panel of Figure 16.12 is a confidence interval for the K-function for the corresponding animal, based on the observed *within-animal variability* in the estimated K-function. Discrepancies between the estimated K-functions for the different animals can be used to infer *between-animal variability* in the K-function (i.e. between animals in the hypothetical population from which the samples were drawn). This is a kind of analysis of variance [66, 232, 68].

16.8.3 Permutation test for difference between groups

In analysing the `pyramidal` data, the main task is to assess evidence for a difference between control, schizoaffective, and schizophrenic subjects in the spatial pattern of pyramidal neurons.

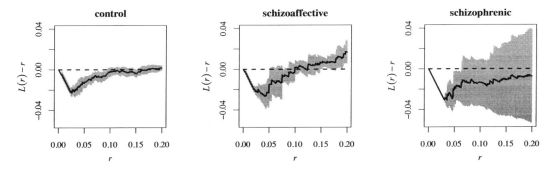

Figure 16.13. *Pooled estimates of centred L-function for each group of subjects in the* pyramidal *data.*

Figure 16.13 shows the pooled estimates of the centred *L*-function for each group of subjects, calculated using equal weight for each subject:

```
> pa <- pyramidal
> pa$L <- with(pa, Lest(Neurons))
> Lsplit <- split(pa$L, pa$group)
> Lpool <- anylapply(Lsplit, pool)
> plot(Lpool, cbind(pooliso,pooltheo,hiiso,loiso) - r ~ r,
      shade=c("hiiso", "loiso"), xlim=c(0, 0.2), equal.scales=TRUE)
```

Variability appears to be smallest in the control group and highest in the schizophrenic group. This is confirmed by Figure 16.14 which shows the individual estimates of the *L*-function for each subject, superimposed on each other within each group of subjects:

```
> Leach <- anylapply(Lsplit, collapse.fv,
                    same="theo", different="iso")
> plot(Leach, legend=FALSE, xlim=c(0, 0.2), ylim=c(0, 0.2))
```

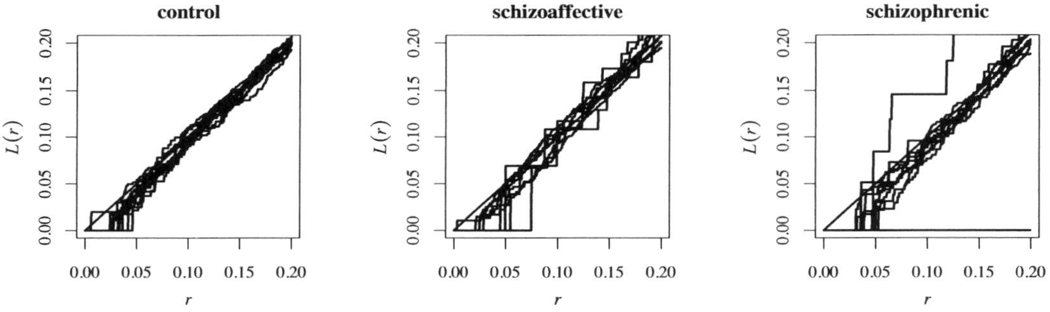

Figure 16.14. *Individual estimates of L-function for each subject in the* pyramidal *data, collected by group.*

The main question is whether the differences between groups are statistically significant. The null hypothesis, of no difference between groups, states that the observed point patterns are independent and identically distributed random patterns, regardless of the group they belong to. The alternative hypothesis of interest is that the point patterns within a group are independent and identically distributed. Diggle *et al.* [232] proposed a test for group differences based on bootstrapping the summary function estimates. Hahn [310] proposed a *permutation test* which is described here.

We gratefully acknowledge Dr Hahn for contributing R code and documentation, which have been incorporated in Sections 16.8.3–16.8.5.

The idea of using permutations to test for differences between groups goes back to Fisher [261] and Pitman [550]. Suppose we have g groups of point patterns, containing $m_1, \ldots, m_g$ patterns, respectively. The patterns in group 1 are named $\mathbf{x}_{11}, \ldots \mathbf{x}_{1m_1}$, etc. The tests proposed in [310] compare group means of estimates $\widehat{K}_{ij}(r)$ of the K-function of pattern $\mathbf{x}_{ij}$ on an interval $[r_0, r_1]$ for r, via the test statistic

$$T = \sum_{1 \le i < j \le g} \int_{r_0}^{r_1} \frac{(\bar{K}_i(r) - \bar{K}_j(r))^2}{\frac{1}{m_i} s_i^2(r) + \frac{1}{m_j} s_j^2(r)} \, dr, \tag{16.7}$$

where $\bar{K}_i(r) = (1/m_i) \sum_{j=1}^{m_i} \widehat{K}_{ij}(r)$ are the group means and

$$s_i^2(r) = \frac{1}{m_i - 1} \sum_{j=1}^{m_i} (\widehat{K}_{ij}(r) - \bar{K}_i(r))^2 \tag{16.8}$$

are the estimated within-group variances of the estimates for a given distance r. The test statistic T is essentially the sum of integrated squared studentized differences between group means, hence the name 'studentized permutation test'. The test is performed by evaluating T for the observed data, and for $N - 1$ random permutations of the list of point patterns, then computing the Monte Carlo p-value.

```
> testpyramidal <- studpermu.test(pyramidal, Neurons ~ group)
> testpyramidal

        Studentized permutation test for grouped point patterns
        Neurons ~ group
        3 groups: control, schizoaffective, schizophrenic
        summary function: Kest, evaluated on r in [0, 0.25]
        test statistic: T, 999 random permutations

data:  pyramidal
T = 2, p-value = 0.03
alternative hypothesis: not the same K-function
```

The first argument of `studpermu.test` contains the data, which may be either a list-of-lists of point patterns, or a hyperframe. The `pyramidal` data are given as a hyperframe, with a column `Neurons` that contains the point patterns, and a column `group` that assigns each pattern to one of the groups. The formula `Neurons ~ group` indicates which columns to use; it may be omitted in the present case since there are no other columns in the dataset.

This test is computationally cheap to perform, because the K-functions only need to be evaluated once. Accordingly a large number of random permutations can be used: the default is 1000 permutations.

Overall, no fewer than 12 patterns should be used in the test; otherwise, the number of possible different outcomes for the test statistic would be too small to yield sensible p-values. Groups should contain at least three patterns to achieve reasonably precise estimates for the within group variances in the denominator of the test statistics.

Instead of the K-function, any other summary function may be used, such as the G-, L-, or J-function. In the following code example, we compare two groups of simulated point patterns, using the pair correlation function for the test, and only 200 random permutations:

```
> set.seed(12345)
> sample1 <- rpoispp(100, nsim=10)
> sample2 <- rMaternII(110, 0.02, nsim=10)
```

```
> patterns <- list(Poisson = sample1, MaternII = sample2)
> studpermu.test(patterns, summaryfunction = pcf, nperm=199,
                 interval=c(0,0.15))
        Studentized permutation test for grouped point patterns
        2 groups: MaternII, Poisson
        summary function: pcf, evaluated on r in [0, 0.25]
        test statistic: T, 199 random permutations

data:  patterns
T = 1, p-value = 0.005
alternative hypothesis: not the same pair correlation function
```

If a variance-stabilized summary function is used, like the *L*-function, the variance estimate in the denominator can be improved by pooling over all *r*-values. This suggests an alternative test statistic of the form

$$\overline{T} = \sum_{1 \le i < j \le g} \int_{r_0}^{r_1} \frac{(\bar{L}_i(r) - \bar{L}_j(r))^2}{\frac{1}{m_i}\overline{s_i^2} + \frac{1}{m_j}\overline{s_j^2}}\,dr \tag{16.9}$$

where

$$\overline{s_i^2} = \frac{1}{r_1 - r_0} \int_{r_0}^{r_1} \frac{1}{m_i - 1} \sum_{j=1}^{m_i} (\widehat{L}_{ij}(t) - \bar{L}_i(t))^2\,dt.$$

Using $\overline{T}$ instead of T improves robustness of the test in the case where the intensity differs widely between the groups, but it can lead to a loss of power for regular point patterns. In studpermu.test the statistic $\overline{T}$ is used if use.Tbar=TRUE:

```
> studpermu.test(pyramidal, summaryfunction = Lest, use.Tbar = TRUE)
        Studentized permutation test for grouped point patterns
        Neurons ~ group
        3 groups: control, schizoaffective, schizophrenic
        summary function: Lest, evaluated on r in [0, 0.25]
        test statistic: Tbar, 999 random permutations

data:  pyramidal
Tbar = 2, p-value = 0.01
alternative hypothesis: not the same L-function
```

16.8.4   Permutation test to compare subsets of a single pattern

The permutation test for differences between groups of point patterns can also be used to test for differences between subsets of a *single* point pattern. The original point pattern is first divided into two or more subsets which we wish to compare. Each subset is then further subdivided into a group of sub-sub-patterns which we will treat as if they were replicated observations within the group. We then test for differences between these groups of sub-sub-patterns.

For example, the dataset waka gives the locations and diameters of trees in a 100 metre square survey plot in Waka National Park, Gabon [74, 547]. We divide the pattern into large and small trees, declaring a tree to be large if its diameter at breast height *dbh* exceeds 20 *cm*:

```
> waka2 <- split(cut(waka, breaks=c(0,20,200),
                 labels = c("small", "large")))
```

See Figure 16.15. Then invoking

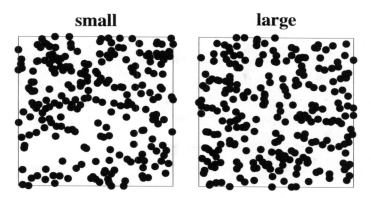

Figure 16.15. *The* waka *dataset split into two size classes of trees. Survey region is 100 m across.*

```
> Lwak <- anylapply(waka2, Lest, correction="iso")
```

we obtain the estimated *L*-functions of the small and large trees, shown in Figure 16.16: the small trees appear to be clustered, while the large trees do not.

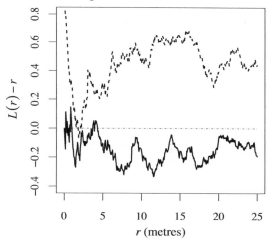

Figure 16.16. *Estimates of the centred L-function $L(r) - r$ for the two subsets in the previous figure. Solid lines: large trees. Dashed lines: small trees. Dotted line: theoretical value for random pattern.*

The discrepancy between the two *L*-functions for $r > 5$ is visually impressive. In order to test whether this difference is significant, we subdivide the large and small trees into sub-patterns defined by a 3×3 grid of quadrats.

```
> subwindows <- quadrats(waka, 3, 3)
> treegroups <- lapply(waka2, split, f = subwindows)
```

Then treegroups is a list of length 2, with first and second elements giving the data from the small and large trees, respectively. Each element of treegroups is a list of 9 point patterns obtained by splitting the relevant points over the 3×3 grid of quadrats. In this special case, the two point patterns to compare are obtained from the same pattern waka. This makes it possible to use the same subdivision into quadrats, here stored as subwindows. We can now perform the permutation test:

```
> wakatest <- studpermu.test(treegroups, summaryfunction = Lest,
                              rinterval=c(5,15))
```

We set a maximum distance of 15 *m* for the comparison of summary functions because the side
length of the quadrats is only 33 *m*.

```
> wakatest
        Studentized permutation test for grouped point patterns
        2 groups: large, small
        summary function: Lest, evaluated on r in [5, 15]
        test statistic: T, 999 random permutations

data:  treegroups
T = 20.4, p-value = 0.17
alternative hypothesis: not the same L-function
```

The resulting *p*-value of 0.17 means the null hypothesis is not rejected.

The validity of the permutation test is questionable here, because the patterns within each group
are not independent random point patterns, unless they come from a Poisson point process. The
dependence is, however, much weaker between the estimates of summary functions obtained from
the individual subpatterns, so that the permutation test may be approximately correct. A simulation
study in [310] suggests that the significance level of the test is approximately correct provided each
of the sub-patterns contains at least 100 points. If the sub-patterns have 25 points each, tests become
slightly anticonservative for strongly clustered point processes, but not for processes that are regular
or close to Poisson.

The general recommendation given in [310] is to split the patterns into a few large quadrats,
rather than into many small ones. For patterns in rectangular windows, a 3×3 grid of quadrats
seems to work well, provided the individual subpatterns contain at least 20 points. Thus, these tests
are reliable for datasets of 200 points or more.

Figure 16.17 shows the result of plot(wakatest, . - r ~ r, lwd.mean=2). Group means
are shown if the argument lwd.mean is set.

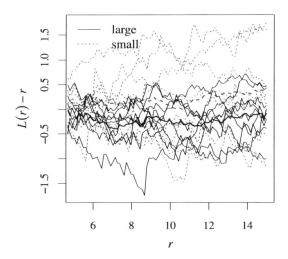

Figure 16.17. *Estimates of the centred L-function $L(r) - r$ on nine quadrats from each of the two
patterns in Figure 16.15, and group means.*

The group means of the estimated *L*-functions of the sub-sub-patterns differ from the *L*-function estimates computed from the patterns of big and small trees. This is partly due to edge correction on the 'internal' edges, and partly due to the fact that the average was computed using equal weights.

A more structured way to perform the same test is to use the function `nestsplit`. If X is a single point pattern, `nestsplit(X, f1, f2)` will split X into sub-patterns according to the splitting factor or splitting tessellation `f1`, then split each of these sub-patterns into sub-sub-patterns according to the splitting factor or splitting tessellation `f2`. The result is a hyperframe containing the sub-sub-patterns, together with two columns `f1` and `f2` indicating the group membership of each sub-sub-pattern. The example above could be accomplished with

```
> sizes <- cut(marks(waka), breaks=c(0,20,200),
             labels = c("small", "large"))
> arbres <- nestsplit(waka, sizes, quadrats(waka, nx=3, ny=3))
```

so that the pattern is split according to the factor `f1=sizes` into two patterns consisting of large and small trees, and then split into sub-sub-patterns according to the tessellation `f2=quadrats(waka, nx=3, ny=3)`. The same result could have been obtained with

```
> arbres <- nestsplit(waka, sizes, nx=3, ny=3)
```

We can perform the test by

```
> epreuveWaka <- studpermu.test(arbres, pts ~ f1)
```

16.8.5 Testing the type of inhomogeneity in a single pattern

When analysing correlation in a single point pattern, a crucial question is whether the pattern can be treated as homogeneous, and if not, what type of inhomogeneity can be assumed. If we used the K-function we would effectively be assuming that the data were generated by a stationary point process. Using the inhomogeneous K-function defined in Section 7.10.2 assumes that the process is correlation stationary. Using the locally scaled K-function defined in Section 7.10.3 assumes that the process is a locally scaled version of a stationary process.

Permutation tests can be used to check which type of homogeneity or inhomogeneity assumption is most appropriate. Suppose we divide the point pattern into sub-patterns using a grid of quadrats, and compute an estimate of a summary function from each quadrat. If the point process is stationary, then estimates of the K-function obtained from different quadrats should be similar. If the process is correlation-stationary, estimates of the inhomogeneous K-function should be similar in each quadrat, and so on.

The bronze filter data (Figure 1.9 on page 8) are clearly inhomogeneous. For simplicity we ignore the profile diameters and consider only the locations. We might contemplate assuming that the underlying process is correlation-stationary or that it is locally scaled. A simple exercise would be to divide the pattern into two halves containing equal numbers of points, and to compare summary function estimates obtained from the two halves:

```
> bro <- unmark(bronzefilter)
> mx <- median(coords(bro)$x)
> halves <- chop.tess(Window(bro), infline(v=mx))
```

Here we used `chop.tess` to divide the window into two halves separated by the vertical line at the median x-coordinate of the data points. The result `halves` is a tessellation consisting of two rectangular tiles, each containing the same number of data points. Figure 16.18 shows the split. (The same result can be obtained using `quantess(bro, "x", 2)` as explained below.)

The locally scaled K-functions for the two subsets can now be computed:

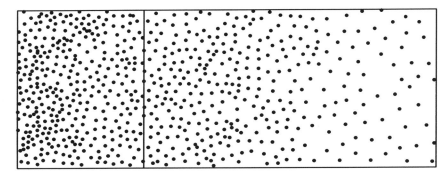

Figure 16.18. *Bronze filter particle locations data, split into two subsets containing equal numbers of points. Outer frame is* 18×7 *mm.*

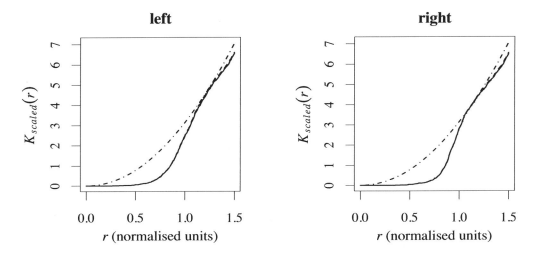

Figure 16.19. *Estimates of the locally scaled K-function obtained from the left and right halves of the bronze filter data.*

```
> Ks <- anylapply(split(bro, halves), Kscaled)
```

The results are shown in Figure 16.19.

For the inhomogeneous K-function we need an estimate of the intensity. It is reasonable to assume the intensity is a function of the x-coordinate, and to estimate it using a relative distribution (Section 6.6.3):

```
> bronzerho <- rhohat(bro, "x", method="tr")
> bronzelambda <- predict(bronzerho)
```

Here we set `method="transform"` to select the transformation smoother (6.23), page 180, which usually works well for locally scaled data. The function `bronzerho` is plotted in Figure 16.20.

Using this intensity estimate we compute the inhomogeneous K-functions for the two halves:

```
> Ki <- anylapply(split(bro, halves), Kinhom, lambda=bronzelambda)
```

The results are shown in Figure 16.21.

The results of `Kinhom` from the two halves of the data are quite different, which suggests that the assumption of correlation-stationarity is not appropriate.

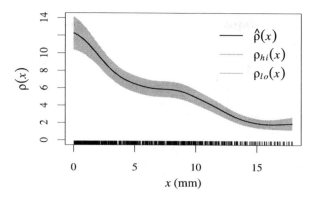

Figure 16.20. *Estimated intensity of* `bronzefilter` *pattern as a function of x-coordinate.*

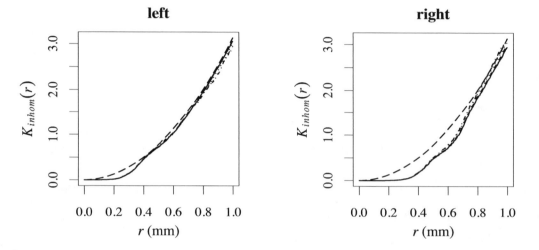

Figure 16.21. *Estimates of the inhomogeneous K-function obtained from the left and right halves of the bronze filter data.*

Hahn and Jensen [311] propose a permutation test of the inhomogeneity assumption. The point pattern is divided into quadrats, and the quadrats are organised into two groups representing high and low intensity, respectively. A permutation test is then applied to the summary function estimates obtained from these groups of patterns.

To apply this test to the bronze filter data, we first need to split the data into *several* quadrats with roughly equal numbers of points. For convenience we can use the function `quantess` which computes a 'quantile tessellation', a division of space into tiles which contain roughly equal numbers of points:

```
> bronze6 <- quantess(bro, "x", 6)
```

The result `bronze6` is a tessellation consisting of 6 tiles, shown in Figure 16.22.
We then apply `nestsplit` to divide each of these tiles into three sub-tiles:

```
> b63 <- nestsplit(bro, bronze6, ny=3)
```

The result is a hyperframe containing the sub-sub-patterns and two factors indicating the splitting history. The first factor `f1` indicates the horizontal position (the sextile of the *x*-coordinate).

692 Spatial Point Patterns: Methodology and Applications with R

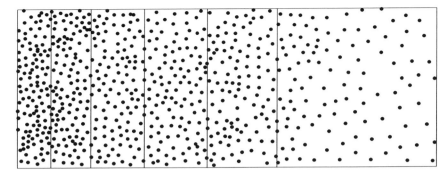

Figure 16.22. *Subdivision of the* `bronzefilter` *pattern into 6 quadrats with approximately equal numbers of points.*

```
> head(b63)
     pts f1 f2
1 (ppp)  1  1
2 (ppp)  1  2
3 (ppp)  1  3
4 (ppp)  2  1
5 (ppp)  2  2
6 (ppp)  2  3
```

We now re-group the 6 sub-sub-patterns into 2 groups of 3 patterns, where the groups contain patterns with high and low intensity, respectively:

```
> b63$inten <- factor(as.integer(b63$f1) <= 3, labels=c("Hi","Lo"))
```

Finally we can apply the permutation test:

```
> locTest <- studpermu.test(b63, pts ~ inten, summaryfunction=Kscaled,
                            rinterval=c(0, 1.5))
> corrTest <- studpermu.test(b63, pts ~ inten, summaryfunction=Kinhom,
                             lambda=bronzelambda, rinterval = c(0, 0.7))
```

Note that the argument `lambda` is passed to `Kinhom`.

```
> locTest

        Studentized permutation test for grouped point patterns
        pts ~ inten
        2 groups: Hi, Lo
        summary function: Kscaled, evaluated on r in [0, 1.5]
        test statistic: T, 999 random permutations

data:  b63
T = 1, p-value = 0.2
alternative hypothesis: not the same locally scaled K-function
> corrTest

        Studentized permutation test for grouped point patterns
        pts ~ inten
        2 groups: Hi, Lo
        summary function: Kinhom, evaluated on r in [0, 0.7]
        test statistic: T, 999 random permutations
```

```
data:  b63
T = 5, p-value = 0.004
alternative hypothesis: not the same inhomogeneous K-function
```

As one might have expected, the null hypothesis of correlation-stationarity is emphatically rejected, while the null hypothesis of local scaling is accepted. A weakness of the test based on `Kinhom` is that the intensity function must be estimated somehow; an inaccurate estimate of intensity could lead to a spuriously significant outcome.

16.9 Poisson models

The command `mppm` fits models to multiple point patterns. Its syntax is very similar to that of `lm`, `glm`, and `ppm`:

```
mppm(formula, data, interaction, ...)
```

where `formula` is a formula describing the systematic trend part of the model, `data` is a hyperframe containing all the data (responses and covariates), and `interaction` determines the stochastic interpoint interaction part of the model.

16.9.1 Trend formula

The trend formula has a left-hand side, which identifies the response. This should be the name of a column of `data`. For example with the `simba` dataset, to fit a model in which all patterns are independent replicates of the same uniform Poisson process, with the same constant intensity:

```
> mppm(Points ~ 1, simba)
Point process model fitted to 10 point patterns
Log trend formula: ~1
Fitted trend coefficients:
(Intercept)
       4.19
Interaction for all patterns:        Poisson process
```

The right side of the `formula` is an expression for the logarithm of the spatial trend. The variables appearing in the right hand side of `formula` should be either the names of columns in `data`, the names of objects in the R global environment (such as `pi` and `log`), or the reserved names x, y (representing Cartesian coordinates), `marks` (representing mark values attached to points), or `id` (a factor representing the row number in the hyperframe).

16.9.1.1 Design covariates

The variables in the trend could be 'design covariates'. To fit a model to the `simba` data in which the two groups of patterns (control and treatment groups) are assumed to consist of independent replicates of a uniform Poisson process, but with possibly different intensity in each group:

```
> mppm(Points ~ group, simba)
Point process model fitted to 10 point patterns
Log trend formula: ~group
Fitted trend coefficients:
```

```
     (Intercept) grouptreatment
           4.268          -0.154
Interaction for all patterns:              Poisson process
```

To fit a uniform Poisson process to each pattern, with a different intensity for each pattern:

```
> mppm(Points ~ id, simba)
Point process model fitted to 10 point patterns
Log trend formula: ~id
Fitted trend coefficients:
(Intercept)         id2         id3         id4         id5         id6
     4.2627      0.1193     -0.1195     -0.0432      0.0548     -0.1518
        id7         id8         id9        id10
    -0.0432     -0.3709     -0.2022     -0.0142
Interaction for all patterns:              Poisson process
```

To reduce the size of output in the rest of this chapter, we shall use a subset of the simba dataset consisting of every second row:

```
> simba2 <- simba[c(FALSE,TRUE), ]
```

16.9.1.2 Spatial covariates

The variables in the trend could be 'spatial covariates'. For example, the demohyper dataset has a column Image containing pixel images.

```
> mppm(Points ~ Image, data=demohyper)
Point process model fitted to 3 point patterns
Log trend formula: ~Image
Fitted trend coefficients:
(Intercept)        Image
      2.971        0.439
Interaction for all patterns:              Poisson process
```

This model postulates that each pattern is a Poisson process with intensity of the form $\lambda(u) = \exp(\beta_0 + \beta_1 Z(u))$ at location u, where β_0, β_1 are coefficients to be estimated, and $Z(u)$ is the value of the pixel image Image at location u.

Interaction terms between covariates were explained in Section 9.3.9 for ppm, and can be used similarly in mppm. To allow the coefficients β_0, β_1 to be different in each group, we could use either of the following:

```
> mppm(Points ~ Group/Image, data=demohyper)
Point process model fitted to 3 point patterns
Log trend formula: ~Group/Image
Fitted trend coefficients:
 (Intercept)       Groupb Groupa:Image Groupb:Image
       2.911        0.151        1.049       -0.705
Interaction for all patterns:              Poisson process
> mppm(Points ~ (Group-1)/Image, data=demohyper)
Point process model fitted to 3 point patterns
Log trend formula: ~(Group - 1)/Image
Fitted trend coefficients:
      Groupa       Groupb Groupa:Image Groupb:Image
       2.911        3.062        1.049       -0.705
Interaction for all patterns:              Poisson process
```

These two commands fit the same model using different parametrisations, as explained in Section 9.3.9.

16.9.2 Random effects

The *effect* of a covariate is the contribution made by the covariate to the experimental outcome. The name originated in agricultural field trials [261] where we think of a fertiliser or soil treatment as having an effect on crop yield. In a Poisson point process model, the effect of a covariate is the term in the log intensity that is associated with that covariate. The coefficients of these terms are also called 'effects'.

In the Poisson point process models described above, the model coefficients are assumed to be fixed numbers, whose unknown values must be estimated. This means that, if were to repeat the experiment, the same model coefficients would apply, so the effect of the covariate is repeatable and predictable. These are known as *fixed effects*. When we fit a fixed effect to data, we estimate the numerical value of the fixed effect, making it possible to predict the outcomes of future experiments.

An alternative is to assume that the effect of a particular covariate on a particular experimental unit is a *random* value, drawn from a population with a certain mean and variance. This is called a *random effect*. When we fit a random effect to data, we estimate only the mean and variance of the random effect, making it possible to predict the amount of variability in future experiments.

A *mixed effects model* is a model involving both fixed and random effects. Mixed effects models make it easy to analyse experiments with multiple sources of variability [549].

A random effect is appropriate when each experimental unit can be regarded as a randomly selected representative of a population, for example, when the observations are replicates.

In the model mppm(Points ~ id, simba) fitted on page 694, we fitted a separate fixed effect for each point pattern. In this simple case, the fitted fixed effects are estimates of the log intensity for each point pattern. There are many coefficients to be estimated — one coefficient for each point pattern, in this example. The individual fitted effects do not really have predictive value: if we were to repeat the experiment, we do not expect these fitted effect values to apply. It would have been more appropriate to treat the point patterns as replicates drawn randomly from a population. We would then fit a random effects model, estimating only the mean and variance of the random effects. This could be fitted by

```
> mppm(Points ~ 1, simba, random = ~ 1 | id)
```

Random-effects models for spatial point processes were first considered in generality by Bell and Grunwald [87].

Random effects are specified in mppm by the argument random. This is a formula, with no left-hand side, that specifies the structure of the random effects. The names in the formula may be any of the covariates supplied by data. Additionally the formula may involve the reserved name id, a factor representing row number in the hyperframe.

In a random effects formula, the operator | is used to represent 'conditioning'. The term 1 | g, where g is a factor, represents a random intercept that is the same for all experimental units that have the same level of g. If x is a numerical covariate, then the term x | g represents a random intercept and slope, that is, a term of the form $\beta_0 + \beta_1 x$, where the values of β_0, β_1 are assigned randomly, with the same values for all experimental units that have the same level of g.

The simplest random effects formula that makes sense for mppm is

$$\sim 1 \mid id$$

which signifies a random intercept term that is different for each point pattern:

```
> H <- hyperframe(P=waterstriders)
> mppm(P ~ 1, data=H, random=~1|id)
Point process model fitted to 3 point patterns
Log trend formula: ~1
Fixed effects:
(Intercept)
    -4.13
```

```
Random effects:
  (Intercept)
1     3.81e-08
2    -1.03e-07
3     6.46e-08
Random effects summary:
Random effects:
 Formula: ~1 | id
         (Intercept) Residual
StdDev:     0.000214        1

Variance function:
 Structure: fixed weights
 Formula: ~invwt
Interaction for all patterns:          Poisson process
```

Each point pattern is modelled as a stationary Poisson process, but the intensity is assumed to take different values in different point patterns. Hence, this is technically a Cox process model (a mixed Poisson process) rather than a Poisson process. The log intensity is assumed to be normally distributed, but this is not strictly a log-Gaussian Cox process, rather it is a degenerate case where the driving intensity is a random constant function. From the printout, the log intensity $\log \Lambda$ has an estimated mean of -4.13 and estimated standard deviation of 0.000214. Note that the mean of the lognormal distribution with parameters μ and σ is $\exp(\mu + \sigma^2/2)$, so that the estimated intensity of the model is $\exp(-4.13 + 0.000214^2/2) = 0.0161$ larvae per square *cm*.

The following mixed model postulates that there is a systematic difference between the three groups of subjects in the pyramidal neurons data (fixed effect), but that individual subjects also vary (random effect) according to a common distribution regardless of their grouping.

```
> mppm(Neurons ~ group, data=pyramidal, random=~1|id)
Point process model fitted to 31 point patterns
Log trend formula: ~group
Fixed effects:
         (Intercept) groupschizoaffective   groupschizophrenic
               3.956               -0.275               -0.545
Random effects:
  (Intercept)
1     -0.1669
2     -0.2472
3      0.2096

 [...]

29     0.1894
30     0.3782
31    -0.6949
Random effects summary:
Random effects:
 Formula: ~1 | id
         (Intercept) Residual
StdDev:        0.368        1

Variance function:
 Structure: fixed weights
 Formula: ~invwt
Interaction for all patterns:          Poisson process
```

The output has been shortened to save space. The mixed model allows us to ask the right questions about the pyramidal neurons data, since we are interested in the differences between groups and the *variability* amongst subjects within each group, rather than the parameters applicable to individual subjects.

The formula `random` should be recognisable to the function `lme` in the R package `nlme`. It can be of the form

$$\sim \texttt{x1} + \ldots + \texttt{xn} \mid \texttt{g1}/\ldots/\texttt{gm}$$

or

$$\sim \texttt{x1} + \ldots + \texttt{xn} \mid \texttt{g}$$

where `x1, ..., xn` are numerical covariates and `g, g1, ..., gm` are factors. Full details are given in the help file for `lme` (under the argument `random`).

16.9.3 Methods for fitted models

The result of `mppm` is a fitted model of class `"mppm"`. Important methods for this class are listed in Table 16.2.

coef	Extract fitted coefficients
fixef	Extract *fixed effect* coefficients
ranef	Extract *random effect* coefficients
anova	Analysis of deviance
print	Print basic information
summary	Print extensive summary information
plot	Plot fitted model
fitted	Compute fitted trend or conditional intensity
predict	Compute predictions for model
vcov	Variance-covariance matrix for coefficient estimates
logLik	log-likelihood
is.poisson	Determine whether model is Poisson
cdf.test	goodness-of-fit test based on spatial CDF
quadrat.test	χ^2 goodness-of-fit test from quadrat counts
residuals	point process residuals

Table 16.2. *Generic operations with methods for class* `"mppm"`.

Note that for a model that includes random effects, the results of `ranef` and `coef` will be data frames with one row for each point pattern, and one column for each effect. The result of `fixef` is a vector.

The methods for `anova`, `vcov`, `logLik`, `cdf.test`, `quadrat.test` effectively pool the contributions from all point patterns to obtain a single result. The methods for `predict` and `residuals` produce a list or hyperframe containing a separate result for each point pattern.

The analysis of deviance for a random-effects model is quite similar to the familiar analysis of variance:

```
> fitpyr <-mppm(Neurons ~ group, data=pyramidal, random=~1|id)
> anova(fitpyr)
            numDF denDF F-value p-value
(Intercept)     1 33237    1473  <.0001
group           2    28       3  0.0783
```

16.10   Gibbs models

The interaction between points in a point process model is specified by the arguments `interaction` and (optionally) `iformula` in

```
mppm(formula, data, interaction, ..., iformula)
```

The argument `interaction` can be either an object of class `"interact"` specifying a point process interaction, or a hyperframe containing such objects. The optional argument `iformula` determines which interactions will apply to each row of data, and determines how the interaction parameters applicable to different rows are connected to each other.

16.10.1   Exactly the same interaction for all patterns

In the simplest case, the argument `interaction` is one of the familiar objects that describe the point process interaction structure. It is an object of class `"interact"` created by calling one of the functions in Tables 13.2 and 13.3 on pages 516 and 519, respectively. In this 'simple' usage of `mppm`, the point process model assumes that all point patterns have exactly the same interpoint interaction, with the same interaction parameters.

```
> mppm(Points ~ id, data=simba2, interaction=Strauss(0.07))
Point process model fitted to 5 point patterns
Log trend formula: ~id
Fitted trend coefficients:
(Intercept)         id2         id3         id4         id5
      4.790      -0.243      -0.370      -0.655      -0.197
Interaction for all patterns:         Strauss(r = 0.07, gamma = 0.6462)
```

The result is a Strauss process with the same interaction radius r and interaction strength γ for each pattern, but with different first order parameters β for each pattern, because of the trend formula.

16.10.2   Same interaction, different interaction parameters

We may specify that the *fitted* interaction parameters (such as the interaction parameter γ in the Strauss process) are different in different rows of the hyperframe, using the argument `iformula`. This is a formula without a left-hand side, involving the reserved names `Interaction` and `id`, and the names of any columns of `data` which are vectors or factors. The `iformula` specifies how the interaction in each row depends on these variables.

To fit a Strauss model with the same interaction range $r = 0.07$ but a different interaction parameter γ in each row:

```
> fiteach <- mppm(Points ~ id, data=simba2, interaction=Strauss(0.07),
                  iformula = ~Interaction:id)
> fiteach
Point process model fitted to 5 point patterns
Log trend formula: ~id
Fitted trend coefficients:
      (Intercept)             id2             id3             id4             id5
           4.8258         -0.8530         -0.1774         -0.4101          0.0872
  id1:Interaction id2:Interaction id3:Interaction id4:Interaction id5:Interaction
          -0.4835          0.2236         -0.9057         -1.3252         -1.0755
Interaction 1:          Strauss(r = 0.07, gamma = 0.6166)
```

```
Interaction 2:          Strauss(r = 0.07, gamma = 1.251)
Interaction 3:          Strauss(r = 0.07, gamma = 0.4043)
Interaction 4:          Strauss(r = 0.07, gamma = 0.2657)
Interaction 5:          Strauss(r = 0.07, gamma = 0.3411)
```

The argument `iformula = ~Interaction:id` specifies (since `id` is a factor) that the interpoint interaction shall have a different coefficient γ for each row of the hyperframe.

The `iformula` allows us to specify an interaction that depends on the experimental design. For example

```
> fitgr <- mppm(Points ~ group, simba2, Strauss(0.07),
                iformula = ~Interaction:group)
```

This model has two different values for the Strauss interaction parameter γ, one for the control group and one for the treatment group. We also allowed the first order trend to depend on the grouping, which is usually sensible, but not obligatory.

The `print` method tries to do 'automatic interpretation' of the fitted model (translating the fitted interaction coefficients into meaningful numbers like γ). This will be successful in *most* cases:

```
> fitgr
Point process model fitted to 5 point patterns
Log trend formula: ~group
Fitted trend coefficients:
              (Intercept)              grouptreatment    groupcontrol:Interaction
                   4.3504                      0.3012                     -0.0459
grouptreatment:Interaction
                  -1.0258
Interaction 1:          Strauss(r = 0.07, gamma = 0.9551)
Interaction 2:          Strauss(r = 0.07, gamma = 0.9551)
Interaction 3:          Strauss(r = 0.07, gamma = 0.3585)
Interaction 4:          Strauss(r = 0.07, gamma = 0.3585)
Interaction 5:          Strauss(r = 0.07, gamma = 0.3585)
```

The fitted model can also be interpreted directly from the fitted canonical coefficients:

```
> coef(fitgr)
              (Intercept)              grouptreatment    groupcontrol:Interaction
                   4.3504                      0.3012                     -0.0459
grouptreatment:Interaction
                  -1.0258
```

The last output shows all the coefficients β_j in the linear predictor for the (log) conditional intensity. The interpretation of the model coefficients, for any fitted model in R, depends on the *contrasts* which were applicable when the model was fitted. See Section 9.3.4.2. In the output above, there is a coefficient for `(Intercept)` and one for `grouptreatment`. These are coefficients related to the group factor. According to the default 'treatment contrasts' rule, the `(Intercept)` coefficient is the estimated effect for the control group, and the `grouptreatment` coefficient is the estimated difference between the treatment and control groups. Thus the fitted first order trend is $\exp(4.35) = 77.51$ for the control group and $\exp(4.35 + 0.3012) = 104.7$ for the treatment group. The correct values used to generate this simulated dataset were 80 and 100.

The coefficients `groupcontrol:Interaction` and `grouptreatment:Interaction` are estimates of the canonical parameter of the interaction in the control and treatment groups respectively. Recall that the Strauss process interaction term is $\gamma^{t(u,\mathbf{x})} = \exp(t(u,\mathbf{x})\log\gamma)$ at a spatial location u, for a point pattern $\mathbf{x}$. The canonical parameter is $\log\gamma$. The estimated Strauss interaction parameter γ is $\exp(-0.0459) = 0.9551$ for the control group and $\exp(-1.026) = 0.3585$ for the treatment group. The correct values of γ were 1 and 0.5.

16.10.3   Hyperframe of interactions

More generally the argument `interaction` can be a hyperframe containing interpoint interactions (objects of class `"interact"`). For example, we might want to fit a Strauss process to each point pattern, but with a different Strauss interaction radius for each pattern.

```
> radii <- with(simba2, mean(nndist(Points)))
```

Then `radii` is a vector of numbers which we could use as the values of the interaction radius for each case. First we need to make the interaction objects:

```
> Rad <- hyperframe(R=radii)
> Str <- with(Rad, Strauss(R))
```

Then we put them into a hyperframe and fit the model:

```
> Int <- hyperframe(str=Str)
> mppm(Points ~ 1, simba2, interaction=Int)
Point process model fitted to 5 point patterns
Log trend formula: ~1
Fitted trend coefficients:
(Intercept)
      4.47
Interaction for each pattern:        Strauss process
Interaction 1:         Strauss(r = 0.06588, gamma = 0.7245)
Interaction 2:         Strauss(r = 0.06511, gamma = 0.7245)
Interaction 3:         Strauss(r = 0.07948, gamma = 0.7245)
Interaction 4:         Strauss(r = 0.09014, gamma = 0.7245)
Interaction 5:         Strauss(r = 0.07684, gamma = 0.7245)
```

Note that the interaction parameter γ is the same on each row, making this a rather unusual model.

An important constraint is that all of the interaction objects in one column must be *instances of the same process* (e.g., Strauss), albeit possibly having different parameter values. For example, you cannot put Poisson and Strauss processes in the same column.

If `interaction` is a hyperframe, then `iformula` may involve the names of columns of this hyperframe. In the example above, to allow different values of γ in each group, and different first-order terms β,

```
> mppm(Points ~ id, simba2, interaction=Int, iformula = ~str:group)
```

where `str` is the name of the column in the hyperframe `Int`.

16.10.3.1   Selecting one column

If the right hand side of `iformula` is a single name, then this identifies the column in `interaction` to be used as the interpoint interaction structure.

```
> h <- hyperframe(Y=waterstriders)
> g <- hyperframe(po=Poisson(), str4 = Strauss(4), str7= Strauss(7))
> mppm(Y ~ 1, data=h, interaction=g, iformula=~str4)
Point process model fitted to 3 point patterns
Log trend formula: ~1
Fitted trend coefficients:
(Intercept)
     -3.52
Interaction formula:        ~str4
Interactions defined for each pattern:
Interaction 'str4':        Strauss(r = 4, gamma = 0.2556)
```

An `iformula` is mandatory if `interaction` is a hyperframe with more than one column.

16.10.3.2 Completely different interactions for different cases

Suppose we now want to fit a model to the `simba2` data which is Poisson in the control group, and Strauss in the treatment group. The Poisson and Strauss interactions can be given as separate columns in a hyperframe of interactions:

```
interaction=hyperframe(po=Poisson(), str=Strauss(0.07))
```

What do we write for the `iformula`? The following *will not* work:

```
iformula=~ifelse(group=="control", po, str)
```

This does not work because the Poisson and Strauss models are 'incompatible' inside such expressions. The canonical sufficient statistics for the Poisson and Strauss processes do not have the same dimension. Internally in `mppm` we translate the symbols `po` and `str` into matrices in which each row is the value of the sufficient statistic at some location; in the example above, these matrices have different numbers of columns, so the `ifelse` expression cannot be evaluated.

Instead we need something like the following:

```
iformula=~I((group=="control")*po) + I((group=="treatment") * str)
```

Here we use `I` to prevent its argument from being interpreted as a formula (thus the symbol `*` is interpreted as multiplication instead of a model interaction). The expression `(group=="control")` is logical, and, when multiplied by the matrix `po`, yields a matrix.

So the following does work:

```
> g <- hyperframe(po=Poisson(), str=Strauss(0.07))
> fit2 <- mppm(Points ~ 1, simba2, g,
               iformula=~I((group=="control")*po)
                         + I((group=="treatment") * str))
> fit2
Point process model fitted to 5 point patterns
Log trend formula: ~1
Fitted trend coefficients:
                    (Intercept)    I((group == "control") * po)
                           4.45                              NA
I((group == "treatment") * str)
                          -0.87
Active interactions:         (5 x 2 matrix)
Interaction 1:        Poisson process
Interaction 2:        Poisson process
Interaction 3:        Strauss(r = 0.07, gamma = 0.419)
Interaction 4:        Strauss(r = 0.07, gamma = 0.419)
Interaction 5:        Strauss(r = 0.07, gamma = 0.419)
```

Note that the coefficient for `I((group=="control")*po)` is NA. This happens for `Poisson`, and for certain other interactions such as `Hardcore`, because the sufficient statistic for the interaction is zero-dimensional. Terms involving the `Poisson` interaction are vacuous and could have been dropped from the interaction formula. The analysis above could have been performed using either of the following:

```
> fit2a <- mppm(Points ~ 1, simba2, g,
                iformula=~I((group=="treatment") * str))
> fit2b <- mppm(Points ~ 1, simba2, Strauss(0.07),
                iformula=~I((group=="treatment") * Interaction))
```

16.11   Model validation

Fitted models produced by `mppm` can be examined and validated in many ways.

16.11.1   Fits for each pattern

16.11.1.1   Subfits

The command `subfits` takes an `"mppm"` object and extracts, for each individual point pattern, the fitted point process model for that pattern *that is implied by the overall fit*. It returns a list of objects of class `"ppm"`.

```
> H <- hyperframe(W=waterstriders)
> fit <- mppm(W ~ 1, H)
> subfits(fit)
1:
Stationary Poisson process
Intensity: 0.0161
2:
Stationary Poisson process
Intensity: 0.0161
3:
Stationary Poisson process
Intensity: 0.0161
```

In this example the result is a list of three `"ppm"` objects representing the implied fits for each of the three point patterns in the `waterstriders` dataset. Notice that *the fitted coefficients are the same* in all three submodels, because the model assumes this.

The main advantage of extracting a separate fitted model of class `"ppm"` for each row of the hyperframe is that it enables us to apply existing diagnostic methods available for `"ppm"` objects.

Note that the objects returned by `subfits` were not actually fitted by ppm; they are artificially constructed to resemble `"ppm"` objects as closely as possible. Some idiosyncrasies remain. In particular, the sum of raw residuals is zero in a true `"ppm"` object, but not in these fake `"ppm"` objects.

16.11.1.2   Fitting separately to each pattern

To fit a *different* uniform Poisson point process to each of the three water striders patterns, we could type

```
> fitI <- mppm(W ~ id, H)
> subI <- subfits(fitI)
```

or alternatively use `with.hyperframe`:

```
> subII <- with(H, ppm(W ~ 1))
```

The result `subII` is again a list of three fitted point process models (objects of class `"ppm"` actually produced by ppm).

16.11.2   Residuals

One standard way to check a fitted model is to examine the residuals.

16.11.2.1 Point process residuals

Residuals for a fitted point process model (fitted to a *single* point pattern) were described in Chapter 11. The command `residuals.mppm` computes the point process residuals for an "mppm" object.

```
> fit <- mppm(P ~ x, hyperframe(P=waterstriders))
> res <- residuals(fit, type="Pearson")
```

The result is a list, with one entry for each of the point pattern datasets. Each list entry contains the point process residuals for the corresponding point pattern dataset. Each entry in the list is a signed measure (object of class "msr") as explained in the help for `msr` or `residuals.ppm`). It can be plotted by typing `plot(res)`. Typically one would prefer the smoothed residual field:

```
> smor <- with(hyperframe(res=res), Smooth(res, sigma=4))
```

which can be plotted by `plot(smor)`. In order to compare the smoothed residual values in different panels we use `plot(smor, equal.ribbon=TRUE)`, shown in Figure 16.23.

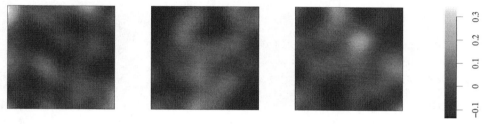

Figure 16.23. *Smoothed residual fields for each* `waterstriders` *pattern.*

16.11.2.2 Sums of residuals

It would be useful to have a residual that is a single numerical value for each point pattern (representing how much that point pattern departs from the model fitted to all the point patterns). That can be computed by *integrating* the residual measures using the function `integral.msr`

```
> sapply(res, integral.msr)
     1      2      3
  6.11 -16.19  10.03
```

In designed experiments we can plot these total residuals against the design covariates.

```
> mod <- mppm(Neurons ~ group * x, data=pyramidal)
> res <- residuals(mod, type="raw")
> df <- as.data.frame(pyramidal, warn=FALSE)
> df$resid <- sapply(res, integral.msr)
> plot(resid ~ group, df)
```

The result is shown in Figure 16.24.

16.11.2.3 Four-panel diagnostic plots

The function `diagnose.ppm` produces a four-panel diagnostic plot based on the point process residuals. It is designed for "ppm" objects, but it can be applied to each of the 'subfits' in an "mppm" object:

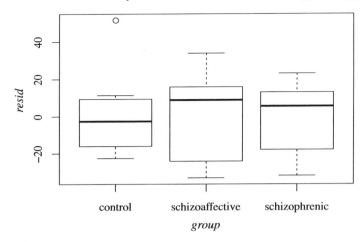

Figure 16.24. *Boxplots of total raw residuals against group for a model fitted to the pyramidal neurons data.*

```
> fit <- mppm(P ~ 1, hyperframe(P=waterstriders))
> subs <- hyperframe(Model=subfits(fit))
> plot(subs, quote(diagnose.ppm(Model)))
```

The result is shown in Figure 16.25. Note that the significance bands for the lurking variable plots are open ended, rather than closed at both ends as they would be for a "ppm" object. The residuals in an "mppm" object do not have to sum to zero for each individual point pattern.

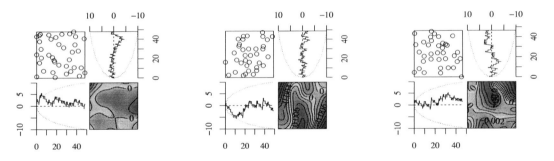

Figure 16.25. *Four-panel diagnostic plots for each of the* waterstriders *patterns.*

16.11.2.4   Residuals of the parameter estimates

We can also compare the parameter estimates obtained by fitting the model simultaneously to all patterns (using mppm) with those obtained by fitting the model separately to each pattern (using ppm).

```
> H <- hyperframe(P = waterstriders)
> fitall <- mppm(P ~ 1, H)
> together <- subfits(fitall)
> separate <- with(H, ppm(P))
> Fits <- hyperframe(Together=together, Separate=separate)
```

```
> dr <- with(Fits, unlist(coef(Separate)) - unlist(coef(Together)))
> dr
      1       2       3
 0.0223 -0.0607  0.0402
> exp(dr)
     1     2     3
 1.023 0.941 1.041
```

16.11.3   Goodness-of-fit tests

16.11.3.1   Quadrat count test

The χ^2 goodness-of-fit test based on quadrat counts (Sections 6.4.2, 6.7.1 and 10.4) is implemented for objects of class "ppm" (in quadrat.test.ppm) and also for objects of class "mppm" (in quadrat.test.mppm).

This is a goodness-of-fit test for a fitted Poisson point process model only. The model could be uniform or non-uniform and the intensity might depend on covariates.

```
> H <- hyperframe(X=waterstriders)
> # Poisson with constant intensity for all patterns
> fit1 <- mppm(X~1, H)
> quadrat.test(fit1, nx=2)
        Chi-squared test of fitted Poisson model 'fit1' using quadrat counts
        Pearson X2 statistic

data:  fit1
X2 = 2, df = 10, p-value = 0.004
alternative hypothesis: two.sided

Pooled test
> # uniform Poisson with different intensity for each pattern
> fit2 <- mppm(X ~ id, H)
> quadrat.test(fit2, nx=2)
        Chi-squared test of fitted Poisson model 'fit2' using quadrat counts
        Pearson X2 statistic

data:  fit2
X2 = 2, df = 9, p-value = 0.01
alternative hypothesis: two.sided

Pooled test
```

See the help for quadrat.test.ppm and quadrat.test.mppm for further details.

16.11.3.2   Spatial CDF test

The Kolmogorov-Smirnov, Cramér-Von Mises, and Anderson-Darling tests of goodness-of-fit of a Poisson point process model compare the observed and predicted distributions of the values of a spatial covariate (Sections 6.7.2 and 10.5). For a Poisson model fitted to multiple point patterns, the same rationale applies. Values of the covariate at the data points, $Z(x_i)$, will be pooled over all point patterns; values of the covariate at each spatial location, $Z(u)$, will likewise be pooled over all windows; then the CDF test will be applied to compare these two distributions.

The spatial CDF test is implemented as cdf.test.mppm, with syntax

```
cdf.test(model, covariate, test)
```

where model is a fitted model (of class "mppm") and covariate is either a function(x,y) making it possible to compute the value of the covariate at any location (x,y); a pixel image containing the covariate values; a list of functions, one for each row of the hyperframe of original data; a list of pixel images, one for each row of the hyperframe of original data; or a hyperframe with one column containing either functions or pixel images. Again the argument test is either "ks", "cvm", or "ad" for the Kolmogorov-Smirnov, Cramér-Von Mises, or Anderson-Darling test, respectively.

16.12   Theory∗

16.12.1   Notation

In keeping with the general notation in experimental design, a 'response' will be denoted by the letter Y, and explanatory variables will be denoted X or Z.

Our experimental design involves n experimental units. The response from unit i is a point pattern $\mathbf{y}^{(i)}$ in a spatial window $W^{(i)}$. For clarity, the unit index i will always appear as a superscript in this section. Thus, the response data consist of n point patterns $\mathbf{y}^{(1)}, \ldots, \mathbf{y}^{(n)}$ in possibly different spatial windows $W^{(1)}, \ldots, W^{(n)}$. The windows $W^{(i)}$ are assumed fixed and known.

16.12.2   Poisson models

We assume that the point patterns $\mathbf{y}^{(i)}$, $i = 1, \ldots, n$ are independent, conditional on the covariates and the random effects. In a Poisson model we assume that for each i the pattern is a Poisson process with intensity function

$$\lambda^{(i)}(u; \beta, \gamma) = \exp\left\{ B^{(i)}(u) + \beta^{\mathsf{T}} \mathbf{x}^{(i)}(u) + \gamma^{\mathsf{T}} \mathbf{z}^{(i)}(u) \right\} \tag{16.10}$$

where β is a vector of fixed coefficients, γ is a vector of random coefficients, $B^{(i)}(u)$ is a known baseline function, and

$$\mathbf{x}^{(i)}(u) = (x_1^{(i)}(u), \ldots, x_p^{(i)}(u))$$
$$\mathbf{z}^{(i)}(u) = (z_1^{(i)}(u), \ldots, z_p^{(i)}(u))$$

are vector-valued functions. The term $\beta^{\mathsf{T}} \mathbf{x}^{(i)}(\mathbf{y})$ is the fixed-effect part of the model, and $\gamma^{\mathsf{T}} \mathbf{z}^{(i)}(\mathbf{y})$ is the random-effect part. The loglikelihood for the entire model *given the random effects* is

$$\log L(\beta, \gamma) = \sum_{i=1}^{n} \left[\sum_{y_j \in \mathbf{y}^{(i)}} \left(B^{(i)}(y_j) + \beta^{\mathsf{T}} \mathbf{x}^{(i)}(y_j) + \gamma^{\mathsf{T}} \mathbf{z}^{(i)}(y_j) \right) \right.$$
$$\left. - \int_{W^{(i)}} \exp\left(B^{(i)}(u) + \beta^{\mathsf{T}} \mathbf{x}^{(i)}(u) + \gamma^{\mathsf{T}} \mathbf{z}^{(i)}(u) \right) \mathrm{d}u \right]. \tag{16.11}$$

This is a sum of Poisson process loglikelihoods; in fact it is the loglikelihood of a single Poisson point process on the disjoint union of the windows $W^{(1)}, \ldots, W^{(n)}$. Existing methods for approximating and maximising the Poisson likelihood described in Chapter 9 can be applied.

When random effects are present, there are several choices for the formation of the likelihood, and several options for fitting the model. These are discussed in [87].

∗ Starred sections contain advanced material, and can be skipped by most readers.

16.12.3 Gibbs models

In a Gibbs model for multiple point patterns, the most general form of the probability density for $\mathbf{y}^{(i)}$ is

$$f_i(\mathbf{y}) = M_i(\boldsymbol{\beta}, \boldsymbol{\gamma}) \exp\{B^{(i)}(\mathbf{y}) + \boldsymbol{\beta}^{\mathsf{T}} \mathbf{x}^{(i)}(\mathbf{y}) + \boldsymbol{\gamma}^{\mathsf{T}} \mathbf{z}^{(i)}(\mathbf{y})\} \tag{16.12}$$

where $M_i(\boldsymbol{\beta}, \boldsymbol{\gamma})$ is a normalising constant, $B^{(i)}(\mathbf{y})$ is a known baseline function, and

$$\mathbf{x}^{(i)}(\mathbf{y}) = (x_1^{(i)}(\mathbf{y}), \dots, x_p^{(i)}(\mathbf{y}))$$
$$\mathbf{z}^{(i)}(\mathbf{y}) = (z_1^{(i)}(\mathbf{y}), \dots, z_p^{(i)}(\mathbf{y}))$$

are vector-valued functions. Conditions must be imposed in order that (16.12) may define a valid point process.

Any function $S(\mathbf{y})$ of a point pattern can be decomposed as $S(\mathbf{y}) = S_{[1]}(\mathbf{y}) + S_{[\geq 2]}(\mathbf{y})$ where

$$S_{[1]}(\mathbf{y}) = \sum_{y_j \in \mathbf{y}} S(\{y_j\})$$

is the first order component and $S_{[\geq 2]}(\mathbf{y}) = S(\mathbf{y}) - S_{[1]}(\mathbf{y})$ is the interaction component. We also write $S_{[1]}(u) = S_{[1]}(\{u\})$. Applying this decomposition to each of the functions $B^{(i)}(\mathbf{y})$, $\mathbf{x}^{(i)}(u)$, and $\mathbf{z}^{(i)}(u)$ we get first-order components $B_{[1]}^{(i)}(u)$, $\mathbf{x}_{[1]}^{(i)}(u)$, and $\mathbf{z}_{[1]}^{(i)}(u)$, and interaction components $B_{[\geq 2]}^{(i)}(\mathbf{y})$, $\mathbf{x}_{[\geq 2]}^{(i)}(\mathbf{y})$, and $\mathbf{z}_{[\geq 2]}^{(i)}(\mathbf{y})$.

In the `spatstat` function `mppm`, the main model formula determines the 'trend' $\mathbf{x}_{[1]}^{(i)}(u)$ and 'offset' $B_{[1]}^{(i)}(u)$, and the random effects formula determines $\mathbf{z}_{[1]}^{(i)}(u)$. It is currently assumed that $\mathbf{z}_{[\geq 2]}^{(i)}(\mathbf{y}) \equiv 0$ so that interactions between points are not dependent on random effects (future versions may remove this constraint). The interaction between points is specified by the arguments `interaction` and `iformula` to `mppm` which jointly specify $\mathbf{x}_{[\geq 2]}^{(i)}(\mathbf{y})$ and $B_{[\geq 2]}^{(i)}(\mathbf{y})$. An example of the latter is the interaction term in the Gibbs hard core model, which has no canonical (loglinear) parameters.

The conditional intensity for the ith pattern is

$$\lambda^{(i)}(u, \mathbf{y}) = \exp\{\Delta_u B(\mathbf{y}) + \boldsymbol{\theta}^{\mathsf{T}} \Delta_u \mathbf{x}^{(i)}(\mathbf{y}) + \boldsymbol{\gamma}^{\mathsf{T}} \Delta_u \mathbf{z}^{(i)}(\mathbf{y})\} \tag{16.13}$$

where Δ_u is the difference operator $\Delta_u S(\mathbf{y}) = S(\mathbf{y} \cup \{u\}) - S(\mathbf{y})$. The log pseudolikelihood given the random effects is

$$\log \mathrm{PL}(\boldsymbol{\theta}, \boldsymbol{\gamma}) = \sum_{i=1}^{n} \sum_{y_j \in \mathbf{y}^{(i)}} \left[\Delta_{y_j} B(\mathbf{y}^{(i)}) + \boldsymbol{\theta}^{\mathsf{T}} \Delta_{y_j} \mathbf{x}^{(i)}(\mathbf{y}^{(i)}) + \boldsymbol{\gamma}^{\mathsf{T}} \Delta_{y_j} \mathbf{z}^{(i)}(\mathbf{y}^{(i)}) \right]$$
$$- \sum_{i=1}^{n} \int_{W^{(i)}} \exp\left(\Delta_u B(\mathbf{y}^{(i)}) + \boldsymbol{\theta}^{\mathsf{T}} \Delta_u \mathbf{x}^{(i)}(\mathbf{y}^{(i)}) + \boldsymbol{\gamma}^{\mathsf{T}} \Delta_u \mathbf{z}^{(i)}(\mathbf{y}^{(i)}) \right) \mathrm{d}u. \tag{16.14}$$

This is the pseudolikelihood of a Gibbs process on the disjoint union of the windows, so again one can apply existing methods to evaluate and maximise the log pseudolikelihood. For further detail see [87].

Arguments about the inefficiency of pseudolikelihood have less force in the setting of replicated patterns, because we have multiple independent observations, and the MPLE normal equations are unbiased estimating equations. However, experiments reported in [87] suggest bias is still present in maximum pseudolikelihood fits for small experiments.

16.13 FAQs

- *When I try to plot several objects in a list or hyperframe, I get an error message stating* `"Figure margins too large"`.

  The figure margins are too large, leaving insufficient space for the graphics in each panel of the plot. Reduce the outer margins by changing the value of the base graphics parameter `mar`, for example `par(mar=rep(0.2, 4))`. Reduce the margins in each figure panel by setting the argument `mar.panel` in the plot command: see the help for `plot.solist` or `plot.anylist`.

- *After encountering problems with a plot, I cannot display any further plots; the error message is* `"Corrupted graphics state"`.

  The internal state of the base graphics code has been corrupted by an error which occurred in the middle of a series of commands. Destroy the current plot device by typing `dev.off()` and start a new plot.

- *Can I perform a MAD test or DCLF test for replicated patterns?*

  Yes. However, there are now many possible choices for the null hypothesis. A null hypothesis of 'no difference between groups' should be handled using the methods of Section 16.8.3 which are the counterparts of the DCLF test. A null hypothesis stating that all patterns are CSR (for example) should be handled by performing a separate test for each point pattern using `dclf.test` or `mad.test`, then combining the *p*-values using standard methods for multiple hypothesis testing.

17

Point Patterns on a Linear Network

This chapter concerns the analysis of spatial patterns of points that lie on a network of lines, such as the pattern of road traffic accidents on a road network.

17.1 Introduction

Figure 17.1 is a record of crimes in the period 25 April to 8 May 2002, in an area of Chicago (Illinois, USA) close to the University of Chicago. The original crime map was published in the university newspaper, *Chicago Weekly News*, in 2002. The graphic was scanned by staff of the Department of Statistics, University of Chicago, and manually digitised by Adrian Baddeley using `clickppp` and `clickjoin` (explained below).

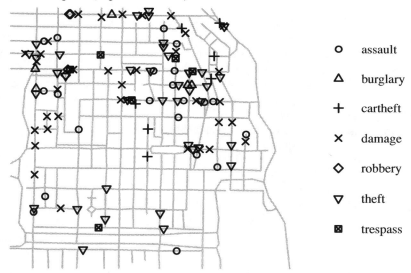

o	assault
△	burglary
+	cartheft
×	damage
◇	robbery
▽	theft
▣	trespass

Figure 17.1. *Chicago crimes data (full version). Locations and types of 116 crimes, recorded near the University of Chicago over a two-week period in 2002. Survey area 1280 feet (390 metres) across. Available in* spatstat *as the dataset* chicago.

Figure 17.2 shows the positions of 48 webs of the urban wall spider *Oecobius navus* on the mortar lines of a brick wall, recorded on graph paper by Voss [676] and manually digitised by M.S. Handcock. The habitat preferences of this species were studied by Voss *et al.* [677]. The main unresolved question is whether there is evidence for interaction between nearby individuals. The mortar line network is relevant because the mortar spaces provide the only opportunity for constructing webs [676].

Figure 17.3 shows the locations of small protrusions called *spines* on the dendrite network of a neuron (nerve cell). The spines are also classified into three types, based on their shape. This

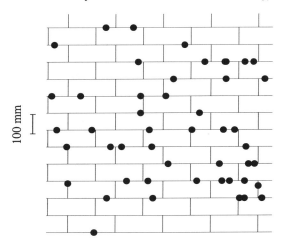

Figure 17.2. *Spider webs on the mortar lines of a brick wall. Recorded by S. Voss [676, 677] and manually digitised by M.S. Handcock. Available in* spatstat *as the dataset* spiders.

neuron was grown in a cell culture *in vitro* so that the dendrite network is almost flat, and is shown in two-dimensional projection in Figure 17.3. Since the dendrites propagate electrical signals, and convey nutrients and molecular genetic signals from the cell body, the network structure is highly relevant.

A three-dimensional image (232 by 168 by 2.6 microns) was obtained with a laser-scanning confocal microscope [361]. NeuronStudio software [585, 695] was used to trace the dendritic network and the spines. The 3D coordinates were read into R and converted to an "lpp" object by A. Jammalamadaka. The third dimension was discarded after resolving the connectivity of the network.

The network shown in Figure 17.3 is one of the ten dendritic trees of this neuron. A dendritic tree consists of all dendrites issuing from a single root branch off the cell body; each neuron typically has 4 to 10 dendritic trees. This example was chosen because it is large enough to demonstrate our techniques clearly, without being too large for graphical purposes.

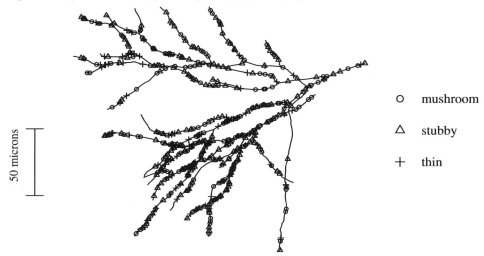

Figure 17.3. *Dendritic spines on a branch of the dendrite network of a neuron. Recorded by the Kosik Lab (UCSB) and A. Jammalamadaka [361, 42]. Available in* spatstat *as the dataset* dendrite.

In each of Figures 17.1–17.3, the points in the spatial point pattern are constrained to lie on a network of lines. Other examples include traffic accidents and vehicle thefts on a road network [435, 705, 706]; roadside trees or invasive species in an agricultural area [623, 209]; and green spaces, retail stores, or roadside kiosks accessible from urban streets [165, 522, 528, 523]. John Snow's pioneering observations of cholera cases in London [621] could also be described as a point pattern on a linear network.

For such data, it is clearly not appropriate to use statistical techniques designed for point patterns in two-dimensional space, such as Ripley's K-function (Chapter 7). The analysis needs to take into account the geometry of the network. In the last decade, substantial research effort has been addressed to this problem by A. Okabe and collaborators [524, 526] and others [120, 121, 245].

The dependence between points in a point process is more difficult to study on a linear network than in the two-dimensional plane. A linear network is not a homogeneous space: different spatial locations on a network are surrounded by different configurations of lines. Recently it was discovered how to 'adjust' for the geometry of the network when defining quantities like the K-function [19, 42].

Spatial analysis on a network of rivers or streams is another active research topic [672, 671]. Special techniques are appropriate because the network is *directed* (water flows in one direction) and *acyclic* (there are no loops). Currently most of this research does not concern point processes.

17.2 Network geometry

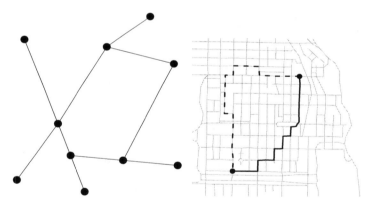

Figure 17.4. *Linear network concepts.* Left: *A linear network is a collection of line segments (—) joining vertices (•).* Right: *A path (dashed line) and a path of shortest length (solid line) between two points (•) in the Chicago street network.*

First we need to define a few basic terms. A linear network consists of *line segments* which end at *vertices* (nodes of the network). See the left panel of Figure 17.4. In our definition, line segments are only allowed to meet at vertices. If two roads intersect, their crossing-point should be a vertex of the network, and each of the roads should be split into segments which terminate at the crossing-point.

The distance between any two points u and v in the network L will be measured by the *shortest-path distance* $d_L(u, v)$ defined as the length of the shortest path in L connecting u and v. See the right panel of Figure 17.4. It will often, but not always, be more appropriate to measure distance using the shortest path than using Euclidean distance.

The *diameter* D of the network is the largest value of shortest-path distance $d_L(u, v)$ between any pair of points u, v in L.

17.3   Data handling

17.3.1   Data classes in `spatstat`

Table 17.1 lists the main classes of spatial data on linear networks. The basic tools for creating these objects are listed in Table 17.2.

`"lpp"`	point pattern on a linear network
`"linnet"`	linear network
`"linim"`	pixel image on a linear network
`"linfun"`	function defined on a linear network

Table 17.1. *Classes of data on a linear network.*

A point pattern on a linear network is represented by an object of class `"lpp"`. The installed datasets `spiders`, `chicago`, and `dendrite` belong to this class. The functions `lpp` and `as.lpp` convert raw data into an object of class `"lpp"`. They require a specification of the underlying network of lines, which is represented by an object of class `"linnet"`.

If covariate data are available, they can be given in any of the usual formats, but it is usually best to convert them to objects of the special classes `"linim"` (image on a linear network) or `"linfun"` (function on a linear network). Objects of class `"linim"` are also returned from many `spatstat` commands when applied to data on a linear network.

`lpp`	create point pattern on linear network
`as.lpp`	convert raw data into a point pattern on a network
`linnet`	create a linear network
`as.linnet`	convert other data to a linear network
`linim`	create pixel image on a network
`as.linim`	convert other data into pixel image on network
`linfun`	create function on linear network

Table 17.2. *Functions for creating data on a network.*

17.3.2   Creating a linear network

An object of class `"linnet"` represents a network of straight line segments in two dimensions. It contains information about each line segment, each vertex or node of the network (i.e. each endpoint, crossing point or meeting point of lines), and about the connectivity of the network — that is, which vertices are joined by line segments, and which line segments have a vertex in common.

The installed dataset `simplenet` (shown in the left panel of Figure 17.4) is a simple example of a linear network. Table 17.3 lists functions which create a linear network.

`linnet`	create linear network from raw data
`as.linnet`	convert other data to linear network
`clickjoin`	interactively add lines to a network
`delaunayNetwork`	network of Delaunay triangulation
`dirichletNetwork`	network of Dirichlet edges

Table 17.3. *Functions which create a linear network.*

Building a network from vertices

The function `linnet` creates a linear network object from the spatial location of each vertex and information about which vertices are joined by an edge. Its syntax is

```
linnet(vertices, m, edges, sparse)
```

where the arguments `m` and `edges` are incompatible, and `sparse` is optional. Here `vertices` is a point pattern (class `"ppp"`) specifying the vertices of the network. The argument `m` is the adjacency matrix: a matrix of logical values equal to TRUE when the corresponding vertices are joined by a line. The argument `edges` is the edge list: a two-column matrix of integers, specifying all pairs of vertices that should be joined by an edge. The argument `sparse` controls the internal format, and is explained below.

For example, the following code creates the letter 'A' as a linear network:

```
> v <- ppp(x=(-2):2, y=3*c(0,1,2,1,0), c(-3,3), c(-1,7))
> edg <- matrix(c(1,2,3,4,2,
                  2,3,4,5,4), ncol=2)
> letterA <- linnet(v, edges=edg)
```

The result is an object of class `linnet`, which can be plotted immediately by `plot(letterA)`.

Building a network from line segments

The generic function `as.linnet` converts other data to a linear network. Many methods for `as.linnet` simply extract the linear network from another class of object, such as `"lpp"` or `"linim"`.

The method `as.linnet.psp` is different: it converts any collection of line segments, given as an object of class `"psp"`, into a linear network by *inferring* the connectivity of the network, using a specified distance threshold `eps`, as in `as.linnet(X, eps=0.01)`. If any segments in X cross over each other (as determined by `selfcross.psp(X)`), the segments are first cut into uninterrupted pieces using `selfcut.psp`. Then any pair of segment endpoints lying closer than `eps` units apart is treated as a single vertex. The linear network is then constructed using `linnet`. It would be wise to check the result graphically, since it is essentially a guess.

In this approach, the line segments in X must portray the geometry of the network accurately, because intersection points will be computed from them. Also, lines which cross each other on the map are assumed to intersect, which is not always the case with road networks or electrical circuit diagrams. Non-intersecting lines can be accommodated using `linnet` or by editing the adjacency matrix. The dendritic spines data (Figure 17.3) required the use of `linnet` rather than `as.linnet` since some dendrites appear to cross over in the two-dimensional projection.

To make a linear network from the boundary edges of a polygonal window `w`, one would type `as.linnet(edges(w))`.

Sparse and non-sparse representations

At the time of writing, the default behaviour of `linnet` and `as.linnet` is to compute additional data whenever a linear network is created. The matrix of shortest-path distances *between all pairs of vertices* in the network is computed and stored as part of the `"linnet"` object. This is computationally efficient because the pre-computed distance values can be used by many other functions. However, it requires a lot of memory, and can only be used for small datasets.

If the argument `sparse=TRUE` is given, then a much more compact representation of the linear network is used, in which the adjacency matrix is stored as a sparse matrix (class `"lgCMatrix"` from the `Matrix` package) and the shortest-path distance matrix is not computed in advance. A network can be converted from one format to the other by `as.linnet`.

The installed dataset `spiders` is stored in the default non-sparse form, while the larger datasets `chicago` and `dendrite` are stored in sparse form. This saves a considerable amount of space:

```
> b <- as.linnet(dendrite, sparse=FALSE)
> print(object.size(dendrite), units="Mb")
0.1 Mb
> print(object.size(b),          units="Mb")
4.7 Mb
```

Computing b takes about 1 second on a laptop, but computation time increases exponentially with the size of the network.

Interactively plotting and editing a network

An object of class `"linnet"` can be plotted by `plot.linnet` or `iplot.linnet`. The `plot` method gives full control over the graphical display while the `iplot` method is particularly useful for zooming in to small details in the network. In `iplot.linnet` the user can choose to display the vertices, the line segments, and the *vertex degree* (number of line segments emanating from each vertex), to check the validity of the network data.

Given a two-dimensional point pattern, the function `clickjoin` provides a point-and-click interface allowing the user to join pairs of selected points by edges. This function makes it easier for the user to create a linear network, given a set of vertices. The result of `clickjoin` is an adjacency matrix that could be used as the argument m in `linnet`. Editing an existing linear network is also possible using `clickjoin`.

The degree of each vertex in a linear network L can be computed directly by `vertexdegree(L)`, so the following idiom is often useful:

```
plot(L); text(vertices(L), labels=vertexdegree(L))
```

17.3.3   Manipulating a linear network

`L; print(L)`	print basic information
`summary(L)`	print summary information
`plot(L)`	plot network
`iplot(L)`	interactively plot network
`vertices(L)`	extract vertices (nodes) of network
`as.psp(L)`	extract line segments of network
`as.owin(L)`	extract window containing network
`L[W]`	sub-network inside window W
`unitname(L)`	extract unit of length
`unitname(L) <- u`	change unit of length
`rescale(L, s)`	convert to different unit of length
`volume(L)`	total length of network
`diameter(L)`	maximum network distance between two locations
`circumradius(L)`	maximum network distance from central location
`vertexdegree(L)`	degree of each vertex
`rotate(L, angle)`	rotate entire network
`shift(L, vec)`	shift (translate) entire network
`scalardilate(L, f)`	apply scale change to entire network
`affine(L, mat, vec)`	apply affine transformation to network

Table 17.4. *Functions for manipulating a* `"linnet"` *object* L.

Table 17.4 shows functions for manipulating a linear network L. The generic function `volume` computes length, area, or volume, depending on the dimension; the method `volume.linnet` calculates the total length of the network. The *diameter* of a network is the maximum possible distance between any two points on the network, measured by the shortest path. It is infinite if the network is not connected. The *circumradius* of a network is the radius of the smallest 'circle' covering the network — that is, the smallest distance d such that there is a centre point c on the network from which every location u on the network can be reached by a path of length at most d.

Additional possibilities are available by extracting the line segments of the network using as.psp and applying methods for a line segment pattern (Table 4.17 on page 120). To compute the average length of uninterrupted stretches of street (i.e. between street intersections) in the Chicago data network,

```
> mean(lengths.psp(as.psp(domain(chicago))))
[1] 61.9
```

17.3.4 Creating a point pattern on a linear network

lpp	create point pattern on linear network from raw data
as.lpp	convert other data to point pattern on linear network
rpoislpp	generate Poisson point process on network
runiflpp	uniformly random points on a network

Table 17.5. *Functions which create a point pattern on a given linear network.*

An object of class "lpp" represents a point pattern on a linear network, such as the datasets in Figures 17.1–17.3. It contains the linear network information, the spatial coordinates (x, y) of the data points, *local coordinates* seg and tp for the data points, and any number of columns of marks. The local coordinate seg is an integer index specifying which line segment of the network contains the point. The local coordinate tp is a real number between 0 and 1 specifying the position of the point along this segment: tp=0 corresponds to the first endpoint and tp=1 to the second endpoint. Every "lpp" object also belongs to class "ppx" (Section 15.4) so that the marks attached to points can be a hyperframe containing any kind of data.

The creator function lpp builds an object of class "lpp":

$$lpp(X,L)$$

where L is a linear network (class "linnet") and X is some kind of data specifying the locations of the points: a point pattern (class "ppp"), or data acceptable to as.ppp, or a matrix or data frame. The function as.lpp will convert data in various other formats to an "lpp" object.

Unless local coordinates are provided explicitly in X, they are computed by projecting the (x, y) locations onto the lines of L, using the function project2segment. This computation will change the (x, y) coordinates if they did not lie exactly on the lines. For example, using the linear network letterA computed above, we can add points manually:

```
> xx <- list(x=c(-1.5,0,0.5,1.5), y=c(1.5,3,4.5,1.5))
> X <- lpp(xx, letterA)
> X
Point pattern on linear network
4 points
Linear network with 5 vertices and 5 lines
Enclosing window: rectangle = [-3, 3] x [-1, 7] units
```

Random point patterns on a linear network can be generated by rpoislpp or runiflpp. Crossing-points between the network and another set of lines can be computed by crossing.psp.

17.3.5 Manipulating a point pattern on a linear network

To manipulate a point pattern on a linear network, we can call on methods for class `"ppx"` as well as methods for class `"lpp"`, because `"lpp"` inherits from `"ppx"`. Basic methods are listed in Table 17.6. Utility functions for manipulating a point pattern on a linear network are listed in Table 17.7. Functions for basic data analysis are shown in Table 17.8.

```
> spiders
Point pattern on linear network
48 points
Linear network with 156 vertices and 203 lines
Enclosing window: rectangle = [0, 1125] x [0, 1125] mm
> spidersm <- rescale(spiders, 1000, c("metre", "metres"))
> summary(spidersm)
Point pattern on linear network
48 points
Linear network with 156 vertices and 203 lines
Total length 20.2 metres
Average intensity 2.37 points per metre
Enclosing window: rectangle = [0, 1.125] x [0, 1.125] metres
```

`X; print(X)`	print basic information
`summary(X)`	print summary information
`plot(X)`	plot point pattern and network
`iplot(X)`	interactively plot point pattern and network
`as.linnet(X); domain(X)`	extract linear network
`as.owin(X); Window(X)`	extract containing window
`as.ppp(X)`	extract points in two dimensions
`as.psp(X)`	extract line segments
`coords(X)`	extract spatial and local coordinates
`coords(X) <- value`	assign new coordinates
`marks(X)`	extract marks
`marks(X) <- value`	assign new marks
`unmark(X)`	remove marks
`is.multitype(X)`	check whether pattern is multitype
`as.hyperframe(X)`	extract coordinates and marks
`as.matrix(X)`	extract coordinates and numerical marks
`as.data.frame(X)`	extract coordinates and marks
`X[W]`	subset inside window `W`
`X[i]`	subset of points selected by index `i`

Table 17.6. *Basic operations on an object* X *of class* `"lpp"` *representing a point pattern on a linear network.*

17.3.6 Covariates on a linear network

By a 'spatial covariate on a linear network' we mean a quantity that could conceivably be measured at any location on the network. It does not necessarily have a meaning at other locations in two-dimensional space. Examples include measurable properties of a road, such as road width; geometrically defined quantities such as the distance to the nearest road intersection, or the sighting distance along the road; and management labels such as the class of road or the speed limit. Covariates can be based on other spatial data, for example, the terrain slope, flooding susceptibility, or distance to the nearest traffic accident.

`anyDuplicated(X)`	determine whether any points are identical
`duplicated(X)`	determine which points are duplicated
`multiplicity(X)`	count number of duplicates of each point
`unique(X)`	remove duplicated points
`subset(X, subset, select)`	subset specified by conditions
`unitname(X)`	extract name of unit of length
`unitname(X) <- value`	change name of unit of length
`superimpose(X,Y,...)`	combine several point patterns
`split(X, f)`	divide point pattern into several point patterns
`rescale(X, s)`	convert to another unit of length
`rotate(X, angle)`	rotate the entire spatial dataset
`shift(X, vec)`	shift the entire spatial dataset
`affine(X, mat, vec)`	apply affine transformation
`scalardilate(X, f)`	multiply spatial coordinates by a factor
`round(X, digits)`	applying rounding to spatial coordinates
`rlabel(X)`	permute the marks

Table 17.7. *Utility functions for manipulating a point pattern* X *on a linear network.*

`npoints(X)`	number of points
`intensity(X)`	number of points per unit length
`pairdist(X)`	shortest-path distances between pairs of points
`nndist(X)`	shortest-path distance to nearest neighbour
`crossdist(X,Y)`	shortest-path distances between pairs of points
`nncross(X, Y)`	shortest-path distance to nearest neighbour
`distfun(X)`	distance function (shortest-path distance)
`nnwhich(X)`	identify which point is nearest neighbour
`nnfun(X)`	nearest-neighbour index function
`rounding(X)`	guess the amount of rounding in spatial coordinates
`envelope(X, Fun)`	simulation envelopes of summary function

Table 17.8. *Basic functions for data analysis of a point pattern* X *on a linear network.*

Spatial covariates on a linear network are represented in `spatstat` by objects of class `"linim"` (pixel image on a linear network) or `"linfun"` (function on a linear network). Geometrically defined quantities such as sighting distance are best represented as functions, while sampled data such as terrain elevation maps will be represented as pixel images. In model-fitting, more accurate fits will be obtained when the covariates can be evaluated exactly at the data points, and this requires a function rather than a pixel image.

Making a `linfun`

A `"linfun"` object is created by calling `g <- linfun(f, L)` where `f` is a `function` in the R language, and `L` is a linear network on which `f` is defined. The function `f` should have formal arguments `f(x,y,seg,tp)` or `f(x,y,seg,tp, ...)` where `x,y` are Cartesian coordinates of locations on the linear network, `seg, tp` are the local coordinates, and `"..."` are optional additional arguments.

The function `f` should be vectorised: that is, if `x,y,seg,tp` are numeric vectors of the same length n, then `v <- f(x,y,seg,tp)` should be a vector of length n.

The result of `g <- linfun(f, L)` can be printed and plotted. It is also recognised as a function, and it can be called in several different ways: as `g(X)` where `X` is an `"lpp"` object, or as `g(x,y)` or `g(x,y,seg,tp)` where `x,y,seg,tp` are coordinates.

For example, in the `spiders` data in Figure 17.2, it appears that almost all the webs lie on horizontal lines rather than vertical lines. To investigate this we could create a covariate which indicates, for each location on the mortar line network, whether the mortar line is vertical or horizontal:

```
> bricks <- domain(spiders)
> alpha <- angles.psp(as.psp(bricks)) * 180/pi
> mortarvert <- (round(alpha/90) == 1)
> f <- function(x, y, seg, tp) { mortarvert[seg] }
> vertical <- linfun(f, bricks)
```

In this calculation, `mortarvert` is a logical vector indicating whether each line segment in the mortar line network is vertical. Given a point on the network, the function `f` extracts the segment index `seg` and looks up the value of `mortarvert` for that segment. The final result `vertical` is an object of class `"linfun"`. It can be plotted simply by `plot(vertical)`, and also serves as a function that can be applied to point patterns or to coordinates:

```
> vertical(spiders[5:8])
[1] FALSE FALSE  TRUE  TRUE
> which(vertical(spiders))
[1] 7 8
> table(vertical(spiders))
FALSE  TRUE
   46     2
```

There are methods for `print` and `plot` for objects of class `"linfun"`. Other methods convert a `"linfun"` to other formats: `as.function.linfun`, `as.linim.linfun`, or extract information about the spatial domain: `as.linnet.linfun`, `as.owin.linfun`. The function `nnfun.lpp` also returns an object of class `"linfun"`.

Making a `linim`

The functions `linim` and `as.linim` create an object of class `"linim"` that represents a pixel image on a linear network. Many other `spatstat` functions return an object of class `"linim"`.

The function `linim` would often be used when the original data were supplied as a pixel image of class `"im"`, such as a terrain elevation map. Its syntax is `linim(L, Z)` where `L` is a linear network and `Z` is a pixel image (class `"im"`) that gives a pixellated approximation of the function values. Covariate functions (of class `"linfun"`) can be converted to `"linim"` by `as.linim`.

Pixel images on a linear network are convenient for computation: mathematical operators on `"linim"` are supported, and there is a function `eval.linim` analogous to `eval.im`.

```
> dna <- distfun(split(chicago)$assault)
> Dna <- as.linim(dna)
> a <- sqrt(Dna) + 3
> b <- eval.linim(pmin(Dna, 250))
```

Expressions involving `pmin` or `ifelse` are two of the few cases where `eval.linim` is needed.

Plotting and manipulating covariates

There are methods for `print` and `plot` for `"linim"` objects. A covariate on a linear network could be plotted using colours or greyscales to represent the different covariate values, as shown in the left panel of Figure 17.5. Alternatively the line segments of the network could be drawn with varying thickness, proportional to the covariate value, as shown in the right panel. The `plot` methods for `"linfun"` and `"linim"` have an argument `style` which determines the plotting style. The figure was produced by

```
> plot(dna, style="colour", ribside="left", box=FALSE)
> plot(dna, style="width", adjust=2.5)
```

Figure 17.5. *Distance to the nearest recorded assault in the Chicago crimes data by the shortest street path, plotted using greyscale (*Left*) and using line thickness (*Right*). Scale markings are distances in feet.*

17.4   Intensity

A point pattern **x** on a linear network L will be treated as a realisation of a *point process* **X** on L. This concept does not require any new technical definitions. In simple terms, it means that each realisation of **X** is a finite set $\mathbf{x} = \{x_1, \ldots, x_n\}$ of points $x_i \in L$, where $n \geq 0$ is not fixed in advance.

The first property that should be investigated is the *intensity* or average density of points along the network. A non-uniform intensity of points can easily be conflated with clustering between points, as we saw in Chapters 7, 8, and 12. It is therefore very important to investigate any evidence of spatial variation in the first-order intensity of points, before conclusions can reliably be drawn about correlations or interactions between points. The dendritic spines data in particular illustrate the need for careful analysis of spatial variation in the intensity of points, before assessing any evidence of clustering [42].

17.4.1   Homogeneous intensity

A point process **X** on L has *homogeneous intensity* λ if the expected number of points of **X** falling in any nominated subset $B \subseteq L$ is equal to λ times the total length of B:

$$\mathbb{E} n(\mathbf{X} \cap B) = \lambda \, \ell(B). \tag{17.1}$$

Notice that λ is now expressed as an expected number per unit *length* of network. That is, λ has dimensions $length^{-1}$. If we convert the spatial coordinates from metres to kilometres, the intensity value in points per metre should be multiplied by 1000 to obtain the intensity in points per kilometre.

Assuming **X** has homogeneous intensity, an unbiased estimate of λ is the empirical intensity

$$\overline{\lambda} = \overline{\lambda}(\mathbf{x}) = \frac{n(\mathbf{x})}{\ell(L)},$$

the total number of points divided by the total length of network. This calculation is performed by the spatstat function intensity.lpp, a method for the generic function intensity:

```
> spidersm <- rescale(spiders, 1000, c("metre", "metres"))
> intensity(spidersm)
[1] 2.37
```

The mortar lines have a total length of 20.2 metres and there are 48 spider webs, giving an average of 2.37 webs per metre. The summary method gives the same information, and explains the units.

```
> summary(spidersm)
Point pattern on linear network
48 points
Linear network with 156 vertices and 203 lines
Total length 20.2 metres
Average intensity 2.37 points per metre
Enclosing window: rectangle = [0, 1.125] x [0, 1.125] metres
```

If we ignored the network structure, the result would be:

```
> intensity(as.ppp(spidersm))
[1] 37.9
```

Observations were made inside a square quadrat of side 1.125 metres, so there are $48/1.125^2 = 37.9$ webs per square metre of wall.

17.4.2 Inhomogeneous intensity function

Alternatively we can assume the intensity is spatially varying over the network according to an *intensity function* $\lambda(u)$ defined at all locations u on L. Loosely speaking $\lambda(u)$ is the expected number of points per unit length of network, in the vicinity of location u. Formally, the point process $\mathbf{X}$ has intensity function $\lambda(u)$ if the expected number of points in a subset $B \subseteq L$ is

$$\mathbb{E}n(\mathbf{X} \cap B) = \int_L \lambda(u)\,\mathrm{d}_1 u \tag{17.2}$$

where $\mathrm{d}_1 u$ denotes integration with respect to arc length. A version of *Campbell's formula* (6.11) holds for linear networks: for any function h,

$$\mathbb{E}\left[\sum_{x_i \in \mathbf{X}} h(x_i)\right] = \int_L h(u)\lambda(u)\,\mathrm{d}_1 u. \tag{17.3}$$

Practical choices for estimation of the intensity function $\lambda(u), u \in L$ include nonparametric methods such as kernel smoothing, and parametric model-fitting. Various kernel smoothing techniques for linear networks have been proposed *ad hoc* [120, 121, 705, 525, 615, 526]. The method of choice is the 'equal-split continuous' method of [525, Section 5] which also has strong theoretical support [461].

Given a point pattern $\mathbf{x} = \{x_1, \ldots, x_n\}$ on a linear network L, the kernel estimate of intensity has the form

$$\widetilde{\lambda}(u) = \sum_i \kappa(x_i, u) \tag{17.4}$$

where $\kappa(v, u)$ is the smoothing kernel. Note that the kernel is now a function of *two* spatial locations u and v, unlike the two-dimensional case (6.7) where the kernel was a function only of the distance between locations, $\kappa_2(u, v) = \kappa_2(\|u - v\|)$. The precise form of the kernel $\kappa(u, v)$ is somewhat complicated and we refer to [525, Section 5], [461] for details. Further mathematical insight and computational advances were contributed by McSwiggan [461].

Figure 17.6 illustrates the behaviour of the smoothing kernel on a linear network. The left panel shows a simple network and a point u which will be smoothed by applying the kernel. The remaining panels show the effect of smoothing this point by a Gaussian kernel with bandwidth $\sigma = 0.1, 0.2, 0.3$ units, respectively. The thickness of the lines is proportional to the kernel value. For $\sigma = 0.1$ the shape of the kernel is very close to the usual 'bell' shape of the Gaussian density on the real number line, confined to the segment containing the original point u. For larger values of bandwidth σ, the kernel spreads to the rest of the network. The panels in Figure 17.6 use equal scales so the total area of black ink is the same in the three rightmost panels.

Figure 17.6. *Smoothing kernel on a network.* Left: *simple example of linear network and one data point u (●). Other panels: Line-thickness display of the smoothing kernel* $\kappa(u, v)$ *as a function of v, with bandwidth* $\sigma = 0.1, 0.2,$ *and* 0.3 *(left to right).*

Kernel estimation of the intensity function is performed in `spatstat` by `density.lpp`. At the time of writing, automatic bandwidth selection is not available, but this will change soon [461].

Visual inspection of Figure 17.1 strongly suggests that the crime rate is spatially nonuniform. From a criminological standpoint, a spatially inhomogeneous crime rate might be attributable to variation in opportunity (such as the population density of victims or property), in surveillance, or in travel distance to crime location. For initial investigation we could ignore the marks (because of the relatively small counts of some types of crime) and try a kernel estimate of intensity, with the arbitrarily chosen bandwidth of 60 feet:

```
> d60 <- density(unmark(chicago), 60)
```

This is plotted in Figure 17.7.

Figure 17.7. *Kernel estimate of intensity for the Chicago crimes (ignoring types of crimes) with smoothing bandwidth* $\sigma = 60$ *feet.* Left: *line thickness presentation.* Right: *greyscale presentation. Scale markings are numbers of crimes per mile of street.*

Figure 17.8 shows a kernel-smoothing estimate of the spatially varying intensity of the dendritic

Figure 17.8. *Kernel smoothing estimates of intensity of dendritic spines. Smoothing bandwidth 10 microns. Intensity value is proportional to ribbon width.*

spines regardless of type, using a Gaussian kernel with arbitrarily chosen standard deviation 10 microns. The ribbon width in Figure 17.8 is proportional to the intensity estimate, which ranges between 0.01 and 0.79 spines per micron.

17.4.3   Intensity depending on covariate

Section 6.6.3 discussed the situation where the point process intensity $\lambda(u)$ depends on a spatial covariate $Z(u)$, through the relationship $\lambda(u) = \rho(Z(u))$ where ρ is an unknown function to be estimated (called the 'resource selection function' by ecologists). The 'relative distribution' techniques for estimating ρ, described in Section 6.6.3, do not depend on the underlying space, apart from a few details, because smoothing is applied to the values of Z on the real number line. Accordingly the same technique can be applied to point patterns on a linear network. The method rhohat.lpp computes a kernel-smoothed estimate of the function ρ from an "lpp" object X. In the command rhohat(X, covariate) the covariate can be a "linim", "linfun", or one of the strings "x" or "y" representing the spatial coordinates. The left panel of Figure 17.9 shows the result of rhohat(unmark(chicago), "y") suggesting crime rates *per unit length of street* are higher in the northern part of the study region. The right panel shows the result of

```
> along <- linfun(function(x,y,seg,tp) { tp }, domain(spiders))
> rhoalong <- rhohat(spiders, along)
```

Since tp is the local coordinate measuring position along the line segment, with 0 and 1 representing the endpoints, this plot suggests tentatively that spiders prefer building webs in the middle of a mortar line, away from intersections with other mortar lines.

Hypothesis tests for the dependence of intensity on a spatial covariate, such as Berman's tests and CDF tests (Sections 6.7.2 and 10.5.2), likewise depend only on the values of the covariate, so they can also be applied to data on a linear network. The functions berman.test and cdf.test have methods for class "lpp".

```
> berman.test(unmark(chicago), "y")
```

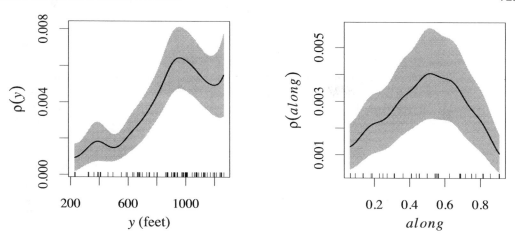

Figure 17.9. *Estimates of the intensity of (Left) the Chicago crimes as a function of y-coordinate, and (Right) the spider data as a function of position along the mortar line.*

```
        Berman Z1 test of CSR in linear network

data:  covariate 'y' evaluated at points of 'unmark(chicago)'
Z1 = 2, p-value = 0.02
alternative hypothesis: two-sided
> cdf.test(spiders, along)
        Spatial Kolmogorov-Smirnov test of CSR in linear network

data:  covariate 'along' evaluated at points of 'spiders'
    and transformed to uniform distribution under CSR
D = 0.2, p-value = 0.1
alternative hypothesis: two-sided
```

There is strong evidence for a trend in the Chicago data, but insufficient evidence that spiders prefer the middle of a mortar line. However, spatial inhomogeneity may have affected the latter conclusion.

If the explanatory covariate is categorical or takes only a discrete set of values, tabulation methods can be employed. For example, using the vector `mortarvert` and function `vertical` defined on page 718,

```
> mortlen <- lengths.psp(as.psp(bricks))
> (totlen <- tapply(mortlen, mortarvert, sum)/1000)
FALSE   TRUE
14.62   5.59
> (totpts <- table(vertical(spiders)))
FALSE   TRUE
   46      2
> totpts/totlen
FALSE   TRUE
3.145  0.358
```

The vertical segments have a total length $\ell_v = 5.59$ metres and contain $n_v = 2$ webs, with average intensity $n_v/\ell_v = 2/5.59 = 0.358$ webs per metre, while the horizontal segments have total length $\ell_h = 14.62$ metres and contain $n_h = 46$ webs, yielding $n_h/\ell_h = 46/14.625 = 3.145$ webs per metre.

17.4.4   Intensity for multitype point patterns

The datasets `chicago` and `dendrite` are examples of *multitype* point patterns on linear networks, where points are classified into different types.

We use the notation for multitype point patterns and point processes that was established in Chapter 14. On a linear network L, a multitype point pattern dataset **y**, with each point belonging to one of M possible categories or types, is a finite set $\mathbf{y} = \{(x_1, m_1), \ldots, (x_n, m_n)\}$ where $x_k \in L$ is the location of the kth point and $m_k \in \{1, 2, \ldots, M\}$ is the type classification of the kth point, $k = 1, 2, \ldots, n$, and the number n of random points is not fixed in advance.

A multitype point process **Y** on a linear network L is a stochastic process whose realisations **y** are multitype point patterns. It can be regarded as a point process on $L \times \{1, 2, \ldots, M\}$. We write $\mathbf{X}^{(\bullet)} = \{x_k : (x_k, m_k) \in \mathbf{Y}\}$ for the 'projected' or 'unmarked' process consisting of the locations of points of **Y** ignoring their types. For each possible type i, we write $\mathbf{X}^{(i)} = \{x_k : (x_k, m_k) \in \mathbf{Y}, \, m_k = i\}$ for the point process of locations of random points of type i. Then we may regard **Y** as equivalent to the multivariate process $(\mathbf{X}^{(1)}, \ldots, \mathbf{X}^{(M)})$.

A multitype point process $\mathbf{Y} = (\mathbf{X}^{(1)}, \ldots, \mathbf{X}^{(M)})$ on the network L has *homogeneous intensity* if the points of each type i have homogeneous intensity, meaning that

$$\mathbb{E}[n(\mathbf{X}^{(i)} \cap B)] = \lambda_i \, \ell(B)$$

for any $B \subseteq L$. Here λ_i is the intensity of points of type i. The process $\mathbf{X}^{(\bullet)}$, of points regardless of type, has intensity $\lambda_\bullet = \sum_i \lambda_i$. The probability that any given random point of **Y** belongs to type i equals $p_i = \lambda_i / \lambda_\bullet$.

For a multitype pattern, `intensity.lpp` estimates the homogeneous intensity of each type of point:

```
> intensity(chicago)
 assault burglary cartheft   damage  robbery    theft trespass
0.000674 0.000161 0.000225 0.001124 0.000128 0.001220 0.000193
```

The `summary` method gives the same information, and explains the units. Let us rescale the coordinates from feet to miles:

```
> chicagomiles <- rescale(chicago, 5280, c("mile","miles"))
> summary(chicagomiles)
Point pattern on linear network
116 points
Linear network with 338 vertices and 503 lines
Total length 5.9 miles
Average intensity 19.7 points per mile
Multitype:
          frequency proportion intensity
assault          21     0.1810     3.560
burglary          5     0.0431     0.848
cartheft          7     0.0603     1.190
damage           35     0.3020     5.930
robbery           4     0.0345     0.678
theft            38     0.3280     6.440
trespass          6     0.0517     1.020
Enclosing window: rectangle = [1e-04, 0.2428] x [0.029, 0.2418] miles
```

The street network has a total length of 31150 feet (5.9 miles, 9.5 km). The average intensity is 3.72 crimes per 1000 feet (19.7 crimes per mile or 12.2 crimes per km).

For an inhomogeneous point process on a network, we would normally assume it has an *intensity function* $\lambda(u, i) = \lambda_i(u)$ so that

$$\mathbb{E}[n(\mathbf{X}^{(i)} \cap B)] = \int_B \lambda_i(u) \, \mathrm{d}_1 u$$

for any type i, and any subset $B \subseteq L$. Kernel smoothing (Section 17.4.2) can be applied to the points of a given type i to estimate the intensity function $\lambda_i(u)$ of the points of type i. For the dendritic spines data, Figure 17.10 shows kernel estimates of intensity for each type of spine, computed by

```
> DSD <- density(split(dendrite), sigma=10)
```

with the arbitrarily chosen bandwidth of 10 microns.

Figure 17.10. *Kernel estimates of intensity for each type of dendritic spine.*

17.5 Poisson models

17.5.1 Poisson processes on a linear network

A Poisson point process on a linear network is defined in the same way as a Poisson process in two dimensions, by properties very similar to (PP1)–(PP4) on page 300. The only substantial change is in the meaning of the intensity, which is now the expected number of points per unit *length*. For a homogeneous Poisson process, property (PP2) is replaced by equation (17.1), and for an inhomogeneous Poisson process, (PP2′) is replaced by (17.2).

To simulate a homogeneous Poisson process with intensity λ on a network L, we simply divide the network into its component line segments. The random points on each segment are determined independently of other segments, by the independence property (PP3). For a segment of length ℓ, we calculate the mean number of random points $\mu = \lambda \ell$, generate a random number N from the Poisson distribution with mean μ, and then place these points independently and uniformly along the segment, by the conditional property (PP4).

For an inhomogeneous Poisson process with intensity function $\lambda(u)$, the simplest strategy is the Lewis-Shedler thinning method described in Section 5.4.2.

The spatstat function rpoislpp generates simulated realisations of homogeneous and inhomogeneous Poisson processes on a linear network. The first argument lambda can be a constant, a function such as a "linfun" object, or a pixel image such as a "linim" object. The second argument L should provide the linear network if this is not inherent in lambda.

```
> XS <- rpoislpp(intensity(spiders), domain(spiders))
> XC <- rpoislpp(d60)
```

Here d60 is the kernel-smoothed intensity estimate for the Chicago data calculated above, belonging to class "linim".

17.5.2   Formulating and fitting a Poisson model

For point pattern data on a linear network, a Poisson point process model can be formulated and fitted in essentially the same manner as for two-dimensional point patterns. Typically we assume that the intensity (points per unit *length*) has a loglinear form

$$\lambda_{\boldsymbol{\theta}}(u) = \exp(B(u) + \boldsymbol{\theta}^{\top}\mathbf{Z}(u)) \tag{17.5}$$

where $B(u)$ is a known 'offset' or 'log baseline' function with numerical values, $\mathbf{Z}(u)$ is a known spatial covariate function with numerical or vector values, and $\boldsymbol{\theta}$ is the parameter vector.

The function lppm fits a point process model to a point pattern on a linear network, using syntax very similar to ppm:

```
lppm(X ~ trend, ...)
lppm(X ~ trend, ..., data)
```

The left-hand side of the model formula must specify a point pattern X on a linear network. It may be the name of a point pattern dataset, or an expression that can be evaluated to yield a point pattern. The right hand side of the model formula determines the form of the *logarithm* of the point process intensity. The terms in the trend on the right-hand side of the model formula can be functions in the R language (including objects of class "linfun" or "funxy"), pixel images (class "linim" or "im"), windows, tessellations, single values, or the reserved names x, y representing the spatial coordinates. If the data are multitype, then the model formula can include the reserved name marks representing the types. Unrecognised variable names in the model formula will be matched to entries of the list argument data, if it is provided, and otherwise to objects in the global environment.

```
> lppm(spiders ~ polynom(x,y,2))
Point process model on linear network
Nonstationary Poisson process
Log intensity:  ~x + y + I(x^2) + I(x * y) + I(y^2)
Fitted trend coefficients:
(Intercept)           x          y      I(x^2)     I(x * y)       I(y^2)
  -7.59e+00    1.22e-03   4.89e-03    6.99e-08    -1.03e-06    -4.00e-06
Original data: spiders
```

The additional arguments "..." can be any of the optional arguments of ppm. The arguments eps and nd will be interpreted differently in the case of a linear network: eps will be the spacing between dummy points along each segment of the network, in length units; nd will be the total number of dummy points.

At the time of writing, only Poisson process models can be fitted, but this will change soon.

For multitype point patterns, the trend formula describes the log intensity $\lambda(u,i) = \lambda_i(u)$ for *each* type of point i. The simplest trend formula, ~ 1, describes a model in which each type i has equal, constant intensity. The reserved name marks can be used to formulate models where the intensity depends on the type of point. The trend formula ~ marks describes a model in which intensity is homogeneous but is different for each type of point. In a model such as

```
> fitca <- lppm(chicago ~ marks + polynom(x,y,2))
```

the intensities of different types of crimes are proportional, $\lambda_i(u) = p_i \lambda_\bullet(u)$. In general the intensities are proportional if there are no *interaction* terms involving marks. The model

```
> fitcx <- lppm(chicago ~ marks * polynom(x,y,2))
```

has a different, unrelated, log-quadratic intensity function for each type of crime. The chicago data may need to be rescaled to a larger unit of length, such as miles, to obtain convergence of the fitting algorithm when the trend includes higher-order polynomials.

17.5.3 Methods for class "lppm"

The result of lppm is an object of class "lppm" representing the fitted model. Important methods for this class are listed in Table 17.9.

fit; print(fit)	print the fitted model
summary(fit)	print a summary of the model
coef(fit)	extract fitted coefficients
vcov(fit)	variance-covariance matrix for coefficient estimates
anova(fit)	perform analysis of deviance
step(fit)	stepwise model selection
predict(fit, ...)	compute fitted intensity of model
plot(fit)	plot fitted intensity of model
simulate(fit)	simulate realisations of model
update(fit, ...)	re-fit the model using new data or new formula
formula(fit)	extract model formula
terms(fit)	extract terms in model formula
as.linnet(fit); domain(fit)	extract linear network
as.owin(fit); Window(fit)	extract enclosing window
nobs(fit)	number of data points ('observations')
logLik(fit)	evaluate maximised log-likelihood of model
extractAIC(fit); AIC(fit)	evaluate Akaike Information Criterion
is.lppm(fit)	check whether model is of class "lppm"
is.poisson(fit)	check whether model is Poisson
is.stationary(fit)	check whether model is stationary
is.multitype(fit)	check whether model is multitype
model.matrix(fit)	extract 'design matrix' of canonical covariates
model.images(fit)	extract canonical covariates as images
berman.test(fit, covariate)	Berman test
cdf.test(fit, covariate)	spatial CDF test
rhohat(fit, covariate)	relative distribution estimate
envelope(fit, Fun)	envelopes of summary function

Table 17.9. *Methods for an object* fit *of class* "lppm".

Figure 17.11 shows the fitted intensity of a Poisson point process with log-quadratic intensity fitted to the unmarked Chicago crimes data, computed by

```
> fit <- lppm(unmark(chicago) ~ polynom(x,y,2))
> lam <- predict(fit, dimyx=512)
```

Note that predict.lppm will allow predictions to be made at any spatial location, if the relevant covariates are available. For example, crime rate on the Chicago street network can be extrapolated into two-dimensional space:

```
> fillwin <- Window(chicago)
> fillmask <- as.mask(fillwin)
> lamfill <- predict(fit, locations=fillmask)
```

The result is shown in the right panel of Figure 17.11. This can be quite useful for visualising a gentle spatial trend such as the log-quadratic trend in the figure. It may also be useful for special purposes such as cross-validation. However, very careful interpretation is required: the predicted intensity values are still expressed in *points per unit length* of network.

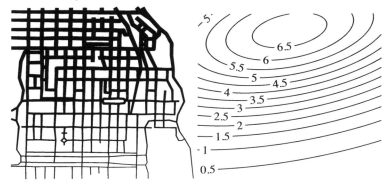

Figure 17.11. *Fitted intensity of inhomogeneous Poisson point process for Chicago crimes (ignoring types) assuming log-quadratic intensity function.* Left: *line thickness proportional to fitted intensity.* Right: *contours of fitted intensity extrapolated into two dimensions (contour levels: crimes per 1000 feet of street network).*

Analysis of deviance for fitted models (Section 9.4.6) is supported by `anova.lppm`. In the spider web data, for example, we may fit a model incorporating both a spatial trend and a preference for horizontal rather than vertical lines:

```
> fit0 <- lppm(spiders ~ polynom(x,y,2) + vertical)
```

then add a preference for different positions along a segment:

```
> fit1 <- update(fit0, . ~ . + polynom(along,2))
```

where `vertical` and `along` are functions computed on pages 718 and 722, respectively. The fitted coefficients (not shown) indicate a strong preference for the middle of a line segment: try plotting `effectfun(fit1, "along", x=0, y=0, vertical=FALSE)`. To assess significance we apply the likelihood ratio test:

```
> anova(fit0, fit1, test="LR")
Analysis of Deviance Table
Model 1: ~x + y + I(x^2) + I(x * y) + I(y^2) + vertical          Poisson
Model 2: ~x + y + I(x^2) + I(x * y) + I(y^2) + vertical + (along + I(along^2))
  Npar Df Deviance Pr(>Chi)
1   7
2   9  2     6.36    0.042 *
---
Signif. codes:  0 '***' 0.001 '**' 0.01 '*' 0.05 '.' 0.1 ' ' 1
```

There is significant evidence for a preference for the middle of a segment. Similarly one can use `anova(fit1, test="LR")` or `step(fit1)` for model selection.

In the Chicago crimes data, some types of crimes were uncommon, so a detailed analysis of each type of crime is not appropriate. A simple strategy is to group the crime types into a smaller number of types of crime:

```
> chicagoG <- mergeLevels(chicago,
              person=c("assault", "robbery"), property=NULL)
```

This idiom means that the crime types "assault" and "robbery" have been mapped to a single type "person", and all other crime types have been mapped to the type "property".

```
> chicagoG
Point pattern on linear network
116 points
Multitype, with possible types: person, property
Linear network with 338 vertices and 503 lines
Enclosing window: rectangle = [0, 1282] x [153, 1277] feet
```

We may then apply the likelihood ratio test:

```
> fitGcx <- lppm(chicagoG ~ marks * polynom(x,y,2))
> fitGca <- lppm(chicagoG ~ marks + polynom(x,y,2))
> anova(fitGca, fitGcx, test="LR")
Analysis of Deviance Table
Model 1: ~marks + (x + y + I(x^2) + I(x * y) + I(y^2))        Poisson
Model 2: ~marks * (x + y + I(x^2) + I(x * y) + I(y^2))        Poisson
   Npar Df Deviance Pr(>Chi)
1   7
2   12  5     1.59      0.9
```

Poisson point process models on a linear network are discussed in detail in [460] for applications to road accident research.

17.6   Intensity on a tree

A network which has no loops is called an acyclic network or 'tree'. The dendrite network in Figure 17.3 is a tree. Other important examples include river and stream networks [672, 671] and capillary networks. A tree-like network can be divided naturally into sub-trees by cutting it at the branch points. This provides new ways of analysing the data. Strong inhomogeneity has been observed in some tree-like branching networks, including the spines data [42]. It is conceivable that strong inhomogeneity is more likely in tree-like networks than in networks with loops, such as road networks.

17.6.1   Branch-splitting analysis of dendrite data

Figure 17.8 strongly suggests that different branches of the dendrite network may have different intensities of spines. In biological terms it is conceivable that a dendrite network may exhibit different structural characteristics in different branches [42]. For simplicity we ignore the marks and analyse only the locations of the spines.

For any proposed split of the network into several subsets $S_1, \ldots, S_K$, we may apply the χ^2 test (or likelihood ratio test) of homogeneous intensity based on the counts $n_k = n(\mathbf{X} \cap S_k)$ using the procedure of Section 6.4.2. The only modification is that the expected counts are proportional to *length*: $e_k = n\ell_k/\ell$ where ℓ_k is the length of dendrite in the kth subset, while $\ell = \sum_k \ell_k$ and $n = \sum_k n_k$ are totals.

Selection of an appropriate split of the network into branches is a problem of model selection; here we can use a recursive splitting approach similar to that used for classification and regression trees [123]. Starting from the cell body as the root of the tree, we can visit each successive branching point of the network, and apply the χ^2 test of uniformity to the subset *beyond* this branch point. The

data will be split if the null hypothesis is rejected at, say, the 5% level. Of course the usual significance interpretation of the tests is not applicable in this context where multiple tests are performed
on the same data.

First we ensure that the dendritic spines data include complete information about distances:

```
> dend <- as.lpp(unmark(dendrite), sparse=FALSE)
> L <- domain(dend)
```

The dendrite network L was originally traced outward from the neuron cell body, so that the first-
listed vertex vertices(L)[1] is also the base of the dendrite network adjacent to the cell body,
and will serve as the root of the tree. We create the distance function to the cell body:

```
> v1 <- lpp(vertices(L)[1], L)
> d <- distfun(v1)
```

We shall attach labels to the network nodes and edges indicating the branch that they belong to. The
root is labelled "", the first branches are labelled "a" and "b", the next branches "aa", "ab", etc.
See Figure 17.12.

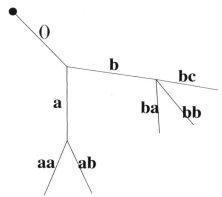

Figure 17.12. *Lexicographical labelling of segments of a tree.*

The spatstat functions treebranchlabels and branchlabelfun perform this labelling.

```
> tb <- treebranchlabels(L, root=1)
> b <- branchlabelfun(L, root=1)
```

For example, to test whether the two branches "a" and "b" have different (but uniform) intensities,
we could fit the model

```
> lppm(dend ~ (substr(b,1,1) == "a"))
```

and compare this with lppm(dend ~ 1), where the condition substr(b,1,1) == "a" catches
every label that *begins* with "a". However we also need to allow for the possibility that the intensity
may depend on distance from the cell body. For convenience we define

```
> g <- function(x,n) factor(substr(x, 1, n))
```

so that g(b, 1) is a factor with levels "a" and "b" identifying the two branches. Our comparison
of branches "a" and "b" is

```
> step(lppm(dend ~ d * g(b, 1)), trace=0)
```

```
Point process model on linear network
Nonstationary Poisson process
Log intensity:  ~d + g(b, 1)
Fitted trend coefficients:
(Intercept)            d     g(b, 1)a      g(b, 1)b
   -1.43722     0.00242      -0.02620      -0.42221
Original data: dend
```

The output has been shortened to save space. The final model does include an effect for the two branches, but none of the associated coefficients is significant, which suggests at most weak evidence of differences between the two branches at this level. Next we apply the same technique for the branches at the next level:

```
> step(lppm(dend ~ d * g(b, 2)), trace=0)
Point process model on linear network
Nonstationary Poisson process
Log intensity:  ~g(b, 2)
Fitted trend coefficients:
(Intercept)      g(b, 2)a    g(b, 2)aa    g(b, 2)ab    g(b, 2)ac    g(b, 2)b
  -1.402456      0.162669     0.453779    -0.210436    -0.998513   -0.825029
   g(b, 2)ba    g(b, 2)bb
  -0.095930    -0.000625
Original data: dend
```

By printing coef(summary(.Last.value)) for this fitted model (not shown) we find a significant effect for the branch labelled "ac".

Once a branch of the tree has been identified for special attention, models can include a term representing this branch. The expression begins(b, "ac") returns TRUE inside the branch labelled "ac":

```
> lppm(dend ~ d * begins(b, "ac"))
Point process model on linear network
Nonstationary Poisson process
Log intensity:  ~d * begins(b, "ac")
Fitted trend coefficients:
         (Intercept)                    d    begins(b, "ac")TRUE
            -1.305162             0.000899             -3.556131
d:begins(b, "ac")TRUE
            0.018700
Original data: dend
```

A branch of the tree can also be extracted as a sub-network using extractbranch, or deleted from the existing network using deletebranch. To extract the branch labelled "ac", while keeping both the dendrite segments and the spine locations in that branch,

```
> dendAC <- extractbranch(dend, code="ac", labels=tb)
```

Similarly deletebranch(dend, "ac", tb) would remove branch *ac* from the dataset dend. In either case, covariates may have to be adjusted to the new network geometry, in order to continue building models.

17.7   Pair correlation function

Chapter 7 introduced ways of measuring correlation between the points in a two-dimensional spatial point process, including Ripley's K-function and the pair correlation function. For a point process on a linear network, these techniques need to be modified to take account of the geometry of the network. This section discusses the pair correlation function on a linear network, which is much easier to think about than the K-function.

17.7.1   Pair correlation function on a linear network

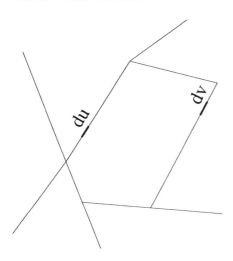

To define the pair correlation we first introduce the *second moment intensity* (or *product density*) $\lambda_2(u,v)$. Heuristically, given any two distinct locations u, v on the network, we consider a very short piece of the network around location u of length $d_1 u$, and similarly a short piece of network around v of length $d_1 v$, sketched in Figure 17.13. The probability $p(u,v)$ that both of these short segments contain at least one random point is assumed to be $p(u,v) = \lambda_2(u,v)\,d_1 u\, d_1 v$. Then for any two line segments $A, B \subset L$ that are disjoint ($A \cap B = \emptyset$) we have

$$\mathbb{E}[n(\mathbf{X} \cap A)\, n(\mathbf{X} \cap B)] == \int_A \int_B \lambda_2(u,v)\,d_1 v\, d_1 u, \quad (17.6)$$

analogously to equation (7.45). For example, the Poisson point process on a linear network, with intensity function $\lambda(u)$, has second moment intensity $\lambda_2(u,v) = \lambda(u)\lambda(v)$.

The (general) *pair correlation function* is then defined as

Figure 17.13. *Heuristic definition of second-order intensity $\lambda_2(u,v)$ for a pair of locations u and v on a linear network.*

$$g_2(u,v) = \frac{\lambda_2(u,v)}{\lambda(u)\lambda(v)} \quad (17.7)$$

where λ is the intensity function. This is analogous to (7.46). For example, any Poisson point process on a linear network has general pair correlation function $g_2(u,v) = 1$.

The general pair correlation function has the same interpretation on a linear network as it does in two dimensions. It is a ratio of two probabilities: the probability of finding two random points at the specified locations, divided by the corresponding probability for a Poisson process with the same intensity.

In analysing two-dimensional point pattern it is common practice to simplify matters by assuming that $g_2(u,v)$ depends only on the Euclidean distance $\|u - v\|$ between the two locations:

$$g_2(u,v) = g(\|u - v\|)$$

where $g(r)$ is the function usually known as *'the'* pair correlation function.

Similarly for point patterns on a linear network, to simplify matters we shall assume that $g_2(u,v)$ depends only on the *shortest-path distance* $d_L(u,v)$:

$$g_2(u,v) = g(d_L(u,v)). \quad (17.8)$$

Then $g(r)$ is the *linear network pair correlation function*. A Poisson point process on a linear network has $g(r) = 1$. A value $g(10) = 0.5$ would mean that, for a pair of locations u and v separated by a shortest-path distance of 10 units, the probability that random points fall in both locations is half as much as it would be for a Poisson process with the same intensity. The pair correlation $g(r)$ is defined for all distances $r \leq D$, where D is the diameter of the network (maximum possible shortest-path distance between any two points in the network).

17.7.2 Estimating the pair correlation function

Recall from Section 7.6 that, to estimate the pair correlation function of a stationary point process in two dimensions, we measure the Euclidean distance $d_{ij} = \|x_i - x_j\|$ between each pair of points x_i and x_j, and apply kernel smoothing to the distance values. Either before or after performing the smoothing, weights must be introduced into the calculation. Either each observed distance d_{ij} is weighted by $w_{ij} = 1/(2\pi d_{ij})$ before smoothing, or the unweighted kernel smoother of the distances d_{ij} is subsequently multiplied by the factor $w(r) = 1/(2\pi r)$. This weighting accounts for the fact that shorter distances are less likely to be observed than longer distances; the denominator $2\pi r$ is the length of the circle of radius r.

Similarly on a linear network, suppose that the point process has constant intensity λ, and that its pair correlation function depends only on shortest-path distance, as in (17.8). To estimate the pair correlation function, we measure the shortest-path distance $d_{ij} = d_L(x_i, x_j)$ between each pair of data points, and apply kernel smoothing with weights. Ang *et al.* [19] showed that the appropriate weight is the reciprocal of the 'perimeter' of a 'circle of radius r' in the network. See Figure 17.14. For any point u on the network, let $m(u, r)$ denote the number of locations on the network which lie exactly r units away from u by the shortest path:

$$m(u, r) = \#\{v : d_L(u, v) = r\}. \tag{17.9}$$

This is the analogue for linear networks of the perimeter of a circle of radius r. Then the appropriate weight for the distance d_{ij} is $1/m(x_i, d_{ij})$.

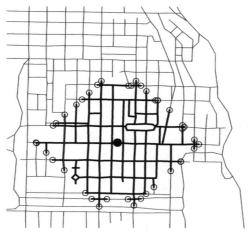

Figure 17.14. *A 'disc' of radius $r = 400$ feet in the Chicago street network. Filled circle: centre point u. Thick lines: disc $b_L(u, r)$ consisting of all locations that can be reached from the centre point u by walking less than 400 feet on the street network. Open circles: the relative boundary $\partial b_L(u, r)$, consisting of locations which are exactly 400 feet away from the centre point, measured by the shortest path. Computed by the* `spatstat` *function* `lineardisc`.

A kernel estimator of the pair correlation function $g(r)$ is

$$\widehat{g}^L(r) = \frac{\ell(L)}{n(n-1)} \sum_{i=1}^{n} \sum_{j \neq i} \frac{\kappa(d_L(x_i, x_j) - r)}{m(x_i, d_L(x_i, x_j))} \tag{17.10}$$

where $\kappa(\cdot)$ is a smoothing kernel on the real line. Standard techniques for one-dimensional kernel smoothing [687] can be used.

We call (17.10) the **geometrically corrected** function because the denominator $m(x_i, d_L(x_i, x_j))$ adjusts for the geometry of the network. As usual, we need to avoid situations where the weight factor $m(x_i, d_L(x_i, x_j))$ is zero. It is shown in [19] that the estimator (17.10) is valid for all $r \leq R$, where R is the largest value such that $m(u, r) \neq 0$ for all locations u and all $r \leq R$. If L is a connected network, then $R = \min_{u \in L} \max_{v \in L} d_L(u, v)$ is the radius of the smallest disc that covers the entire network, which we call the *circumradius* of the network. If L is not connected, then R is the minimum of the circumradii of the connected components of L.

The estimator (17.10) is computed in `spatstat` by the function `linearpcf`. If X is a point pattern on a linear network (class "`lpp`") then `linearpcf(X)` returns a function object (class "`fv`") giving the values of (17.10) and the theoretical value $g(r) = 1$ for a sequence of values of r. It uses the fixed-bandwidth kernel smoother `density.default`, with a standard bias correction at each end of the interval of r values (because distances outside this range are impossible). To switch off the bias correction, set `endcorrect=FALSE`. Arguments can be passed to `density.default` to select the smoothing kernel and the bandwidth. By default the bandwidth is chosen by Silverman's rule of thumb [617, eq (3.31), p. 48], applied to the observed shortest-path distances; this rule was adequate in all examples we have studied.

The left panel of Figure 17.15 shows the result of `plot(linearpcf(spiders))`. The smoothing bandwidth can be recovered from

```
> attr(linearpcf(spiders), "bw")
[1] 63
```

If we were confident that the spiders data have constant intensity, then this graph might suggest mild clustering between spider webs at distances of 200 *mm*, and possibly inhibition at very short distances.

For the spiders data, envelopes of the pair correlation function computed from 39 simulations were found to be too irregular, due to the weak correlation between pcf estimates at different distances r. To reduce sampling variability, we have increased the number of simulations to $m = 199$, and computed the order statistics of rank $k = 5$ and $m - k + 1 = 195$, yielding a Monte Carlo test of size $2(k+1)/(m+1) = 0.05$. The right panel of Figure 17.15, generated by

```
> plot(envelope(spiders, linearpcf, nsim=199, nrank=5))
```

shows 5% significance bands for the estimated pair correlation function for a homogeneous Poisson process on the mortar lines network with the same average intensity of points. The envelopes suggest there is no significant departure from randomness.

The functions `dclf.test` and `mad.test` can also be applied to point pattern data on a linear network, to perform the DCLF and MAD tests based on any appropriate summary function.

17.7.3 Pair correlation function for inhomogeneous process

Instead of assuming that the point process has constant intensity λ, we could allow it to have a non-constant intensity function $\lambda(u)$. Assuming the pair correlation function depends only on the shortest-path distance as in (17.8), a kernel estimator of the pair correlation function $g(r)$ is

$$\widehat{g}^{L,\text{ih}}(r) = \frac{1}{\sum_i 1/\hat{\lambda}(x_i)} \sum_{i=1}^n \sum_{j \neq i} \frac{\kappa(d_L(x_i, x_j) - r)}{\hat{\lambda}(x_i)\hat{\lambda}(x_j)m(x_i, d_L(x_i, x_j))} \qquad (17.11)$$

where $\hat{\lambda}(x_i)$ is an estimate of $\lambda(x_i)$, and other terms are defined in (17.10).

The function `linearpcfinhom` calculates the estimate (17.11). The argument `lambda` should

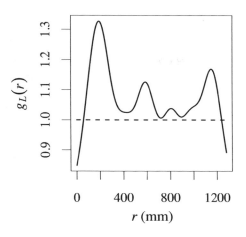 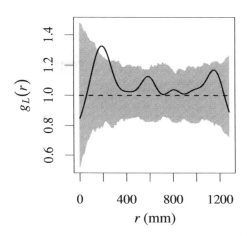

Figure 17.15. *Pair correlation function analysis for spider webs data.* Left: *estimated pair correlation function.* Right: *estimate and pointwise 5% significance bands from 199 simulations of homogeneous Poisson process on the same network.*

provide values of $\hat{\lambda}(x_i)$ and can be given as a numeric vector, a `function(x,y)`, a pixel image, or a fitted point process model of class `"lppm"`. In the latter case, $\hat{\lambda}(u)$ is the fitted intensity of the point process model; by default, the model will be re-fitted to the point pattern dataset before the fitted intensities $\hat{\lambda}(x_i)$ are computed. This ensures that a Monte Carlo test based on the inhomogeneous pair correlation function remains valid. This behaviour can be disabled by setting `update=FALSE`.

```
> fit <- lppm(spiders ~ polynom(x,y,2))
> ghat <- linearpcfinhom(spiders, fit)
> genv <- envelope(spiders, linearpcfinhom, lambda=fit,
                   nsim=199, nrank=5)
```

The results are shown in Figure 17.16. Any suggestion of clustering seems to have been removed by accounting for inhomogeneous intensity. There is still a suggestion of inhibition at short distances.

17.8 *K*-function

We turn now to the counterpart of the *K*-function for a point pattern on a linear network.

17.8.1 Danger in using the two-dimensional *K*-function

To measure correlation between points on a linear network, it is clearly not appropriate to apply Ripley's *K*-function to the two-dimensional spatial locations of the points. At least, it would be fallacious to take a point pattern on a linear network, forget the linear network and retain only the spatial (x, y) coordinates, compute the empirical Ripley *K*-function of these points, and compare this with the theoretical *K*-function for a completely random pattern *in two dimensions*.

For points which are constrained to lie along a linear network, this comparison could produce

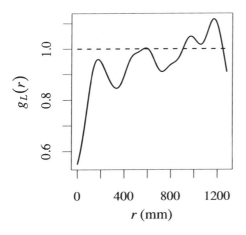 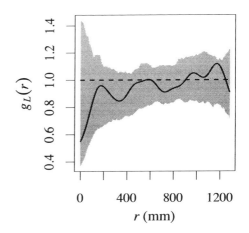

Figure 17.16. *Inhomogeneous pair correlation analysis of spider webs data.* Left: *estimated inhomogeneous pair correlation function based on a fitted log-quadratic intensity model, and using a Gaussian smoothing kernel with standard deviation 63 mm.* Right: *estimate and pointwise 5% critical envelopes based on 199 simulations of a uniform Poisson process on mortar lines.*

spurious evidence of short-range clustering (due to concentration of points on a road) and long-range regularity (due to spatial separation of different roads). This has been observed in real data on traffic accidents [706] and urban crime [435].

The fallacy is demonstrated in the left panel of Figure 17.17, which shows the result of applying the two-dimensional K-function to a synthetic point pattern generated by a homogeneous Poisson process on the Chicago street network:

```
> X <- rpoislpp(0.004, domain(chicago))
> envKestCSR <- envelope(as.ppp(X), Kest, nsim=39)
```

The apparent discrepancy is an artefact, arising because we have chosen the wrong null hypothesis: the envelopes are computed from simulations of CSR in two dimensions, while the appropriate null hypothesis is a Poisson process on the linear network.

A possible pragmatic solution is to generate envelopes of the (two-dimensional) K-function based on simulations from a Poisson point process *on the linear network*:

```
> L <- domain(chicago)
> envKestL <- envelope(as.ppp(X), Kest, nsim=39,
                  simulate=expression(as.ppp(rpoislpp(0.004, L))))
```

The result is shown in the right panel of Figure 17.17. This suggests (correctly) that the empirical K-function of the data is consistent with the empirical K-functions of realisations of a Poisson process on the network.

The appropriate 'theoretical' curve — the expected value of the two-dimensional empirical K-function of a homogeneous Poisson process *on the network* — can be computed in spatstat in the case where $\widehat{K}(r)$ is calculated without edge correction. Then the theoretical value is $K_0(r) = |W|H_L(r)$ where $H_L(r)$ is the cumulative distribution function of the Euclidean distance between two independent uniformly distributed random points on the network. For the Chicago street network:

```
> L <- domain(chicago)
> HL <- distcdf(as.owin(L), dW=pixellate(L))
```

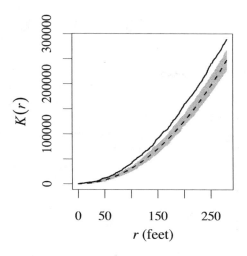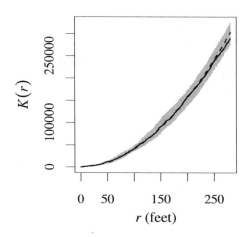

Figure 17.17. *The two-dimensional K-function of a random point pattern on a linear network (solid lines) and the envelopes based on 39 simulations of CSR in two dimensions (Left) and a homogeneous Poisson process on the linear network (Right).*

```
> A <- area(as.owin(L))
> KO <- eval.fv(A * HL)
```

The result of `pixellate(L)` is a pixel image giving for each pixel the length of the intersection between the pixel and the network. Then `distcdf` uses the Fast Fourier Transform to compute the CDF of the distance between a pair of pixels chosen randomly with probabilities proportional to these pixel values; this is a good approximation to the desired CDF. The resulting function `KO` could be superimposed on the graphs in Figure 17.17 to give the 'theoretical' curve.

While this approach is defensible for hypothesis tests, it cannot be used for other purposes, because the usual interpretation of the K-function does not apply. The empirical two-dimensional K-function of a point pattern on a linear network (computed by forgetting the network and using only the spatial coordinates) does not have a straightforward meaning; it is affected by the network geometry; *we cannot compare two point patterns on different networks* using this function.

17.8.2 *K*-function for shortest-path distance

In most (but not all) applications it is sensible to measure the distance between two points u, v on a network by the shortest-path distance $d_L(u, v)$. Suppose we are given a linear network L on which events have been observed to occur at the locations $x_1, \ldots, x_n$. Okabe and Yamada [527] define the (empirical) 'network K function' by

$$\widehat{K}_{\text{net}}(r) = \frac{\ell(L)}{n(n-1)} \sum_{i=1}^{n} \sum_{j \neq i} \mathbf{1} \left\{ d_L(x_i, x_j) \leq r \right\}, \tag{17.12}$$

where $\ell(L)$ denotes the total length of the linear network. This is a modification of the estimator (7.3) of Ripley's K-function of a two-dimensional point process, modified by replacing the Euclidean distance $\|x_i - x_j\|$ by the shortest-path distance $d_L(x_i, x_j)$, replacing the two-dimensional area of the observation window by the one-dimensional length of the network, and avoiding edge correction. Thus it is a renormalised version of the empirical cumulative distribution function of the shortest-path distances between all pairs of data points.

A problem with this definition is that the corresponding 'theoretical' curve — the expected value of $\widehat{K}_{net}(r)$ for a homogeneous Poisson process on the network — is not a simple function of r, and depends on the network geometry (essentially, on how much road length is present within a given travelling distance of any point).

Given a point u on the network, consider all locations in the network that can be reached from u by a path of length shorter than r. The set of such locations deserves to be called the 'disc of radius r' in the shortest-path metric, and is written $b_L(u,r)$. See Figure 17.14. In a homogeneous Poisson process on the network, the expected number of points which fall within a distance r of the location u (measured by the shortest path) is proportional to the total length of all line segments within distance r — that is, the total length of the segments making up the disc $b_L(u,r)$. The 'theoretical' expected value of $\widehat{K}_{net}(r)$ is the average, over all locations u on the network, of the length of the disc of radius r centred at u:

$$\frac{\mathbb{E}\left[N(N-1)\widehat{K}_{net}(r)\right]}{\mathbb{E}[N(N-1)]} = \frac{1}{\ell(L)}\int_L \ell(b_L(u,r))\,\mathrm{d}_1 u. \tag{17.13}$$

Calculation of this 'theoretical' curve is a complicated task in itself. Network K-functions obtained from different networks are not directly comparable. For example, it is difficult to compare the spatial patterns of crime in two different cities using the respective network K-functions.

The `spatstat` function `linearK` computes the Okabe-Yamada function $\widehat{K}_{net}(r)$ when the argument `correction="none"` is given.

```
> KN <- linearK(spiders, correction="none")
```

This computes only the empirical estimate $\widehat{K}_{net}(r)$ and not the 'theoretical' curve described above.

The generic `envelope` command has a method `envelope.lpp` for point patterns on a linear network. By default, simulations are generated from the homogeneous Poisson process on the network, with intensity equal to the estimated intensity of the data pattern. Figure 17.18 shows the result of plotting

```
> envelope(spiders, linearK, correction="none", nsim=39)
```

showing pointwise envelopes of the Okabe-Yamada network K-function based on 39 simulations of a homogeneous Poisson process on the same network and with the same average intensity as the `spiders` data. Equivalently the total number of points in each simulated pattern is Poisson distributed with mean equal to 48, with the points distributed uniformly on the mortar lines.

Note that Figure 17.18 does not include a benchmark curve for the expected value (17.13) under a uniform Poisson process; this value is expensive to compute exactly. Instead, the Figure shows the sample mean of the simulations, which gives a rough estimate of the reference value (17.13).

Strictly speaking, $\widehat{K}_{net}$ should be described as an 'empirical' or 'estimated' network K-function. As originally defined for two-dimensional point processes [571], the K-function is a theoretical expectation for the point process that is assumed to have generated our data, while the empirical K-function of the point pattern data is an empirical estimator of the K-function of the point process.

17.8.3   Assumptions implicit in the K-function

To make further progress, we need to return to basic principles discussed in Chapter 7. An important message there was that computing the K-function or pair correlation from point pattern data involves making implicit assumptions.

In two dimensions, the K-function and pair correlation function are defined under the assumption that the point process is stationary, at least to first and second order. Roughly speaking, the probability $p(u,v)$ that there are random points at two specified locations u and v is assumed to depend only on the Euclidean distance $\|u-v\|$ between them.

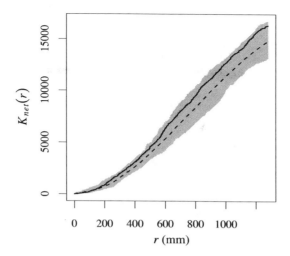

Figure 17.18. *Okabe-Yamada network K-function for the spider web data (solid black lines) and the sample mean (dashed lines) and pointwise envelopes (boundary of shaded region) of the same function for 39 simulations of a homogeneous Poisson process on the same network.*

On a linear network, we will need to make analogous assumptions in order to define the *K*-function. Unfortunately the concept of a (fully) stationary point process is meaningless on a linear network, because the network itself is not homogeneous: different locations in the network are surrounded by different configurations of line segments. However, a *second-order stationary* point process on a linear network is meaningful, and we can define its *K*-function and pair correlation function.

17.8.4 Geometrically corrected *K*-function

One of the good properties of Ripley's *K*-function in two dimensions is that it is normalised to enable comparison between different point processes, with different intensities, observed in different windows. This comparison is not possible with the Okabe-Yamada network *K*-function because it depends on the network geometry.

To correct this problem, Ang *et al.* [19] proposed a 'geometrically corrected *K*-function' in which each pair of points u, v is weighted by the factor $1/m(u, d_L(u, v))$ described above. This compensates for the varying geometry of the network. For a point process with constant intensity λ and pair correlation function satisfying (17.8), the geometrically corrected *K*-function is defined by

$$K^L(r) = \frac{1}{\lambda} \mathbb{E}\left[\sum_j \frac{1\{0 < d_L(u, x_j) \le r\}}{m(u, d_L(u, x_j))} \,\Big|\, u \in \mathbf{X}\right] \tag{17.14}$$

for all $r \le R$, where u is any location on the network. Under the given assumptions, the right-hand side of (17.14) does not depend on u.

For a Poisson process with constant intensity λ, the geometrically corrected *K*-function is

$$K^L(r) = r \tag{17.15}$$

for all $0 \le r < R$. This provides a simple benchmark for completely random point patterns on a linear network. It also permits comparison between the corrected *K*-functions obtained from different point patterns on different networks.

For the estimation of spatial interaction, the geometrically corrected K-function and pair correlation function have the strong advantage that they permit the range of interaction to be identified.

In two dimensions, the K-function and pair correlation function are intimately connected. A similar relationship holds in a linear network. For a point process with constant intensity λ and pair correlation function satisfying (17.8), the geometrically corrected K-function is related to the pair correlation function by

$$K^L(r) = \int_0^r g(s)\,ds \tag{17.16}$$

analogous to (7.25).

This geometrical correction restores many natural and desirable properties of K, including its direct relationship to the pair correlation function. For a completely random point pattern, on any network, the corrected network K-function is the identity. The corrected estimator $\hat{K}(r)$ has approximately constant variance.

Given a point pattern $\mathbf{x} = \{x_1, \ldots, x_n\}$ on the network L, the natural estimator of $K^L(r)$ is

$$\hat{K}^L(r) = \frac{\ell(L)}{n(n-1)} \sum_{i=1}^n \sum_{j \neq i} \frac{1\{d_L(x_i, x_j) \leq r\}}{m(x_i, d_L(x_i, x_j))} \tag{17.17}$$

for $0 \leq r < R(L)$. That is, the contribution to (17.17) from each pair of points (x_i, x_j) is weighted by the reciprocal of the number of points $u \in L$ that are situated at the same distance from x_i as x_j is. This weighting is analogous to the isotropic edge correction of Ripley [572], since $m(u, t)$ is a measure of the size of the boundary of the ball of radius t centred at u. Thus our corrected K-function can be viewed as the natural analogue of the edge-corrected Ripley K-function. The weighting is also intrinsic to the geometry of the linear network.

The geometrically corrected empirical K-function (17.17) is computed in `spatstat` by the function `linearK` with `correction="Ang"` (which is also the default). For the spider webs data:

```
> kls <- linearK(spiders)
> ekls <- envelope(spiders, linearK, nsim=39)
```

The results are plotted in Figure 17.19, showing our geometrically corrected network K-function for the spider data, with pointwise 5% critical envelopes based on 39 simulations of the uniform Poisson process with the same estimated intensity of 2.37 spiders per metre.

Note that the estimator variance is approximately constant in this example. Figure 17.19 suggests no departure from complete randomness. This is a pointwise Monte Carlo test; the simultaneous Monte Carlo test (not shown) also suggests no departure.

17.8.5   Inhomogeneous K-function

A counterpart of the geometrically corrected K-function is available when the point process is inhomogeneous with intensity function $\lambda(u)$. Assume the pair correlation function depends only on the shortest-path distance as in (17.8). Define the inhomogeneous, geometrically corrected K-function

$$K^{L,\text{ih}}(r) = \mathbb{E}\left[\sum_j \frac{1\{d_L(u, x_j) \leq r\}}{\lambda(x_j)\, m(u, d_L(u, x_j))} \;\middle|\; u \in X\right] \tag{17.18}$$

where u is an arbitrary location on the network. It is shown in [19] that the right-hand side does not depend on u, and that

$$K^{L,\text{ih}}(r) = \int_0^r g(t)\,dt. \tag{17.19}$$

The usual plug-in estimator of $K^{L,\text{ih}}(r)$ is

$$\hat{K}^{L,\text{ih}}(r) = \frac{1}{\ell(L)} \sum_{i=1}^n \sum_{j \neq i} \frac{1\{d_L(x_i, x_j) \leq r\}}{\hat{\lambda}(x_i)\hat{\lambda}(x_j) m(x_i, d_L(x_i, x_j))} \tag{17.20}$$

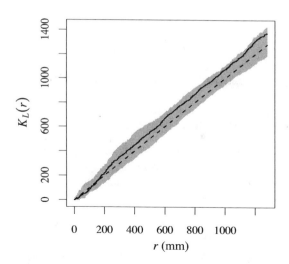

Figure 17.19. *Geometrically corrected linear K function estimate for spider pattern (solid lines), theoretical K-function for completely random pattern on mortar lines (dashed line), and pointwise 5% critical envelopes based on 39 simulations of a uniform Poisson process on mortar lines (grey shading).*

where $\hat{\lambda}(\cdot)$ is an estimate of the intensity function obtained by some chosen method. Although (17.20) is asymptotically unbiased for $K^{L,\text{ih}}(r)$ under suitable conditions, it has relatively high variance, attributable to variability in the number of data points. Campbell's formula (17.3) gives

$$\mathbb{E}\left[\sum_i \frac{1}{\lambda(x_i)}\right] = \ell(L)$$

so that if we replace (17.20) by

$$\widehat{K}^{L,\text{ih}}(r) = \frac{1}{\sum_i 1/\hat{\lambda}(x_i)} \quad \sum_{i=1}^{n}\sum_{j\neq i} \frac{1\{d_L(x_i,x_j) \leq r\}}{\hat{\lambda}(x_i)\hat{\lambda}(x_j)m(x_i,d_L(x_i,x_j))}; \tag{17.21}$$

then since the numerator and denominator are positively correlated, we obtain an approximately unbiased estimator of $K^{L,\text{ih}}(r)$ with smaller variance than (17.20). See [68, 640].

The `spatstat` function `linearKinhom` computes the estimates (17.21) and (17.20). It requires the intensity estimates $\hat{\lambda}(x_l)$ which can be supplied as a numeric vector, a function, a pixel image, or a fitted model of class `"lppm"`. In the latter case the model is re-fitted to the data before the fitted intensity values are computed, so that simulation envelopes based on `linearKinhom` are valid. By default the normalised, geometrically corrected estimate (17.21) is computed. Setting `normalise=FALSE` will select the unnormalised version (17.20), and setting `correction="none"` will remove the geometrical correction weights $m(u,r)$.

A very simple illustration is the model for the `spiders` data in which the intensity has a different constant value on vertical and horizontal segments, respectively. Using the `vertical` covariate constructed on page 718, this model can be fitted by:

```
> (fithv <- lppm(spiders ~ vertical))
Point process model on linear network
Nonstationary Poisson process
Log intensity: ~vertical
```

```
Fitted trend coefficients:
 (Intercept) verticalTRUE
       -5.76         -2.17
Original data: spiders
```

If we then compute the inhomogeneous K-function

```
> Khv <- linearKinhom(spiders, fithv)
> Khv <- linearKinhom(spiders, predict(fithv))  # equivalent
```

the weights $1/\lambda(x_i)$ in (17.18) take only two values, depending on whether the data point is situated on a vertical or horizontal segment. In this example, the estimated vertical intensity is low, so vertical points are up-weighted and contribute more to the final estimate.

For a more complete analysis of the spider data we add a log-quadratic trend term:

```
> fitv <- lppm(spiders ~ polynom(x,y,2) + vertical)
> KiS <- linearKinhom(spiders, fitv)
> EKiS <- envelope(spiders, linearKinhom, lambda=fit, nsim=39)
> EKiSwrong <- envelope(spiders, linearKinhom, lambda=fit,
                        nsim=39, update=FALSE)
```

The results are shown in Figure 17.20. The left panel is the correct version EKiS: it suggests the spider web pattern is consistent with an inhomogeneous Poisson process.

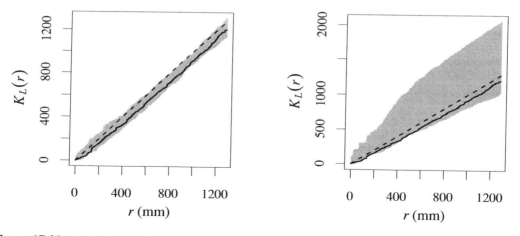

Figure 17.20. *Analysis of spider webs data using the inhomogeneous linear K-function. Solid lines: estimate $\widehat{K}^{L,\mathrm{ih}}(r)$ using fitted intensity which is a log-quadratic function of the spatial coordinates. Grey shading: region bounded by envelopes of $\widehat{K}^{L,\mathrm{ih}}(r)$ obtained from 39 simulations from the fitted model.* Left: *correct envelopes: intensity estimated by refitting model to each simulated pattern.* Right: *incorrect envelopes: intensity obtained from model fitted to original data.*

From this analysis we would conclude that the spider data do not show evidence of clustering after allowing for spatial inhomogeneity.

Second-order analysis can be highly sensitive to the fitted intensity, especially in a tree-like network [42].

17.8.6 Correlation in multitype point patterns

The multitype K-function, multitype pair correlation function, and other second moment summary functions (Section 14.6) have counterparts for multitype point patterns on a linear network, developed in [42].

linearKcross	K-function, type i to type j
linearKdot	K-function, type i to any type
linearKmulti	K-function, group I to group J
linearpcfcross	pair correlation, type i to type j
linearpcfdot	pair correlation, type i to any type
linearpcfmulti	pair correlation, group I to group J
linearmarkconnect	mark connection function
linearmarkequal	mark equality function
linearKcross.inhom	inhomogeneous K-function, type i to type j
linearKdot.inhom	inhomogeneous K-function, type i to any type
linearKmulti.inhom	inhomogeneous K-function, group I to group J
linearpcfcross.inhom	inhomogeneous pair correlation, type i to type j
linearpcfdot.inhom	inhomogeneous pair correlation, type i to any type
linearpcfmulti.inhom	inhomogeneous pair correlation, group I to group J

Table 17.10. *Summary functions for multitype point patterns on a linear network.*

The spatstat implementations of these functions are listed in Table 17.10. The names should be self-explanatory: for example, linearKcross is the counterpart of Kcross for point patterns on a linear network, with typical syntax

linearKcross(X, i, j)

where X is a point pattern on a linear network (class "lpp"). The arguments i and j are interpreted as levels of the factor marks(X).

The inhomogeneous summary functions (with names ending in inhom) require estimates of the relevant intensity function. For example linearKcross.inhom has typical syntax

linearKcross.inhom(X, i, j, lambdaI, lambdaJ)

where lambdaI and lambdaJ are estimates of the intensity for points of type i and j, respectively. Each of these can be either a numeric vector, a function, a pixel image (object of class "im" or "linim"), or a fitted point process model (object of class "ppm" or "lppm"). If lambdaI or lambdaJ is a fitted point process model, the default behaviour is to update the model by re-fitting it to the data, before computing the fitted intensity. This can be disabled by setting update=FALSE.

Any of the summary functions listed in Table 17.10 can be used in the function alltypes. For example,

```
> plot(alltypes(dendrite, linearKcross))
```

would display the bivariate K-function for each pair of types in the dendritic spines data.

The function rlabel also works for "lpp" objects and can be used to support tests of random labelling.

17.9 FAQs

- *I have a large shapefile of roads and want to convert this to a pixel image in which the pixel value is the total length of road in that pixel.*

Convert your shapefile to a "psp" object and use pixellate.psp.

- *What does* density.psp *do, and how is it different from* pixellate.psp?

  Roughly speaking density.psp is the same as applying pixellate.psp followed by blur to blur the image.

- *I have data specifying a road network and the locations of all road signs, including speed limit signs. How can I produce a spatial covariate which gives the speed limit that is in force at each place on the road network?*

  A code snippet is available at www.spatstat.org.

Index of Notation

745

SYMBOL	BRIEF DEFINITION	PAGES
$\widehat{K}_{iso*}(r)$	Ohser-Stoyan estimator of $K(r)$	217
$\widehat{K}_{rigid}(r)$	rigid motion correction estimator of $K(r)$	219
$\widehat{K}_{trans}(r)$	translation correction estimator of $K(r)$	219
$\widehat{K}_{un}(r)$	uncorrected estimate of $K(r)$	213
$\widehat{K}(r,x_i)$	local K function	247
$\mathscr{K}(A)$	reduced second moment measure	240
$\widehat{K}_{\mathrm{net}}(r)$	empirical network K-function	737
$K^L(r)$	geometrically corrected K-function on a network	739
$K^{L,\mathrm{ih}}(r)$	inhomogeneous, geometrically corrected K-function	740
$\ell(B)$	length of B	719
$\log L_P(\theta)$	log Palm likelihood	485
$\log \mathrm{PL}(\theta,\mathbf{x})$	log pseudolikelihood	554
L	linear network	719
$L_-(r)$ and $L_+(r)$	confidence limits for $L(r)$	231
$L(r)$	Besag's L-function	207
$L^*(r)$	resampled bootstrap L function	231
$L_{ij}(r), L_{i\bullet}(r)$	multitype L functions	596, 600
$L_{inhom}(r)$	inhomogeneous L-function	245
$L(r), U(r)$	global envelopes for a summary function	393
$L(\theta) = L(\theta,\mathbf{x})$	likelihood	343
$m(u,r)$	analogue of perimeter of circle	733
$\widetilde{m}(u)$	Nadaraya-Watson mark smoother	642
$\bar{m}_f$	nearest-neighbour correlation index	650
M	number of possible types	565
M	stability constant	552
M	Morisita index	200
$M(B)$	expected sum of covariate values	332
$\mathscr{M}$	all possible marks	627
$n(\mathbf{x})$	number of points in point pattern $\mathbf{x}$	129
$n(\mathbf{x}\cap B), n(\mathbf{X}\cap B)$	number of points falling in B	129
$O_{r_1,r_2}(\phi)$	point pair orientation distribution	238
$p(m\mid u)$	conditional probability of type m at u	577
$p(x,d)$	fraction of length of circle $\partial b(x,d)$ within W	216
$p_{ij}(r)$	mark connection function	596
p_j	proportion of points of type j	600, 626
$\mathrm{pL}(\psi)$	profile likelihood	360
$\hat{p}_j$	simulated p-value for the Dao-Genton test	406
$\mathbb{P}\{\cdot\}$	probability	130
$\mathbb{P}^u\{\cdot\}$	Palm probability	293
$\mathrm{PPL}(\phi)$	profile log pseudolikelihood	518
$Q(u), Q(u,v)$	weight function	485
R, r	interaction distance parameter in Gibbs models	499, 515, 516, 520, 522, 524
R	Matérn cluster radius	139
R	Clark-Evans index	258
$\mathbb{R}^2$	two-dimensional Euclidean plane	128
$r(u)$	relative intensity	413
$\tilde{r}(t)$	ordinate of added variable plot	430

SYMBOL	BRIEF DEFINITION	PAGES
$X_{\oplus r}$	dilation of X by $b(0,r)$	266
$y = (x^{(1)}, \ldots, x^{(M)})$	multivariate form of multitype pattern	565
$y^{(i)}$	i-th replicate pattern	706
y_i	parent point of cluster	459
$Y = (X^{(1)}, \ldots, X^{(M)})$	multivariate form of multitype process	574
$Y(u)$	additional covariate function	430
z_{im}	indicator whether point i has type m	579
Z_1, Z_2	Berman's test statistics	376, 376
$Z(u)$	spatial covariate	66, 304
$Z(u)$	p-dimensional vector-valued covariate	304
$Z(u,m)$	m-dimensional vector-valued covariate	590
$Z(u,x)$	score difference	553
α	significance level	371
α	normalising constant	554
α_m	coefficients in loglinear model	588, 589
β	probability of Type II error	371
β	abundance parameter of Gibbs model	499, 522, 524
β	vector of fixed coefficients	706
$\beta_m(u)$	first-order trend for type m	612
$\beta_i(u)$	cluster intensity function	466
$\beta(u)$	varying intensity	302
γ	interaction parameter of Gibbs models	499, 519, 522, 524
γ	vector of random coefficients	706
$\gamma(r)$	mark variogram	647
Γ	likelihood ratio test statistic	372
$\Gamma(u,r)$	likelihood ratio test statistic, scan test	190
Γ^*	scan statistic	191
Γ^*	composite likelihood ratio statistic	513
$\varepsilon_w(x)$	Epanechnikov kernel	228
$\theta = (\eta, \varphi)$	trend and interaction parameters	505
$\theta = (\varphi, \psi)$	regular and irregular parameters	360
$\widehat{\theta}$	maximum likelihood estimate of θ	360
$\widehat{\theta}_{MPL}$	maximum pseudolikelihood estimate	554, 556
$\kappa, \kappa(u)$	parent intensity	139, 465
κ	proportionality constant in intensity model	320
κ	decay rate parameter in Fiksel model	516
κ	range parameter in Diggle-Gratton model	516
κ	first-order parameter of area-interaction model	519
κ	shape parameter of soft core model	515
$\kappa(u)$	smoothing kernel	168, 720
λ	intensity	132, 719
$\lambda(u)$	intensity function	137
$\lambda(u \mid x)$	conditional intensity	489
$\lambda(u,m)$	multitype intensity function	575
$\lambda((u,m) \mid y)$	multitype conditional intensity	612
$\lambda(x,y), \lambda((x,y))$	intensity function of Cartesian coordinates	162, 319
$\lambda_P(u), \lambda_P(u \mid v; \theta)$	Palm intensity	485, 485
$\lambda_0(u)$	null-hypothesized intensity	382
$\lambda_2(u,v)$	second moment intensity function	242, 729

Bibliography

[1] E.A. Abbott. *Flatland: A Romance of Many Dimensions*. Blackwell, revised 6th edition, 1950.

[2] G. Adelfio and F.P. Schoenberg. Point process diagnostics based on weighted second-order statistics and their asymptotic properties. *Annals of the Institute of Statistical Mathematics*, 61:929–948, 2009.

[3] A. Agresti and B.A. Coull. Approximate is better than "Exact" for interval estimation of binomial proportions. *American Statistician*, 52:119–126, 1998.

[4] F.P. Agterberg. Automatic contouring of geological maps to detect target areas for mineral exploration. *Journal of the International Association for Mathematical Geology*, 6:373–395, 1974.

[5] F.P. Agterberg. *Geomathematics: Theoretical Foundations, Applications, and Future Developments*. Number 18 in Quantitative Geology and Geostatistics. Springer, Cham, 2014.

[6] N. Ahuja. Dot pattern processing using Voronoi neighbourhoods. *IEEE Transactions on Pattern Analysis and Machine Intelligence*, 4:336–343, 1982.

[7] M. Aitkin, D.A. Anderson, B.J. Francis, and J.P. Hinde. *Statistical Modelling in GLIM*. Oxford University Press, 1989.

[8] M. Aitkin and D. Clayton. The fitting of exponential, Weibull and extreme value distributions to complex censored survival data using GLIM. *Applied Statistics*, 29:156–163, 1980.

[9] H. Akaike. A new look at the statistical model identification. *IEEE Transactions on Automatic Control*, 19(6):716–723, 1974.

[10] A. Albert and J.A. Anderson. On the existence of maximum likelihood estimates in logistic regression models. *Biometrika*, 71:1–10, 1984.

[11] D. Allard and C. Fraley. Nonparametric maximum likelihood estimation of features in spatial point processes using Voronoi tessellation. *Journal of the American Statistical Association*, 92:1485–1493, 1997.

[12] S.E. Alm. Approximation and simulation of the distributions of scan statistics for Poisson processes in higher dimensions. *Extremes*, 1(1):111–126, 1998.

[13] J.E. Alt, G. King, and C.S. Signorino. Aggregation among binary, count and duration models: estimating the same quantities from different levels of data. *Political Analysis*, 9:21–44, 2001.

[14] G.K. Ambler. *Dominated coupling from the past and some extensions of the area-interaction process*. PhD thesis, Department of Mathematics, University of Bristol, September 2002.

[15] G.K. Ambler and B.W. Silverman. Perfect simulation of spatial point processes using dominated coupling from the past with application to a multiscale area-interaction process. Technical report, Department of Mathematics, University of Bristol, 2004. arXiv: 0903.2651v1 [stat.ME].

[16] T.W. Anderson and D.A. Darling. Asymptotic theory of certain 'goodness-of-fit' criteria based on stochastic processes. *Annals of Mathematical Statistics*, 23:193–212, 1952.

[17] T.W. Anderson and D.A. Darling. A test of goodness of fit. *Journal of the American Statistical Association*, 49:765–769, 1954.

[18] R.S. Anderssen, A. Baddeley, F.R. DeHoog, and G.M. Nair. Solution of an integral equation arising in spatial point process theory. *Journal of Integral Equations and Applications*, 26(4):437–453, 2014. In press. Published online November 2014.

[19] Q.W. Ang, A. Baddeley, and G. Nair. Geometrically corrected second order analysis of events on a linear network, with applications to ecology and criminology. *Scandinavian Journal of Statistics*, 39:591–617, 2012.

[20] L. Anselin. Local indicators of spatial association – LISA. *Geographical Analysis*, 27:93–115, 1995.

[21] R. Assunção. *Robust estimation in point processes*. PhD thesis, University of Washington, Seattle, 1994.

[22] R. Assunção. Score tests for pairwise interaction parameters of Gibbs point processes. *Brazilian Journal of Probability and Statistics*, 17:169–178, 2003.

[23] R. Assunção and P. Guttorp. Robustness for inhomogeneous Poisson point processes. *Annals of the Institute of Statistical Mathematics*, 51:657–678, 1999.

[24] R.M. Assunção and P.A. Ferrari. Independence of thinned processes characterizes the Poisson process: An elementary proof and a statistical application. *Test*, 16:333–345, 2007.

[25] A.C. Atkinson. Diagnostics, regression analysis and shifted power transformations. *Technometrics*, 25:23–34, 1983.

[26] A.C. Atkinson. *Plots, Transformations and Regression.* Number 1 in Oxford Statistical Science Series. Oxford University Press/ Clarendon, 1985.

[27] A. Baddeley. Spatial point processes and their applications. In A. Baddeley, I. Bárány, R. Schneider, and W. Weil, editors, *Stochastic Geometry: Lectures given at the C.I.M.E. Summer School held in Martina Franca, Italy, September 13–18, 2004*, Lecture Notes in Mathematics 1892 (subseries: Fondazione C.I.M.E., Firenze), pages 1–75. Springer-Verlag, 2006. ISBN 3-540-38174-0.

[28] A. Baddeley. Validation of statistical models for spatial point patterns. In J.G. Babu and E.D. Feigelson, editors, *Statistical Challenges in Modern Astronomy IV*, volume 371 of *Astronomical Society of the Pacific, Conference Series*, pages 22–38, San Francisco, California, USA, 2007. Astronomical Society of the Pacific. Proceedings of a conference held at Penn State University, 12–15 June 2006. www.astrosociety.org.

[29] A. Baddeley. Analysing spatial point patterns in R. Technical report, CSIRO, 2010. Version 4. Available at www.csiro.au/resources/pf16h.html.

[30] A. Baddeley. Modelling strategies. In A.E. Gelfand, P.J. Diggle, M. Fuentes, and P. Guttorp, editors, *Handbook of Spatial Statistics*, chapter 20, pages 339–369. CRC Press, Boca Raton, 2010.

[31] A. Baddeley. Multivariate and marked point processes. In A.E. Gelfand, P.J. Diggle, M. Fuentes, and P. Guttorp, editors, *Handbook of Spatial Statistics*, chapter 21, pages 371–402. CRC Press, Boca Raton, 2010.

[32] A. Baddeley. Spatial point patterns — models and statistics. In E. Spodarev, editor, *Lectures on Stochastic Geometry, Spatial Statistics and Random Fields. Asymptotic Methods*, number 2068 in Lecture Notes in Mathematics, chapter 4. Springer-Verlag, 2012.

[33] A. Baddeley. Local composite likelihood for spatial point processes. Submitted for publication, 2015.

[34] A. Baddeley, M. Berman, N.I. Fisher, A. Hardegen, R.K. Milne, D. Schuhmacher, and R. Turner. Spatial logistic regression and change-of-support for Poisson point processes. *Electronic Journal of Statistics*, 4:1151–1201, 2010.

[35] A. Baddeley, Y.-M. Chang, Y. Song, and R. Turner. Residual diagnostics for covariate effects in spatial point process models. *Journal of Computational and Graphical Statistics*, 22:886–905, 2013.

[36] A. Baddeley, Y.M. Chang, and Y. Song. Leverage and influence diagnostics for spatial point processes. *Scandinavian Journal of Statistics*, 40:86–104, 2013.

[37] A. Baddeley, Y.M. Chang, Y. Song, and R. Turner. Nonparametric estimation of the dependence of a spatial point process on a spatial covariate. *Statistics and its Interface*, 5:221–236, 2012.

[38] A. Baddeley, J.-F. Coeurjolly, E. Rubak, and R. Waagepetersen. Logistic regression for spatial Gibbs point processes. *Biometrika*, 101(2):377–392, 2014.

[39] A. Baddeley and D. Dereudre. Variational estimators for the parameters of Gibbs point process models. *Bernoulli*, 19:905–930, 2013.

[40] A. Baddeley, P.J. Diggle, A. Hardegen, T. Lawrence, R.K. Milne, and G. Nair. On tests of spatial pattern based on simulation envelopes. *Ecological Monographs*, 84(3):477–489, 2014.

[41] A. Baddeley, A. Hardegen, T. Lawrence, R. Milne, G.M. Nair, and S. Rakshit. Pushing the envelope: extensions of graphical Monte Carlo tests. Submitted for publication.

[42] A. Baddeley, A. Jammalamadaka, and G. Nair. Multitype point process analysis of spines on the dendrite network of a neuron. *Applied Statistics (Journal of the Royal Statistical Society, Series C)*, 63(5):673–694, 2014.

[43] A. Baddeley and E.B. Vedel Jensen. *Stereology for Statisticians*. Chapman and Hall/CRC, Boca Raton, 2005. ISBN 1-58488-405-3.

[44] A. Baddeley, M. Kerscher, K. Schladitz, and B.T. Scott. Estimating the J function without edge correction. *Statistica Neerlandica*, 54(3):315–328, November 2000.

[45] A. Baddeley, J. Møller, and A.G. Pakes. Properties of residuals for spatial point processes. *Annals of the Institute of Statistical Mathematics*, 60:627–649, 2008.

[46] A. Baddeley, J. Møller, and R. Waagepetersen. Non- and semiparametric estimation of interaction in inhomogeneous point patterns. *Statistica Neerlandica*, 54(3):329–350, 2000.

[47] A. Baddeley and G. Nair. Approximating the moments of a spatial point process. *Stat*, 1(1):18–30, 2012.

[48] A. Baddeley and G. Nair. Fast approximation of the intensity of Gibbs point processes. *Electronic Journal of Statistics*, 6:1155–1169, 2012.

[49] A. Baddeley, E. Rubak, and J. Møller. Score, pseudo-score and residual diagnostics for spatial point process models. *Statistical Science*, 26:613–646, 2011.

[50] A. Baddeley and R. Turner. Practical maximum pseudolikelihood for spatial point patterns (with discussion). *Australian and New Zealand Journal of Statistics*, 42(3):283–322, 2000.

[51] A. Baddeley and R. Turner. Spatstat: an R package for analyzing spatial point patterns. *Journal of Statistical Software*, 12(6):1–42, 2005. URL: `www.jstatsoft.org`, ISSN: 1548-7660.

[52] A. Baddeley and R. Turner. Modelling spatial point patterns in R. In A. Baddeley, P. Gregori, J. Mateu, R. Stoica, and D. Stoyan, editors, *Case Studies in Spatial Point Pattern Modelling*, number 185 in Lecture Notes in Statistics, pages 23–74. Springer-Verlag, New York, 2006. ISBN: 0-387-28311-0.

[53] A. Baddeley and R. Turner. Bias correction for parameter estimates of spatial point process models. *Journal of Statistical Computation and Simulation*, 84:1621–1643, 2014.

[54] A. Baddeley, R. Turner, J. Mateu, and A. Bevan. Hybrids of Gibbs point process models and their implementation. *Journal of Statistical Software*, 55(11):1–43, 2013.

[55] A. Baddeley, R. Turner, J. Møller, and M. Hazelton. Residual analysis for spatial point processes (with discussion). *Journal of the Royal Statistical Society, Series B*, 67(5):617–666, 2005.

[56] A. Baddeley, R. Turner, and E. Rubak. Adjusted composite likelihood ratio test for spatial Gibbs point processes. *Journal of Statistical Computation and Simulation*, 2015. In press.

[57] A.J. Baddeley. A limit theorem for statistics of spatial data. *Advances in Applied Probability*, 12:447–461, 1980.

[58] A.J. Baddeley. Stereology and survey sampling theory. *Bulletin of the International Statistical Institute*, 50, book 2:435–449, 1993.

[59] A.J. Baddeley. Spatial sampling and censoring. In O.E. Barndorff-Nielsen, W.S. Kendall, and M.N.M. van Lieshout, editors, *Stochastic Geometry: Likelihood and Computation*, chapter 2, pages 37–78. Chapman and Hall, London, 1999.

[60] A.J. Baddeley. Time-invariance estimating equations. *Bernoulli*, 6(5):783–808, 2000.

[61] A.J. Baddeley and L.M. Cruz-Orive. The Rao-Blackwell theorem in stereology and some counterexamples. *Advances in Applied Probability*, 27:2–19, 1995.

[62] A.J. Baddeley and R.D. Gill. Kaplan-Meier estimators for interpoint distance distributions of spatial point processes. Research Report BS-R9315, Centrum voor Wiskunde en Informatica, July 1993.

[63] A.J. Baddeley and R.D. Gill. The empty space hazard of a spatial pattern. Preprint 845, Department of Mathematics, University of Utrecht, 1994.

[64] A.J. Baddeley and R.D. Gill. Kaplan-Meier estimators of interpoint distance distributions for spatial point processes. *Annals of Statistics*, 25:263–292, 1997.

[65] A.J. Baddeley and H.J.A.M. Heijmans. Incidence and lattice calculus with applications to stochastic geometry and image analysis. *Applicable Algebra in Engineering, Communication, and Computing*, 6(3):129–146, 1995.

[66] A.J. Baddeley, C. V. Howard, A. Boyde, and S. Reid. Three-dimensional analysis of the spatial distribution of particles using the tandem-scanning reflected light microscope. *Acta Stereologica*, 6 (supplement II):87–100, 1987.

[67] A.J. Baddeley and J. Møller. Nearest-neighbour Markov point processes and random sets. *International Statistical Review*, 57:89–121, 1989.

[68] A.J. Baddeley, R.A. Moyeed, C.V. Howard, and A. Boyde. Analysis of a three-dimensional point pattern with replication. *Applied Statistics*, 42(4):641–668, 1993.

[69] A.J. Baddeley and B.W. Silverman. A cautionary example on the use of second-order methods for analyzing point patterns. *Biometrics*, 40:1089–1094, 1984.

[70] A.J. Baddeley and M.N.M. van Lieshout. Area-interaction point processes. *Annals of the Institute of Statistical Mathematics*, 47:601–619, 1995.

[71] A.J. Baddeley, M.N.M. van Lieshout, and J. Møller. Markov properties of cluster processes. *Advances in Applied Probability*, 28:346–355, 1996.

[72] A. Baillo, A. Cuevas, and A. Justel. Set estimation and nonparametric detection. *Canadian Journal of Statistics*, 28:765–782, 2008.

[73] D. Bakan. The test of significance in psychological research. *Psychological Bulletin*, 66:423–437, 1966.

[74] M. Balinga, T. Sunderland, G. Walters, Y. Issembé, S. Asaha, and E. Fombod. A vegetation assessment of the Waka national park, Gabon. CARPE report, Herbier National du Gabon, LBG, MBG, WCS, FRP and Simthsonian Institution, Libreville, Gabon, 2006. URL http://carpe.umd.edu/resources/Documents/.

[75] S. Banerjee. Spatial gradients and wombling. In A.E. Gelfand, P.J. Diggle, M. Fuentes, and P. Guttorp, editors, *Handbook of Spatial Statistics*, chapter 31. CRC Press, Boca Raton, FL, 2010.

[76] S. Banerjee and A.E. Gelfand. Prediction, interpolation and regression for spatially misaligned data. *Sanhkya A*, 64:227–245, 2002.

[77] S. Banerjee, A.E. Gelfand, and B.P. Carlin. *Hierarchical modeling and analysis for spatial data*. Chapman and Hall/CRC, Boca Raton, FL, 2005.

[78] J.D. Banfield and A.E. Raftery. Model-based Gaussian and non-Gaussian clustering. *Biometrics*, 49:803–821, 1993.

[79] G. Barnard. Contribution to discussion of "The spectral analysis of point processes" by M.S. Bartlett. *Journal of the Royal Statistical Society, Series B*, 25:294, 1963.

[80] O.E. Barndorff-Nielsen. *Information and Exponential Families in Statistical Theory*. Wiley, Chichester, New York, Brisbane, Toronto, 1978.

[81] C. Barr and F.P. Schoenberg. On the Voronoi estimator for the intensity of an inhomogeneous planar Poisson process. *Biometrika*, 97:977–984, 2010.

[82] M.S. Bartlett. The spectral analysis of point processes. *Journal of the Royal Statistical Society, Series B*, 29:264–296, 1963.

[83] M.S. Bartlett. A note on spatial pattern. *Biometrics*, 20:891–892, 1964.

[84] F. Bavaud. Equilibrium properties of the Vlasov functional: The generalized Poisson-Boltzmann-Emden equation. *Review of Modern Physics*, 63:129–149, 1991.

[85] M. J. Bayarri and J.O. Berger. P values for composite null models. *Journal of the American Statistical Association*, 95(452):1127–1142, December 2000.

[86] T. Bedford and J. van den Berg. A remark on the Van Lieshout and Baddeley J-function for point processes. *Advances in Applied Probability*, 29:19–25, 1997.

[87] M. Bell and G. Grunwald. Mixed models for the analysis of replicated spatial point patterns. *Biostatistics*, 5:633–648, 2004.

[88] M. Berman. Testing for spatial association between a point process and another stochastic process. *Applied Statistics*, 35:54–62, 1986.

[89] M. Berman and P. Diggle. Estimating weighted integrals of the second-order intensity of a spatial point process. *Journal of the Royal Statistical Society, Series B*, 51:81–92, 1989.

[90] M. Berman and T.R. Turner. Approximating point process likelihoods with GLIM. *Applied Statistics*, 41:31–38, 1992.

[91] R. Bernhardt, F. Meyer-Olbersleben, and B. Kieback. Fundamental investigation on the preparation of gradient structures by sedimentation of different powder fractions under gravity. In David Hui, editor, *Proceedings of the 4th International Conference on Composite Engineering, ICCE/4, July 6–12 1997*, pages 147–148, Hawaii, 1997.

[92] K. K. Berthelsen and J. Møller. Bayesian analysis of Markov point processes. In A. Baddeley, P. Gregori, J. Mateu, R. Stoica, and D. Stoyan, editors, *Case Studies in Spatial Point Process Modeling*, pages 85–97. Springer Lecture Notes in Statistics 185, Springer-Verlag, New York, 2006.

[93] K.K. Berthelsen and J. Møller. A primer on perfect simulation for spatial point processes. *Bulletin of the Brazilian Mathematical Society*, 33:351–367, 2002.

[94] K.K. Berthelsen and J. Møller. Likelihood and non-parametric Bayesian MCMC inference for spatial point processes based on perfect simulation and path sampling. *Scandinavian Journal of Statistics*, 30:549–564, 2003.

[95] E. Bertin, J.-M. Billiot, and R. Drouilhet. Existence of "nearest-neighbour" type spatial Gibbs models. *Advances in Applied Probability*, 31:895–909, 1999.

[96] E. Bertin, J.-M. Billiot, and R. Drouilhet. Spatial Delaunay Gibbs point processes. *Stochastic Models*, 15:181–199, 1999.

[97] J. Besag. Statistical analysis of non-lattice data. *The Statistician*, 24:179–195, 1975.

[98] J. Besag. Some methods of statistical analysis for spatial data. *Bulletin of the International Statistical Institute*, 47:77–91, 1977.

[99] J. Besag and P.J. Diggle. Simple Monte Carlo tests for spatial pattern. *Applied Statistics*, 26:327–333, 1977.

[100] J. Besag and P. J. Green. Spatial statistics and Bayesian computation (with discussion). *Journal of the Royal Statistical Society, Series B*, 55:25–37, 1993.

[101] J. Besag, R. Milne, and S. Zachary. Point process limits of lattice processes. *Journal of Applied Probability*, 19:210–216, 1982.

[102] J. Besag and J. Newell. The detection of clusters in rare diseases. *Journal of the Royal Statistical Society, Series A*, 154:143–155, 1991.

[103] J. E. Besag. Contribution to the discussion of the paper by Ripley (1977). *Journal of the Royal Statistical Society, Series B*, 39:193–195, 1977.

[104] J.E. Besag and P. Clifford. Generalized Monte Carlo significance tests. *Biometrika*, 76:633–642, 1989.

[105] J.-M. Billiot, J.-F. Coeurjolly, and R. Drouilhet. Maximum pseudolikelihood estimator for exponential family models of marked Gibbs processes. *Electronic Journal of Statistics*, 2:234–264, 2008.

[106] J.F. Bithell. An application of density estimation to geographical epidemiology. *Statistics in Medicine*, 9:691–701, 1990.

[107] J.F. Bithell. The choice of test for detecting a raised disease risk near a point source. *Statistics in Medicine*, 14:2309–2322, 1995.

[108] J.F. Bithell, S.J. Dutton, G.J. Draper, and N.M. Neary. Distribution of childhood leukaemias and non-Hodgkin's lymphomas near nuclear installations in England and Wales. *British Medical Journal*, 309:501–505, 1994.

[109] R. Bivand and A. Gebhardt. Implementing functions for spatial statistical analysis using the R language. *Journal of Geographical Systems*, 2:307–317, 2000.

[110] R. Bivand, J. Hauke, and T. Kossowski. Computing the Jacobian in Gaussian spatial autoregressive models: An illustrated comparison of available methods. *Geographical Analysis*, 45(2):150–179, 2013.

[111] R. Bivand, E.J. Pebesma, and V. Gómez-Rubio. *Applied spatial data analysis with R*. Springer, 2008.

[112] R. Bivand and G. Piras. Comparing implementations of estimation methods for spatial econometrics. *Journal of Statistical Software*, 63(18):1–36, 2015.

[113] G.E. Blackman. A study by statistical methods of the distribution of species in grassland associations. *Annals of Botany*, os-49(4):749–777, 1935.

[114] M.A. Bognar. Bayesian inference for spatially inhomogeneous pairwise interacting point processes. *Computational Statistics and Data Analysis*, 49:1–18, 2005.

[115] L. Bondesson and J. Fahlén. Mean and variance of vacancy for hard-core disc processes and applications. *Scandinavian Journal of Statistics*, 30:797–816, 2003.

[116] G. Bonham-Carter. *Geographic Information Systems for Geoscientists: Modelling with GIS*. Number 13 in Computer Methods in the Geosciences. Pergamon Press/ Elsevier, Kidlington, Oxford, UK, 1995.

[117] G. Borgefors. Distance transformations in arbitrary dimensions. *Computer Vision, Graphics and Image Processing*, 27:321–345, 1984.

[118] G. Borgefors. Distance transformations in digital images. *Computer Vision, Graphics and Image Processing*, 34:344–371, 1986.

[119] D. Borland and R.M. Taylor. Rainbow color map (still) considered harmful. *IEEE Computer Graphics and Applications*, 27(2):14–17, 2007.

[120] G. Borruso. Network density estimation: Analysis of point patterns over a network. In O. Gervasi, M.L. Gavrilova, V. Kumar, A. Laganà, H.P. Lee, Y. Mun, D. Taniar, and C.J.K. Tan, editors, *Computational Science and its Applications — ICCSA 2005*, number 3482 in Lecture Notes in Computer Science, pages 126–132. Springer, Berlin/Heidelberg, 2005.

[121] G. Borruso. Network density estimation: a GIS approach for analysing point patterns in a network space. *Transactions in GIS*, 12:377–402, 2008.

[122] A. W. Bowman and A. Azzalini. *Applied Smoothing Techniques for Data Analysis: the Kernel Approach with S-Plus Illustrations*. Oxford University Press, Oxford, 1997.

[123] L. Breiman, J. Friedman, C.J. Stone, and R.A. Olshen. *Classification and Regression Trees*. Chapman and Hall/CRC, 1984.

[124] D.R. Brillinger. Comparative aspects of the study of ordinary time series and of point processes. In P.R. Krishnaiah, editor, *Developments in Statistics*, pages 33–133. Academic Press, New York, London, 1978.

[125] D.R. Brillinger and H.K. Preisler. Two examples of quantal data analysis: a) multivariate point process, b) pure death process in an experimental design. In *Proceedings, XIII International Biometric Conference, Seattle*, pages 94–113. International Biometric Society, 1986.

[126] D.R. Brillinger and J.P. Segundo. Empirical examination of the threshold model of neuron firing. *Biological Cybernetics*, 35:213–220, 1979.

[127] D.R. Brillinger and B.S. Stewart. Elephant seal movements: modelling migration. *Canadian Journal of Statistics*, 26:431–443, 1998.

[128] A. Brix. Generalized gamma measures and shot-noise Cox processes. *Advances in Applied Probability*, 31:929–953, 1999.

[129] A. Brix and J. Chadoeuf. Spatio-temporal modeling of weeds and shot-noise G Cox processes. *Biometrical Journal*, 44:83–99, 2002.

[130] A. Brix and P. J. Diggle. Spatio-temporal prediction for log-Gaussian Cox processes. *Journal of the Royal Statistical Society, Series B*, 63:823–841, 2001.

[131] A. Brix and J. Møller. Space-time multi type log Gaussian Cox processes with a view to modeling weed data. Research Report R-98-2012, Department of Mathematics, Aalborg University, 1998. Submitted.

[132] A. Brix and J. Møller. Space-time multitype log Gaussian Cox processes with a view to modelling weed data. *Scandinavian Journal of Statistics*, 28:471–488, 2001.

[133] S.P. Brooks, B.J.T. Morgan, M.S. Ridout, and S.E. Pack. Finite mixture models for proportions. *Biometrics*, 53:1097–1115, 1997.

[134] W.M. Brown, T.D. Gedeon, A.J. Baddeley, and D.I. Groves. Bivariate J-function and other graphical statistical methods help select the best predictor variables as inputs for a neural network method of mineral prospectivity mapping. In U. Bayer, H. Burger, and W. Skala, editors, *IAMG 2002: 8th Annual Conference of the International Association for Mathematical Geology*, volume 1, pages 257–268. International Association of Mathematical Geology, 2002.

[135] S. Byers and A.E. Raftery. Nearest-neighbour clutter removal for estimating features in spatial point processes. *Journal of the American Statistical Association*, 93:577–584, 1998.

[136] K. Byth. θ-stationary point processes and their second-order analysis. *Journal of Applied Probability*, 18:864–878, 1981.

[137] K. Byth. On robust distance-based intensity estimators. *Biometrics*, 38:127–135, 1982.

[138] C. Calenge. The package adehabitat for the R software: a tool for the analysis of space and habitat use by animals. *Ecological Modelling*, 197:516–519, 2006.

[139] E.J.M. Carranza. *Geochemical anomaly and mineral prospectivity mapping in GIS*, volume 11 of *Handbook of Exploration and Environmental Geochemistry*. Elsevier, 2008.

[140] E.J.M. Carranza. Controls on mineral deposit occurrence inferred from analysis of their spatial pattern and spatial association with geological features. *Ore Geology Reviews*, 35:383–400, 2009.

[141] S. Chainey and J. Ratcliffe. *GIS and Crime Mapping*. Wiley, Chichester, 2005.

[142] R.M. Chambers and T. Hastie, editors. *Statistical models in S*. Wadsworth and Brooks/Cole, Monterey, 1992.

[143] H.P. Chan. Detection of spatial clustering with average likelihood ratio test statistics. *American Statistician*, 37:3985–4010, 2009.

[144] R.E. Chandler and E.M. Scott. *Statistical Methods for Trend Detection and Analysis in the Environmental Sciences*. Wiley, Chichester, 2011.

[145] Y.M. Chang, A. Baddeley, J. Wallace, and M. Canci. Spatial statistical analysis of tree deaths using airborne digital imagery. *International Journal of Applied Earth Observation and Geoinformation*, 21:418–426, 2012.

[146] B.J. Chen, G.P. Leser, D. Jackson, and R.A. Lamb. The influenza virus M2 protein cytoplasmic tail interacts with the M1 protein and influences virus assembly at the site of virus budding. *Journal of Virology*, 82:10059–10070, 2008.

[147] J. Chen and A.K. Gupta. *Parametric Statistical Change Point Analysis*. Birkhauser, 2000.

[148] Y.C. Chin and A.J. Baddeley. Markov interacting component processes. *Advances in Applied Probability*, 32:597–619, 2000.

[149] S.N. Chiu. Spatial point pattern analysis by using Voronoi diagrams and Dirichlet tessellations — a comparative study. *Biometrical Journal*, 45:367–376, 2003.

[150] S.N. Chiu and D. Stoyan. Estimators of distance distributions for spatial patterns. *Statistica Neerlandica*, 52:239–246, 1998.

[151] E. Choi and P. Hall. Nonparametric analysis of earthquake point-process data. In M. de Gunst, C. Klaassen, and A. van der Vaart, editors, *State of the Art in Probability and Statistics: Festschrift for Willem R. van Zwet*, pages 324–344. Institute of Mathematical Statistics, Beachwood, Ohio, 2001.

[152] C.F. Chung and F.P. Agterberg. Regression models for estimating mineral resources from geological map data. *Mathematical Geology*, 12:473–488, 1980.

[153] C.F. Chung and A.G. Fabbri. Probabilistic prediction models for landslide hazard mapping. *Photogrammetric Engineering and Remote Sensing*, 62(12):1389–1399, 1999.

[154] G. Claeskens and N.L. Hjort. *Model Selection and Model Averaging*. Cambridge University Press, 2008.

[155] P.J. Clark and F.C. Evans. Distance to nearest neighbor as a measure of spatial relationships in populations. *Ecology*, 35:445–453, 1954.

[156] M. Clausel, J.-F. Coeurjolly, and J. Lelong. Stein estimation of the intensity of a spatial homogeneous Poisson process. `arXiv:1407.4372`, 2014.

[157] R.A. Clements, F.P. Schoenberg, and D. Schorlemmer. Residual analysis for space-time point processes with applications to earthquake forecast models in California. *Annals of Applied Statistics*, 5:2549–2571, 2011.

[158] M. Clyde and D. Strauss. Logistic regression for spatial pair-potential models. In A. Possolo, editor, *Spatial Statistics and Imaging*, volume 20 of *Lecture Notes - Monograph series*, chapter II, pages 14–30. Institute of Mathematical Statistics, Hayward, CA, 1991. ISBN 0-940600-27-7.

[159] W.G. Cochran. *Sampling Techniques*. John Wiley and Sons, third edition, 1977.

[160] J.F. Coeurjolly, D. Dereudre, R. Drouilhet, and F. Lavancier. Takacs-Fiksel method for stationary marked Gibbs point processes. Preprint, 2010.

[161] J.F. Coeurjolly and F. Lavancier. Residuals and goodness-of-fit tests for stationary marked Gibbs point processes. *Journal of the Royal Statistical Society, Series B*, 75:247–276, 2013.

[162] J.F. Coeurjolly and E. Rubak. Fast covariance estimation for innovations computed from a spatial Gibbs point process. *Scandinavian Journal of Statistics*, 40:669–684, 2013.

[163] J. Cohen. The earth is round ($p < .05$). *American Psychologist*, 49:997–1003, 1994.

[164] D. Collett. *Modelling Binary Data*. Chapman and Hall, London, 1991.

[165] A. Comber, C. Brunsdon, and E. Green. Using a GIS-based network analysis to determine urban greenspace accessibility for different ethnic and religious groups. *Landscape and Urban Planning*, 86:103–114, 2008.

[166] R. Condit. *Tropical Forest Census Plots*. Springer-Verlag, 1998.

[167] R. Condit, S.P. Hubbell, and R.B. Foster. Changes in tree species abundance in a neotropical forest: impact of climate change. *Journal of Tropical Ecology*, 12:231–256, 1996.

[168] R.D. Cook. Added-variable plots and curvature in linear regression. *Technometrics*, 38:275–278, 1996.

[169] R.D. Cook and S. Weisberg. *Residuals and Influence in Regression*. Chapman and Hall, London, 1982.

[170] R.D. Cook and S. Weisberg. *Applied Regression, Including Computing and Graphics*. John Wiley and Sons, 1999.

[171] J.W. Cooley and J.W. Tukey. An algorithm for the machine calculation of complex Fourier series. *Mathematics of Computation*, 19:297–301, 1965.

[172] J.B. Copas. Plotting p against x. *Applied Statistics*, 32:25–31, 1983.

[173] C.B. Cordy. An extension of the Horvitz-Thompson theorem to point sampling from a continuous universe. *Statistics and Probability Letters*, 18:353–362, 1993.

[174] D. R. Cox and V. Isham. *Point Processes*. Chapman and Hall, London, 1980.

[175] D.R. Cox. Some statistical models related with series of events (with discussion). *Journal of the Royal Statistical Society, Series B*, 17:129–164, 1955.

[176] D.R. Cox. *Planning of Experiments*. Wiley, New York, 1958.

[177] D.R. Cox. The statistical analysis of dependencies in point processes. In P.A.W. Lewis, editor, *Stochastic Point Processes*, pages 55–66. Wiley, New York, 1972.

[178] D.R. Cox. Discussion contribution. *Journal of the Royal Statistical Society, Series B*, 39:206, 1977.

[179] D.R. Cox. The role of significance tests. *Scandinavian Journal of Statistics*, 4:49–70, 1977.

[180] D.R. Cox and C.A. Donnelly. *Principles of Applied Statistics*. Cambridge University Press, Cambridge, UK, 2011.

[181] D.R. Cox and E.J. Snell. *Applied Statistics: Principles and Examples*. Chapman and Hall, 1981.

[182] T.F. Cox. A method for mapping the dense and sparse regions of a forest stand. *Applied Statistics*, pages 14–19, 1979.

[183] H. Cramér. On the composition of elementary errors: II, Statistical applications. *Skandinavisk Aktuarietidskrift*, 11:141–180, 1928.

[184] N. Cressie. Change of support and the modifiable area unit problem. *Geographical Systems*, 3:159–180, 1996.

[185] N. Cressie and L.B. Collins. Analysis of spatial point patterns using bundles of product density LISA functions. *Journal of Agricultural, Biological and Environmental Statistics*, 6:118–135, 2001.

[186] N. Cressie and L.B. Collins. Patterns in spatial point locations: local indicators of spatial association in a minefield with clutter. *Naval Research Logistics*, 48:333–347, 2001.

[187] N. Cressie and F.L. Hulting. A spatial statistical analysis of tumor growth. *Journal of the American Statistical Association*, 87:272–283, 1992.

[188] N. Cressie and G.M. Laslett. Random set theory and problems of modeling. *SIAM Review*, 28:557–574, 1987.

[189] N. Cressie and A. Lawson. Hierarchical probability models and Bayesian analysis of minefield locations. *Advances in Applied Probability*, 32:315–330, 2000.

[190] N.A.C. Cressie. *Statistics for Spatial Data*. John Wiley and Sons, New York, 1991.

[191] N.A.C. Cressie. *Statistics for Spatial Data*. John Wiley and Sons, New York, second edition, 1993.

[192] N.A.C. Cressie and C.K. Wikle. *Statistics for Spatio-Temporal Data*. Wiley, Hoboken, NJ, 2011.

[193] L.M. Cruz-Orive. Best linear unbiased estimators for stereology. *Biometrics*, 36:595–605, 1980.

[194] S. Csörgő and J.J. Faraway. The exact and asymptotic distributions of Cramér-von mises statistics. *Journal of the Royal Statistical Society, Series B*, 58:221–234, 1996.

[195] J. Cuzick and R. Edwards. Spatial clustering for inhomogeneous populations (with discussion). *Journal of the Royal Statistical Society, Series B*, 52:73–104, 1990.

[196] R.B. D'Agostino and M.A. Stephens. *Goodness-of-Fit Techniques*. Marcel Dekker, New York, 1986.

[197] D.J. Daley and D. Vere-Jones. *An Introduction to the Theory of Point Processes*. Springer-Verlag, New York, 1988.

[198] D.J. Daley and D. Vere-Jones. *An Introduction to the Theory of Point Processes. Volume I: Elementary Theory and Methods*. Springer-Verlag, New York, second edition, 2003.

[199] D.J. Daley and D. Vere-Jones. *An Introduction to the Theory of Point Processes. Volume II: General Theory and Structure*. Springer-Verlag, New York, second edition, 2008.

[200] N.A. Dao and M. Genton. A Monte Carlo adjusted goodness-of-fit test for parametric models describing spatial point patterns. *Journal of Graphical and Computational Statistics*, 23:497–517, 2014.

[201] A. Dasgupta and A. E. Raftery. Detecting features in spatial point processes with clutter via model-based clustering. *Journal of the American Statistical Association*, 93:294–302, 1998.

[202] T. M. Davies, M. L. Hazelton, and J. C. Marshall. sparr: Analyzing spatial relative risk using fixed and adaptive kernel density estimation in R. *Journal of Statistical Software*, 39(1):1–14, 2011.

[203] T.M. Davies and M.L. Hazelton. Adaptive kernel estimation of spatial relative risk. *Statistics in Medicine*, 29:2423–2437, 2010.

[204] M. Davis and P.J.E. Peebles. A survey of galaxy redshifts, V. The two-point position and velocity correlations. *Astrophysical Journal*, 267:465–482, 1983.

[205] A. Davison. *Statistical Models*. Cambridge Series in Statistical and Probabilistic Mathematics. Cambridge University Press, 2003.

[206] M. de la Cruz Rot. *Metodos para analizar datos puntuales*, chapter 3, pages 76–127. Asociacion Espanola de Ecologia Terrestre, Universidad Rey Juan Carlos and Caja de Ahorros del Mediterraneo, 2008.

[207] C.B. Dean and R. Balshaw. Efficiency lost by analyzing counts rather than exact times in Poisson and overdispersed Poisson regression. *Journal of the American Statistical Association*, 92:1387–1398, 1997.

[208] C.B. Dean and J.F. Lawless. Tests for detecting overdispersion in Poisson regression models. *Journal of the American Statistical Association*, 84(406):467–472, 1989.

[209] B. Deckers, K. Verheyen, M. Hermy, and B. Muys. Effects of landscape structure on the invasive spread of black cherry *Prunus serotina* in an agricultural landscape in Flanders, Belgium. *Ecography*, 28:99–109, 2005.

[210] C. Demattei, N. Molinari, and J.P. Daurès. Arbitrarily shaped multiple spatial cluster detection for case event data. *Computational Statistics and Data Analysis*, 51:3931–3945, 2007.

[211] D. Dereudre, R. Drouilhet, and H.-O. Georgii. Existence of Gibbsian point processes with geometry-dependent interactions. *Probability Theory and Related Fields*, 153:643–670, 2012.

[212] D. Dereudre and F. Lavancier. Campbell equilibrium equation and pseudo-likelihood estimation for non-hereditary Gibbs point processes. *Bernoulli*, 15(4):1368–1396, 2009.

[213] D. Dereudre and F. Lavancier. Practical simulation and estimation for Gibbs Delaunay-Voronoi tessellations with geometric hardcore interaction. *Computational Statistics and Data Analysis*, 55:498–519, 2011.

[214] P. Diggle and P.J. Ribeiro. *Model-based Geostatistics*. Number 692 in Springer Series in Statistics. Springer, New York, 2007.

[215] P. J. Diggle, J. A. Tawn, and R. A. Moyeed. Model-based geostatistics (with discussion). *Appl. Statist.*, 47(3):299–350, 1998.

[216] P.J. Diggle. A note on robust density estimation for point patterns. *Biometrika*, 64:91–95, 1977.

[217] P.J. Diggle. On parameter estimation for spatial point processes. *Journal of the Royal Statistical Society, Series B*, 40:178–181, 1978.

[218] P.J. Diggle. On parameter estimation and goodness-of-fit testing for spatial point patterns. *Biometrika*, 35:87–101, 1979.

[219] P.J. Diggle. Statistical methods for spatial point patterns in ecology. In R.M. Cormack and J.K. Ord, editors, *Spatial and Temporal Analysis in Ecology*, pages 95–150. International Co-operative Publishing House, Fairland, 1979.

[220] P.J. Diggle. Binary mosaics and the spatial pattern of heather. *Biometrics*, 37:531–539, 1981.

[221] P.J. Diggle. *Statistical Analysis of Spatial Point Patterns*. Academic Press, London, 1983.

[222] P.J. Diggle. A kernel method for smoothing point process data. *Journal of the Royal Statistical Society, Series C (Applied Statistics)*, 34:138–147, 1985.

[223] P.J. Diggle. Displaced amacrine cells in the retina of a rabbit: analysis of a bivariate spatial point pattern. *Journal of Neuroscience Methods*, 18:115–125, 1986.

[224] P.J. Diggle. A point process modelling approach to raised incidence of a rare phenomenon in the vicinity of a prespecified point. *Journal of the Royal Statistical Society, Series A*, 153:349–362, 1990.

[225] P.J. Diggle. *Statistical Analysis of Spatial Point Patterns*. Hodder Arnold, London, second edition, 2003.

[226] P.J. Diggle. *Statistical Analysis of Spatial and Spatio-Temporal Point Patterns*. Chapman and Hall/CRC, Boca Raton, FL, third edition, 2014.

[227] P.J. Diggle, S.J. Eglen, and J.B. Troy. Modelling the bivariate spatial distribution of amacrine cells. In A. Baddeley, P. Gregori, J. Mateu, R. Stoica, and D. Stoyan, editors, *Case Studies in Spatial Point Process Modeling*, number 185 in Lecture Notes in Statistics, pages 215–233. Springer, New York, 2005.

[228] P.J. Diggle, T. Fiksel, P. Grabarnik, Y. Ogata, D. Stoyan, and M. Tanemura. On parameter estimation for pairwise interaction processes. *International Statistical Review*, 62:99–117, 1994.

[229] P.J. Diggle, D.J. Gates, and A. Stibbard. A nonparametric estimator for pairwise-interaction point processes. *Biometrika*, 74:763–770, 1987.

[230] P.J. Diggle, V. Gómez-Rubio, P.E. Brown, A.G. Chetwynd, and S. Gooding. Second-order analysis of inhomogeneous spatial point processes using case-control data. *Biometrics*, 63:550–557, 2007.

[231] P.J. Diggle and R.J. Gratton. Monte Carlo methods of inference for implicit statistical models (with discussion). *Journal of the Royal Statistical Society, Series B*, 46:193–227, 1984.

[232] P.J. Diggle, N. Lange, and F. M. Benes. Analysis of variance for replicated spatial point patterns in clinical neuroanatomy. *Journal of the American Statistical Association*, 86:618–625, 1991.

[233] P.J. Diggle and B. Matérn. On sampling designs for the estimation of point-event nearest neighbour distributions. *Scandinavian Journal of Statistics*, 7:80–84, 1981.

[234] P.J. Diggle, J. Mateu, and H.E. Clough. A comparison between parametric and non-parametric approaches to the analysis of replicated spatial point patterns. *Advances in Applied Probability (SGSA)*, 32:331–343, 2000.

[235] P.J. Diggle and R.K. Milne. Bivariate Cox processes: some models for bivariate spatial point patterns. *Journal of the Royal Statistical Society, Series B*, 45:11–21, 1983.

[236] P.J. Diggle and B. Rowlingson. A conditional approach to point process modelling of elevated risk. *Journal of the Royal Statistical Society, Series A*, 157(3):433–440, 1994.

[237] P.J. Diggle, P. Zheng, and P. Durr. Non-parametric estimation of spatial segregation in a multivariate point process: bovine tuberculosis in Cornwall, UK. *Applied Statistics*, 54:645–658, 2005.

[238] P. Dixon. Testing spatial segregation using a nearest-neighbour contingency table. *Ecology*, 75(7):1940–1948, 1994.

[239] P.M. Dixon. Nearest-neighbor contingency table analysis of spatial segregation for several species. *Écoscience*, 9(2):142–151, 2002.

[240] A.J. Dobson and A.G. Barnett. *An Introduction to Generalized Linear Models*. CRC Press, third edition, 2008.

[241] S.I. Doguwa. A comparative study of the edge-corrected kernel-based nearest neighbour density estimators for point processes. *Journal of Statistical Computation and Simulation*, 33:83–100, 1989.

[242] S.I. Doguwa. On edge-corrected kernel-based pair correlation function estimators for point processes. *Biometrical Journal*, 32:95–106, 1990.

[243] S.I. Doguwa and D.N. Choji. On edge-corrected probability density function estimators for point processes. *Biometrical Journal*, 33:623–637, 1991.

[244] K. Donnelly. Simulations to determine the variance and edge-effect of total nearest neighbour distance. In I. Hodder, editor, *Simulation Studies in Archaeology*, pages 91–95. Cambridge University Press, Cambridge; New York, 1978.

[245] J.A. Downs and M.W. Horner. Characterising linear point patterns. In A.C. Winstanley, editor, *Proceedings of the GIScience Research UK Conference (GISRUK), Maynooth, Ireland*, pages 421–424, County Kildare, Ireland, 2007. National University of Ireland Maynooth.

[246] M. Dudík, S. J. Phillips, and R.E. Schapire. Maximum entropy density estimation with generalized regularization and an application to species distribution modeling. *Journal of Machine Learning Research*, 8:1217–1260, 2007.

[247] G. Duranton and H.G. Overman. Testing for localisation using microgeographic data. *Review of Economic Studies*, 72:1077–1106, 2005.

[248] M. Dutta. On maximum (information-theoretic) entropy estimation. *Sankhya: The Indian Journal of Statistics, Series A*, 28:319–328, 1966.

[249] M. Dwass. Modified randomization tests for nonparametric hypotheses. *Annals of Mathematical Statistics*, 28:181–187, 1957.

[250] B. Efron and R. J. Tibshirani. *An Introduction to the Bootstrap*, volume 57 of *Monographs on Statistics and Applied Probability*. Chapman and Hall, London, 1993.

[251] H. El Barmi and J.S. Simonoff. Transformation based density estimation for weighted distributions. *Journal of Nonparametric Statistics*, 12:861–878, 2000.

[252] J. Elith, Steven J. Phillips, T. Hastie, M. Dudík, Yung E. Chee, and Colin J. Yates. A statistical explanation of MaxEnt for ecologists. *Diversity and Distributions*, 17:43–57, 2011.

[253] P. Elliott, J. Wakefield, N. Best, and D. Briggs, editors. *Spatial Epidemiology: Methods and Applications*. Oxford University Press, Oxford, 2000.

[254] ESRI. ESRI Shapefile Technical Description. ESRI white paper, Environmental Systems Research Institute, Inc, July 1998.

[255] M. Ezekiel. A method for handling curvilinear correlation for any number of variables. *Journal of the American Statistical Association*, 19:431–453, 1924.

[256] T. Fiksel. Estimation of parameterized pair potentials of marked and non-marked Gibbsian point processes. *Elektronische Informationsverarbeitung u. Kybernetika*, 20:270–278, 1984.

[257] T. Fiksel. Edge-corrected density estimators for point processes. *Statistics*, 19:67–75, 1988.

[258] T. Fiksel. Estimation of interaction potentials of Gibbsian point processes. *Statistics*, 19:77–86, 1988.

[259] R.A. Fisher. Frequency distribution of the values of the correlation coefficient in samples of an indefinitely large population. *Biometrika*, 10:507–521, 1915.

[260] R.A. Fisher. *Statistical Methods for Research Workers*. Oliver and Boyd, Edinburgh, 1925.

[261] R.A. Fisher. *Design of Experiments*. Oliver and Boyd, Edinburgh, 1935.

[262] R.A. Fisher. *Statistical Methods and Scientific Inference*. Oliver and Boyd, Edinburgh, second edition, 1959.

[263] F. Fleischer, M. Beil, M. Kazda, and V. Schmidt. Analysis of spatial point patterns in microscopic and macroscopic biological image data. In A. Baddeley, P. Gregori, J. Mateu, R. Stoica, and D. Stoyan, editors, *Case Studies in Spatial Point Process Modeling*, number 185 in Lecture Notes in Statistics, pages 235–260. Springer, New York, 2005.

[264] O. Flores. *Déterminisme de la régénération chez quinze espèces d'arbres tropicaux en forêt guyanaise: les effets de l'environnement et de la limitation par la dispersion*. PhD thesis, University of Montpellier 2, Montpellier, France, 2005.

[265] E.M. Floresroux and M.L. Stein. A new method of edge correction for estimating the nearest neighbor distribution. *J. Statist. Planning and Inference*, 50:353–371, 1996.

[266] E.D. Ford, P.A. Mason, and J. Pelham. Spatial patterns of sporophore distribution around a young birch tree in three successive years. *Transactions of the British Mycological Society*, 75:287–296, 1980.

[267] A.S. Fotheringham, C. Brunsdon, and M. Charlton. *Geographically Weighted Regression: The Analysis of Spatially Varying Relationships*. Wiley, 2003.

[268] J. Fox. *Applied Regression, Linear Models and Related Methods*. Sage, 1997.

[269] R. Foxall and A. Baddeley. Nonparametric measures of association between a spatial point process and a random set, with geological applications. *Applied Statistics*, 51(2):165–182, 2002.

[270] E.L. Frome. The analysis of rates using Poisson regression models. *Biometrics*, 39:665–674, 1983.

[271] N. Fry. Random point distributions and strain measurement in rocks. *Tectonophysics*, 60:89–105, 1979.

[272] N. Funwi-Gabga. A pastoralist survey and fire impact assessment in the Kagwene Gorilla Sanctuary, Cameroon. Master's thesis, Geology and Environmental Science, University of Buea, Cameroon, 2008.

[273] N. Funwi-Gabga and J. Mateu. Understanding the nesting spatial behaviour of gorillas in the Kagwene Sanctuary, Cameroon. *Stochastic Environmental Research and Risk Assessment*, 26(6):793–811, 2012.

[274] E. Gabriel and P.J. Diggle. Second-order analysis of inhomogeneous spatio-temporal point process data. *Statistica Neerlandica*, 63:43–51, 2009.

[275] C. Gaetan and X. Guyon. *Spatial statistics and modeling*. Springer, 2009. Translated by Kevin Bleakley.

[276] R. Gangnon and M. Clayton. A weighted average likelihood ratio test for spatial clustering of disease. *Statistics in Medicine*, 20:2977–2987, 2001.

[277] A.E. Gelfand, P.J. Diggle, M. Fuentes, and P. Guttorp, editors. *Handbook of Spatial Statistics*. CRC Press, Boca Raton, FL, 2010.

[278] H.-O. Georgii. *Canonical Gibbs Measures*. Number 760 in Lecture Notes in Mathematics. Springer, Berlin, 1979.

[279] H.-O. Georgii. *Gibbs Measures and Phase Transitions*, volume 9 of *Studies in Mathematics*. Walter de Gruyter, Berlin, second edition, 2011.

[280] D.J. Gerrard. Competition quotient: a new measure of the competition affecting individual forest trees. Research Bulletin 20, Agricultural Experiment Station, Michigan State University, 1969.

[281] A. Getis and J. Franklin. Second-order neighbourhood analysis of mapped point patterns. *Ecology*, 68:473–477, 1987.

[282] C.J. Geyer. Likelihood inference for spatial point processes. In O.E. Barndorff-Nielsen, W.S. Kendall, and M.N.M. van Lieshout, editors, *Stochastic Geometry: Likelihood and Computation*, number 80 in Monographs on Statistics and Applied Probability, chapter 3, pages 79–140. Chapman and Hall / CRC, Boca Raton, FL, 1999.

[283] C.J. Geyer and J. Møller. Simulation procedures and likelihood inference for spatial point processes. *Scandinavian Journal of Statistics*, 21(4):359–373, 1994.

[284] R.D. Gill. Lectures on survival analysis. In P. Bernard, editor, *22e Ecole d'Eté de Probabilités de Saint-Flour 1992*, number 1581 in Lecture Notes in Mathematics. Springer, 1994.

[285] L. Glass and W.R. Tobler. Uniform distribution of objects on a homogeneous field: Cities on a plain. *Nature*, 233:67–68, 1971.

[286] R. Gnanadesikan and M.B. Wilk. A probability plotting procedure for general analysis of variance. *Journal of the Royal Statistical Society, Series B*, 32:88–101, 1970.

[287] V. Gómez-Rubio, J. Ferrándiz-Ferragud, and A. Lopez-Quílez. Detecting clusters of disease with R. *Journal of Geographical Systems*, 7(2):189–206, 2005.

[288] P.V. Gorsevski, P.E. Gessler, R.B. Folz, and W.J. Elliott. Spatial prediction of landslide hazard using logistic regression and ROC analysis. *Transactions in GIS*, 10:395–415, 2006.

[289] C.A. Gotway and L.J. Young. Combining incompatible spatial data. *Journal of the American Statistical Association*, 97:632–648, 2002.

[290] M. Goulard, A. Särkkä, and P. Grabarnik. Parameter estimation for marked Gibbs point processes through the maximum pseudolikelihood method. *Scandinavian Journal of Statistics*, 23:365–379, 1996.

[291] Gnu General Public License, version 2. URL: `www.gnu.org/licenses/old--licenses/gpl-2.0.html`, 2007. Last retrieved 2015-05-01.

[292] P. Grabarnik, M. Myllymäki, and D. Stoyan. Correct testing of mark independence for marked point patterns. *Ecological Modelling*, 222:3888–3894, 2011.

[293] P. Grabarnik and A. Särkkä. Modelling the spatial structure of forest stands by multivariate point processes with hierarchical interactions. *Ecological Modelling*, 220:1232–1240, 2009.

[294] J. Grandell. *Doubly Stochastic Poisson Processes*. Number 529 in Lecture Notes in Mathematics. Springer-Verlag, Berlin, 1976.

[295] J. Grandell. *Mixed Poisson Processes*. Chapman and Hall, London, 1997.

[296] P. Greig-Smith. The use of random and contiguous quadrats in the study of the structure of plant communities. *Annals of Botany*, 16:293–316, 1952.

[297] P. Greig-Smith. Pattern in vegetation. *Journal of Ecology*, 67:755–779, 1979.

[298] D.I. Groves, R.J. Goldfarb, C.M. Knox-Robinson, J. Ojala, S. Gardoll, G.Y. Yun, and P. Holyland. Late-kinematic timing of orogenic gold deposits and significance for computer-based exploration techniques with emphasis on the Yilgarn Block, Western Australia. *Ore Geology Reviews*, 17:1–38, 2000.

[299] Y. Guan. A composite likelihood approach in fitting spatial point process models. *Journal of the American Statistical Association*, 101:1502–1512, 2006.

[300] Y. Guan. A least-squares cross-validation bandwidth selection approach in pair correlation function estimation. *Statistics and Probability Letters*, 77(18):1722–1729, 2007.

[301] Y. Guan. On consistent nonparametric intensity estimation for inhomogeneous spatial point processes. *Journal of the American Statistical Association*, 103:1238–1247, 2008.

[302] Y. Guan and D.R. Afshartous. Test for independence between marks and points of marked point processes: a subsampling approach. *Environmental and Ecological Statistics*, 14:101–111, 2007.

[303] Y. Guan, A. Jalilian, and R. Waagepetersen. Quasi-likelihood for spatial point processes. *Journal of the Royal Statistical Society, Series B*, 2015. In press. Published online 4 Sep 2014.

[304] Y. Guan and Y. Shen. A weighted estimating equation approach for inhomogeneous spatial point processes. *Biometrika*, 97:867–880, 2010.

[305] D. Guo. Regionalization with dynamically constrained agglomerative clustering and partitioning (REDCAP). *International Journal of Geographical Information Science*, 22(7):801–823, 2008.

[306] D. Guo and H. Wang. Automatic region building for spatial analysis. *Transactions in GIS*, 15:29–45, 2011.

[307] O. Häggström, M.N.M. van Lieshout, and J. Møller. Characterisation results and Markov chain Monte Carlo algorithms including exact simulation for some spatial point processes. *Bernoulli*, 5:641–659, 1999.

[308] U. Hahn. Scale families in spatial point processes. Conference poster presented at the 12th Workshop on Stochastic Geometry, Stereology and Image Analysis, Prague, Czech Republic, 2003.

[309] U. Hahn. *Global and Local Scaling in the Statistics of Spatial Point Processes*. Habilitationsschrift, Universitaet Augsburg, 2007.

[310] U. Hahn. A studentized permutation test for the comparison of spatial point patterns. *Journal of the American Statistical Association*, 107(498):754–764, 2012.

[311] U. Hahn and E.B.V. Jensen. Inhomogeneous spatial point processes with hidden second-order stationarity. CSGB Research Report 2013–07, University of Aarhus, 2013.

[312] U. Hahn, E.B.V. Jensen, M.-C. van Lieshout, and L.S. Nielsen. Inhomogeneous spatial point processes by location-dependent scaling. *Advances in Applied Probability (SGSA)*, 35:319–336, 2003.

[313] P. Hall. Resampling a coverage pattern. *Stochastic Processes and Their Applications*, 20:231–246, 1985.

[314] P. Hall. *An Introduction to the Theory of Coverage Processes*. John Wiley and Sons, New York, 1988.

[315] P. Hall. The effect of bias estimation on coverage accuracy of bootstrap confidence intervals for a probability density. *Annals of Statistics*, 20:675–694, 1992.

[316] P. Hall. *The Bootstrap and Edgeworth Expansion*. Springer, 2002.

[317] A.J.S. Hamilton. Toward better ways to measure the galaxy correlation function. *Astrophysical Journal*, 417:19–35, 1993.

[318] M.S. Handcock and M. Morris. *Relative Distribution Methods in the Social Sciences.* Springer-Verlag, New York, 1999.

[319] K.-H. Hanisch. On Palm and second-order quantities of point processes and germ-grain models. Wissenschaftliche Sitzung WSS–01/84, Akademie der Wissenschaften der DDR, Berlin, 1984.

[320] K.-H. Hanisch. Some remarks on estimators of the distribution function of nearest neighbour distance in stationary spatial point patterns. *Mathematische Operationsforschung und Statistik, series Statistics*, 15:409–412, 1984.

[321] S.S. Hanna and N. Fry. A comparison of methods of strain determination in rocks from southwest Dyfed (Pembrokeshire) and adjacent areas. *Journal of Structural Geology*, 1:155–162, 1979.

[322] M.B. Hansen, A.J. Baddeley, and R.D. Gill. First contact distributions for spatial patterns: regularity and estimation. *Advances in Applied Probability*, 31:15–33, 1999.

[323] M.B. Hansen, R.D. Gill, and A.J. Baddeley. Kaplan-Meier type estimators for linear contact distributions. *Scandinavian Journal of Statistics*, 23:129–155, 1996.

[324] R.D. Harkness and V. Isham. A bivariate spatial point pattern of ants' nests. *Applied Statistics*, 32:293–303, 1983.

[325] F. Harrell. *Regression Modeling Strategies*. Springer, New York, 2001.

[326] R.J. Hasenstab. A preliminary cultural resource sensitivity analysis for the proposed flood control facilities construction in the Passaic River basin of New Jersey. Technical report, Soil Systems, Inc, New York, 1983. Submitted to the Passaic River Basin Special Studies Branch, Department of the Army, USA. New York District Army Corps of Engineers.

[327] T.J. Hastie and R.J. Tibshirani. *Generalized Additive Models*. Chapman and Hall/CRC, 1990.

[328] T.J. Hastie, R.J. Tibshirani, and J.H. Friedman. *The Elements of Statistical Learning.* Springer-Verlag, second edition, 2009.

[329] W.K. Hastings. Monte Carlo sampling methods using Markov chains and their applications. *Biometrika*, 57:97–109, 1970.

[330] W.W. Hauck, Jr. and A. Donner. Wald's test as applied to hypotheses in logit analysis. *Journal of the American Statistical Association*, 72(360, part 1):851–853, 1977.

[331] M.L. Hazelton and T.M. Davies. Inference based on kernel estimates of the relative risk function in geographical epidemiology. *Biometrical Journal*, 51:98–109, 2009.

[332] J. Heikkinen and E. Arjas. Non-parametric Bayesian estimation of a spatial Poisson intensity. *Scandinavian Journal of Statistics*, 25:435–450, 1998.

[333] J. Heikkinen and A. Penttinen. Bayesian smoothing in the estimation of the pair potential function of Gibbs point processes. *Bernoulli*, 5:1119–1136, 1999.

[334] N. Henze. A multivariate two-sample test based on the number of nearest neighbour type coincidences. *Annals of Statistics*, 16:772–783, 1988.

[335] N.L. Hjort and M.C. Jones. Locally parametric density estimation. *Ann. Statist.*, 24:1619–1649, 1996.

[336] Y. Hochberg and A. Tamhane. *Multiple Comparison Procedures*. Wiley, New York, 1987.

[337] H. Högmander and A. Särkkä. Multitype spatial point patterns with hierarchical interactions. *Biometrics*, 55:1051–1058, 1999.

[338] A.C.A. Hope. A simplified Monte Carlo significance test procedure. *Journal of the Royal Statistical Society, Series B*, 30:582–598, 1968.

[339] B. Hopkins. A new method of determining the type of distribution of plant individuals. *Annals of Botany*, 18:213–227, 1954. With an appendix by J.G. Skellam.

[340] J.L. Horowitz. The bootstrap. In J.J. Heckman and E. Leamer, editors, *Handbook of Econometrics*, volume 5, pages 3159–3228. North-Holland, Amsterdam, 2001.

[341] D.G. Horvitz and D.J. Thompson. A generalization of sampling without replacement from a finite universe. *Journal of the American Statistical Association*, 47:663–685, 1952.

[342] D.W. Hosmer, T. Hosmer, S. le Cessie, and S. Lemeshow. A comparison of goodness-of-fit tests for the logistic regression model. *Statistics in Medicine*, 16:965–980, 1997.

[343] D.W. Hosmer and S. Lemeshow. A goodness-of-fit test for the multiple logistic regression model. *Communications in Statistics*, 10:1043–1069, 1980.

[344] D.W. Hosmer and S. Lemeshow. *Applied Logistic Regression*. Wiley, second edition, 2000.

[345] J.B. Hough, M. Krishnapur, Y. Peres, and B. Viràg. Determinantal processes and independence. *Probability Surveys*, 3:206–229, 2006.

[346] C.V. Howard, S. Reid, A.J. Baddeley, and A. Boyde. Unbiased estimation of particle density in the tandem-scanning reflected light microscope. *Journal of Microscopy*, 138:203–212, 1985.

[347] J.C. Hsu. *Multiple Comparisons: Theory and Methods*. Chapman and Hall, 1996.

[348] F. Huang and Y. Ogata. Improvements of the maximum pseudo-likelihood estimators in various spatial statistical models. *Journal of Computational and Graphical Statistics*, 8(3):510–530, 1999.

[349] S.P. Hubbell and R.B. Foster. Diversity of canopy trees in a neotropical forest and implications for conservation. In S.L. Sutton, T.C. Whitmore, and A.C. Chadwick, editors, *Tropical Rain Forest: Ecology and Management*, pages 25–41. Blackwell Scientific Publications, Oxford, 1983.

[350] D. Hug and G. Last. On support measures in Minkowski spaces and contact distributions in stochastic geometry. *Annals of Probability*, 28:796–850, 2000.

[351] D. Hug, G. Last, and W. Weil. A local Steiner-type formula for general closed sets and applications. *Mathematische Zeitschrift*, 246(1-2):237–272, 2004.

[352] M.J. Hutchings. Standing crop and pattern in pure stands of *Mercurialis perennis* and *Rubus fruticosus* in mixed deciduous woodland. *Oikos*, 31:351–357, 1979.

[353] R. Ihaka and R. Gentleman. R: A language for data analysis and graphics. *Journal of Computational and Graphical Statistics*, 5(3):299–314, 1996.

[354] J. Illian, E. Benson, J. Crawford, and H. Staines. Principal component analysis for spatial point processes — assessing the appropriateness of the approach in an ecological context. In A. Baddeley, P. Gregori, J. Mateu, R. Stoica, and D. Stoyan, editors, *Case Studies in Spatial Point Process Modeling*, number 185 in Lecture Notes in Statistics, pages 135–150. Springer, New York, 2005.

[355] J. Illian, A. Penttinen, H. Stoyan, and D. Stoyan. *Statistical Analysis and Modelling of Spatial Point Patterns*. John Wiley and Sons, Chichester, 2008.

[356] J.B. Illian, S. Martino, S.H. Sørbye, J.B. Gallego-Fernández, M. Zunzunegui, M.P. Esquivias, and J.M.J. Travis. Fitting complex ecological point process models with integrated nested Laplace approximation. *Methods in Ecology and Evolution*, 4(4):305–315, 2013.

[357] J.B. Illian, J. Møller, and R.P. Waagepetersen. Hierarchical spatial point process analysis for a plant community with high biodiversity. *Environmental and Ecological Statistics*, 16:389–405, 2009.

[358] J.B. Illian, S.H. Sørbye, and H. Rue. A toolbox for fitting complex spatial point process models using Integrated Nested Laplace Approximation (INLA). *The Annals of Applied Statistics*, 6(4):1499–1530, 2012.

[359] V.S. Isham. Multitype Markov point processes: some approximations. *Proceedings of the Royal Society of London, Series A*, 391:39–53, 1984.

[360] A. Jalilian, Y. Guan, and R. Waagepetersen. Decomposition of variance for spatial Cox processes. *Scandinavian Journal of Statistics*, 40(1):119–137, 2013.

[361] A. Jammalamadaka, S. Banerjee, B.S. Manjunath, and K. Kosik. Statistical analysis of dendritic spine distributions in rat hippocampal cultures. *BMC Bioinformatics*, 14:287, 2013.

[362] E. B. V. Jensen and L. S. Nielsen. Inhomogeneous Markov point processes by transformation. *Bernoulli*, 6:761–782, 2000.

[363] E.B.V. Jensen and L.S. Nielsen. Statistical inference for transformation inhomogeneous point processes. *Scandinavian Journal of Statistics*, 31:131–142, 2004.

[364] J.L. Jensen and H.R. Künsch. On asymptotic normality of pseudo likelihood estimates for pairwise interaction processes. *Annals of the Institute of Statistical Mathematics*, 46:475–486, 1994.

[365] J.L. Jensen and J. Møller. Pseudolikelihood for exponential family models of spatial point processes. *Annals of Applied Probability*, 1:445–461, 1991.

[366] E. Jolivet. Central limit theorem and convergence of empirical processes for stationary point processes. In P. Bastfai and J. Tomko, editors, *Point Processes and Queueing Problems*, pages 117–161. North-Holland, Amsterdam, 1980.

[367] M.C. Jones. Kernel density estimation for length-biased data. *Biometrika*, 78:511–519, 1991.

[368] O. Kallenberg. An informal guide to the theory of conditioning in point processes. *International Statistical Review*, 52:151–164, 1984.

[369] E.L. Kaplan and P. Meier. Nonparametric estimation from incomplete observations. *Journal of the American Statistical Association*, 53:457–481, 1958.

[370] A.F. Karr. *Point Processes and Their Statistical Inference*. Dekker, New York, 1985.

[371] F.P. Kelly and B.D. Ripley. A note on Strauss's model for clustering. *Biometrika*, 63:357–360, 1976.

[372] J.E. Kelsall and P.J. Diggle. Kernel estimation of relative risk. *Bernoulli*, 1:3–16, 1995.

[373] J.E. Kelsall and P.J. Diggle. Spatial variation in risk of disease: a nonparametric binary regression approach. *Applied Statistics*, 47:559–573, 1998.

[374] M. Kendall. Hiawatha designs an experiment. *The American Statistician*, 13:23–24, 1959.

[375] W. S. Kendall and J. Møller. Perfect simulation using dominating processes on ordered spaces, with application to locally stable point processes. *Advances in Applied Probability*, 32:844–865, 2000.

[376] W.S. Kendall, M.N.M. van Lieshout, and A.J. Baddeley. Quermass-interaction processes: conditions for stability. *Advances in Applied Probability*, 31:315–342, 1999.

[377] M. Kerscher. Regularity in the distribution of superclusters? *Astronomy and Astrophysics*, 336:29–34, 1998.

[378] M. Kerscher. Statistical analysis of large-scale structure in the Universe. In K.R. Mecke and D. Stoyan, editors, *Statistical Physics and Spatial Statistics*, Lecture Notes in Physics, pages 36–71. Springer-Verlag, Berlin, 2000.

[379] M. Kerscher, M. J. Pons-Borderia, J. Schmalzing, R. Trasarti-Battistoni, T. Buchert, V. J. Martinez, and R. Valdarnini. A global descriptor of spatial pattern interaction in the galaxy distribution. *Astrophysical Journal*, 513:543–548, 1999.

[380] M. Kerscher, J. Schmalzing, T. Buchert, and H. Wagner. Fluctuations in the IRAS 1.2 Jy catalogue. *Astronomy and Astrophysics*, 333:1–12, 1998.

[381] S.S. Kind. *The Scientific Investigation of Crime*. Forensic Science Services, Harrogate, UK, 1987.

[382] J.F.C. Kingman. Remarks on the spatial distribution of a reproducing population. *Journal of Applied Probability*, 14:577–583, 1977.

[383] J.F.C. Kingman. *Poisson Processes*. Oxford University Press, New York, 1993.

[384] J.G. Kirkwood and E. Monroe. Statistical mechanics of fusion. *Journal of Chemical Physics*, pages 514–526, 1941.

[385] W. Klein. Potts-model formulation of continuum percolation. *Physical Review B*, 26:2677–2678, 1982.

[386] C.M. Knox-Robinson and D.I. Groves. Gold prospectivity mapping using a geographic information system (GIS), with examples from the Yilgarn Block of Western Australia. *Chronique de la Recherche Minière*, 529:127–138, 1997.

[387] D.E. Knuth. *Literate Programming*. Stanford University Center for the Study of Language and Information, Stanford, California, 1992.

[388] A. Kolmogorov. Sulla determinazione empirica di una legge di distribuzione. *Giornale dell'Istituto Italiano degli Attuari*, 4:83–91, 1933.

[389] S.A.L.M. Kooijman. The description of point patterns. In R.M. Cormack and J.K. Ord, editors, *Spatial and Temporal Analysis in Ecology*, pages 305–332. International Co-operative Publication House, Fairland, Maryland, 1979.

[390] S.A.L.M. Kooijman. Inference about dispersal patterns. *Acta Biotheoretica*, 28:149–189, 1979.

[391] J. Kornak, M.E. Irwin, and N. Cressie. Spatial point process models of defensive strategies: detecting changes. *Statistical Inference for Stochastic Processes*, 9:31–46, 2006.

[392] P. Kovesi. Designing colour maps with uniform perceptual contrast. Submitted for publication.

[393] K. Krickeberg. Processus ponctuels en statistique. In P.L. Hennequin, editor, *Ecole d'Eté de Probabilités de Saint-Flour X*, volume 929 of *Lecture Notes in Mathematics*, pages 205–313. Springer, 1982.

[394] M. Kulldorff. Spatial scan statistics: models, calculations, and applications. In J. Glaz and N. Balakrishnan, editors, *Recent Advances on Scan Statistics*, pages 303–322. Birkhauser, Boston, 1999.

[395] M. Kulldorff and N. Nagarwalla. Spatial disease clusters: detection and inference. *Statistics in Medicine*, 14:799–810, 1995.

[396] Y.A. Kutoyants. *Statistical Inference for Spatial Poisson Processes*. Number 134 in Lecture Notes in Statistics. Springer, New York, 1998.

[397] K.L. Kvamme. Computer processing techniques for regional modeling of archaeological site locations. *Advances in Computer Archaeology*, 1:26–52, 1983.

[398] K.L. Kvamme. A view from across the water: the North American experience in archeological GIS. In G.R. Lock and Z. Stančič, editors, *Archaeology and Geographical Information Systems: a European Perspective*, pages 1–14. CRC Press, 1995.

[399] K.L. Kvamme. There and back again: revisiting archeological locational modeling. In M.W. Mehrer and K.L.Wescott, editors, *GIS and Archaeological Site Modelling*, pages 3–40. CRC Press, 2006.

[400] J.M. Landwehr, D. Pregibon, and A.C. Shoemaker. Graphical methods for assessing logistic regression models. *Journal of the American Statistical Association*, 79:61–83, 1984.

[401] S.L. Landy and A.S. Szalay. Bias and variance of angular correlation functions. *Astrophysical Journal*, 412:64–71, 1993.

[402] C. Lantuéjoul. On the estimation of mean values in individual analysis of particles. *Microscopica Acta*, Suppl. 4:266–273, 1980.

[403] W.A. Larsen and S.J. McCleary. The use of partial residual plots in regression analysis. *Technometrics*, 14:781–790, 1972.

[404] N.A. Laskurain. *Dinámica espacio-temporal de un bosque secundario en el Parque Natural de Urkiola (Bizkaia)*. PhD thesis, Universidad del País Vasco /Euskal Herriko Unibertsitatea, 2008.

[405] G.M. Laslett. Censoring and edge effects in areal and line transect sampling of rock joint traces. *Mathematical Geology*, 14:125–140, 1982.

[406] G.M. Laslett. The survival curve under monotone density constraints with applications to two-dimensional line segment processes. *Biometrika*, 69:153–160, 1982.

[407] G. Last and M. Holtmann. On the empty space function of some germ-grain models. *Pattern Recognition*, 32(9):1587–1600, 1999.

[408] G. Last and R. Schassberger. On the distribution of the spherical contact vector of stationary germ-grain models. *Advances in Applied Probability*, pages 36–52, 1998.

[409] G. Last and R. Szekli. Comparisons and asymptotics for empty space hazard functions of germ-grain models. *Advances in Applied Probability*, 43(4):943–962, 2011.

[410] F. Lavancier, J. Møller, and E. Rubak. Determinantal point process models and statistical inference. *Journal of Royal Statistical Society: Series B (Statistical Methodology)*, 2015. In press.

[411] R. Law, J. Illian, D.F.R.P. Burslem, G. Gratzer, C.V.S. Gunatilleke, and I.A.U.N. Gunatilleke. Ecological information from spatial patterns of plants: insights from point process theory. *Journal of Ecology*, 97:616–628, 2009.

[412] A. Lawson. GLIM and normalising constant models in spatial and directional data analysis. *Computational Statistics and Data Analysis*, 13:331–348, 1992.

[413] A.B. Lawson. On the analysis of mortality events around a prespecified fixed point. *Journal of the Royal Statistical Society, Series A*, 156(3):363–377, 1993.

[414] A.B. Lawson. *Bayesian Disease Mapping: Hierarchical Modeling in Spatial Epodemiology*. CRC Press, second edition, 2013.

[415] A.B. Lawson and D.G.T. Denison, editors. *Spatial Cluster Modelling 4*. Chapman and Hall/CRC Press, Boca Raton, 2002. ISBN 1-58488-266-2.

[416] E.L. Lehmann. *Theory of Point Estimation*. John Wiley and Sons, New York, 1983.

[417] E.L. Lehmann and J.P. Romano. *Testing Statistical Hypotheses*. Springer, New York, third edition, 2005.

[418] J.E. Lennard-Jones. On the determination of molecular fields. *Proc. Royal Soc. London A*, 106:463–477, 1924.

[419] P.A.W. Lewis. Recent results in the statistical analysis of univariate point processes. In P.A.W. Lewis, editor, *Stochastic Point Processes*, pages 1–54. Wiley, New York, 1972.

[420] P.A.W. Lewis and G.S. Shedler. Simulation of nonhomogeneous Poisson processes with log linear rate function. *Biometrika*, 63:501–505, 1976.

[421] P.A.W. Lewis and G.S. Shedler. Simulation of non-homogeneous Poisson processes by thinning. *Naval Logistics Quarterly*, 26:406–413, 1979.

[422] P.A.W. Lewis and G.S. Shedler. Simulation of nonhomogeneous Poisson processes with degree-two exponential polynomial rate function. *Operations Research*, 27:1026–1040, 1979.

[423] S. Liang, S. Banerjee, and B. Carlin. Bayesian wombling for spatial point processes. *Biometrics*, 65:1243–1253, 2009.

[424] J.K. Lindsey. *The Analysis of Stochastic Processes using GLIM*. Springer, Berlin, 1992.

[425] J.K. Lindsey. *Modelling Frequency and Count Data*. Oxford University Press, 1995.

[426] J.K. Lindsey. *Applying Generalized Linear Models*. Springer, 1997.

[427] J.K. Lindsey and G. Mersch. Fitting and comparing probability distributions with log linear models. *Computational Statistics and Data Analysis*, 13:373–384, 1992.

[428] C. Loader. *Local Regression and Likelihood*. Springer, New York, 1999.

[429] C.R. Loader. Large-deviation approximations to the distribution of scan statistics. *Advances in Applied Probability*, 23:751–771, 1991.

[430] J.M. Loh. A fast and valid spatial bootstrap for correlation functions. *Astrophysical Journal*, 681:726–734, 2008.

[431] N.B. Loosmore and E.D. Ford. Statistical inference using the G or K point pattern spatial statistics. *Ecology*, 87:1925–1931, 2006.

[432] H. W. Lotwick. Simulation of some spatial hard core models, and the complete packing problem. *Journal of Statistical Computation and Simulation*, 15:295–314, 1982.

[433] H. W. Lotwick and B. W. Silverman. Methods for analysing spatial processes of several types of points. *Journal of the Royal Statistical Society, Series B*, 44:406–413, 1982.

[434] H.W. Lotwick and B.W. Silverman. Convergence of spatial birth-and-death processes. *Mathematical Proceedings of the Cambridge Philosophical Society*, 90:155–165, 1981.

[435] Y. Lu and X. Chen. On the false alarm of planar K-function when analyzing urban crime distributed along streets. *Social Science Research*, 36:611–632, 2007.

[436] J. Lund, A. Penttinen, and M. Rudemo. Bayesian analysis of spatial point patterns from noisy observations. Preprint 1999:57, Department of Mathematical Statistics, Chalmers University of Technology, 1999.

[437] J. Lund and M. Rudemo. Models for point processes observed with noise. *Biometrika*, 87:235–249, 2000.

[438] E. Mammen and A.B. Tsybakov. Asymptotical minimax recovery of sets with smooth boundaries. *Annals of Statistics*, 23:502–524, 1995.

[439] B.J.F. Manly, L.L. McDonald, and D.L. Thomas. *Resource Selection by Animals: Statistical Design and Analysis for Field Studies*. Chapman and Hall, London, 1993.

[440] M. Månsson and M. Rudemo. Random patterns of nonoverlapping convex grains. *Advances in Applied Probability*, 34:718–738, 2002.

[441] A.F. Mark and A.E. Esler. An assessment of the point-centred quarter method of plotless sampling in some New Zealand forests. *Proceedings of the New Zealand Ecological Society*, 17:106–110, 1970.

[442] F.H.C Marriott. Barnard's Monte Carlo tests: how many simulations? *Applied Statistics*, 28:75–77, 1979.

[443] G. Marsaglia and J. Marsaglia. Evaluating the Anderson-Darling distribution. *Journal of Statistical Software*, 9(2):1–5, 2004. URL: http://www.jstatsoft.org/v09/i02.

[444] R. Marshall. A review of methods for the statistical analysis of spatial patterns of disease. *Journal of the Royal Statistical Society, Series A*, 154(3):421–441, 1991.

[445] V. Martinez and E. Saar. *Statistics of the Galaxy Distribution*. Chapman and Hall/CRC, 2002.

[446] S. Mase. Mean characteristics of Gibbsian point processes. *Annals of the Institute of Statistical Mathematics*, 42:203–220, 1990.

[447] S. Mase. Uniform LAN condition of planar Gibbsian point processes and optimality of maximum likelihood estimators of soft-core potential functions. *Probability Theory and Related Fields*, 92:51–67, 1992.

[448] S. Mase. Consistency of the maximum pseudo-likelihood estimator of continuous state space Gibbsian processes. *Annals of Applied Probability*, 5:603–612, 1995.

[449] S. Mase. The threshold method for estimating annual rainfall. *Annals of the Institute of Statistical Mathematics*, 48:201–213, 1996.

[450] S. Mase. Marked Gibbs processes and asymptotic normality of maximum pseudo-likelihood estimators. *Mathematische Nachrichten*, 209:151–169, 1999.

[451] S. Mase, J. Møller, D. Stoyan, R.P. Waagepetersen, and G. Döge. Packing densities and simulated tempering for hard core Gibbs point processes. *Annals of the Institute of Statistical Mathematics*, 53:661–680, 2001.

[452] B. Matérn. Spatial variation: stochastic models and their application to some problems in forest surveys and other sampling investigations. *Meddelanden från Statens Skogsforskningsinstitut*, 49(5):1–144, 1960.

[453] B. Matérn. Doubly stochastic Poisson processes in the plane. In G.P. Patil, E.C. Pielou, and W. E. Waters, editors, *Statistical Ecology, Volume 1: Spatial Patterns and Statistical Distributions*, pages 195–213. Pennsylvania State University Press, University Park, 1971.

[454] B. Matérn. *Spatial Variation*. Number 36 in Lecture Notes in Statistics. Springer Verlag, New York, 1986.

[455] J. Mateu. Parametric procedures in the analysis of replicated spatial point patterns. *Biometrical Journal*, 43:375–394, 2001.

[456] G. Matheron. *Random Sets and Integral Geometry*. John Wiley and Sons, New York, 1975.

[457] K. Matthes, J. Kerstan, and J. Mecke. *Infinitely Divisible Point Processes*. Wiley, Chichester, 1978.

[458] P. McCullagh and J.A. Nelder. *Generalized Linear Models*. Chapman and Hall, second edition, 1989.

[459] G. McSwiggan. Personal communication, 2013.

[460] G. McSwiggan, A. Baddeley, and G. Nair. Fitting point process models to road accident data. Submitted for publication, 2015.

[461] G. McSwiggan, A. Baddeley, and G. Nair. Kernel smoothing on a linear network. In preparation, 2015.

[462] R. Mead. A test for spatial pattern at several scales using data from a grid of contiguous quadrats. *Biometrics*, 30:295–307, 1974.

[463] J. Mecke. Stationäre zufällige maße auf lokalkompakten abelschen gruppen. *Zeitschrift für Wahrscheinlichkeitstheorie und verwandte Gebiete*, 9:36–58, 1967.

[464] J. Mecke, R.G. Schneider, D. Stoyan, and W.R.R. Weil. *Stochastische Geometrie*. DMV Seminar Band 16. Birkhäuser, Basel, 1990.

[465] N. Metropolis, A.W. Rosenbluth, M.N. Rosenbluth, A.H. Teller, and E. Teller. Equation of state calculations by fast computing machines. *Journal of Chemical Physics*, 21:1087–1092, 1953.

[466] R.E. Miles. On the homogeneous planar Poisson point process. *Mathematical Biosciences*, 6:85–127, 1970.

[467] R.E. Miles. The fundamental formula of Blaschke in integral geometry and its iteration, for domains with fixed orientation. *Australian Journal of Statistics*, 16(2):111–118, 1974.

[468] R.E. Miles. On the elimination of edge effects in planar sampling. In E.F. Harding and D.G. Kendall, editors, *Stochastic Geometry: a Tribute to the Memory of Rollo Davidson*, pages 228–247. John Wiley and Sons, London-New York-Sydney-Toronto, 1974.

[469] R.E. Miles. A synopsis of 'Poisson flats in Euclidean spaces'. In E.F. Harding and D.G. Kendall, editors, *Stochastic Geometry*, pages 202–227. John Wiley and Sons, 1974.

[470] A.J. Miller. *Subset Selection in Regression*. Chapman and Hall, London, 1990.

[471] E. Miranda, D. Jiménez, J. Suñé, E. O'Connor, S. Monaghan, I. Povey, K. Cherkaoui, and P.K. Hurley. Nonhomogeneous spatial distribution of filamentary leakage current paths in circular area Pt/HfO2/Pt capacitors. *J. Vac. Sci. Technol. B*, 31:01A107, 2013.

[472] E. Miranda, E. O'Connor, and P.K. Hurley. Simulation of the breakdown spots spatial distribution in high-K dielectrics and model validation using the spatstat package for R language. *ECS Transactions*, 33(3):557–562, 2010.

[473] I. Molchanov. *Statistics of the Boolean Model for Practitioners and Mathematicians*. Wiley Series in Probability and Statistics. Wiley, Chichester, 1997.

[474] J. Møller. On the rate of convergence of spatial birth-and-death processes. *Annals of the Institute of Statistical Mathematics*, 41:565–581, 1989.

[475] J. Møller. Markov chain Monte Carlo and spatial point processes. In O.E. Barndorff-Nielsen, W.S. Kendall, and M.N.M. van Lieshout, editors, *Stochastic Geometry: Likelihood and Computation*, number 80 in Monographs on Statistics and Applied Probability, chapter 4, pages 141–172. Chapman and Hall / CRC, Boca Raton, FL, 1999.

[476] J. Møller. Shot noise Cox processes. *Advances in Applied Probability*, 35:614–640, 2003.

[477] J. Møller, editor. *Spatial Statistics and Computational Methods*, number 173 in Lecture Notes in Statistics, New York, 2003. Springer-Verlag.

[478] J. Møller. Parametric methods. In A.E. Gelfand, P.J. Diggle, M. Fuentes, and P. Guttorp, editors, *Handbook of Spatial Statistics*, chapter 19, pages 317–337. CRC Press, Boca Raton, FL, 2010.

[479] J. Møller and C. Diaz-Avalos. Structured spatio-temporal shot-noise cox point process models, with a view to modelling forest fires. Technical Report R-2008-07, Department of Mathematical Sciences, Aalborg University, 2008. Submitted.

[480] J. Møller and M. Sørensen. Parametric models of spatial birth-and-death processes with a view to modelling linear dune fields. *Scandinavian Journal of Statistics*, 21:1–19, 1994.

[481] J. Møller, A.R. Syversveen, and R. Waagepetersen. Log Gaussian Cox processes: A statistical model for analyzing stand structural heterogeneity in forestry. In H. Kure, I. Thysen, and A. R. Kristensen, editors, *Proc. First European Conference for Information Technology in Agriculture*, pages 339–342. Department of Mathematics and Physics, The Royal Veterinary and Agricultural University, Denmark, 1997.

[482] J. Møller and G.L. Torrisi. Generalised shot noise Cox processes. *Advances in Applied Probability*, 37:48–74, 2005.

[483] J. Møller and R. P. Waagepetersen. Modern spatial point process modelling and inference (with discussion). *Scandinavian Journal of Statistics*, 34:643–711, 2007.

[484] J. Møller and R.P. Waagepetersen. *Statistical Inference and Simulation for Spatial Point Processes*. Chapman and Hall/CRC, Boca Raton, FL, 2004.

[485] J. Møller and S. Zuyev. Gamma-type results and other related properties of Poisson processes. *Adv. in Appl. Probab.*, 28(3):662–673, 1996.

[486] M. Moore. On the estimation of a convex set. *Annals of Statistics*, 12:1090–1099, 1984.

[487] P.A.P. Moran. The interpretation of statistical maps. *Journal of the Royal Statistical Society, Series B*, 10:243–251, 1948.

[488] M. Morisita. Measuring of the dispersion of individuals and analysis of the distributional patterns. Memoirs of the faculty of science, E 2, Kyushu Univ. Ser., Kyushu University, 1959.

[489] M.J. Moroney. *Facts from Figures*. Penguin, London, 1951. Penguin reprint, 1990.

[490] R.A. Moyeed and A.J. Baddeley. Stochastic approximation of the MLE for a spatial point pattern. *Scandinavian Journal of Statistics*, 18:39–50, 1991.

[491] T. Mrkvička. Distinguishing different types of inhomogeneity in Neyman-Scott point processes. *Methodology and Computing in Applied Probability*, 16:385–395, 2014.

[492] T. Mrkvička and I. Molchanov. Optimisation of linear unbiased intensity estimators for point processes. *Annals of the Institute of Statistical Mathematics*, 57:71–81, 2005.

[493] T. Mrkvička, M. Muška, and J. Kubečka. Two step estimation for Neyman-Scott point process with inhomogeneous cluster centers. *Statistics and Computing*, 24(1):91–100, 2014.

[494] E.A. Nadaraya. On estimating regression. *Theory of Probability and its Applications*, 9:141–142, 1964.

[495] E.A. Nadaraya. *Nonparametric Estimation of Probability Densities and Regression Curves*. Kluwer, Dordrecht, 1989.

[496] J.I. Naus. The distribution of the size of the maximum cluster of points on the line. *Journal of the American Statistical Association*, 60:532–538, 1965.

[497] J.I. Naus. Approximations for distributions of scan statistics. *Journal of the American Statistical Association*, 77:177–183, 1982.

[498] J. Neyman and E. Pearson. On the problem of the most efficient tests of statistical hypotheses. *Philosophical Transactions of the Royal Society of London, Series A*, 231:289–337, 1933.

[499] J. Neyman and E.L. Scott. A statistical approach to problems of cosmology. *Journal of the Royal Statistical Society, Series B*, 20:1–43, 1958.

[500] J. Neyman and E.L. Scott. Processes of clustering and applications. In P.A.W. Lewis, editor, *Stochastic Point Processes*, pages 646–681. John Wiley and Sons, 1972.

[501] X.X. Nguyen and H. Zessin. Integral and differential characterizations of the Gibbs process. *Mathematische Nachrichten*, 88:105–115, 1979.

[502] L.S. Nielsen and E.B.V. Jensen. Statistical inference for transformation inhomogeneous Markov point processes. *Scandinavian Journal of Statistics*, 31:131–142, 2004.

[503] M. Numata. Forest vegetation, particularly pine stands in the vicinity of Choshi — flora and vegetation in Choshi, Chiba prefecture, VI (in Japanese). *Bulletin of the Choshi Marine Laboratory*, (6):27–37, 1964. Chiba University.

[504] Y. Ogata. Statistical models for earthquake occurrences and residual analysis for point processes. *Journal of the American Statistical Association*, 83:9–27, 1988.

[505] Y. Ogata. Statistical model for standard seismicity and detection of anomalies by residual analysis. *Tectonophysics*, 169:159–174, 1989.

[506] Y. Ogata. Detection of precursory relative quiescence before great earthquakes through a statistical model. *Journal of Geophysical Research*, 97:19845–19871, 1992.

[507] Y. Ogata. Space-time point-process models for earthquake occurrences. *Annals of the Institute of Statistical Mathematics*, 50:379–402, 1998.

[508] Y. Ogata. Increased probability of large earthquakes near aftershock regions with relative quiescence. *Journal of Geophysical Research*, 106(B5):8729–8744, 2001.

[509] Y. Ogata. Significant improvements of the space-time ETAS model for forecasting of accurate baseline seismicity. *Earth Planets Space*, 63:217–229, 2011.

[510] Y. Ogata and K. Katsura. Likelihood analysis of spatial inhomogeneity for marked point patterns. *Annals of the Institute of Statistical Mathematics*, 40:29–39, 1988.

[511] Y. Ogata and K. Katsura. Maximum likelihood estimates of the fractal dimension for random spatial patterns. *Biometrika*, 78:463–474, 1991.

[512] Y. Ogata, K. Katsura, and M. Tanemura. Modelling heterogeneous space-time occurrences of earthquakes and its residual analysis. *Applied Statistics*, 52(4):499–509, 2003.

[513] Y. Ogata and M. Tanemura. Estimation of interaction potentials of spatial point patterns through the maximum likelihood procedure. *Annals of the Institute of Statistical Mathematics*, B 33:315–338, 1981.

[514] Y. Ogata and M. Tanemura. Likelihood analysis of spatial point patterns. *Journal of the Royal Statistical Society, Series B*, 46:496–518, 1984.

[515] Y. Ogata and M. Tanemura. Estimation of interaction potentials of marked spatial point processes through the maximum likelihood method. *Biometrics*, 41:421–433, 1985.

[516] Y. Ogata and M. Tanemura. Likelihood estimation of interaction potentials and external fields of inhomogeneous spatial point patterns. In I.S. Francis, B.J.F. Manly, and F.C. Lam, editors, *Pacific Statistical Congress*, pages 150–154. Elsevier, Amsterdam, 1986.

[517] Y. Ogata and M. Tanemura. Likelihood estimation of soft-core interaction potentials for Gibbsian point patterns. *Annals of the Institute of Statistical Mathematics*, 41:583–600, 1989.

[518] J. Ohser. On estimators for the reduced second moment measure of point processes. *Mathematische Operationsforschung und Statistik, series Statistics*, 14:63–71, 1983.

[519] J. Ohser and F. Mücklich. *Statistical Analysis of Microstructures in Materials Science*. John Wiley and Sons, Chichester, 2000.

[520] J. Ohser and D. Stoyan. On the second-order and orientation analysis of planar stationary point processes. *Biometrical Journal*, 23:523–533, 1981.

[521] A. Okabe, B. Boots, and K. Sugihara. *Spatial tessellations: concepts and applications of Voronoi diagrams*. Wiley series in probability and mathematical statistics. John Wiley and Sons, Chichester, England; New York, 1992.

[522] A. Okabe and M. Kitamura. A computational method for market area analysis on a network. *Geographical Analysis*, 28:330–349, 1996.

[523] A. Okabe and K. Okunuki. A computational method for estimating the demand of retail stores on a street network and its implementation in GIS. *Transactions in GIS*, 5:209–220, 2001.

[524] A. Okabe and T. Satoh. Spatial analysis on a network. In A.S. Fotheringham and P.A. Rogers, editors, *The SAGE Handbook on Spatial Analysis*, chapter 23, pages 443–464. SAGE Publications, London, 2009.

[525] A. Okabe, T. Satoh, and K. Sugihara. A kernel density estimation method for networks, its computational method and a GIS-based tool. *International Journal of Geographical Information Science*, 23:7–32, 2009.

[526] A. Okabe and K. Sugihara. *Spatial Analysis Along Networks*. Wiley, 2012.

[527] A. Okabe and I. Yamada. The K-function method on a network and its computational implementation. *Geographical Analysis*, 33:271–290, 2001.

[528] K. Okunuki and A. Okabe. Solving the Huff-based competitive location model on a network with link-based demand. *Annals of Operations Research*, 111:239–252, 2003.

[529] S. Openshaw. *The Modifiable Area Unit Problem*. Geo Books, Norwich, 1984.

[530] S. Openshaw, M.G. Charlton, C. Wymer, and A.W. Craft. A mark 1 geographical analysis machine for the automated analysis of point data sets. *International Journal on Geographical Information Systems*, 1:335–358, 1987.

[531] L.S. Ornstein and F. Zernike. Accidental deviations of density and opalesence at the critical point of a single substance. *Proceedings of the Section of Sciences, Royal Academy of Sciences/ Koninklijke Akademie van Wetenschappen, Amsterdam*, 17:793–806, 1914.

[532] L. Pace, A. Salvan, and N. Sartori. Adjusting composite likelihood ratio statistics. *Statistica Sinica*, 21:129–148, 2011.

[533] C. Palm. Intensitätsschwankungen im Fernsprechverkehr. *Ericsson Technics*, 44:1–189, 1943.

[534] A.L. Patterson. A Fourier series method for the determination of the components of interatomic distances in crystals. *Physical Review*, 46:372–376, 1934.

[535] A.L. Patterson. A direct method for the determination of the components of inter-atomic distances in crystals. *Zeitschrift fuer Krystallographie*, 90:517–554, 1935.

[536] M.J. Paulo, A. Stein, and M.Tomé. A spatial statistical analysis of cork oak competition in two Portuguese silvapastoral systems. *Canadian Journal of Forest Research*, 32:1893–1903, 2002.

[537] P.J.E. Peebles and E.J. Groth. Statistical analysis of extragalactic objects. V: three-point correlation function for the galaxy distribution in the Zwicky catalog. *Astrophysical Journal*, 196:1–11, 1975.

[538] R. Pélissier and F. Goreaud. ads package for R: A fast unbiased implementation of the K-function family for studying spatial point patterns in irregular-shaped sampling windows. *Journal of Statistical Software*, 63(6):1–18, 2015.

[539] A. Penttinen. *Modelling Interaction in Spatial Point Patterns: Parameter Estimation by the Maximum Likelihood Method*. Number 7 in Jyväskylä Studies in Computer Science, Economics and Statistics. University of Jyväskylä, 1984.

[540] A. Penttinen, D. Stoyan, and H. M. Henttonen. Marked point processes in forest statistics. *Forest Science*, 38:806–824, 1992.

[541] J.K. Percus and G.J. Yevick. Analysis of classical statistical mechanics by means of collective coordinates. *Physical Review*, 110:1–13, 1958.

[542] H. Persson. Root dynamics in a young Scots pine stand in Central Sweden. *Oikos*, 30:508–519, 1978.

[543] O. Persson. The robustness of estimating density by distance measurements. In G. Patil, E.C. Pielou, and W.E. Waters, editors, *Statistical Ecology*, volume II, pages 175–187. Penn State University Press, University Park, PA, 1971.

[544] J. Pfanzagl. On the measurability and consistency of minimum contrast estimates. *Metrika*, 14:249–276, 1969.

[545] D.U. Pfeiffer, T. Robinson, M. Stevenson, K. Stevens, D. Rogers, and A. Clements. *Spatial Analysis in Epidemiology*. Oxford University Press, Oxford, UK, 2008.

[546] S. J. Phillips, R. P. Anderson, and R. E. Schapire. Maximum entropy modeling of species geographic distributions. *Ecological Modelling*, 190:231–259, 2006.

[547] N. Picard, A. Bar-Hen, F. Mortier, and J. Chadoeuf. The multi-scale marked area-interaction point process: a model for the spatial pattern of trees. *Scandinavian Journal of Statistics*, 36:23–41, 2009.

[548] E.S. Pielou. Segregation and symmetry in two-species populations as studied by nearest neighbour relations. *Journal of Ecology*, 49:255–269, 1961.

[549] J.C. Pinheiro and D.M. Bates. *Mixed-Effects Models in S and S-PLUS*. Springer, 2000.

[550] E.J.G. Pitman. Significance tests which may be applied to samples from any populations. *Supplement to the Journal of the Royal Statistical Society*, 4(1):119–130, 1937.

[551] W. J. Platt, G. W. Evans, and S. L. Rathbun. The population dynamics of a long-lived Conifer (Pinus palustris). *The American Naturalist*, 131:491–525, 1988.

[552] M.J. Pons-Bordería, V.J. Martínez, D. Stoyan, H. Stoyan, and E. Saar. Comparing estimators of the galaxy correlation function. *Astrophysical Journal*, 523:480–491, 1999.

[553] D. Pregibon. Logistic regression diagnostics. *Annals of Statistics*, 9:705–724, 1981.

[554] C.J. Preston. Spatial birth-and-death processes. *Bulletin of the International Statistical Institute*, 46:371–391, 1977.

[555] M. Prokešová, J. Dvořák, and E.B.V. Jensen. Two-step estimation procedures for inhomogeneous shot-noise Cox processes. Submitted for publication, 2015.

[556] M. Prokešová, U. Hahn, and E.B. Vedel Jensen. Statistics for locally scaled point processes. In A. Baddeley, P. Gregori, J. Mateu, R. Stoica, and D. Stoyan, editors, *Case Studies in Spatial Point Process Modeling*, number 185 in Lecture Notes in Statistics, pages 99–123. Springer, New York, 2005.

[557] M. Prokešová and E.B. Vedel Jensen. Asymptotic Palm likelihood theory for stationary point processes. *Annals of the Institute of Statistical Mathematics*, 65:387–412, 2013.

[558] M. Prokešová. Inhomogeneity in spatial Cox point processes – location dependent thinning is not the only option. *Image Analysis and Stereology*, 29(3):133–141, 2010.

[559] M. A. Proschan and B. Presnell. Expect the unexpected from conditional expectation. *The American Statistician*, 52(3):248–252, 1998.

[560] S. Raha. A critique of statistical hypothesis testing in clinical research. *Journal of Ayurveda and Integrative Medicine*, 2(3):105–114, 2011.

[561] T. Rajala. A note on Bayesian logistic regression for spatial exponential family Gibbs point processes. *ArXiv e-prints*, 2014.

[562] J.O. Ramsay and B.W. Silverman. *Functional Data Analysis*. Springer-Verlag, 1997.

[563] C.R. Rao. Large sample tests of statistical hypotheses concerning several parameters with applications to problems of estimation. *Proceedings of the Cambridge Philosophical Society*, 44:50–57, 1948.

[564] S. L. Rathbun and N. Cressie. A space-time survival point process for a longleaf pine forest in southern Georgia. *Journal of the American Statistical Association*, 89:1164–1173, 1994.

[565] S.L. Rathbun and N. Cressie. Asymptotic properties of estimators of the parameters of spatial inhomogeneous Poisson point processes. *Advances in Applied Probability*, 26:122–154, 1994.

[566] T.R.C. Read and N.A.C. Cressie. *Goodness-of-Fit Statistics for Multivariate Data*. Springer-Verlag, New York, 1988.

[567] I.W. Renner, J. Elith, A. Baddeley, W. Fithian, T. Hastie, S.J. Phillips, G. Popovic, and D.I. Warton. Point process models for presence-only analysis. *Methods in Ecology and Evolution*, 6(4):366–379, 2015.

[568] I.W. Renner and D.I. Warton. Equivalence of MAXENT and Poisson point process models for species distribution modeling in ecology. *Biometrics*, 69:274–281, 2013.

[569] B. D. Ripley. The foundation of stochastic geometry. *Annals of Probability*, 4(6):995–998, 1976.

[570] B. D. Ripley and A. I. Sutherland. Finding spiral structures in images of galaxies. *Philosophical Transactions of the Royal Society of London, Series A*, 332:477–485, 1990.

[571] B.D. Ripley. The second-order analysis of stationary point processes. *Journal of Applied Probability*, 13:255–266, 1976.

[572] B.D. Ripley. Modelling spatial patterns (with discussion). *Journal of the Royal Statistical Society, Series B*, 39:172–212, 1977.

[573] B.D. Ripley. Simulating spatial patterns: dependent samples from a multivariate density. *Applied Statistics*, 28:109–112, 1979.

[574] B.D. Ripley. Tests of 'randomness' for spatial point patterns. *Journal of the Royal Statistical Society, Series B*, 41:368–374, 1979.

[575] B.D. Ripley. *Spatial Statistics*. John Wiley and Sons, New York, 1981.

[576] B.D. Ripley. *Statistical Inference for Spatial Processes*. Cambridge University Press, 1988.

[577] B.D. Ripley. Gibbsian interaction models. In D.A. Griffiths, editor, *Spatial Statistics: Past, Present and Future*, pages 1–19. Image, New York, 1989.

[578] B.D. Ripley and K. Hornik. Date-time classes. *R News*, 1(2):8–11, June 2001.

[579] B.D. Ripley and F.P. Kelly. Markov point processes. *Journal of the London Mathematical Society*, 15:188–192, 1977.

[580] B.D. Ripley and J.-P. Rasson. Finding the edge of a Poisson forest. *Journal of Applied Probability*, 14:483–491, 1977.

[581] B.D. Ripley and B.W. Silverman. Quick tests for spatial interaction. *Biometrika*, 65:641–642, 1978.

[582] J.M. Robins, A. van der Vaart, and V. Ventura. Asymptotic distribution of p values in composite null models. *Journal of the American Statistical Association*, 95(452):1143–1156, December 2000.

[583] W.S. Robinson. Ecological correlations and the behavior of individuals. *American Sociological Review*, 15:351–357, 1950.

[584] R.T. Rockafellar. *Convex Analysis*. Princeton University Press, Princeton, NJ, 1972.

[585] A. Rodriguez, D.B. Ehlenberger, D.L. Dickstein, P.R. Hof, and S.L. Wearne. Automated three-dimensional detection and shape classification of dendritic spines from fluorescence microscopy images. *PLoS ONE*, 3(4):234–778, 2008. doi:10.1371/journal.pone.0001997.

[586] A. Rosenfeld and J. L. Pfalz. Sequential operations in digital picture processing. *Journal of the Association for Computing Machinery*, 13:471, 1966.

[587] A. Rosenfeld and J. L. Pfalz. Distance functions on digital pictures. *Pattern Recognition*, 1:33–61, 1968.

[588] D. Rossiter. *Introduction to the R Project for Statistical Computing for Use at the ITC*. Available at r-project.org under manuals > contributed documentation.

[589] J.S. Rossman, X.H. Jing, G.P. Leser, and R.A. Lamb. Influenza virus M2 protein mediates ESCRT-independent membrane scission. *Cell*, 142:902–913, 2010.

[590] B. Rowlingson and P. Diggle. Splancs: spatial point pattern analysis code in S-PLUS. *Computers and Geosciences*, 19:627–655, 1993.

[591] J.S. Rowlinson. Penetrable sphere models of liquid-vapor equilibrium. *Advances in Chemical Physics*, 41:1–57, 1980.

[592] H. Rue, S. Martino, and N. Chopin. Approximate Bayesian inference for latent Gaussian models by using integrated nested Laplace approximations. *Journal of the Royal Statistical Society: Series B (Statistical Methodology)*, 71(2):319–392, 2009.

[593] K. Safi and B. Kranstauber. *Analysis and Mapping of Animal Movement in R*. Chapman and Hall/CRC, 2015.

[594] A. Särkkä. *Pseudo-likelihood approach for pair potential estimation of Gibbs processes.* Number 22 in Jyväskylä Studies in Computer Science, Economics and Statistics. University of Jyväskylä, 1993.

[595] A. Särkkä. Pseudo-likelihood approach for Gibbs point processes in connection with field observations. *Statistics*, 26(1):89–97, 1995.

[596] X. Saura, D. Moix, J. Suñé, P.K. Hurley, and E. Miranda. Direct observation of the generation of breakdown spots in MIM structures under constant voltage stress. *Microelectronics Reliability*, 53:1257–1260, 2013.

[597] X. Saura, S. Monaghan, P.K. Hurley, J. Suñé, and E. Miranda. Failure analysis of MIM and MIS structures using point-to-event distance and angular probability distributions. *IEEE Transactions on Devices and Materials Reliability*, 14(4):1080–1090, 2014.

[598] X. Saura, J. Suñé, S. Monaghan, P.K. Hurley, and E. Miranda. Analysis of the breakdown spot spatial distribution in Pt/HfO2/Pt capacitors using nearest neighbor statistics. *Journal of Applied Physics*, 114:154112, 2013.

[599] K. Schladitz. Surprising optimal estimators for the area fraction. *Advances in Applied Probability*, 31:995–1001, 1999.

[600] K. Schladitz and A.J. Baddeley. A third order point process characteristic. *Scandinavian Journal of Statistics*, 27:657–671, 2000.

[601] M. Schlather, P. Riberio, and P.J. Diggle. Detecting dependence between marks and locations of marked point processes. *Journal of Royal Statistical Society, Series B*, 66:79–93, 2004.

[602] F.P. Schoenberg. Multi-dimensional residual analysis of point process models for earthquake occurrences. *JASA*, 98:789–795, 2003.

[603] F.P. Schoenberg. Consistent parametric estimation of the intensity of a spatial-temporal point process. *Journal of Statistical Planning and Inference*, 128:79–93, 2005.

[604] S.C. Scholtz. Location choice models in Sparta. In R. Lafferty III, J.L. Ottinger, S.C. Scholtz, W.F. Limp, B. Watkins, and R. D. Jones, editors, *Settlement Predictions in Sparta: A Locational Analysis and Cultural Resource Assessment on the Uplands of Calhoun County, Arkansas*, number 14 in Arkansas Archaeological Survey Research Series, pages 207–222. Arkansas Archaeological Survey, Fayetteville, Arkansas, USA, 1981.

[605] W. Schottky. Über spontane Stromschwankungen in verschiedenen Elektrizitätsleitern. *Annalen der Physik*, 57:541–567, 1918. In German.

[606] D. Schuhmacher and K. Stucki. Gibbs point process approximation: Total variation bounds using Stein's method. *Annals of Probability*, 42:1911–1951, 2014.

[607] A.J. Scott and M.J. Symons. Clustering methods based on likelihood ratio criteria. *Biometrics*, 27:387–397, 1971.

[608] D.W. Scott. *Multivariate Density Estimation. Theory, Practice and Visualization.* Wiley, New York, 1992.

[609] K. Seefeld and E. Linder. *Statistics Using R with Biological Examples.* Available at r-project.org under manuals > contributed documentation.

[610] T.G. Seidler and J.B. Plotkin. Seed dispersal and spatial pattern in tropical trees. *PLoS Biology*, 4:e344, 2006.

[611] J. Serra. *Image analysis and mathematical morphology.* Academic Press, London, 1982.

[612] J. P. Shaffer. Multiple hypothesis testing. *Annual Review of Psychology*, 46:561–584, 1995.

[613] A. Särkkä. A note on robust intensity estimation for point processes. *Biometrical Journal*, 34:757–764, 1992.

[614] G. Shen, F. He, R. Waagepetersen, I.-F. Sun, Z. Hao, Z.-S. Chen, and M. Yu. Quantifying effects of habitat heterogeneity and other clustering processes on spatial distributions of tree species. *Ecology*, 94(11):2436–2443, November 2013.

[615] S. Shiode and N. Shiode. Detection of hierarchical point agglomerations by the network-based variable clumping method. *International Journal of Geographical Information Science*, 23:75–92, 2009.

[616] M. Silvapulle. On the existence and uniqueness of the maximum likelihood estimates for the binomial response models. *Journal of the Royal Statistical Society, Series B*, 43:310–313, 1981.

[617] B.W. Silverman. *Density Estimation for Statistics and Data Analysis.* Chapman and Hall, London, 1986.

[618] J.G. Skellam. Appendix to article by Hopkins (1954). *Annals of Botany*, 18:226–227, 1954.

[619] N. Smirnov. Table for estimating the goodness of fit of empirical distributions. *Annals of Mathematical Statistics*, 19:279–281, 1948.

[620] G.K. Smyth. *Pearson's goodness of fit statistic as a score test statistic*, volume 40 of *Lecture Notes–Monograph Series*, pages 115–126. Institute of Mathematical Statistics, Beachwood, Ohio, USA, 2003.

[621] J. Snow. *On the Mode of Communication of Cholera.* John Churchill, New Burlington Street, London, England, 1855.

[622] P. Soille. *Morphological Image Analysis: Principles and Applications.* Springer, New York, 2003.

[623] P.G. Spooner, I.D. Lunt, A. Okabe, and S. Shiode. Spatial analysis of roadside Acacia populations on a road network using the network K-function. *Landscape Ecology*, 19:491–499, 2004.

[624] A. Stein, M. N. M. van Lieshout, and H. W. G. Booltink. Spatial interaction of methylene blue stained soil pores. *Geoderma*, 102:101–121, 2001.

[625] M.L. Stein. A new class of estimators for the reduced second moment measure of point processes. *Biometrika*, 78:281–286, 1991.

[626] M.L. Stein. Asymptotically optimal estimation for the reduced second moment measure of point processes. *Biometrika*, 80:443–449, 1993.

[627] M.L. Stein. *Interpolation of Spatial Data: Some Theory for Kriging.* Springer, New York, 1999.

[628] R.A. Stone. Investigations of excess environmental risks around putative sources: statistical problems and a proposed test. *Statistics in Medicine,* 7:649–660, 1988.

[629] D. Stoyan. Thinnings of point processes and their use in the statistical analysis of a settlement pattern with deserted villages. *Statistics,* pages 45–56, 1988.

[630] D. Stoyan. A spatial statistical analysis of a work of art: Did Hans Arp make a "truly random" collage? *Statistics,* 24:71–80, 1993.

[631] D. Stoyan. On estimators of the nearest neighbour distance distribution function for stationary point processes. *Metrika,* 64:139–150, 2006.

[632] D. Stoyan, U. Bertram, and H. Wendrock. Estimation variances for estimators of product densities and pair correlation functions of planar point processes. *Annals of the Institute of Statistical Mathematics,* 45:211–221, 1993.

[633] D. Stoyan and P. Grabarnik. Second-order characteristics for stochastic structures connected with Gibbs point processes. *Mathematische Nachrichten,* 151:95–100, 1991.

[634] D. Stoyan and P. Grabarnik. Statistics for the stationary Strauss model by the cusp point method. *Statistics,* 22:283–289, 1991.

[635] D. Stoyan, W.S. Kendall, and J. Mecke. *Stochastic Geometry and its Applications.* John Wiley and Sons, Chichester, 1987.

[636] D. Stoyan, W.S. Kendall, and J. Mecke. *Stochastic Geometry and its Applications.* John Wiley and Sons, Chichester, second edition, 1995.

[637] D. Stoyan and H.-D. Schnabel. Description of relations between spatial variability of microstructure and mechanical strength of alumina ceramics. *Ceramics International,* 16:11–18, 1990.

[638] D. Stoyan and H. Stoyan. *Fractals, Random Shapes and Point Fields.* John Wiley and Sons, Chichester, 1995.

[639] D. Stoyan and H. Stoyan. Non-homogeneous Gibbs process models for forestry — a case study. *Biometrical Journal,* 40:521–531, 1998.

[640] D. Stoyan and H. Stoyan. Improving ratio estimators of second order point process characteristics. *Scandinavian Journal of Statistics,* 27:641–656, 2000.

[641] D. Stoyan and O. Wälder. On variograms in point process statistics II: models of markings and ecological interpretation. *Biometrical Journal,* 42:171–187, 2000.

[642] L. Strand. A model for stand growth. In *IUFRO Third Conference Advisory Group of Forest Statisticians,* pages 207–216, Paris, 1972. INRA, Institut National de la Recherche Agronomique.

[643] D.J. Strauss. A model for clustering. *Biometrika,* 63:467–475, 1975.

[644] D.R. Strong, Jr. Null hypotheses in ecology. *Synthese,* 43:271–285, 1980.

[645] K. Stucki and D. Schuhmacher. Bounds for the probability generating functional of a Gibbs point process. *Advances in Applied Probability,* 46(1):21–34, 2014.

[646] R. Takacs. Estimator for the pair-potential of a Gibbsian point process. Institutsbericht 238, Institut für Mathematik, Johannes Kepler Universität Linz, Austria, 1983.

[647] R. Takacs. Estimator for the pair potential of a Gibbsian point process. *Statistics*, 17:429–433, 1986.

[648] R. Takacs and T. Fiksel. Interaction pair-potentials for a system of ants' nests. *Biometrical Journal*, 28:1007–1013, 1986.

[649] U. Tanaka, Y. Ogata, and D. Stoyan. Parameter estimation and model selection for Neyman-Scott point processes. *Biometrical Journal*, 50:43–57, 2008.

[650] B. M. Taylor, T. M. Davies, B. S. Rowlingson, and P. J. Diggle. lgcp: An R package for inference with spatial and spatio-temporal log-Gaussian Cox processes. *Journal of Statistical Software*, 52(4):1–40, 2013.

[651] B. M. Taylor, T. M. Davies, B. S. Rowlingson, and P. J. Diggle. Bayesian inference and data augmentation schemes for spatial, spatiotemporal and multivariate log-Gaussian Cox processes in R. *Journal of Statistical Software*, 63(7):1–48, 2015.

[652] B.M. Taylor and P.J. Diggle. INLA or MCMC? A tutorial and comparative evaluation for spatial prediction in log-Gaussian Cox processes. *Journal of Statistical Computation and Simulation*, 84(10):1–19, 2014.

[653] M. Thomas. A generalisation of Poisson's binomial limit for use in ecology. *Biometrika*, 36:18–25, 1949.

[654] A.L. Thurman, R. Fu, Y. Guan, and J. Zhu. Regularized estimating equations for model selection of clustered spatial point processes. *Statistica Sinica*, 25:173–188, 2015.

[655] R. Tibshirani. Regression shrinkage and selection via the lasso. *Journal of the Royal Statistical Society, Series B*, 58(1):267–288, 1996.

[656] A. Tscheschel and S.N. Chiu. Quasi-plus sampling edge correction for spatial point patterns. *Computational Statistics and Data Analysis*, 52:5287–5295, 2008.

[657] E.R. Tufte. *The Visual Display of Quantitative Information*. Graphics Press, second edition, 2001.

[658] J.W. Tukey. Discussion of paper by F.P. Agterberg and S.C. Robinson. *Bulletin of the International Statistical Institute*, 44(1):596, 1972. Proceedings, 38th Congress, International Statistical Institute.

[659] B. Turnbull, E.J. Iwano, W.S. Burnett, H.L. Howe, and L.C. Clark. Monitoring for clusters of disease: application to leukaemia incidence in upstate New York. *Amer. Jour. Epidem.*, 132, supplement:136–143, 1990.

[660] G.J.G. Upton and B. Fingleton. *Spatial Data Analysis by Example. Volume I: Point Pattern and Quantitative Data*. John Wiley and Sons, Chichester, 1985.

[661] M.N.M. van Lieshout. *Markov Point Processes and Their Applications*. Imperial College Press, London, 2000.

[662] M.N.M. van Lieshout. A J-function for inhomogeneous point patterns. *Statistica Neerlandica*, 65:183–201, 2011.

[663] M.N.M. van Lieshout and A.J. Baddeley. A nonparametric measure of spatial interaction in point patterns. *Statistica Neerlandica*, 50:344–361, 1996.

[664] M.N.M. van Lieshout and A.J. Baddeley. Indices of dependence between types in multivariate point patterns. *Scandinavian Journal of Statistics*, 26:511–532, 1999.

[665] M.N.M. van Lieshout and I.S. Molchanov. Shot noise weighted processes: a new family of spatial point processes. *Stochastic Models*, 14:715–734, 1998.

[666] Ashlee Vance. Data analysts are mesmerised by the power of program R. *The New York Times*, page B6, 7 January 2009. New York Edition.

[667] C. Varin and P. Vidoni. A note on composite likelihood inference and model selection. *Biometrika*, 92:519–528, 2005.

[668] A. Veen and F.P. Schoenberg. Assessing spatial point process models using weighted K-functions: analysis of California earthquakes. In A. Baddeley, P. Gregori, J. Mateu, R. Stoica, and D. Stoyan, editors, *Case Studies in Spatial Point Process Modeling*, number 185 in Lecture Notes in Statistics, pages 293–306. Springer, New York, 2005.

[669] W.N. Venables and B.D. Ripley. *Modern Applied Statistics with S-Plus*. Springer, 1994.

[670] W.N. Venables and B.D. Ripley. *Modern Applied Statistics with S-Plus*. Springer, fourth edition, 2002.

[671] J. Ver Hoef and E. Peterson. A moving average approach for spatial statistical models of stream networks. *Journal of the American Statistical Association*, 105:6–18, 2010.

[672] J.M. Ver Hoef, E. Peterson, and D. Theobald. Spatial statistical models that use flow and stream distance. *Environmental and Ecological Statistics*, 13(4):449–464, 2006.

[673] D. Vere-Jones. Stochastic models for earthquake occurrence (with discussion). *Journal of the Royal Statistical Society, Series B*, 32:1–62, 1970.

[674] D. Vere-Jones. Space-time correlations for microearthquakes — a pilot study. *Supplement to Advances in Applied Probability*, 10:73–87, 1978.

[675] R. von Mises. *Wahscheinlichkeitsrechnung und Ihre Anwendung in der Statistik und Theoretischen Physik*. Deuticke, Leipzig, 1931.

[676] S. Voss. Habitat preferences and spatial dynamics of the urban wall spider: Oecobius annulipes Lucas. Honours thesis, Department of Zoology, University of Western Australia, 1999.

[677] S. Voss, B.Y. Main, and I.R. Dadour. Habitat preferences of the urban wall spider *Oecobius navus* (Araneae, Oecobiidae). *Australian Journal of Entomology*, 46:261–268, 2007.

[678] R. Waagepetersen. An estimating function approach to inference for inhomogeneous Neyman-Scott processes. *Biometrics*, 63:252–258, 2007.

[679] R. Waagepetersen. Estimating functions for inhomogeneous spatial point processes with incomplete covariate data. *Biometrika*, 95:351–363, 2008.

[680] R. Waagepetersen and Y. Guan. Two-step estimation for inhomogeneous spatial point processes. *Journal of the Royal Statistical Society, Series B*, 71:685–702, 2009.

[681] J. Wakefield. A critique of statistical aspects of ecological studies in spatial epidemiology. *Environmental and Ecological Statistics*, 11:31–54, 2004.

[682] J. Wakefield. Disease mapping and spatial regression with count data. *Biostatistics*, 8:158–183, 2007.

[683] O. Wälder and D. Stoyan. On variograms in point process statistics. *Biometrical Journal*, 38:895–905, 1996.

[684] J.F. Wallace, M. Canci, X. Wu, and A. Baddeley. Monitoring native vegetation on an urban groundwater supply mound using airborne digital imagery. *Journal of Spatial Science*, 53:63–73, 2008. ISSN 1449-8596.

[685] L. Waller, B. Turnbull, L.C. Clark, and P. Nasca. Chronic disease surveillance and testing of clustering of disease and exposure: Application to leukaemia incidence and TCE-contaminated dumpsites in upstate New York. *Environmetrics*, 3:281–300, 1992.

[686] L.A. Waller and C.A. Gotway. *Applied Spatial Statistics for Public Health Data*. Wiley, 2004.

[687] M.P. Wand and M.C. Jones. *Kernel Smoothing*. Chapman and Hall, 1995.

[688] P.C. Wang. Adding a variable in generalized linear models. *Technometrics*, 27:273–276, 1985.

[689] P.C. Wang. Residual plots for detecting nonlinearity in generalized linear models. *Technometrics*, 29:435–438, 1987.

[690] D. Wartenberg. Exploratory spatial analyses: outliers, leverage points, and influence functions. In D.A. Griffith, editor, *Spatial Statistics: Past, Present and Future*, pages 133–162. Institute of Mathematical Geography, Ann Arbor, Michigan, USA, 1990.

[691] D.I. Warton and L.C. Shepherd. Poisson point process models solve the "pseudo-absence problem" for presence-only data in ecology. *Annals of Applied Statistics*, 4:1383–1402, 2010.

[692] H. Wässle, B.B. Boycott, and R.-B Illing. Morphology and mosaic of on- and off-beta cells in the cat retina and some functional considerations. *Proceedings of the Royal Society of London, Series B*, 212:177–195, 1981.

[693] K.P. Watkins and A.H. Hickman. Geological evolution and mineralization of the Murchison Province, Western Australia. Bulletin 137, Geological Survey of Western Australia, 1990. Published by Department of Mines, Western Australia, 1990. Available online from Department of Industry and Resources, State Government of Western Australia, www.doir.wa.gov.au.

[694] G.S. Watson. Smooth regression analysis. *Sankhya A*, 26:359–372, 1964.

[695] S.L. Wearne, A. Rodriguez, D.B. Ehlenberger, A.B. Rocher, S.C. Henderson, and P.R. Hof. New techniques for imaging, digitization and analysis of three-dimensional neural morphology on multiple scales. *Neuroscience*, 136:661–680, 2005.

[696] S. Webster, P.J. Diggle, H.E. Clough, R.B. Green, and N.P. French. Strain-typing transmissible spongiform encephalopathies using replicated spatial data. In A. Baddeley, P. Gregori, J. Mateu, R. Stoica, and D. Stoyan, editors, *Case Studies in Spatial Point Process Modeling*, number 185 in Lecture Notes in Statistics, pages 197–214. Springer, New York, 2005.

[697] R.W.M. Wedderburn. On the existence and uniqueness of maximum likelihood estimates for certain generalized linear models. *Biometrika*, 63:27–32, 1976.

[698] D. Wheatley and M. Gillings. *Spatial Technology and Archaeology: the Archaeological Applications of GIS*. Taylor and Francis, 2002.

[699] B. Widom and J.S. Rowlinson. New model for the study of liquid-vapor phase transitions. *The Journal of Chemical Physics*, 52:1670–1684, 1970.

[700] T. Wiegand and K.A. Moloney. Rings, circles and null-models for point pattern analysis in ecology. *Oikos*, 104:209–229, 2004.

[701] M.B. Wilk and R. Gnanadesikan. Probability plotting methods for the analysis of data. *Biometrika*, 55:1–17, 1968.

[702] G. N. Wilkinson and C. E. Rogers. Symbolic description of factorial models for analysis of variance. *Applied Statistics*, 22:392–399, 1973.

[703] D.A. Williams. Generalised linear model diagnostics using the deviance and single case deletions. *Applied Statistics*, 36:181–191, 1987.

[704] J.H. Wolfe. Pattern clustering by multivariate mixture analysis. *Multivariate Behavioral Research*, 5:329–350, 1970.

[705] Z. Xie and J. Yan. Kernel density estimation of traffic accidents in a network space. *Computers, Environment and Urban Systems*, 32:396–406, 2008.

[706] I. Yamada and J.-C. Thill. Comparison of planar and network K-functions in traffic accident analysis. *Journal of Transport Geography*, 12:149–158, 2004.

[707] F. Yates. The influence of 'Statistical Methods for Research Workers' on the development of the science of statistics. *Journal of the American Statistical Association*, 46:19–34, 1951.

[708] T. Yoshida and T. Hayashi. On the robust estimation in Poisson processes with periodic intensities. *Annals of the Institute of Statistical Mathematics*, 42:489–507, 1990.

[709] Y.R. Yue and J.M. Loh. Variable selection for inhomogeneous spatial point process models. *Canadian Journal of Statistics*, 2015. Published online 13 Feb 2015.

[710] J. Zhuang. Second-order residual analysis of spatiotemporal point processes and applications in model evaluation. *Journal of the Royal Statistical Society, Series B*, 68:635–653, 2006.

[711] J. Zhuang, P. Chang, Y. Ogata, and Y. Chen. A study on the background and clustering seismicity in the Taiwan region based on a point process model. *Journal of Geophysical Research Solid Earth*, 110(B05S18), 2005. doi: 10.1029/2004JB003157.

[712] J. Zhuang, Y. Ogata, and D. Vere-Jones. Diagnostic analysis of space-time branching processes for earthquakes. In A. Baddeley, P. Gregori, J. Mateu, R. Stoica, and D. Stoyan, editors, *Case Studies in Spatial Point Process Modelling*, number 185 in Lecture Notes in Statistics, chapter 15. Springer, New York, 2005.

[713] S. Zuyev. Strong Markov property of Poisson processes and Slivnyak formula. In A. Baddeley, P. Gregori, J. Mateu, R. Stoica, and D. Stoyan, editors, *Case studies in spatial point process modeling*, number 185 in Lecture Notes in Statistics, pages 78–84. Springer, New York, 2005.

Index